17th International Conference on Textures of Materials 2014

(ICOTOM 17)

IOP Conference Series: Materials Science and Engineering Volume 82

Dresden, Germany
24 – 29 August 2014

Editors:

W. Skrotzki
C. Oertel

ISBN: 978-1-5108-0296-4
ISSN: 1757-8981

Printed from e-media with permission by:

Curran Associates, Inc.
57 Morehouse Lane
Red Hook, NY 12571

Printed by Curran Associates, Inc. (2015)

For permission requests, please contact the Institute of Physics
at the address below.

Institute of Physics
Dirac House, Temple Back
Bristol BS1 6BE UK

Phone: 44 1 17 929 7481
Fax: 44 1 17 920 0979

techtracking@iop.org

Additional copies of this publication are available from:

Curran Associates, Inc.
57 Morehouse Lane
Red Hook, NY 12571 USA
Phone: 845-758-0400
Fax: 845-758-2634
Email: curran@proceedings.com
Web: www.proceedings.com

TABLE OF CONTENTS

CRYSTALLIZATION, RECRYSTALLIZATION AND GROWTH TEXTURES

TRANSFORMATION TEXTURES

TEXTURES IN FUNCTIONAL MATERIALS

TEXTURES IN ADVANCED MATERIALS

TEXTURES IN ROCKS

TEXTURE RELATED RESEARCH IN MICROSTRUCTURES

TEXTURE-INDUCED ANISOTROPY

INSIGHT THROUGH NEW EXPERIMENTAL METHODS

TECHNOLOGICAL APPLICATIONS OF TEXTURE STUDIES

Preface

The 17th International Conference on Textures of Materials (ICOTOM 17) took place in Dresden, Germany, August 24-29, 2014. It belongs to the "triennial" series of ICOTOM meetings with a long tradition, starting in 1969 - Clausthal, 1971 - Cracow, 1973 - Pont-à-Mousson, 1975 - Cambridge, 1978 - Aachen, 1981 - Tokyo, 1984 - Noordwijkerhout, 1987 - Santa Fe, 1990 - Avignon, 1993 - Clausthal, 1996 - Xian, 1999 - Montreal, 2002 - Seoul, 2005 - Leuven, 2008 - Pittsburgh, 2011 - Mumbai, 2014 - Dresden. ICOTOM 17 was hosted by the Dresden University of Technology, Institute of Structural Physics.

Following the tradition of the ICOTOM conferences, the main focus of ICOTOM-17 was to promote and strengthen the fundamental understanding of the basic processes that govern the formation of texture and its relation to the properties of polycrystalline materials. Nonetheless, it was the aim to forge links between basic research on model materials and applied research on engineering materials of technical importance. Thus, ICOTOM 17 provided a forum for the presentation and discussion of recent progress in research of texture and related anisotropy of mechanical and functional properties of all kinds of polycrystalline materials including natural materials like rocks. Particular attention was paid to recent advances in texture measurement and analysis as well as modeling of texture development for all kinds of processes like solidification, plastic deformation, recrystallization and grain growth, phase transformations, thin film deposition, etc. Hence, ICOTOM 17 was of great interest to materials scientists, engineers from many different areas and geoscientists.

The topics covered by ICOTOM 17 were:
1. Mathematical, numerical and statistical methods of texture analysis
2. Deformation textures
3. Crystallization, recrystallization and growth textures
4. Transformation textures
5. Textures in functional materials
6. Textures in advanced materials
7. Textures in rocks
8. Texture related research on microstructures
9. Texture-induced anisotropy
10. Insight through new experimental methods
11. Technological applications of texture studies
12. Other new developments and future trends related to the field

While there was large interest in the topics 2, 3 and 8, contributions to topic 7 were much less than expected.

ICOTOM 17 attracted 266 scientists from 34 countries with about one third of the participants being students. This is a very good ratio showing that we could attract the young generation. There have been 216 oral and 76 poster presentations, three of which received a poster award. It is our pleasure to thank the members of the International ICOTOM

Committee for their valuable help, especially for proposing and choosing the 15 plenary speakers as well as the distinguished scientist of the texture community for the "Bunge Award". 130 papers were submitted for publication in the proceedings, 116 were accepted after reviewing. We would like to express our thanks to all referees for their efficient and prompt efforts. We acknowledge particularly support from the German Research Society (DFG) and the City of Dresden. We are also grateful for industrial support from Bruker Nano GmbH, Oxford Instruments GmbH, Ametek GmbH / EDAX, Labosoft S.C., PANalytical GmbH and IOP Publishing. Finally we thank all members of the National Organizing Committee, Intercom Dresden and Conwerk / Laboratory Ten for the excellent organization of ICOTOM 17 and the very pleasant collaboration.

On the first day of the conference three tutorials have been offered. Each of them has been attended by about 30 participants.

1. Texture-aided residual stress identification system (TARSIuS)
   (organized by Prof. Dr. J. Bonarski and Mr. B. Kania)
2. MTEX – MATLAB toolbox for quantitative texture analysis
   (organized by Dr. R. Hielscher and Mr. F. Bachmann)
3. Grain boundary engineering
   (organized by Prof. N. Bozzolo and Prof. Dr. A.D. Rollett)

A highlight of ICOTOM 17 was the ceremony honoring Prof. Dr. Claude Esling with the Bunge Award for his distinguished contributions to the field of Textures of Materials and his continuous effort to pass on his knowledge to future generations of texture experts. The Bunge Award is named after Professor Hans Bunge († 2004), to whom the world's texture community is very much indebted not only for his magisterial work on the Mathematical Theory of Texture, but also for his lifelong promotion of the field of Textures of Materials.
To the great delight of all participants, Helga Bunge and her son Prof. Hans-Peter Bunge, to whom many of the older generation have a personal relationship, attended the ceremony, s. Fig. 1).

Figure 1: Bunge award ceremony: from left to right Prof. Hans-Peter Bunge, Helga Bunge, Prof. Claude Esling (recipient of Bunge Award), Prof. Werner Skrotzki, Prof. Paul van Houtte

Following the award ceremony Prof. Dr. Claude Esling gave an in memoriam tribute to Prof. Dr. Richard Penelle, who was an internationally recognized texture specialist. Details can be found in the proceedings paper by Esling et al. [this issue].

During the conference the International ICSMA Committee decided to convene the next conference in St. George, USA, in 2017. We wish the organizers of ICOTOM 18 great success and look forward to meeting you in St. George.

Werner Skrotzki* (Chairman of ICOTOM 17, Dresden University of Technology)

Carl-Georg Oertel (Dresden University of Technology)

Guest Editors

Dresden, March, 2015

(* Corresponding author; e-mail address: werner.skrotzki@tu-dresden.de)

IOP Conf. Series: Materials Science and Engineering **82** (2015) 011002 doi:10.1088/1757-899X/82/1/011002

In memoriam tribute to Richard Penelle (1936-2012)

C Esling[1, 2], T Baudin[3] and A-L Helbert[3]
[1] Laboratoire LEM3, CNRS UMR 7239, Université de Lorraine, 57045 Metz, France
[2] Laboratory of Excellence DAMAS, Université de Lorraine, 57045 Metz, France
[3] Université Paris-Sud, ICMMO, SP2M, UMR CNRS 8182, F-91405 Orsay, France
E-mail: claude.esling@univ-lorraine.fr

Abstract. The present paper is dedicated to Richard Penelle who was an internationally recognized texture specialist. The first part quickly describes his scientific carrier and the second one presents his last paper as first author and his last research projects on the Invar and 1050 aluminium recrystallization, respectively. Most of his last studies were linked to the understanding of the Cube texture development in fcc materials.

1. The scientific carrier of Richard Penelle

Richard Penelle was an immutable figure of texture specialist at all national, European and international texture conferences, except the last ICOTOM 16 in December 2011, Mumbai. At that time, Richard was already severely ill of a disease which took him away just before Christmas 2012.

After a solid scientific education at the historical Sorbonne university, Richard Penelle prepared his thesis at the Centre de Recherches Métallurgiques (CRM) of the Ecole des Mines de Paris, under the supervision of Professor Paul Lacombe (see figure 1).

Figure 1: Meeting at the "Maison de la Chimie", Paris, September 1989. From the left to the right: Dr. T. Okada, Prof. J. Philibert, Prof. P. Lacombe, Dr. R. Penelle and Dr. M. Aucouturier (Photo: courtesy of Prof. J. Philibert)

The title of his thesis, defended in 1967, was "Relations between deformation mechanisms in rolled iron single crystals and their deformation and primary recrystallization textures". In 1983, Professor Lacombe received the sword of Academician with the specialty metallurgy, a specialty which is only exceptionally honored by the Academy of Sciences. All along his career at the French CNRS, Richard Penelle continued to keep metallurgy as a common thread in his wide ranging research on metallic materials from iron and low alloy steels, to alloys of iron and nickel, and even to zirconium and titanium and their alloys. While conducting a fundamental scientific approach, he became also interested in the technological aspects involved, e.g. his contribution to deep drawing.

Even after the scientific meeting held in 2009 in his honor at Orsay at the occasion of his "official" retirement as a Research Director of the CNRS, Richard Penelle continued his scientific activity and loved to share his knowledge with the younger scientists of his former small research group, researchers who pursue studies on the topics he had initiated.
Richard Penelle had also an intense international cooperation with many foreign countries during his entire career. When he fell ill, he was just returned from his last mission to Brazil, where he gave a series of lectures at the Mining School of Ouro Preto.
Despite his dense scientific activity, Richard also enjoyed an intense and harmonious family life with his wife, Gabrielle, who was his faithful companion and stood on his side at most of his missions, notably the ICOTOM conferences. She offered him three daughters Anne, Isabelle and Aude, who gave him five grandchildren and his family accompanied him to the end.

2. Some last scientific achievements of Richard Penelle

From Web of Science™, it appears that Richard Penelle has published his last paper as a first author in 2010 on recrystallization of the Fe-36%Ni (Invar) alloy [1]. However, until the end, he continued to supervise research works of the Orsay team. The last one concerned the recrystallization study of 1050 aluminium alloy. Richard Penelle has co-published two last papers in international journals in 2012 on this subject [2,3].
Let us note that in 2012 he has also published three other papers in the Proceedings of ICOTOM 16 (Bombay, 2011) [4-6] and a book chapter on the influence of textures on the physical properties [7].
In the present paper dedicated to Richard Penelle, instead of presenting broadly the results obtained during his long scientific carrier, it was decided to select some results of his last publication [1] and last research project [2-4]. These two studies are linked to the understanding of the Cube texture development during recrystallization of cold rolled sheets, of Invar and 1050 aluminium alloy.

2.1. Primary recrystallization of Invar

The French German "Texture and Anisotropy" symposium, organized by Claude Esling and Richard Penelle, for the SF2M-DGM societies on April 3rd 2009, was preceded by a scientific day to honour Richard Penelle at the occasion of his "official" retirement. During this day, organized by Thierry Baudin and Claude Esling, Richard Penelle presented a communication entitled "On primary recrystallization of Invar". The corresponding paper was then published in a special issue of Advanced Engineering Materials. It was his last paper as first author.
This paper was a short review of results obtained with his team on recrystallization of the Fe-36%Ni (Invar) alloy. The Fe-36%Ni austenitic alloy is used in electronic devices because of its low thermal expansion coefficient and its good magnetic properties [8]. For that purpose it is beneficial to have a high fraction of Cube orientation. These works have started with the two postdoctoral years of Stefan Zaefferer (10-1995-10-1997), and then with the PhD theses of Frank Julliard (2001) [9] and Sorphal Chhann (2007) [10] plus some additional contributions of his close collaborators.
The Cube orientation nucleation, notably its development after high deformation by cold rolling and subsequent annealing, was discussed from multiscale analyses based on TEM, SEM/EBSD, X-ray and neutron diffraction measurements. It was explained why the Cube texture sharpness increased with the deformation amount.
These experimental observations were also coupled with a quite original Monte-Carlo simulation, since the calculation was performed from a TEM crystallographic orientation map and stored energy values estimated by neutron diffraction for all the components of the deformation texture [11].
From these studies, which were also completed by additional work on Fe-Ni alloys supervised by Richard Penelle (see [12] for example), some results have been highlighted.
After 95% cold rolling, the Copper texture is developed. It is composed of the C {112}<111>, B {110}<112> and S {123}<634> components [13]. In addition, a minor Cube {001}<100> orientation is also observed. On the contrary, after complete recrystallization (60 min at

600°C), a sharp Cube texture is observed (about 80%). At the same time, its twin {122}<221> orientation appears (about 10%) and a random part remains (about 10%).
The Cube orientation nucleates preferentially because of its cellular dislocation substructure which presents a low stored energy. On the contrary, the other deformation texture components have a lamellar substructure with a high stored energy. In fact, the most important parameter is the stored energy difference between the Cube and the non-Cube orientations which increases with the deformation amount and explains why the Cube orientation is the main recrystallization component for the highest cold rolling reduction. For low reductions (<70%), this parameter is quite low and all the deformation texture components can nucleate. For very low reduction, because of twinning, the recrystallization texture tends to become random [1].
So, after dislocation cell growth or coalescence to attain a critical size, a Cube subgrain (named A in figure 2) adjacent to a grain boundary (high misorientation), can grow by Strain Induced Boundary Migration (SIBM) as shown in figure 2.

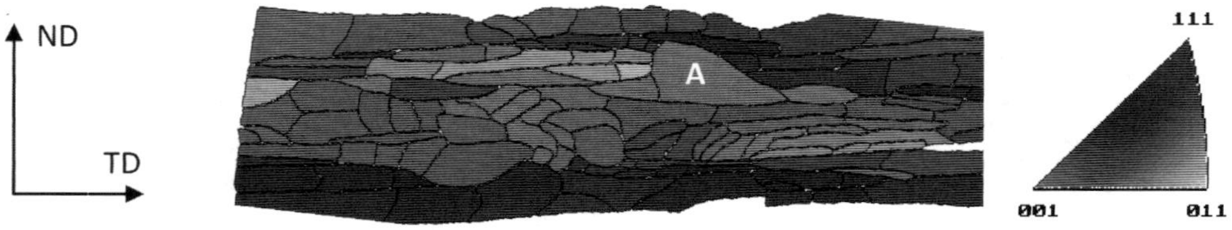

Figure 2: TEM orientation map ({hkl} distribution) showing a Cube band in a 95% cold rolled and 2 min. at 600°C annealed sample [9].

This research work was initiated by Richard Penelle in the years 1994-1995 but a lot of work still remains to improve the Cube texture development. Indeed, despite of a better understanding of the metallurgical evolution, the Orsay team currently continues to work on this topic and particularly on the Fe-48%Ni recrystallization but for other applications such as photovoltaic cells. In this case, it is necessary to develop 100% of Cube texture with a low dispersion (lower than 5°) [14] to lead to an epitaxial deposition of silicon on the Fe-Ni substrate. In order to succeed, the cold rolling reduction is increased up to 99% and the thermal treatment must be optimized as well as the material chemical composition. As an example, figure 3 shows the adverse effect of niobium amount on the Cube fraction.

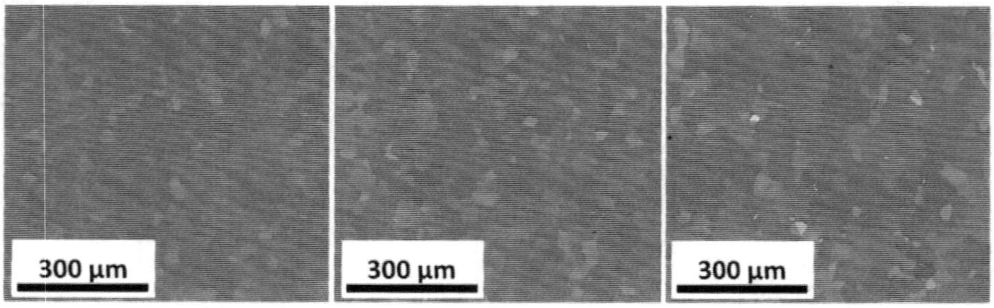

(a) (b) (c)
Figure 3: EBSD maps of Fe-48%Ni alloy with (a) 0%, (b) 0,035% and (c) 0,050% of Nb (weight %) - ({hkl} distribution) [14]

This figure shows that the Cube fraction decreases (99.9 %, 98.8 % and 97.2 %, respectively, with a 10° misorientation from the ideal Cube orientation) when the percentage of niobium increases because niobium is a recrystallization inhibitor. On the contrary, it was shown that sulphur favours the Cube development through MnS precipitation. Indeed, they are the origin of hardening that increases the stored energy gap between Cube and the other components. Thus, this difference favours the Cube development [14].

2.2. Primary recrystallization of 1050 aluminium alloy

In the year 2010, Richard Penelle and his close colleague Anne-Laure Helbert started to work on the formation and the development of a strong Cube recrystallization texture in aluminium of commercial purity, in collaboration with François Boutin, previously research director of the Alcan society. Wei Wang worked on this topic in 2010 as part of his master thesis [15] before starting a PhD project in the same research team on recrystallization of NiCrW alloys [16].

The manufacture of electrolytic capacitors requires the development of a very pronounced Cube texture. So, Boutin [17] proposed an industrial process to develop a predominant Cube texture in high purity aluminium. However in the case of commercial purity aluminium of low cost, the Cube fraction remains lower than about 30%. Knowing that silicon and iron contents in solid solution slow down the Cube texture development, the first idea was to introduce a pre-annealing for 5h at 300°C to precipitate Si and Fe aluminides to limit solute dragging effect in the 1050 aluminium of commercial purity. Moreover, since the energy stored during deformation provides the driving force for recrystallization, the authors proposed to introduce an additional 10% cold rolling before the final annealing after a partial recrystallization of the alloy at moderate temperature (200°C). Because of the geometry of the slip systems, the Cube grains are less affected by strain hardening than the non-Cube grains. Thus, the stored energy difference is increased and the Cube grain nucleation is favoured. Under these conditions, the optimized thermo-mechanical process, described in figure 4, allows to develop about 65% of Cube grains, in contrast to 30% using a more traditional route.

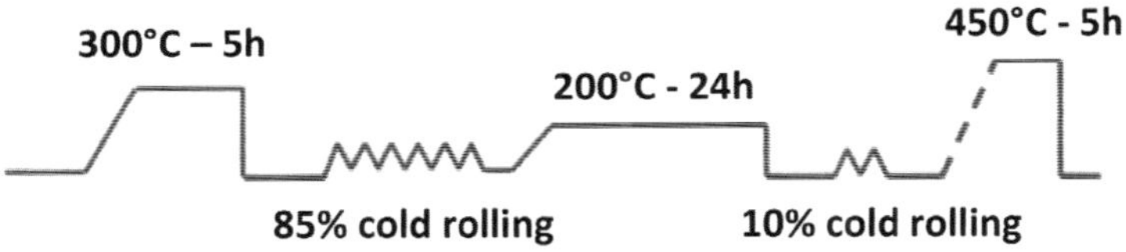

Figure 4: Processing route established to develop 65% of Cube texture after final recrystallization [2-4].

Obviously, this fraction remains low with regard to that developed in high purity aluminium. However, this result is encouraging, but to still increase the Cube fraction, the nucleation and growth mechanisms of Cube grains have been studied [2-4].

To illustrate the necessity to introduce a pre-annealing for 5h at 300°C, figure 5 shows that this annealing allows to form Si and Fe aluminide precipitates. According to the study of Zhang et al. [18], most of them are AlFeSi and $FeSiAl_5$ intermetallic compounds. This pre-annealing allows doubling the Cube fraction [2-4].

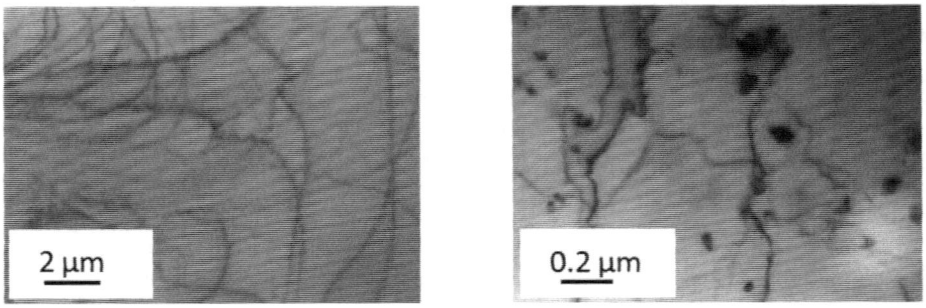

(a) (b)

Figure 5: TEM micrographs of the 1050 aluminium alloy (a) in the as-received state without precipitates and (b) after pre-annealing, with precipitates [2-4].

As observed in the Invar alloy, the dislocation substructure in Cube grains is composed of equiaxed dislocation cells. On the contrary, the other deformation texture components present a lamellar or an irregular substructure. Moreover, it has been observed that the Cube grains recover quickly compared to the other components [19-21]. Thus, due to its dislocation cell shape and the stored energy difference, the Cube nucleation is favoured. The 10% cold rolling before the final annealing allows to further increase this stored energy difference and then favours the Cube grain growth by strain induced boundary migration (SIBM) [2-4]. Besides, a second recrystallization mechanism is observed. During annealing, the sub-structure of polygonized Cube grains is removed by in-situ recrystallization (or generalized recovery). As an example, figure 6 shows the sub-structure evolution in a Cube grain observed during in-situ annealing in a FEG-SEM from room temperature up to 450°C with a heating rate of 55°C/h on a sample partially recrystallized at 200°C and then 10% cold-rolled (see figure 4). This mechanism promotes the development of the Cube orientation in the texture.

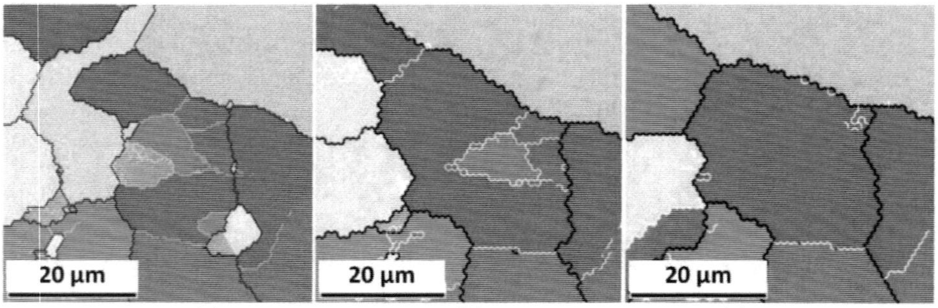

(a) (b) (c)

Figure 6: EBSD maps showing the sub-structure evolution of a Cube grain at (a) 290°C, (b) 310°C and (c) 350°C observed during in-situ annealing in a FEG-SEM (from [2-4]).

Apart from this mechanism, that clearly shows that the Cube development is linked to the Cube dislocation cell morphology, it was also evidenced that the stored energy difference between Cube grains (low energy) and non-Cube grains (high energy) promotes the Cube development, as well as in the FeNi alloys.
However, even if the Cube nucleation and growth mechanisms are now quite well understood, it still remains to increase the Cube volume fraction by optimizing the thermo-mechanical process.

3. Conclusion

During his long carrier mainly dedicated to the understanding of the correlation between microstructure, texture and physical properties, Richard Penelle has contributed to a better understanding of the deformation, recovery, primary recrystallization and (normal or abnormal) grain growth mechanisms in different metallic or geological materials of various crystal structures.
In this paper, as an example, the understanding of the Cube texture development was particularly discussed in Invar and 1050 aluminium alloys. The Cube texture fraction was increased notably thanks to Richard Penelle's research. However, it still remains a lot of work to do for the young co-workers to carry on the studies initiated by him.

Acknowledgements

The authors would like to acknowledge Gabrielle and Aude Penelle for their help in documentation.

References

[1] Penelle R and Baudin T 2010 *Advanced Engineering Materials* **12** 1047
[2] Wang W, Helbert A-L, Baudin T, Brisset F and Penelle R 2012 *Materials Characterization* **64** 1

[3] Helbert A-L, Wang W, Brisset F, Baudin T and Penelle 2012 R *Advanced Engineering Materials* **14** 39
[4] Helbert A-L, Wang W, Baudin T, Brisset F and Penelle R 2012 *Materials Science Forum* **702-703** 391
[5] Cruz-Gandarilla F, Salcedo Garrido AM, Baudin T, Mendoza-León H and Penelle R 2012 *Materials Science Forum* **702-703** 519
[6] Maazi N and Penelle R 2012 *Materials Science Forum* **702-703** 599
[7] Penelle R and Solas D 2012 Influence des textures sur les propriétés physiques", Chapitre 6.4 de l'ouvrage, Rayonnement synchrotron, rayons X et neutrons au service des matériaux – Analyse des contraintes et des textures, Eds Lodini A and Baudin T, *EDP Sciences* 343
[8] The Iron Nickel alloys, Eds. Béranger G, Duffaut F, Morlet J, Tiers J-F, Lavoisier Technique & Documentation (1996)
[9] Julliard F 2001 Etude des mécanismes de recristallisation dans l'Invar, alliage Fe-36%Ni *PhD Thesis*, Université de Paris-Sud, France
[10] Chhann S 2007 Etude de la recristallisation de l'Invar (alliage Fe-36%Ni) déformé par laminage asymétrique *PhD Thesis*, Université de Paris-Sud, France
[11] Baudin T, Solas D, Etter A-L, Ceccaldi D and Penelle R 2004 *Scripta Materialia* **51** 427
[12] Caleyo F, Baudin T and Penelle R 2002 *Scripta Materialia* **46** 829
[13] Zaefferer S, Baudin T and Penelle R 2001 *Acta Materialia* **49** 1105
[14] Ateba-Betanda Y 2012 Hypertexture Cube {100}<001> de feuillards Fer-Nickel et Nickel-Tungstène pour substrats d'épitaxie *Rapport de stage de master 2*, Université de Paris-Sud, France
[15] Wang W 2010 Développement de la texture cubique dans l'aluminium de pureté commerciale *Rapport de stage de master 2*, Université de Paris-Sud, France
[16] Wang W 2014 Etude des mécanismes de recristallisation de nouveaux alliages à base de Ni *PhD Thesis*, Université de Paris-Sud, France
[17] Boutin FR 1979 *Japanese Patent* No B2-45-11242
[18] Zhang J, Pan F, Zuo R and Bai C 2008 *Journal of Materials Processing Technology* **206** 382
[19] Umezawa A and Inagaki H 2006 Zeitschrift für Metallkunde **97** 39
[20] Barou F 2009 Etude de la croissance des sous-grains dans des alliages binaires aluminium-manganèse *PhD Thesis*, Ecole des Mines de St Etienne, France
[21] Penelle R 2009 *International Journal of Materials Research* **100** 1420

IOP Conf. Series: Materials Science and Engineering **82** (2015) 011003 doi:10.1088/1757-899X/82/1/011003

Peer review statement

All papers published in this volume of *IOP Conference Series: Materials Science and Engineering* have been peer reviewed through processes administered by the proceedings Editors. Reviews were conducted by expert referees to the professional and scientific standards expected of a proceedings journal published by IOP Publishing.

IOP Conf. Series: Materials Science and Engineering **82** (2015) 012001 doi:10.1088/1757-899X/82/1/012001

On the development of a model predicting the recrystallization texture of aluminum using the Taylor model for rolling textures and the coincidence lattice site theory

Morimoto T[1], Yoshida F[2], Yanagida A[3] and Yanagimoto J[4]

[1] Nakayama Amorphous,Osaka,Japan
[2] Nakayama Steel,Osaka,Japan
[3] Tokyo Denki University,Tokyo,Japan
[4] University of Tokyo,Tokyo,Japan
E-mail: morimoto_t@nakayama-amorphous.co.jp

Abstract. First, hardening model in f.c.c. metals was formulated with collinear interactions slips, Hirth slips and Lomer-Cottrell slips. Using the Taylor and the Sachs rolling texture prediction model, the residual dislocation densities of cold-rolled commercial pure aluminum were estimated. Then, coincidence site lattice grains were investigated from observed cold rolling texture. Finally, on the basis of oriented nucleation theory and coincidence site lattice theory, the recrystallization texture of commercial pure aluminum after low-temperature annealing was predicted.

1. Introduction

Polycrystalline metals generally recrystallize both from grain boundaries, called strain-induced boundary migration, and from microbands or shear bands. A general mechanism of recrystallization from stored energy has been proposed. On the basis of a nucleation theory, cube orientated grains recrystallize from a bandlike structure formed during rolling[1]. The orientation of the recrystallized grain must exist within the spread of orientations present in the deformed state[2].Recrystallization texture prediction models of b.c.c. metals have been proposed using the Taylor factors and total amount of microscopic slips based on the oriented nucleation model[3]-[4]. However, few recrystallization texture prediction models of f.c.c. metals have been proposed[5]-[6]. Many studies have been carried out on the formation of recrystallization textures in aluminum alloys and rexrystallization textures of aluminum have not been analyzed as the Taylor factors. Furthermore, after primary recrystallization, two types of grain growth models are developed, one is based on a grain boundary character theory[7]-[8] and the other is based on a <111> 40° rotation theory[9]-[10]. Thus, hardening model considering collinear interaction slips[11]-[12] and sessile slips was formulated to estimate the density of accumulated dislocations in f.c.c. metals. For a while, many precise models have been proposed to predict rolling texture of f.c.c. metals[13]-[15]. Then, using the Taylor and Sachs deformation texture prediction model, dislocations at grain boundaries of pure aluminum was quantified. Next, coincidence site lattice grains were investigated from observed cold rolling texture. Recrystallization prediction model that depended on strain-induced boundary migration and coincidence lattice theory was developed.

2. Recrystallization texture prediction model

2.1. Hardening model of f.c.c. metals

Many hardening model of f.c.c. metals were proposed[16]-[17]. Inoko proposed tensile and compression cross slip models of aluminum[18]. We guess that this phenomena is the same as collinear interaction slips[11].When a single aluminum crystal is stretched in the <100> direction, two slip systems cross in the same direction. This slip is called repulsive collinear interaction slip. On the other hand, when a single aluminum crystal is stretched in the <111> direction, two slip systems cross in the opposite directions. This cross slip is called attractive collinear interaction slip. Figure 1 shows slip system analysis of <001>-direction-stretched and <111>-direction-stretched single crystals of f.c.c. metals. By adding attractive collinear interaction slips, the total number of collinear interaction slips was estimated as

$$\Gamma_{collinear} = \sum d\gamma^{attractiv} \tag{1}$$

Adding collinear interaction slips ($\Gamma_{collinear}$), Lomer Cottrell slips (Γ_{lomer}) and Hirth lock slips (Γ_{hirth}),the number of accumulated dislocations was estimated using equation (2). Coefficient of equation (2) was referred from Devincrea's research[12].

$$\Gamma=\Gamma_{collinear}+\Gamma_{lomer}/5+\Gamma_{hirth}/10 \tag{2}$$

$$\tau_y = 13.0\sqrt{\Gamma} \tag{3}$$

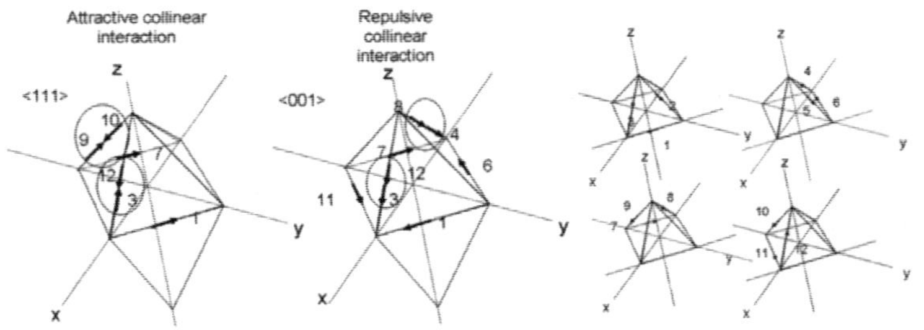

Figure 1:Attractive and repulsive collinear interaction slip of f.c.c. metals

Obeying equation (3), critical resolved shear stress (τ_y) was calculated. Figure 2 shows stress-strain curve prediction accuracy of a single f.c.c. crystal.

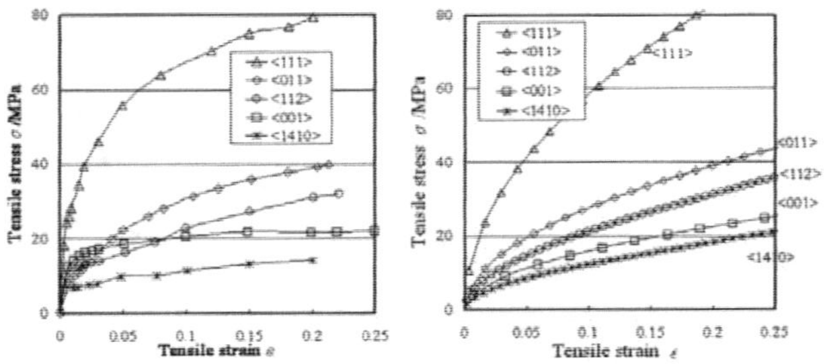

Figure 2: Stress-strain curve prediction accuracy of a single f.c.c. crystal
(Left :Inoko observed[18], Right:Predicted)

2.2. Rolling texture analysis

The plastic strain rate $\boldsymbol{D}^P$ and plastic spin $\boldsymbol{W}^P$ can be defined using Eqs.(4) and (5).

$$D_{ij}^{P} = \dot{\gamma}(a_i b_j + a_j b_i)/2 = P_{ij}\dot{\gamma} \,. \tag{4}$$

$$W_{ij}^{P} = \dot{\gamma}(a_i b_j - a_j b_i)/2 = Q_{ij}\dot{\gamma} \,, \tag{5}$$

where $\dot{\gamma}$, $\boldsymbol{a}$, $\boldsymbol{b}$ and are the slip rate, the unit vector normal to the slip plane and the unit vector in the slip direction, respectively. The f.c.c. crystal lattice has 12 slip systems. The strain equilibrium, the resolved shear stress of each slip system, the rate-dependent rule and boundary condition are expressed using Eqs.(6)-(9), respectively. Equation (6) implies that the sum of the slip values $\gamma^{(k)}$ multiplied by $P_{ij}^{(k)}$ for each slip system in a crystal is equal to the sum of the microstrains $\boldsymbol{\varepsilon}$. Equation (7) gives the relationship between the microstress $\boldsymbol{\sigma}$ and the resolved shear stress $\tau^{(k)}$ acting on the slip plane. Equation (8) is the Asaro rate-dependent rule and m is a material parameter. For the boundary conditions, the microstrain $\boldsymbol{\varepsilon}$ is assumed to be equal to the macrostrain $\boldsymbol{E}$ in accordance with the Taylor theory, as shown in Eq.(9). Here, k is number of slip system.

$$d\varepsilon_{ij} = \sum_k P_{ij}^{(k)} d\gamma^{(k)} \tag{6}$$

$$\tau^{(k)} = \sum_i \sum_j P_{ij}^{(k)} \sigma_{ij} \quad (i,j=1,2,3) \tag{7}$$

$$d\gamma^{(k)} = \dot{\gamma}_0 * dt * \left(\frac{\tau^{(k)}}{\tau_y} \right)^{1/m} * sign(\tau^{(k)}) \tag{8}$$

$$\boldsymbol{\varepsilon} = \boldsymbol{E} \tag{9}$$

where, $\dot{\gamma}_0$, τ_y and dt are the reference slip strain rate, the critical resolved shear stress and the time increment, respectively. The critical resolved shear stress can be expressed as

$$\tau_y = S/M \tag{10}$$

IOP Conf. Series: Materials Science and Engineering **82** (2015) 012001 doi:10.1088/1757-899X/82/1/012001

where M is the average Taylor Factor for all orientations and $\bar{s}$ is equivalent flow stress. The Newton Raphson method was used to resolve equations (6)-(9) as the microstress tensor $\boldsymbol{\sigma}$ was unknouwn and the initial microstress tensor $\boldsymbol{\sigma}_0$ is assumed to be equal to the macrostress tensor $\boldsymbol{S}$. However, when the microstress was not resolved, the Sachs boundary condition was used. For the boundary conditions, the microstress $\boldsymbol{\sigma}$ is assumed to be equal to the macrostress $\boldsymbol{S}$ in accordance with the Sachs theory, as shown in Eq.(11).

$$\boldsymbol{\sigma} = \boldsymbol{S} \tag{11}$$

Because using combined the Taylor model and the Sachs model, rolling texture prediction accuracy was almost good as shown in Figure 3.

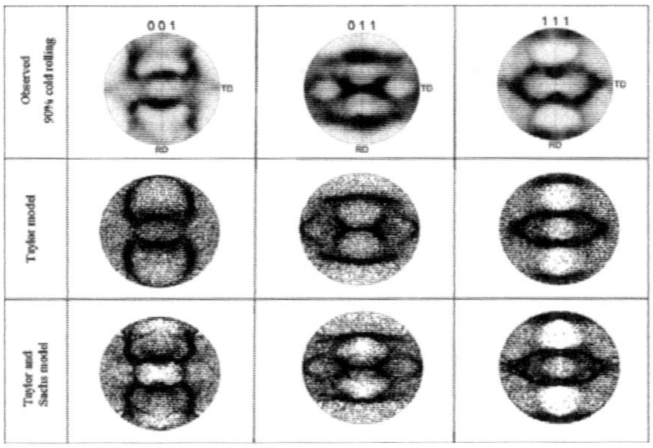

Figure 3:Rolling texture prediction accuracy of f.c.c. metals

2.3. Recrystallization texture analysis

Next, coincidence site lattice grains were investigated from observed cold rolling texture. Figure 4 shows coincidence site lattice analysis[19]. Grains in black had coincidence site lattice boundaries such as $\Sigma 5, \Sigma 7, \Sigma 13$. Figure 5 shows that selection result of grains which had coincidence site lattice boundaries from rolling grains measured by EBSD. Recrystallization texture of aluminum was not predicted only by the CSL theory.

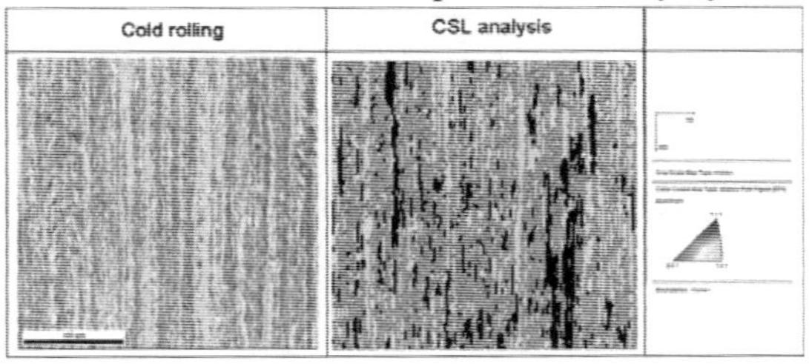

Figure 4:Coincidence site lattice (CSL) analysis (Left :observed, Right: analyzed)

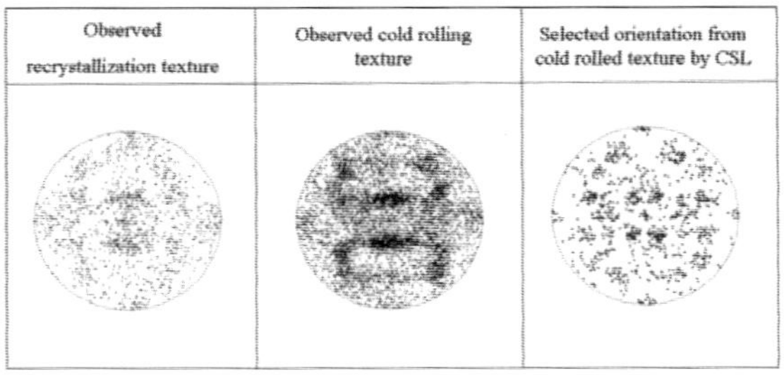

Figure 5:Observed and predicted recrystallized aluminum texture by CSL model

Strain-induced grain boundary migration starts from the boundaries of grains with a large dislocation density difference[20]-[21]. In accordance with the number of dislocations given by equation (2), grains with high dislocation densities were encroached on grains with low dislocation densities. Succeeding the oriented nucleation theory, selective growth theory was introduced. First, grains with a small number of accumulated dislocations were selected using the oriented nucleation model. Next, grains with the special coincident lattice shown were selected in seven types of low grain boundary energies, namely <100> and <111>[22].

Figure 6 shows the observed and predicted recrystallization textures. In our predicted recrystallized texture in Fig.6, each grain is drawn as a dot and all grains are assumed to be of the same size. Employing the universal recovery theory[23], our recrystallization texture prediction accuracy will be improved. However, the recrystallization texture observed by EBSD was almost in agreement with our predicted recrystallization texture.

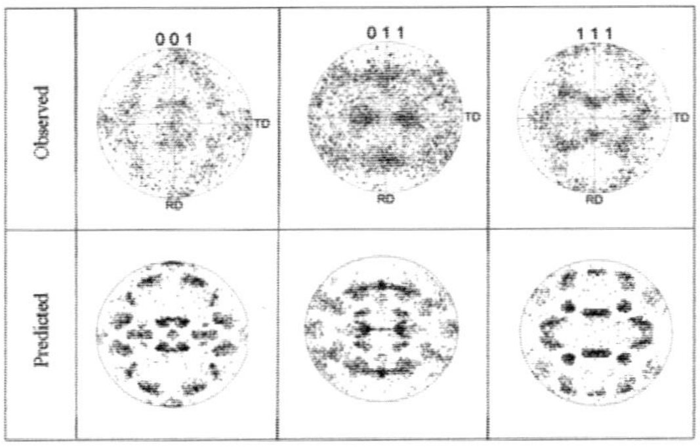

Figure 6: Observed and predicted recrystallized aluminum texture

3. Discussion of recrystallization texture

Many researchers have indicated that grains with <111> 40° rotation from a deformed matrix have high mobility. From the S orientation to the Cube orientation, the axis and angle of rotation are <1.08,1.00,1.23> and 48.6°, respectively and from the S orientation to the R orientation, <1.08,1.00,1.19> and 169.1°.Coincidence site lattice theory is ruled by the Brandon allowance. For example, one of 7 coincident lattices is <111> 38.2° and the Brandon allowance of an ideal coincident lattice was 5.7°, as calculated from

$$Vm = 15 \times \Sigma^{-0.5} \tag{12}$$

Sabin suggested that a combination of oriented nucleation and growth occurs in growth selection[24]. Engler observed the characteristics of both oriented nucleation and growth in obtained textures[25-26]. Our prediction model showed the same manner of grain selection with a small number of accumulated dislocations and the selective growth of grains with a coincident lattice.

4. Conclusions

The recrystallization texture of commercial pure aluminum was proposed on the basis of the number of accumulated dislocations calculated using the Taylor and the Sachs rolling texture prediction model with collinear interaction slips and sessile slips and using grain boundary characteristics. The prediction accuracy for the texture of cold-rolled and annealed aluminum was reasonable. To improve the prediction accuracy of recrystallization texture, grain-hardening model and rolling texture prediction model must be modified. And grain migration rule will be more precisely quantified.

References

[1]Humphreys F J 2004 Materials Science Forum **467-474** 107
[2]Hutchinson W B 1974 Metal Science **8** 185
[3]Morimoto T Yoshida F Kusumoto Y and Yanagida A 2012 ISIJ Int. **52** 592
[4]Morimoto T Yoshida F Kusumoto Y Oda M and Yanagimoto J 2012 Materials Transactions **53** 1837
[5]Engler O 1999 Texture and Microstructure **32** 197
[6]Sidor J Petrov P and Kestens L 2012 Material Science Forum **702-703** 611
[7] Liu Y C and Hibbard W R 1955 Journal of Metals 381
[8] Matsumoto K Shibayanagi T and Umakoshi Y 1997 Acta Mater. **45** 439
[9] Ito K 1993 J. Japan Inst. Light Metals **43** 285
[10]Huang Y and Humphereys F J 1999 Acta Mater. **47** 2259
[11]Madec R 2004 Science **301** 1879
[12]Devincrea B Hoc T and Kubin L P 2005 Materials Science and Engineering A Volumes **400–401** 182
[13]Raabe D 1998 Computational Materials Science WILEY –VCH New York 267
[14]Houtte P V Kanjarla A K Bael A V Seefeldt M and Delannay L 2000 European Journal of Mechanics A/ Solids **25** 634
[15]Tadano Y Kuroda M and Noguchi H 2012 Computational Materials Science **51** 290
[16]Franciosi P and Zaoui A 1982 Acta Metall **30** 2141
[17]Asaro R J and Needleman A 1985 Acta Metall **33** 923
[18] Inoko F Kashihara K Tagami M and Okada T 2011 J. Japan Inst. Light Metals **61** 410
[19]K.Ikeda,K.Tsumagari,F.Yoshida,E.Nakashima and H.Abe 2001 J. JILM **51** 119
[20]Koike Y 1959 J. Japan Inst. Light Metals **1** 3
[21]Beck P A and Sperry P 1950 J.Appl. Phys. **21** 150
[22]Kogtenkcova O Stranmel B Protasovo S Tsurekawa S and Watanabe T 2005 Z. Metallkd **96** 216
[23]Humphreys F J 1997 Acta mater. **45** 4231
[24]Sabin T J Winther G and Jensen D J 2003 Acta Materiailia **51** 3999
[25]Engler O Hirsch J and Lucke K 1995 Acta metal. mater. **43** 121
[26]Engler O and Lucke K 1992 Scripta Metallurgica **27** 1527
[27]Alanker A and Mastorakos N Field D P 2012 J.Engineering Materials and Techonology **134** 02018

Changes of inclusion, texture and magnetic property of non-oriented Si steel treated by Ca alloy

X Lv, F Zhang and X Chen
Si Steel Department, Baoshan Iron & Steel Co., Ltd, Shanghai, 200941, China
E-mail: zhangfengbaosteel@163.com

Abstract: Based on the industrial production of non-oriented Si steel, Ca treatment by Ca alloy adding during the RH refining process was studied. The changes of inclusion, crystal texture and microstructure, and its effect on magnetic properties of final steel sheets were analyzed. The results showed that, in present work, Ca treatment can improve the texture proportion of {110} and {111} significantly, and the formation of MnS and AlN inclusions were restrained. Meanwhile, the recrystallization effects of hot rolled strip get bad and the fiber structure enhanced obviously. The grain size of finished steel sheets increased as the increase of Ca alloy adding amount quickly, and then decreased. Compared with the non-Ca treatment charge, the numbers of inclusions whose size below 1.0μm will decrease by 68.06%, 87.50% and 94.94%, the texture proportion of {110} and {111} was 30.3%, 39.1%, 17.6% and 2.8%, 5.5%, 20.6%, while the correspondent Ca alloy adding amount is 0.67 kg/t steel, 1.00 kg/t steel and 1.67 kg/t steel, respectively. In addition, the core loss gradually decreases to a stable level as the increasing of Ca added, and the magnetic induction decreases quickly after slow increasing, respectively. The optimal Ca treatment mode depends on the chemical compositions of steel grades.

1 Introduction

The demands for energy saving and environment protection are more and more increasing recent years. Various kinds of EI core, motor, compressor are being developed toward smaller scale, and higher efficiency [1]. Thus, as raw material of the cores, the non-oriented Si steel with excellent magnetic properties [2], is required to fulfill the strong demand for energy saving and high efficiency of the electric power consuming products. The magnetic properties of non-oriented Si steel mainly depend on the grain size of ferrite, crystallographic texture, and the inclusions in the steel [3-4].

The present work studied the Ca treatment by adding Ca alloy in the liquid steel during refining of Si steel in industrial production process. The morphology and size distribution of inclusions were observed. The behavior change of inclusions in Si steel after Ca treatment and its influence on the magnetic properties of finished product was discussed.

2 Experimental

2.1 Production Process

The Ca treatment was carried out by adding Ca alloy to the liquid steel during RH refining. The main composition of Ca alloy is CaSi which size is 10～50mm. The requirement of some technique parameters during LD tapping and RH refining is listed in Table 1. The main chemical compositions of steel used in present work are shown in Table 2.

Table 1 Requirement of technique parameters during LD tapping and RH refining

Tapping			RH refining	
Temperature	Free oxygen	T·Fe + MnO	Temperature	Ca adding
1640±20 ℃	0.04%-0.12%	≤5%	1600±20℃	0-2.0 kg/t steel

Table 2 Main chemical compositions of steel used in present work, mass%

C	Si	Mn	P	S	Al	Ca
0-0.005	0-2.1	0.2-1.0	0-0.2	0-0.005	0.2-1.0	0-0.01

2.2 Experimental Methods
Steel samples were taken in steelmaking process. The correspondent steel sheets of the same batch were taken. The microstructures were observed by SEM and the grain size of finished sheets was measured by intersection method. The morphologies, chemical compositions, quantities, size distribution of non-metallic inclusions were investigated by SEM and EDS. Each specimen was observed for 10 connected fields of view, an image analysis software was used to obtain the quantity and the size distribution of the non-metallic inclusions. The magnetic properties of finished steel sheet were measured by the Epstein frame method.

3 Results and discussion

3.1 Quantity and Size Distribution of Extracted Non-metallic Inclusions
The results showed that the size and numbers of non-metallic inclusion were increased in 1000 times, with the increase of Ca alloy adding amount. When Ca alloy adding amount reached 1.67 kg/t steel, the average size of non-metallic inclusions was largest and the numbers was most, and the numbers of non-metallic inclusions decreased at first and then increased in 5000 times, and some of the non-metallic inclusions was relative larger. The size distribution of these non-metallic inclusions was obtained from SEM photos by imagine analysis software, as shown in Fig.1. The size of most inclusions in finished Si steel sheet without Ca treatment is that whose size was smaller than 1.0 μm, the numbers was $1.6\times10^7/mm^3$; and the numbers of inclusions in finished Si steel sheet with Ca treatment sharply decreased, especially the fine inclusions whose size was smaller than 1.0 μm. Furthermore, the fine inclusion was still decreased with the increase of Ca alloy adding amount, but the change was rather small. Compared with the charge without Ca treatment, the numbers of inclusions whose size below 1.0μm decrease by 68.06%, 87.50% and 94.94%, while the correspondent Ca alloy adding amount is 0.67 kg/t steel, 1.00 kg/t steel and 1.67 kg/t steel, respectively.

3.2 Types of Non-metallic Inclusions
The fine inclusions MnS and Cu_xS will not disappear even the Ca alloy adding amount reached to 1.67 kg/t steel and is relate to the lower Ca element, and the weak combine of Ca and S. The precipitation of fine MnS and Cu_xS is either single, or precipitation as the core of AlN and combined to the component inclusions [5-6]. The fine inclusions of MnS and Cu_xS exist single, or combined with AlN and CaS under different Ca treat condition in the range of 1-5 μ m, and the CaS inclusion has been found after Ca treat in the above size range. That showed that some Ca will combine with S, and the S can't be floated and removed immediately after the Ca alloy adding into the liquid steel. When the Ca alloy adding amount reached 0.67 kg/t steel, there are still some MnS and AlN appeared in the finished steel sheets; when the Ca alloy adding amount reached 1.00 kg/t steel or 1.67 kg/t steel , the main type of inclusions is oxysulfide of Ca in the finished steel sheets. Because of the distortion and extension of MnS in the processes of cold rolling and annealing, the recrystallization of finished steel will be retard seriously, and decrease the magnetic properties of finished steel sheets. Meanwhile, according to the Zener formula [7], the inhibitor force on the grain

boundary is proportional to the volume fraction of inclusions, and inverse to the average radius of inclusions. Thus, from the viewpoint of improving the magnetic properties, in order to obtain the excellent controlling effects of inclusions, it is better to choose the suitable Ca treat method to decrease the numbers and increase the size of inclusions [8].

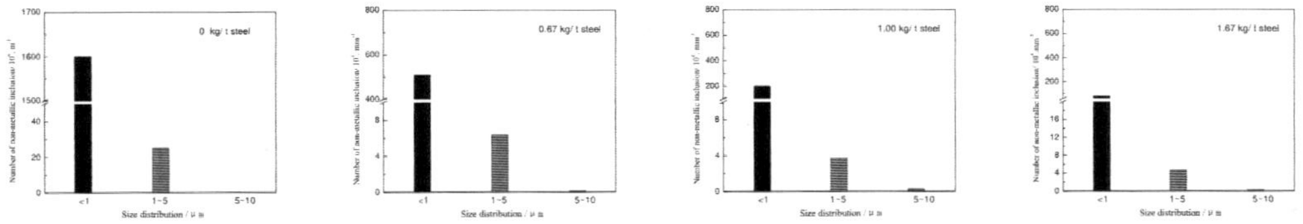

Fig.1 Size distribution of inclusions from Si steel sheets with and without Ca treatment

3.3 Microstructures Heredity

The relation between grain size before and after SRA and the Ca concentration is shown in Fig 2. The grain size of steels with Ca alloy adding amount was 0.67 kg/t steel and 1.00 kg/t steel are almost same before SRA (31.5 μm and 34.3 μm, respectively), but the later was more easy to grow after SRA (38.4 μm and 48.7 μm, respectively). But when Ca concentration reached 1.67 kg/t steel, the grain size of finished product are almost same as that without Ca treatment probably due to the grain refinement effect of CaS.

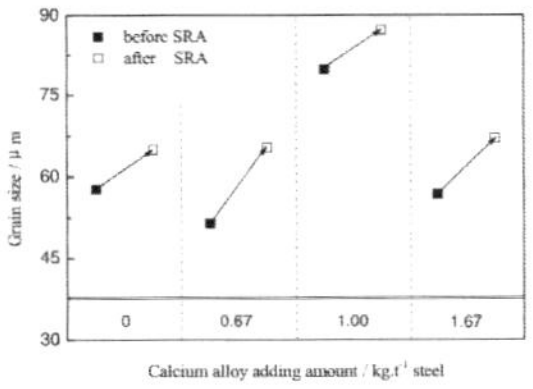

Fig.3 Relation between grain size before and after SRA and the Ca adding amount

3.4 Crystal Texture of Finished Steel Sheets

It is harmful to the magnetic properties, due to the {111} plane of more grain is parallel to the rolling plane. So, in order to improve the magnetic properties of finished steel sheets, the heat treating should be adopted to improve the proportion of the grain whose {100} and {110} plane paralleled to the rolling plane, and replace the {111} and {112} plane by the {100} and {110} plane. In present work, we can see that, the proportion of {100} is decreased, and the proportion of {111} is increased with the increase of Ca alloy adding amount, and the proportion of {110} is increased slowly at first, and then decreased sharply with the increase of Ca alloy adding amount. The texture proportion of {110} and {111} was 30.3%, 39.1%, 17.6% and 2.8%, 5.5%, 20.6%, while the correspondent Ca alloy adding amount is 0.67 kg/t steel, 1.00 kg/t steel and 1.67 kg/t steel, respectively. The ODF graphs are showed in Fig 3.

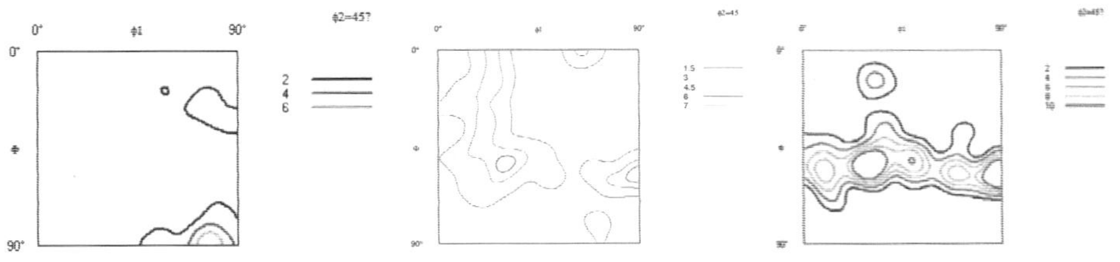

Fig. 3 Change of texture proportion of Si steel with Ca adding amount

3.5 Magnetic Property of Finished Steel Sheets

The core loss and magnetic induction flux density were measured for Si steel sheets with different Ca treatment condition before and after SRA as shown in Fig. 4. From Fig. 4, we can see that the core loss decreases with the increase of Ca concentration and tends to stable, the magnetic induction flux density increases with the increase of Ca concentration and then decreases. This is because that the number of fine inclusions decreased by Ca treatment so the grains grown easily during SRA, resulted the decrease of core loss. But the mechanism of magnetic induction flux density change with Ca treatment seems complicated, it needs further studies [9]. Moreover, it is clear that the core loss after SRA decreased quickly and attained a minimum value when Ca concentration was at 1.00 kg/t steel then increased. This can be explained according with the relation between grain size and Ca concentration, as shown in Fig.2.

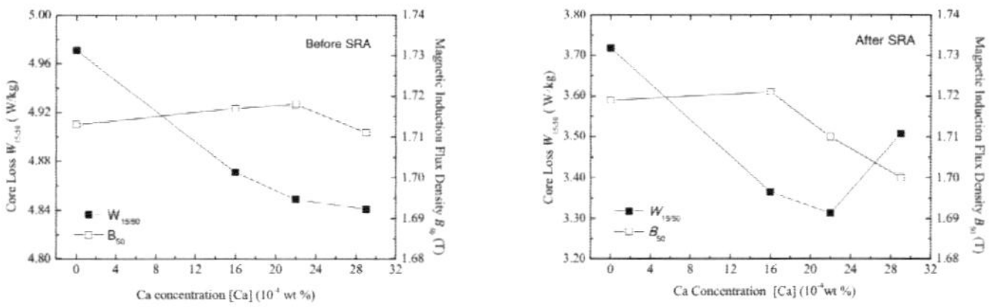

(a) Before SRA (b) After SRA

Fig. 4 Change of magnetic properties of Si steel with Ca adding amount before and after SRA

4. Conclusions

Following conclusions can be summarized: (1) Ca treatment can improve the texture proportion of {110} and {111} significantly, and the formation of MnS and AlN inclusions was restrained. Meanwhile, the grain size of finished steel sheets increased as the increase of Ca alloy adding amount quickly, and then decreased. (2) Compared with the non-Ca treatment charge, the numbers of inclusions whose size below 1.0μm will decrease by 68.06%, 87.50% and 94.94%, the texture proportion of {110} and {111} was 30.3%, 39.1%, 17.6% and 2.8%, 5.5%, 20.6%, while the correspondent Ca alloy adding amount is 0.67 kg/t steel, 1.00 kg/t steel and 1.67 kg/t steel, respectively. (3) In addition, the core loss gradually decreases to a stable level as the increasing of Ca added, and the magnetic induction decreases quickly after slow increasing, respectively. The optimal Ca treatment mode depends on the chemical compositions of steel grades.

References

[1] Yoshihiko O, et al. 2003 *J. Mag. Mat.* **254-255** pp. 361-363.
[2] Takeshi K. 2003 *CAMP-ISIJ* **16** p. 626.
[3] Ben X, et al. 2005 *Mater. Sci. Forum* **475-479** pp. 2279-2282.
[4] Shiozaki M and Kurosaki Y 1989 *J. Mater. Eng.* **11** pp. 37-43.
[5] Hashi W, et al. 2008 *United States Patent* US2008/0118389 A1, 5, 22.
[6] Nakayama T, et al. 2001 *J. Mag. Mag. Mat.* **234** pp. 55 - 61.
[7] Akamatsu S, et al. 1992 *ISIJ Int.* **32** pp. 75-282.
[8] Xuejun L, et al. 2012 *Special Steel* **33** pp. 22-25.
[9] He Z, "Electric Steel" Metallurgical Industry Press, Beijing, 1996.

IOP Conf. Series: Materials Science and Engineering **82** (2015) 012003 doi:10.1088/1757-899X/82/1/012003

Quantifying texture evolution during hot rolling of AZ31 Twin Roll Cast strip

S Gorelova[1,2] and H Schaeben[3]
[1] Leibniz-Institut für Polymerforschung Dresden e. V., Dresden, Germany
[2] TU Bergakademie Freiberg, Institut of Metal Forming, Freiberg, Germany
[3] TU Bergakademie Freiberg, Geophysics and Geoinformatics, Freiberg, Germany
E-mail: gorelova@ipfdd.de and schaeben@tu-freiberg.de

Abstract. Multi-pass rolling experiments with an AZ31 Twin Roll Cast (TRC) alloy were performed on an industrial scaled four-high rolling mill. Within the rolling with an intermediate annealing the evolution of texture was investigated. To quantify the extent of preferred crystallographic orientation experimental X-ray pole figures were measured after different process steps and analyzed using the free and open Matlab® toolbox MTEX for texture analysis. The development of the fiber texture was observed and analyzed in dependence on rolling conditions. In the initial state the specimen exhibits a texture composed of a weak basal texture and a cast texture with {0001}-planes oriented across the rolling direction. During the following rolling process a fiber texture was developed. The expected strength increment of the fiber texture was quantitatively confirmed in terms of volume portions of the orientation density function around the fiber and in terms of the canonical parameters of fitted pseudo Bingham distributions. On the results of this work a model for prediction of the texture evolution during the strip rolling of magnesium in the examined parameter range was developed.

1. Introduction

The reducing of mass plays an important role in a development of modern construction technologies. Magnesium is the lightest metal among the construction materials because of its low density. The disadvantage of this material is its low formability at room temperature due to a limited number of effective deformation systems. Basal slip is the dominant slip system during the deformation, resulting in the formation of a basal texture during deformation and reducing the resulting mechanical properties. At the Institute of Metal Forming of TU Bergakademie Freiberg a new technology process chain of Twin Roll Casting (TRC) with subsequent rolling of Mg-alloys has been developed [1]. The TRC-plant with a subsequent four-high reverse strip rolling mill in an industrial scale allows us to investigate the material properties in a production like process.

In this paper the texture evolution of the rolled TRC AZ31 strip aiming at a quantification of the evolving fiber texture were investigated. Understanding of the texture and microstructure evolution is essential to predict and control it during rolling of TRC strips of Mg alloys in the future.

Definitions, conventions, and the notation with respect to texture analysis closely follow the publication introducing MTEX [2].

All computations were done with the free and open source Matlab® toolbox MTEX 3.3.2 for texture analysis, which can be downloaded from http://mtex.googlecode.com. It employs fast Fourier transforms for S^2, $S^2 \times S^2$, and SO (3) as encoded in the public domain software NFFT, cf. http://www-user.tu-chemnitz.de/~potts/nfft/.

2. Problem

The subject of this paper is the hypothesized systematically increasing extent of preferred orientation of a fiber texture during rolling of TRC-material in a production like process. The laboratory experiments cannot always be transferred to production processes, therefore, the first objective of the investigation is to demonstrate the texture evolution in the production like process. The second objective is to quantify the texture evolution. The texture was represented in terms of orientation density functions (ODFs) derived from the experimental diffraction pole figures, and the hypothesized evolution of a fiber component due to progressive deformation was analyzed in terms of characteristics of the ODFs and in particular in terms of the canonical parameters of the fitted pseudo Bingham orientation density functions.

3. Experiments

The initial material was Twin-Roll Cast strip of AZ31 (3% Al, 1% Zn, Mg) alloy with a thickness of 5 mm. The Twin Roll casted coil with a width of 600 mm was used for the rolling experiment. Before rolling the TRC coil was homogenized above 400 °C in a circulation air furnace and roughed down in two passes to a thickness of 2.45 mm. Then it was intermediately annealed at 360 °C and finish-rolled in two passes to the final thickness of 1.5 mm. The rolling speed was 80–120 m/min with a rolling temperature in the range of 270–360 °C. The deformation degree for every pass was 20–35% for rough down and for finish-rolling steps. The roughed and finish-rolling steps were carried out on coil material. The coil was set on the coiler and reversible rolled on the four-high rolling mill. The texture evolution was investigated in the chosen process conditions because of importance and relevancy for the safety of reversing rolling. It provides the favorable combination of rolling temperature, rolling speed, deformation degree and its effect of the microstructure and geometrical properties of the final strip, which is important for the product quality [1, 3].
The texture was measured at the Institute of Materials Science (TU Bergakademie Freiberg) with X-ray goniometry at the middle layer on RD-TD cross sections of the rolled specimens. Due to the coarse grained microstructure of initial stage stack of four ND-RD cross sections were measured to improve the statistics. The measured data was analyzed with the free and open Matlab® MTEX for texture analysis [2].

4. Descriptive texture analysis

The fitted pole figures from the experiment are shown in Fig. 1.The pole figures of the texture of initial material are shown in Fig. 1a. The {0001}-pole figures have two different maximums, indicating a weak basal texture within the deformed parts with the other maximum transversal to the rolling direction, showing the undeformed parts of the material cast state [4]. During roughing down the basal texture increased due to an increase in deformation (see Fig. 1b and c). {10-10}-pole figures show beginning of a fiber texture development along the (0001)-axis, which is more pronounced with a higher deformation degree.
During deformation the dislocation slips along the basal slip systems primarily, although other slip systems are activated at this temperature [5]. It leads to evolution of basal texture with the increasing of deformation degree. The effect is shown on the pole figures after the 1st and 2nd passes. After the 2nd pass the coil was intermediate annealed. During this process static recrystallization was activated. After the heat treatment the basal texture is more distinct with the increase of the fiber portion. This sharpening depends on the incomplete dynamic recrystallization after two roughing passes. Some not recrystallized grains saved a dislocation portion after the deformation. The activation of the static recrystallization during the intermediate annealing leads to the initial grain growth and to rotation and coalescence of new grains with subgrains [6, 7]. It provides more intensive maximum of basal texture after heat treatment (see Fig. 1d). The same mechanism of the evolution of the basal texture with increasing of fiber portion was activated during 3rd and 4th finish-rolling passes (see Fig. 1e and f). It provides more intensive basal texture with homogeneous fiber distribution.

IOP Conf. Series: Materials Science and Engineering **82** (2015) 012003 doi:10.1088/1757-899X/82/1/012003

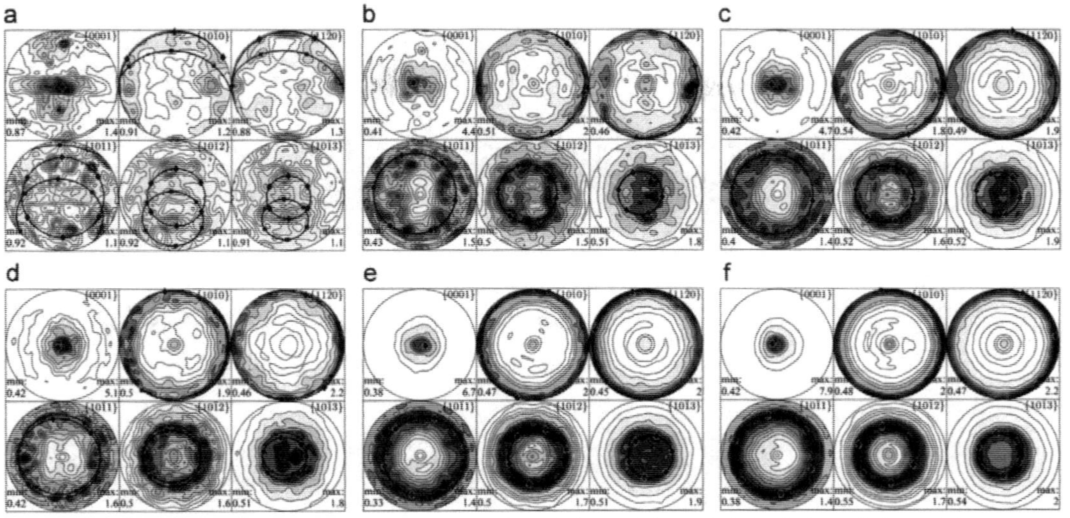

Figure 1 : Fitted X-ray pole figures of six crystal forms h={0001}, {10-10}, {11-20}, {10-11}, {10-12}, {10-13} displaying (a) texture in homogenized TRC-strip; (b) texture after the 1st pass; (c) texture after the 2nd pass; (d) texture after the intermediate annealing; (e) texture after the 3rd pass; (f) texture after the 4th pass.

5. Quantitative texture analysis

The experimental X-ray pole figures of the six specimens reveal the systematically increasing strength of a fiber texture as to be expected from the successive deformation steps in this process window. Next, the texture evolution, in particular the increasing strength of the fiber portion, is confirmed by means of quantitative texture analysis with MTEX. First, an orientation density function (ODF) is estimated from the experimental pole figures and characterized by a few quantities like uniform portion, texture index and entropy. The extent of an increasingly strong fiber texture is measured in terms of volume portions of the ODF around the fitted fiber. Then a model ODF, here a fiber pseudo Bingham orientation density, is fitted. The pseudo Bingham distribution is the most versatile model distribution, which able to present several different patterns of preferred orientation including asymmetric unimodal textures and incomplete and asymmetric fiber textures. Its four canonical shape parameters measure the extent of anisotropy κ, and of a bipolar B, of a fiber C, and of a spherical proportion S, respectively. The absolute extent of a fiber texture is captured by $2\kappa C$ and confirms again the increasing strength of a fiber texture along the successive deformation.

5.3. Practical application to rolled AZ31 TRC strip

The fitted canonical Bingham parameters κ, B, C, S for the six sets of experimental pole figures are shown in Table 1. Their systematic evolution, especially the systematic increasing of $2kC=k_3-k_2$ capturing the extent of a fiber texture, is consistent with the increasing volume portions around the fibers with respect to both the orientation density estimated with the experimental pole intensities and the fitted pseudo Bingham orientation density for all sizes of neighborhoods from 2° to 32°. The quantitative texture analysis as summarized in Table 1 is consistent in all its figures and clearly confirms the initial hypothesis of a fiber texture of increasing strength. The fitted pole figures are shown if Fig.2.

The details for canonical parametrization of the Bingham orientation density and fitting a pseudo Bingham ODF are described in [8].

All calculations were done with MTEX's defaults wherever they apply.

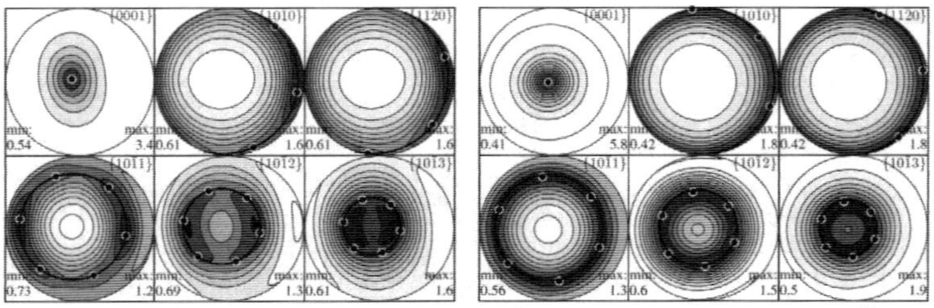

Figure 2 : Pseudo Bingham fitted h-pole densities of specimen 1st pass (left) and of specimen 4th pass (right) augmented with their modal directions (black dots) and the fitted ((0001), $r_{mode(0001)}$)-fiber (black circle).

Quantitative texture analysis in terms of fitted Bingham parameters and volume portions						
process stage	homogenized TRC-strip	1st pass	2nd pass	intermediate annealing	3rd pass	4th pass
anisotropy κ	7.4129444e+000	8.6070332e+000	1.1570152e+001	1.0603488e+001	1.4995928e+001	2.2429046e+001
bimodal B	1.8575467e-003	2.0642562e-003	5.6262756e-003	1.1471018e-002	2.7158914e-003	1.6682365e-003
fiber C	2.2812029e-003	1.4202273e-001	1.4013803e-001	1.6351103e-001	1.3179895e-001	1.0482021e-001
spherical S	9.9586125e-001	8.5591302e-001	8.5423570e-001	8.2501796e-001	8.6548515e-001	8.9351156e-001
$2\kappa C = \kappa_3 - \kappa_2$	**3.3820860e-002**	**2.4447886e+000**	**3.2428365e+000**	**3.4675745e+000**	**3.9528952e+000**	**4.7020344e+000**
vol($\mathcal{U}_2(\mathbf{c},\mathbf{r})$) epf–fit	8.5897016e-004	2.6038779e-003	2.8832333e-003	3.0296859e-003	3.9655463e-003	4.7405316e-003
vol($\mathcal{U}_2(\mathbf{c},\mathbf{r})$) Bingham–fit	7.6937037e-004	2.0062491e-003	2.5932498e-003	2.5872391e-003	3.3573408e-003	4.3855930e-003
vol($\mathcal{U}_4(\mathbf{c},\mathbf{r})$) epf–fit	3.5288945e-003	1.0371037e-002	1.1574777e-002	1.2061598e-002	1.5784445e-002	1.8985669e-002
vol($\mathcal{U}_4(\mathbf{c},\mathbf{r})$) Bingham–fit	3.1914850e-003	8.2812889e-003	1.0674593e-002	1.0658407e-002	1.3774106e-002	1.7873499e-002
vol($\mathcal{U}_8(\mathbf{c},\mathbf{r})$) epf–fit	1.3085129e-002	3.5107493e-002	4.0336222e-002	4.1352302e-002	5.3651277e-002	6.5375058e-002
vol($\mathcal{U}_8(\mathbf{c},\mathbf{r})$) Bingham–fit	1.2179298e-002	3.1032508e-002	3.9602588e-002	3.9655117e-002	5.0502181e-002	6.4041872e-002
vol($\mathcal{U}_{16}(\mathbf{c},\mathbf{r})$) epf–fit	4.9241187e-002	1.1603805e-001	1.4264877e-001	1.4132609e-001	1.7682212e-001	2.0248692e-001
vol($\mathcal{U}_{16}(\mathbf{c},\mathbf{r})$) Bingham–fit	4.8781337e-002	1.1553517e-001	1.4215937e-001	1.4371419e-001	1.7400054e-001	2.0555045e-001
vol($\mathcal{U}_{32}(\mathbf{c},\mathbf{r})$) epf–fit	1.6883367e-001	3.1859113e-001	3.8515509e-001	3.9087263e-001	4.3814811e-001	4.4050111e-001
vol($\mathcal{U}_{32}(\mathbf{c},\mathbf{r})$) Bingham–fit	1.7161651e-001	3.2756035e-001	3.7287372e-001	3.8191330e-001	4.2306807e-001	4.5791790e-001
uniform portion	8.2317539e-001	3.4492614e-001	3.4154289e-001	3.3959066e-001	2.9149666e-001	3.2342474e-001
texture index epf–fit	1.0157083e+000	1.5029599e+000	1.7064079e+000	1.7536885e+000	2.0234547e+000	2.1647929e+000
texture index Bingham–fit	1.0051204e+000	1.3282550e+000	1.5303073e+000	1.5689386e+000	1.8269390e+000	2.0691460e+000
entropy epf–fit	-6.3188062e-003	-1.9561308e-001	-2.5222319e-001	-2.6416862e-001	-3.3428893e-001	-3.4637583e-001
entropy Bingham–fit	-2.4223380e-003	-1.2991566e-001	-1.9599677e-001	-2.1148514e-001	-2.8447396e-001	-3.7027462e-001

Table 1 : Summary of texture analysis of specimens homogenized TRC strip, 1st pass, 2nd pass, intermediate annealing, 3rd pass, and 4th pass, comprising the L^1-fitted canonical parameters (i) extent of anisotropy κ, and proportional (ii) bimodal B, (iii) fiber C, and (iv) spherical S , and, most instructively, the extent of a fiber preferred orientation $2\kappa C$ (bold) of the pseudo Bingham orientation den- sity function, and then for both the orientation density function estimated from the experimental pole figures (epf–fit) and the fitted pseudo Bingham orientation density function (Bingham–fit) volume proportions around the (c, $r_{mode(c)}$)-fiber, and uniform portion, texture index and entropy.

6. Conclusions

The evolution of texture was investigated and quantified during the rolling of TRC AZ31 strip in the production-like process. The rolling process resulted in an increasingly strong $(\mathbf{c},\mathbf{r_o})$-fiber texture with $\mathbf{r_o} \approx \mathbf{z}$. Texture analysis of six sets of experimental X-ray pole figures referring to consecutive rolling passes actually reveals systematically increasing volume portions around the $(\mathbf{c},\mathbf{r_o})$- fiber with respect to the orientation densities estimated with the sets of experimental pole intensities. Moreover, fitting pseudo Bingham orientation densities the systematic evolution of their canonical parameters κ, B, C, S and especially of $2kC=k_3-k_2$ is consistent with the evolution of the volume portions around the $(\mathbf{c},\mathbf{r_o})$- fiber. Thus the initially hypothesized evolution towards an increasingly pronounced $(\mathbf{c},\mathbf{r_o})$- fiber texture is confirmed twofold, (i) by the volume portions about the $(\mathbf{c},\mathbf{r_o})$- fiber of the orientation density functions fitted to the six sets of experimental pole figures as well as the fitted pseudo Bingham orientation densities, and (ii) by the extent $2kC=k_3-k_2$ of preferred crystallographic fiber orientation quantified in terms of canonical parameters κ, B, C, S of the fitted pseudo

IOP Conf. Series: Materials Science and Engineering **82** (2015) 012003 doi:10.1088/1757-899X/82/1/012003

Bingham orientation distribution. Thus, beyond visualization the method enables us to define and characterize a fiber texture quantitatively in terms of its parameters. The achieved knowledge enables the prediction of the texture evolution in the production-like process, which is important for the optimization of the rolling process in the future.

References

[1] Ullmann M, Oswald M, Gorelova S, Kawalla R, Vogt H-P 2012 *Steel Res. J.: Metal Form.* 855–858
[2] Hielscher R, Schaeben H 2008 *J. Appl. Cryst.* **41**, 1024–1037
[3] Gorelova S, Ullmann M, Oswald M, Kawalla R 2013, *KEM* **554-557**, 274-279
[4] Chalmers B 1968 Principles of Solodification (Metallugriya, Moscow)
[5] Schmidt C 2011 Textur- und Gefügeentwicklung bei der Umformung von Magnesium-Gießwalzband (PhD-Thesis, Freiberg)
[6] Li J C M 1962 *J.Appl.Phys.* **33**, 2958-2965
[7] Rios P R, Siciliano F, Sandim H R Z, Plaut R L, Padilha A F 2005 *Mater. Res.* **8**. 225-238
[8] Gorelova S, Schaeben H, Kawalla R 2014 *Mater. Sc. & Eng.* A **602** 134–142

IOP Conf. Series: Materials Science and Engineering **82** (2015) 012004 doi:10.1088/1757-899X/82/1/012004

Measuring twin dependent triple junctions from a single section plane

GB Hardy[1] and DP Field[2]
School of Mechanical and Materials Engineering, Washington State University PO BOX 642920, Pullman, WA 99164-2920, USA
[1]gradenhardy@gmail.com, [2]dfield@wsu.edu

Abstract. Given that polycrystalline triple junctions are significant contributors to material properties, they are frequently becoming the focus of emerging research. Despite this interest, the tools to quickly and quantitatively analyze triple junction textures remain severely limited. To enable characterization of triple junctions on a large scale, the parameters, space, and conventions of twin dependent triple junction distributions have been developed. In addition, by adopting grain boundary stereological techniques, triple junction distributions have been generated from a single section plane for triple junctions containing a coherent twin boundary. This methodology has been validated using simulated microstructures and, with further experimental development, will provide insight into actual triple junction structures. This technique also establishes the foundation for a generalized non-twin dependent approach in the future.

1. Introduction

A triple junction is a polycrystalline microstructural feature formed by the common intersection of three grains. As defined in this work, a fully characterized triple junction consists of three crystal lattices, three grain boundaries, and a triple line. Numerous studies indicate that triple junctions possess a host of junction-specific properties. These properties include exerting a drag effect on grain boundaries during grain growth [1], serving as prime locations for heterogeneous solute segregation [2], providing locations of rapid diffusion [3], and acting as favorable sites for heterogeneous nucleation [4]. This list is not all encompassing but provides a glimpse of the influence of triple junctions on material properties. These properties also appear to be character dependent, though in most cases the relationship is not well quantified. Triple junction distributions also demonstrate unique traits from grain boundary character distributions. It has been shown that the triple junction distribution is capable of describing properties of the grain boundary network, whereas the more common grain boundary character distribution is not [5]. It has been further shown that the triple junction distribution accounts for approximately 75% of the correlations within the grain boundary network [6]. The importance of these network descriptive properties is underscored by the consideration that many material properties are likely the result of the grain boundary network as opposed to isolated grain boundaries.

Triple junctions are three-dimensional structures, which complicates experimental characterization. Triple junctions are also composed of several geometrical features which requires a preponderance of measurements for minimal statistical reliability of distribution results [7]. Currently, this dimensionality requires a quantity of measurements beyond what is experimentally practical from three dimensional characterization methods. Many times a simplifying model is utilized to reduce the junction dimensionality. While these models have at times been beneficial, they lack resolution and frequently rely heavily on poor assumptions such as the coincident site lattice model [8]. Clearly, if integrated materials engineering is to be performed, there is a distinct need for a relatively rapid and inexpensive method for fully characterizing triple junctions on a large scale.

2. Triple junction parameters and space

2.1. Reducing the triple junction parameters

As explained, a fundamental difficulty of characterizing triple junction distributions is their high number of parameters. These parameters require too many measurements for practicality. A triple junction may be parameterized by the following macroscopic parameters: the three crystal lattices by the orientations g_A, g_B, and g_C, the three grain boundaries by the normal vectors $\boldsymbol{n}_1$, $\boldsymbol{n}_2$, and $\boldsymbol{n}_3$, and the triple line by the vector $\boldsymbol{t}$. The crystal orientations are given by the Euler angles φ_1, Φ, and φ_2 in the Bunge convention [9]. The vectors are given in spherical coordinates by the polar and azimuthal angles θ and ϕ. This results in 17 macroscopic parameters, 11 of which are independent. In order to reduce the number of parameters to a more measurable number, a coherent twin boundary may be introduced into the measured junction as demonstrated in Fig. 1. This distribution will be measured in FCC materials. As a coherent twin within FCC materials always points in a $\langle 111 \rangle$ direction, a reduction in parameters is made by rotating the coherent twin to the [111] direction. The twin boundary then becomes known. In addition, the symmetry across the twin boundary allows the three orientations to be determined by a single misorientation g_{AB}. The plane parameters are referenced from the internal coordinate system within Grain *A*, and consist of the $\boldsymbol{n}_{1A}(\phi_{1A}, \theta_{1A})$ Plane 1 normal and the dihedral angle α, between Planes 2 and 3. This reduces the number of independent junction parameters from 11 to six, which is within the capabilities of two-dimensional electron backscatter diffraction (EBSD) measurements.

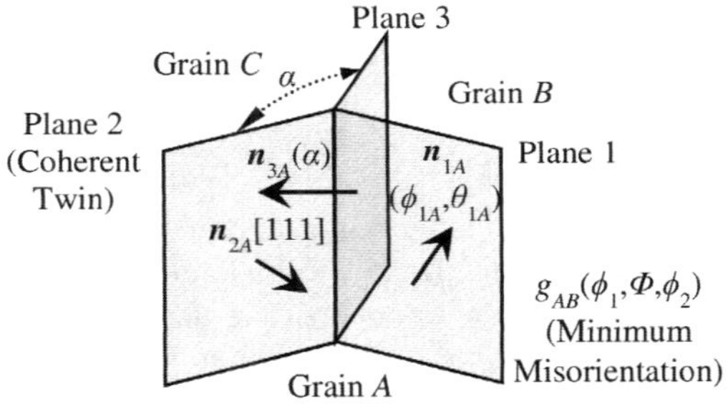

Fig. 1 All utilized parameters and conventions for a triple junction with a coherent twin boundary

2.2. The distribution space

The twin dependent triple junction distribution space is formed by the domains of the parameters and is depicted in Fig. 2. The six dimensional space has been broken into two three-dimensional spaces to enable visual inspection. It also conveniently separates the misorientation parameters from the grain boundary plane parameters. Each parameter is divided into equal discrete units forming unit volume cells within the distribution

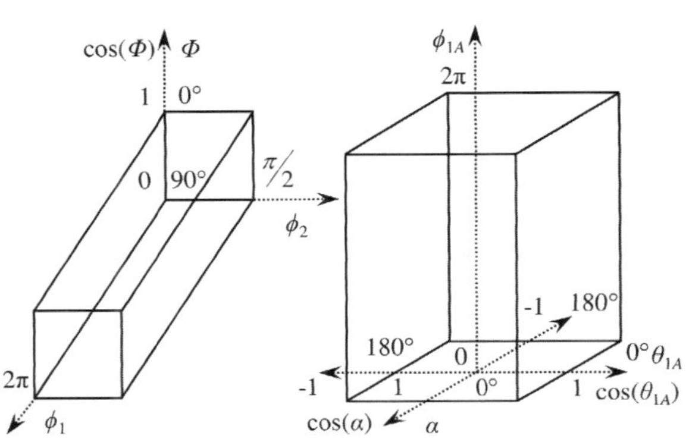

Fig. 2 The six-dimensional space of the twin dependent triple junction distribution

space. As apparent in the figure, some of the parameters are implemented with a cosine function which forms equal volume cells within the space. This has been the common practice for discrete grain boundary distributions [10]. For the simulations presented here, a resolution of approximately 10 degrees is utilized and the parameters are divided accordingly.

Unlike grain boundary distributions, the triple junction distribution does not fill the plane parameter space. There are many variations of $\boldsymbol{n}_{1A}(\phi_{1A}, \cos(\theta_{1A}))$ and $\cos(\alpha)$ which correspond to triple junctions with dihedral angles greater than 180 degrees. These are thermodynamically unfavorable and will not exist in actual microstructures. They are therefore treated as an empty region within the space. Inversion symmetries are removed from the plane parameters by always selecting the normal vectors such that they point in the same direction about the triple line. Grain *A* is identified as the grain between the twin and the boundary with the minimum misorientation angle or g_{AB}. The coherent twin boundary, or $\boldsymbol{n}_{2A}$, is selected such that it always points into Grain *A*. The other normal vectors then point in the same direction about the triple line as $\boldsymbol{n}_{2A}$. A fundamental zone is a region of the parameter space where each unique triple junction is represented only once. Given crystal symmetries within Grains *A* and *B*, there are many fundamental zones with equivalent triple junctions within the parameter space. A single fundamental zone must be designated in order to select a single set of parameters to represent each unique junction. A single fundamental zone is chosen by selecting the symmetrically equivalent misorientation with minimum misorientation angle and misorientation axis lying in an identified zone within the standard [111] stereographic projection. This is an adaptation of the method of selecting a fundamental zone for grain boundary planes [11].

3. Plotting triple junction distributions

A triple junction distribution formed from 500,000 simulated twin dependent junctions at a fixed misorientation is shown in Fig. 3a. The figure consists of slices through the plane parameter space at $\cos(\alpha)$ values moving into the page. Each stereographic projection is a projection of the $\boldsymbol{n}_{1A}(\phi_{1A}, \cos(\theta_{1A}))$ parameter values. These figures are the southern hemispheres of the stereographic projections. The empty region within the parameter space is apparent in the projections. The raw distribution is plotted and the visible cells result in a pixelated appearance. $\cos(\alpha)$ values were selected at both ends of the distribution, at the approximate distribution center, and at the distribution peak which is located at $\boldsymbol{n}_{1A} = [1\bar{2}\bar{1}]$ and $\cos(\alpha) = -0.73$ or $\boldsymbol{n}_{3A} = [\bar{3}0\bar{1}]$. After generating a triple junction distribution it must be weighted by the inverse of a uniform random distribution (due to a sampling bias). This weights all values to their representative values in terms of multiples of a random distribution (MRD).

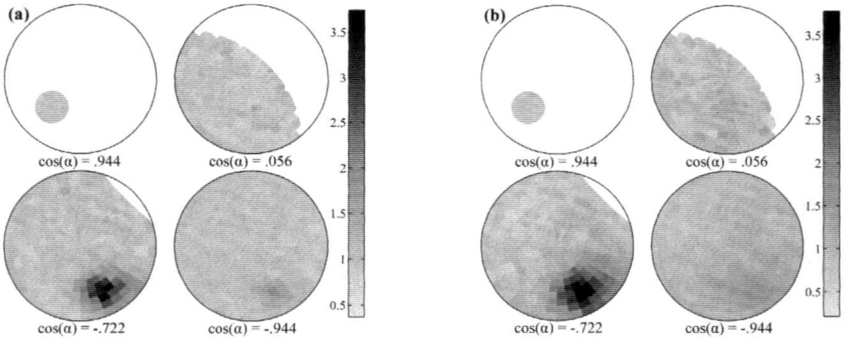

Fig 3 (a) Southern hemispheres of a simulated twin dependent triple junction distribution. (b) A distribution from the adopted stereological method using experimental data produced from the simulated distribution.

4. Measuring the distribution

In the section plane, the triple junction parameter $\cos(\theta_{1A})$ is immeasurable. Given this value, all other values may be determined from EBSD measurements. By adopting Saylor's stereological method for measuring five-parameter grain boundary distributions from a single section plane, $\cos(\theta_{1A})$ may be determined and the twin dependent distribution generated [12]. To validate this application, a simulated distribution was developed as shown in Fig. 3a. 500,000 simulated triple junction measurements, as generated from EBSD on a section plane, were then produced from this known simulated distribution and the stereological method applied. The distribution as appears in Fig. 3b was that generated from the simulated section plane data at a fixed misorientation. While the peak shows some mild broadening, otherwise the technique produced a similar distribution. 500,000 measurements for a fixed misorientation are an order of magnitude larger than what may be practically measured experimentally, but this number clearly demonstrates the success of the technique. Successful results are obtained with less data but the distributions do not reach this same level of convergence.

5. Conclusion

The developed method will allow twin dependent triple junction distributions to be generated from EBSD measurements on a single section plane. No serial sectioning will be required. This work also establishes necessary foundational work for developing triple junction distributions. Not only will this work further efforts in integrated materials engineering but it prepares the way for a non-twin dependent approach in the future.

Acknowledgements

This work was funded by NSF under Award DMR-1003004. The stereographic projections were generated by MTEX.

References

[1] Czubayko U, Sursaeva V G, Gottstein G and Shvindlerman L S 1998 Influence of triple junctions on grain boundary motion *Acta Mater.* **46** 5863–71
[2] Yin K-M, King A H, Hsieh T E, Chen F-R, Kai J J and Chang L 1997 Segregation of bismuth to triple junctions in copper *Microsc. Microanal.* **3** 417–22
[3] Bokstein B, Ivanov V, Oreshina O, Peteline A and Peteline S 2001 Direct experimental observation of accelerated Zn diffusion along triple junctions in Al *Mater. Sci. Eng. A* **302** 151–3
[4] King A H 2010 Triple lines in materials science and engineering *Scr. Mater.* **62** 889–93
[5] Schuh C A, Kumar M and King W E 2003 Analysis of grain boundary networks and their evolution during grain boundary engineering *Acta Mater.* **51** 687–700
[6] Frary M and Schuh C A 2005 Connectivity and percolation behaviour of grain boundary networks in three dimensions *Philos. Mag.* **85** 1123–43
[7] Bunge H-J 1994 Statistical crystallography of the polycrystal *Mater. Sci. Forum* **157-162** 13–30
[8] Rohrer G S 2011 Grain boundary energy anisotropy: a review *J. Mater. Sci.* **46** 5881–95
[9] Bunge H-J 1982 *Texture Analysis in Materials Science* (London: Butterworths) chapter 2 p 4
[10] Saylor D M, Morawiec A and Rohrer G S 2003 Distribution of grain boundaries in magnesia as a function of five macroscopic parameters *Acta Mater.* **51** 3663–74
[11] Morawiec A 2004 *Orientations and Rotations: Computations in Crystallographic Textures* (Berlin: Springer-Verlag) chapter 6 pp 130-131
[12] Saylor D M, El-Dasher B S, Adams B L and Rohrer G S 2004 Measuring the five-parameter grain-boundary distribution from observations of planar sections *Metall. Mater. Trans. A* **35** 1981–9

On the possibilities and limitations of modeling the elastic properties of textured multi-phase materials

S Matthies
Leading Scientist, Joint Institute of Nuclear Research, Dubna, RU
Permanent address: Müller-Berset-Str. 3, 01309 Dresden, Germany
E-mail: matthies.vinel@gmail.com

Abstract. The two ways of theoretical determination and modeling of the elastic properties of geological or modern technological materials are discussed. Such samples are characterized by a whole string of parameters or distributions (volume parts of the phases, sizes, shapes of the grains, or voids and their orientations etc.). The direct way uses first principles, averaging the strain field calculated by FEM techniques. The indirect way, without absolute spatial characteristics, uses distributions averaging the single-crystalline elastic properties, supposing infinite samples with ideal disorder and non-overlapping. space filling units. Overlapping can effectively been simulated by the DEM technique. The (GeoMixSelf) GMS-approximation able to consider ellipsoidal units, including voids, combines elements of the geometric mean and the self-consistent approximation. One example demonstrates the potentials, limitations and possible combinations of both ways .

1. Introduction

The description of the elastic properties of macroscopic samples (metals, alloys, compounds, minerals and rocks, cf. Fig. 1) is one of the first and typical application of quantitative texture analysis using

- the single-crystalline elastic parameters of their microscopic constituents (grains), and
- distribution characteristics, describing
- the texture of single-phase materials by the well-known ODF f(g),
- the volume fractions v_ν in case of multi-phase compounds (ν = 1, 2, ...,N),
- grain shapes, grain sizes, shape orientations $F(\Omega)$,
- the orientation of the crystal lattice inside a given shape, or
- the elastic properties of an effective homogeneous matrix, in order to determine its response to inclusions that model single grains [1-3].

Fig. 1. Microstructure of a biotite gneiss [4] (40 v.% quartz, 37% plagioclase, 23% biotite) with grains of different shapes, sizes and orientations, including a small amount of pores.

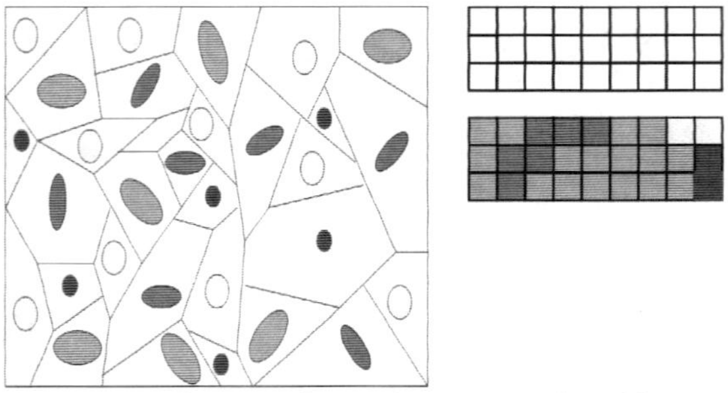

Fig. 2. 2D-scheme of covering a sample with a FE-net, getting its 'regular' representation. This representation will be loaded by a outer stress $\underline{\sigma}$ and using first principle equations the strain field $\boldsymbol{\varepsilon}(\mathbf{r})$ will be calculated.

Starting with Hook's law that combines the stress $\boldsymbol{\sigma}$ or strain $\boldsymbol{\varepsilon}$ by the stiffness **C** or compliance **S** on the microscopic (single-crystalline '^{o}') level

$$\boldsymbol{\sigma} = \mathbf{C}^{o} \bullet \boldsymbol{\varepsilon}\,,\ \boldsymbol{\varepsilon} = \mathbf{S}^{o} \bullet \boldsymbol{\sigma}\,,\ \mathbf{C}^{o} = (\mathbf{S}^{o})^{-1}, \tag{1}$$

the resulting effective elastic parameters $\underline{\mathbf{C}}$ or $\underline{\mathbf{S}}$ of a macroscopic sample follow as

$$\underline{\boldsymbol{\sigma}} = \underline{\mathbf{C}} \bullet \underline{\boldsymbol{\varepsilon}}\ ,\ \underline{\boldsymbol{\varepsilon}} = \underline{\mathbf{S}} \bullet \underline{\boldsymbol{\sigma}}\ ,\ \underline{\mathbf{C}} = (\underline{\mathbf{S}})^{-1}. \tag{2}$$

The $\underline{\boldsymbol{\sigma}}$ and $\underline{\boldsymbol{\varepsilon}}$ are applied or describe the response of the macroscopic sample. It holds

$$\underline{\boldsymbol{\sigma}} = \langle \boldsymbol{\sigma} \rangle\ ,\ \underline{\boldsymbol{\varepsilon}} = \langle \boldsymbol{\varepsilon} \rangle\ , \tag{3}$$

where the brackets represent an average over all microscopic values in the whole sample. There are a 'direct' and an 'indirect' way to determine the asked $\underline{\mathbf{C}}$ or $\underline{\mathbf{S}}$.

2. The direct way

The direct way uses the first principle equations

$$\mathbf{u}(\mathbf{r}) \approx \boldsymbol{\sigma} \approx \mathbf{C}^{o}_{V}\,\boldsymbol{\varepsilon} \approx \mathbf{C}^{o}_{V} \quad \mathbf{u}(\mathbf{r}) \tag{4}$$

(that combine the displacement $\mathbf{u}(\mathbf{r})$ with the corresponding stresses and single-crystal data C^{o}_{V} in the finite elements 'FE' of the regular representation of the sample, cf. Fig. 2) and the corresponding boundary conditions (continuity of stress and strain at the interfaces between the FE) in order to determine $\boldsymbol{\varepsilon}(\mathbf{r})$. Then the asked $\underline{\mathbf{C}}$ follows from

$$\underline{\boldsymbol{\sigma}}\,(\text{loaded, known}) = \underline{\mathbf{C}} < \boldsymbol{\varepsilon}(\mathbf{r}) > \tag{5}$$

Obviously a concrete calculation can only consider a 'mesoscopic' unit of a sample to be described. The mesoscopic unit should statistically well represent the macroscopic homogenous sample in order to represent it by multiplying the mesoscopic units. The finer the FE mesh (in order to describe details of the grains) and the more phases with non-trivial lattice orientations or shape orientations have to be described, the number of the necessary FE (and correspondingly the number of equations to be resolved) will drastically rise. Moreover, the calculation of a unit represents only one 'experiment' and the repetition of the experiments (considering another equivalent units) gives an impression about the reached accuracy of the resulting $\underline{\mathbf{C}}$. However, this aspect of the problem (i.e. information on real numbers for the accuracy or the computing expenses) is quite scarcely reported in literature.
A convincing example was already published in (1997) by Mullen et al. [5]. N centers of 'grains' have been statistically determined within a 2D box. Then the grains have been constructed by Voronoi tessellation. The orientation of the cubic crystal lattice in each grain was also statistically chosen in order to represent an isotropic texture (f(g) = 1). The number of FE was determined by the criterion, that a narrowing of the mesh gives only a minor gain of accuracy. Each experiment with N grains have been repeated 100 times. The accuracy of the calculated Youngs modulus E can be described by the spread around the corresponding mean value (for 100 experiments) coded as (N, ΔE [%]) : (5, 26 %), (10, 19 %), (100, 6 %), (1000, 2 %) for Si with A=1.65 ; and (5, 108 %), (10, 83 %), (100,28 %), (1000, 14 %) for Cs with A=6.98; $A = 2\,C^{o}_{44} / (C^{o}_{11} - C^{o}_{12})$.

3. The indirect way

The indirect way considers infinite samples with space filling, non-overlapping units, without any explicit correlations, i.e. realizing an 'ideal disorder'. All absolute spatial characteristics

IOP Conf. Series: Materials Science and Engineering **82** (2015) 012005 doi:10.1088/1757-899X/82/1/012005

(of meaning in the direct way) are lost, becoming relative quantities. Describing e.g. the half-axes {a,b,c} of a ellipsoidal grain, effectively all grains of the kind {a/c,b/c,1} will be meant. Distributions are used, characterizing the sample, like those mentioned in the introduction. Instead of the averaging relations (3) and the resulting connection (5), the asked $\underline{\mathbf{C}}$ or $\underline{\mathbf{S}}$ are determined using averages $<\mathbf{C}^o{}_v>$ or $<\mathbf{S}^o{}_v>$ by variants of 'micromechanical models'. There are three levels of approximations.

The first level considers non-interacting point-like grains (voids cannot be considered because of '$\mathbf{S}^o{}_v$'>> 1) with phase-specific weights v_v and orientation-probabilities described by the corresponding phase-specific ODF $f_v(g)$. The simplest average variant is the arithmetic mean 'a'. The well-known VOIGT $<\mathbf{C}^o{}_v>^a$ or REUSS $<\mathbf{S}^o{}_v>^a$ approximations commonly differ, leading to two different 'bounds', with

$$\mathbf{C}^V = <\mathbf{C}^o{}_v>^a \neq \mathbf{C}^R \,;\quad \mathbf{C}^R = (\mathbf{S}^R)^{-1} \;;\ \mathbf{S}^R = <\mathbf{S}^o{}_v>^a \neq \mathbf{S}^V = (\mathbf{C}^V)^{-1}\,. \tag{6}$$

Therefore the 'inversion relation' $\mathbf{C} = \mathbf{S}^{-1}$, that works on the microscopic (1) and macroscopic (2) levels, is not obeyed by $< >^a$ (6). So long the bounds differ by percents only, there are no problems. But the more anisotropic the $\mathbf{C}^o{}_v$ are, the more $\mathbf{C}^V$ and $\mathbf{C}^R$ differ. E.g. for biotite with isotropic oriented grains the VOIGT and REUSS values of the Young's modulus ($1/\underline{S}_{33}$) differ by about 300% [6]. There are several attempts to 'narrow' the bounds (Hashin & Shtrikman bounds, really applicable to cases with $f_v(g)$=1 only [2]) or using 'means' of them, like HILL's approximations ${}^{HILL}\mathbf{C}^C = (\mathbf{C}^V + \mathbf{C}^R)/2 \neq {}^{HILL}\mathbf{C}^S = ((\mathbf{S}^V + \mathbf{S}^R)/2)^{-1} = ({}^{HILL}\mathbf{S}^S)^{-1}$. The 'SuperHILL' approximation [6,7] ${}^{SUPER}\mathbf{C} = \mathrm{SQRT}^{sym}(\mathbf{C}^V\bullet\mathbf{C}^R) = [\mathrm{SQRT}^{sym}(\mathbf{S}^V\bullet\mathbf{S}^R)]^{-1} = ({}^{SUPER}\mathbf{S})^{-1}$ obeys the inversion relation. The geometric mean 'GEO' $\mathbf{C}^{GEO} = \exp(<\mathrm{LN}\mathbf{C}^o{}_v>^a)$ obeys the inversion relation as well the 'group' principle' [7], that should be obeyed by systems with non-interacting units. Consequently the geometric mean leads to results quite narrow to experimental ones (even for multi-phase cases) if the sample corresponds to the conditions: macroscopic homogeneous, almost spherical grain shapes, minimum correlations between the sort and orientations of the grains. The first level approximations explicitly neglect any interactions between the in reality 'hard-wired' to each other grains?! Already about 100 years there are manifold ('**second level'**) attempts to include elements of interaction into the consideration, finally using in any way the Green-function of a real or working matrix of the sample [2]. A very constructive step was done by Eshelby (1957) who succeeded to describe in a closed form the strain field $\boldsymbol{\varepsilon}(\mathbf{r})$ for a system of a loaded anisotropic homogeneous matrix that contains an anisotropic ellipsoidal ({a,b,c}) (not point-like!) inclusion. The key element is the dimension-less Eshelby-tensor $\mathbf{S}^{Esh}$ that depends on $\mathbf{C}$ of the matrix and {a,b,c}. Basing on that result several kinds of self-consistent 'SC' schemes have been developed. The SC-idea is: the $\mathbf{C}$ or $\mathbf{S}$ of a 'system' consisting of a 'matrix' and inclusions are iteratively determined. The next iteration step (i+1) uses the ${}^i\mathbf{C}$ or ${}^i\mathbf{S}$ of the ystem as the elastic properties of the matrix up to the moment when the properties of the system and the matrix coincide. In the variant of Williams (1970) it reads:

'p-branch': ${}^{i+1}\mathbf{C} = {}^i\mathbf{C} + < ({}^i\mathbf{C} - {}^i\mathbf{E}^{-1})\bullet{}^i\mathbf{u} > \equiv {}^i\mathbf{C} + {}^i\mathbf{p}$; *'q-branch'*: ${}^{i+1}\mathbf{S} = {}^i\mathbf{S} + < {}^i\mathbf{u}\bullet{}^i\mathbf{S} > \equiv {}^i\mathbf{S} + {}^i\mathbf{q}$, (7)

$\mathbf{u} = [\mathbf{C}^o - (\mathbf{C} - \mathbf{E}^{-1})]^{-1}\cdot\mathbf{E}^{-1} - \mathbf{I}$; $\mathbf{E} = \mathbf{S}^{Esh}(\mathbf{C},\{a_1,a_2,a_3\})\cdot\mathbf{S}$; ${}^p\underline{\mathbf{C}}^{SC} \equiv {}^i\mathbf{C}$, ${}^q\underline{\mathbf{S}}^{SC} \equiv {}^i\mathbf{S}$ for $(i \rightarrow \infty)$. (8)

Commonly the results of the two branches ${}^p\underline{\mathbf{C}}^{SC}$ and ${}^q\underline{\mathbf{C}}^{SC} = ({}^q\underline{\mathbf{S}}^{SC})^{-1}$ do not coincide. In order to obey the inversion relation the GeoMixSelf ('GMS') approximation was introduced [7], that applies the SUPER-relation (with the character of a geometric mean) at each iteration step i :

$${}^{i+1}\mathbf{C} = {}^i\underline{\mathbf{C}} + {}^i\underline{\mathbf{p}}\,,\ {}^{i+1}\mathbf{S} = {}^i\underline{\mathbf{S}} + {}^i\underline{\mathbf{q}}\,,\ {}^i\underline{\mathbf{S}} = ({}^i\underline{\mathbf{C}})^{-1},\ {}^i\underline{\mathbf{C}} = \mathrm{SQRT}^{sym}({}^i\mathbf{C}\cdot({}^i\mathbf{S})^{-1}),\ \underline{\mathbf{C}}^{GMS} = {}^i\underline{\mathbf{C}}\ (i \rightarrow \infty) \equiv (\underline{\mathbf{S}}^{GMS})^{-1}. \tag{9}$$

IOP Conf. Series: Materials Science and Engineering **82** (2015) 012005 doi:10.1088/1757-899X/82/1/012005

The averaging procedure < > in (7) can be generalized without difficulties to the case of mixed phases, with their own v_v , $f_v(g)$, $F_v(\Omega)$, and $\{a_1,a_2,a_3\}_v$-sets. The GMS approximation is also able to consider pores, because the critical $\mathbf{S}^{\circ}_v$ are never asked in (8). The second level can describe non-overlapping ellipsoidal grains or voids only. In porous geologic samples or ceramics 'nets' of voids can exist that can be imagined as overlapped ellipsoids. In **third level** approximations such overlaps can effectively be considered by the differential element method ('DEM'). A small amount of ellipsoidal pores (with may be other orientation) is added to a working matrix (may be calculated by GMS) with a starting porosity of non-overlapped ellipsoidal pores. The resulting system (used as the matrix for the next iteration step, up to the final porosity) effectively contains overlapping effects. E.g. considering porous materials (cf. Fig. 3) by non-overlapping pores only, the critical porosity (decay of the sample) will earlier be reached than for overlapped pores with a more stable skeleton of the embedding matrix. Sample specific correlations between constituents of the sample (like in graphite or shale's) can also effectively be modeled ([8], [9]) combining e.g. the $F_v(\Omega)$ of flat ellipsoidal grains with those of flat pores or cracks.

4. Example, comparisons

Because of the limited space, only one example can be given. Fig. 3 demonstrates in a sufficient manner the discussed points of interest.

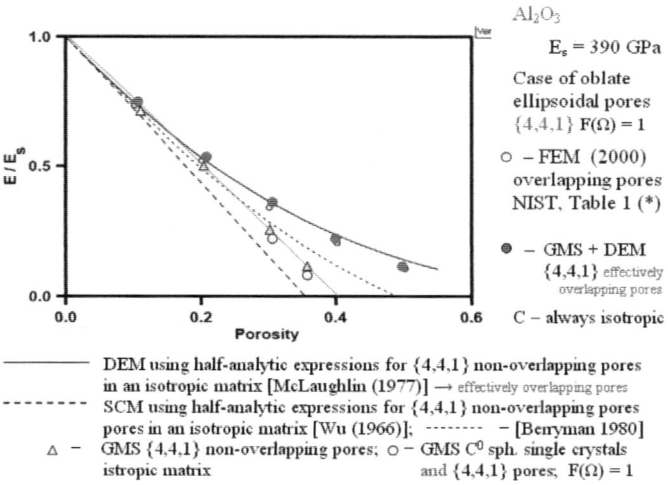

Fig. 3 Dependence of Young' modulus on porosity for Al_2O_3 modeled and calculated by FEM [10] and the corresponding results applying SC-, GMS- and DEM- techniques.

5. Conclusion

The direct way is of uni- versal charater limited by computing expenses only. The 'simpler' indirect way can successfully be applied, especially investigating the dependence of properties on sample parameters in cases, if the to be modeled samples belong to the class of the simplifying assumptions.

A general problem is to get the necessary amount of reliable microscopic (experi- mental or in any way estimated) 'INPUT'-data.

References

[1] Watt J P et al 1976 Rev. Geophys. **14** 541
[2] Mura T 1991 *Micromechanics of Defects in Solids* (Kluwer Academic Publishers)
[3] Kocks U F et al 2000 *Preferred orientations in polycrystals* (Cambridge University Press)
[4] Kern H et al 2008 *Tectonophysics* **457** 143
[5] Mullen R L et al 1997 *Acta mater.* **45** 2247
[6] Matthies S and Humbert M 1995 *J. Appl. Cryst.* **28** 254
[7] Matthies S 2010 *Solid State Phenom.* **160** 87
[8] Matthies S 2012 *J. Appl. Cryst.* **45** 1
[9] Vasin R N et al 2013 Journal *of Geophysical Research: Solid Earth.* **118** 3931
[10] Roberts A P et al [NIST] 2000 in *J. Am. Ceramic Soc.*

IOP Conf. Series: Materials Science and Engineering **82** (2015) 012006 doi:10.1088/1757-899X/82/1/012006

Boundary Fractal Analysis of Two Cube-oriented Grains in Partly Recrystallized Copper

J. Sun[1], Y.B. Zhang[1], A.B. Dahl[2], K. Conradsen[2] and D. Juul Jensen[1]
[1] Danish-Chinese Center for Nanometals, Section for Materials Science and Advanced Characterization, Department of Wind Energy, Technical University of Denmark, Risø Campus, 4000 Roskilde, Denmark.
[2] Section for Image Analysis and Computer Graphics, Department of Applied Mathematics and Computer Science, Technical University of Denmark, Richard Petersens Plads, 2800, Kgs. Lyngby, Denmark.
E-mail: jusu@dtu.dk

Abstract. The protrusions and retrusions observed on the recrystallizing boundaries affect the migration kinetics during recrystallization. Characterization of the boundary roughness is necessary in order to evaluate their effects. This roughness has a structure that can be characterized by fractal analysis, and in this study the so-called "Minkowski sausage" method is adopted. Hereby, two cube-oriented grains in partly recrystallized microstructures are analyzed and quantitative information regarding the dimensions of protrusions/retrusions is obtained.

1. Introduction

During recrystallization of deformed metals, new almost perfect nuclei form in the deformed matrix and grow by boundary migration until the entire deformed matrix is replaced by recrystallized grains. Traditionally it is considered that the nuclei form and grow in a relatively homogeneous manner. Modern techniques including synchrotron X-rays and advanced electron microscopy have however revealed the heterogeneity of recrystallization and the importance of local structural variations for the boundary migration. The migrating recrystallization boundaries are not smooth and do not move in a homogeneous manner with constant speed as described in the classic models, but rather occur with local variations: most segments move in a jerky stop-go manner, with local protrusions and retrusions forming as the boundary migrate [1-3]. Phase-field modeling has shown that the protrusions and retrusions observed on the migrating boundaries may contribute an additional driving force as a result of the boundary curvature and change the migration kinetics on both the local and global scale [4]. Accurate quantitative description of the boundary roughness is necessary in order to evaluate effects hereof on the boundary migration kinetics. Such a description should at least incorporate information about the frequency of protrusions and retrusions as well as their dimensions. One or more parameters that describe this roughness should be developed to allow reasonable comparison among various boundaries.

Mandelbrot has shown that self-similar objects are common in nature and their accurate characterization is of great importance to the field of analytical morphology [5]. Fractals can describe mathematical sets that strictly obey the scaling invariant principle, and fractals are also adopted to describe geometrical objects that exhibit self-similarity. Fractal is generally to be interpreted as a geometrical property describing objects that resembles themselves, exactly or statistically, at different observation scales. Since Mandelbrot made the first trial to correlate the fractal dimension of fracture surface roughness with the impact energy absorbed in fracture specimens [6], fractal analysis has been applied in many fields regarding characterizing irregular and rough morphological features [7]. Numerous methods have been proposed and validated in obtaining the fractal dimension [7,8]. These methods basically follow the same algorithm: (1) Measure the quantities of the objects using various step sizes; (2) Plot the measured quantities versus step sizes on logarithm scale and fit a least-squares regression line through the data points; (3) Estimate the fractal dimension as the slope of the regression line. It should be noted that the calculated fractal dimension cannot be expected to be one single value for features that are not strictly self-similar as the Mandelbrot set. Also, the finite resolution of the image data gives a limitation to how small the step sizes can be, so self-similarities can only be observed within a range of scales. However, by investigating the variation of measuring quantities with various step sizes, which is termed fractal analysis in the present work, information is provided of the geometrical features in complex morphologies.

2. Experimental and Fractal Analysis Method

Two cube oriented recrystallizing grains (Figure 1 (a) and (b)) in partly recrystallized microstructures are investigated. Both are from Oxygen Free High Conductivity (OFHC) copper (99.9% purity) 90% cold rolled and annealed at 150 °C. The interior twin boundaries are ignored in the analysis and the boundaries between the recrystallized and the deformed regions are extracted (Figure 1 c and d). The boundary shown in Figure 1(c) is designated as B1 and the boundary shown in Figure 1(d) is designated as B2. Both B1 and B2 have protrusions/retrusions on certain scales, but B2 appears smoother as B1 has more long, narrow branch-like features. Koch snowflake curves [9] of various iterations are included here as a reference for fractal analysis of the two boundaries. The Koch snowflake curve is constructed by iteration, and the geometrical structures are constructed following the same scaling law, whereby it obtains a fractal dimension of around 1.26. Snowflakes with 2 and 3 iterations are shown in Figure 2 and the characteristic edge length of the finest repeating motif is indicated in Figure 2 (c). This length is 675, 225, 75 and 25 pixels for Koch snowflake curves constructed with 2, 3, 4 and 5 iterations, respectively.

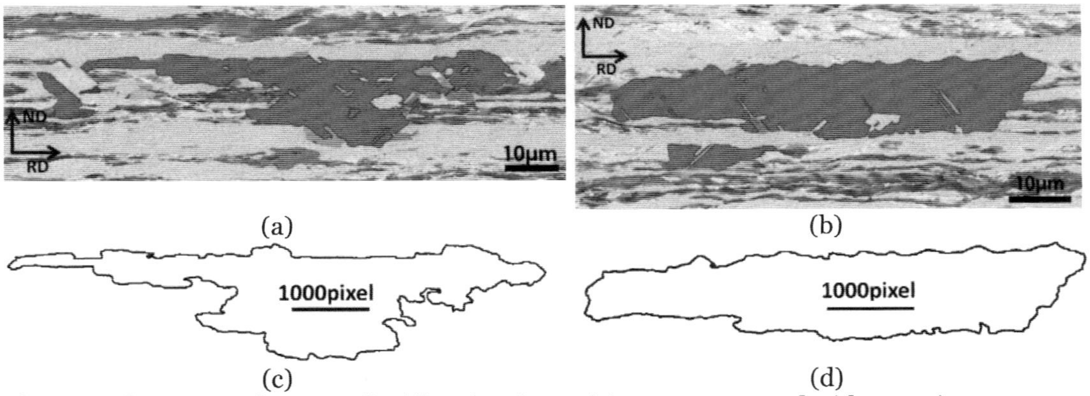

Figure 1. Electron Backscattered Diffraction (EBSD) images measured with step size 0.1 μm of two copper grains. The boundaries surrounding the two grains are shown in (c) and (d), with a pixel bar indicating the scale of digital images; interior twin boundaries are ignored.

The "Minkowski sausage" fractal analysis method [8] follows the algorithm mentioned in the introduction and is used to estimate the fractal behavior in current study. The boundary is dilated with varying widths, and the perimeter is obtained by dividing the area after dilation by the corresponding width. The fractal dimension (FD) is calculated as: $\log(perimeter) = (1 - FD)\log(width) + C$, where FD is the fractal dimension and C is a constant. To obtain more information from the data, as a new idea, the slopes from each set of 2 adjacent points are also calculated referred to as '2-point fractal dimension (FD)' in the following text. Further information about the method will be published later.

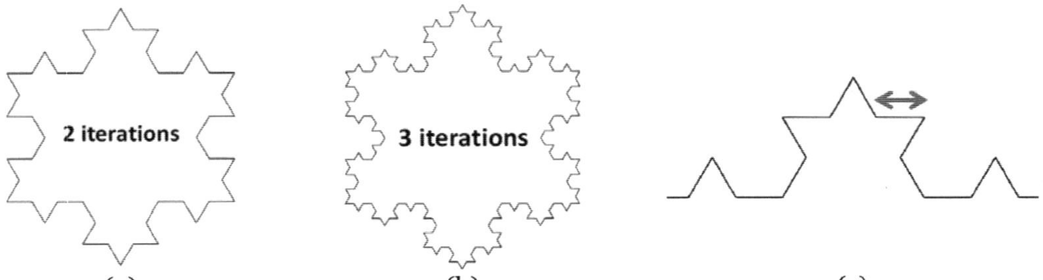

(a) (b) (c)

Figure 2. Koch snowflake curve with (a) 2 and (b) 3 iterations. (c) The red arrow indicates the characteristic edge length of the finest motif.

3. Results and Discussion

Results for the Koch snowflake curves with 2, 3, 4 and 5 iterations are shown in Figure 3. Two main points should be noted: (1) For the ideal Koch snowflake curve, the expected log(*perimeter*) versus log(*width*) plot is a straight line with slope of about -0.26 corresponding to a fractal dimension of 1.26. But with limited iterations, the snowflakes are not ideal fractal geometries and the whole shape is constructed with non-continuous motifs. This result in the inflections of slopes at low log(*width*) values in Figure 3(a) and a saw-tooth shape in Figure 3(b); (2) The transitive points as indicated by the arrows ①-④ in Figure 3(b) reveal the positions where the finest motif is resolved by the measurement with increasing width. When a smaller motif is constructed (i.e. more iteration), the transitive point moves towards the smaller width that corresponds with characteristic edge length of the finest motif. This correspondence provides a correlation between the dimension of the features, e.g. finest edge length and the transitive points in the plots.

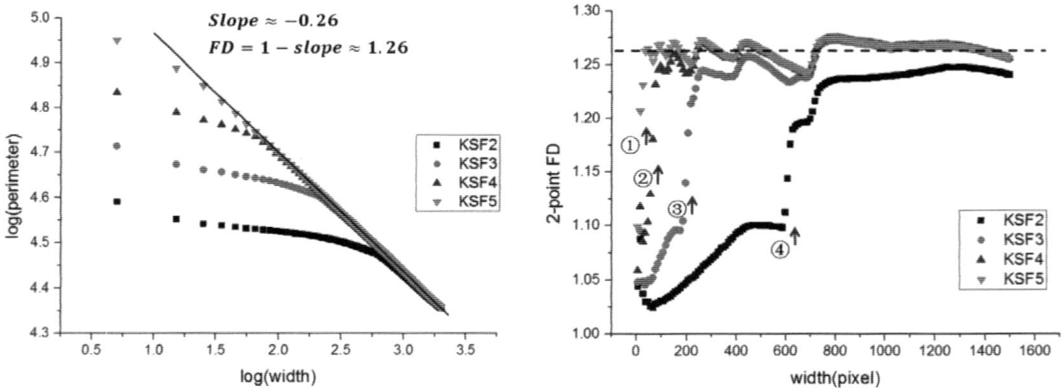

(a) (b)

Figure 3 (a) log(*perimeter*) versus log(*width*) plots and (b) 2-point fractal dimension plot for Koch snowflake curves with 2, 3, 4 and 5 iterations. Arrows ①-④ in (b) mark the transitive points which corresponds to the finest motif of each Koch snowflake curves.

The calculated results for B1 and B2 are shown in Figure 4. It is found that the slope varies in a complicated manner for both B1 and B2. It is not possible to fit a single straight line through the data points and a single fractal dimension value corresponding to the entire boundaries is therefore not found (which agrees with expectation). The arrows in Figure 4(b) mark major transitive points of B1 and B2 curves. For B1, there is an almost steady increase of the 2-point FD values between arrow ⑤ and ⑥. Unlike the sharp increase or saw-tooth behavior in Koch snowflake curves, the increase seen in B1 indicates that there are protrusions/retrusions distributed over the entire length scale between ⑤ and ⑥, which agrees with the visual impression of the morphology of B1 (see Figure 2(c)). The transitive point ⑥, at the width of about 300 pixels, corresponds to the approximate width of the narrow branches in B1. For B2, two distinct plateaus (between ⑤ and ⑧, ⑨ and ⑩) are seen in Figure 4(b) and they indicate that the protrusions/retrusions are mostly on two scales. Referring to the morphology of B2 (see Figure 2(d)), relative small protrusions/retrusions are mostly on the lower boundary (bottom right) and their dimensions are about 50 pixels corresponding to transitive points marked by ⑤, while the larger protrusions/retrusions are more frequently observed on the upper boundary and their dimensions are about 250 pixels corresponding to ⑨. The 2-point FD value of B1 and B2 shows an almost linear drop at positions ⑦ (width at about 660 pixels) and ⑪ (width at about 500 pixels) respectively, and both widths correspond to the approximate overall geometrical width along the normal direction of B1 and B2, respectively.

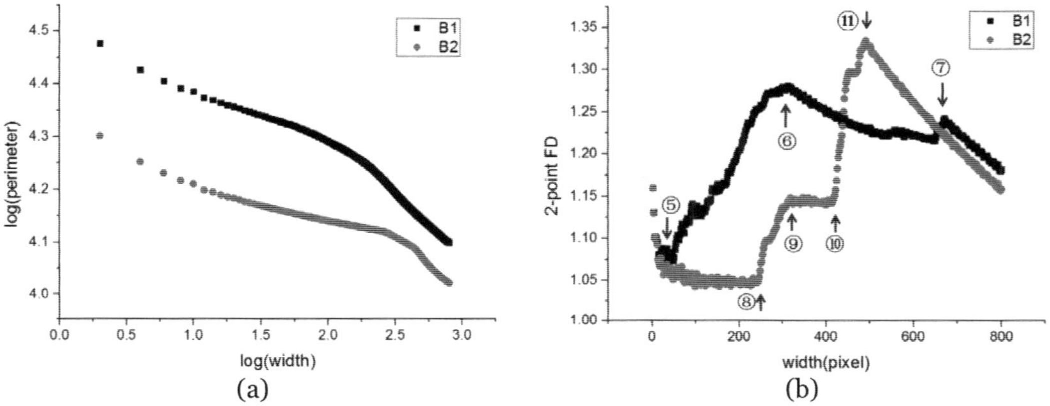

Figure 4. (a) log(*perimeter*) versus log(*width*) plot and (b) 2-point fractal dimension plot for B1 and B2. Arrows ⑤-⑪ mark the major transitive points for curves of B1 and B2.

4. Conclusions and Outlook

The fractal analysis using the "Minkowski sausage" method provides a way to reveal the characteristics of geometrical shapes, with measurement results reflecting the morphological roughness. The method is adequate to analyze grain boundary roughness but more accurate quantitative description is required and will be discussed in future work.

5. References

[1] Juul Jensen, D.; Lin, F.; Zhang, Y.; Zhang, Y. *Mat. Sci. Forum.* 2013, 753, 37-41.
[2] Schmidt, S.; Nielsen, S.; Gundlach, C.; Margulies, L.; Huang, X.; Juul Jensen, D. *Science.* 2004, 305, 229-232.
[3] Zhang, Y.; Juul Jensen, D.; Godfrey, A. *Mat. Sci.Forum* 2012, 715-716, 329-332.
[4] Moelans, N.; Godfrey, A.; Zhang, Y.; Juul Jensen, D. *Phys. Rev. B* 2013, 88, 054103.
[5] Mandelbrot, B, "The Fractal Geometry of Nature", *W. H. Freeman and Company.* 1983.
[6] Mandelbrot, B.; Passoja, D.; Paullay, A. *Nature* 1984.
[7] Li, J, "Image-based Fractal Description of Microstructures", Kluwer Academic Publisher. 2003.
[8] Bérubé, D.; Jébrak, M. *Comp. & Geos.* 1999, 25, 1059-1071.
[9] Weisstein, E. "Koch Snowflake." From *MathWorld--A Wolfram Web Resource.*

IOP Conf. Series: Materials Science and Engineering **82** (2015) 012007 doi:10.1088/1757-899X/82/1/012007

Internal optimization of the texture component approximation method

D. Nikolayev[1], T. Lychagina[1], M. Rusetsky[2], A. Ulyanenkov[3] and A. Sasaki[4]
[1] Frank Laboratory of Neutron Physics, Joint Institute for Nuclear Research, Dubna, Russia
[2] Belarusian State University, Minsk, Belarus
[3] Rigaku Europe SE, Ettlingen, Germany
[4] Rigaku Corporation, Tokyo, Japan
E-mail: dmitry@nf.jinr.ru

Abstract. The component approximation method for the reconstruction of orientation distribution function (ODF) is based on the assumption that the texture could be presented as a weighted linear combination of distributions depending on the parameters, which are related to the position of bell shaped function in orientation space and to the dispersion. The method uses a minimization procedure to obtain the values of ODF parameters. Traditionally, the mean-square deviation of the measured and recalculated pole figures is minimized. However, the quantitative measure of the fit is RP value which differs from the mean-square deviation. In the present work it is suggested to minimize the RP value to obtain ODF parameters. We are using Trust Region method for solving a non-linear optimization problem. The convergences of the proposed method for different minimized functional are compared. We also illustrate a usage of the different objective function on modeling data for the cubic crystalline symmetry. This study is fulfilled using new RIGAKU software for quantitative texture analysis.

1. Introduction

The core problem of the quantitative texture analysis is the ODF reconstruction from the measured pole figures. The component approximation method [1-3] is recently developed for the ODF reconstruction. One of the advantages of the method is a clear physical meaning of the computed parameters. The central difficulty of this method is the nonlinear minimization problem which should be solved for the parameters in question.

2. Component approximation method

Before introduction of the ODF [4] the texture had been characterized by the analysis of pole figures, which assigned one or several preferred orientations explaining the bell shaped peaks on the measured pole figures. The general idea of the component approximation method is a formalization and quantification of such approach. The texture is assumed to be caused by a linear combination of the components centered at the positions of the preferred orientations in the orientation space or rotation group SO(3). Formally this means the ODF can be presented as follows:

$$f(g) = FON + \sum_{n=1}^{NUMB} A_n f^p(g, g_{0n}, \varepsilon_n) + \sum_{k=1}^{NUMF} A_k f^f(g, \vec{n}_t, g_o\vec{n}_t, \varepsilon_k) \qquad (1)$$

Here $f^p(g, g_{0n}, \varepsilon_n)$ and $f^f(g, \vec{n}_t, g_o\vec{n}_t, \varepsilon_k)$ are the peak and the fiber component, respectively. Components are non-negative functions integrating to 1 on SO3 (to be probability density) The values A_n and A_k are the weights, ε_n and ε_k are the parameters describing the width of the components, and $g_{0n} = (\alpha_n, \beta_n, \gamma_n)$ are parameters describing the peak position or position of the ideal orientation on the SO(3) group, *NUMB* is the number of peak components. $\vec{n}_t = (\theta_y, \varphi_y)$, $g_o\vec{n}_t = (\theta_h, \varphi_h)$ are the parameters describing the fiber component position, *NUMF* is the number of fiber components, and *FON* is the background value or isotropic component weight. Weights are non-negative and sum up to 1

$\left(FON+\sum_{n=1}^{NUMB}A_n+\sum_{k=1}^{NUMF}A_k=1\right)$. It is important to note that this is a constrained to the optimization problem.

The recalculated pole figures for the ODF in the form (1) are expressed in the following way:

$$P^r{}_{\vec{h}_i}(\vec{y})=FON+\sum_{n=1}^{NUMB}A_nP_{\vec{h}_i}{}^p(\vec{y}_j,g_{0n},\varepsilon_n)+\sum_{k=1}^{NUMF}A_kP_{\vec{h}_i}{}^f(\vec{y}_j,\vec{n}_t,g_o\vec{n}_t,\varepsilon_k) \qquad (2)$$

For a special appearance of the components, the expression for the recalculated pole figures could also be presented in the forms given below:

• Brownian motion distribution

$$P_{\vec{h}}^p(\vec{y})=\sum_{l=0}^{\infty}(2l+1)\exp(-l(l+1)\varepsilon^2)P_l(\cos\theta_0) \qquad \cos\theta_0=\left(\vec{y},g_0^{-1}\vec{h}\right) \qquad (3)$$

$$P_{\vec{h}}^f(\vec{y},\vec{n}_t)=\sum_{l=0}^{\infty}(2l+1)\exp(-l(l+1)\varepsilon^2)P_l(\cos\theta_1)P_l(\cos\theta_2) \quad \cos\theta_1=\left(\vec{h},g_0^{-1}\vec{n}_t\right) \quad \cos\theta_2=\left(\vec{y},\vec{n}_t\right)$$

$$f^p(g,g_0,\varepsilon)=\sum_{l=0}^{\infty}(2l+1)\exp(-l(l+1)\varepsilon^2)\frac{\sin(l+\frac{1}{2})t}{\sin\frac{t}{2}} \qquad \cos t=[Tr(gg_0{}^{-1})-1]/2$$

$$f^f(g,g_0,\varepsilon,n_t)=\sum_{l=0}^{\infty}(2l+1)\exp(-l(l+1)\varepsilon)P_l(\cos\tau) \qquad \cos\tau=(g\vec{n}_t,g_o\vec{n}_t)$$

• Fisher distribution

$$P_{\vec{h}}^p(\vec{y})=\frac{1}{I_0(S)-I_1(S)}I_0(S(1+\cos\theta_0)/2)\exp(S(\cos\theta_0-1)/2) \qquad \cos\theta_0=\left(\vec{y},g_0^{-1}\vec{h}\right) \qquad (4)$$

$$P_{\vec{h}}^f(\vec{y},\vec{n}_t)=\frac{1}{I_0(S)-I_1(S)}I_0(S\sin\theta_1\sin\theta_2)\exp(S\cos\theta_1\cos\theta_2)$$

$$f^p(g,g_0,S)=\frac{1}{I_0(S)-I_1(S)}\exp(S\cos t) \quad \cos t=[Tr(gg_0{}^{-1})-1]/2$$

$$f^f(g,g_0,S,\vec{n}_t)=\frac{S}{\sinh S}\exp(S\vec{h}\cdot g\vec{n}_t)$$

These both distributions have a similar nature as normal distribution. There are two more distributions, namely Lorentz [5] and the hyperspherical de la Vallée Poussin kernel [6], which also can be used as texture components. Finally, the following distribution approximates [7] the normal one, too:

$$P_{\vec{h}}^p(\vec{y})=\frac{\sigma}{\cos\theta_0}\left(\frac{1}{\cos^2\theta_0}+\frac{1}{2\sigma}\right)\exp\left(-\sigma\tan^2(\theta_0/2)\right) \qquad \cos\theta_0=\left(\vec{y},g_0^{-1}\vec{h}\right) \qquad (5)$$

$$f^p(g,g_0,\sigma)=\frac{\sigma\ \ \pi\sigma}{\cos^4(t/2)}\exp(-\sigma\tan^2(t/2))$$

The distributions with a normal nature do not allow the isotropic generalization, and only the last one has a simple form, which does not contain the series or integrals. The main problem of the quantitative texture analysis thus can be formulated and resolved as follows. For the existing NPF measured pole figures, we can imply the texture is described by *NUMF* fiber components and *NUMB* peak components. The certain suggestions can be made about the initial values of the components parameters using the pole figure pictures. By using the form (2) of the recalculated pole figures, the component method is applied to find the corresponding parameters of peak and fiber components, which are found from the minimum of the functional:

$$\Phi(P^{\exp}_{\vec{h}_i}(\vec{y}_j), P^{r}{}_{\vec{h}_i}(\vec{y})) \to min\,.$$

The total number of the parameters to minimize is $5 \cdot NUMB + 6 \cdot NUMF + NPF + 1$. There are several approaches for the choice of the functional. The old approach is based on the least squares minimization with the functional:

$$\sum_{i=1}^{NPF}\left(P^{\exp}_{\vec{h}_i}(\vec{y}_j) - N_i \times \left(\sum_{n=1}^{NUMB} A_n P_{\vec{h}_i}(\vec{y}_j, g_{0n}, \varepsilon_n) + \sum_{k=1}^{NUMF} A_k P_{\vec{h}_i}(\vec{y}_j, \vec{n}_t, g_o\vec{n}_t, \varepsilon_k)\right) - FON\right)^2 \to min \qquad (6)$$

In [8-10] a set of alternative functionals had been proposed. We consider in this work the new approach with the functional in the form of RP value. The RP values are defined as follows [5]:

$$RP_i = \sum_{j=1}^{J} 100\% * \frac{P_{\vec{h}_i}(\vec{y}_j) - P^{M}_{\vec{h}_i}(\vec{y}_j)}{P_{\vec{h}_i}(\vec{y}_j)} \Theta(\zeta, P_{\vec{h}_i}(\vec{y}_j)) \Bigg/ \sum_{j=1}^{J} \Theta(\zeta, P_{\vec{h}_i}(\vec{y}_j)), \quad \Theta(\zeta, x) = \begin{cases} 0, & x < \zeta \\ 1, & x \geq \zeta \end{cases} \qquad (7)$$

Here ζ is a sensitivity level, and the minimization is fulfilled by the TRUST REGION method [11], which has been recently developed for finding of global maxima in the nonlinear optimization problems.

3. Numerical examples

The solution of the nonlinear optimization problem requires an essential number of the recalculated pole figures, computed with different parameters. To reach a reliable result in a reasonable time, the component function must be computed as fast as possible. Despite the distributions (3) and (4) have very useful features (entropy maximization, fundamental solution of the diffusion equation, etc.), the distribution (5) is much easier and faster for computation. The figure 1 shows the correspondence between all these distributions.

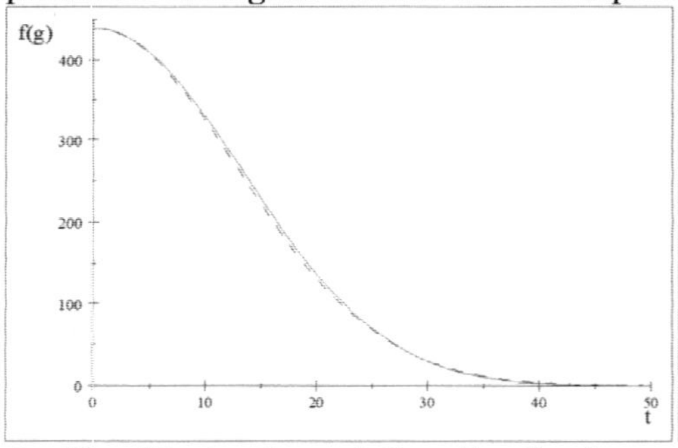

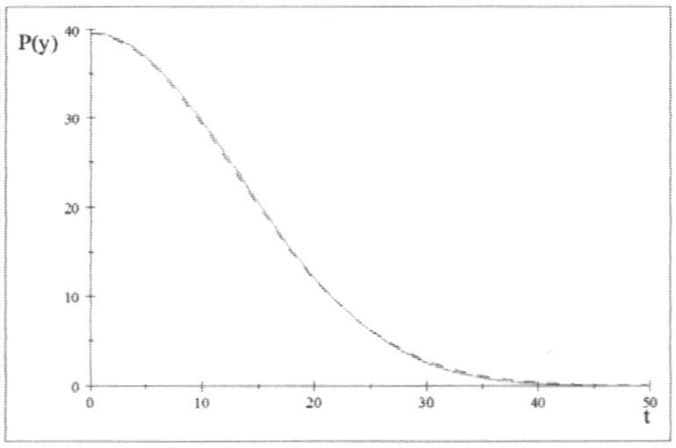

Fig 1. The illustration of the distributions used in calculations. The parameters of the distributions are S = 20, σ = 39.483, ε^2 =2.5435×10^{-2}. All the parameters correspond to a full width at the half maximum b = 30.17°. The dashed line is for distribution (3) and (4), the solid line is for (5).

For a smaller value of width, these distributions become indistinguishable, and thus the distribution with a fastest computation time can be used, which is the distribution (5). Because of all these tree distributions approximate each other quite well, we can use any of them for the further analysis after a proper adjustment of parameters. For example, the simplest form of C coefficients has the distribution (3), and we used formula (3) for the peak components and (4) for the fiber components to generate numerical example. The connection between the parameters describing the width of the components ε and b is presented in [12-13].

Table 1. Summary of the numerical experiment

Parameter	model	initial assumption	optimized (old approach)	optimized (new approach)
$g_0(\alpha,\beta,\gamma)$	$\{45^\circ,45^\circ,45^\circ\}$	$\{46^\circ,46^\circ,46^\circ\}$	$\{45.01^\circ,45.04^\circ,44.96^\circ\}$	$\{45.16^\circ,45.20^\circ,45.16^\circ\}$
b_{peak}	15°	15°	14.97°	14.92°
A_{peak}	0.450	0.450	0.447	0.449
$\vec{n}_t(\theta_y,\varphi_y)$	$\{10^\circ,10^\circ\}$	$\{11^\circ,11^\circ\}$	$\{10.34^\circ,11.00^\circ\}$	$\{10.30^\circ,10.99^\circ\}$
$g_0\vec{n}_t(\theta_h,\varphi_h)$	$\{10^\circ,10^\circ\}$	$\{11^\circ,11^\circ\}$	$\{10.18^\circ,10.18^\circ\}$	$\{10.21^\circ,10.21^\circ\}$
b_{fiber}	15°	15.66°	15.67°	15.67°
A_{fiber}	0.450	0.450	0.454	0.453
FON	0.10	0.10	0.096	0.101

The corresponding pole figures are shown in the figure 2. The functionals (6) and (7) have been used for nonlinear minimization procedure, and the distribution (5) is applied during the fitting procedure. The results of the fit are collected in the table 1.

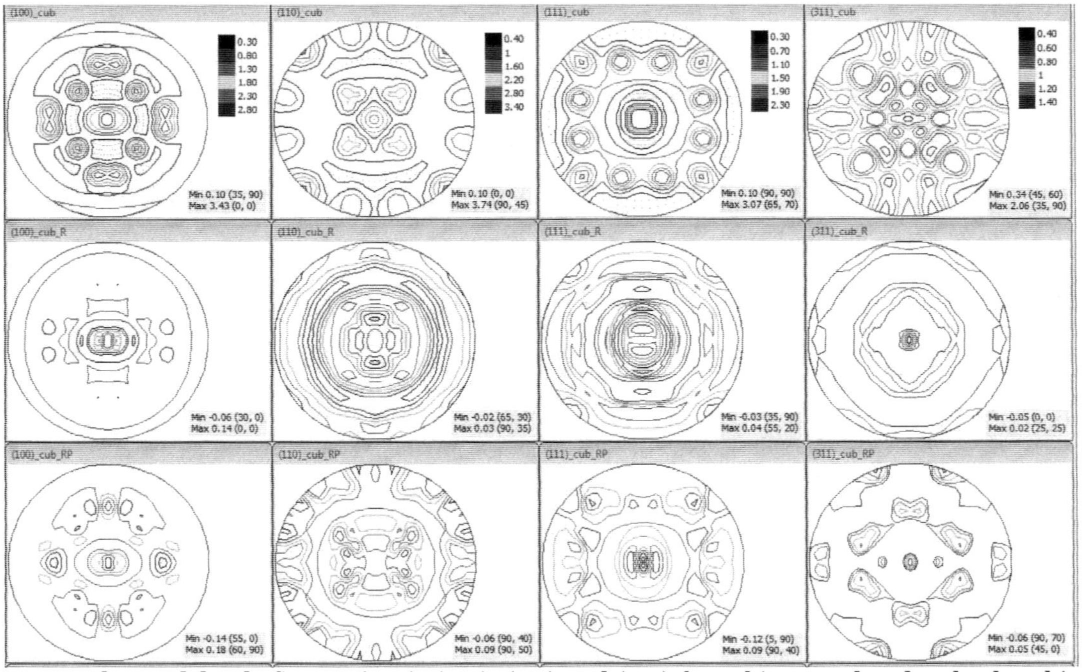

Fig 2. The model pole figures (100), (110), (111) and (311) for cubic crystal and orthorhombic sample symmetry simulated for one peak and one fiber component with parameters presented in the table 1 (1st row). The spatial distribution of the recalculation error (2nd-3rd rows for the old and new approaches respectively).

Both procedures converge to the similar correct results. The minimization procedure with functional (7) converges with 33 iterations and mean RP=1.93, whereas ones with functional (6) converges with 18 iterations and mean RP=1.05. The recalculated pole figures are indistinguishable with the modeled ones. The functional (6) evidently converges faster due to a less number of iterations required, and the RP mean value over all pole figures is slightly smaller.

4. Discussion

The component approximation method has a number of evident advantages, which are demonstrated in the paper. The reconstructed parameters have a clear physical meaning. The number of pole figures required for reconstruction of ODF is relatively small. However, the component approximation method has some disadvantages. The method is not automatic, i.e. the researcher needs to choose the number and the type (peak or fiber) of components. In the case of not distinguishable peak and fiber types of components, the method is not recommended for use to avoid a misinterpretation of the texture character. The method is sensitive to a self-consistency of the measured pole figures. The measured pole figures must be self-consistent, but due to experimental errors this condition can be violated. Another challenging factor influencing the application of the method is a use of non-linear minimization problem. The results of the ODF reconstruction by component approximation method depend on the specific component choice and the functional to be minimized, which both are the key levers to optimize the analysis. The RP value is found to be more stable but less sensitive parameter for minimization procedure and the functional (6) seems to be more sensitive one. As a consequence, for example, for the noisy data the using of the RP value is preferable.

Acknowledgment
The present work has been initiated and supported by Rigaku Corporation.

References

[1] Nikolayev D I, Savyolova T I and Feldman K 1992 *Textures and Microstructures* **19** 9
[2] Bucharova T I and Savyolova T I 1993 *Textures and Microstructures* **21** 161
[3] Bermig G, Tobisch J, Richter K and Helming K 1993 *Material Science Forum* **133-136** 163
[4] Wassermann G and Greven J 1962 *Texturen metallischer Werkstoffe* (Berlin: Springer Verlag)
[5] Matthies S, Vinel G W and Helming K 1987 *Standard Distributions in Texture Analysis* (Berlin: Akademieverlag)
[6] Schaeben H 1997 *Phisica Status Solidi (b)* **200** 367
[7] Ivanova T M and Nikolayev D I 2001. *Physica Status Solidi (b)* **228** 3 825
[8] Barton N R, Boyce D E, Dawson P R, 2002 *Textures and Microstructures*, **35(2)** 113
[9] Bernier J V, Miller M P and Boyce D E 2006 *J. Appl. Cryst.* **39** 69
[10] Hielscher R and Schaeben H 2008 *J. Appl. Cryst.* **41** 1024
[11] Conn A R, Gould N I and Toint P L 2000 Trust-region Methods. SIAM Society for Industrial & Applied Mathematics, Englewood Cliffs, New Jersey, MPS-SIAM Series on Optimization edition.
[12] Lychagina T A, Nikolayev D I and Wagner F 2009 *Texture, Stress and Microstructure*, 237485
[13] Lychagina T A and Nikolayev D I 2003 *Physica Status Solidi (a)*, **195(2)** 322

IOP Conf. Series: Materials Science and Engineering **82** (2015) 012008 doi:10.1088/1757-899X/82/1/012008

Determination of orientations from monochromatic diffraction patterns as the constellation problem

A Morawiec

Institute of Metallurgy and Materials Science, Polish Academy of Sciences, Kraków, Poland

E-mail: nmmorawi@cyf-kr.edu.pl

Abstract. Determination of crystal orientations from diffraction patterns is directly linked to pattern indexing. The problem of indexing can be seen as matching scattering vectors to vectors of the crystal reciprocal lattice. With known crystal structure, the simplest version of indexing is formulated as the (constellation) problem of matching vectors under rotations: given two sets X and Y of unit vectors, determine a rotation carrying the largest subset of X to a position approximating a subset of Y. It is shown that algorithms for solving the constellation problem establish a framework for several orientation determination methods. A class of these algorithms is based on accumulating contributions in the rotation space. A rotation with the largest accumulation is considered to solve the problem. The contributions can be made by n-tuples of vectors with n starting from 1. Formulas for the points of accumulation are given for arbitrary $n \geq 1$. Particularly simple turns out to be the case of 2-tuples. It has a potential of being robust, and it is easy to implement.

1. Introduction

Over the last two decades, automatic orientation imaging has become one of the most significant methods of investigating polycrystalline materials. That occurred thanks to the progress in orientation mappings carried out using mainly electron backscatter diffraction (EBSD) [1] and similar systems relying on transmission electron microscopy [2] or high-energy X-ray diffraction [3]. The orientation imaging is directly linked to indexing of diffraction patterns; orientation determination and indexing of patterns originating from known crystal structures are practically equivalent. As numerous patterns are being analyzed, the indexing algorithms need to be time efficient. Equally important is their robustness; indexing is expected to function properly despite errors in geometric parameters of the patterns and/or spurious reflections. Robustness is crucial in diffraction-based non-destructive grain shape reconstruction [3] when multiple crystallites contribute to recorded patterns. Indexing and orientation determination algorithms are as good as their ability to cope with imperfections of input data. For devising optimal algorithms, it is essential to express the problem in a formal way based on simple principles.

In pattern indexing and crystal orientation determination, the point is to match measured scattering vectors (differences between reflected and incident wave vectors) to low-index vectors of the crystal reciprocal lattice. From a computational perspective, indexing can be formulated as the so-called *largest common point set* (LCPS) problem; cf. [4]. In the simplest case, e.g., when EBSD bands are indexed, the magnitudes of the scattering vectors are unknown, and all involved vectors are assumed to be of unit magnitude. The corresponding restricted case of LCPS is referred to as the *constellation problem* [5]: given two sets X and Y of unit vectors, determine a rotation carrying the largest subset of X to a position approximating a subset of Y. Solving this problem involves both combinatorial matching (assignment) and continuous alignment. In a strict formulation of the constellation problem, a given maximum distance between the matched vectors is allowed. In indexing, such a distance is difficult to specify, or it may vary widely because experimental errors depend on the scattering vector and the direction in space. The heuristic procedures considered below are suitable for solving cases with unspecified distance thresholds, but they do not provide any theoretical guarantee of the quality of the output. In that respect, they are similar to the softassign matching algorithm proposed in [6].

This paper describes a common framework for a number of indexing algorithms. Approaches which at first sight may look different, turn out to be based on similar foundations. We demonstrate that a number of these algorithms can be seen as an accumulation of contributions, i.e., a form of the generalized Hough transform with the abstract rotation space (or the symmetry induced fundamental region [7]) as the parameter space. The contributions can be made by individual vectors (one vector from each set), and in this case accumulation takes place along continuous curves in the rotation space; with two or more vectors from each set, accumulation takes place at discrete points in the rotation space. At the end, the rotation with the largest total accumulation is considered to solve the problem. We list straightforward formulas for the points of accumulation, and we check optimality of these points. The advantage of the considered methods lies in their robustness and conceptual simplicity. The algorithms are relatively easy to implement. We show that particularly simple is the case when pairs of vectors are matched.

Further on, X and Y contain unit vectors bound to a fixed center of rotation. The vectors are represented by column matrices with Cartesian components. Given a non-zero square matrix A, the symbol $\mathcal{O}(A)$ denotes the special orthogonal matrix closest to A in the sense that it minimizes the distance $\|A - \mathcal{O}(A)\|^2$; see [7]. For a non-zero vector x, the symbol $O(x, \omega)$ denotes the special orthogonal matrix representing the rotation by the angle ω about the axis determined by x.

2. Accumulation in the rotation space

One of the strategies for indexing of bands in EBSD patterns is known as 'triplet indexing' [1]. Inter-band angles are compared to angles following from known crystal structure. Based on each band triplet, potential crystal orientations are determined, and each of these orientations gets a vote. The orientation with the largest number of votes is considered to solve the problem. For a more formal description, instead of the angles, we use vectors. If a triplet of linearly independent vectors x_i of X ($i = 1, 2, 3$) is rotated to form the triplet y_i ($= Rx_i$) of Y, the special orthogonal matrix R representing the rotation can be expressed as

$$R = [\,y_1\ y_2\ y_3\,]\,[\,x_1\ x_2\ x_3\,]^{-1}\ . \tag{1}$$

In indexing, the (normalized scattering) vectors x_i of X are affected by errors, the match between (normalized reciprocal lattice) vectors y_i and the rotated x_i is only approximate, and the matrix $[\,y_1\ y_2\ y_3\,]\,[\,x_1\ x_2\ x_3\,]^{-1}$ needs to be adjusted to satisfy orthogonality conditions; e.g., one can use $R'^{(3)} = \mathcal{O}\left([\,y_1\ y_2\ y_3\,]\,[\,x_1\ x_2\ x_3\,]^{-1}\right)$. Now, similarly to [1], to determine the rotation carrying the largest subset of X to a subset of Y, all triplets from X and from Y are considered, and each pair of triplets contributes to a point in the rotation space. Assuming exact congruence between the largest matching subsets of X and Y, the rotation relating them gets the largest number of contributions, and the problem is reduced to locating a maximum of the contributions.

This 'triplet indexing' can be generalized to higher order procedures by employing n-tuples of vectors. To make such a generalization, one needs a formula calculating the rotation best matching large vector sets (and a clear understanding of what the "best matching" really means). Let $n \geq 3$, and let the rank of the involved matrices be equal to 3. In analogy to eq.(1) and definition of $R'^{(3)}$, the rotation matrix can be obtained from

$$R'^{(n)} = \mathcal{O}\left([\,y_1\; y_2\; \dots\; y_n\,]\,[\,x_1\; x_2\; \dots\; x_n\,]^{+}\right) , \tag{2}$$

where the superscript + denotes the generalized (Moore–Penrose) inverse. This solution is optimal in the sense that $R'^{(n)}$ is the special orthogonal matrix closest to the matrix $A = [\,y_1\; y_2\; \dots\; y_n\,]\,[\,x_1\; x_2\; \dots\; x_n\,]^{+}$ which in turn minimizes $\sum_{i=1}^{n}(y_i - Ax_i)^2$; cf. [8]. Less involved and more suitable is the solution given by

$$R^{(n)} = \mathcal{O}\left([\,y_1\; y_2\; \dots\; y_n\,]\,[\,x_1\; x_2\; \dots\; x_n\,]^{T}\right) ; \tag{3}$$

it is optimal in the sense of spherical regression, i.e., one can show that the above $R^{(n)}$ is the special orthogonal matrix minimizing the sum $\sum_{i=1}^{n}(y_i - R^{(n)}x_i)^2$.

In matching based on large n, additional vectors sharpen the criteria for the congruence of given subsets. On the other hand, using higher order n-tuples increases the computational costs. To avoid combinatorial explosion, n needs to be kept small. Moreover, numerical experiments show that with error affected data, higher order n-tuples do not really improve the quality of indexing; cf. [9]. This draws attention to indexing based on $n < 3$, i.e., on individual vectors ($n = 1$) or pairs of vectors ($n = 2$).

The procedure for accumulating contributions based on individual vectors ($n = 1$) was described in [4]. Given two vectors x_1 and y_1 in X and Y, respectively, we look for all rotations transforming x_1 on y_1. Assuming that $x_1 \neq -y_1$, these rotations are represented by the matrices $R^{(1)}(\omega) = O(y_1, \omega)\; O(x_1 + y_1, \pi)$, and they constitute an ω-parameterized geodesic in the rotation space. Each pair from $X \times Y$ contributes to rotations located on one of these lines. As above, the rotation relating the largest matching subsets of X and Y gets the largest number of contributions, and again, the problem is reduced to locating a maximum of the contributions. A variant of this approach was implemented by Schmidt [10].

The unexplored case of $n = 2$ turns out to be particularly interesting. It is in a sense elemental among those providing discrete solutions. Eqs.(2–3) cannot be directly extended to get rotations matching 2–tuples. (When the rank of $[\,x_1\; x_2\; \dots\; x_n\,]$ is smaller than 3, the Moore–Penrose inverse gives A satisfying $y_i = Ax_i$ and minimizing $\|A\|^2$ [8], but $\mathcal{O}(A)$ is generally distant from the best rotation matrix carrying x_i on y_i. Also eq.(3) is inconvenient because of complications with calculating the special orthogonal matrix closest to a singular matrix.) The simplest way of circumventing this problem is to use eqs.(2–3) for $n = 3$ with the third linearly independent vector, say x, formed from x_1, x_2, and a corresponding vector y analogously constructed from y_1, y_2. The simplest choice is to use the cross products $x = x_1 \times x_2$ and $y = y_1 \times y_2$; with these vectors, the matrices $[\,x_1\; x_2\; x\,]^T$ and $[\,x_1\; x_2\; x\,]^{-1}$ are generally different, but eqs. (2) and (3) lead to the same result

$$R^{(2)} = \mathcal{O}\left([\,y_1\; y_2\; y\,]\,[\,x_1\; x_2\; x\,]^{T}\right) = \mathcal{O}\left([\,y_1\; y_2\; y\,]\,[\,x_1\; x_2\; x\,]^{-1}\right) = [\,r_1^y\; r_2^y\; r_3^y\,]\,[\,r_1^x\; r_2^x\; r_3^x\,]^T , \tag{4}$$

where r_1^x, r_2^x and r_3^x represent normalized versions of x, $x_1 + x_2$ and $x_1 - x_2$, respectively, and the r_i^y vectors are defined in analogous way. The essential point is that $R^{(2)}$ given by eq.(4) also transforms the pair x_1, x_2 to the position closest to that of y_1, y_2 in the sense of spherical regression. Moreover, with the last part of eq.(4), the computation of this optimal rotation matrix involves only elementary steps.

As in all accumulation-based methods, at the end, one needs to determine the locations of maxima in the accumulator space. Since the match between vectors is only approximate, some tolerance must be allowed. A straightforward way to evaluate the contributions is by partitioning the rotation space into equivolume bins of size linked to the accuracy of experimental data and the resolution of resulting orientations. The center of the bin with the largest accumulation is the sought rotation matching the largest subsets of X and Y. With this approach, extra measures are needed to take account of contributions distributed in neighboring bins.

Instead of binning, cluster analysis can be implemented. One of the methods is to use a list of potential solutions: Every new rotation obtained from eqs.(2–4) is compared to already saved potential solutions. If the rotation deviates from a given solution by a misorientation angle smaller than a threshold, a ranking number of that solution is increased, and the solution is corrected by taking a weighted average [11] of the solution and the new rotation. Otherwise, the rotation is appended to the list as a new potential solution. Clearly, this approach is more suitable for resolving the discrete cases ($n \geq 2$) than the continuous one ($n = 1$).

With $n \geq 2$, some of the matchings can be omitted by assuming that the vectors x_i can match y_i only if the angles between vectors of the n-tuples differ by less than a threshold; if not, the n-tuples are rejected as a possible match. (This is particularly simple when $n = 2$: the vectors x_1 and x_2 can match y_1 and y_2 only if the angle between x_1 and x_2 is close to that between y_1 and y_2.) This approach links the indexing methods described above to methods in which measured angles between scattering vectors are compared to angles between reciprocal lattice vectors with potential solutions scored according to the number of good matches. Thus, in essence, these methods also use a form of accumulation, but the accumulator space and the way of contributing are not explicitly identified.

3. Final remarks

An inspection of the field (including ab initio indexing) shows that robust solutions to the indexing problem are based on various forms of accumulation. They differ by the nature of contributions, the parameter space, and the methods of collecting and counting the contributions. Depending on demands, one may use various combinations of these forms; in particular, to improve reliability of indexing, one may apply multiple contributions of different types and various methods of counting the contributions. Moreover, as the orientation determination via accumulation can be seen as the generalized Hough transform, its reliability can also be improved by adapting known enhancements of that transform [12].

For brevity, we considered only the constellation problem (LCPS problem limited to rotations), with all vectors having the same magnitude, but it is worth noting that the described procedures can be relatively easily generalized to matching vectors with arbitrary magnitudes.

Acknowledgments

The work was supported by National Science Center based on DEC-2012/06/M/ST8/00449.

References

[1] Adams B L Wright S I and Kunze K 1993 *Metall. Trans.* A **24** 819
[2] Fundenberger J J Morawiec A Bouzy E Lecomte J S 2003 *Ultramicroscopy* **96** 127
[3] Poulsen H F 2004 *Three-Dimensional X-Ray Diffraction Microscopy* (Berlin: Springer)
[4] Morawiec A and Bieda M 2005 *Arch. Metall. Mater.* **50** 47
[5] Cardoze D E and Schulman L J 1998 *Proc. 39th Ann. Symp. on Foundations of Computer Science* (Los Alamitos, California: IEEE) p 156
[6] Rangarajan A Chui H and Bookstein F L 1997 *Information Processing in Medical Imaging*, ed J Duncan and G Gindi (Berlin: Springer) p 29
[7] Morawiec A 2004 *Orientations and Rotations, Computations in Crystallographic Textures* (Berlin: Springer)
[8] Golub G H and Reinsch C 1971 *Handbook for Automatic Computation. II. Linear Algebra* ed J M Wilkinson and C Reinsch (Berlin: Springer) p 134
[9] *EDAX Technical Note – EBSD* 2013 EDAX Inc. at www.edax.com/download/Triplet_Indexing_LR.pdf
[10] Schmidt S 2014 *J. Appl. Cryst.* **47** 276
[11] Morawiec A 1998 *J. Appl. Cryst.* **31** 818
[12] Leavers V F 1993 *CVGIP: Image Understanding* **58** 250

Loading and texture bias on the competitive slip activity for basal and prismatic slip systems in HCP alloys

A K Saxena [1,2], A Tewari[3] and P Pant[1]
[1]Indian Institute of Technology, Bombay, Department of Metallurgical Engineering and Materials Science - 400076, India
[2]National Centre for Aerospace Innovation and Research, IIT Bombay - 400076, India
[3]Indian Institute of Technology, Bombay, Department of Mechanical Engineering - 400076, India
E-mail: ashishkumar2911@gmail.com

Abstract: Asymmetry in hexagonal crystal structure makes the occurrence of slip strongly dependent on the texture of sample. In titanium, which has a c/a ratio less than ideal, slip occurs preferentially on prismatic slip system. However other slip systems may get activated depending on the resolved shear stresses. In this paper we present results from plane strain compression experiments where the same area of the sample was imaged before and after deformation to document changes in microstructure. We then compare these results with a simple calculation of plastic strain based on activation of various slip systems depending on their respective critical resolved shear stresses. We show that incorporation of a strain rate dependent hardening parameter provides a reasonable match with the experimentally observed deformation behaviour of various grain orientations.

1. Introduction:

Titanium and its alloys high specific strength, high oxidation and corrosion resistance which makes them suitable for aerospace, biomedical and other high performance applications. Among all Ti alloys, Ti-6Al-4V (Ti64) has biggest market share. In the aerospace industry Ti64 is used largely in fuselage, nacelles, landing gear beams, fan disc, low pressure compressor discs and components of engine[1].

Ti64 is a two phase alloy with the predominant phase being α (HCP crystallite) along with a small fraction of β (BCC structure) dispersed along the grain boundaries of α phase. The α phase deforms by both slip and twinning, depending on the stress state, loading rate, and grain shape and size distribution. Of the three possible slip systems: basal, prismatic and pyramidal, prismatic is the most preferred since the c/a ratio for Ti64 is less than ideal [2–7]. In this paper we present results from plane strain compression (PSC) experiments for fine grain Ti64 samples. The average grain size was about 2 μm, thus making slip the only feasible mode of deformation[8]. Further the experiments were carried out at room temperature and at strain rates of about 10^{-3} s^{-1}, both conditions favouring slip activity. So these experiments allow a detailed study of the influence of grain orientation on slip activity. The results are then compared with a simple crystal plasticity calculation for various orientations. We show that the incorporation of strain rate dependent plasticity allows us to capture the trends observed during successive deformation of near-basal and near-prismatic grains in the polycrystalline sample.

2 Experimental method

PSC was carried out using the setup shown in figure 1(a). Load is applied along the Z direction, and the sample is free to deform along the Y axis, while deformation along the X axis is restricted by the die (figure 1(b)). Figure 1c is the backscattered image of the sample before PSC showing microindent marks to identify the same area after deformation for sequential analysis of the microstructure. Figure 1d shows the inverse pole figure map (IPF map) of undeformed sample. Grains are randomly oriented and the average grain size is about 1.9μm. It has been reported that twinning becomes more difficult with decreasing grain size and is unlikely to occur in this sample [9]. So the changes in microstructure observed after successive deformation steps can be attributed only to the slip activity in Ti64.

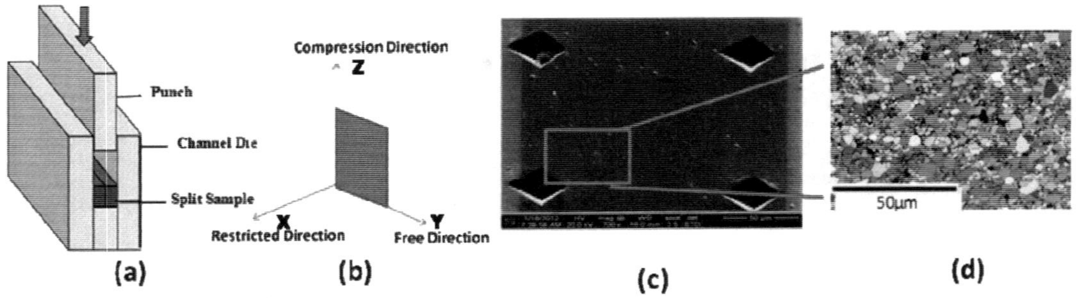

Figure.1 (a) Set up used for PSC experiment (b) loading direction, restricted direction and free to deform direction (c) SEM image of undeformed Ti64 sample with microindent mark for imaging the same area after each deformation step (d) EBSD image of selected region.

Electron back scattered diffraction (EBSD) imaging was done for the marked region in the undeformed, 5% deformed and 7% deformed sample. From EBSD data Kernel Average Misorientation (KAM) of grain is calculated. KAM of a grain is used to quantifying the local strain accumulated in a grain due to plastic deformation [10]. In figure 2 average KAM for near basal and near prismatic ($10\bar{1}0$ and $2\bar{1}\bar{1}0$) oriented grains is plotted as a function of % deformation.KAM increases with deformation for all orientations, however the rate of increase is higher for near-prismatic grains as compared to the near basal grains.

It can be seen that even for the "undeformed" sample, KAM values range from about 0.5 for near-basal grains to 0.3 for near prismatic orientations.

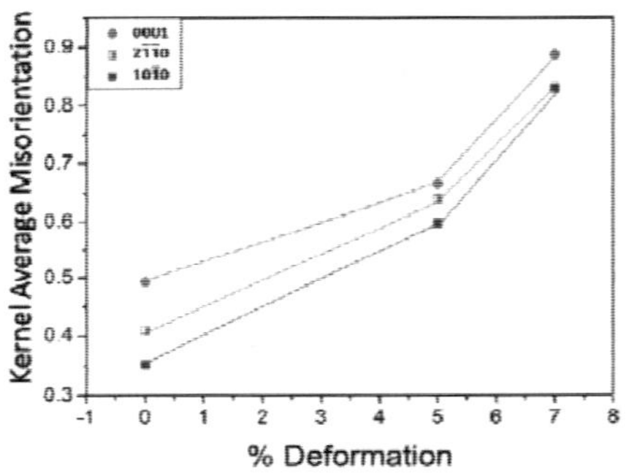

Figure 2 KAM as a function of deformation for differently oriented grains

However this spread ΔKAM decreases with increasing deformation. Since misorientation development is related to dislocation accumulation, reduction in ΔKAM is indicates less slip activity in near basal orientations relative to near prismatic oriented grains.

3. Modelling framework

In this section we present a calculation of average plastic strain in hexagonal Ti64 by calculating resolved shear stresses on all slip systems and using CRSS value as threshold for slip activation.

The resolved shear stress (τ) on the slip plane and in the slip direction is the primary driving force for slip. For each slip system α, with slip direction m^α and slip plane normal n^α

$$\tau^\alpha = \sigma^c m^\alpha \otimes n^\alpha, \quad (1)$$

where σ^c is the stress tensor in crystal coordinate system (depending on the grain orientation).

The total plastic strain tensor D^p [11] can be calculated by summation over all slip systems,

$$D^p = \sum_\alpha \dot{\gamma}^\alpha P^\alpha, \quad (2)$$

where $P^\alpha = 0.5(m^\alpha n^\alpha \otimes n^\alpha m^\alpha)$.

Here, the shearing rate $\dot{\gamma}^\alpha$ is given by

$$\dot{\gamma}^\alpha = \dot{\gamma}^0 \left(\frac{|\tau^\alpha|}{S^\alpha}\right)^{\frac{1}{m}} sign(\tau^\alpha), \qquad (3)$$

where $\tau^\alpha = \sigma'.P^\alpha$ and σ' is the deviatoric component of applied stress tensor
Finally the plastic strain tensor has the form

$$D^p = \sum_\alpha \dot{\gamma}^0 \left(\frac{|\sigma'.P^\alpha|}{S^\alpha}\right)^{\frac{1}{m}} sign(\sigma'.P^\alpha)\, P^\alpha \qquad (4)$$

In equation 3, $\dot{\gamma}^0$ is the rate of strain in PSC experiments, taken as 0.001 sec^{-1}. S^α is the critical resolved shear stress for α^{th} slip system. For titanium, CRSS values for basal, prismatic and pyramidal slip systems are in the ratio 1:0.7:3[2,12–14].

As seen from equation 3, the rate of shear depends on the strain rate sensitivity parameter, *m*, in addition to the resolved shear stress. Typically *m* ranges from 0.01 to 0.02 for most metals [11] In our calculations, basal oriented grains show higher KAM than prismatic grains in the undeformed sample, *m* is taken equal to 0.01 for basal and 0.005 for other orientations for 5% deformation. for next step of deformation to 7%, *m* is 0.02 for basal and 0.018 for other orientations.

Finally from the plastic strain tensor obtained for each orientation, the effective Von Mises strain[15] can be calculated by

$$\bar{\varepsilon}^p = \sqrt{\sum_{i,j}^{3} \frac{2}{3} \varepsilon_{ij}{}^p \varepsilon_{ij}{}^p}, \qquad (5)$$

4.Results and discussion

In figure 3, plastic strain due to each slip system is shown separately, along with the total effective plastic strain for m values 0.005, 0.015 and 0.02. It is clear that plastic strain decreases significantly with increasing *m*. In addition, differences between various slip systems also decrease as *m* increases.

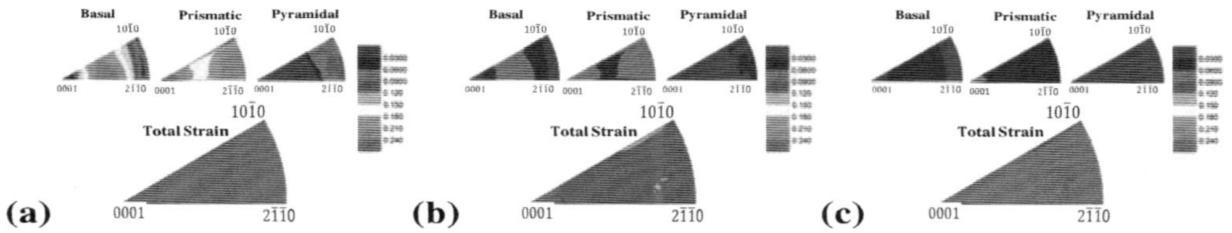

Figure 3: The slip system wise and total effective strain with (a) m=0.005,(b) m= 0.015 and (c) m=0.02. (color online)

Figure 4(a) shows the effective plastic strain values calculated by using *m* = 0.01 for basal and *m* = 0.005 for other orientations. In our model this corresponds to 5% deformation of the specimen. The difference in the effective plastic strain between basal and prismatic orientations is 0.06. Results for 7% deformation are shown in figure 4(b) obtained by setting *m* = 0.018 for basal and 0.02 for other orientations. These values are selected because the 5% deformed sample is expected to be more strain hardened than the starting sample, so *m* is higher than the previous case. At the same time ΔKAM is less (see figure 4(c)) so we expect less difference in the hardening response of basal and non-basal grains. The difference in the effective plastic strains obtained from this calculation for near-basal and near-prismatic orientations is 0.02.

A comparison of differences in effective plastic strain in various orientations obtained from the rate-dependent plasticity model, and ΔKAM values shows that we are able to capture the trend of decreasing ΔKAM with increasing % deformation. When the sample is deformed to 5%, more slip occurs in the near-prismatic grains due to a combination of orientation effect and less prior-hardening. This can be seen from calculated effective strain values and matched with the decrease in ΔKAM – from 0.14 in the undeformed sample to 0.07 after 5% deformation. In the next step of deformation to 7%, both orientations are harder and difference in effective plastic strain decreases to 0.02, while corresponding ΔKAM decreases to 0.06

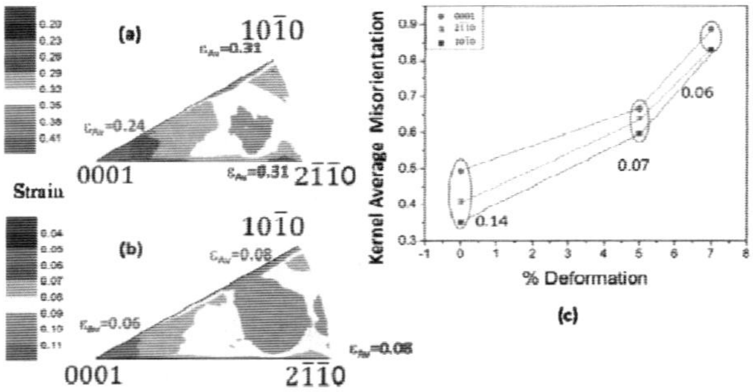

Figure 4: The total effective strain calculated for (a) first deformation step (b) for 2nd deformation step of plane strain compression (d)experimentally obtained KAM as function of deformation.

5. Conclusions

- Two key observations from experiments: 1) the rate of KAM increase with increasing deformation is less for basal grains and more for prismatic; 2) spread of KAM values decreases with deformation – more deformed orientations deform less and vice versa.
- Slip activity was modeled by taking into account slip activation based on relative CRSS values for various slip system. Slip was assumed to be rate dependent. The strain rate sensitivity parameter *m* was varied from 0.005 to 0.02 for 5% and 7% deformation.
- By selecting appropriate *m* values depending on the KAM values of orientations, we could capture differences in the deformation response of near-basal and near-prismatic orientations, even though grain interactions have been ignored in the current model.
- Results show that the strain rate sensitivity, *m*, plays a critical role in determining the hardening response.

Acknowledgements
Authors gratefully acknowledge the support provided for this work by National Centre for Aerospace Innovation and Research, IIT-Bombay, Dept. of Science and Technology- Government of India, Boeing Company and IIT Bombay Collaboration. EBSD was carried out at the National Facility of Texture and OIM – DST-IPHRA facility at IIT Bombay.

References

[1] Boyer R 1996 *Materials Science and Engineering A* **213** 103
[2] Guven G, Karaman I 2006 *Acta Materialia* **54** 3755
[3] Bao L, Schuman C, Lecomte J 2010 *CMC* **15** 113
[4] Ahn SH et al 2010 *Materials Science and Engineering:* A **528** 165
[5] Some T S, The D, Ii T 1981 *Metallurgical Transactions A* **12A** 409
[6] Munroe N, Tan X, 1997 *Acta Metallurgica* **36** 1383
[7] Chichili DR, Ramesh KT, Hemker KJ1998 *Acta Metallurgica* **46** 1025
[8] Barnett M R 2008 *Scripta Materialia* **59** 696
[9] Ghaderi A, Barnett MR 2011 Acta Materialia 59 7824
[10] Chen P et al 2013 *Materials Science and Engineering A* **580** 114
[11] Asaro R J, Needleman A 1985 *Acta Metallurgica* **33** 923
[12] Bieler TR, Glavicic MG, Semiatin SL 2002 *JOM* **54** 31
[13] Warwick JLW et al, 2012 *Acta Materialia* **60** 4117
[14] Betsofen SY et al 2007 *Russian Metallurgy Metally* **2007** 387
[15] Colby RB, *Equivalent Plastic Strain for the Hill's Yield Criterion under General Three-Dimensional*, 2013 Massachusetts Institute of Technology

IOP Conf. Series: Materials Science and Engineering **82** (2015) 012010 doi:10.1088/1757-899X/82/1/012010

Simulation of texture formation of body-centered-cubic metals by three kinds of intersections of two {110} slip planes

H Masui
Professor emeritus, Teikyo University, Utsunomiya, Japan
E-mail: masuihiro@jcom.home.ne.jp

Abstract. Based on the rotational symmetry of the principal axes of X[100], Y[010] and Z[001] ,72 possible combinations of the five slips on {110} planes based on Taylor's formidable restriction rule of the five slips are calculated among three kinds of intersections of two {110} planes on ⟨111⟩ direction in bcc metal. Crystal rotation is carried out by only one solution among the 72 by the minimum total slip at every strain and simulates properly lengthy of accumulated researcher's experimental results such as the three stable orientations of bcc metal in rolling {112}⟨110⟩, {11 11 8}⟨4 4 11⟩ and {100}⟨011⟩.

1. Introduction

The pencil glide theory combining {110},{112} and {123} planes with common direction ⟨111⟩ gave a very clear account of prediction about crystal rotation of body-centered-cubic (bcc) metal.[1-4] Barrett et al found that high silicon steel slips by only {110}⟨111⟩ and is less ductile than mild steel. [5] But they also reported in other experiment that the high silicon steel forms finally almost same cold rolling textures as those of low carbon mild steel, although the latter are supposed to have been formed by the pencil glide. [6] This led directly to a motivation in this study to investigate whether crystal rotation could take place by only slip on {110} plane on direction ⟨111⟩ in bcc metal.

2. Model

The principal axes of X[100], Y[010] and Z[001] are perpendicular to each other as the three orbits of $\{\pm X\},\{\pm Y\}$ and $\{\pm Z\}$ by a rotational symmetry of mathematical group theory in such way that component X is not related to Y or Z one another whichever.[7] There is a conservation quantity in the symmetry.[8] For example, a small ball at the top of Mexican hat may drop down on the bottom everywhere equally and symmetrically within 360° around the top due to existence of potential energy of the ball as a conservation quantity. Once the hat inclined, the ball would lose both the quantity and symmetry. As Taylor proved, crystal rotates so that slips occurs associating themselves with the minimum total slip amount[9]. The minimum total slip amount in crystal by Taylor corresponds to the conservation quantity in the rotational symmetry of cubic crystal.

There are three kinds of {110} planes having a common glide axis ⟨111⟩ in bcc crystal such as those illustrated on Fig.1 in case around Z[001] axis.

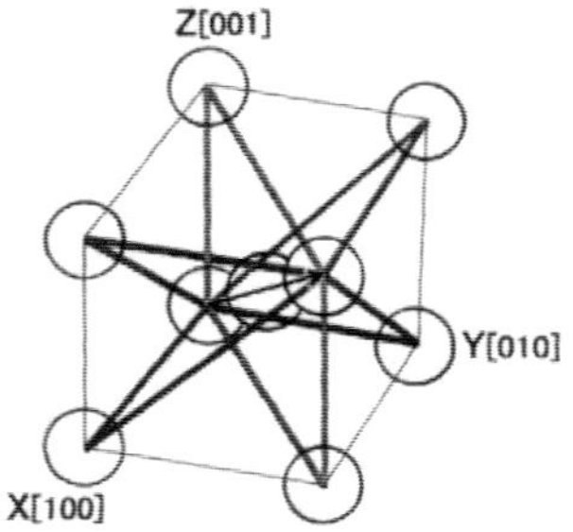

Fig.1 Three kinds of {110} planes composed of {110}, {101} and {011} ones around the principal Z[001] axis with a common slip direction ⟨111⟩ in bcc crystal.

Calculation method is exemplified as follows.

Firstly, an intersection of two kinds of {110} planes from the three ones composed of {110}, {101} and {011} as illustrated on Fig.1 is chosen. As example, an intersection between {101} and {011} in Fig.1 is chosen and shears on {101} group planes on four ⟨111⟩ directions in case around Z[001] axis are named by turns dX_1, dX_2, dX_3 and dX_4 and those on {011} group planes are done dY_1, dY_2, dY_3 and dY_4 correspondently as shown in Fig.2. These are geometrically correlated to strain components $d\varepsilon_{11}$, $d\varepsilon_{22}$, $d\varepsilon_{33}$, $d\varepsilon_{12}$, $d\varepsilon_{13}$ and $d\varepsilon_{23}$ and crystal rotation $d\phi_1$, $d\phi_2$ and $d\phi_3$ around the principal axes X, Y and Z respectively and accordingly a crystal rotation and texture formation can be calculated for the bcc crystal during any of compression, elongation, rolling and others.

Second the eight slips, however, must be reduced to five shears according to Taylor's formidable combination rule of the five slips as follows.[9] The geometrical condition rule for a given strain cannot be satisfied if the five shears are chosen so that two are taken from one group, *i.e.* one slip plane, and the remaining three are chosen one from each of the three remaining groups, in other words, all of which must be chosen so that two shears occur on each of two planes, one on the third and none on the fourth. This rule was properly applied to the present model as exemplified in Fig.2. By the Taylor's combination rule of the five, Fig.2 in case of $dX_1 = dY_1 = 0$ shows two possible combinations (○) of the five slips and two impossible ones (×). This phenomenon of two possible combinations on Fig.2 appears in each case of $dXi = dYi = 0$ for four ⟨111⟩ directions (i=1~4), and furthermore every for each group around X, Y and Z axes.

This shows 24(=2×4×3) possible combinations in the intersection of two groups of {110} planes (in this example case {101} and {011}) from the three ones. There are three kinds of intersections of two {110} planes as illustrated on Fig.1 and the model accordingly provides with 72(=24×3) possible combinations of the five slips as a whole.

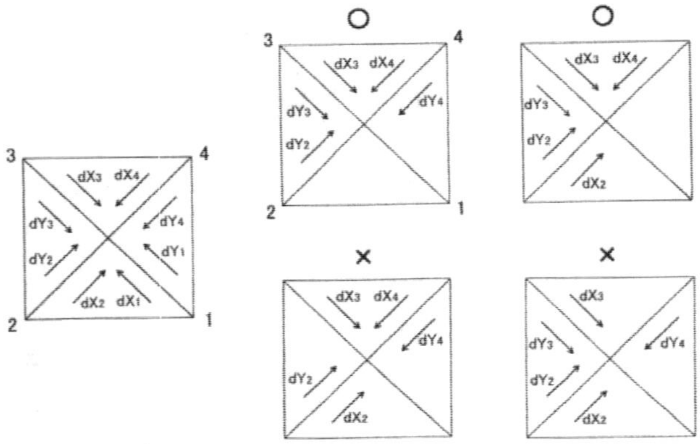

(1) (2)

Fig.2 Example of an intersection between {101} and {011} in Fig.1.

(1) Eight slips in bcc cubic metal wherefrom five slips are needed for deformation and crystal rotation.

(2) Two possible combinations of the five slips (○) and two impossible ones (×) in the model by the rule of the Taylor's formidable restriction of the five slips, "all of which must be chosen so that two shears occur on each of two planes, one on the third and none on the fourth", are exemplified in case of $dX_1=dY_1=0$ ($dX_2=0$ or $dY_4=0$ is possible).

In this case of the intersection example between {101} and {011} from the three kinds of intersections around Z[001] axis in Fig.1, there are geometrical relations as demonstrated in equation (1) among the shears dX_1,dX_2,dX_3 and dX_4 on {101} group planes on four ⟨111⟩ directions as well as dY_1,dY_2,dY_3 and dY_4 on {011} group planes, and strain components, crystal rotations. Similar equations exist also in each case around X[100] and Y[010] cases. Further, applied to equation (1), a calculation is carried out as example for crystal rotation in the possible case(○) of $dX_1=dY_1=0$, $dX_2=0$ on Fig.2 according to the Taylor's combination rule of the five slips and its solution is introduced as in equation (2). The above example is only one of total 72 possible combinations of the five slips by the three kinds of intersections of two {110} planes on ⟨111⟩ directions in bcc metal and each of the 72 has respectively similar equations as equation (2) .

Third an actual crystal rotation by equation (2) proceeds only for the case of which the total slip amount Γ(gamma) defined by equation (3) is minimum among the 72 cases. Calculation of equation (3) is performed by inserting both equation (1) and equation (2) to equation (3). Suppose one of the 72 cases is selected in the model.

$$dX_1 = (\sqrt{6}/4)(d\varepsilon_{11} + d\varepsilon_{12} + d\varepsilon_{13} - 2d\varepsilon_{23} + d\phi_2 - d\phi_3)$$
$$dX_2 = (\sqrt{6}/4)(-d\varepsilon_{11} + d\varepsilon_{12} - d\varepsilon_{13} - 2d\varepsilon_{23} - d\phi_2 - d\phi_3) \quad (1)$$
....
$$dY_4 = (\sqrt{6}/4)(-d\varepsilon_{22} + d\varepsilon_{12} - 2d\varepsilon_{13} - d\varepsilon_{23} + d\phi_1 + d\phi_3)$$

$$d\phi_1 = d\varepsilon_{22} + 2d\varepsilon_{12} - 2d\varepsilon_{13} - d\varepsilon_{23} \ , \quad d\phi_2 = -d\varepsilon_{11} - d\varepsilon_{13} \ , \quad d\phi_3 = d\varepsilon_{12} - 2d\varepsilon_{23} \quad (2)$$

$$\Gamma = \sqrt{dX_1 + dY_1} + \sqrt{dX_2 + dY_2} + \sqrt{dX_3 + dY_3} + \sqrt{dX_4 + dY_4} \quad (3)$$

3. Experimental results

According to the model Table 1 demonstrates how final stable rolling orientation in bcc $(112)[\bar{1}10]$ is derived from initial orientation $(111)[\bar{1}10]$ with increase of strain. At each strain the model selects one solution among the 72 cases composed of 24 ones each belonging to any one of X, Y and Z group. In this Table, it may be noted i) how the model gains the orientation with strain by way of selecting one of the partitions $\{\pm X\}, \{\pm Y\}$ and $\{\pm Z\}$ independently in the rotational symmetrical system and ii) how it holds a continuous value of Γ(gamma), the minimum total slip by way of equation (3) throughout X,Y and Z group at each strain so smoothly as not to give a fatal discontinuity in the value of Γ(gamma). This is in reasonable accord with an expectation that amount of the external work by applied force to material shall be continuously changed with strain.

Table 1. Simulation results by the model how a final stable rolling orientation $(112)[\bar{1}10]$of bcc metal is derived from initial orientation $(111)[\bar{1}10]$ with increase of strain.

initial			phi2 45.00	Phi 54.74	phi1 180.00	0.58	ND 0.58	0.58	-0.71	RD 0.71	0.00
strain	mode	gamma	phi2	Phi	phi1	ND			RD		
0.069	Y	3.31	45.00	52.87	181.15	0.56	0.56	0.60	-0.70	0.72	-0.02
0.139	X	3.28	45.12	51.21	180.19	0.55	0.55	0.63	-0.70	0.71	0.00
0.208	X	3.28	45.18	49.74	179.40	0.54	0.54	0.65	-0.71	0.70	0.01
0.277	Y	3.26	45.03	48.43	180.12	0.53	0.53	0.66	-0.71	0.71	0.00
0.347	X	3.25	45.11	47.25	179.55	0.52	0.52	0.68	-0.71	0.70	0.01
0.416	Y	3.24	44.98	46.20	180.06	0.51	0.51	0.69	-0.71	0.71	0.00
0.485	X	3.23	45.07	45.26	179.63	0.50	0.50	0.70	-0.71	0.70	0.00
0.555	Y	3.22	44.96	44.41	180.00	0.49	0.50	0.71	-0.71	0.71	0.00
0.624	Y	3.21	44.90	43.65	180.29	0.49	0.49	0.72	-0.71	0.71	0.00
0.693	X	3.20	45.00	42.96	180.02	0.48	0.48	0.73	-0.71	0.71	0.00
0.762	X	3.19	45.06	42.33	179.80	0.48	0.48	0.74	-0.71	0.71	0.00
0.832	Y	3.18	44.99	41.76	179.99	0.47	0.47	0.75	-0.71	0.71	0.00
0.901	Y	3.17	44.94	41.24	180.15	0.47	0.47	0.75	-0.71	0.71	0.00
0.970	Z	3.16	45.00	40.77	180.00	0.46	0.46	0.76	-0.71	0.71	0.00
1.040	X	3.16	45.04	40.34	179.89	0.46	0.46	0.76	-0.71	0.71	0.00
1.109	Y	3.15	44.99	39.94	179.99	0.45	0.45	0.77	-0.71	0.71	0.00
1.178	Y	3.14	44.97	39.58	180.08	0.45	0.45	0.77	-0.71	0.71	0.00
1.248	X	3.13	45.00	39.25	180.00	0.45	0.45	0.77	-0.71	0.71	0.00
1.317	Y	3.13	44.98	38.95	180.06	0.44	0.44	0.78	-0.71	0.71	0.00
1.386	X	3.12	45.00	38.67	180.01	0.44	0.44	0.78	-0.71	0.71	0.00
1.456	X	3.11	45.02	38.42	179.96	0.44	0.44	0.78	-0.71	0.71	0.00
1.525	Y	3.11	45.00	38.18	180.00	0.44	0.44	0.79	-0.71	0.71	0.00
1.594	X	3.10	45.01	37.97	179.97	0.44	0.43	0.79	-0.71	0.71	0.00
1.664	Y	3.10	45.00	37.77	180.00	0.43	0.43	0.79	-0.71	0.71	0.00
1.733	X	3.09	45.01	37.58	179.97	0.43	0.43	0.79	-0.71	0.71	0.00
1.802	Y	3.09	45.00	37.41	180.00	0.43	0.43	0.79	-0.71	0.71	0.00
1.872	Y	3.08	44.99	37.26	180.02	0.43	0.43	0.80	-0.71	0.71	0.00
1.941	Z	3.08	45.00	37.11	180.00	0.43	0.43	0.80	-0.71	0.71	0.00
2.010	Y	3.08	44.99	36.98	180.01	0.43	0.43	0.80	-0.71	0.71	0.00
2.079	X	3.07	45.00	36.85	180.00	0.42	0.42	0.80	-0.71	0.71	0.00
2.149	Y	3.07	45.00	36.74	180.01	0.42	0.42	0.80	-0.71	0.71	0.00
2.218	X	3.07	45.00	36.63	180.00	0.42	0.42	0.80	-0.71	0.71	0.00
2.287	X	3.06	45.00	36.54	179.99	0.42	0.42	0.80	-0.71	0.71	0.00
2.357	Y	3.06	45.00	36.44	180.00	0.42	0.42	0.80	-0.71	0.71	0.00
2.426	Y	3.06	45.00	36.36	180.01	0.42	0.42	0.81	-0.71	0.71	0.00
2.495	X	3.06	45.00	36.28	180.00	0.42	0.42	0.81	-0.71	0.71	0.00
2.565	X	3.05	45.00	36.21	180.00	0.42	0.42	0.81	-0.71	0.71	0.00
2.634	Y	3.05	45.00	36.14	180.00	0.42	0.42	0.81	-0.71	0.71	0.00
2.703	Y	3.05	45.00	36.08	180.00	0.42	0.42	0.81	-0.71	0.71	0.00
2.773	Z	3.05	45.00	36.02	180.00	0.42	0.42	0.81	-0.71	0.71	0.00
2.842	X	3.05	45.00	35.97	180.00	0.42	0.42	0.81	-0.71	0.71	0.00
2.911	Y	3.05	45.00	35.92	180.00	0.41	0.41	0.81	-0.71	0.71	0.00
2.981	Y	3.05	45.00	35.87	180.00	0.41	0.41	0.81	-0.71	0.71	0.00
3.050	X	3.04	45.00	35.83	180.00	0.41	0.41	0.81	-0.71	0.71	0.00

Fig.3 illustrates by the model dynamically how orientations in ODFs map at rolling ratio of (a)6%,(b)38%,(c)62% and (d)95% are rigorously accumulated from initial random ones with strain in rolling by the present model and consequently

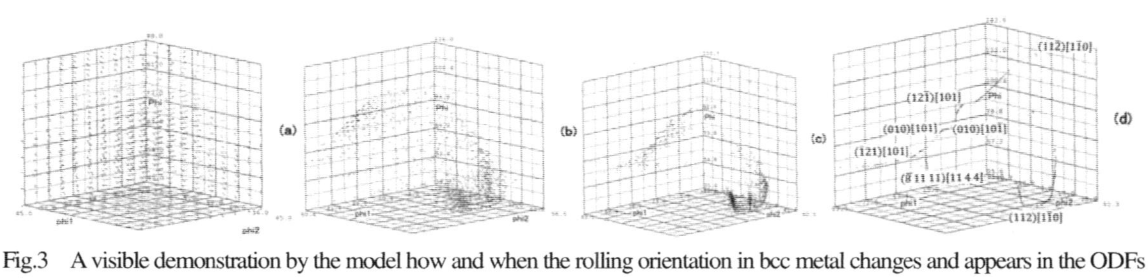

Fig.3 A visible demonstration by the model how and when the rolling orientation in bcc metal changes and appears in the ODFs map which shows the orientations at rolling ratio of (a)6%,(b)38%,(c)62% and (d)95%.

the three stable orientations of bcc metal in rolling[3,10,11] such as $\{112\}\langle 110\rangle, \{11\ 11\ 8\}\langle 4\ 4\ 11\rangle$ and $\{100\}\langle 011\rangle$ are attained as drawn on Fig.3(d).

4. Discussion

According to the model Table 1 illustrates how final stable rolling orientation in bcc $(112)[\bar{1}10]$ is derived from initial orientation $(111)[\bar{1}10]$ with increase of strain sequentially selecting one among the 72 cases composed of 24 ones each belonging to any one of X, Y and Z group. As in Table 1, Γ(gamma), the minimum total slip by way of equation (3) in the three X,Y and Z group, is not constant and changes but so gradually as not to give a fatal discontinuity in the value of Γ (gamma) so that the model may hold the symmetry in the system. As generally known, even in the stable system composed of extreme symmetry, it can lose the symmetry immediately and transiently when exposed to external forces or other physical energy. Even in this case, however, as it is still the orthodoxy accepted by the majority, if a break of symmetry is so small where the symmetry recovers immediately and sustains still continuously the original symmetric state that the symmetry may still give birth to forceful analytical means to the phenomenon .[12] Besides example of Table 1, it is supposed that such phenomena by the model may happen throughout the whole orientation as shown in Fig.3.

Similar model on face-centered-cubic(fcc) metal already reported by the author requires 24 cases in total for one solution where two combinations of five slips are accepted by the Taylor's formidable combination rule of the five slips as similarly as in Fig.2 on $\{111\}$ planes on every four directions of $\langle 110\rangle$ for each group of X, Y and Z principal axes of fcc metal.[13-15] A ratio of the 24(=2×4×3) of fcc metal to the 72 of bcc metal which has three intersections of $\{110\}$ planes, implies a ratio of one $\{111\}$ slip plane on direction $\langle 110\rangle$ of fcc metal to at least three slip planes $\{110\}, \{112\}$ and $\{123\}$ on direction of $\langle 111\rangle$ in the pencil glide theory of bcc metal. The idea which utilized the rotational symmetry among X, Y and Z principal axes was also applied for texture formation of NaCl structure.[16]

5. Conclusion

There are 72 combinations of the five slips on $\{110\}$ planes based on Taylor's formidable restriction rule of the five slips among three kinds of intersections of two $\{110\}$ planes on $\langle 111\rangle$ direction in bcc metal based on the rotational symmetry of the principal axes of X[100], Y[010] and Z[001]. One solution of crystal rotation among the 72 is selected by the minimum total slip at every strain and simulates properly lengthy of accumulated researcher's experimental results such as the three stable orientations of bcc metal in rolling $\{112\}\langle 110\rangle$, $\{11\ 11\ 8\}\langle 4\ 4\ 11\rangle$ and $\{100\}\langle 011\rangle$.

References

[1] Calnan E A and Clews C J B 1951 *Phil.Mag.* **42** 616
[2] Opinsky A J and Smoluchowski R 1951 *J.App.Phys.* **22** No.12 p1488
[3] Dillamore I L and Katoh H 1974 *Metal Science* **8** 73
[4] Rollet A D and Kocks U F 1988 *Proc.ICOTOM* **8** p 375
[5] Barrett C S, Ansel G and Mehl R F 1937 *Trans.Am.Soc.Metals* **25** 702
[6] Barrett C S, Ansel G and Mehl R F 1937 *Trans AIME* **125** 516
[7] Armstrong M A 2010 *Groups and Symmetry* (Springer) pp 104-165
[8] Noether E 1918 *Nachr.Gesellsch.Wiss.Goettingen* **2** 235
[9] Taylor G I 1938 *J.Inst.Met* **62** 307
[10]Gensamer M and Mehl R F 1936 *Trans. AIME* **120** 277
[11]Rollet A D and Wright S I 1998 *Texture and Anistropy* (Cambridge University Press) pp 179-201
[12] Kazama Y 2008 *Symmetry and conservation quantity* (Science Ltd, Japan) pp 26-32
[13] Masui H 1999 *Acta mater* **47** No.17 p 4283
[14] Masui H 2005 *Materals Science Forum* **495-497** p 971
[15] Masui H 2008 *Ceramics Trans* **201** 489
[16] Masui H 2008 *Ceramics Trans* **201** 239

Quantitative comparison between simulated and experimental FCC rolling textures

M Wroński[1], K Wierzbanowski[1] and T Leffers[2]
[1] *AGH University of Science and Technology, Faculty of Physics and Applied Computer Science, al. Mickiewicza 30, 30-059 Kraków, Poland*
[2] *Danish-Chinese Center for Nanometals, Department of Wind Energy, Technical University of Denmark, Risø Campus, DK 4000 Roskilde, Denmark*
E-mail: mwronskii@gmail.com, wierzbanowski@fis.agh.edu.pl, tole@dtu.dk

Abstract. The degree of similarity between simulated and experimental fcc rolling textures is characterized by a single scalar parameter. The textures are simulated with a relatively simple and efficient 1-point model which allows us to vary the strength of the interaction between the grains and the surrounding matrix and the scheme for the calculation of the lattice rotation. For the copper-type texture the best agreement between simulation and experiment is obtained by {111}<110> slip combined with relaxed constraints according to a procedure different from the traditional scheme for relaxed constraints. A Sachs-type model without any volume effect of deformation twins provides a very high degree of similarity between simulated and experimental brass texture. Addition of volume fractions of deformation twins corresponding to those observed experimentally has practically no effect on the simulated textures.

1. Introduction

The existence of two different types of fcc rolling texture, the copper-type texture and the brass-type texture, is *the* classical problem within the field of deformation texture. The history of the investigations of the two types of fcc rolling texture has been reviewed by Leffers and Ray [1]. The development of the copper-type texture is favoured by high stacking fault energy, high rolling temperature and low strain rate while the development of the brass-type texture is favoured by the opposite set of conditions. As also reviewed by Leffers and Ray there is by now general agreement that the development of the copper-type texture can be simulated by models with {111}<110> slip. There is not a corresponding general agreement about the simulation of the development of the brass-type texture. According to the predominant ideas it may be simulated by a combination of {111}<110> slip and a volume effect of deformation twinning as originally suggested by Wassermann [2] or by {111}<110> slip only with a special pattern of deformation and lattice rotation, e.g. [1,3,4]. In the present work we do strictly quantitative comparisons between the simulated and experimental textures via a single scalar correlation factor obtained from the procedure developed by Tarasiuk and Wierzbanowski [5]. This correlation factor involves all components/aspects of the textures considered. We also show the simulated and experimental ODFs [6].
Our present simulations refer to {111}<110> slip without and with deformation twinning. Our model is a rather simple 1-point model which includes the possibility to vary the strength of the interaction between the grains – from very weak to very strong – and to make a choice between two procedures for the calculation of the lattice rotations. The simulated and experimental textures correspond to the reduction of 60%.

2. Computational procedure

2.1. The polycrystal deformation model

The model for plastic deformation of polycrystals proposed by Leffers [7,8] and further developed by Wierzbanowski et al. [9,10] is used to simulate the texture formation during rolling deformation. The elastic-plastic isotropic interaction between a grain and its environment is expressed as:

$$\dot{\sigma}_{ij} = \dot{\Sigma}_{ij} + L(\dot{E}^{p}_{ij} - \dot{\varepsilon}^{p}_{ij}) \tag{1}$$

where σ_{ij} and ε_{ij}^{p} are the stress and plastic strain tensors of a grain, Σ_{ij} is the stress applied to the material, E_{ij}^{p} is the plastic deformation of the sample and L is a scalar interaction parameter. Equation (1) was used by Berveiller and Zaoui [11] with $L = \alpha\mu$ where μ is the shear modulus and α is an elastic-plastic accommodation factor suggested to be in the range 0.01-0.1. This equation was successfully appliedd for the interpretation of residual stress measurements [12, 13]. For the materials we are going to simulate, copper and brass, this corresponds to an approximate L range of 500-5000 MPa. We examined a continuous transition from Sachs to Taylor type models by doing the simulations with a range of L values: 10, 30, 100, 300, 1000, 3000 and 5000 MPa. Other equations define the deformation model (cf. [7-11]), e.g., active of slip systems are selected on the basis of the Schmid law, the linear hardening law is used and the plastic strain increment of a grain is:

$$\delta\varepsilon_{ij}^{p} = \frac{1}{2}(m_i n_j + m_j n_i)\delta\gamma \tag{2}$$

where $\boldsymbol{m}$ and $\boldsymbol{n}$ are unit vectors along the *[uvw]* slip direction and the *(hkl)* slip plane normal and $\delta\gamma$ is a slip shear increment. The sample strain is the average of grain strains.

2.2. Lattice rotations

As first pointed out by Hosford [14] and recently recapitulated by the present authors [1,3,4] there is an ambiguity in the calculation of the lattice rotation resulting from a given slip event. For plane strain/rolling as described with 1-point models there are two possible lattice-rotation schemes which we here, in accordance with the designations used in [3,4], refer to as "Classical Condition", *CL*, and "Preservation Condition", *PR* (the same quantities were designed as: *MA* and *PSA* in [15]). Let us take the total sample rotation and the rotations of the individual grains to be zero. Then, according to *CL*, the grain lattice rotation $d\omega_{ij}^{latt(CL)}$ must be added in order to compensate for the plastic rigid body rotation:

$$\delta\omega_{ij}^{latt(CL)} = -\delta\omega_{ij}^{p} = -\frac{1}{2}(m_i n_j - m_j n_i)\ \delta\gamma \tag{3}$$

CL is the classical lattice-rotation scheme used in solid mechanics.

On the other hand, during the *rolling process* initially equiaxed grains get flattened parallel to the rolling plane and elongated parallel to the rolling direction. In this situation it is logical to assume that a plate of material initially parallel to the rolling plane ($x_1\, x_2$) preserves this orientation after deformation and that a string of material initially parallel to the rolling direction (x_1) preserves this orientation. This leads to the following lattice rotation (*PR*):

$$\begin{aligned} \delta\omega_{21}^{latt(PR)} &= -\delta e_{21}^{p} = -m_2 n_1 \delta\gamma \\ \delta\omega_{31}^{latt(PR)} &= -\delta e_{31}^{p} = -m_3 n_1 \delta\gamma \qquad \text{with: } \delta\omega_{ij}^{latt(PR)} = -\delta\omega_{ji}^{latt(PR)}. \\ \delta\omega_{32}^{latt(PR)} &= -\delta e_{32}^{p} = -m_3 n_2 \delta\gamma \end{aligned} \tag{4}$$

2.3. The correlation factor

For the comparison of the simulated textures with experimental ones we use a correlation factor R which expresses the degree of similarity between two textures. $R = 1$ means that the two textures are identical, and decreasing R values reflect a decreasing degree of similarity. R is calculated - from pole figures or orientation distribution functions, ODFs – with a linear-regression procedure developed by Tarasiuk and Wierzbanowski [5]. The resulting R values are particularly sensitive to the positions in pole figure/Euler space. It is considered that for $R \geq 0.8$ there is a reasonable agreement between simulated and experimental textures.

3. Results

In this section we present our results for the various combinations of lattice-rotation scheme and interaction strength (L values). We refer to the reduction of 60%.

3.1. Simple <110>{111} slip

The correlation factor between the textures simulated for <110>{111} slip and the copper-type experimental texture, R_C, and the correlation factor between the textures simulated with <110>{111} slip and the brass-type experimental texture, R_B, as functions of *L* are shown in Fig. 1 for *CL* and *PR* lattice rotation definitions. For the highest *L* value one notices that the *R* values (both R_C and R_B) are (practically) identical for *CL* and *PR* in accordance with the statement from [3] that *CL* and *PR* lead to the same lattice rotations for high *L* values.

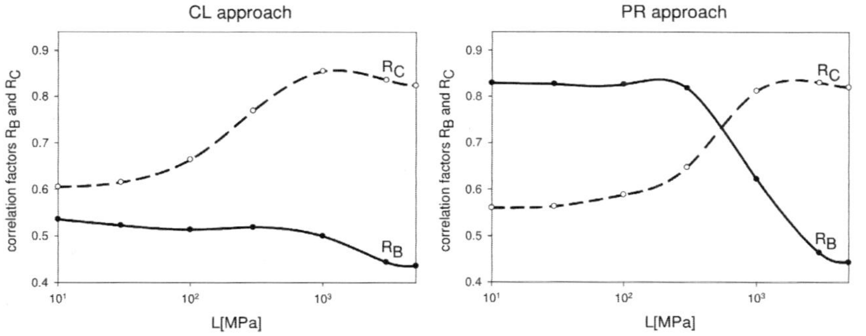

Fig. 1. *Correlation factors R_C and R_B versus L for CL and PR lattice rotation with <110>{111} slip mode.*

For the higher values of *L* there is agreement between the textures simulated with both lattice-rotation schemes and the experimental copper-type texture ($R_C > 0.8$). The agreement is the best for L = 1000 MPa combined with *CL*: $R_C = 0.86$, approaching "really good agreement". For the lower values of *L* the textures simulated with *PR* agree reasonably well with the experimental brass-type texture: $R_B = 0.83$.

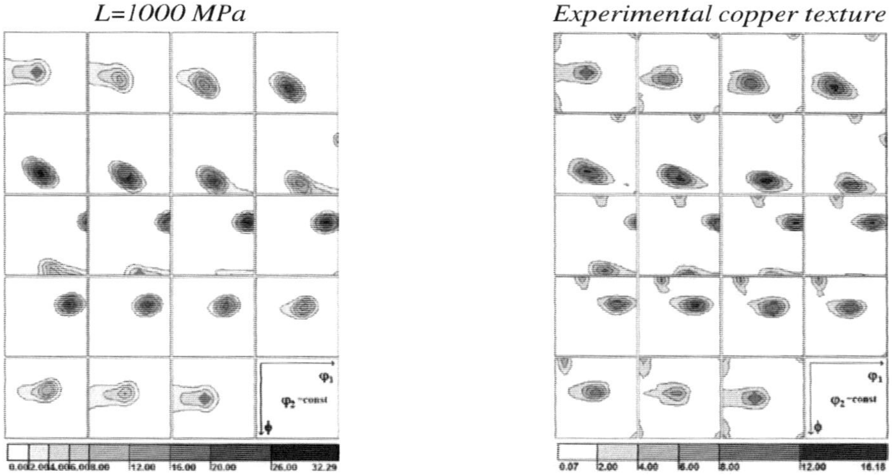

Fig. 2. *ODFs simulated with CL lattice rotation and <110>{111} slip mode for and L=1000 MPa together with the experimental copper ODF. The copper component C (•) and the brass component B (♦) are marked.*

Fig. 2 shows the ODFs simulated with *CL* lattice rotation for an *L* value of 1000 MPa together with the experimental copper ODF. The figure confirms the above statement that *CL* lattice rotation leads to a copper-type simulated texture for L = 1000 MPa. Fig. 3 shows the ODFs simulated with *PR* lattice rotation for an *L* value of 10 MPa together with the experimental brass ODF. The figure confirms the statement that *PR* lattice rotation leads to a brass-type simulated texture for low L values. In both cases the intensities of simulated textures are higher than those of experimental ones but their shapes and characters agrees very well.

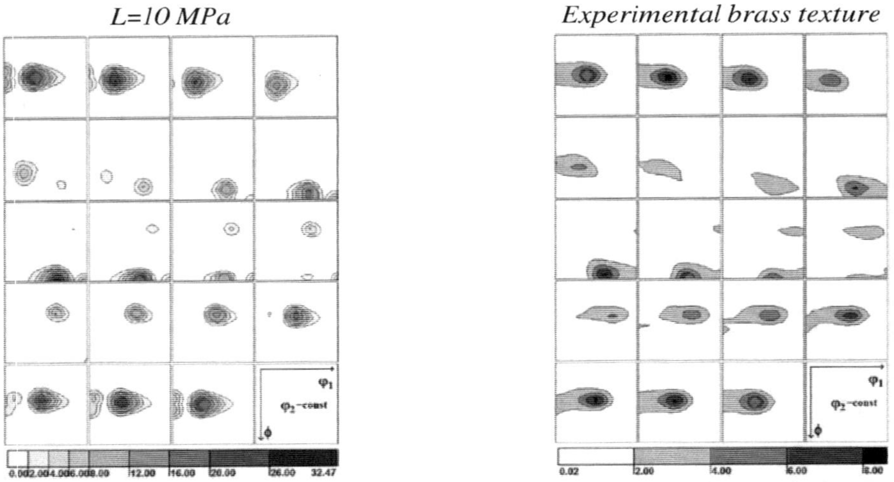

Fig. 3. *ODFs simulated with PR lattice rotation and <110>{111} slip mode for L=10 MPa together with the experimental brass ODF. The brass component B (♦) is marked.*

3.2. <110>{111} slip with random stresses added

Leffers [16] suggested a "modified Sachs model" for the simulation of the brass-type texture. In our terminology the modified Sachs model corresponds to a combination of <110>{111} slip, *PR* lattice rotation and low *L* values with *random stresses* added [7,8]. We add six components of random stress each with a random value between ± a maximum value which is 35% of the deterministic applied stress Σ_{11} (Eq. 1). A given random-stress component is allowed to operate in a number of steps randomly selected between zero and 30 % of the average number of steps needed to reach 60% reduction. The random stresses were originally introduced with the specific purpose of mimicking the interaction with individual neighbouring grains which, in 1-point models, is a random effect. Their magnitude was established by trial-and-error. Fig. 4 shows R_C and R_B as functions of *L* for *PR* lattice rotation calculated for <110>{111} slip with random stresses added. The overall picture is similar to that in Fig.1, but some of the actual values of R_C and R_B have changed. The best agreement with the experimental copper texture (for moderately high L values combined with CL) is unchanged as compared with Fig.1 (highest R_C value is 0.86 in both figures). The best agreement with the experimental brass texture (for low *L* combined with *PR*) has improved very significantly as compared with Fig. 1 (highest R_B value increased from 0.83 to 0.92). This is nearly a perfect agreement.

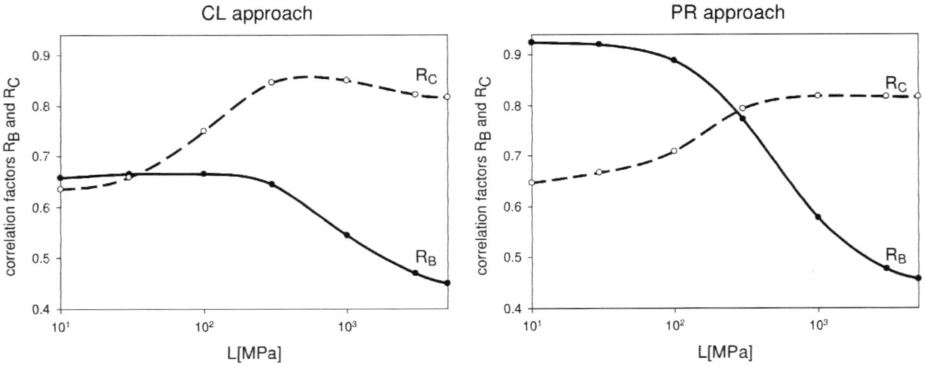

Fig. 4. *Correlation factors R_B and R_C versus L for CL and PR lattice rotation calculated for <110>{111} slip with random stresses added.*

Fig. 5 shows the ODF simulated with *L* value of 10 MPa, *PR* lattice rotation and <110>{111} slip with random stresses added together with the experimental brass ODF. As compared with Fig. 3 there are certain changes. For instance the simulated ODF in Fig.5 is less sharp than this in Fig. 3. As already reflected in Fig. 4 there is a very good agreement between simulated and experimental textures.

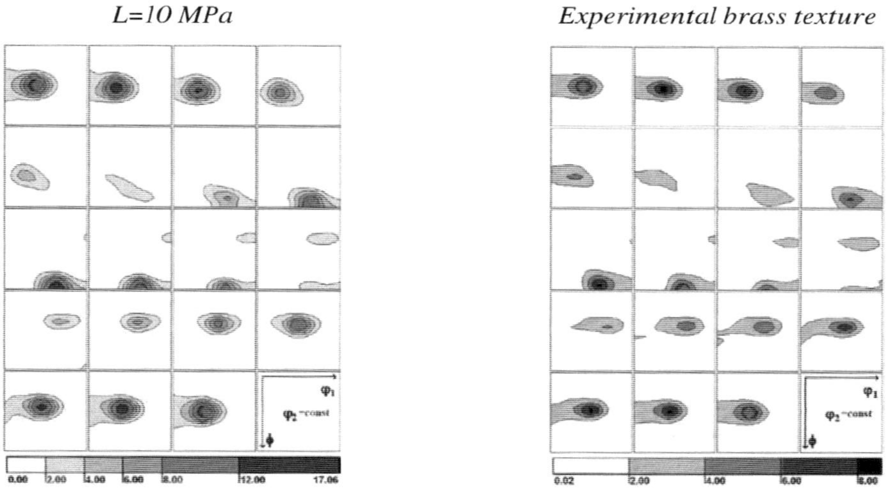

Fig. 5. *ODF simulated with L = 10 MPa combined with PR lattice rotation for <110>{111} slip with random stresses added together with the experimental brass ODF. The brass component B (♦) is marked.*

3.3. <110>{111} slip with <112>{111} twinning added

It is obvious that a possible volume effect of deformation twins on the formation of the brass-type texture must be seen together with the experimentally observed volume fraction of deformation twins in materials developing a brass-type texture. According to [1] there are only two reliable series of observations: In brass with 15% zinc rolled to reductions of about 40% the volume fraction of deformation twins is only few percent, and in brass with 30% zinc rolled to 76% reduction the volume fraction of deformation twins is less than 25%. By interpolation we derive a volume fraction of deformation twins of ~15% for our reduction of 60%, and this is basically the volume fraction that we use in our simulations.

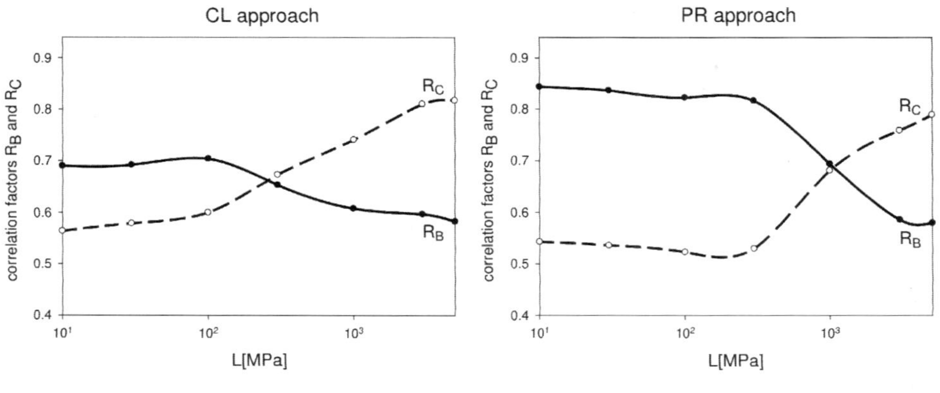

Fig. 6. *Correlation factors versus L for CL and PR lattice rotation with 15% volume fraction of deformation twins.*

We only consider twinning in the matrix material. We do not consider detwinning in the twins. Deformation twinning is introduced in our simulations with the procedure suggested by Tomé et al. [17]. Fig. 6 shows the correlation factors versus *L* for *CL* and *PR* with a 15% volume fraction of deformation twins. Comparison with Fig. 1 shows that the addition of 15% twins has a negligible effect on the correlation factors (and hence on the textures) in the case of *PR* lattice rotation. In the case of *CL* rotation twinning slightly increases values of R_B – but not enough to obtain a satisfactory agreement with experimental brass texture, particularly so for the higher values of *L* (R_B smaller than 0.6). This lacking effect of the 15% twins goes beyond earlier observations by Kalidindi [18]. He found that the addition of a 22% twin volume fraction had a "very minor" effect on the Taylor-simulated texture.
It should be mentioned here that the 'modified Sachs model, which gives a nearly perfect prediction of the brass type texture, can be explained by a formation of thin twin lamellae (not volume twins !) parallel to primary slip systems planes [1]. This favors a long action of primary slip systems by blocking a passage of dislocations to secondary slip systems, hence the 'Sachs type' behavior is realized.

4. Conclusions

Our results lead to the following conclusions:
- Use of correlation factor, based on linear regression, permits to compare precisely the predicted and experimental textures,
- There is no problem with explaining the copper type texture formation: strong interactions (*L* of the order of 1000 MPa) and preferably *CL* lattice rotation,
- The best agreement with brass type texture is obtained for weak interactions (*L* of the order of 10 MPa) and *PR* lattice rotation,
- Adding of *volume* twinning in realistic amount does not explain the brass texture formation when using strong interaction and *CL* rotation definition.

Acknowledgements

The work was partially financed by the Polish National Centre for Science - NCN (Grant No. DEC-2011/01/B/ST8/07394). TL acknowledges the support from the Danish National Research Foundation, the National Natural Science Foundation of China (Grant No. 50911130230) and the Danish-Chinese Center for Nanometals.

References

[1] Leffers T and Ray R K 2009 *Prog. Mater. Sci.* **54** 351
[2] Wassermann G 1963 *Zeit. Metallkd.* **54** 61
[3] Wierzbanowski K et al 2014 *Crit. Rev. Sol. State & Mater. Sci.* **39** 391-422
[4] Wronski M et al. 2013 *Mater. Sci. Technol.* **29** 129
[5] Tarasiuk J and Wierzbanowski K 1996 *Phil. Mag. A* **73** 1083
[6] Tarasiuk J et al. 1998 *Cryst. Res. Technol.* **33** 101
[7] Leffers T 1968 *phys. stat. sol.* **25** 337
[8] Leffers T 1968 *Risø Report No. 184* (Risø National Laboratory, Denmark)
[9] Baczmanski A et al. 1995 *Zeit. Metallkd.* **86** 507
[10] Wierzbanowski K et al. 2007 *Arch. Metall. Mater.* **52** 77
[11] Berveiller M and Zaoui A 1979 *J. Mech. Phys. Solids* **26** 325
[12] Baczmanski A et al. 1997 *Revue de Metallurgie - Cahiers d'Informations Techn.* **94** 1467
[13] Baczmanski A et al. 2006 *Mat. Sci. Forum* **524-525** 235-240
[14] Hosford W F 1977 *Text. Cryst. Solids* **2** 175
[15] Lebensohn R A and Leffers T 1999 *Text. Microstruct.* **31** 217
[16] Leffers T 1979 *Strength of Metals and Alloys, Eds.: P. Haasen et al.* (Pergamon Press, Oxford) p 769
[17] Tomé C N et al. 1991 *Acta Metall. Mater.* **39** 2667
[18] Kalidindi S R 2001 *Int. J. Plasticity* **17** 837

IOP Conf. Series: Materials Science and Engineering **82** (2015) 012012 doi:10.1088/1757-899X/82/1/012012

Asymmetric rolling textures of aluminium studied with crystalline model implemented into FEM

M Wronski[1,2], K Wierzbanowski[1], B Bacroix[2] and P Lipinski[3]

[1] *AGH University of Science and Technology, Faculty of Physics and Applied Computer Science, al. Mickiewicza 30, 30-059 Kraków, Poland*

[2] *LSPM-CNRS.* Université Paris 13, 99, av. J.B. Clément, 93 430 Villetaneuse, France

[3] *LaBPS, ENIM, 1, Route d'Ars Laquenexy CS 65820, 57078 Metz Cedex 3, France*

E-mails: mwronskii@gmail.com, wierzbanowski@fis.agh.edu.pl, brigitte.bacroix@univ-paris13.fr, lipinski@enim.fr

Abstract. The goal of this work was to study the asymmetric rolling process using the Finite Element Method (FEM) coupled with the deformation model of polycrystalline material. The Leffers-Wierzbanowski (LW) model was selected to be implemented into FEM. This implementation enables a study of heterogeneous plastic deformation process, like asymmetric rolling, taking into account its crystallographic nature. Our aim was to examine the crystallographic texture and mechanical properties of asymmetrically rolled aluminium 6061. The simulation results are compared with experimental textures determined by X-ray diffraction. The advantages of asymmetrical rolling over symmetrical rolling are reduced rolling normal forces and rolling torques, improvement of microstructure and producing the homogeneous crystallographic texture.

1. Introduction

Asymmetric rolling (AR) is a promising forming technique offering numerous possibilities of material properties modification and the improvement of technological process parameters. The geometry of this process can be realized using existing rolling mills, commonly used in symmetric rolling (SR). AR can provide a large volume of a deformed material which is its advantage. Moreover, AR can be useful for industrial applications because it reduces the applied rolling forces and torques, changes the rolled plate shape and modifies the microstructure [1-4]. AR is a rolling process in which outer velocities of the upper and lower rolls are different. It can be achieved in different ways. AR studied in this work was realized using two identical rolls, of 180 mm diameter, driven by independent motors and rotating with different angular velocities ω_1 and ω_2. This ensured a wide range of rolling asymmetry. Asymmetry of the process is described by the parameter $A=\omega_1/\omega_2$. In the present work the modifications of crystallographic texture and mechanical characteristics of aluminium samples, caused by AR, were studied using the Finite Element Method (FEM) coupled with a polycrystalline deformation model. Theoretical results were checked by comparison of predicted and experimental textures.

2. Implementation of Crystallographic Model into Finite Element Method

The implementation of a crystallographic model into FEM code is an important issue for the description of anisotropic behavior of metals, like during AR process. In this work the model of polycrystalline plastic deformation proposed by Leffers [5] and further developed by Wierzbanowski et al. [6-8] was used (LW model). This model was selected, because it is fast and relatively simple in its formulation. The elasto-plastic isotropic interaction between a grain and a surrounding material is expressed by the equation:

$$\dot{\sigma}_{ij} = \dot{\Sigma}_{ij} + L(\dot{E}^p_{ij} - \dot{\varepsilon}^p_{ij}) \tag{1}$$

where σ_{ij} and ε^p_{ij} are the stress and plastic strain tensors of a grain, Σ_{ij} is the stress tensor applied to the sample, E^p_{ij} is the plastic strain tensor of the sample and L is a scalar interaction parameter. This equation is a simplified version of the equation derived by Hill [9]; the total strain was replaced by the plastic strain (correct for the ranges of deformation considered), and the four-rank interaction tensor was replaced by the scalar parameter L. A given slip system *[uvw](hkl)* can be activated if the resolved shear stress $\sigma_{[uvw](hkl)}$ reaches a critical value τ_c (the Schmid law): $\tau_c = \sigma_{[uvw](hkl)}$. During deformation τ_c increases due to the work hardening. The latter process was described by a hardening matrix with a magnitude obeying a nonlinear saturating law [10]. For each elementary act of slip system activation, with a small shear strain $\delta\gamma$, the increments of plastic strain: $\delta\varepsilon^p_{ij} = \frac{1}{2}(m_i n_j + m_j n_i)\delta\gamma$ and of lattice rotation: $\delta\omega^{latt}_{ij} = -\frac{1}{2}(m_i n_j - m_j n_i)\,\delta\gamma$ were calculated (**m** and **n** are unit vectors along the *[uvw]* slip direction and the *(hkl)* slip plane normal).

In the present work the FEM calculations were done using ABAQUS/Explicit software [11]. The crystallographic model (LW) was implemented into ABAQUS via the user subroutine VUMAT. A representative polycrystalline model sample was attached to each finite element. Calculation in ABAQUS software are carried out incrementally. The calculation sequence in VUMAT subroutine (where LW model is implemented), in each Gauss point is as follows:
- at the beginning of each calculation step the ABAQUS program transfers to the VUMAT procedure a total strain increment ΔE_{ij} to be attained in this step (it is calculated basing on a stress state from the end of the preceding step and is called Σ_{ij}^{old} in the present step),
- LW model performs the imposed deformation ΔE_{ij} and calculates a new resulting stress state, Σ_{ij}^{new}, which is transferred to ABAQUS (it will become Σ_{ij}^{old} at the beginning of the next step).

The logical scheme of these operations is shown in Fig. 1.

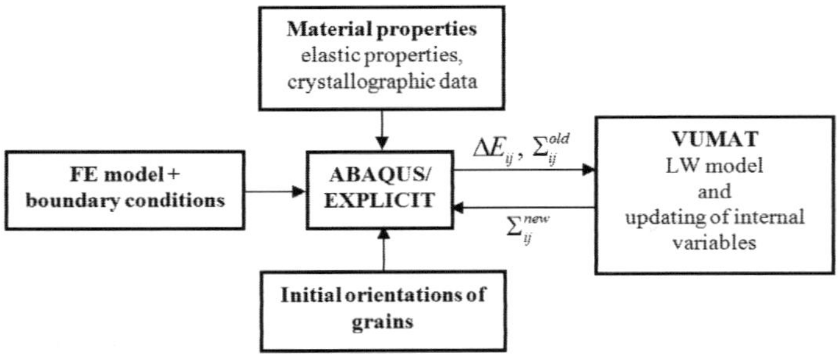

Fig. 1. *Interaction of LW model (included in VUMAT subroutine) with ABAQUS program.*

In order to decrease the number of finite elements the model of *plane strain* was used. The FEM sample contained 128 finite elements: 8 elements along ND, one element along TD and 16 elements along RD (8x1x16=128 elements). The initial textures of aluminium sample, determined by X-ray diffraction, was represented by 19200 grains - Fig. 4. A polycrystalline sample containing 150 grains was attached to each finite element of the C3D8R type (in each of 128 finite elements was 150 crystal grains - this gives the total number of 19200 grains representing a texture). When modelling AR, the Coulomb friction model was applied to describe the interaction of contacting surfaces (rolls and deformed material). The friction coefficient was estimated on the basis of the capture angle [12] and the value μ=0.30 was used in the calculations. The rolls were modelled as rigid ones, using the *RIGID SURFACE option. In order to decrease the calculation time the process was artificially accelerated by increasing the rate of deformation. The total calculation time of the joint model was 0.275 s. The Mass Scaling factor equal to 5 was applied. The condition for quasi-static process in *Explicit mode* [11] was checked and fulfilled (the kinetic energy was negligible compared with the internal one).

3. Mechanical characteristics of asymmetrically rolled aluminium

The results presented in this section were obtained for the aluminium rolled to reduction of 36% with initial texture shown in Fig. 2. The calculated distributions of Σ_{13} and Σ_{33} components of the *internal stresses* in SR (*A*=1.0) and AR samples (*A*=1.05 and 1.3) are shown in Fig. 3. One can note that during SR the shear stresses component, Σ_{13}, has the opposite sequence of signs on two surface layers (Fig. 3a-left). In contrast, when the degree of rolling process asymmetry increases, i.e., for A=1.3, there is no change of sign and the distribution of the Σ_{13} shear stress component becomes nearly homogeneous across the sample thickness (Fig. 3c-left). This results is in qualitative agreement with the calculations for isotropic material [13,14].

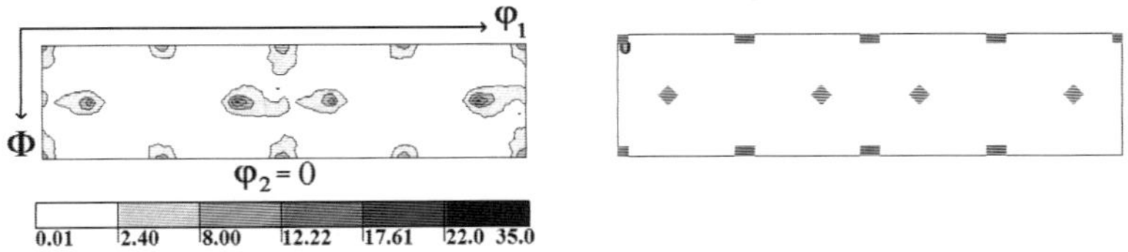

***Fig. 2**. The crystallographic texture of the initial aluminium sample determined by X-ray diffraction. φ_2=0° sections are shown and orientations W (■) and B (◆) are marked.*

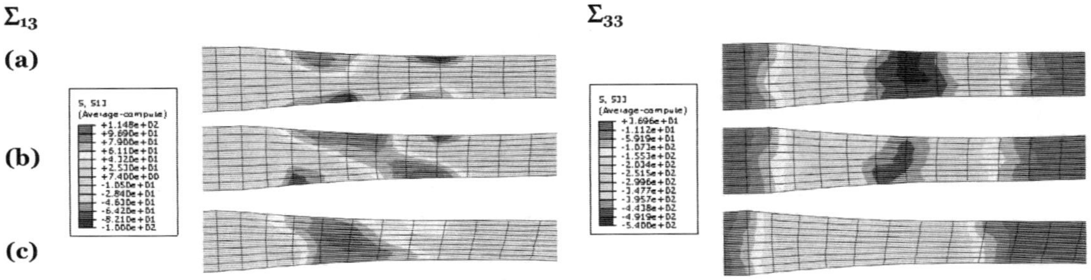

***Fig.3**. Calculated distributions of Σ_{11}, Σ_{33} internal stress components during rolling with asymmetry ratios $A=\omega_1/\omega_2$ equal to: (a) 1, (b) 1.05 and (c) 1.3. Calculation was for polycrystalline aluminium rolled to 36% reduction in one pass (FEM+LW model).*

The distribution of the Σ_{33} normal component of *internal stress* for SR and AR is shown in Fig. 3-right. It is nearly homogeneous across the sample thickness and its magnitude decreases vs. rolling asymmetry, i.e., it is lower for AR (A=1.05 and A=1.3) than for SR (A=1.0). This effect explains the appearance of lower *residual stresses* in AR materials, as determined by diffraction measurements (cf. [15] and also [16-18])).
Also the external characteristics of the rolling process were examined. One of the advantages of AR, predicted by FEM calculations, is the reduction of the average normal force (F_{AR}) exerted on the sample surface. This is a consequence of reduction of the magnitude of the normal component of internal stress, Σ_{33}, exerted on the material by rolls (cf., Fig. 3-right). The predicted normal force (presented as the relative value F_{AR}/F_{SR}) vs. A, during rolling of polycrystalline aluminium, is shown in Fig. 4a. Also the *bend magnitude* versus A was calculated and it is shown in Fig. 3b. We note that sample bending can be reduced by an appropriate choice of rolling anisotropy (A). In our case its magnitude is strongly reduced starting from A=1.3. As expected, no bending was predicted for symmetric rolling, A=1.0. Another important variable is the total deformation work, W, supplied by the rolling mill to obtain a given deformation. It is presented as the ratio W_{AR}/W_{SR} , i.e., as the relative deformation work. This work increases with the degree of rolling asymmetry, A (Fig. 4c).

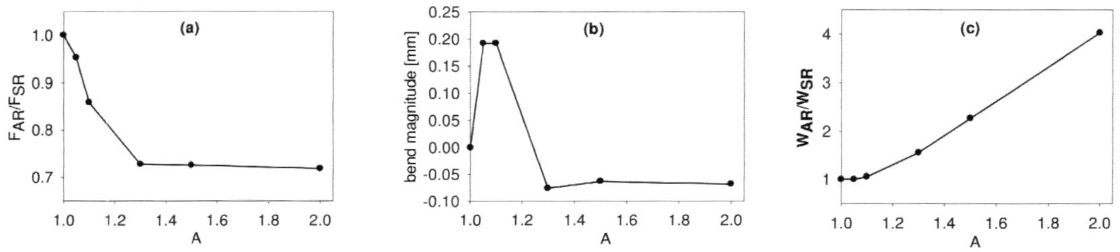

***Fig.4**. Mechanical characteristics versus $A=\omega_1/\omega_2$ predicted by FEM+LW model for polycrystalline aluminium (rolling reduction 36% in one pass): a) relative normal force (F_{AR}/F_{SR}), b) bend magnitude defined as deflection (in mm) at the end of 200 mm rolled plate, c) relative deformation work (W_{AR}/W_{SR}).*

3. Texture variation

Pole figures were determined in three layers of the rolled samples: in the top, bottom and central layers, using X-ray diffraction and corresponding ODFs were calculated [19]. Texture of the initial material is presented in $\varphi_2=0^o$ section of the Euler space (according to Bunge convention) and two main orientations are visible in this section: Brass (*B*) and Cubic (*W*) ones - Fig. 2. This material was next rolled symmetrically (*A*=1) and asymmetrically (*A*=1.5). The predicted rolling textures (by FEM+LW model) for three sample layers are compared with experimental textures of 6061 aluminium in Figs. 5, 6. Again only $\varphi_2=0^o$ sections, representative for texture modification, are shown in these figures. In the predicted SR texture (Fig.5a) we observe subtle shifts of texture maxima. Let us consider B orientation: in the top material layer the first two maxima of B component are shifted up and the two next ones - are shifted down (with respect to the red horizontal line drawn at $\Phi=45^o$), while the opposite sequence of shifts is observed in the bottom layer. These shifts do not occur in the centre layer of SR material: all equivalent B component maxima lie on the horizontal red line (Fig 5a-centre). The observed effect can be explained by the distribution of Σ_{13} shear stress component - Fig. 2a-left. In the top surface layer we observe the variation from positive to negative values of Σ_{13} and the opposite sequence appears in the bottom layer surface, while this stress component do not practically appear in the centre layer. Corresponding shifts of B component appear also in the experimental texture but with a lower magnitude - Fig. 5b. It can be shown that the discussed shifts of texture maxima are caused by texture rotations around transverse direction (TD) - cf. [20].

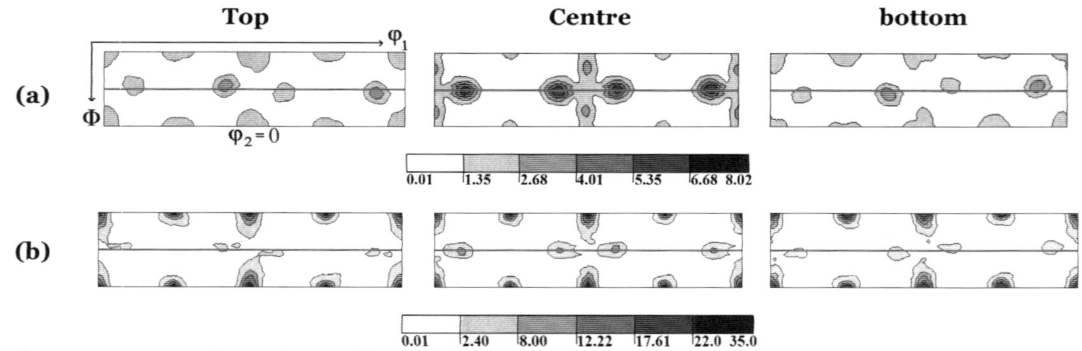

***Fig. 5**. Textures of symmetrically rolled aluminium ($A=\omega_1/\omega_2$ =1.0) in top, centre and bottom material layers: a) predicted by FEM+LW model, b) determined by X-ray diffraction. $\varphi_2=0^o$ sections are shown and red line is located at $\Phi=45^o$.*

Another behaviour is observed in predicted textures of AR material (Fig.6) - the shifts of *B* maxima have the same character in all three examined material layers (the first two maxima are shifted up and two next ones - are shifted down) - Fig. 6a. The same effect appears is experimental textures, though with a smaller magnitude - Fig. 6b. Therefore, we conclude that texture is much more homogeneous after AR process than after SR one. This behaviour can be explained by a nearly homogeneous distribution of the Σ_{13} shear stress component across the sample thickness- Fig. 2c-left.

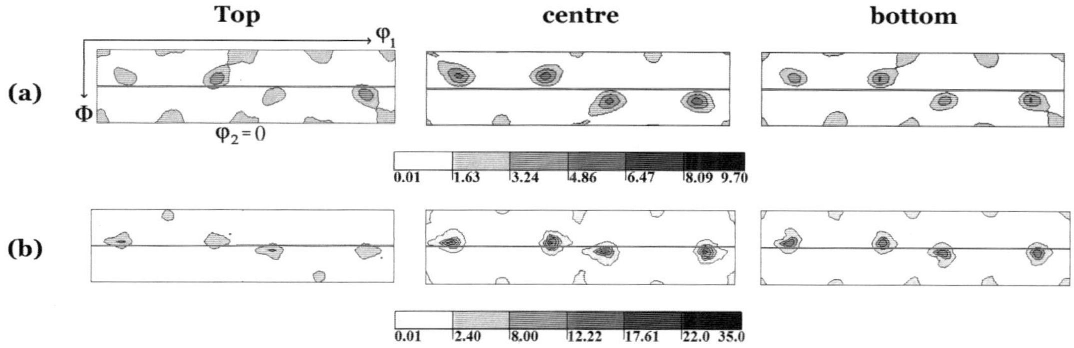

Fig. 6. *Textures of asymmetrically rolled aluminium ($A=\omega_1/\omega_2$ =1.5) in top, centre and bottom material layers: a) predicted by FEM+LW model, b) determined by X-ray diffraction. $\varphi_2=0^o$ sections are shown and red line is located at $\Phi=45^o$.*

The variation of texture across the material thickness was analyzed in function of distance from the top surface. The material studied by FEM+LW model was divided into eight layers, as shown in Fig. 7.

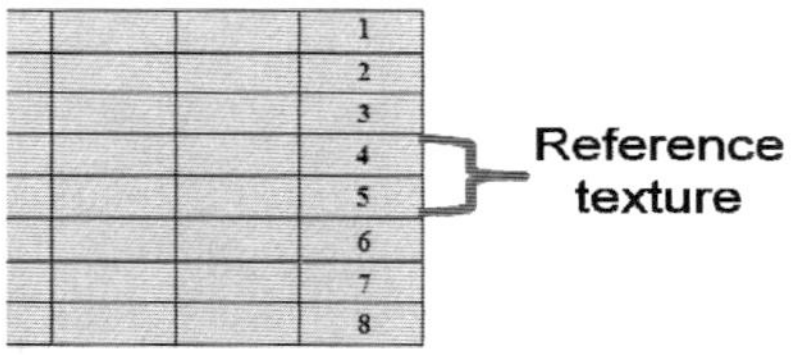

Fig.7. *Division of the rolled material into eight layers used in FEM+LW calculations.*

The average texture of two central layers (No. 4 and 5) served as the reference one. Next, to compare a texture from a given layer with the reference one we used the correlation factor R (according to the procedure developed in [21]) which expresses the degree of similarity between two textures. R = 1 means that two textures being compared are identical, and R=0 mean that textures are totally different. The calculated correlation factor R versus depth for different degrees of rolling asymmetry (A) is shown in Fig. 8. We note that the strongest variation of R appears in the case of SR, i.e., A=1.0. This is consistent with the character of shifts visible in Fig. 5. For higher degrees of asymmetry (A=1.5 and A=2.0) the variation of R becomes much more flat and R reaches the values around 0.9. This confirms the effect of texture homogenization across the sample thickness during AR.

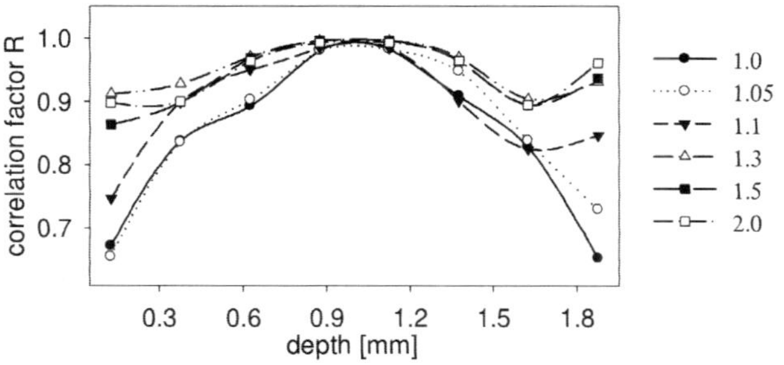

Fig.8. *Correlation factor R expressing similarity of textures from different layers of aluminium sample with the reference SR texture of the centre layer, calculated by FEM+LW model. Results for $A=\omega_1/\omega_2$ equal to 1.0, 1.05, 1.1, 1.3, 1.5 and 2.0 are shown.*

Conclusions

The obtained theoretical and experimental results for rolled polycrystalline aluminium confirm that AR leads to:

- Texture modification- rolling texture are more homogeneous across the sample thickness after AR (they are homogeneously rotated around TD),
- Decrease of the applied rolling force due to a decrease of Σ_{33} internal stress component vs. rolling asymmetry (which is favorable for rolling mill durability),
- Modification of bend magnitude (bending can be reduced by appropriate choice of rolling asymmetry).

Acknowledgements

This study was financed by the Polish National Centre for Science (NCN) under decision numbers: DEC-2011/01/B/ST8/07394 and DEC-2011/01/D/ST8/07399.

References

[1] Gao H and Chen G 1998 Asymmetrical cold rolling realized on plain mill for steel sheet by laser-textured rolls, *Iron and Steel* **33** 63
[2] Lee S H and Lee G N 2001 Analysis of deformation textures of asymmetrically rolled steel sheets, *Int. J. Mech. Sci.* **43** 1997
[3] Wronski S et al. 2011 Analysis of textures heterogeneity in cold and warm asymmetrically rolled aluminium, *Mater. Character.* **62** 22
[4] Wronski S and Bacroix B 2014 Microstructure evolution and grain refinement in asymmetrically rolled aluminium, *Acta Mat.* **76** 404
[5] Leffers T 1968 Computer Simulation of the Plastic Deformation in Face-Centred Cubic polycrystals and the Rolling Texture Derived, *phys. stat. solidi* **25** 337
[6] Wierzbanowski K et al. 2007 Elasto-plastic models of polycrystalline material deformation and their applications, *Arch. Metall. Mater.* **52** 77
[7] Baczmanski A et al. 1995 Models of Plastic Deformation Used for Internal Stress Measurements, *Zeit. Metallkd.* **86** 507
[8] Wierzbanowski K et al., 2014 FCC rolling textures reviewed in the light of quantitative comparisons between simulated and experimental textures, *Crit. Rev. Solid State & Mater. Sci.* **39** 391
[9] Hill R 1965 A Self-Consistent Mechanics of Composite Materials, *J. Mech. Phys. Solids*, **13** 213
[10] Wierzbanowski K et al. 1999 Deformation Characteristics Important for Nucleation Process. Case of Low-Carbon Steels, *Archives of Metallurgy* **44** 183
[11] Hibbitt H.D, Karlson B.I, Sorensen D, 2004 *ABAQUS Example Problems Manual. Rolling of Thick Plates* (1.3.6) (*Dassault Systèmes)* pp 485-502
[12] Gudur P et al. 2008 A theoretical study on the application of asymmetric rolling for the estimation of friction, *Int. J. Mech. Sci.* **50** 315
[13] Wronski S et al. 2009 Texture heterogeneity of asymmetrically rolled low carbon steel, *Arch. Met. Mater* **54** 89
[14] Wronski S et al. 2010 Crystallographic Textures Variation in Asymmetrically Rolled Steel, *Mater. Sci. Forum* **638-642** 2811
[15] Wronski M et al. 2014 Modification of stress and texture distributions in asymmetrically rolled titanium, *Adv. Mater. Res.* **996** 688
[16] Baczmanski A et al. 1997 Anisotropy of Micro-Stresses Measured by Diffraction, *Revue de Metallurgie - Cahiers d'Informations Techniques* **94** 1467
[17] Baczmanski A et al. 1999 Internal Stresses in Two Phase Polycrystalline Materials, *Archives of Metallurgy* **44** 39
[18] Baczmanski A et al. 2006 New Type of Diffraction Elastic Constants for Stress Determination, *Mat. Sci. Forum* **524-525** 235
[19] Tarasiuk J et al. 1998 New Algorithm of Direct Method of Texture Analysis, *Cryst. Res. Technol.* **33** 101
[20] Wronski M et al. 2014 Texture variation in asymmetrically rolled titanium. Study by Finite Element Method with implemented crystalline model, *Int. J. Mech. Sci.* **87** 258
[21] Tarasiuk J and Wierzbanowski K 1996 Application of the Linear Regression Method for Comparison of Crystallographic Textures, *Phil. Mag. A* **73** 1083

IOP Conf. Series: Materials Science and Engineering **82** (2015) 012013 doi:10.1088/1757-899X/82/1/012013

Local shear texture formation in adiabatic shear bands by high rate compression of high manganese TRIP steels

J Li[1], P Yang[1], WM Mao[1] and FE Cui[2]
[1] School of Materials Science and Engineering, University of Science and Technology Beijing, Beijing, 100083
[2] Research Institute of Advanced Materials, University of Science and Technology Beijing, Beijing, 100083
E-mail: yangp@mater.ustb.edu.cn

Abstract. Local shear textures in ASBs of high manganese TRIP steels under high rate straining are determined and the influences of initial microstructure is analyzed using EBSD technique. It is seen that even at the presence of majority of two types of martensite before deformation, ASB is preferred to evolve in austenite, rather than in martenite, due to reverse transformation. Ultrafine grains of thress phases due to dynamic recrystallization are formed and all show shear textures. The less ε-martensite in ASB is distributed as islands and its preferred orientation can be found to originate from the variants in matrix. The grain orientation rotation around ASB in multi-phase alloy reveals significant influence of α′-martensite on texture in ASB. The mechanism of local texture formation in ASB of high manganese TRIP steel is proposed in terms of the interaction of early TRIP and later reverse transformation.

1. Introduction

Adiabatic shear bands (ASBs) are commonly seen in materials subjected to high rate compression or shocking and are related with instability of stress-strain behavior. Although it is well known that shear texture develops in ASBs which accompanies with dynamic recrystallization, the concurrent TRIP effect in high manganese steels with three different phases which have different behaviors from the ASB formation to the appearance of cracks makes the formation of such shear texture in ASB quite complicated. Our previous works show that γ (austenite), ε (hexagonal martensite), α′ (BCC martensite) or even ferrite may be present in ASBs in different volume fractions, but γ is always the dominant phase [1-3] which may partially relate with reverse transformation of martensite. It is further observed that TRIP effect occurs before ASB formation and thus delays its formation [4] and the cracks in ASBs and outside ASB are all surrounded by α′–M. Hence, similar shear texture in ASB may relate to quite different formation processes due to concurrent TRIP effect. The objective of this study is to elucidate the features of microtextures within and around ASBs in a high manganese TRIP steel.

2. Experimental

The forged high manganese steel with composition of Fe-17.22Mn-2.87Si-0.48Al-0.022C mass% was annealed for 2h at 1100°C followed by water quenching. Cylindrical samples of Φ11mm×15mm were machined and subjected to a compression of 15% and 30% at a strain rate of 10^{-3}/s. EBSD was used for phase and orientation determination. The pre-compressed samples were machined into hat-shaped specimens along the compression direction, and were impacted by modified split-Hopkinson pressure bar (SHPB) at strain rate of 10^5/s. After testing, the specimens were sectioned parallel to the compression axis, electrolytic polished in 5% $HClO_4$, and etched in 4% Nital solution.

3. Results and discussion

Fig. 1a shows the Euler angle map of ε-M around an ASB after high rate compression in the sample containing initially 26.9%γ, 52.2%ε-M and 8.2%α′-M. By comparison of orientations of γ, ε-M and α′-M grains on both sides of ASB, the same grains on both sides of ASB can be found, as labeled by the numbers in Fig. 1a. It is seen that the ASB penetrates the grains interior rather than along grain boundaries. The shifts of matrix on both sides by shearing along ASB are ~100μm. Considering the width of ASB (10μm), the true shear strain is about 10. The ε-M grains neighboring ASB rotate their orientations for over 30° as shown in Fig. 1b. Fig.1c shows the phase distribution in ASB region (dashed lines) indicating dominant ultrafine γ (red color) (Fig. 1d) within ASB and lots of strain induced α′-M (blue color) in the surrounding matrix. Fig.1e shows the phase distribution in ASB of the sample subject to 15% pre-compression before high rate compression depicting more or wider γ-regions in matrix neighboring ASB. The γ matrix outside ASB shows grains with lath morphology and many Σ3 and Σ11 relationships (green and yellow lines) with large angular deviation indicating the reverse transformation of α′-M→γ in matrix due to adiabatic heating. As the temperature in ASB is higher than that outside ASB, we can assume that reverse transformation of α′→γ should proceed in ASB too. In addition it is seen that more γ-phase located around ASB in sample containing more α′-M before high rate compression. The reason for this is ascribed to stronger reverse transformation when more α′-M existed before high rate compression.

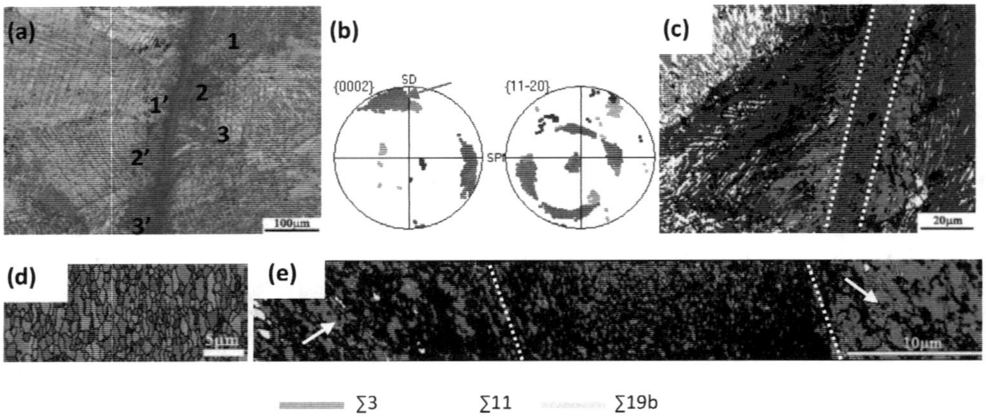

Fig.1 General feature of ASB formation with concurrent TRIP. (a) Euler angle map of ε-M in sample without pre-compression (the numbers show corresponding initial grains); (b) pole figures of ε-M of grain 1 in (a) (the red arrows denote the rotating direction, and the red circle shows the rotation axis); (c) phase distribution around ASB in the sample without pre-compression; (d) Euler angle map of γ in ASB of (c); (e) phase distribution of ASB region subject to 15% pre-compression before high rate straining. (red: γ; yellow: ε-M; blue: α′-M; black: un-indexed region; the arrows show the places where martensite reversion transformation happens)

Fig.2 shows the local textures within ASB containing ultra-fined γ with less ε-M (yellow color) distributed as islands. Fig.2b and 2c show the orientation distribution of two phases indicating a uniform $\{111\}<110>_{\gamma}$ and weak $\sim<11\text{-}20>_{\varepsilon}$ shear textures. Fig.2d to 2f exhibit the ε-M orientations within ASB and on both sides of ASB indicating that the ε-M in ASB originated from the right side ε-M as a variant.

Fig.3a shows the EBSD map of ASB in Fig.1e. Again shear texture of {111}<110>$_\gamma$ developed during ASB formation. No other phases than γ is detected here which is at least partially due to reverse transformation because Σ3 and Σ11 can be found in matrix neighboring this ASB.

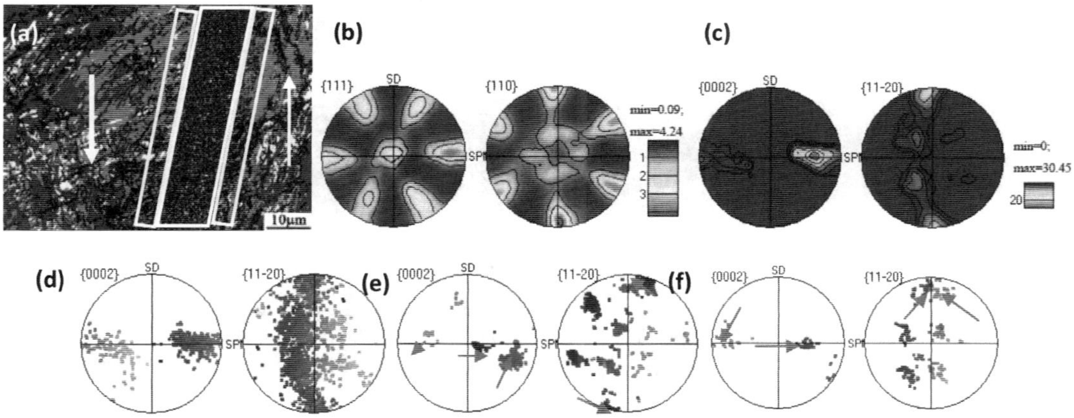

Fig.2 EBSD maps of the sample without pre-compression. The white arrows depict shear direction of matrix; (a) phase distribution around ASB (the white box denotes ASB); (b) orientation of γ in ASB, (c), (d) orientation of ε-M in ASB; (e) orientation of ε-M on the left side of ASB, (f) orientation of ε-M on the right of ASB. (SD: shear direction; SPN: the normal of shear plane)

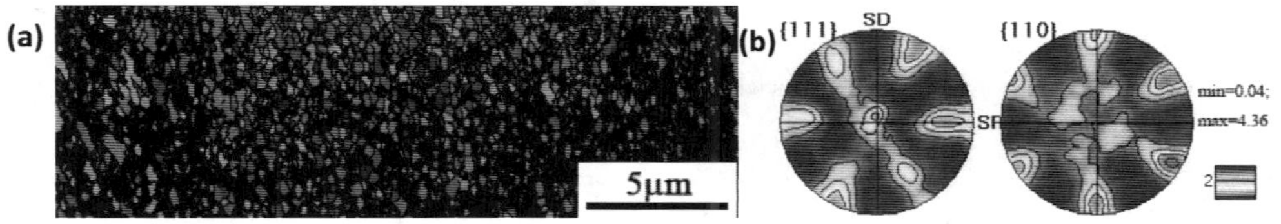

Fig.3 (a): the Euler angle map of γ in ASB of sample with 15% pre-compression; (b): the pole figures of γ in (a).

After 30% pre-compression, γ, ε and α′-M turn to be 0.3%, 26.3% and 54.0% (the rest is no-indexing region). In this α′-M dominant initial microstructure, cracks initiated firstly within ASB and then grow both along ASB and extend from ASB out into matrix. Fig.4 shows the EBSD maps in ASB containing dominant γ and a crack. The ASB shows a sharp boundary to the matrix on the right side, but a rather diffusive boundary with matrix on the left. From the pole figures of γ on each side (Fig. 4g and 4h), it is clear that the orientation on the left is close to shear texture. Judging from the elongated and fragmentized grains, the left side is more likely to be sheared which results in a typical shear texture. On the contrary, the right side reveals clear boundary which means difficult to shear and corresponds to the large deviation to shear texture. High resolution EBSD map of Fig.4b reveals the α′-M preferentially located around crack and fine ε-M distributed nearby. Fig. 4d-4f show local shear textures of three phases within ASB. Whereas γ shows a weak {111}<110> shear texture and ε-M shows a <11-20> fiber texture, the α′-M presents nearly {110}<111>$_{\alpha'}$ shear texture. However, we may explain it in other way, namely, two orientations with preferred <111>||SD are merely two variants with Σ3 initially. It is clear that during ASB formation the ideal shear direction is firstly achieved, with the shear plane rotating slowly. The α′-M around cracks extending from ASB into matrix did not show strong deformation.

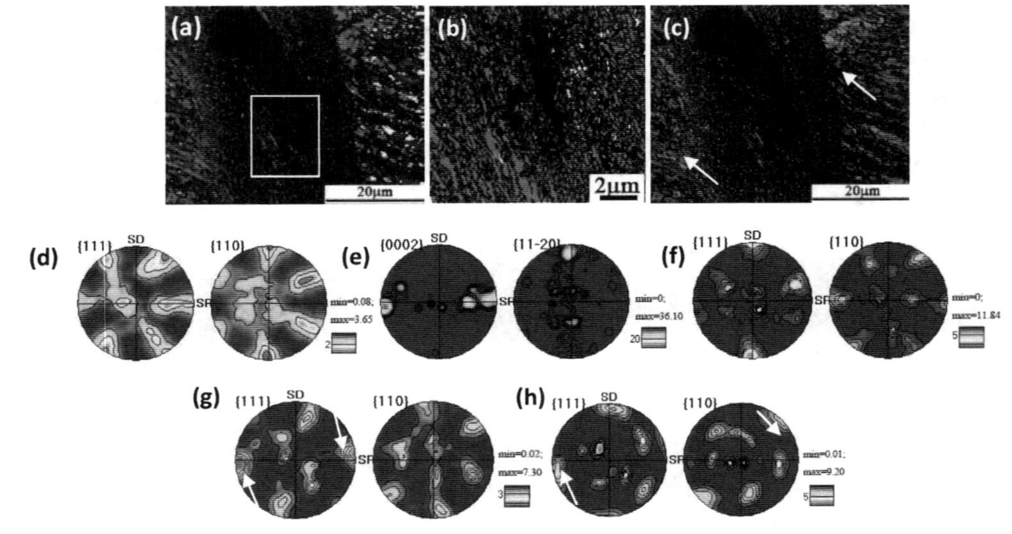

Fig.4 Local textures of three phases in an ASB containing cracks in the sample with 30% pre-compression. (a) phase distribution of crack in ASB at low magnification; (b) high resolution EBSD map showing phase distribution in the white box in (a); (c) Euler angle map of γ corresponding to (a); The pole figures of γ (d), ε-M (e) and α'-M (f); (g) the pole figures of γ on the left (g) and right (h) sides of ASB.

4. Conclusions

TRIP occurs smoothly before the ASB formation and happens around the ASB during its development. Reverse transformation of martensite occurs in and around ASB leading to the dominate phase of γ in ASB. The microtextures in ASB with concurrent TRIP effect have some characteristics as follows:

1) The reverse transformation in ASB leads to dominant γ shear texture {111}<110>. The more α'-M present before high rate compression, the stronger the reverse transformation in ASB and its surrounding. Whereas TRIP is dominant at early stage of deformation around ASB, reverse transformation is dominant at later stage in ASB. The minor ε- and α'-M in ASB show shear textures of <11-20> and {110}<111> too with ultra-fine grains before they transform back into γ due to adiabatic heating.
2) Fine grained and textured α'-M was observed to locate preferentially around cracks in ASB, which is the main location for crack formation.
3) The origin of ultra-fine ε-M grains in ASB can be found in matrix neighboring ASB according to the orientation features, and the ε-M retains the characteristic of initial variants. Additionally, the rotation of ε-M is less apparent than γ.

Acknowledgments

This work was supported by the Natural Science Foundation of China (Grant No. 51271028).

References

[1] Wang H Z, Sun X R, Yang P and Mao W M 2013 *Mater. Sci. Forum* **753** pp 72-75
[2] Sun X R, Wang H Z and Yang P 2014 *Steel Res. Int.* http://dx.doi.org/10.1002/srin.201300381
[3] Wang H Z, Sun X R, Yang P, Mao W and Meng L 2014 *J. Mater. Sci. Technol.* http://dx.doi.org/10.1016/j.jmst.2014.03.027
[4] Sun X R, Wang H Z, Yang P and Mao W M 2014 *Acta Metall. Sin.* **50** pp 387-394

IOP Conf. Series: Materials Science and Engineering **82** (2015) 012014 doi:10.1088/1757-899X/82/1/012014

Deformation-Induced Recrystallization of Magnesium Single Crystals at Ambient Temperature

KD Molodov, T Al-Samman, DA Molodov
Institute of Physical Metallurgy and Metal Physics, RWTH Aachen University, Germany
E-mail: kmolodov@imm.rwth-aachen.de

Abstract. Specially oriented magnesium single crystals were subjected to plane strain compression along the <11$\bar{2}$0> direction in *c*-axis extension at ambient temperature. The samples exhibited outstanding formability deforming up to a logarithmic final strain of -1. Investigations by optical and orientation imaging microscopy revealed that massive {10$\bar{1}$2} extension twinning at low strains consumed the whole sample and resulted in new soft orientations for slip. Observations also indicated that additional twinning took place in the completely twinned matrix by secondary and tertiary twinning events. At advanced stages of deformation newly formed, equiaxed small grains were observed within numerous bands related to former deformation twins. These "recrystallized" grains characterized by a low grain orientation spread of less than 1° generated new orientations, which led to a substantial weakening and randomization of the texture during deformation up to very large strains. The reported results in this paper are discussed with regard to the microstructure evolution arising from multiple twinning and continuous dynamic recrystallization at room temperature.

1. Introduction

Due to their hexagonal closed packed (HCP) crystal structure with a limited number of independent slip systems, the ductility of wrought polycrystalline magnesium is usually poor at ambient temperature and the material shows strong anisotropy in mechanical behavior. For ambient temperature deformation the primary deformation modes available are usually basal slip and extension twinning. Both modes have critical resolved shear stress values of less than 5 MPa, and are hence very easy to operate at low temperatures [1-3]. In Mg alloys, the most commonly observed twin types are the {10$\bar{1}$2} extension and {10$\bar{1}$1} contraction twins, which can accommodate extension or contraction along the hexagonal (*c*) - axis. In practice, unlike extension twins, contraction twins in magnesium are rather scant. Hence due to their low volume fraction, they do not generally cause significant changes in the global texture during deformation but in certain cases they can give rise to continuous dynamic recrystallization (CDRX) [4] owing to crystallographic softening within the twinned material. Continuous DRX (also denoted rotation DRX [5, 6]) is based on progressive misorientation of subgrains during recovery until high angle grain boundaries are formed.

In terms of formability, one very important characteristic of deformation twinning is that grains unfavorably oriented for basal slip can be reoriented into a more favorable position upon twinning. In addition, by the onset of numerous double twinning events [7], the total plastic strain can be increased considerably provided that the latter does not lead to void nucleation, as reported by Hartt and Reed-Hill [8], who investigated the significance of deformation and fracture along {10$\bar{1}$1}-{10$\bar{1}$2} double twins in polycrystalline Mg. Recent research has, however, indicated that not all types and variants of double twins provide equivalent void nucleation sites. It has been actually observed that voids are formed preferentially at boundaries between primary and secondary twins rotated around the same <11$\bar{2}$0> axis [9]. It is thus pointed out that the formation of such voids, in principle, can be avoided if the factors responsible for twin variant selection were understood. For HCP crystal plasticity this criterion can be extremely important, as was observed in [10], where it was demonstrated that the variant selection of extension twins can greatly harden the material under certain loading conditions, leading to high fracture strength and limited ductility or soften it under others, giving rise to CDRX and very large ductility. The present work focuses on the deformation and recrystallization behavior of a magnesium single crystal compressed along the <11$\bar{2}$0> axis in a channel-die at ambient temperature, so as to form {10$\bar{1}$2} extension twins. The deformation experiments were terminated at predefined strains so that the microstructure and texture evolution could be tracked throughout the deformation as a function of imposed strain. The obtained results are discussed with respect to the multiplicity of deformation twinning and its consequent crystallographic softening effect that gave rise to deformation-induced recrystallization at ambient temperature.

2. Experiments

High-purity magnesium single crystals (99.95%), oriented with the basal plane parallel to the loading direction, were deformed in plane strain compression (PSC) at ambient temperature, such that the stress axis was parallel to the <11$\bar{2}$0> crystal direction and extension was confined to the *c*-axis. The misalignments between the crystallographic axes and the sample reference frame were less than 1°. All PSC tests were performed at a constant strain rate of 10^{-3} s^{-1} using a conventional mechanical testing machine equipped with an automated data acquisition system. For minimization of friction between the sample and channel-die, lubrication oil was used. The microstructure in the mid-layer of the compression plane was characterized by means of optical microscopy and electron backscatter diffraction (EBSD). EBSD measurements were performed with a scanning electron microscope equipped with a field emission gun and an HKL-Nordlys II EBSD detector. X-ray pole figure (PF) measurements were conducted using a Bruker D8 Advance diffractometer, equipped with a HI-STAR area detector. The MTEX toolbox [11] was employed to analyze and visualize the texture data collected by X-ray diffraction and EBSD methods.

3. Results and discussion

The mechanical response of the investigated monocrystalline specimen during room temperature PSC along the <11$\bar{2}$0> direction (*c*-axis extension) is reported elsewhere [12]. In summary, the specimen achieved a remarkable logarithmic strain of -1 without mechanical failure. During initial deformation the flow curve showed a sigmoidal work hardening regime consistent with predominant {10$\bar{1}$2} extension twinning at low strains. After reaching a peak stress of 220 MPa at 11% strain, there was an unexpected drop of stress with strain (flow softening) down to 200 MPa. During further straining the stress-strain curve depicted a nearly steady-state behavior consistent with the formation of new recrystallized grains. Fig. 1 shows the microstructure development of the originally monocrystalline specimen as a function of different strain levels from 3% to 100%. The EBSD maps are represented in terms of the Schmid Factor (SF) for basal slip considering a stress of σ and $\sigma/2$ in the compression and transverse direction, respectively. The latter passive stress component takes into account the contribution of back stress due to the channel-die constraints. At 3% strain, extensive twinning was evident throughout the whole specimen. The twin area fraction was approx. 53 pct. Two opposite sets of {10$\bar{1}$2} twins were observed to emanate from one TD edge of the specimen (or from a preexisting twin boundary), and extend diagonally to the opposite side, occasionally penetrating other twins (Fig. 1a). The twinned areas exhibited characteristic orientations corresponding to crystallographic rotation of 86° about <11$\bar{2}$0> from the initial single crystal (parent) orientation. This rendered a large portion of the microstructure favorable for basal slip, as indicated by a significant increase of the SF for basal slip from zero to approx. 0.2. Within the large twins that seemed to form first (primary twins), secondary small twins of the {10$\bar{1}$2} extension type (for details see [12]) possessed even higher SF values of about 0.3. At 8% strain (Fig. 1b), the remaining parent microstructure was completely consumed by the {10$\bar{1}$2} primary twins. The only twins observed were the secondary twins that appeared isolated within a "new" parent matrix. Additionally, there were other small twins with very high SF values of approx. 0.45 that seemed to form inside the secondary twins (i.e. third generation {10$\bar{1}$2} twins). At 40% strain, the second generation of {10$\bar{1}$2} twins appeared to have undergone significant growth compared to what was observed at low strains (Fig. 1c). When reaching the final strain of 100%, the fraction of recrystallized regions, characterized by fine equiaxed grains with grain orientation spread (GOS) values below 1°, was very prominent throughout the whole sample (Fig. 1d). Although not shown in Fig. 1, it is noted that the formation of recrystallized grains was already seen at 11% strain [12].

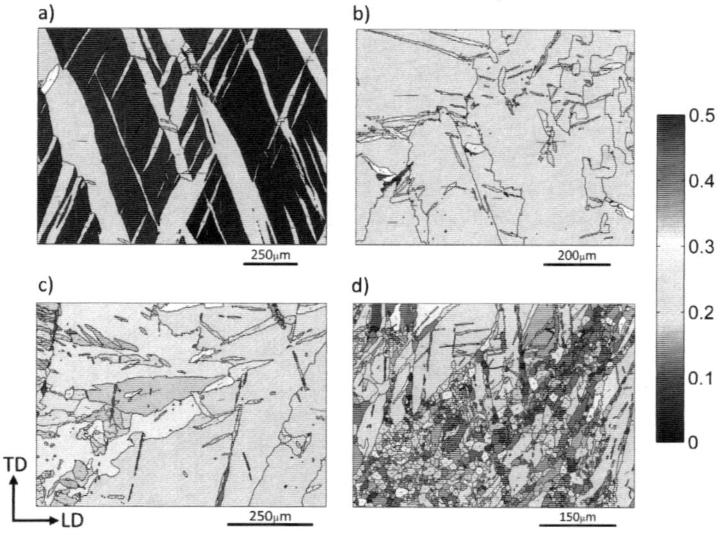

Fig. 1: EBSD maps of the deformed magnesium crystal colored according to the Schmid factor for basal slip. (a): 3%, (b): 8%, (c): 40%, (d): 100% strain.

Fig. 2 shows recalculated (0002) and $\{10\bar{1}0\}$ pole figures that present the macrotexture development in correspondence with the microstructures given in Fig. 1. The initial texture of the single crystal prior to deformation corresponded to the orientation (φ_1, Φ, φ_2) = (90°, 90°, 0°) in Euler space (Bunge notation). After 3% strain, primary $\{10\bar{1}2\}$ twinning imparted significant texture changes, seen by two new texture components (0, 30, 0) and (180, 30, 0) with an angle of 30° between the direction of their respective *c*-axes and the compression direction (Fig. 2a). After 8% deformation the initial orientation of the monocrystalline matrix was completely converted into the new twinning orientations with soft slip conditions (Fig. 2b), which was crucial for the operation of further deformation mechanisms beyond primary twinning. With increasing strain up to 40% the twinning components appeared to accumulate some scatter and were shifted from the characteristic (0/180, 30, 0) orientations (Fig. 2c).

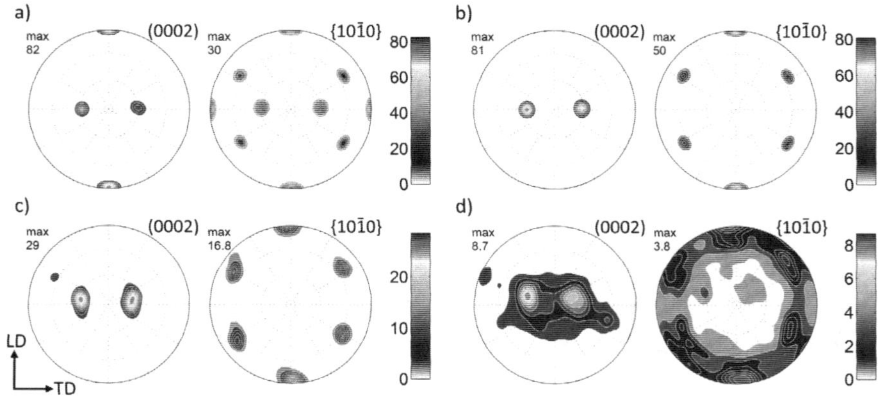

Fig. 2: X-ray texture development in the deformed magnesium single crystal as a function of strain. (a): 3%, (b): 8%, (c): 40%, (d): 100%. ODF calculations have been performed with a kernel half-width of 5°.

The significant growth of secondary $\{10\bar{1}2\}$ twins at this strain level gave rise to an additional weak texture component further away from the center of the basal pole figure. At the final strain of 100% the advanced state of DRX resulted in a massive decrease in texture intensity associated with significant texture spread around the twin orientations (Fig. 2d). Fig. 3 shows a representative example of a recrystallized region at 100% deformation, represented in terms of EBSD microstructures. From the SF map in Fig. 3a it is evident that the majority of recrystallized grains had a favorable basal slip condition. Fig. 3b depicts additionally, that these grains possessed a broad orientation spectrum that contributed to the observed weakening of the overall macrotexture. As revealed by the results reported in the current paper and elsewhere [10, 12], the crucial mechanism of the deformation-induced DRX was based on dynamic recovery characterized by high slip activity of dislocations in the soft twinned regions, and the formation of subgrains by dislocation rearrangement. These subgrains, as indicated in Fig. 3b, continuously transform into recrystallized grains with high-angle boundaries by means of subgrain coarsening and coalescence.

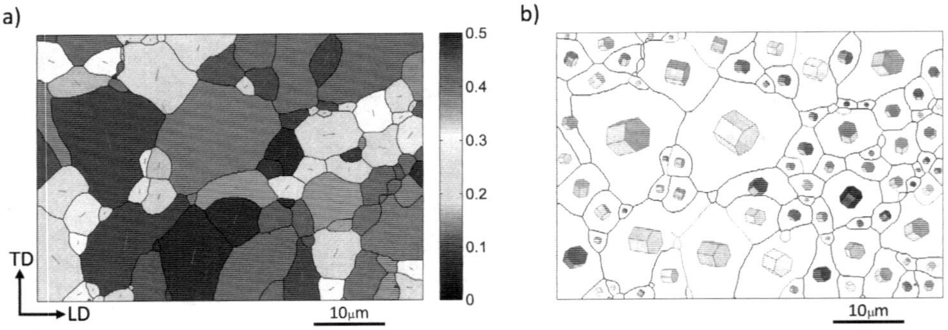

Fig. 3: (a) Schmid factor map for basal slip with slip trace overlay. (b) Grain boundary map with crystal lattice overlay. The color key corresponds to inverse pole figure coloring with respect to the compression axis. Light-colored boundaries 5°-15°, dark-colored boundaries > 15°.

4. Conclusions

Although easy slip modes were suppressed for the initial single crystal orientation investigated in this study, it was still possible to deform the specimens at room temperature up to 100% without failure. Initial deformation proceeded by profuse $\{10\bar{1}2\}$ extension twinning, converting the whole sample into softer orientations for slip and further twinning. This gave rise to deformation-induced dynamic recrystallization that led to a randomization of the overall texture, which was vital for enhancing the room temperature ductility.

References

[1] Reed-Hill RE and Robertson W 1957 *Acta Metall.* **5** 728.
[2] Akhtar A and Teghtsoonian E 1969 *Acta Metall.* **17** 1339.
[3] Wonsiewicz BC and Backofen WA 1967 *Trans. Metall. Soc. AIME* **239** 1422.
[4] Al-Samman T, Molodov KD, Molodov DA, Gottstein G and Suwas S 2012 *Acta Mater.* **60** 537.
[5] Guillopé M and Poirier JP 1979 *J. Geophys. Res.* **84** 5557.
[6] Burrows SE, Humphreys FJ and White SH 1979 Proc. 5th Int. Conf. on Strength of Metals and Alloys, edited by Haasen P, Gerald V and Kostorz G, Pergamon Press p 607.
[7] Couling SL, Pashak JF and Sturkey L 1959 *Trans. ASM* **51** 94.
[8] Hartt WH and Reed-Hill RE. 1967 *Trans. Metall. Soc. AIME* **239** 1511.
[9] Ando D, Koike J and Sutou Y 2010 *Acta Mater.* **58** 4316.
[10] Molodov KD, Al-Samman T, Molodov DA and Gottstein G 2014 *Metall. Mater. Trans. A* **45** 3275.
[11] Hielscher R and Schaeben H 2008 *J. Appl. Cryst.* **41** 1024.
[12] Molodov KD, Al-Samman T, Molodov DA and Gottstein G 2014 *Acta Mater.* **76** 314.

IOP Conf. Series: Materials Science and Engineering **82** (2015) 012015 doi:10.1088/1757-899X/82/1/012015

A new cluster-type statistical model for the prediction of deformation textures

P Van Houtte[1], Q Xie[2], A Van Bael[1], J Sidor[3] and J Moerman[4]
[1] Department MTM, KU Leuven, Leuven, Belgium
[2] University of Science and Technology, Beijing, People's Republic of China
[3] Department of Materials Science and Engineering, Ghent University, Belgium
[4] Tata Steel, IJmuiden Technology Centre, IJmuiden, The Netherlands
E-mail: paul.vanhoutte@mtm.kuleuven.be

Abstract. An attempt was done to improve the quality of deformation texture predictions by statistical models through the introduction of "clusters" of *N* grains thus defining a third, intermediate length scale. The interaction between each cluster and the macroscopic length scale is of the Taylor type, whereas inside each cluster a VPSC scheme is used. Predictions of cold rolling deformation textures were quantitatively compared with experimental results for a steel alloy. The results are encouraging.

1. Introduction

During the plastic deformation of polycrystalline materials, the boundary conditions (forces, displacements) applied at the outer borders of the work piece are somehow transmitted to the individual grains (stresses, strains). This process is quite complex, and can be captured by CPFE [1] or CP-FFT [2] models, which could be used for deformation texture predictions. However, these models are so far too computation-intensive for certain applications, such as FE simulations of industrial deep drawing processes in which the evolution of texture and plastic anisotropy in every integration point is tracked and accounted for *during* the FE simulation. Statistical models may still be preferred in that case [3]. Note that only single-phase materials are considered in what follows.

2. Boundary conditions as used in statistical models for deformation textures

These models operate on discrete sets of crystals, each modelled by at least their crystal orientation and weight factor. Such set must be statistically representative for the initial condition of the material of which a metal forming operation is to be simulated. Statistical models use simple assumptions about the transmission of external forces and displacements throughout the polycrystal. They are usually formulated in terms of 'interactions': the long range interaction is the relationship between the stresses and velocity gradients of the RVE (Representative Volume Element) at the macroscopic length scale (see Fig. 1) and those of the RVEs at smaller length scales, usually grains (at what is called here 'mesoscale', Fig. 1). Short-range interactions are those between directly neighbouring grains. The well-known Taylor model adopts for example the following assumption about a long range interaction: the plastic velocity gradient at macroscale must be equal to the one at the mesoscale (grain scale). This is a very strong interaction. As a result of this, all grains are supposed to have the same plastic strain; this assumption constitutes strong short-range interaction applied by the model. In contrast to this, no assumptions about short-range stress interactions (between directly neighbouring grains) are adopted. Concerning the long range stress interaction, it is merely assumed that the macroscopic stress is the volume average of the stresses in the mesocopic RVEs (grains). The model also assumes homogeneous distributions of stresses and plastic strains inside the grains. The equally well-known VPSC model [4-5] also assumes homogeneity of stress and plastic strain rate inside the grains (called 'inclusions'). As for the long-range interaction, it (i) requires that the stresses and plastic strain rates of the macroscopic RVE are the volume averages of those of the inclusions and (ii) treats each grain as an inclusion embedded in the macroscopic RVE (assumed to be homogenous) and solves for the local stresses and (plastic) strains as would be done in an Eshelby-type inclusion model. In this, the constitutive model used for the inclusions is visco-plasticity, which can be seen as an approximation of the Generalized Schmid Law. The classical Eshelby solution, which uses linear elasticity, had of course to be adapted to this [4]. Note that this procedure leads to a much more subtle long-range interaction between the mesoscopic RVEs (grains) and the macroscopic RVE. On the other hand, short range interactions (which take place at grain boundaries) are not directly modelled. Compared to this, the Taylor model strictly models both the long-range and the short range strain-rate interaction, but no short range stress interaction at all. In this, it must be pointed out that in truth, all grains interact to different extents with all other ones; short range and long range interactions are just extreme cases. The CPFE (see for example [1]) and CP-FFT (see for example [2]) models can in principle capture the entire range of interactions, however at a rather high computational cost. "Cluster models" try to make a compromise between statistical models and CPFE / CP-FFT models by introducing a third length scale which is intermediate between the macroscopic and the mesoscopic length scales. This is done by subdividing the polycrystal into *M* "clusters" of *N* grains. Such cluster can then be seen as the RVE of a third intermediate length scale, here called "mesoscopic-2". The interaction between each cluster and the macroscopic length scale is of the Taylor type, whereas inside each cluster a VPSC [4-5] scheme is used.

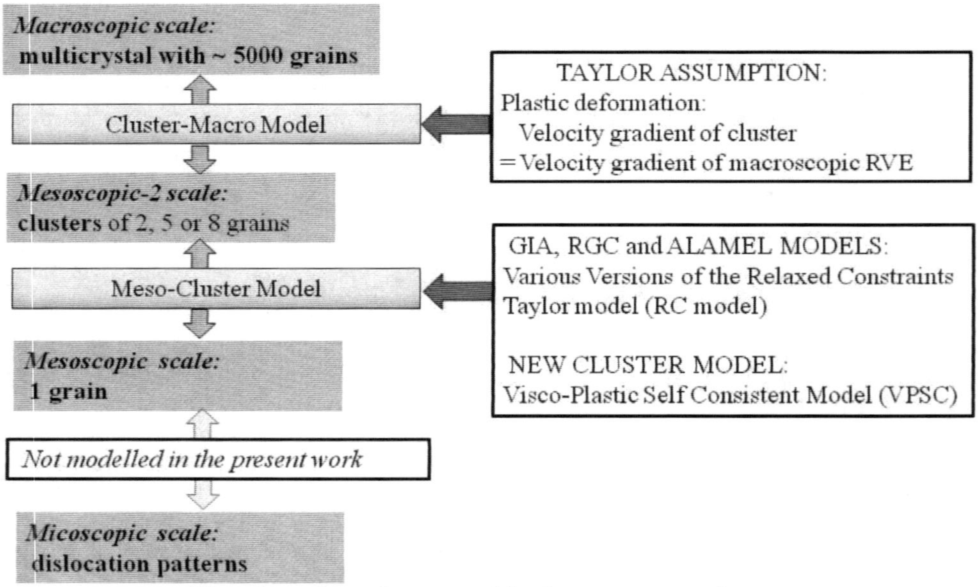

Fig. 1 Length scales as used in the present work.

3. Cluster Models

3.1. Existing Models

The LAMEL model [6] adopts the Taylor-interaction between a cluster (N=2) of two grains and the macroscopic RVE. It focuses on short-range interaction. It is of limited use, as has been designed to simulate plane strain deformation only (the strain mode at mid-plane in rolling). The internal grain boundary of the cluster needs to be parallel with the rolling plane. The velocity gradient of the two grains must be equal to each other, except for the two simple shears for which the shear plane is parallel to the internal grain boundary. The average plastic velocity gradient of the two grains needs to be equal to the macroscopic velocity gradient (Taylor assumption for the cluster). The solution for the slip rates in the

IOP Conf. Series: Materials Science and Engineering **82** (2015) 012015 doi:10.1088/1757-899X/82/1/012015

two grains is sought by minimizing the total rate of plastic work dissipated in the cluster taken as a whole, using the Generalized Schmid Law for each grain but respecting the conditions described above. This results in (i) geometrical compatibility at the internal grain boundary and (ii) equilibrium of shear stresses at the grain boundary [6].
The GIA model [7] (which is in fact older than the LAMEL model) and the closely related RGC model [8] can be seen as a sort of generalization of the LAMEL model. These models look at a brick-shaped cluster subdivided into 8 grains by 3 planes, which are normal to ND, RD and TD, respectively. This creates 12 internal grain boundaries, as opposed to a single one in the case of LAMEL. Each internal grain boundary is treated in a similar way as the one of LAMEL. This leads to 24 possible shear-type relaxations. An extra complication is that these may cause voids or overlaps inside the cluster, which are kept under control by the adoption of a penalty term in the overall minimization procedure. The ALAMEL model is another generalisation of the LAMEL model [9]. It is very similar to it as it also works with a cluster of two grains (N=2). It no longer assumes that the internal grain boundary is parallel to a rolling plane. Its initial orientation is chosen at random (for each cluster separately) from a 'grain boundary orientation distribution function'. The cluster is supposed to represent two thin layers of material draped at both sides of a grain boundary segment. The orientation changes of these internal grain boundaries are tracked during the simulation of a large strain. The application of this model is not restricted to plane strain.

3.2. New Model

Fig. 1 describes how the new cluster model [10] works. The interaction between the cluster and the macroscopic RVE is of the Taylor type, as in LAMEL, ALAMEL, GIA and RGC. The interaction between each grain and the cluster as a whole is modelled by the "Neffect-VPSC"-software [10] from the Los Alamos National Lab, with n^{eff}=10. In this, the cluster, i.e. the mesoscopic-2 RVE, was used as the "macroscopic" RVE of the VPSC software. So far, tests have been done with N=2 and N=5.

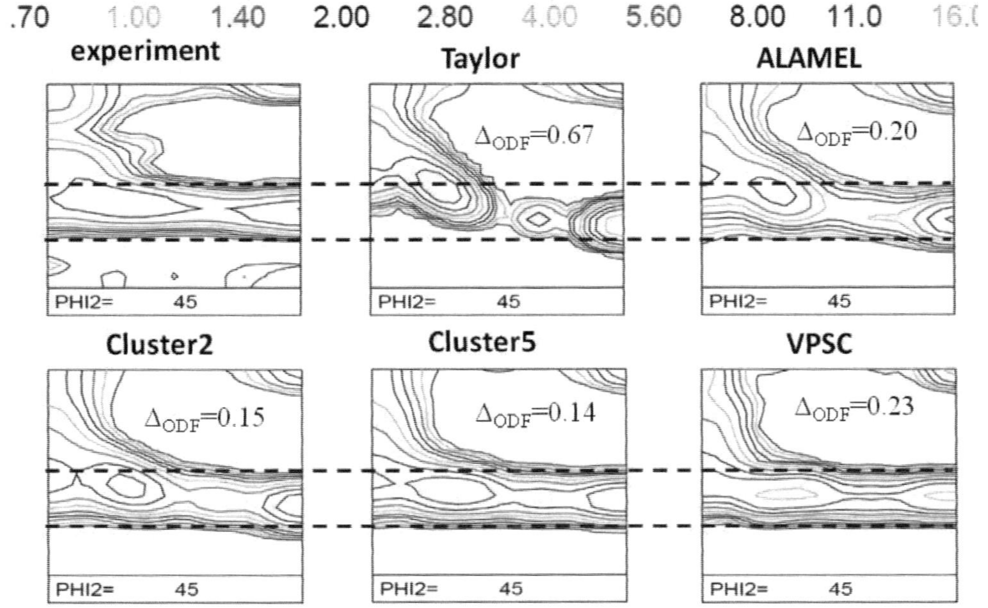

Fig.2 φ_2=45° sections of the ODFs of the experimental cold rolling texture and model predictions for the T61 cup drawing steel. Δ_{ODF} is the normalized error integral.

4. Results

Experimental validations for cold rolling applied to two steel qualities (DC06 and T61) and one aluminium alloy (AA6016) have been described in [10], as well as results for simple shear tests on DC06. Fig. 2 reproduces some of the results for T61, which is a low carbon steel with 0.04 wt% C for beverage cans. Sheets of it have been cold rolled to 0.2 mm thickness followed by continuous annealing, after which the material was further cold rolled (60 % reduction) in the laboratory. The figure also gives the normalized error integral Δ_{ODF}, i.e. the square of the difference between the values of the ODF of the cold rolling texture as predicted by the models and the experimentally measured one, integrated over entire Euler space, and then divided by the texture index of the experimental ODF.

5. Conclusions

The results reported in Fig. 2 and in [10] all indicate, that (i) the quantitative agreement between the measured and predicted ODFs after rolling and shearing are much better for the ALAMEL, VPSC and the new cluster models with N=2 or 5, than for the Taylor model; (ii) the two New Cluster models perform better than ALAMEL and VPSC and (iii), the New Cluster model with N=5 performs slightly better than the one with N=2. Further research should indicate whether even better results might be obtained with N>5. Note also that in case N=5 would indeed be the best choice, it could be concluded that the simple Taylor assumption of homogeneous plastic strain would very well model the interaction between the mesoscopic-2 RVE corresponding to N=5 and the macroscopic RVE.

Acknowledgement

This research was carried out under the project numbers M41.10.08307b and M22.1.10394 in the framework of the Research Program of the Materials innovation institute M2i (www.m2i.nl). The authors gratefully acknowledge the financial support of the Interuniversity Attraction Poles Program (P7/21) from the Federal Government of Belgium. The authors also want to thank P. Eyckens at KU Leuven, H. Vegter and Y. An at Tata Steel in the Netherlands for providing some of the measured texture data and samples used in this work.

The authors wish to express their gratitude to Carlos Tomé and Ricardo Lebensohn of the Los Alamos National Laboratory for the permission to use their VPSC software.

References

[1] Roters F, Eisenlohr P, Hantcherli L, Tjahjanto DD, Bieler TR and Raabe D 2010 *Acta Mater.* **58** 1152
[2] Lebensohn RA, Rollet A and Suquet P 2011 *JOM* **63-3** 13
[3] Van Houtte P, Gawad J, Eyckens P, Van Bael B, Samaey G and Roose D 2011 *JOM* **63-11** 37
[4] Molinari A, Canova GR and Ahzi S 1987 *Acta Metall.* **35** 2983
[5] Lebensohn RA and Tomé CN 1993 *Acta Metal. Mater.* **41** 2611
[6] Van Houtte P, Delannay L and Samajdar I 1999 *Textures and Microstructures* **31** 109
[7] Engler O, Crumbach M and Li S 2005 *Acta Mater.* **53** 2241
[8] Eisenlohr P, Tjahjanto DD, Hochrainer T, Roters F and Raabe D 2009 *Int. J. Mat. Res.* **100** 500
[9] Van Houtte P, Li S, Seefeldt M and Delannay L 2005 *Int. J. Plasticity* **21** 589
[10] Xie Q, Van Bael A, Sidor J, Moerman J and Van Houtte P 2014 *Acta Mater.* **69** 175
[11] Tomé CN 1999 *Modell Simul Mater Sci Eng* 7 723

IOP Conf. Series: Materials Science and Engineering **82** (2015) 012016 doi:10.1088/1757-899X/82/1/012016

Texture comparison between cold rolled and cryogenically rolled pure copper

L Lapeire[1], J Sidor[1], E Martinez Lombardia[2], K Verbeken[1], I De Graeve[2], H Terryn[2,3], LAI Kestens[1,3]
[1] Department of Materials Science and Engineering, Ghent University, Technologiepark 903, B-9052 Zwijnaarde (Ghent), Belgium
[2] Research Group Electrochemical and Surface Engineering, Vrije Universiteit Brussel, Pleinlaan 2, B-1050 Etterbeek (Brussels), Belgium
[3] Department of Materials and Engineering, Delft University of Technology, Mekelweg 22628 CD Delft, The Netherlands
E-mail: Linsey.Lapeire@ugent.be

Abstract. Nowadays, there is a considerable scientific interest in bulk ultrafine grained materials, due to their potential for superior mechanical properties. One of the possible formation methods of nano-grained materials is cryogenic rolling. The influence of rolling at cryogenic temperatures has been investigated. Significant differences in the textures and the microstructures can be observed between the cryogenically rolled copper and conventionally cold rolled copper, reduced to the same thickness.

1. Introduction
Nowadays, there is a considerable interest in the deformation of metals at very low absolute temperatures, which is known as cryogenic deformation [1]. It is assumed that rolling at cryogenic temperature can result in enhanced grain refinement and associated strengthening when compared to rolling at room temperature (RT) [2]. It is believed that very low temperatures suppress the processes of dynamic recovery, promote an increase in the dislocation density, activate mechanical twinning [3, 4] and thereby promote the formation of extremely fine-grained structures. Another potential of cryogenic rolling is the formation of a very weak, almost random texture after annealing in Cu and Al alloys [5, 6]. There is evidence that this behavior can be linked with the occurrence of shear bands in the deformation texture.
In the present work the aim is to get a better understanding of the texture development in the cryogenic rolling process of pure copper. Therefore, the texture and microstructures of cold rolled and cryogenic rolled material, starting from an identical batch of pure copper, is studied in detail.

2. Experimental
The investigated material in this study is high purity cast copper. A hot rolling reduction (400°C) of 50% allows destroying the cast microstructure. After an annealing step of 30 min at 400°C a completely recrystallized material is obtained, called the *initial* material. Part of the initial material is cryogenic rolled to 50%, 75% and 90% overall thickness reduction. The other part is rolled at RT to the same thickness reductions.
Electron backscatter diffraction (EBSD) observations on both the cryogenically rolled and the cold rolled copper are made on the longitudinal plane containing the rolling and the normal directions. The textural observations are made on the mid-thickness section of the rolled samples. For the deformed materials, textures are calculated without imposed orthorhombic symmetry.

3. Results
3.1 Initial material
As the texture of the initial material can influence the texture of the deformed material [7], a detailed texture analysis of the material prior to rolling is done at the mid-thickness section. The texture (Fig 1) has an almost random distribution with the strongest peak being 2.4 x random.

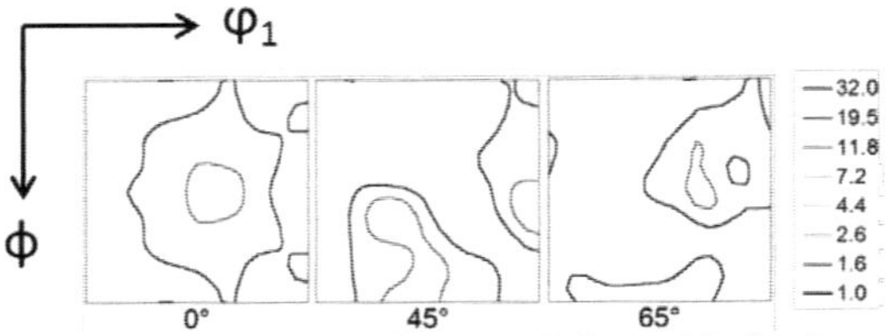

Figure 1: ODF sections (with φ_2 = 0°, 45° and 65°) of the mid-thickness of the copper sample prior to subsequent rolling.

3.2 Deformed state after 50% deformation

After 50% cold deformation, the texture in the mid-thickness (Fig 2a) remains weak (no peaks higher than 4 x random) and the observed texture is obviously different from the conventional β-fibre rolling texture. The ND-rotated cube component is present. After 50% cryogenic reduction (Fig 3b), also a weak texture is observed in the mid-thickness of the sample, with the same intensity as in the cold rolled copper. The same texture components are observed as in the 50% cold rolled material, namely the ND-rotated cube component. Both textures have a high degree of symmetry.

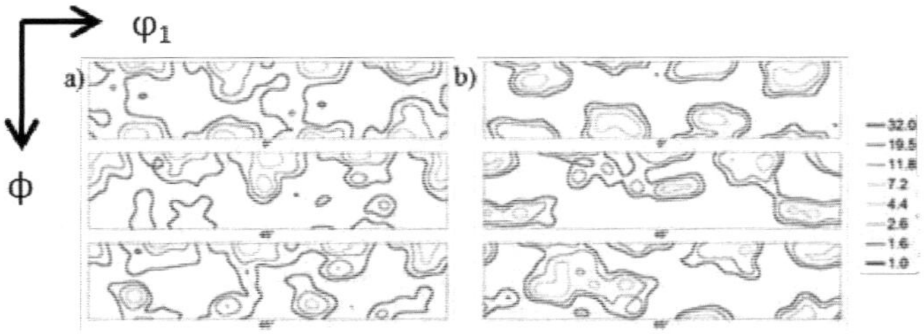

Figure 2: ODF sections (with φ_2 = 0°, 45° and 65°) of the mid-thickness section after 50% deformation (a) cold rolled, (b) cryogenic rolled.

3.3 Deformed state after 75% deformation

In the material cold rolled to 75% reduction, the overall texture intensity is strengthened and both α- and β-fibers become evident in the orientation distribution function (ODF) (Fig 3a). Within the α-fiber, Goss and Brass components are noticed, whereas the copper and S components are observed within the β-fiber. The brass component is prevalent over the other components, with a maximum intensity of 13.8 x random. A high degree of symmetry is observed. After 75% cryogenic rolling (Fig 3b), the texture looks similar to the cold rolled material rolled to the same thickness reduction. However, some differences are observed. After cryogenic rolling the texture looks less symmetric. Also a weaker texture is observed, compared to the cold rolled copper, with a maximum intensity of 10.6 x random.

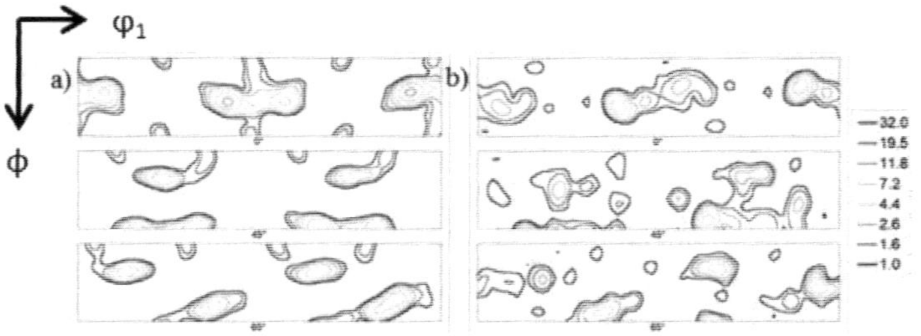

Figure 3: ODF sections (with φ_2 = 0°, 45° and 65°) of the mid-thickness after 75% deformation (a) cold rolled, (b) cryogenic rolled.

Figure 4 shows the image quality (IQ) maps of the cold rolled (a) and cryogenically rolled (b) copper after 75% deformation. Although the deformation texture of these two materials is similar, a large difference is observed in their microstructures. In the cryogenic rolled material, in contrast to the cold rolled material, numerous shear bands are observed. These bands form thin planar sheets that are parallel to the transverse direction and inclined at an angle of ~35° to the rolling plane. A secondary minor set of shear bands are observed inclined at an angle of ~20° to the rolling plane.

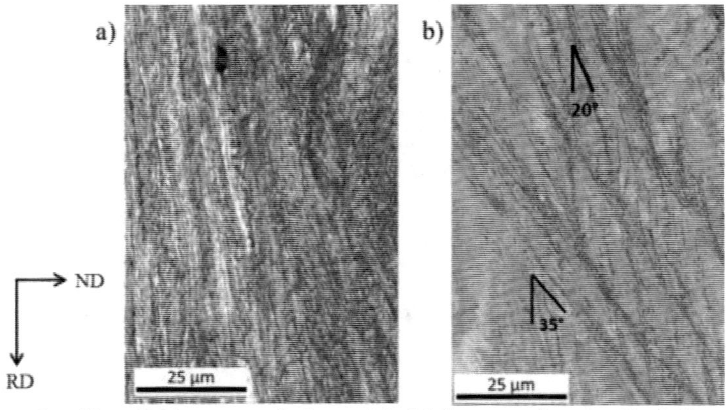

Figure 4: Image Quality (IQ) map of the mid-thickness after 75% deformation (a) cold rolled (b) cryogenic rolled.

3.3 Deformed state after 90% deformation

After 90% cold reduction the texture becomes much sharper (Fig 5a), with the strongest peak being 19 x random. The homogeneity along the α-fiber further degenerates and develops into the single brass component, although the β-fiber remains present. Another remarkable observation is the strengthening of the cube component compared to the 75% cold rolled sample.

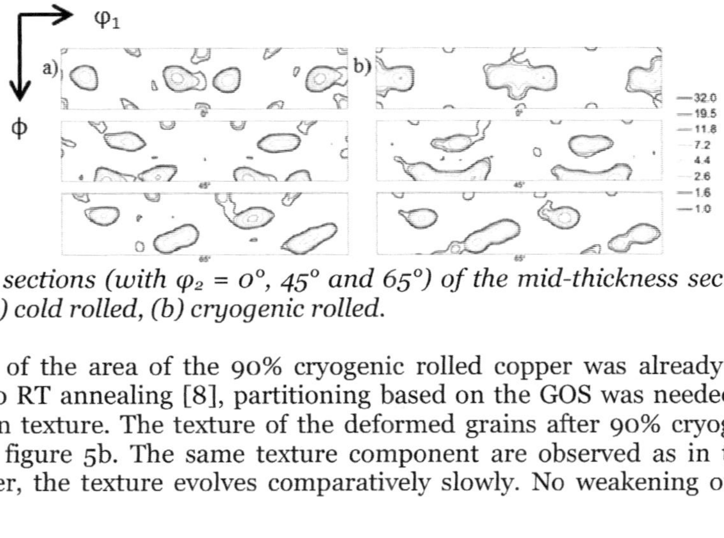

Figure 5: ODF sections (with φ_2 = 0°, 45° and 65°) of the mid-thickness section after 90% deformation (a) cold rolled, (b) cryogenic rolled.

As almost half of the area of the 90% cryogenic rolled copper was already recrystallized, probably due to RT annealing [8], partitioning based on the GOS was needed to determine the deformation texture. The texture of the deformed grains after 90% cryogenic rolling is represented in figure 5b. The same texture component are observed as in the cold rolled copper, however, the texture evolves comparatively slowly. No weakening of the α-fiber is observed.

4. Discussion

Since the initial material is for both thermo-mechanical treatments identical and the texture of this initial copper is completely random (Fig. 1), a thorough comparison can be conducted.

After 50% cold and cryogenic reduction (Fig 2a and 2b), no large difference can be observed between the two textures. Both textures are fairly weak, nearly symmetric and none of the typical fcc rolling fibers are already present. The observed ND-rotated cube orientation is a typical shear texture component which may be the result of the roll gap geometry in this experiment.
Comparing the 75% cold and cryogenically rolled (Fig. 3) and the 90% cold and cryogenically rolled copper (Fig. 5), some difference can be perceived. The most notable difference is the slower texture evolution of the cryogenically rolled copper compared to the cold rolled one. It appears that after the cryogenic deformation the texture resembles an alloy type of texture, in contrast to the pure metal type of texture formed after cold rolling of the pure copper. A possible explanation for the alloy type of texture is the occurrence of deformation twinning [9], which is observed experimentally. Another explanation is that the development of a weaker texture after cryogenic rolling is due to the presence of shear bands. It is reasonable to assume that the shear bands retard the development of the typical rolling textures in the copper matrix.

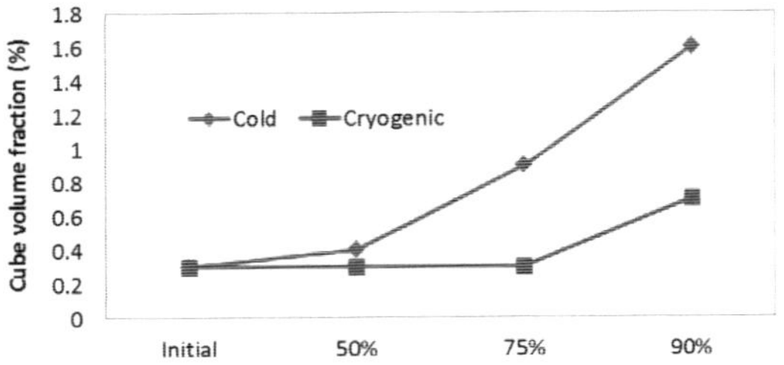

Figure 6: Cube volume fraction of the cold rolled and cryogenically rolled pure copper as a function of the applied reductions.

Another remarkable observation is done in the texture of the cold rolled pure copper, and to lesser extent in the cryogenically deformed. Although the initial material has an almost fully random texture, after higher deformations, the cube component is appearing, albeit in minute strength. This indicates that the cube component is part of the deformation texture. Figure 6 shows the percentage of cube volume fraction of both cold rolled and cryogenic rolled pure copper as a function of the different reductions. This figure shows that cube is formed during the deformation process, where higher deformations stimulate a higher intensity of the cube component. The reason why this cube deformation component forms at higher reductions in the cryogenically rolled material than in the cold rolled material can be explained by the following suppositions. Since cube is very sensitive to shear and the higher friction in the cryogenically rolled material leads to more shear, no cube is formed at lower reductions. At higher reductions, the rolled material becomes thinner, so more heating up of the samples between every rolling pass is possible. In this way, the deformation texture will be more closely related to the deformation textures formed at RT.

5. Conclusion

The texture development in cryogenically rolled and cold rolled pure copper is compared after different rolling reductions, starting from the same randomly textured material. Whereas the cold rolled copper gives rise to a pure metal texture, the texture after cryogenic rolling resembles an alloy type of texture. The cube component is observed in both cold and cryogenic rolled copper as part of the deformation texture.

Acknowledgments

The authors wish to thank the Agency for Innovation by Science and Technology in Flanders (IWT) and the fund for Scientific Research in Flanders (FWO) for financial support (FWO10/PRJ/336). Aurubis (Belgium) is thanked for the high purity cast Cu.

References

[1] Konkova T N et al 2013 *Dokl. Phys.* **58** 240-3
[2] Wang Y M, Chen M W, Sheng H W and Ma E 2002 *J Mater. Res.* **17** 3004-7
[3] Li Y S, Tao N R and Lu K 2008 *Acta Mater.* **56** 230-41
[4] Zhang Y, Tao N R and Lu K 2008 *Acta Mater.* **56** 2429-40
[5] Ridha A A and Hutchinson W B 1982 *Acta Metall.* **30** 1929-39
[6] Duckham A, Engler O and Knutsen R D 2002 *Acta Mater.* **50** 2881-93
[7] Humphreys F J and Hatherly M 2004 *Recrystallization and Related Annealing Phenomena* (Oxford: Elsevier Ltd.)
[8] Konkova T et al 2011 *Mater. Sci. Eng. A* **528** 7432-43
[9] Dillamore I L and Roberts W T 1964 *Acta Metall.* **12** 281-93

IOP Conf. Series: Materials Science and Engineering **82** (2015) 012017 doi:10.1088/1757-899X/82/1/012017

Development of materials model based on microstructure evolution for formability analysis of low-Ni austenitic stainless steel

C Shanta [1*], N Harsh[2], M Bhargava[2], P Prita[1] and M Sushil[2]
[1] Department of Metallurgical Engineering & Materials Science, IIT Bombay, India
[2] Department of Mechanical Engineering, IIT Bombay, India
E-mail: chakrabarty.shanta@gmail.com

Abstract. Low Ni austenitic stainless steels exhibit outstanding mechanical properties due to the occurrence of two deformation mechanisms: dislocations glide (slip) and phase transformation (austenite fcc phase to strain induced α martensite (bcc)). Both mechanisms have strong dependence on strain and strain paths. The local misorientation development due to slip (grain average misorientation (GAM)) is more in biaxial strain-path (BS) followed by plane strain (PS) and uniaxial strain-path (US); whereas martensite volume fraction followed an opposite trend (US > PS > BS) as compared to GAM. In the present study a constitutive material model is developed based on phase transformation and GAM which has been implemented to describe the strain hardening behaviour of material for various strains and strain paths. Variation of normal anisotropy ($\bar{r}$) with stain and strain paths was also considered. Finally the constitutive model and instantaneous $\bar{r}$ were incorporated in Finite Element (FE) based simulation to improve the strain path predictability for this material.

1. Introduction

For high strength and formability applications, such as transport industries, austenitic steels are promising material. They are attractive due to their favorable mechanical properties and good corrosion resistance properties. As Ni cost is very high relative to other alloying elements, efforts are ongoing to substitute Ni with Mn [1]. Ni and Mn both act as austenite stabilizing elements. Stability of austenite depends on the grade of alloying (compositions). The highly alloyed grades are generally stable and form martensite at very low temperature, but the lean grade (low Ni) austenite is considered to be metastable. It forms deformation induced martensite at room temperature. This deformation induced martensite transformation along with slip, changes the properties of austenitic steel and leads to very high strain hardening [2, 3] with increase in plastic strain. Most of the industrial metal-forming processes are characterized by a complex strain path history. A change in strain path has a significant effect on the mechanical behavior of materials [4, 5]. Physical origin for this strain path dependence may be due to evolution of microstructural features such as twinning, phase transformation (α martensite), orientation changes, and local strain development. In the present study a constitutive material model was developed that includes the effect of these microstructure (phase fraction and GAM) developments and implemented to predict the strain hardening behaviour of material at different strain for different strain paths.

2. Experimental procedure

Low Ni austenitic 201 stainless steel (Mn: 9.24, Ni: 1.02, Cr: 15.15, Cu: 1.69, Si: 0.33, N: 0.13, C: 0.06) selected for this study was deformed by limit dome height (LDH) tests up to various strains (low strain (LS), intermediate strain (IS) and high strain (HS)) and under different strain paths (uniaxial strain (US), plane strain (PS) and biaxial strain (BS)). To find the correlation between microstructural development during deformation and mechanical properties tensile tests were carried for 201 SS samples fabricated at different orientation relative to the rolling direction: 0°, 45° and 90°. Mechanical properties were determined by tensile tests and microstructural characterizations were carried out using electron back scattered diffraction (EBSD) and X-ray diffraction (XRD) techniques. EBSD data were used to measure the development of local misorientation and fraction of martesite formed during deformation. From the X-ray profile (2θ, d-spacing) different phases and bulk crystallographic texture were identified. X-ray ODFs (Orientation Distribution functions) and volume fractions for individual ideal texture components were estimated by 3-4 pole figures of corresponding phases by using MTM-FHM programme [6]. SPD simulations were carried out using finite element based software (PAMSTAMP 2G) using the mechanical properties obtained from tensile tests. Strain paths predicted from simulations were compared with experiments and a constitutive model was developed that improves the match between predicted and measured Strain paths.

3. Results and discussion

Variation of work hardening coefficient (n) for 0°, 45° and 90° to RD (rolling direction) samples is shown in figure 1(a). Since this material undergoes both slip and phase transformation during deformation, the variation in n could be attributed to differences in the relative contribution of these two mechanisms at various stages of deformation.

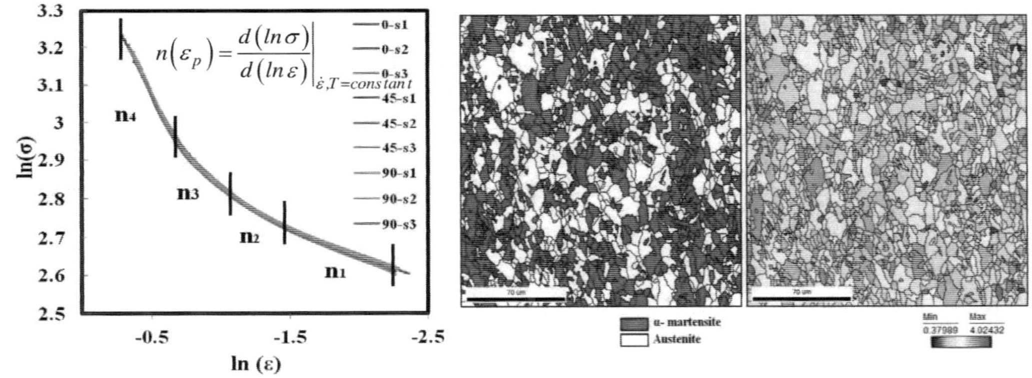

(a) (b) (c)

Fig.1 Effect of microstructural changes on strain hardening coefficient (n) (a) log-log plot of true stress v/s true strain obtained from tensile test of 201SS in different degree (0°,45° and 90°) to rolling direction (b) phase map of US specimen (c) GAM plot of US specimen

The phase transformation and local misorientation development due to slip (GAM) for US specimen were shown in figure 1(b) and 1(c) respectively. The evolution of microstructural features during deformation for all strain paths (US, PS and BS) at different strain levels (LS, IS and HS) are tabulated in table 1. Based on these features (phase fraction and GAM) the strain hardening coefficient (n) was calculated. Figure 2 shows the algorithm for prediction of strain hadening coefficient based on microstructural features for different strain paths and to incorporate in FE based simulation for better SPD prediction. Correaltion of 'n' based on phase fraction and GAM for US is shown on figure 3(a). It shows the increase in 'n' value takes place as

Table1: Microstructural evolution of low Ni-austenitic steel

Strain Path	Strain levels	GAM (°)	phase fraction	r_0	r_{45}	r_{90}
US	LS	1.35	0.083	0.861	1.163	0.643
PS		1.33	0.032	0.985	1.099	1.010
BS		1.43	0.042	0.997	1.114	1.038
US	IS	1.96	0.383	1.075	1.520	1.473
PS		1.58	0.166	1.428	1.747	1.051
BS		2.64	0.186	1.093	1.542	1.050
US	HS	2.27	0.467	1.318	1.635	0.990
PS		1.84	0.265	1.340	2.139	1.023
BS		2.72	0.249	2.098	2.103	1.337

GAM and phase fraction increases. This correlation was assumed to be true for othe strain paths also.Based on this correlation 'n' value for other strain paths were calculated and plotted with respect to the corresponding effective strain as shown in figure 3(b).

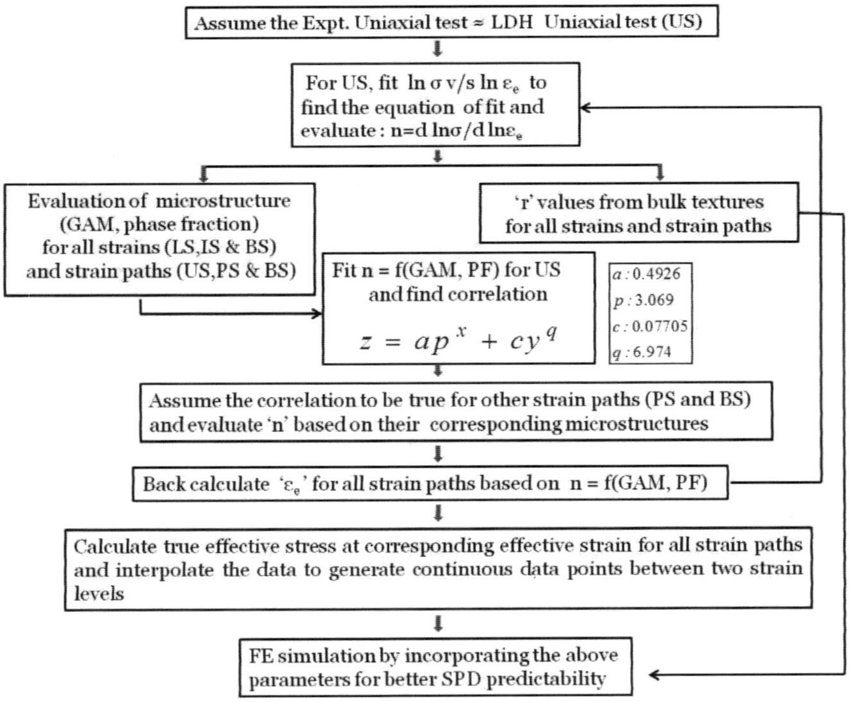

Fig. 2 Algorithm for prediction of strain hadening coefficient based on microstructural features for different strain paths and to incorporate in FE based simulation for better SPD prediction

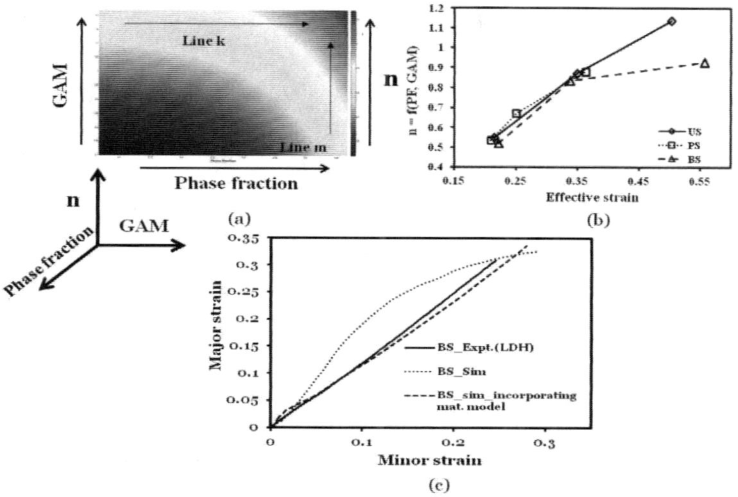

Fig. 3 (a) Variation in n with GAM and phase fraction of martensite for uniaxial strain path (US) (b) Calculated values of n for various strain paths (c) Comparison of experimental and simulated BS strain path before and after incorporating empirical model in FE simulation.

Figure 3 (a) shows that the rate of increase of n with either GAM (line m) or with phase fraction of martensite (line k) is higher when the other quantity – phase fraction and GAM respectively - is high. This is because both martensite and dislocations act as barriers for further plasticity thus increasing strain hardening. Figure 3(b) shows that the increase in strain hardening coefficient with strain is more in case of US than that of PS and BS. After incorporating this empirical model in FE simulation the predictability of strain path is improved and this is shown in figure 3(c).

4. Conclusions

It was observed that strain and strain path during deformation strongly influence microstructure developments, which in turn influence the strain hardening behaviour of material and normal anisotropy ($\bar{r}$). In this case the important microstructural parameters influencing strain hardening coefficient (n) are phase transformation and GAM. Martensite phase fraction in uniaxial strain-path (US) is more compared to biaxial strain-path (BS) whereas GAM is more in BS. n value increases with increasing martensite phase fraction and GAM values. A constitutive materials model is developed based on the evolved microstructural features (phase fraction and GAM) to predict n. The predictability of strain-path has improved significantly after incorporating proposed model and instantaneous ($\bar{r}$) in FE simulation.

References

[1] Cunat P J 2004 *Alloying elements in stainless steels and other Cr containing alloys* (Technical report Euro Inox)
[2] Byun T S, Hashimoto N, Farrell K 2004 Acta *Materialia,Vol* **52** p 3889
[3] Fang X F and Dahl W 1991 *Mater. Sci. Engg A,* Vol **141** p 189
[4] Zhibing Z, Yuqi L, Ting D, Zhigang L 2008 J. *Mater. Process. Tech.* Vol **205** p 425
[5] Abedrabbo N, Pourboghrat F, Carsley J 2007 *Int. J. Plasticity, Vol* **23** p 841
[6] Van-Houtte P 1995 *MTM-FHM Software Manual*, Belgium

IOP Conf. Series: Materials Science and Engineering **82** (2015) 012018 doi:10.1088/1757-899X/82/1/012018

Characterization of the deformation texture after tensile test and cold rolling of a Ti-6Al-4V sheet alloy

B Mehdi[1+], H Azzeddine[2], R Badji[1], V Ji[3], B Alili[2] and D. Bradai[2]
[1]Welding and NDT Research Center, BP 64, Route de Dely Brahim, Chéraga, Algiers, Algeria
[2] Faculty of Physics, USTHB, BP 32 El-Alia, Dar El Beida, Algiers, Algeria
[3] ICMMO/SP2M UMR CNRS_8182, bat 410 University-Paris-Sud, Orsay Cedex, French
E-mail: b.mehdi@csc.dz

Abstract. The deformation texture after cold rolling and tensile test of an industrial Ti-6Al-4V sheet alloy was studied using X-ray diffraction. The alloy was subjected to a cold rolling to different thickness reductions (from 20% to 60%) and then tensile tests have been carried out along three directions relatively to the rolling direction (0°, 45° and 90°). The experimental results were compared to the existing literature and discussed in terms of active plastic deformation mechanisms.

1. Introduction

Titanium alloys containing both aluminium and vanadium (Ti-6Al-4V alloy) are widely used in aerospace applications due to their high strength and good corrosion resistance. It is well known that titanium alloys exhibit different types of texture depending on the deformation conditions (deformation mode, temperature and strain rate) and the alloy composition [1, 2]. Tailoring the properties of the industrial alloys has as prerequisite the knowledge of the simultaneous evolution of their texture and microstructure. In this framework, several works were devoted to titanium alloys in general and Ti-6Al-4V alloy in particular [2–6]. Francillette et al. [5, 6] have presented results of a detailed analysis of anisotropic behaviour of a Ti–6Al–4V sheet during cold rolling in correlation with crystalline texture evolution with a specific initial texture. The present work was undertaken in order to clarify details of the plastic deformation texture evolution for Ti-6Al-4V alloy sample with a slight different initial texture.

2. Experimental

The material used in the present study was a commercial titanium based Ti-6Al-4V alloy provided by Air Algérie. The rolling experiments were conducted at room temperature. The total reductions in thickness were of 20, 40, 50 and 60% with a 20% reduction per pass.
The tensile tests were performed using Zwick Roell test machine. Samples were taken from sheets having three orientations: rolling direction (0°RD), 45°RD and transverse direction (90°RD). The samples were strained to failure. The anisotropy parameter (R-value) was determined by tensile test measurements as the ratio between the transverse to longitudinal strain change. The average Lankford factor $\overline{R}$ and its variation ΔR were calculated using following equations [7] :

$$\overline{R} = \frac{(R(0) + 2R(45) + R(90))}{4} \qquad (1)$$

$$\Delta R = \frac{(R(0) - 2R(45) + R(90))}{2} \qquad (2)$$

The texture was determined in the mid-plane of the specimen by measuring incomplete pole figures (0°≤ α ≤ 70°) in back reflection mode using Cu-Kα radiation. A set of three pole figures ($\{10\bar{1}0\}$, (0002) and $\{11\bar{2}0\}$ was used to calculate the orientation distribution function (ODF) using Bunge series-expansion method, with an expansion degree of l_{max} =22, using MTex software [8].

3. Results and discussion

Figure 1 presents the initial texture of Ti-6Al-4V sheet alloy represented by means of a (0002) recalculated pole figure. As mentioned by Francillette et al. [5], the pole figures associated with initial samples after 45° and 90° rotation around ND can be easily deduced. The as-received alloy sheets present a texture somewhat different from conventionally presented ones and consists of a (0002) pole figure of two components the c axes of which are almost in TD and one central component with two peaks at 20° to ND in the ND-RD plane (all these components result from prior hot and cold rolling) [3, 5, 6]. Here, in Fig. 1, besides the two components with c axis in TD (here referre to as A component), five other texture components can be identified:
$\{01\bar{1}3\}<2\bar{1}\bar{1}0>$, $\{01\bar{1}3\}<0\bar{3}32>$, $\{11\bar{2}3\}<\bar{1}\bar{1}22>$, $\{11\bar{2}2\}<1\bar{2}12>$, $\{11\bar{2}2\}<\bar{1}3\bar{2}\bar{3}>$.
The main ideal texture component positions are also presented in Fig. 1 as well as their characteristics. Such a texture may be due to the variation and heterogeneity induced by the combination of plastic deformation, recrystallization and phase transformation during elaboration [3, 6].

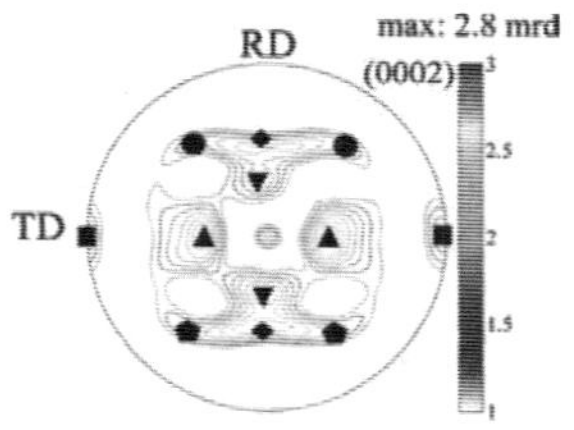

	Component	{hkil}<uvtw>	Euler Angle		
			φ_1	ϕ	φ_2
●	N	$\{11\bar{2}2\}<1\bar{2}12>$	55/125°	60°	30°
⬟	K	$\{11\bar{2}2\}<\bar{1}3\bar{2}\bar{3}>$	240/300°	60°	30°
◆	M	$\{11\bar{2}3\}<\bar{1}\bar{1}22>$	90/270°	45°	30°
▲	N_1	$\{01\bar{1}3\}<2\bar{1}\bar{1}0>$	0/180/360°	30°	0°
▼	N_2	$\{01\bar{1}3\}<0\bar{3}32>$	90/270°	30°	0°
■	*A*	$\{11\bar{2}0\}<10\bar{1}0>$	0°	90°	0/60°

Figure 1. Recalculated (0002) pole figure of the as-received Ti-6Al-4V sheet alloy (left) and texture characteristics (right).

The textures after cold rolling of Ti-6Al-4V sheets are shown in Fig. 2. After 20% thickness reduction, the A ($\{11\bar{2}0\}<10\bar{1}0>$) component disappeared reflecting the early stage of basal texture formation. The intensity of the $\{11\bar{2}3\}<\bar{1}\bar{1}22>$ component decreased after 40% of thickness reduction. The basal poles were shifted from ND towards TD direction (i.e. T-type texture) during cold rolling. Such a texture is also termed as split rolling direction texture (STD) observed for strains above 40% of thickness reduction while below this strain level, a split rolling direction texture (SRD) is often observed [3– 6].

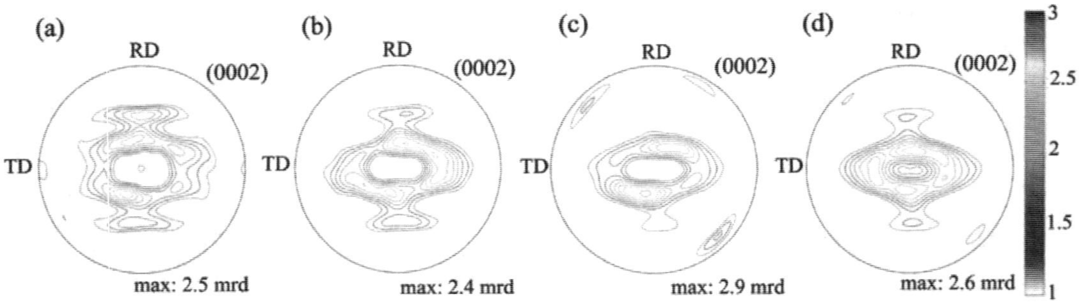

Figure 2. Recalculated (0002) pole figures of Ti-6Al-4V sheets after cold rolling at: a) 20%, b) 40%, c) 50% and d) 60% of thickness reduction.

Rolling up to 50% of thickness reduction caused the development of a T-type texture and the M ($\{11\bar{2}3\}<\bar{1}\bar{1}22>$) component disappeared. The ($\{11\bar{2}3\}<\bar{1}\bar{1}22>$) component was formed also again after rolling to 60% of thickness reduction. The texture simulation made by Philippe et al. [9] predicted that the T-type texture was caused by the activation of $\{1010\}<11\bar{2}0>$ and $\{1011\}<11\bar{2}0>$ glide systems, $\{1012\}<10\bar{1}1>$ tension and $\{1122\}<11\bar{2}3>$ compression twinning.

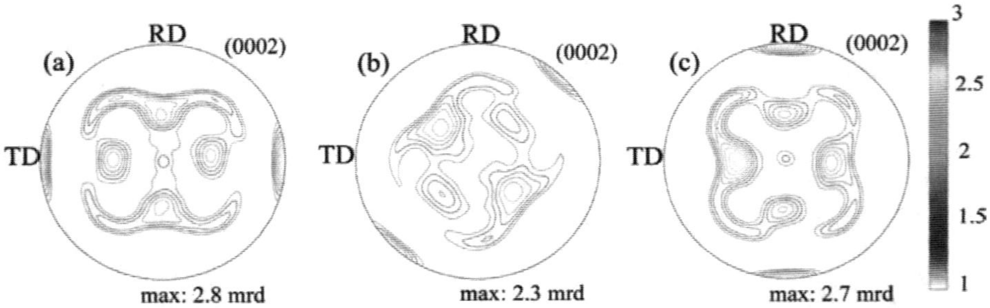

Figure 3. Recalculated {0002} pole figures of the Ti-6Al-4V sheet after cold rolling and after tensile tests for a) 0°RD, b) 45°RD and c) 90°RD.

The textures of tensile tested samples of the Ti-6Al-4V sheet are shown in Fig. 2. Tensile tests in the three different directions do not cause strong changes in the texture characteristics evolution except some intensity sharpening around the $\{11\bar{2}0\}<10\bar{1}0>$, $\{11\bar{1}0\}<2\bar{1}\bar{1}0>$ and $\{01\bar{1}3\}<0\bar{3}32>$ components. Typical stress–strain curves from tensile tests for cold rolled samples are shown in Fig. 4a for 0°RD, 45°RD and 90°RD samples. The deduced mechanical properties are summarized in Fig. 4a. The Young's modulus varied very slightly. The highest yield strength was obtained along 90°RD and the lowest along 45°RD. Our results are in rather good agreement with those of [5, 6]. The elongation-to-failure was slightly higher in 90°RD than in other orientations. 0°RD alloy sample showed the lower value. Additionally, the yield point was much more pronounced in the 90°RD than in any other orientation. The planar anisotropy, or R value, shown in Fig. 4b was lowest along RD (3.14) and increased towards TD (4.75). These values seem to be higher than those presented by Kuwabara et al. [10] (1.8-5.2) or Revil-Baudard et al. [11] (2.3-5.1).

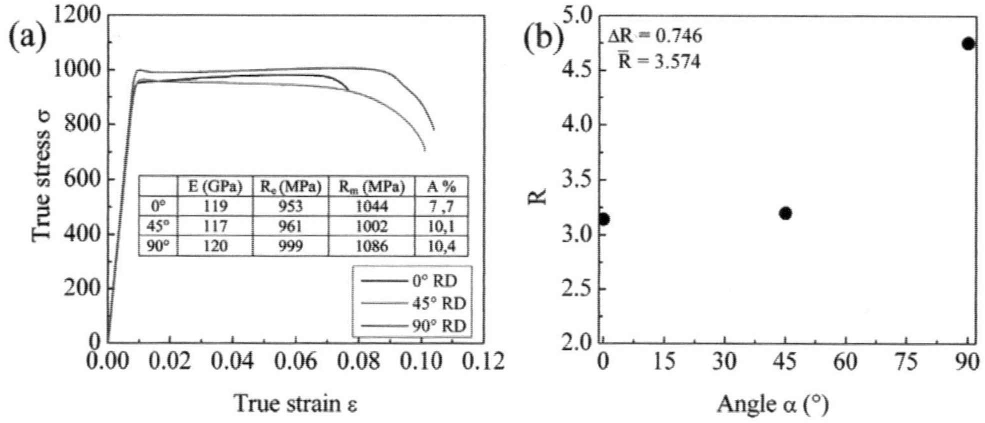

Figure 4. Stress-strain curves of Ti-6Al-4V sheet alloy and b) evolution of R value versus angle to rolling direction.

4. Conclusion

The crystalline texture evolution formed after cold rolling and tensile tests along three different directions (0°, 45° and 90° to the rolling direction) of Ti-6Al-4V sheets was investigated by X-ray diffraction. It was found that cold rolling of the Ti-6Al-4V sheet with a complex initial texture (with six components: $\{11\bar{1}3\}<2\bar{1}\bar{1}0>$, $\{01\bar{1}3\}<0\bar{3}32>$, $\{11\bar{2}3\}<\bar{1}\bar{1}22>$, $\{11\bar{2}2\}<1\bar{2}12>$, $\{11\bar{2}2\}<\bar{1}3\bar{2}\bar{3}>$, $\{11\bar{2}0\}<10\bar{1}0>$) resulted in a split transverse direction (STD) texture. The texture after tensile tests was not considerably modified and was quite similar to the initial texture in the 3 studied directions.

References

[1] Larson FR, Zarkades A and Avery DH 1973 *Titanium Science and Technology* **2**, Plenum Press, London, p. 1169.
[2] Singh AK and Schwarzer RA 2000 *Z. Metallkd.* **91** 702–716.
[3] Phillipe MJ, Serghat M, van Houtte P and Esling C 1995 *Acta Metall. Mater.* **43** 1619-1630.
[4] Nourbakhsh S and O'Brien TD 1988 *Mater. Sci. Eng.* **100** 109-114.
[5] Francillette H and Benmaouche M 2006, *Proc. Materiaux 2006,* Dijon, France.
[6] Francillette H, Benmaouche M and Gauquelin N 2008 *J. Mater. Proc. Technol.* **198** 86–92.
[7] Kocks UE 1998 Texture and anisotropy: preferred orientations in polycrystals and their effect on material properties. Cambridge University Press, 427–130
[8] Bachmann F, Hielscher R and Schaeben H 2010 *Solid State Phenom.* **160** 63–68.
[9] Philippe MJ, Esling C and Hocheid B 1988 *Textures Microstruct.* **7** 265.
[10] Gilles G, Hammami W, Libertiaux V, Cazacu O, Yoon JH, Kuwabara T, Habraken AM and Duchêne L 2011 *Int. J. Solids Struct.* **48** 1277–1289.
[11] Knezevic M, Lebensohn RA, Cazacu O, Revil-Baudard B, Proust G, Vogel SC and Nixon ME 2013 *Mater. Sci. Eng. A* **564** 116–126.

IOP Conf. Series: Materials Science and Engineering **82** (2015) 012019 doi:10.1088/1757-899X/82/1/012019

Quantitative correlation between slip patterning and microstructure during tensile elongation in 6xxx series aluminum alloy

S Ghodrat[1] H Pirgazi[2] and LAI Kestens[1,2]
[1] Delft University of Technology, Materials Science and Engineering Department, Mekelweg 2, 2628CD Delft, The Netherlands.
[2] Ghent University, Department of Materials Science and Engineering, Technologiepark 903, 9052 Zwijnaarde-Gent, Belgium
E-mail: s.ghodrat@tudelft.nl, hadi.pirgazi@ugent.be, leo.kestens@ugent.be

Abstract. To the purpose of evaluating the effect of deformation on the microstructure, aluminum structures were analyzed on tensile strained samples extended to 25% elongation. In the substructure of these deformed samples linear slip patterns were observed, generally confined to the bulk of the grain. In order to study the crystallographic aspect of these slip patterns, two methods were applied based on orientation contrast microscopy (EBSD). The first method is the statistical analysis of stereological nature, which allows us to determine the incidence of certain crystallographic planes with the slip patterns. In other to corroborate the statistical method, also a 3D analysis was carried out on two perpendicular planes of observation (TD and ND sections). The results of both methods were in a very good agreement. It was found that the linear features are predominantly parallel to the {111} crystal planes, although the frequency of {111} planes was not exclusive; also other crystal planes such as {112} and {110} are involved. These observations give a stronger statistical basis for similar observations earlier made by TEM on much smaller fields of observation.

1. Introduction

The evolution of crystal misorientation is an indication of local strains induced in the microstructure by plastic deformation involving dislocation glide in the metal matrix. The dislocations remaining in the microstructure configure themselves in cells or subgrains and more aggregate slip patterns, which will give rise to small orientation gradients. The microstructural effect of plasticity can be gauged by monitoring the crystallographic features of such slip patterns. Dislocation induced structures in aluminum have been extensively studied in the last decades [1-4]. Crystals and grains during deformation are subdivided on a macroscopic scale into deformation bands and on a more microscopic scale into cell blocks and cells. Parallelogram-shaped Cell Blocks (CBs), which are delineated by planar dislocation walls, contain cells surrounded by incidental dislocation boundaries [4]. Dense Dislocation Walls (DDWs) are of a thin planar nature, whereas Microbands (MBs) are thicker and can be distinguished as double boundaries in the TEM. There is a lot of controversy on the crystallographic nature of these phenomena. According to Delannay [4] the DDWs are formed parallel to the most stressed {111} plane. However, Hansen [2,3] also observed DDWs not showing any obvious alignment with the {111} planes.

Analyzing the crystallographic orientation of slip patterns will contribute to the understanding of mechanism by which they are generated and hence it will provide an improved insight in the deformation microstructure. Most of the previous investigations are base on TEM observations, which unfortunately lack statistical reliability. Even most EBSD mappings are done on a rather small scale. Therefore the aim of this work is to give a better statistical basis for the crystallographic aspects involved in the formation of the deformation substructures combining wide field data gathering with high spatial resolution.

2. Experimental Procedure

The material used in the current study is a conventionally 6016 aluminum alloy for automotive applications. The sheet from which the samples were extracted was in cold rolled and annealed T4 condition. Tensile specimens were cut from the sheet with the tensile access parallel to the rolling direction. The specimens were elongated to 25% strain. Samples for EBSD observations were cut from the middle of the gauge length of the tensile specimens. The planes of the observation were TD and ND planes (*i.e.* planes normal to TD and ND, respectively). Sample preparation for this work was very critical because we wanted to be sure that no extra deformation be induced to the samples. Therefore, after conventional mechanical polishing up to 0.25 μm with OPS, electropolishing was applied by using the Electrolyte A2 Struers electrolyte with a voltage of 39 V, flow rate 9 for 15 s. It should be noted that we mostly prepared a non-deformed reference sample jointly with the deformed sample. The EBSD system of type EDAX-TSL® that was used in the current study was mounted on a FEI quanta 450® SEM with Field Emission Gun (FEG) filament. The orientation contrast scans were collected and analyzed by the commercial OIM-TSL® software. Wide filed EBSD scans, covering an area of ~9 mm^2, in combination with good spatial resolution (step sizes between 0.5 and 1 μm), gave rise to huge data sets with as much as 10 Mpixels.

3. Results

Figure 1 shows the orientation contrast map of the sample extended in tension to an elongation of 25%. We are interested to analyze the linear slip patterns observed in a subset of grains as shown in figure 1 (cf. white markers). In reality, these linear slip patterns in 3D are of a planar nature. The question that we would like to address here is whether these slip patterns are parallel to specific crystallographic planes, or in other words, we want to quantify the frequency of the coincidence of crystal planes {hkl} with the planar slip patterns. This question relates to the nature and the origin of the slip patterns.

Figure 1 IPF map overlaid by IQ map on a specimen loaded to 25% elongation. White markers indicate linear slip patterns.

A very large dataset, with more than 2000 grains, was analyzed covering a total scanned area of various square millimeters. The line segments corresponding to the linear slip patterns were considered as intersections of an unknown crystal plane with the plane of observation. These unknown planes can be represented as pole traces on unit triangle of standard stereographic projection. Figure 2a shows the distribution of these pole traces after appropriate background correction, which considers a random distribution of slip segments for the specific texture of the sample. The method used here to process the data was based on the stereological line segment analysis proposed by Rohrer *et al.* [5, 6]. It can be seen that the trace distribution exhibits the maximum on the [111] pole but it is quite a broad maximum extending to the [112] pole. Also in the vicinity of [110] a local maximum can be observed.

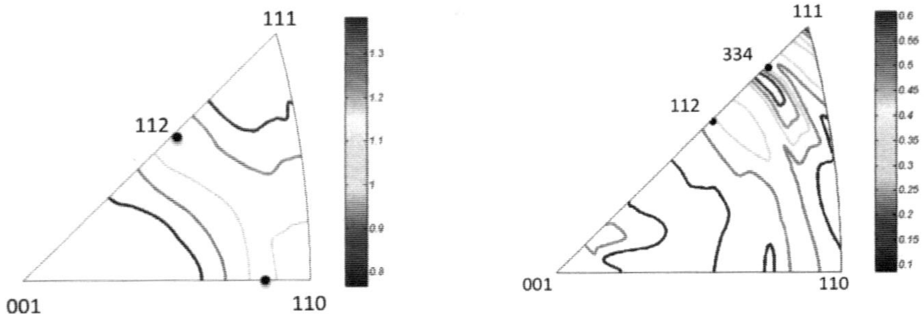

(a) (b)

Figure 2 (a) Distribution of the normal of the planes parallel to the slip patterns obtained by the stereological method. (b) Distribution of the plane normals defined by the linear slip segments (observed on the perpendicular sections in 3D). The colorbar shows the intensity.

Alternative to the statistical method, a 3D analysis was carried out on two perpendicular planes of observation (TD and ND sections). In this method we identified slip patterns on 20 corresponding grains at the common edge of two sections. Figure 2b shows the distribution of the normals of the planes that were defined by both slip segments. It can be seen that the pole distribution of planes exhibits the maximum on the [111] pole, in the vicinity of the [112] and the [334] poles. Also in the neighborhood of the [110] pole there is a small maximum. The planar pole distribution obtained from 3D analysis (figure 2b) is in reasonable correspondence with the trace distribution produced by the statistical method (figure 2a), even though only 20 grains were considered in 3D analysis.

4. Discussion

If we compare the results of this work with the observations of Hansen (2001), we may conclude that what we have observed are cell blocks or cell block boundaries. The reason why we come to this conclusion is that our observed slip features, just as cell blocks, exhibit a linear character and in many cases they are parallel to the presumed {111} octahedral slip planes. It has to be mentioned, though, that the alignment of the planar slip features with the {111} planes is not restricted to these planes only, as also cell blocks are observed, which are aligned to other crystal planes such as {112} and ~{110}.
In an alternative interpretation, the slip patterns may be related to in-grain shear bands. However, shear bands (SB) have rather a fixed angle (*e.g.* ≈35°) with the main axis of strain, whereas the linear slip segments observed here, exhibit scattered angles with respect to the TA, ranging from 10 to 60°. Additionally, SBs would give rise to saw-tooth features when

intersecting grain boundaries, which was not observed here. Moreover, it would be expected to observe shear bands in high Taylor factor grains, as they would be the most benefiting from geometric softening. In the present case, though, the linear slip segments are also observed in grains with low Taylor factor (*cf.* red grains in Figure 1). An additional argument in favor of the cell blocks hypothesis is that the slip patterns analyzed in this work have an appropriate width (≈5 μm), which very well correspond to the reported width in the literature [2, 3]. Based on these facts, it is unlikely that these patterns can be associated with shear bands, but their characteristics better correspond to cell blocks features [1-4].

5. Conclusions

To the purpose of analyzing the linear slip patterns observed in tensile elongated (25%) aluminum samples two methods were applied based on orientation contrast microscopy (EBSD): (i) a statistical analysis of stereological nature, and (ii) a 3D analysis.
The results of both methods were in reasonable agreement. It was found that the linear features are predominantly parallel to the {111} crystal planes, although the frequency of {111} planes was not exclusive; also other crystal planes such as {112} and {110} are involved. These observations give a stronger statistical basis for similar observations earlier made by TEM on much smaller fields of observation. In the discussion of the present data we argue that the observed slip segments may be related to the presence of cell blocks in the deformed structure.

Acknowledgments

The authors wish to thank Dr. A.C. Riemslag for helping us to carry out tensile tests at Delft University of Technology (TU Delft) in the Netherlands. Also many thanks are due to Dr. J. Sidor for providing the aluminium sheets for this research.

References

[1] Bay, B., Hansen, N., Hughes, D.A., Kuhlmann-Wilsdorf, D. (1992), Evolution of FCC deformation structures in polyslip, *Acta Mater.,* **40**, 205-219.
[2] Hansen, N. (2001), New discoveries in deformed metals, *Metall Mater Trans A*, **32A**, 2917-2935.
[3] Hansen, N. Juul Jensen D. (1999), Development of microstructure in FCC metals during cold work, *Phil Trans. R. Soc. Lond. A*, **357**, 1447-1469.
[4] Delannay, L. (2001), Observation and modelling of grain interactions and grain subdivision in rolled cubic polycrystals, *PhD Thesis*, K.U.Leuven, Mai 2001.
[5] Rohrer, G.S. et al. (2004), The distribution of internal interfaces in polycrystals, *International Journal of Materials Research and Advances Techniques, Z. Metallkd.* ***95*** (2004) 4, 1-18.
[6] Saylor, D.M. et al. (2004), Measuring the Five-Parameter Grain-Boundary Distribution from Observations of Planar Sections, *Metall Mater Trans A,* **35A**, JULY 2004, 1981-1989.

IOP Conf. Series: Materials Science and Engineering **82** (2015) 012020 doi:10.1088/1757-899X/82/1/012020

Evolution of deformation texture of high-pressure ω-phases in pure Ti and Zr during high-pressure torsion straining

N Adachi,[1] Y Todaka[1,*], H Suzuki[2] and M Umemoto[1]
[1] Department of Mechanical Engineering, Toyohashi University of Technology, Aichi, JP
[2] Quantum Beam Science Center, Japan Atomic Energy Agency, Ibaraki, JP
[*] Corresponding author (E-mail: todaka@me.tut.ac.jp)

Abstract.
High-pressure torsion process is the only way to obtain bulk-shaped samples of nearly 100% of high-pressure ω-phase of pure Ti and Zr at ambient condition. The neutron diffraction experiment of HPT-processed pure Ti and Zr revealed that the deformation of high-pressure ω-phase mainly occurs at the prismatic plane, which is completely different as α-phase. This paper also reports the correlation between deformation textures and mechanical property of ω-phase obtained by HPT process.

1. Introduction

Group IV transition metals such as Ti and Zr display three phases, depending on the temperature and pressure. At ambient temperature and pressure, these metals have a hexagonal close-packed (hcp) structure (α-phase), which transforms into a body-centered cubic (bcc) structure (β-phase) with increasing temperature. The α-phase structure also transforms into a simple hexagonal structure (ω-phase) under high pressure (3.4 GPa [1]); this pressure-induced α-to-ω transformation was first observed by Jamieson [2]. The high-pressure ω-phase essentially exhibits a reverse phase transformation upon pressure reduction or cooling. Therefore, the application of these phases as structural material could not be achieved so far. However, recent studies have revealed that, by using the high-pressure torsion (HPT) process, the high-pressure ω-phases in both pure Ti and Zr can be stabilized under ambient conditions [3–5]. The HPT process can realize a severe plastic deformation and involves the application of an intense shear strain under high pseudohydrostatic pressure on the gigapascal scale. This process is based on a procedure developed by Bridgman [6], in which a disk-shaped specimen is placed between two anvils, and a torsional shear strain is applied to the specimen by rotating the lower anvil under high pressure. The number of revolutions, N (accompanied by an amount of shear strain, γ) and the compression pressure, P, that the specimen is subjected to are the critical parameters. The fraction of the stabilized ω-phase after the HPT process is dependent on both N and P [3,7]. The HPT process is the only method that can result in nearly 100 vol% of the ω-phase in pure group IV metals. This fact enables the determination of the mechanical properties of single high-pressure ω-phase, such as elastic constants [8] and slip systems.

In this study, we investigated the deformation behavior of high-pressure ω-phase of pure Ti and Zr during the HPT process by using the neutron diffraction technique. The correlation between the texture of ω-phase and the mechanical properties was also investigated.

2. Experimental procedures

The materials used in this study are commercially pure Zr (99.2 %) and Ti (99.9 %). First, the Ti and Zr disks—with 20 and 10 mm diameters, respectively, and 0.85 mm thickness—were annealed at 1073 K for 3.6 ks in vacuum to homogenize their microstructures. The disks were individually placed between two anvils with a depression of 10 mm in diameter and 0.25 mm in depth, and a torsional strain was applied under high pressure. A compression pressure, P, of 5 GPa was used for each sample, and N was up to 20 revolutions. The rotating speed was 0.2 rpm, and the HPT process was performed at room temperature. The γ introduced by the HPT process was calculated using the expression $\gamma = 2\pi rN/t$, where r is the distance from the disk's center and t is the specimen thickness.

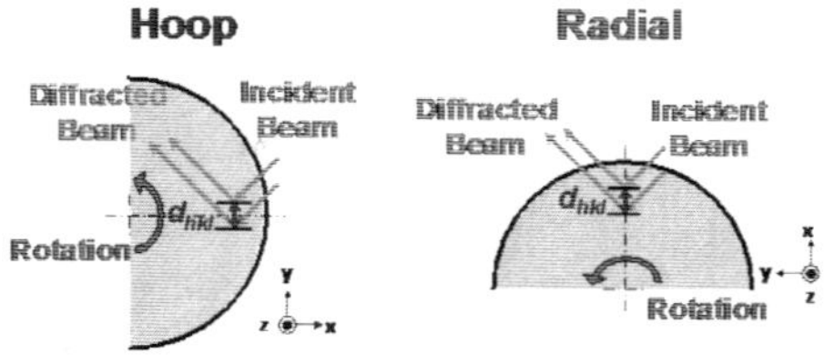

Fig.1 Schematic illustrations of the beam incident direction in the neutron diffraction experiment.

The neutron diffraction (ND) experiment was performed at the Japan Research Reactor No. 3 (JRR-3) of the Japan Atomic Energy Agency. The wavelength of the neutron beam was 0.164 nm and the slit size was 2 (width) × 15 (height) mm. The experiment was performed under ambient conditions, using RESA-1 (diffractometer for the residual stress analysis). The ND samples consisted of stacks of the HPT-processed disks with a total thickness of 5 mm and neutron beam is irradiated from the lateral side of the disks. The diffraction patterns were measured along two directions of the incident beam—hoop and radial—at each diffraction angle. The diffraction plane of the sample in the hoop (or radial) direction was perpendicular (parallel) to the direction of shear applied during the HPT process (Fig.1).

3. Results and discussion

The XRD profiles of pure Ti and Zr after the HPT process measured at ambient condition were shown in Figs.2(a) and (b). Note that measured diffraction planes are parallel to disk surface (i.e., perpendicular to hoop and radial). The stabilization of high-pressure ω-phase was confirmed. The peak intensities of stabilized ω-phase increases dependent on the introduced strain (*i.e.*, N) and the α-phase peaks almost disappeared after the HPT process of $N = 5$, these results consistent with the previous results [3]. The P required to obtain ω-phase has threshold value; when the applied P is 3 GPa or lower, the ω-phase could not be observed. Zhilyaev *et al.* reported that the threshold value in pure Zr (99.8%) is 1 GPa, which is much lower value compared with the present study. This difference is probably caused by the purity of material. The effect of impurities are still unclear, but it can be expected that the oxygen content affects threshold value, as oxygen is α-phase stabilization element in Zr and Ti and increases the energy barrier for α-to-ω transformation [9]. For the ND experiment, the $N = 10$ samples were used to investigate the deformation behavior of single ω-phase.

Using scanning transmission electron microscopy, we could confirm that both the β-phase and the ω-phase existed in high-purity Zr after the HPT process; this result will be published elsewhere. During the ND experiment, however, the β-phase and the ω-phase cannot be distinguished, even when a long-wavelength beam is used. This is because the ω-phase's superlattice structure includes a bcc structure in its interior, and all the diffraction peaks of the β-phase overlap with those of the ω-phase [1,10]. Therefore, the existence of the β-phase was not considered in this study.

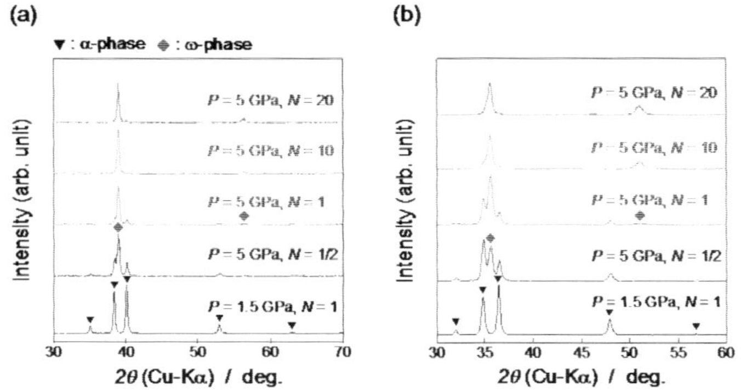

Fig. 2 XRD profiles of the HPT-processed ($P = 1.5$, 5 GPa, $N = 1/2 \sim 20$) (a) pure Ti and (b) pure Zr samples measured at ambient condition.

We previously reported the results of the ND experiment in pure Ti after the HPT process [11]. The obtained results in pure Zr have almost same tendency. By employing pure Zr instead of pure Ti for ND experiment, however, more detailed information of deformation texture of ω-phase was obtained since Zr have larger lattice constant compared with Ti and thus diffraction peaks which can not be separated in Ti can be separated in Zr. The observed diffraction planes in both the hoop and radial directions were summarized in Fig. 3. The ω-phase after the HPT process has quite strong texture. For instance, in the radial direction (parallel to shear direction applied in the HPT process), the {11-20}ω deformation texture were gradually formed with increasing γ (and r), and basal plane could not observed. In contrast, basal plane was mainly observed in the hoop direction (perpendicular to shear direction). These results suggest that the active slip plane of ω-phase during the HPT process is prismatic plane. It is well known that, in the case of α-phase, the main slip system is close-packed plane and direction: {0001}<11-20>. After α- to ω-phase transformation, {11-20}ω prismatic plane becomes close-packed plane and close packed direction is <0001>ω. Therefore, the main slip system changed to {11-20}<0001>ω and aligned parallel to shear direction applied in the HPT process.

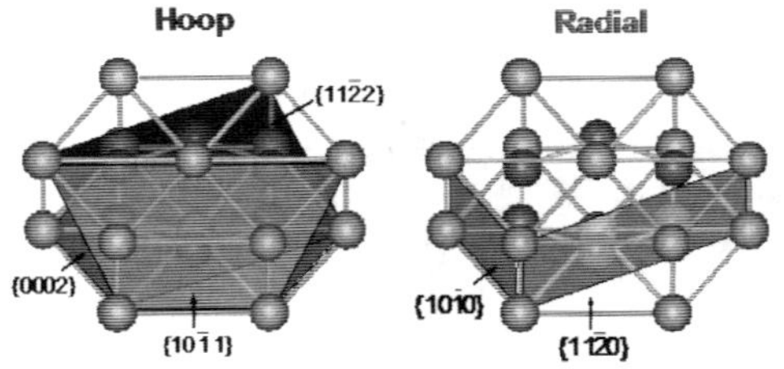

Fig.3 Observed diffraction planes in the neutron diffraction experiments of the HPT-processed pure Ti and Zr.

In the hoop direction, pyramidal planes were also observed in addition to basal planes, indicate that the c-axis of some part of ω-phase tilted to the shear direction. The observed lattice plane of ω-phase in the ND experiment and the interplanar angles between observed lattice planes and reference planes (basal plane, prismatic planes) were listed in Table 1. As seen in table, diffraction peaks were observed when the interplanar angle between basal or prismatic plane is below 36°; if an interplanar angle between basal (prismatic) plane is below 36°, a diffraction peak is observed in hoop (radial) direction. In the case of {11-21}ω, every interplanar angles are over 36°, thus this plane could not observed in both the hoop and radial directions. The pyramidal plane {30-31}ω could not observed in spite of low interplanar angle. This is probably owing to the quite small structure factor of this plane.

In order to investigate the effect of texture on the mechanical properties of ω-phase, the samples for the tensile tests were cut out from the HPT processed pure Ti as shown in Fig.4. Two types of tensile sample (parallel area: 1.0 mm width, 0.5 mm thickness, and 3.0 mm length) were prepared as shown in Fig.5(a). The tensile direction of one sample is parallel to shear direction applied in the HPT process (sample-A) and thaxht of the other sample is perpendicular to shear direction (sample-B). The obtained tensile stress – strain curves are shown in Fig.5(b). The sample-A has the high strength of about 1250 MPa, but fractured with brittle manner after yielding. In contrast, the sample-B showed the tensile elongation of about 1 %. Such difference in the deformation behavior is probably caused by the

Table 1 List of the observed diffraction planes in the neutron diffraction experiment, and interplanar angle between the observed and basal or prismatic planes.

Lattice plane	Observed direction	Interplanar angle [°]		
		(0001)	(10-10)	(11-20)
(10-11)	Hoop	35.7	54.3	59.6
(11-21)	×	51.2	47.5	38.8
(11-22)	Hoop	31.9	62.8	58.1
(20-21)	Radial	55.2	34.8	44.7
(31-41)	Radial	68.9	25.1	26.3
(30-31)	×	65.1	24.9	38.2
(21-31)	Radial	62.3	33.2	29.6

deformation texture formed by the HPT process. In the case of sample-A, as observed in the ND experiment, main slip plane of ω-phase {11-20}ω is parallel to tensile direction (Schmidt factor = 0) or close to parallel. Therefore, slip deformation could not be activated during the tensile test and fractured with brittle manner. In the case of sample-B, the slip plane of most part of material is perpendicular to tensile direction. However, as seen in Table 1, observed plane in radial direction {20-21}ω have angle of nearly 45° between slip plane {11-20}ω. It is suggested that the slip deformation of grains with such orientation is easy to activated during the tensile test. Consequently, slight plastic deformation was observed in the sample B.

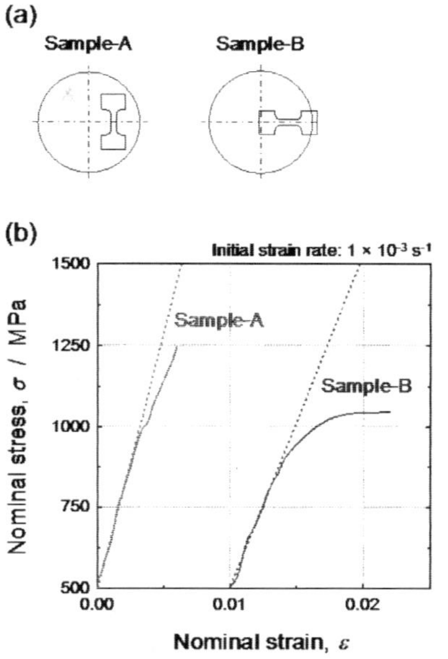

Fig. 4 (a) Geometry of the tensile test samples of the HPT-processed pure Ti used in this study. The tensile direction is parallel and perpendicular to the shear direction by the HPT process in the sample-A and sample-B, respectively. (b) The obtained stress-strain curves of the sample-A and sample-B.

4. Summary

By investigating the deformation texture of single high-pressure ω-phase by using the neutron diffraction technique, {11-20}<0001>ω seems main slip system of ω-phase. The high-pressure ω-phase exhibits slight plastic deformation when the tensile direction is perpendicular to shear direction applied in HPT process. This is probably the effect of texture; the slip deformation of some grains easy to activated upon loading.

Acknowlegments

The neutron diffraction experiments were performed at the RESA-1 of JRR-3 with the approval of the Japan Atomic Energy Agency (proposal No. 2008B-A08). This work was supported by a Grant-in-Aid for Scientific Research on Innovative Area, "Bulk Nanostructured Metals," (Contract No. 22102002) through MEXT, Japan.

References

[1] Zhang J, Zhao Y, Pantea C, Qian J, Daemen L L, Rigg P A, Hixson R S, Greeff C W, Gray III G T, Yang Y, Wang L, Wang Y, Uchida T 2005 J. Phys. Chem. Sol. **66** 1213
[2] Jamieson J C, 1963 Science **140** 72
[3] Todaka Y, Sasaki J, Moto T, Umemoto M 2008 Scripta Mater. **59** 615
[4] Pérez-Prado M T, Gimazov A A, Ruano O A, Kassner M E, Zhilyaev A P 2008 Scripta Mater. **58** 219
[5] Pérez-Prado M T, Zhilyaev A P 2009 Phys. Rev. Lett. **102** 1775504
[6] Bridgeman P W 1952 Studies in Large Plastic Flow and Fracture (McGraw-Hill, New York)
[7] Zhilyaev A P, Galvez F, Sharafutdinov A, Pérez-Prado M T 2010 Mater. Sci. Eng. A **527** 3918
[8] Tane M, Okuda Y, Todaka Y, Ogi H, Nagakubo A 2013 Acta Mater. **61** 7543
[9] Hennig G K, Trinkle D R, Bouchet J, Srinivasan S G, Albers R C, Wilkins J W 2005 Nat. Mater **4** 129
[10] Dey G K, Tewari R, Banerjee S, Jyoti G, Gupta S C, Joshi K D, Sikka S K 2004 Acta Mater. **52** 5243
[11] Todaka Y, Azuma H, Ohnishi Y, Suzuki H, Umemoto M 2010 J. Phys.: Conf. Series. **240** 012113

IOP Conf. Series: Materials Science and Engineering **82** (2015) 012021 doi:10.1088/1757-899X/82/1/012021

Variant selection of twins with low Schmid factors in cross grain boundary twin pairs in a magnesium alloy

Z-Z Shi[1], YD Zhang[1], F Wagner[1], P-A Juan[2], S Berbenni[1], L Capolungo[2], J-S Lecomte[1] and T Richeton[1]

[1] Laboratoire d'Etude des Microstructures et de Mécanique des Matériaux, LEM3, CNRS 7239, Université de Lorraine U.L. – Metz, Ile du Saulcy, 57045 Metz, France

[2] George Woodruff School of Mechanical Engineering, Georgia Institute of Technology Lorraine, UMI 2958 Georgia Tech–CNRS, 57070 Metz, France

E-mail: ryansterne@gmail.com (Zhangzhi Shi)

Abstract. Samples of magnesium AZ31 alloys are deformed in compression at room temperature under a strain rate of 1×10^{-3} s^{-1}. The initial texture with respect to the loading direction is favorable for {10–12}<-1011> extension twinning during the deformation. At an engineering strain of 2.75%, many extension twins are found to be connected with each other at grain boundaries, forming cross grain boundary twin pairs. Some have low positive or even negative Schmid factors (SFs). The variant selection of them are interpreted in terms of shear accommodations. The observed twin variants require the least or no accommodation through deformation modes with high CRSSs, but the most or more accommodation through those with low CRSSs.

1. Introduction

Mg and its alloys have drawn much attention in recent years mainly because of their low densities, which renders them high specific yield strength critical for aviation and great potentiality of fuel saving and CO_2 emission reduction for car industry. Extended civil application of Mg alloy as light-weighted structure material depends on achieving ductility, simpler manufacturing and high processing qualities [1]. Twinning is an important deformation mode of wrought Mg alloys at low temperatures [2, 3], the study of which is necessary to predict their mechanical responses, and thereby benefits computational alloy design. When formed abundantly, {10–12}<-1011> extension twins (ETWs) contribute to significant strain softening [2, 4]. After they introduce sufficient interfaces, they lead to strain hardening due to impediment of continuous dislocation slip by the additional interfaces [4, 5].

ETWs are easy to form "cross grain boundary" twin pairs [6, 7]. A statistical study reveals that the minimum misorientation angle between two twins in a cross grain boundary pair is nearly equal to that between their host grains [6], which is an empirical rule. A geometrical factor (m) is widely used to evaluate the shear compatibility between two twins in a cross grain boundary pair. This m-factor is described by $m = \cos\alpha\cdot\cos\beta$, where α and β are the angles between the twinning planes and the shear directions, respectively [7, 8]. Previous studies find that the majority ETWs in cross grain boundary twin pairs have m factors close to 1, indicating high shear compatibilities [7, 8]. This paper studies on ETWs with low Schmid factors (SFs) in cross grain boundary pairs, which cannot be interpreted neither by m-factor nor by the empirical rule of misorientation angles.

2. Experimental procedure

The material is a hot rolled sheet of commercial Mg AZ31 alloy (Mg-3Al-1Zn, wt.%). Its rolling, transverse and normal directions are designated as RD, TD and ND, respectively. Uni-axial compressions along RD were performed on cubic specimens from the sheet, at room temperature and a strain rate of 1×10^{-3}/s. The microstructures of the specimens were measured by electron backscatter diffraction (EBSD). EBSD samples were grinded using SiC papers with grits from 2400 to 4000, and then electrolytically polished in an electrolyte of 62.5% phosphoric acid and 37.5% ethanol at 3 V for 30 s followed by at 1.5 V for 2 min, both at -15 °C. The instrument used was JEOL 6500F FEG SEM with Channel 5 analysis system (Oxford HKL).

3. Results and discussion

The rolled sheet exhibits a typical fiber texture with grains' c-axes around ND. The as-received material has been heat-treated without twins. After the compression along RD, ETWs are abundantly intrigued, as shown in Fig 1. The (macro) SF of an ETW can be calculated, which is the external stress resolved along the shear direction on the twinning plane. The ETWs are then classified into two groups, i.e., low and high SF twins. Here, a low SF twin is defined by two criteria. One is that its SF ≤ 0.3, the other is its SF ratio ≤ 0.6 when its SF > 0. The SF ratio (designated as SF_{ratio}) is the ratio of the SF of a twin to the possible highest SF of the twins in a particular grain. SF_{ratio} is only applied to a twin with SF > 0, because a negative SF renders the appearance of a twin abnormal but a positive low SF may be the possible highest one in a particular grain so that its associated twin should be activated preferentially. The above criteria select a twin with not only a SF absolutely low, but also a SF relatively lower in a particular grain. Oppositely, a high SF twin can be defined. It is intuitive that high SF ETWs are easy to be activated, which indeed constitute the observed majority.

Fig. 1. An EBSD microstructure showing ETWs in the compressed sample with inverse pole figure (IPF) coloring component and band contrast (BC) component. Boundaries between the ETWs and matrices are outlined in red. Other boundaries with misorientation angles larger than 5° are outlined in black.

However, low SF twins are also detected. Fig. 2a shows a cross grain boundary pair of twins consisting of Twin 1 in Grain 1 and Twin 2 in Grain 2. Fig. 2b and 2c provide {0001} pole figures of the grains and the twins in Fig. 2a. From these figures, it is seen that Twin 1 has a negative SF of -0.08, while Twin 2 has a positive SF of 0.22 and the highest SF_{ratio} of 1. Therefore, the cross grain boundary twin pair contains low SF Twin 1 and high SF Twin 2. The m-factor mentioned in the introduction section is first applied to the twins in order to evaluate their shear compatibility. It can be calculated that their m-factor is 0.32, indicating a low shear compatibility between the twinning systems of Twin 1 and Twin 2. The minimum misorientation angle between Grain 1 and Grain 2, and that between Twin 1 and Twin 2 are calculated to be 26.2° and 35.4°, respectively. The difference between them is 9.2° so that the empirical rule of misorientation angles mentioned above cannot be applied to this case. Next, we further explore the rule of variant selection governing low SF Twin 1.

IOP Conf. Series: Materials Science and Engineering **82** (2015) 012021 doi:10.1088/1757-899X/82/1/012021

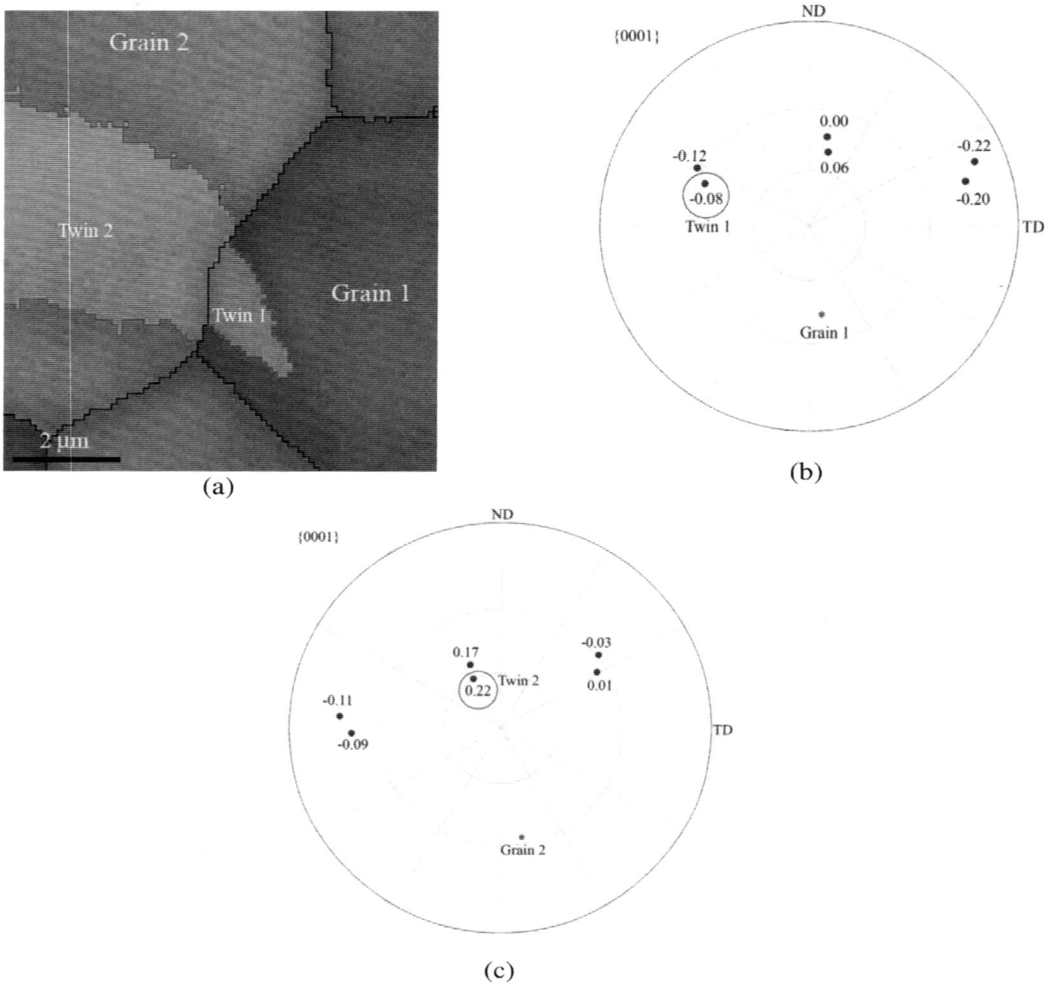

Fig. 2. (a) An EBSD microstructure showing a cross grain boudanry twin pair, Twin 1 and Twin 2, with IPF coloring and BC components as contained in Fig. 1. (b) {0001} pole figure of Grain 1 and its 6 possible ETW variants with their associated SFs. (c) {0001} pole figure of Grain 2 and its 6 possible ETW variants with their associated SFs.

Since twinning produces shear deformation, the growth of a twin requires shear accommodations in its neighboring crystals, which could happen in its host grain, neighboring grain or twin. The accommodation abilities of the shear induced by one twin through different deformation modes in another twin can be evaluated by a method developed in reference [9]. For Mg alloys, 5 common deformation modes for accommodation are chosen to be studied, which are basal, prismatic and <c+a> pyramidal slips as well as {10-12}<-1011> and {10-11}<10-1-2> twinnings. The shear magnitude of Twin 1 along the shear direction on the twinning plane is 0.129 when the c/a ratio is 1.624 [3]. It can be resolved along the slip direction on the slip plane of a slip mode or along the shear direction on the twinning plane of a twinning mode in Twin 2, with the method provided in reference [9]. The resulted shear in Twin 2 is thus called as a geometric accommodation shear for Twin 1. There exists several variants of a deformation mode, e.g. basal slip mode has 3 variants, and thereby several corresponding geometric accommodation shears. The largest one of a deformation mode "i", designated as e^{imax}, sheds light on twin variant selection.

Table 1. The e^{imax} values associated with the 6 possible twin variants in Grain 1, where i = BS (basal slip), PR (prismatic slip), PY (<c+a> pyramidal slip), ETW ({10-12}<-1011> twinning) or CTW ({10-11}<10-1-2> twinning). The observed Twin 1 is variant 3.

ETW variant in Grain 1	SF	BS	PR	PY	ETW	CTW
1	-0.20	0.0242	0.0685	0.0815	0.0280	0.0968
2	0.06	0.0278	0.0426	0.1145	0.0091	0.1226
3	**-0.08**	**0.0049**	**0.1138**	**0.0533**	**0.0367**	**0.0473**
4	-0.22	0.0823	0.0871	0.0672	0.0290	0.0682
5	0.00	0.0918	0.0434	0.0859	0.0095	0.0915
6	-0.12	0.0431	0.1064	0.0379	0.0352	0.0292

Table 1 lists the e^{imax} values associated with the 6 possible ETW variants in Grain 1. Comparing the e^{imax} values in each column of the table, one can find that variant 3 has the highest e^{PRmax} and e^{ETWmax} values, but the second lowest e^{PYmax} and e^{CTWmax} values, in the 6 ETW variants in Grain 1. In AZ31 Mg alloy, the critical resolved shear stresses (CRSSs) of the 5 deformation modes can be ordered increasingly as BS < ETW < PR < RY & CTW [10]. Therefore, variant 3 requires the most shear accommodations through prismatic slip and extension twinning with low CRSSs, but less those through pyramidal slip and contraction twinning with high CRSSs. This indicates that Twin 2 is readier to accommodate the strain induced by variant 3, which corresponds to the observed Twin 1 with negative SF. Analogous to Twin 1, statistical analysis reveals that over 70% of the low SF twins in 66 cross grain boundary twin pairs require the most or more shear accommodations through prismatic slip and/or extension twinning with low CRSSs, but the least or less through pyramidal slip and/or contraction twinning with high CRSSs. Due to the length of the paper, the statistical data is not included here.

4. Conclusions

This paper focuses on the ETWs with low SFs in cross boundary twin pairs. The shear compatibility between two twins in such a twin pair is low, which is evaluated by the widely used m-factor. A newly developed method is used to calculate the shear accommodation of a deformation mode in one twin, which is generated by the shear induced by another twin. With this method, it is found that a low SF twin requires the most or more shear accommodations through prismatic slip and/or extension twinning with low CRSSs, but the least or less those through pyramidal slip and/or contraction twinning with high CRSSs, in the other twin in the twin pair.

Acknowledgements The authors acknowledge the support of the French National Research Agency under the project MAGTWIN (referenced as ANR-12-BS09-0010-02).

References

[1] Friedrich HE and Mordike B 2006 *Magnesium Technology*, in *Metallurgy, Design Data,Applications* (Germany: Springer-Verlag Berlin Heidelberg)
[2] Barnett MR 2007 *Mater. Sci. Eng.* A **464** 1
[3] Christian JW and Mahajan S 1995 *Prog. Mater. Sci.* **39** 1
[4] Jiang L, Jonas JJ, Luo AA, Sachdev AK and Godet S 2006 *Scr. Mater.* **54** 771
[5] Barnett MR, Keshavarz Z, Beer AG and Atwell D 2004 *Acta Mater.* **52** 5093
[6] Beyerlein IJ, Capolungo L, Marshall PE, McCabe RJ and Tomé CN 2010 *Phil. Mag.* **90** 2161
[7] Barnett MR, Nave MD and Ghaderi A 2012 *Acta Mater.* **60** 1433
[8] Xin R, Guo CF, Xu ZR, Liu GD, Huang XX and Liu Q 2014 *Scr. Mater.* **74** 96
[9] Shi Z-Z, Zhang YD, Wagner F, Juan P-A, Berbenni S, Capolungo L, Lecomte J-S and Richeton T *paper in publication*
[10] Hutchinson WB and Barnett MR 2010 *Scr. Mater.* **63** 737

IOP Conf. Series: Materials Science and Engineering **82** (2015) 012022 doi:10.1088/1757-899X/82/1/012022

Texture analysis of zirconium samples deformed by uniaxial tension using neutron and X-ray diffraction

M Kucerakova[1], S Vratislav[1], L Kalvoda[1] and Z Trojanova[2]
[1] Department of Solid State Engineering, Faculty of Nuclear Sciences and Physical Engineering, Czech Technical University in Prague, Trojanova 13, 120 00 Prague 2, Czech Republic
[2] Department of Physics of Materials, Faculty of Mathematics and Physics, Charles University in Prague, Ke Karlovu 5, 121 16 Prague 2, Czech Republic
E-mail: monika.kucerakova@fjfi.cvut.cz

Abstract. Seven zirconium samples were studied by neutron and X-ray diffraction after deformation on uniaxial tensile machine INSTRON 5882 from strain 5% to strain 30% (strain step was 5%). Preferred orientation parameters were determined by using pole figures and inverse pole figures. The X-ray measurements were performed at theta/theta X'Pert PRO diffractometer with Cr X-ray tube. Observed data were processed by software packages GSAS and X'Pert Texture. Our results can be summarized as follows: (i) Samples prefer orientation of planes (100) and (110) perpendicular to rolling direction. (ii) The position of the basal poles is tilted by 30° from the normal direction toward the transverse direction. (iii) Samples prefer orientation of planes (102) and (103) perpendicular to normal direction. (iv) Level of resulting texture increases with deformation. The obtained results are characteristic for zirconium.

1. Introduction

The textured material usually exhibits anisotropic mechanical, physical and chemical properties. Preferred orientation occurs in both powder samples (shape texture) and compact samples (caused by mechanical or heat treatment of material) [1]. Looking on diffraction methods to measure textures mainly three different radiations are in use. These are X-rays produced by conventional X-ray tube or synchrotron, electrons and thermal neutrons. Basic differences in analytical use of these radiations follow from their different interaction with matter [2].

This article deals with the use of neutron an X-ray diffraction in texture analysis of alpha zirconium.

Zirconium has very low absorption cross-section of thermal neutrons, high hardness, ductility and corrosion resistance. Therefore, zirconium and its alloys are used in the nuclear industry as fuel rod cladding, especially in water reactors.

In the alpha zirconium slip takes place usually on the $\{10\bar{1}0\}$ first order prism planes along the $\langle 12\bar{1}0\rangle$ direction. Slip can occur also on the (0002) basal plane in the same direction. In regions of high stress concentration, the {10-11} slip is observed. Activation of slip with $\langle c\rangle$ component was noticed on first and second-order pyramidal planes$\{10\bar{1}1\}$ and $\{11\bar{2}1\}$ in $\langle c+a\rangle$ direction [3].

Most common twinning modes at room temperature are $\{10\bar{1}2\}\langle\bar{1}011\rangle$ tensile twins and $\{11\bar{2}2\}\langle\bar{1}\bar{1}23\rangle$ compressive twins [3].

2. Experimental

2.1. Samples

The alpha zirconium plate was forged and hot rolled to the thickness 3.3 mm. Then it was annealed at 664 °C and the tensile specimens were made from the plate. Fig. 1 shows the shape, dimensions and sample coordinate system (ND means Normal Direction, RD is Rolling Direction and TD is Transverse Direction). Tab. 1 shows the amount of the admixtures in the samples.

Table 1. Chemical composition of investigated material.

Zr [at%]	Hf [at%]	Ca [at%]	Mg [at%]	O [at%]	H [at%]
99.39	0.41	0.044	0.096	0.055	0.0011

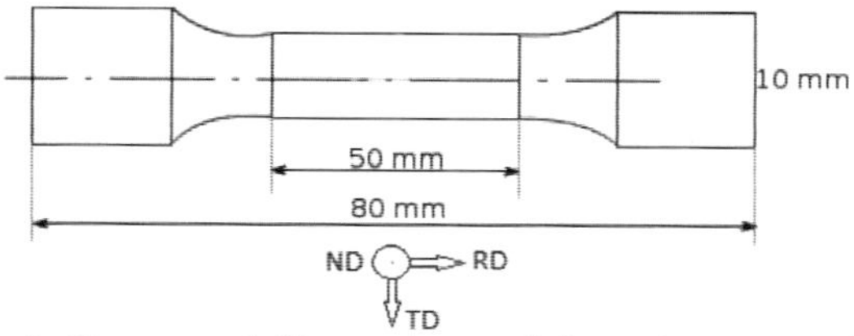

Figure 1. Shape and dimensions of zirconium samples.

2.2. Deformation experiments

The samples were deformed on uniaxial tensile machine INSTRON 5882 from strain 5% to strain 30% (strain step was 5%) at room temperature. The structure of the initial (non-deformed by uniaxial tension) sample observed by using light microscope Zeiss Axio Imager ZM1 is in Fig. 2. The grain size is about 20 µm.

Figure 2. Structure of non-deformed sample.

IOP Conf. Series: Materials Science and Engineering **82** (2015) 012022 doi:10.1088/1757-899X/82/1/012022

2.3. Diffraction experiments

The neutronographic texture measurements were performed on the KSN-2 neutron diffractometer situated at the horizontal channel of the research reactor LVR-15 in the Nuclear Research Institute, plc. Rez, Czech Republic. The monochromatic neutrons having wavelength 0.1362 nm were used. The single-crystal Cu(200) was used as monochromator. The KSN-2 diffraction device offers good intensity and the best resolution value of $\Delta d/d$ = 0.007 in the region d ~ 1.0÷0.1 nm (d is interplanar spacing). All the obtained neutronographic patterns were corrected for non-linear background and then evaluated using the Rietveld method implemented in the software package GSAS [Von Dreele, 2000]. The samples were mounted in reflection geometry with its normal parallel to the direction of interest - usually RD, TD, or ND - and the reflected intensities were measured depending on Bragg angle. The resulting peak intensities were normalized to the intensities of a standard sample with random texture by using Mueller formula [4]:

$$p_{hkl,q} = \frac{\frac{I_{hkl,q}}{I_{R,hkl}}}{\frac{1}{n}\sum_n \frac{I_{hkl,q}}{I_{R,hkl}}},$$

where $I_{hkl,q}$ is measured intensity of hkl reflection for directions q = TD, ND and RD, $I_{R,hkl}$ is intensity of hkl reflection for non-textured sample, n is number of reflections measured.

The X-ray texture measurements were performed at theta/theta X'Pert PRO diffractometer with Cr X-ray tube. Full pole figures for planes $(10\bar{1}0)$, (0002), $(10\bar{1}1)$ and $(11\bar{2}0)$ were calculated from incomplete pole figures using orientation distribution function (ODF). Observed data were processed by software package X'Pert Texture, PANalytical.

3. Results

The intensity ratios $p_{hkl,q}$ calculated (for q=ND, RD and TD) by using Mueller formula from neutronographic data are in Tab. 2. Pole figures measured by X-ray diffraction for planes $(10\bar{1}0)$, (0002), $(10\bar{1}1)$ and $(11\bar{2}0)$ for samples deformed at 5%, 15% and 30% are given in Fig. 3.

Table 2. Inverse pole figures calculated from neutronographic measurements of deformed zirconium samples.

	$p_{002, ND}$	$p_{002, TD}$	$p_{100, RD}$	$p_{101,TD}$	$p_{110, RD}$	$p_{102, RD}$	$p_{103, RD}$
0%	2.82	1.27	1.43	1.49	1.72	1.74	1.65
5%	2.67	1.73	1.51	1.10	1.45	1.58	1.41
10%	2.60	1.76	2.19	1.22	1.42	1.62	1.55
15%	2.77	2.01	2.49	1.01	1.45	1.36	1.34
20%	2.91	2.12	1.58	0.83	0.72	0.81	1.20
25%	2.85	2.27	2.34	1.31	0.83	0.34	1.52
30%	2.77	3.30	2.67	1.22	0.87	0.47	1.42

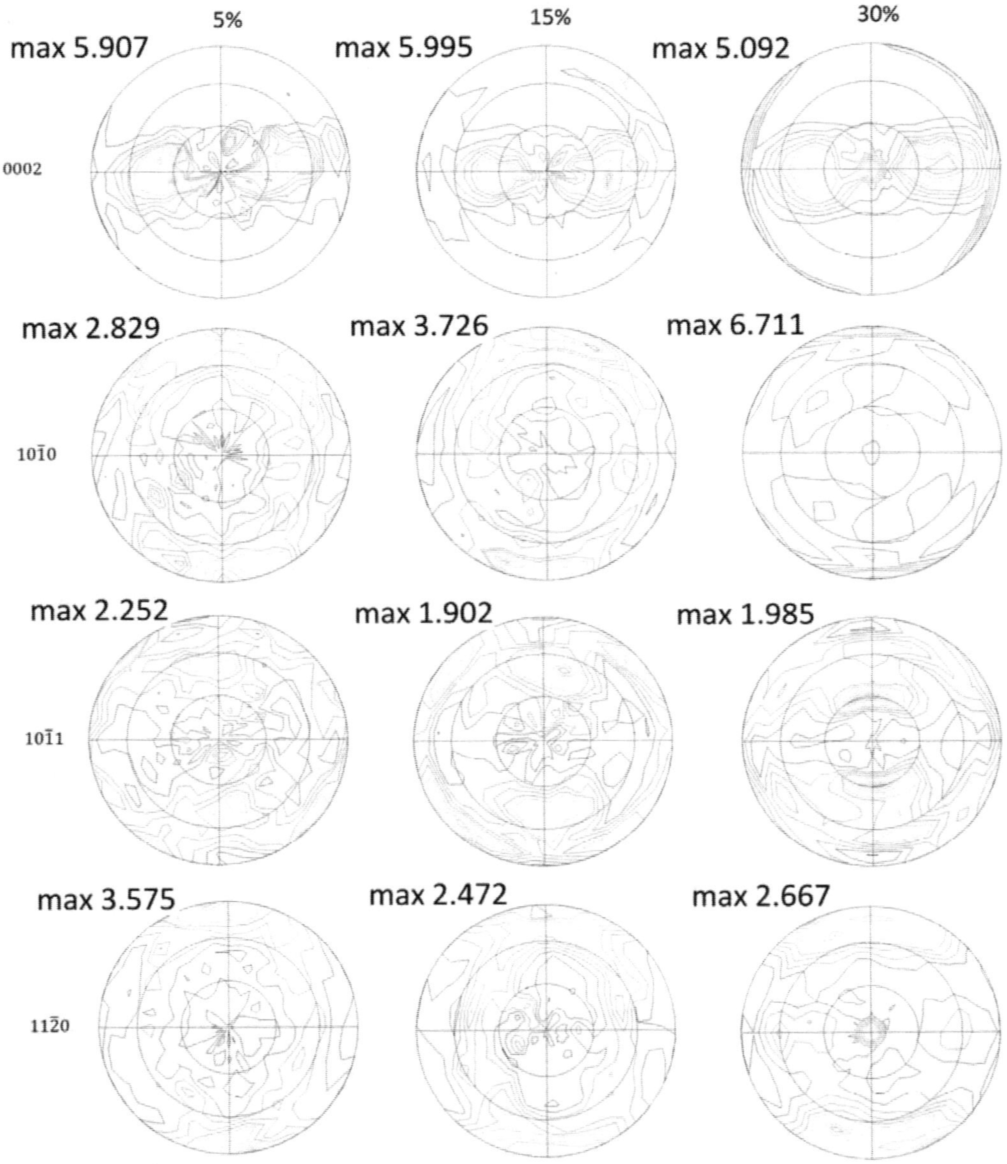

Figure 3. Pole figures measured by XRD for planes (10$\bar{1}$0), (0002), (10$\bar{1}$1) and (11$\bar{2}$0) for samples deformed at 5%, 15% and 30%.

4. Discussion

The position of basal poles is tilted from the normal direction toward the transverse direction by ±20° for 5% deformation and by ±45° for 30% deformation samples (Fig. 3). This rotation indicates twinning. Similar results can be seen from calculated intensity ratios in Tab. 2. The value of pole density $p_{002,ND}$ does not significantly change during the deformation. On the other hand, the value of pole density $p_{002,TD}$ shows a remarkable growth (from 1.27 to 3.30, see Tab. 2) during the deformation.

Planes (10$\bar{1}$0) are oriented perpendicularly to the rolling direction. An analogous situation is observed for planes (11$\bar{2}$0) - see Fig. 3 and Tab. 2. From (11$\bar{2}$0) pole figure (Fig. 3) obtained

for deformation 30% it is obvious that it shows maxima along the normal direction. Such texture component was not observed in the inverse pole figures.
Planes $(10\bar{1}2)$ and $(10\bar{1}3)$ are mostly oriented perpendicular to the normal direction.

5. Conclusions

From the results reported above it can be seen that samples prefer orientation of planes $(10\bar{1}0)$ and $(11\bar{2}0)$ perpendicular to the rolling direction.
The results show that, even at low deformations, twinning causes the basal poles orient from the normal direction toward the transverse direction. The split of basal poles toward the transverse direction can be explained by pyramidal slip with a $(c + a)$ Burgers vector on $\{11\bar{2}1\}$ and $\{10\bar{1}1\}$ planes [5].
Furthermore, samples prefer orientation of planes $(10\bar{1}2)$ and $(10\bar{1}3)$ perpendicular to the normal direction. Texture become stronger with increasing the level of deformation (Tab. 2). The obtained results are characteristic for zirconium [5,6].

Acknowledgement
This work was supported by the Grant Agency of the Czech Technical University in Prague, grant No. SGS13/219/OHK4/3T/14 and the Czech Science Foundation, grant No. GACR 14-36566G.

References

[1] Hsun H 1974 Texture of metals, Technical report (United States: Steel Corporation Research Laboratory)
[2] Bacon G E 1975 *Neutron Diffraction* (Oxford: Clarendon Press)
[3] Tenckhoff E 2005 *Journal of ASTM International, **2***
[4] Harris G B 1952 *Philos Mag, **43*** p 113
[5] Allen V M et al 2008 *Journal of ASTM International* **5**
[6] Santisteban J R et al 2012 *Journal of Nuclear Materials* **425** pp 218-227

IOP Conf. Series: Materials Science and Engineering **82** (2015) 012023 doi:10.1088/1757-899X/82/1/012023

Shear banding and its contribution to texture evolution in rotated Goss orientations of BCC structured materials

T Nguyen-Minh[1], JJ Sidor[1], RH Petrov[1,2] and LAI Kestens[1,2]
[1] Department of Materials Science and Engineering, Ghent University, B9052 Ghent, Belgium
[2] Department of Materials Science and Engineering, Delft University of Technology, Mekelweg 2, 2628 CD Delft, The Netherlands
E-mail: Tuan.NguyenMinh@ugent.be

Abstract. Due to progressive deformation, the dislocation densities in crystals are accumulated and the resistance of grains to further deformation increases. Homogeneous deformation becomes energetically less favorable, which may result for some orientations in strain localization. In-grain shear banding, a typical kind of localized deformation in metals with BCC crystal structure, has been accounted for by the geometric softening of crystals. In this study, the occurrence of shear bands in rotated Goss ({110}<110>) orientations of Fe-Si steel is predicted by crystal plasticity simulations and validated by EBSD measurements. It was observed and confirmed by crystal plasticity modeling that such shear bands exhibit stable cube orientations The orientation evolution of crystals in shear bands and its impact on annealing texture of materials are also described.

1. Introduction

Plasticity of structural materials, in general, is induced by dislocation slip, mechanical twinning and grain boundary sliding. For BCC structured materials, which have high stacking fault energy, the plastic deformation of mechanical twinning is not energetically favored. In addition, super-plasticity by grain boundary sliding hardly occurs during conventional deformation processes of polycrystalline materials. Therefore, dislocation slip is by far the dominant way to induce plasticity in BCC structured materials.

During plastic deformation, dislocations are generated and accumulated in the grains of the polycrystal. The interactions and accumulations of dislocations result in an increase in the resolved shear stresses of the slip systems, and hence inhibit grains from further plastic deformation. In an aggregate, because of the difference in slip activities, the accumulation of dislocations in various grains is not the same. The total density of dislocations stored in a grain depends on its crystallographic orientation, the strain mode applied and the manner in which this strain mode is accommodated locally. For grains of higher slip resistance, i.e. less favorable for plastic deformation, the density of accumulated dislocations is usually higher. On the contrary, grains, which have higher susceptibility to deformation, trap fewer dislocations, and thus have lower dislocation densities.

In the first approximation of crystal plasticity, the accumulation of dislocation density in a crystal grain can be represented by the Taylor factor (M). This factor is defined by the ratio of the internally dissipated friction work (per time unit) and the plastic work (per time unit) done by the applied deformation:

$$M = \frac{\sum_{s} \tau_s^c \left|\dot{\gamma}_s\right|}{\tau^c \dot{\varepsilon}_{vM}} \tag{1}$$

in which τ_s^c and $\dot{\gamma}_s$ are the resolved shear stress and the slip rate of active slip system (s), τ^c and $\dot{\varepsilon}_{vM}$ are the equivalent shear stress and strain rate of external deformation. The sum in eq. (1) is taken over all active slip systems, regardless of their glide directions. To simplify, a unique value is usually assigned to τ_s^c and τ^c which accounts for the very beginning or the steady state of deformation where hardening is saturated and all slip systems have nearly the same resolved shear stress. Providing the same strain rate is applied to all grains in an aggregate, the Taylor factor is directly proportional to the total slip rates of crystallites.

For BCC structured materials, the Taylor factors of the preferred orientations in plane strain compression of conventional rolling can be observed in the φ_2 =45° section of Euler space. From this Taylor factor map, it is expected that the resistance to slip and the dislocation densities of crystallites oriented along the α-fiber (<110>//TD) are lower than for crystals along the γ-fiber (<111>//ND). These differences in behavior of α- and γ-fiber oriented grains can be well demonstrated by orientation images of the deformation microstructures. In an inverse pole figure (IPF) map of any BCC structured material after conventional rolling the grains which have <100>-axes parallel to ND mostly appear as long and continuous bands. On the contrary, grains having <110>- and <111>-axes close to ND are usually fragmented into inclined strips and contain smaller grains. The grain fragmentation in high plastic resistant crystallites (i.e. with high Taylor factors) is due to the inhomogeneous deformation at grain scale.

Another effect of in-grain inhomogeneous deformation is the occurrence of micro shear banding. Micro shear bands appear in grains as narrow and inclined strips of heavily deformed crystallites. They are not crystallographic features, since the band inclination angles are not related to any specific crystallographic orientation of the grains. However, the occurrence of micro shear bands is greatly influenced by the grain orientations. They are more frequently observed in high Taylor factor grains, rather than in those with low Taylor factor. Inside bands, crystallites are severely strained by shear, whereas outside they are hardly deformed.

2. Geometric softening

Strain localization in the form of shear bands can be accounted for by geometric softening [1] of heavily deformed polycrystals. According to this theory, the accumulation of dislocations in grains during progressive deformation increases the resolved shear stresses for slip activities. Thus, homogeneous deformation of the entire grain volume becomes less energetically favorable than local straining of a specific sub-volume. When the applied deformation is better accommodated in local bands, shear-banding occurs. The transition from homogeneous to local plastic flow of materials is usually marked by a plastic instability point ($d\sigma/d\varepsilon$ = 0) on the stress-strain (σ-ε) curve. Beyond this point, geometric softening ($d\sigma/d\varepsilon$ < 0) due to local strain accommodation is dominant, and thus grains in shear bands are deformed.

Geometric softening is not a reduction of dislocation density in crystals by thermally activated processes, like recovery or recrystallization. It is simply a change of the deformation geometry to *soften* the high hardening stage of crystals. The condition for this change of deformation configurations is proposed by Dillamore et al [1], based on crystal plasticity theory:

$$\frac{d\sigma}{d\varepsilon} \propto \frac{dM}{d\varepsilon} \leq 0 \tag{2}$$

Accordingly, shear banding will occur, if the rotation of crystals in localized bands is towards lower Taylor factor orientations.

3. Shear banding in BCC structured materials

In BCC structured materials after conventional rolling, micro shear banding is most frequently observed in {111}<112> oriented grains. Previous studies [2-4] showed that the presence of shear bands in such grains indeed exhibits the geometric softening effect. In addition, calculation results confirmed the formation of the Goss ({110}<001>) orientation in these shear bands. These Goss oriented crystals are believed to have important effects on the nucleation of recrystallized grains during subsequent annealing.

Among all preferred orientations in BCC structured materials after conventional rolling, the {111}<112> orientation has a Taylor factor of 3.5, which is one of the highest values. However, the maximum value in the Taylor factor map of BCC structured materials in plane strain compression is 4.24 at the rotated Goss ({110}<110>) orientation. An interesting question is why geometric softening by shear banding occurs in this orientation and what is the preferred orientation in these shear bands.

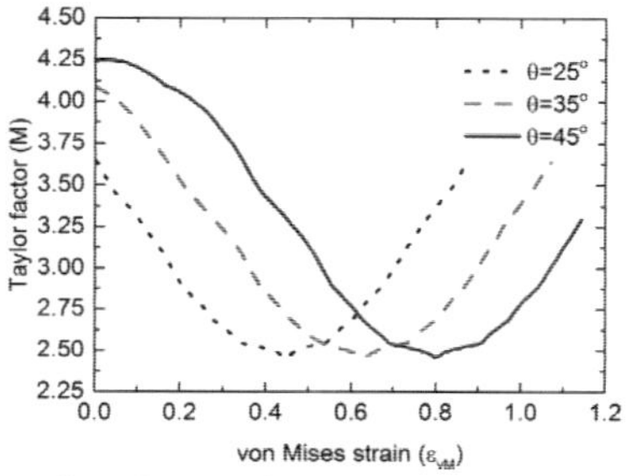

Figure 1. Strain evolution of the Taylor factor for a rotated Goss oriented crystallite in shear bands of different inclination angles (θ = 25°, 35° and 45°).

To investigate shear banding in the rotated Goss orientation, simulation and experiment have been carried out. The full constraints Taylor model was used to investigate the geometric softening effect and the orientation evolution of crystallites in shear bands. Deformation of crystallites in shear bands inclined by 25°, 35° and 45° with respect to the rolling plane were simulated. In all three simulations, a strong geometric softening effect can be observed (Figure 1). The softening effect is strongest in bands of 45° inclination. It starts at the highest Taylor factor value (4.24) and finishes after the highest strain (0.8). Furthermore, the formation of shear bands is supported by the maximum shear stress of conventional rolling, which is on the plane 45° inclined to the normal direction. Shear bands with 45° inclination angle, therefore, should have the highest potential to form in the rotated Goss oriented grains.

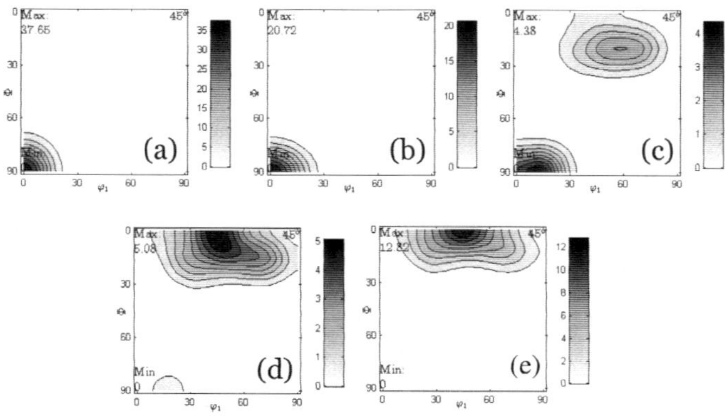

Figure 2. Texture evolution in the 45° inclined shear band of a rotated Goss grain, after straining (ε_{vM}) of (a) 0, (b) 0.2, (c) 0.4, (d) 0.6 and (e) 0.8.

The texture evolution in a shear band with 45° inclination angle was simulated after strain steps of 0.2 from 0 to 0.8. A strong rotated Goss texture discretized by a set of 2000 orientations was used as initial texture for crystal plasticity calculations. Deformations and rotations of these orientations in the shear band are calculated by the full constraints Taylor model. The interpolation of continuous ODFs from discrete orientation sets for each deformation step was then calculated by the MTEX toolbox of the MATLAB software. The φ_2

= 45° sections of these ODFs are shown in Figure 2. With increasing strain from 0 to 0.8 the intensity of the rotated Goss orientation decreases, while above a strain of 0.4 the intensity of the Cube ({001}<100>) orientation appears and strengthens. When the equivalent strain increases from 0.6 to 0.8 the intensity of the Cube orientation increases by a factor of 2.5, i.e. from 5.1 to 12.8 random. Since the geometric softening in the 45° inclined shear bands is exhausted at the equivalent strain of 0.8 (Figure 1), the in-band texture may not further evolve after this deformation. As a result, the Cube oriented crystals are the most stable in shear bands of the rotated Goss grains.

To validate simulation results, the microstructure of deformed materials has been investigated by the electron backscatter diffraction (EBSD) technique. A Fe-1.2 wt.% Si alloy after thickness reduction of 75% by conventional cold rolling was used as material. In the rolled steel sheet, the rotated Goss oriented grains are not very frequently observed, as rotated Goss only represents a minor component of the rolling texture. Although the stability of these grains in plane strain compression is very high, the {110}<110> oriented grains are virtually absent in the texture because even before deformation the volume fraction of this component is quite low. However, by chance it is possible to find some rotated Goss grains in the samples studied. In Figure 3, shear bands in a rotated Goss grain can be clearly observed. Orientations in these bands, indeed, are very close to the Cube orientation.

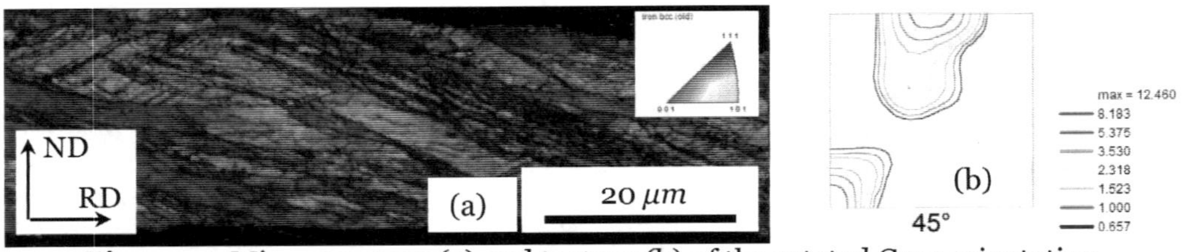

Figure 3. Microstructure (a) and texture (b) of the rotated Goss orientation

4. Summary

During plastic deformation, the accumulation and interaction of dislocations increase the hardening of crystallites. The homogeneous deformation of certain crystallites becomes energetically less favored compared to local plastic flow in micro shear-bands. Micro shear banding in heavily deformed crystals can be accounted for by geometric softening of crystal plasticity. In BCC structured materials, rotated Goss oriented grains have the highest resistance to plane strain compression, therefore they exhibit the most prominent potential for micro shear banding. The most preferred orientation in shear bands of rotated Goss oriented grains is the Cube orientation.

Acknowledgement

The authors acknowledge the financial support from the Belgium Federal Science Policy Office for the Interuniversity Attraction Poles (IAP) program via the project P7/21 INTEMATE.

References

[1] Dillamore IL, Roberts JG, Bush AC 1979 *Met. Sci.* **13** 73.
[2] Haratani T, Hutchinson WB, Dillamore IL, Bate P 1984 *Met. Sci.* **18** 57.
[3] Ushioda K, Hutchinson WB 1989 *ISIJ Int.* **29** 862.
[4] Barnett MR 1998 *ISIJ Int.* **38** 78.

IOP Conf. Series: Materials Science and Engineering **82** (2015) 012024 doi:10.1088/1757-899X/82/1/012024

Evolution of deformation heterogeneity at multiple length scales in a strongly textured zinc layer on galvanized steel

A Ghosh[1] and NP Gurao[2]
[1] Department of Materials Science and Engineering, Indian Institute of Technology Kanpur, Kanpur 208 016
Indian and National Institute of Technology Durgapur, India
[2] Department of Materials Science and Engineering, Indian Institute of Technology Kanpur, Kanpur 208 016, India
E-mail: npgurao@iitk.ac.in

Abstract. The evolution of heterogeneity of plastic deformation in a zinc layer has been probed at multiple length scales using a battery of characterization tools like X-ray diffraction, electron back scatter diffraction (EBSD) and digital image correlation. The experimental results indicate that plastic deformation is heterogeneous at different length scales and the value of micro, meso and macro strain by different characterization techniques shows a different value. The value of strain determined at the meso and micro length scale from EBSD and X-ray diffraction was negligible, however, the macro-strain as determined from X-ray peak shift was significant. EBSD results showed evidence of profuse $\{10\bar{1}2\}<10\bar{1}1>$ contraction twinning in the zinc layer with higher intragranular misorientation in the twin compared to the matrix. It is therefore, inferred that the evolution of higher intergranular (between matrix and twin) strain due to prolific contraction twinning contributes to the failure of zinc layer on galvanized steel.

1. Introduction

Hexagonal close packed (HCP) metals and their alloys find extensive applications in automobile, aerospace and biomedical industries. HCP metals and alloys are characterized with respect to the ideal c/a ratio of 1.633 as less than ideal c/a ratio materials like titanium and zirconium and more than ideal ratio like zinc and cadmium. They undergo deformation by different slip systems namely; $\{0002\}<11\bar{2}0>$ basal, $\{10\bar{1}0\}<11\bar{2}0>$ prismatic and $\{10\bar{1}1\}<11\bar{2}0>$ pyramidal <a> slip systems. However, these slip systems do not provide five independent slip systems to accommodate imposed strain according to von Mises criterion. These slip systems are insufficient to provide strain along <c> direction and additional mechanisms like $\{10\bar{1}1\}<\bar{1}\bar{1}23>$ pyramidal <c+a> and $\{10\bar{1}2\}<\bar{1}011>$ extension and $\{2\bar{1}\bar{1}2\}<2\bar{1}\bar{1}\bar{3}>$ contraction twinning systems are also operative [1]. Twinning, unlike slip causes sudden reorientation and a significant characteristic strain is associated with it. Twinning contributes to strain hardening thereby aiding in ductility but at the same time, sudden reorientation can also cause initiation of cracks in the material at the twin-matrix interface. However, the twin shear (s) for most twin systems in HCP materials is lower (for 10-12 twinning, s = 0.12 for ideal c/a ratio) compared to that in face centre cubic (FCC) materials (s = 0.707) [2]. Hence the twins in HCP materials are wider compared to lenticular twins in FCC materials. This in turn leads to reorientation of large volume of grains to the twinned orientation in hcp materials compared to FCC materials. Reorientation due to twinning can aid in activity of favourable slip system with low critical resolved stress as observed in titanium and zirconium and lead to sufficient ductility. However, in case of brittle HCP materials like magnesium and zinc, reorientation does not lead to easy slip activity and twinning can contribute to failure. In order to decipher the reasons of failure, the evolution of stress at different length scale has to be monitored to elucidate the causes of failure. This is achieved by monitoring the evolution of strain at different length scales using X- ray peak shift and broadening, Electron Back Scatter Diffraction (EBSD) and Digital Image Correlation (DIC) techniques. In order to achieve simple stress and strain state, two dimensional polycrystal offer a model material [3]. In the present investigation, strain

evolution at different length scales was monitored in a zinc coating of hot-dip galvanized steel.
The X-ray peak broadening of the samples after and prior to tensile test have been compared and the strain evolution was measured at the angstrom scale [4]. EBSD was performed on the deformed sample and strain evolved at the meso-scale was estimated by analyzing the orientation maps obtained [5]. DIC technique provided an estimate of strain at the meso length scale using optical method to measure strain from speckle pattern on the specimen using mathematical correlation analysis from digital images taken during the mechanical test. The sample was prepared by application of a random dot pattern (speckle pattern) then consecutive images are captured during the deformation to understand the change in the sample behaviour and surface characteristics when subjected to incremental loads in the millimeter or continuum scale. Such investigations have been carried out separately but in this work we aim to correlate them together and study the deformation occurring in the zinc coating in galvanized steel sheets.

2. Experimental

Commercially available hot-dip galvanized plain carbon steel sheets of about 0.8 mm thickness were studied in the present investigation. Initial characterizations of the coating were performed on the as received sample by optical microscopy, X-ray diffraction (XRD) and Scanning Electron Microscope (SEM) with an Energy Dispersive Spectroscopy and Electron Back Scatter Diffraction facility. The as-received galvanized sheet was cut into small sections, mounted and fine polished by 5µm alumina slurry followed by 1µm alumina slurry. It was etched with 5 ml HCl in 100ml water for 20 seconds and observations were made by an Zeiss Axioskop 2 MAT microscope equipped with a digital camera Axiocam ERc5s. Rectangular sheet of dimensions 6.8cm x 2cm was cut and a tensile specimen was punched out from it. The tensile specimen was etched with 5ml HCl in 100 ml water and a tensile test was performed with it and the fracture was recorded using a video camera placed over a magnifier. Another set of samples with 100 mm length and 15 mm width were provided with a collar and were coated with a speckle pattern to carry out tensile test using DIC.

Fig. 1: Initial microstructure of the zinc layer on the steel sample (left) and snapshots from video during the tensile test.

3. Results and discussion

The initial characterization of the zinc coating under optical microscope reveals very large grains of Zinc of about 500 micron size (Fig. 1). SEM results indicate that the thickness of the coating was 13.8 ±1.8 µm indicating that the zinc coating on a galvanized steel sheet can be considered as a classical case of a two dimensional microstructure. Cross section Energy Dispersive Spectroscopy analysis indicated variation in zinc content from the surface with 100% Zinc to the centre with 100% iron. It is to be mentioned here that the steel layer and the zinc layer is separated by a layer of complex Fe-Zn intermetallics. The presence of

IOP Conf. Series: Materials Science and Engineering **82** (2015) 012024 doi:10.1088/1757-899X/82/1/012024

elemental zinc is confirmed by X-ray diffraction patterns (Fig. 2) that shows only zinc peaks though some minor peaks are also observed. The X-ray diffraction spectra shows a very high intensity (0002) peak compared to the other peaks of Zinc which suggests that the sample has a very strong texture with almost all the grains having their <c> axes pointing out perpendicular to the plane of the sheet. The tensile tested sample showed brittle behaviour (not included in the manuscript) and the engineering stress-strain curve started dropping after yield. It is to be mentioned here that the engineering stress-strain curve is for the entire galvanized sheet and showed ultimate tensile strength of 132 MPa. Multiple tests carried out led to failure near the grip section and design modification to ensure failure in the grip section using a dumbell shape specimen is in progress. Snapshots of the video during the tensile test shown in Fig. 1 indicate intergranular nature of fracture in the zinc layer similar to the one reported in literature [6-7].

In the deformed sample, the intensity of (0002) peak decreased and (10-10) and (10-11) peaks gained prominence as seen in Fig. 2. A quantitative estimate of texture is obtained by plotting the ratio of intensities of different peaks (Table I). The decrease in intensity ratio of 0002 peak to 10-10 peak indicates a transition from basal (c-axis out of plane) to prismatic texture (in-plane c-axis). In addition, there was shift in 0002 peak position at lower angle (Fig. 2 and Table I) indicating presence of Type III tensile residual stress. Thus, there is an evolution of macroscopic strain of 0.0032 in the zinc layer. However, microscopic strain evolution as manifested in terms of peak broadening as seen is negligible. The crystal orientation map obtained from EBSD (Fig. 3a) clearly shows the presence of contraction twins on the surface of the zinc layer. The reorientation of grains due to twinning can explain the change in intensity ratio of different XRD peaks indicating change in crystallographic texture. Different parameters like Minimum Angular Deviation which is deviation of angle between Kikuchi bands from the ideal relationship for a EBSD scan, Local Average Misorientation which is the average misorientation in a grain and Strain contouring which is obtained by considering misorientation between larger regions of microstructure are not significant (Table II) and are in the same range as that of undeformed sample. This indicates that failure is not occurring due to processes at this length scale. The DIC map shows that strain evolution is concentrated in the form of bands perpendicular to the direction of loading and that high strain evolves at the area close to the final failure (Fig. 3b). However, it is the macroscopic tensile stresses which are expected to contribute to failure. This is counter intuitive as compressive stresses are expected normal to the sheet plane during tensile test along the axis. This can be explained on the basis of contraction twinning in the zinc layer due to compressive stresses which is compensated by tensile stress in the matrix and is reflected as shift in the 0002 peak. The incompatibility within the grain and stronger character of the twin boundary ensures that crack propagates in a intergranular manner as shown in Fig. 4. Thus macroscopic stress evolution in the grains contributes to the failure of zinc layer on galvanized steel sample.

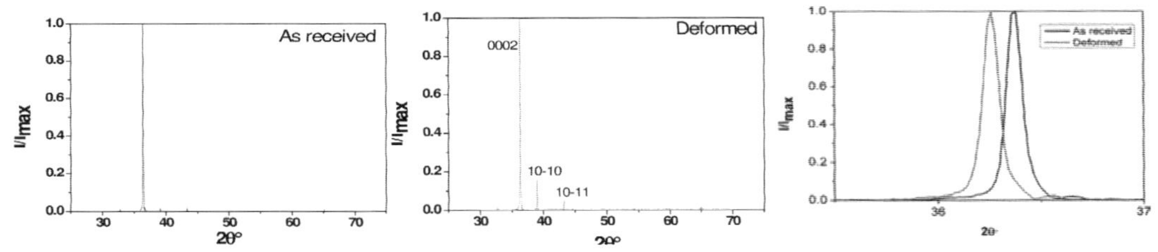

Fig. 2: X-ray diffraction spectra of the as received and deformed sample with close view of 0002 peak.

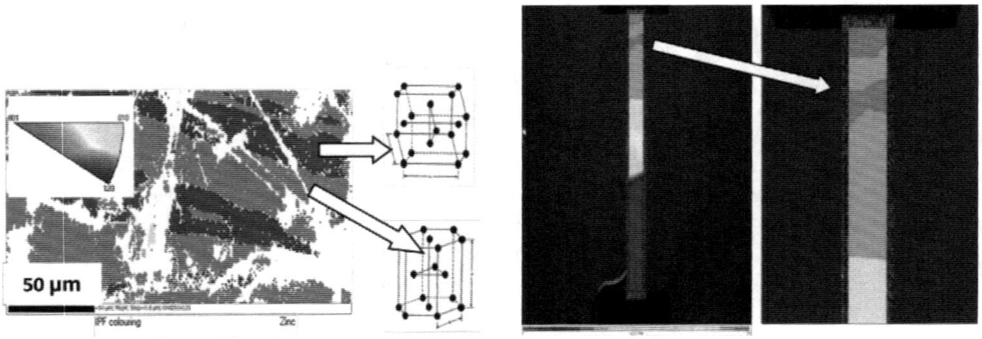

(a) (b)

Fig. 3: (a) Crystal Orientation map from the deformed sample showing contraction twinning and (b) DIC image of the deformed sample with RGB corresponding to strain from 0 to 0.18 of von Mises strain.

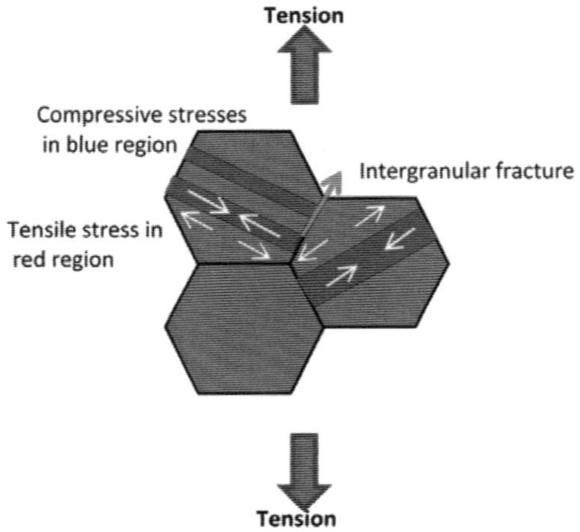

Fig. 4: Schematic showing the mechanism of failure in the zinc layer on galvanized steel.

Table 1: Texture and microstructural parameters obtained from X-Ray diffraction.

Parameter	I_{002} / I_{100}	I_{002} / I_{101}	Crystallite size	Micro strain	Macro strain
Initial	60.49	53.85	836.2 nm	0.00052	---
Deformed	6.39	20.44	491.9 nm	0.00070	0.0032

Table 2: Strain estimation at different length scales.

Length scale	**Strain**	**Technique**
Continuum (mm scale)	0.18	DIC
Macro	0.0032	X-Ray
Meso	LAM : 0.1662°MAD: 0.2699° SC: 0.2699°	EBSD
Micro	0.0007	X-Ray

4. Conclusions

To summarize, it can be inferred that the micro-strain evolution in the zinc layer on steel was negligible, the meso-strain was substantial while the macro strain was the most significant. Thus, the failure of zinc coating on galvanized steel is facilitated by evolution of macro strain accompanied with contraction twinning.

Acknowledgments

The authors would like to thank Prof. S. Sangal (MSE IITK) and Prof. P. Venkitnarayanan (ME IITK) for extending various experimental facilities available in their labs. AG thanks IITK SURGE fellowship that made his stay at IITK possible.

References

[1] Partridge PG 1967 *Metall. Rev.* **12** 169
[2] Christian JW and Mahajan S 1995 *Prog. Mater. Sci.* **39** 1
[3] Parisot R, Forest S, Pineau A, Nguyen F, Demonet X and Mataigne J-M 2004 *Metall. Mater. Trans A* **35** 813
[4] Cullity BD 1966 *Elements of X-ray diffraction* Addison Wesley Metallurgy Series
[5] Kamaya M, Wilkinson AJ and Titchmarsh JM 2005 *Journal of Nucl. Mater.* **235** 713
[6] Parisot R, Forest S, Gourgues AF, Pineau A and Mareuse D 2000 *Comp. Mater. Sci.* **19** 189
[7] Bell RL and Cahn RW 1956 *Proc. Roy. Soc. A* **239** 494

IOP Conf. Series: Materials Science and Engineering **82** (2015) 012025 doi:10.1088/1757-899X/82/1/012025

Effect of strain path change on precipitation behaviour of Al-Cu-Mg-Si alloy

S Mishra[1], K Kulkarni[2], N P Gurao[2]
[1]Department of Materials Science and Engineering, Indian Institute of Technology Kanpur, Kanpur-208016, India
[2]Department of Materials Science and Engineering, Indian Institute of Technology Kanpur, Kanpur-208016, India
E-mail: npgurao@iitk.ac.in

Abstract. The effect of strain path change on precipitation behaviour of Al-Cu-Mg-Si alloy was investigated. Two different types of crystallographic textures were produced by changing the strain path during rolling. The deformed samples were subjected to a short recrystallization treatment and ageing to identify the effect of strain path change manifested in terms of crystallographic texture on precipitation behaviour. Preliminary characterization indicates that ageing kinetics as well as precipitate morphology vary depending upon the mode of rolling. The coherency strains associated with a coherent interface is relieved in a unlike manner for differently rolled samples.

1. Introduction

Precipitation hardening is produced by solutionizing and quenching an alloy in which the second phase precipitates out upon ageing at intermediate temperature. The requirement of a precipitation-hardening alloy is that the phase diagram should show a solvus viz. solid solubility limit should decrease with decrease in temperature. The driving force behind rejection of excess solutes in the form of the precipitates is supersaturation [1, 2]. The sluggish diffusion kinetics at low temperatures ensures that the solute atoms move through only a few tens of interatomic distance [3, 4], hence giving rise to precipitation on a very fine scale. The precipitates of the second phase are generally coherent in nature. If the mismatch between the lattice of matrix and precipitate is large, then considerable amount of coherency strain is associated with a coherent interface, giving rise to internal strain hardening. The main contribution to hardening comes from the interaction between the moving dislocations with finely dispersed precipitates [5, 6]. The Al 6XXX alloys are precipitation hardenable alloys and precipitation hardening is the key aspect behind high strength to weight ratio of these alloys. The decomposition of supersaturated solid solution occurs is several steps involving transition precipitates before the equilibrium precipitate forms [7, 8]. Generally the nucleation barrier for the formation of equilibrium precipitate is too high. The prerequisite of surface energy to create surfaces of critical-sized precipitates dominates over volume Gibbs free energy decrease during the initial stages of precipitation. Usually a sheet of age-hardenable material undergoes some sought of thermo-mechanical processing involving deformation at room temperature and ageing at slightly elevated temperature [9, 10]. Since the recrystallization temperature is higher than the ageing temperature, the alloy retains the polygonised structure, which has high density of dislocations. Thus, precipitation is more fine and homogeneous due to their precipitation on dislocation substructure. Thus, a proper amalgamation between various deformation techniques and ageing cycle is needed to develop attractive mechanical properties. In addition to introduction of dislocations and other lattice defects during deformation, there is also evolution of strong texture in the material. Thus proper scrutiny of texture is required to develop such beneficial properties. Previous investigations indicate that by changing the strain path during room temperature plastic deformation, characteristically different texture can be obtained in the same material [11, 12]. Essentially the deformation substructure is destabilized. Physical phenomena's like recrystallization and phase transformation are also affected by a change in strain path during plastic deformation [13, 14]. Thus, it is worth paying attention on the effect of different deformation texture on precipitation behaviour in age-hardenable alloys. The key is to separate out the effect of crystallographic texture from other microstructural parameters like grain size, dislocation density, micro-strain etc. Thus, the intention behind the present investigation is to examine the effect of different initial textures obtained via strain path change on precipitation behaviour in Al-Cu-Mg-Si alloy.

2. Experimental

2.1. Thermo-mechanical-processing

Hot rolled plates of Al-Mg-Si-Cu alloy in T6 temper was used as starting material for present investigation. To wipe out the previous plastic deformation history the hot rolled plate was subjected to solutionizing treatment in a muffle furnace for 24 hrs, followed by quenching in water. The solutionized plates were rolled in multiple passes at room temperature in two different ways with the true strain being constant for each pass. In uni-directional rolling (UDR) the samples are rolled down to final thickness viz. 1 mm without changing the strain path or the rolling direction. In multi-step cross rolling (MSCR) the sample is rotated by 90° after each pass [11, 12]. A brisk recrystallization technique consisting of holding the rolled samples at 500°C for two minutes in a salt bath furnace, followed by quenching in water was devised to detach the effect of crystallographic texture from other microstructural parameters. The recrystallized samples were subjected to ageing at 160°C in an oven to appreciate the effect of texture on precipitation behaviour in an age-hardenable alloy.

2.2. Characterization

The pole figures were measured at half thickness of the RD-TD plane using Bruker D-8 Discover texture goniometer. For bulk texture determination of the matrix (111), (200), (220) and (311) pole figures were measured assuming triclinic sample symmetry. Zeisss EVO 50 scanning electron microscope (SEM) attached with an electron back scatter diffraction (EBSD) unit from Oxford Instruments was used to carry out the microstructural investigation. It was ensured that step size and binning were kept at 0.5 μm and 4*4 respectively. The electron transparent region required for Transmission electron microscopy (TEM) was produced using PIPS (precision ion milling system). The samples were analysed in a Tecnai-G2-Ultra Twin TEM.

3. Results and discussion

3.1. Texture

The (111) pole figures for the UDR and MSCR samples, after recrystallization at 500°C for 2 minutes (here after referred as UDRX and MSCRX) and for UDRX and MSCRX samples aged at 160°C upto peak ageing condition (UDRXA and MSCRXA) are shown in Fig.1. The bulk texture measurements clearly underline the fact that texture is noticeably different for differently processed samples, while there is no distinctive difference in texture upon peak ageing. The texture of UDRX sample is outlined by the presence of {100} <001> cube component along with a substantial amount of {100} <110> and {110} <112> Brass component. On the other hand the texture for MSCRX samples is appreciably different from the UDRX samples. The MSCRX samples show a dominant cube component with few traces of Brass component. The important point is that there is absence of {100} <110> component in MSCRX samples. From the pole figures it can be deduced that there is no characteristic change in texture during ageing of recrystallized samples, apart from strengthening and weakening of few components. The UDRXA sample shows an augment in Brass component while the cube component decreases. The key difference is that cube component is still the dominant component in MSCRXA with hardly any traces of Brass component. Thus, change in strain path during rolling causes an appreciable difference in texture for UDRX and

MSCRX samples. The texture is relatively weaker for MSCRX samples [11, 12]. This can be attributed to the fact that there is continual destabilization of deformation substructure along with a change in the frame of reference for MSCRX samples [11, 12]. The characteristic deformation texture evolved during rolling leads to a distinct recrystallization texture on annealing for the differently rolled samples.

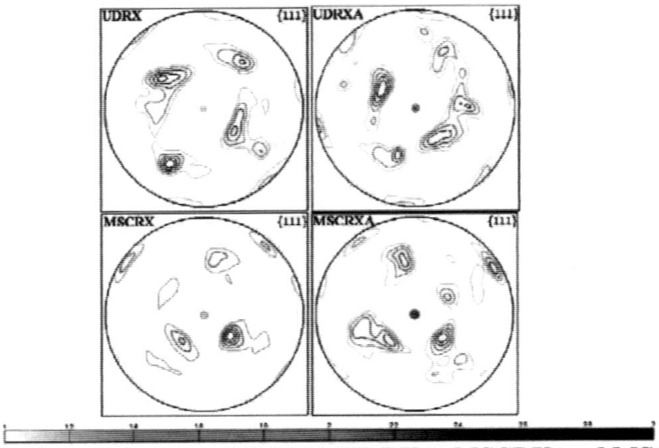

Fig. 1 (111) pole figures of UDRX, UDRXA, MSCRX and MSCRXA.

3.2. Microstructure

Figure.2 represents the EBSD analysis of UDRX, UDRXA, MSCRX and MSCRXA samples. As evident from Figure.2, the microstructure consists of well recrystallized grains with minimal variation in grain size for UDRX (~9 ± 7 μm) and UDRXA (~ 9.5 ± 8μm) samples. A similar trend is observed for MSCRX (~8 ± 7μm) and MSCRXA (~ 12 ± 10μm) samples. Also there is a minimal difference in grain size for UDRX and MSCRX samples. Thus, grain sizes are comparable and no appreciable amount of grain growth has taken place during ageing.

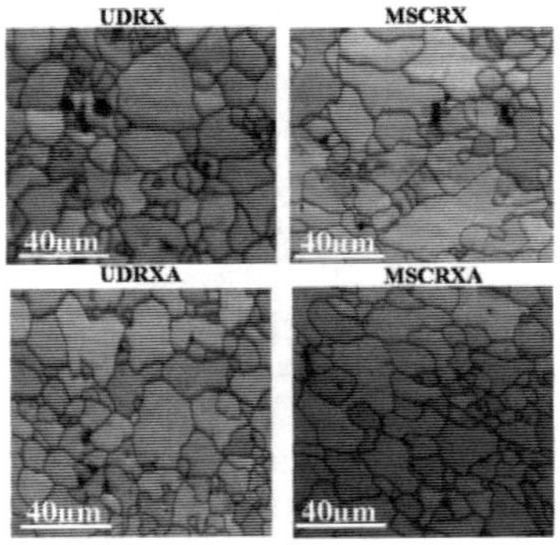

Fig. 2 EBSD micrographs of UDRX, UDRXA, MSCRX and MSCRXA.

The local average misorientation map (LAM) for UDRX, UDRXA, MSCRX and MSCRXA is given in Figure.3. This component calculates small orientation changes on the map, thus highlighting regions of higher deformation. It calculates the misorientation between every pixel and its neighbouring pixel and assigns an average value to that pixel. Figure.3 clearly brings out the fact that recrystallization is more or less complete for UDRX and MSCRX. The lower values of local misorientation angle within the grains suggest that complete recrystallization has taken place with few remaining traces of deformed regions. As expected, the average local misorientation angle is on the higher side near the grain boundaries. During

IOP Conf. Series: Materials Science and Engineering **82** (2015) 012025 doi:10.1088/1757-899X/82/1/012025

ageing of the recrystallized samples coherent and semi-coherent metastable precipitates form within the matrix. This produces elastic lattice strains called 'coherency strains' across the boundary where the lattice planes must be bent to achieve this one to one matching. Coherent precipitates have low interfacial energy, but they are associated with coherency strains. A semi-coherent precipitate will have high interfacial energy but the coherency strains will be relieved. If the misfit between the matrix and the precipitate is on the higher side then it becomes energetically favourable to relieve the coherency strains by introducing geometrically necessary dislocations or misfit dislocations within the matrix. Thus, a semi-coherent boundary consists alternately of region of coherency and region of disregistry. As evident from Figure.3 the average local misorientation angle is on the higher for MSCRXA compared to UDRXA. This suggests sluggish ageing kinetics for MSCRXA. The β' precipitates are taking more time to lose their coherent character and thus the magnitude of coherency strains is on the higher side for MSCRXA compared to UDRXA. Thus, more number of misfit dislocations are introduced in to the matrix for MSCRXA. This causes the average local misorientation angle to be on the higher side for MSCRXA.

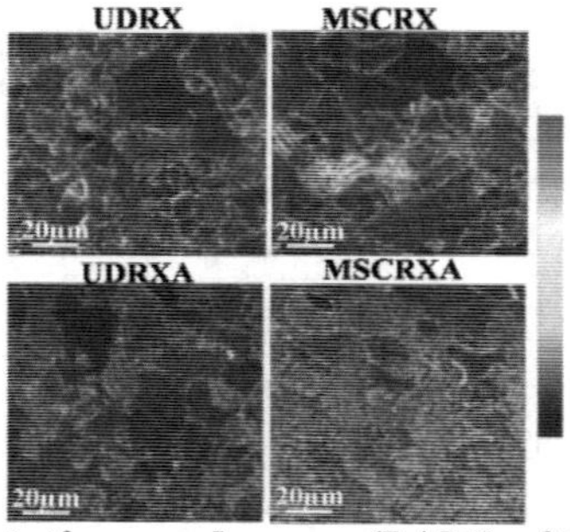

Fig. 3 Local average misorientation angle map (LAM) of UDRX, UDRXA, MSCRX and MSCRXA.

3.3. TEM-studies

The TEM bright field micrograph for UDRXA and MSCRXA samples are shown in Fig.4. The presence of intragranular β' precipitate within the matrix of aluminium aged upto peak ageing condition is evident from the representative TEM micrographs. The vital point is that the aspect ratios of precipitates are different for both the cases. The aspect ratio is on the higher side for UDRXA samples i.e. (~3:1) compared to MSCRXA (~1.2:1). This morphological anisotropy can be attributed to difference in ageing kinetics for UDRX and MSCRX. The faster ageing kinetics for UDRX suggests that precipitates are losing their coherent character earlier than MSCRX. For plate shaped precipitates it is possible to nucleate dislocations at the interface by developing stresses higher than the theoretical strength of the matrix at the edges of the plate [15]. As the plate grows the same process is repeated in order to maintain constant spacing between the dislocations. It has already been shown that plate like precipitates lose their coherency by forming a dislocation loop within the precipitate [16]. The quenched-in mono-vacancies and di-vacancies get attracted towards the coherent interface and then condense to form a dislocation loop within the precipitate. The dislocation loop can expand within the precipitate and cause a simultaneous loss in coherent character. The effect of strain path is manifested in terms of different interface characteristics for UDRXA and MSCRXA, hence different precipitate morphology.

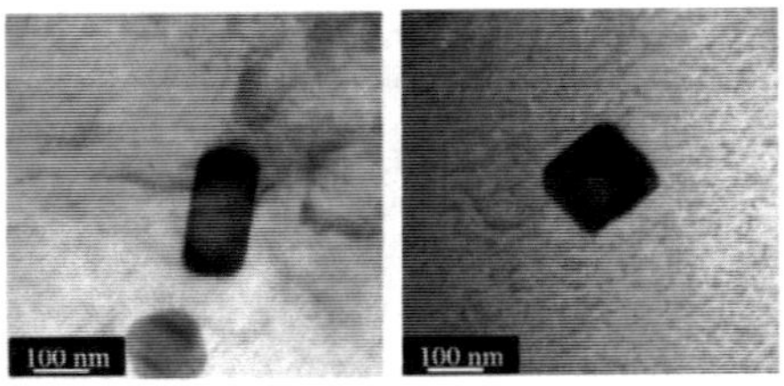

Fig. 4 TEM micrograph of UDRXA and MSCRXA.

Further investigations are needed for better understanding of the governing mechanism but it is quite clear that changing the strain path during rolling plays a key role in controlling the morphology of the precipitates and in the hindsight might also control the precipitation kinetics.

Conclusions

The present investigation clearly shows the effect of strain path change on precipitation behaviour of Al-Cu-Mg-Si alloy. In summary Al-Cu-Mg-Si alloys with different initial texture shows different precipitate morphology and varying texture can give a handle on developing isotropic and anisotropic properties in Al sheets. Crystallographic texture plays a key role in controlling the interface characteristics. The sluggish ageing kinetics for MSCRXA is manifested in terms of higher value of average misorientation angle. The coherency strains are relieved at a much smaller pace for MSCRXA compared to UDRXA. Thus, the growth of the precipitates is almost isotropic for MSCRXA compared to UDRXA where the coherent character of the interface is lost at a much higher pace, thus, leading to anisotropic growth.

References

[1] Smallman RE, Ngan AHW 2014, *Modern Physical Metallurgy* (Elsevier) 8th Edition, p 499
[2] Tilak RV, Morris JG 1985, *Mater. Sci. Eng.* **73**, 139
[3] Soffa WA, Laughlin D E 2014, *Physical Metallurgy* (Elsevier) 5th Edition, p 851
[4] Nie JF 2014, *Physical Metallurgy of Light Alloys* (Elsevier) 5th Edition, p 2009
[5] Ashby M 1966, *Acta Metall.*, **14**, 679
[6] Yang W, Ji S, Huang L, Sheng X, Li Z, Wang M 2014, *Mater. Charact.*, **94**, 170
[7] Son SK, Matsumura S, Fukui K, Takeda M 2011, *J. Alloy Compd.*, **509**, 241
[8] Fukui K, Takeda M, Endo T 2005, *Mater. Lett.*, **59**, 1444
[9] Sun DL, Yang DZ, Hong Y, Lei TC 1989, Proceedings of *ICSMA 8*, **2**, 591
[10] Ghosh SK 2011, *J. Mater. Sci. Technol.*, **27**, 193
[11] Suwas S, Singh AK 2003, *Mater. Sci. Engg. A*, **356**, 368
[12] Suwas S, Gurao NP 2014, *Comprehensive Materials Processing*, (Elsevier) vol. 3, p 81
[13] Garg R, Gurao NP, Ranganathan S, Suwas S 2011, *Philos. Mag.*, **91**, 4089
[14] Gurao NP, Ali A, Suwas S 2009, *Mater. Sci. Eng. A*, **504**, 24
[15] Porter DA, Easterling KE 1992, *Phase Transformation in Metals and Alloys*, (CRC Press London) 3rd Edition, p 160
[16] Sankaran R, Laird C 1974, *Philos. Mag.*, **29**, 179

IOP Conf. Series: Materials Science and Engineering **82** (2015) 012026 doi:10.1088/1757-899X/82/1/012026

Influence of hydrostatic pressure on texture evolution in HPT deformed NiAl

C Tränkner[1], R Chulist[1,2], W Skrotzki[1], T Lippmann[3], J Horky[4] and M Zehetbauer[4]

[1] Institut für Strukturphysik, Technische Universität Dresden, D-01062 Dresden, Germany

[2] Instytut Metalurgii i Inżynierii Materiałowej, Polskiej Akademii Nauk, PL-30-059 Kraków, Poland

[3] Institut für Werkstoffforschung, Helmholtz-Zentrum Geesthacht, D-21502 Geesthacht, Germany

[4] Physik Nanostrukturierter Materialien, Fakultät für Physik, Universität Wien, A-1090 Wien, Austria

E-mail: christine.traenkner@tu-dresden.de

Abstract. NiAl is an intermetallic compound with a brittle-to-ductile transition temperature of about 300°C at ambient pressure. At standard conditions, it is very difficult to deform, but fracture stress and fracture strain are increased under hydrostatic pressure (HP). On account of this, deformation at low temperatures is only possible at high HP, as for instance used in high pressure torsion (HPT). In order to study the influence of HP on texture evolution, small discs of polycrystalline NiAl were deformed by HPT at different temperatures ranging from room temperature to 500°C and different HPs. The influence of HP is presented for deformation at room temperature and 500°C. It is found that HP affects the formability of the samples as well as texture and microstructure.

1. Introduction

Nickel-aluminium (NiAl) is an intermetallic compound with B2 structure. It has a low density of 5.85 g/cm^3 and a high melting point of 1638°C at stochiometric composition as well as a low coefficient of thermal expansion and good oxidation resistance [1]. These characteristics would make it a good material for high temperature applications, but it is hardly used in industry due to its low temperature brittleness. At ambient pressure, the brittle-to-ductile transition temperature (BDTT) lies at about 300°C. According to the von Mises criterion, five independent slip systems are needed for a homogeneous deformation of polycrystals, that means, the three primary {110}<100> slip systems in NiAl are not enough. Below the BDTT, the critical resolved shear stress for the necessary activation of the secondary slip systems {112}<111> or {110}<110> to fulfil the von Mises criterion is higher than the fracture stress and therefore the material breaks before plastically deforming. For temperatures above the BDTT, secondary slip on {110}<110> becomes easy, turning the material ductile. In order to achieve high strains in NiAl at temperatures below 300°C without fracture, high pressure torsion (HPT) [2] was used. In [3] it was shown that the fracture stress and fracture strain of polycrystalline NiAl drastically increase when applying a high hydrostatic pressure (HP) during tensile testing at room temperature (RT). Thus, the BDTT can be shifted below RT by application of a sufficient HP.

In the present work the influence of HP on texture evolution during HPT of NiAl at different temperatures is analyzed by evaluating the volume fraction of the texture components found.

2. Experimental

The polycrystalline NiAl investigated had a nearly stoichiometric composition ((50.2 ± 0.1) at%Ni). In the initial extruded sample, a homogeneous globular microstructure with an average grain size of about 50 μm and a weak <111> fiber texture was found [4, 5]. Small discs of the material with a height of 0.8 mm and a diameter of 8 mm were deformed at RT and 500°C by HPT at different pressures. According to [2], the shear strain γ in torsion is calculated by the relationship

$$\gamma = \frac{2\pi N r}{h}, \quad (1)$$

with N = number of rotations, r = radius and h = height of the sample. Two rotations were used adding up to a maximum shear strain of 60 at the edge of the sample. The rotation rate of 0.2 rotations/min leads to a maximum shear strain rate of 0.1 s^{-1}.
Texture measurements were done on small pins cut along the radial direction of the samples by spark erosion [see Fig. 1]. The investigated part considered here is a small volume at the

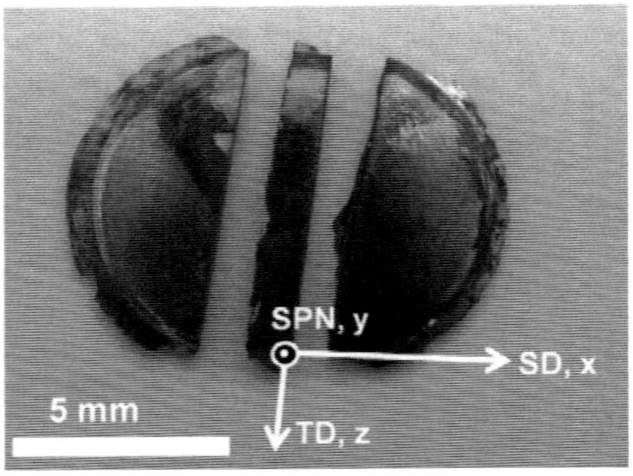

Fig. 1: Sample and crystal coordinate system of HPT deformed sample with cut pin. (SD = shear direction, SPN = shear plane normal, TD = radial direction, x = [100], y = [010], z = [001])

edge of the pin corresponding to γ ≈ 50. The experiments were performed at beamlines HARWI II and HEMS at DESY in Hamburg, Germany. For evaluation of the texture, the Debye-Scherrer rings corresponding to the {100}, {110} and {111} lattice planes were used. Pole figures (PFs) were calculated using the software StressTex [6]. For the calculation of the orientation distribution function (ODF), the software LaboTex [7] was employed. The volume fraction of the texture components was calculated with a Gaussian spread of 10°. More details on texture measurements with synchrotron radiation and data processing can be found in [8].

3. Results and discussion

Depending on temperature, respectively strength, a certain HP is needed to get a smooth HPT deformation without slipping of the sample. While it was possible to deform all samples at a HP of 8 GPa, at 2 GPa and RT the disc slipped. At a temperature of 500°C, HPT at all HPs was performed without problems. Thus, formability is enhanced by high HP. In accordance with the findings of Margevicius and Lewandowski [2], the fracture stress is increased by applying a high enough HP. Thus, activation of the secondary slip systems is possible.
In Fig. 2, PFs of the samples deformed with different HPs at RT and 500°C are shown. For the samples deformed at RT a shear texture with components typical of bcc metals is found (see key figure in Fig. 2b), while for the sample deformed at 500°C a weak F and strong oblique cube component developed. The differences in texture caused by different deformation temperatures are described in [9].

IOP Conf. Series: Materials Science and Engineering **82** (2015) 012026 doi:10.1088/1757-899X/82/1/012026

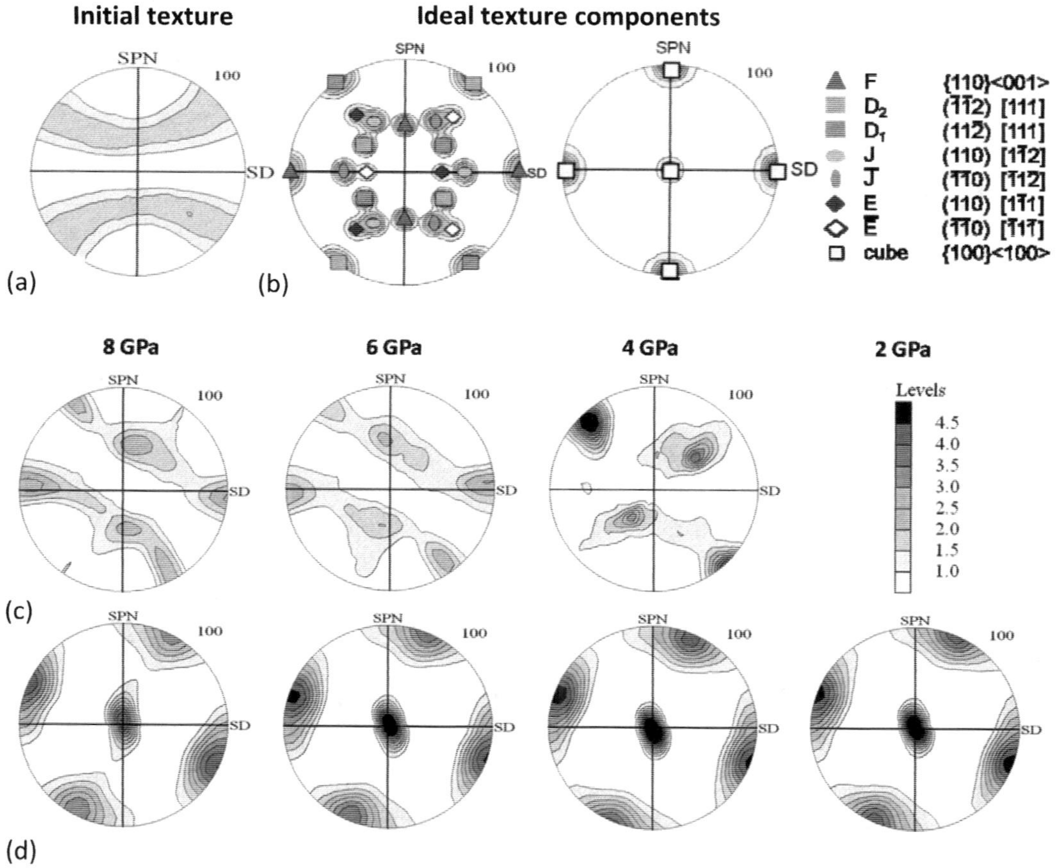

Fig. 2: {100} PFs of the initial texture (a), ideal texture (b) and samples HPT deformed at RT (c) and 500°C (d) at different HPs.

To analyze the texture changes quantitatively, the volume fractions of the texture components were calculated with LaboTex (Fig. 3). The ideal texture components for bcc metals deformed by simple shear can be found in [10].

For the samples deformed at RT shear texture components of different intensities are found. With increasing HP, the volume fraction of the components increases apart from the D_2 component, which shows the opposite effect. The change in volume fraction of the shear texture components may be due to a change in activity of the different secondary slip systems. Simulations to check this are planned.

At 500°C, only a weak F component and a strong oblique cube component are found. The volume fraction of the F component increases with pressure while that of the oblique cube component, which usually develops by discontinuous dynamic recrystallization, decreases. Thus, HP enhances the formation of the shear texture by impeding recrystallization and grain growth due to restrained diffusion [see e.g. 11,12].

These results can be linked to microstructure and microhardness measurements not shown here. According to these measurements, the grain size decreases with increasing HP, whereas microhardness increases which also indicates a slower recrystallization and grain growth process.

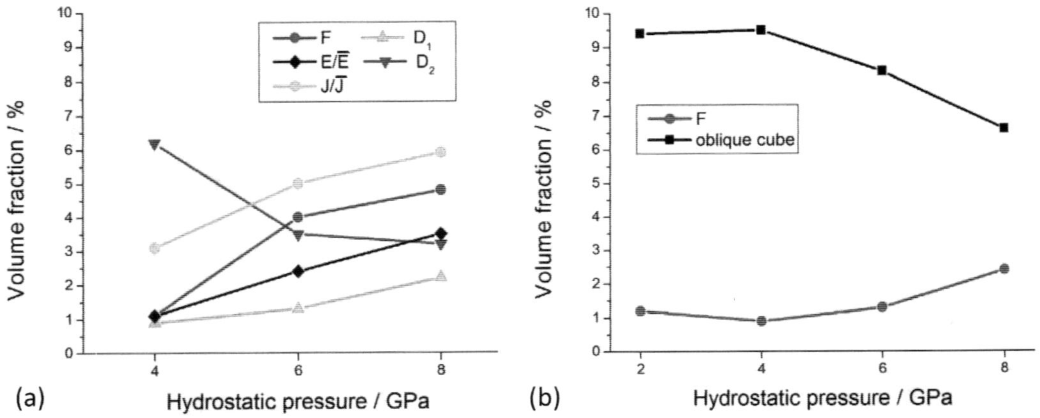

Fig. 3: Volume fraction of texture components for the samples HPT-processed at RT (a) and 500°C (b)

Acknowledgment
The help of H.-T. Reiter with sample cutting is gratefully acknowledged.

References

[1] Bürgel R, Maier HJ and Niendorf T 2011 (4th edition) *Handbuch Hochtemperatur-Werkstofftechnik* (Wiesbaden: Vieweg+Teubner) ch. 6 p 463
[2] Pippan R 2009 *High-pressure Torsion – Features and Applications*, ch. 9 p 217 in: eds. Zehetbauer MJ and Zhu YT 2009 *Bulk Nanostructured Materials* (Weinheim, Germany: Wiley VCH)
[3] Margevicius RW and Lewandowski JJ 1994 *Metall. Mater. Trans.* **A 25** pp 1457-1470
[4] Klöden B 2006 *PhD-Thesis* Institut für Strukturphysik, TU Dresden
[5] Klöden B, Oertel C-G, Skrotzki W and Rybacki E 2009 *J. Eng. Mater. Technol.* **131** 011102
[6] Randau C, Garbe U and Brokmeier H-G 2011 *J. Appl. Crystallogr.* **44** pp 641-646
[7] LaboTex Version 3.0.24
[8] Wcislak L, Klein H, Bunge HJ, Garbe U, Tschentscher T and Schneider JR 2002 *J. Appl. Crystallogr.* **35** pp 82-95
[9] Tränkner C, Chulist R, Skrotzki W, Beausir B, Lippmann T, Horky J and Zehetbauer M 2014 *IOP Conf. Ser.: Mater. Sci. Eng.* **63** 012154
[10] Li S, Beyerlein IJ and Bourke MAM 2005 *Mater. Sci. Eng. A* **934** p 75
[11] Zehetbauer MJ, Stüwe HP, Vorhauer A, Schafler E and Kohout 2003 *Adv. Eng. Mater.* **5** pp 330-337
[12] Kühlein W and Stüwe HP 1988 *Acta metall.* **36** pp. 3055-3059

Mechanical behavior and texture prediction of Ti-6Al-4V based on elastic viscoplastic self-consistent modelling

J Galan-Lopez[1], S Naghdy[1], P Verleysen[1], LAI Kestens[1], F Coghe[2], L Rabet[2] and J Degrieck[1]

[1] Dept. Materials Science and Engineering, Ghent University, Sint-Pietersnieuwstraat 41, 9000 Ghent, Belgium

[2] Dept. COBO, Royal Military Academy, Renaissance Avenue 30, 1000 Brussels, Belgium

Abstract. The goal of this study is to apply an elastic viscoplastic self-consistent crystal plasticity model to predict the texture evolution in a Ti-6Al-4V alloy which has a (mainly) hexagonal crystal structure. The model under consideration is an extension of the viscoplastic self-consistent model proposed by Lebensohn and Tome [1993] which has been adapted to account for elasticity and has been integrated with a new algorithm, making it more computationally efficient within an implicit FE scheme. The flow behavior of Ti-6Al-4V is strongly dependent on strain rate and temperature. To estimate the model parameters, the flow behavior of quasi-static experiments is used. A temperature sensitivity term has been introduced to correct the effects of temperature increase during the dynamic experiments. In order to have a meaningful rate sensitivity exponent, a value is calculated based on valid experimental data, rather than choosing an arbitrary large numerical value. In this way the behavior of Ti-6Al-4V is captured at different strain rates. Predictions of the model are compared to experimental data.

1. Introduction

Polycrystal models are a family of constitutive models that can be used both as a constitutive model and a tool for performing virtual experiments for other constitutive models. VPSC and ALAMEL are examples of polycrystal models, which have been successfully used for texture and anisotropy estimation of HCP metals [1, 2].

Ti-6Al-4V accounts for more than 50% of total titanium usage thanks to its bio-compatibility, high strength-to-weight ratio and good corrosion resistance [3]. HCP titanium alloy exhibits four slip systems: basal {0 0 0 1} <1 1 2 0>, prismatic {1 0 1 0} <1 1 2 0>, first order pyramidal {1 0 1 1} <1 1 2 0>, and second order pyramidal {1 0 1 1} <1 1 2 3>. Plastic deformation is mostly accommodated by basal or prismatic slip along burgers vector <a>, whereas slip is more difficult on the pyramidal slip systems. Besides slip, twinning can be also active in tensile {1 0 1 2} <1 0 1 1> or compressive {1 1 2 2} <1 1 2 3> mode. To accommodate strain parallel to the c axis of the hexagon, second order pyramidal or twin systems should be activated [4]. Although pure titanium is known to exhibit extensive primary and secondary twinning, the high amount of aluminum in Ti-6Al-4V prohibits formation of extensive twins. In monotonic modes of deformation at room temperature, only very low volume fractions (less than 1%) of tension twins have been observed. According to Hammami et al. [2] and Jones and Hutchinson [5] cross-slip occurs along the <c+a> vector on pyramidal planes during uniaxial compression, while it is absent during uniaxial tension.

In this work, the parameters needed for the VPSC90 model [6] (updated version of VPSC) are derived using standard tensile tests at different quasi-static strain rates and an improved fitting procedure, which involves VPSC90. Subsequently, these parameters are used to predict the mechanical behavior and texture development in during quasi-static and dynamic loading.

2. Experimental setup

2.1. Material

The Ti-6Al-4V sheet has a thickness of 0.6 mm. The mill-annealed condition results in equiaxed α-phase grains (96% vol. fraction and average grain size of 6 μm) and bcc β-grains (4% vol, average grain size of less than 1 μm).

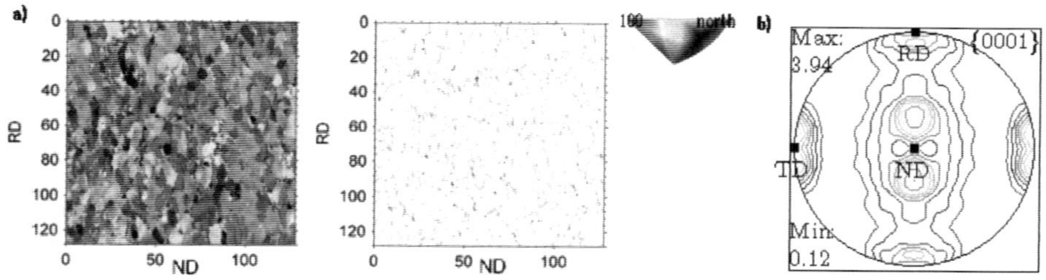

Figure 1: a) microstructure of sheet in RD-ND plane for α and β-phases, respectively, b) {0001} pole-figure of the α-phase, which shows the orthorhombic sample symmetry.

Because of its low volume fraction, the β phase is neglected in the model. The α-phase texture is shown in figure 1-b, which clearly shows a typical hexagonal texture. Two symmetric components along the ND-RD plane with basal poles oriented in the TD direction can be seen, which are attributed to remaining effects of the hot rolling process. Two other symmetric components along ND-TD plane with deviation from the ND pole towards RD are also noticeable. The latter components are attributed to the residual effects of cold rolling.

2.2. Mechanical Testing

Quasi-static and dynamic tension experiments were performed. Samples were cut along the rolling direction of the sheet. Quasi-static experiments at moderate strain rates were used to fit the model parameters. To guarantee the quasi-static conditions and to obtain isothermal flow curves, thermal effects of strain rates were taken in to account using a thermo-mechanical model [7]. Dynamic tests were performed using the Split Hopkinson Tensile Bar (SHTB) method. Subsequently, using the parameters obtained from quasi-static tests, the mechanical behavior of the material at high strain rates was investigated.

3. Modelling

Since the effects of temperature and strain rate can be separated, the parameters obtained by fitting the mechanical behavior at quasi-static conditions should be able to describe the material behavior both at quasi-static and dynamic loading situations. Non-isothermal flow curves of quasi-static experiments ($8*10^{-5}$, $6.7*10^{-4}\,s^{-1}$, $5*10^{-2}$ and $5*10^{-1}\,s^{-1}$) were transformed to isothermal curves using a thermomechanical model based on the Johnson-Cook hardening model [7]. The different slip systems of α titanium considered here are prismatic, basal, pyramidal <a> and pyramidal <c+a>. Since the twinning fraction observed both in literature and in this work for tensile deformation is very low, twinning systems were excluded from the model. The ODF of the initial material, derived from the orientation contrast scans, is discretized in 1000 orientations for the purpose of the simulation. Voce hardening parameters obtained by fitting are shown in table 2.

IOP Conf. Series: Materials Science and Engineering **82** (2015) 012027 doi:10.1088/1757-899X/82/1/012027

Table 2: Voce hardening parameters for α titanium; Latent hardening for all slip systems is equal to 1.

Deformation mode	n_s	τ_s^0	τ_s^1	Θ_s^0	Θ_s^1
Prismatic	88	390	---	0	0
Basal	81	427	46	170	0
Pyramidal <a>	124	457	852	350	0
Pyramidal <c+a>	76	549	657	1040	0

The simulated tensile curves shown in figure 2 are the results of the fitting procedure. From this figure it is clear that an excellent agreement is obtained between the simulated and experimental curves: the model reproduces both the hardening behavior and the strain rate sensitivity.

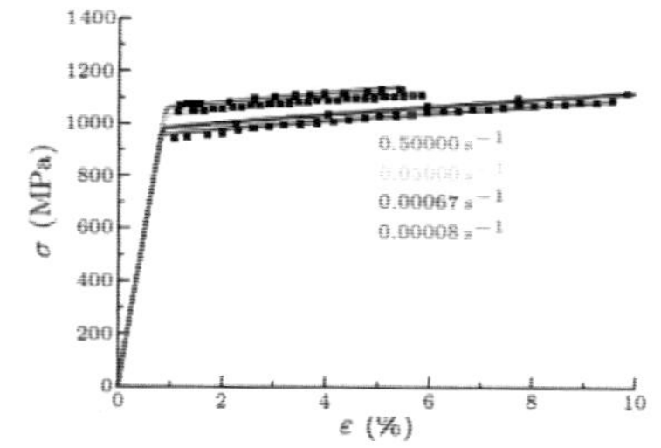

Figure 2: comparison of experimental (black points) and simulated (colour lines) tensile curves.

To assess the capability of the model to predict the mechanical behavior, post-necking local strain measurements obtained by high speed imaging and a DIC technique for a quasi-static ($6.7*10^{-4}\,s^{-1}$) and a dynamic test (average 425 s^{-1}) were investigated.

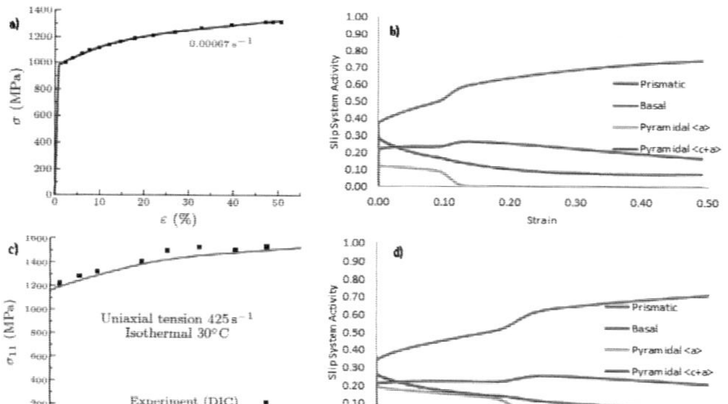

Figure 3: simulated flow behavior for $0.00067s^{-1}$ and $425s^{-1}$ together with the corresponding slip system activity.

4. Results and discussion

The simulation results of the tensile experiments, together with the slip systems activities, are shown in figure 3. The hardening behavior is very well reproduced. Even though the tensile curve obtained by dynamic behavior displays some oscillations, and the strain rate is not constant during the experiment, a good agreement is obtained. The slip systems activity diagrams show that most of the deformation is accommodated by prismatic slip, followed by the basal slip, while the two pyramidal slip systems have minor activity, which even further decreases at higher strains. Although activation of pyramidal <c+a> is more difficult than pyramidal <a>, its contribution to deformation is higher, thus compensating for deformation in the <c> direction of the hexagonal unit cell. No considerable amount of twinning was observed by EBSD inspection both for quasi-static and dynamic cases, and the right prediction of hardening behavior by only considering dislocation glide is thus confirmed. An EBSD measurement of the material adjacent to the fracture zone of a quasi-static tensile sample ($6.7*10^{-4}$ s^{-1}) with almost 30% deformation, together with the simulated texture prediction is represented, cf. figure 4. In the experimental pole figure (Fig. 4a), weakening of the initial rolled components is observed, while new components develop which exhibit the basal plane normal parallel to the rolling direction. The simulated texture (Fig. 4b) shows that without considering twinning the most important features of the experimental texture can be reproduced.

The results represented here show that the VPSC model can be successfully applied for constitutive modelling of Ti-6Al-4V. It is of relevant importance to observe that in spite of all the simplifications, such as considering only one phase and only 1000 grains, the mechanical behavior can still be captured for a wide range of strain rates. Although texture changes are more severe in large deformation processes and in the presence of compressive forces, still the VPSC polycrystal model is capable of tracking texture changes.

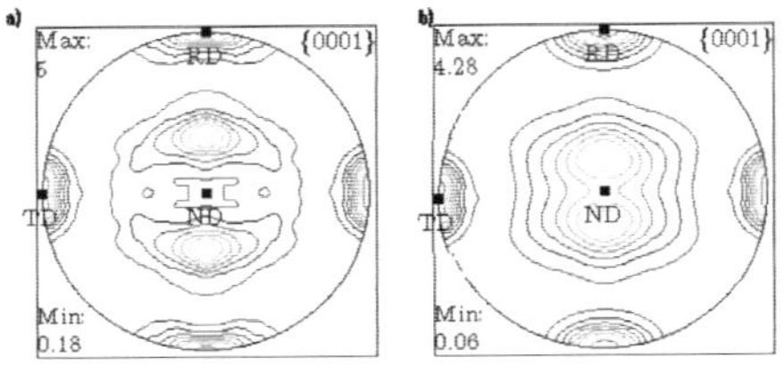

Figure 4: texture evolution in quasi-static deformation at 6.7 10^{-4} s^{-1} after 30% deformation. (a) The experimental pole figure and (b) the simulated one.

5. Conclusion

Quasi-static and dynamic tensile tests were conducted on Ti-6Al-4V. Using the quasi-static data and taking into account the initial texture, the VPSC model parameters were determined by a series of tensile tests. These parameters were successfully applied to simulate the large deformation behavior of material at quasi-static and dynamic cases. No considerable amount of twinning was observed; both texture prediction and hardening behavior of the material are supporting this idea.

Acknowledgments

This work was funded by the InterUniversity Attraction Poles (IUAP) program of the Belgian Science Policy under contract number P7/21.

References

[1] Lebensohn R A and Tomé C N, 1993, *Acta Metall Mater*, 41, 2611-2624.
[2] Hammami W, Gilles G, Habraken A M, Duchene L, 2011, *Int J Mat Form*, 2, 205-215.
[3] Bedinger G M, Titanium Minerals Yearbook, 2013, United States Geological Survey.
[4] Cazacu O, Plunkett B, Barlat F, 2006, *International Journal of Plasticity*, 7, 1171-1194.
[5] Jones I, Hutchinson W, 1981, *Acta Metallurgica*, 6, 951-968.
[6] Galán J, Verleysen P, Lebensohn R A, 2014, Modelling Simul. Mater. Sci. Eng, 22, 5, 055023
[7] Galán J, Verleysen P, Degrieck J, 2013, Strain, 49, 4, 354-365.

IOP Conf. Series: Materials Science and Engineering **82** (2015) 012028 doi:10.1088/1757-899X/82/1/012028

Textures and Microstructures Formed in WE43 and AZ31 Magnesium Alloys during High Speed Rolling and Their Formation Mechanisms

Y Onuki[1], K Hara[2], H Utsunomiya[2] and J A Szpunar[3]
[1] Institute of Multidisciplinary Research for Advanced Materials, Tohoku University, Japan
[2] Division of Materials and Manufacturing Science, Graduate School of Engineering, Osaka University, Japan
[3] Department of Mechanical Engineering, University of Saskatchewan, Canada

E-mail: onuki@tagen.tohoku.ac.jp

Abstract. High speed rolling is recognized as the process that can produce sheets of magnesium alloys having RD-split basal texture without or with minimum preheating. However, the mechanism of the texture formation during high speed rolling has not been fully clarified yet. In this study, conventional AZ31 and a rare earth - yttrium added alloy, WE43 were rolled with high rolling speed. The specimens having different textures were prepared by changing the cutting geometry to initially textured sheets. It is seen that the crack, microstructure and texture formations are strongly influenced by the initial textures in AZ31. These features are strongly related to the extension twinning, {10-12}<-1011>. In the case of WE43, cracks are formed more often than in AZ31, despite of the weak initial textures. It is proposed that the activities of the contraction and double twinning systems give more chance of stress concentration, resulting in the narrow shear banding and subsequent cracking. In addition to the experimental analysis, results of the numerical simulation using VPSC model are also used to discuss the texture formation mechanism.

1. Introduction

Magnesium and its alloys have low density and excellent specific strength. From the viewpoint of reducing carbon emission, the application of this light structural material to larger parts is demanded. However, the applications are currently limited to small parts produced by machining or die-casting due to the poor room temperature ductility.

High speed rolling is one of the solutions for the above problem [1-3]. When rolling deformation is conducted with high rolling speed (> 1000 m/min.), the sample is adiabatically heated by surface and internal friction. With the aid of the friction heat, non-basal slip systems are activated and large thickness reduction can be achieved without or with minimum preheating. In addition, the texture formed by high speed rolling is RD-split basal, which has concentrations of (0001) poles ±10~20° inclined from ND toward RD [1, 2]. The frequent existence of the inclined basal planes enhances the workability of the magnesium sheet [3].

The authors have already presented that the formation of the RD-split basal texture in AZ31 (Mg - 3 mass% Al – 1 mass% Zn) alloy is due to the enhanced activation of <c + a> pyramidal slip systems [1]. In a separate paper, we suggest that the strain hardening could promote the activation of the <c + a> slip as well as the thermal assistance. In addition, we showed that the extension twinning, {10-12}<-1011>, strongly affected the texture formation and cracking [2]. In the present paper, additional results of the high speed rolling test of AZ31 are presented and our conclusion is reinforced. In addition, the results of high speed rolling of WE43 are shown.

On the basis of the comparison between these two alloys, mechanisms of high strain rate deformation and texture formation in magnesium alloys are discussed.

2. Experimental Procedures

The tested materials were AZ31 (Mg-3 mass% Al-1 mass% Zn) and WE43 (Mg-4 mass% Y-3 mass% RE-0.4 mass% Zr) hot-rolled sheets provided by Magnesium Elektron. Specimens were cut with different geometry, so that each of them had different initial texture as shown in Figure 1.

Rolling deformation was applied aiming 40% thickness reduction with a rolling speed of 1000 m/min. The rolling was conducted without preheating and with preheating at 473K for both AZ31 and WE43. The deformation with preheating at 573K was conducted only for AZ31. Textures of the samples at the thickness center were measured by X-ray diffractmeter, Bruker D8 Discover, using Cr Kα radiation and the area detection system [4]. Based on the measurement, the ODF was calculated with ADC method [5] by using Resmat TexTools. EBSD measurement was also conducted by using Hitachi SU6600 equipped with Oxford EBSD system. More details of the experiments can be found in our paper [2].

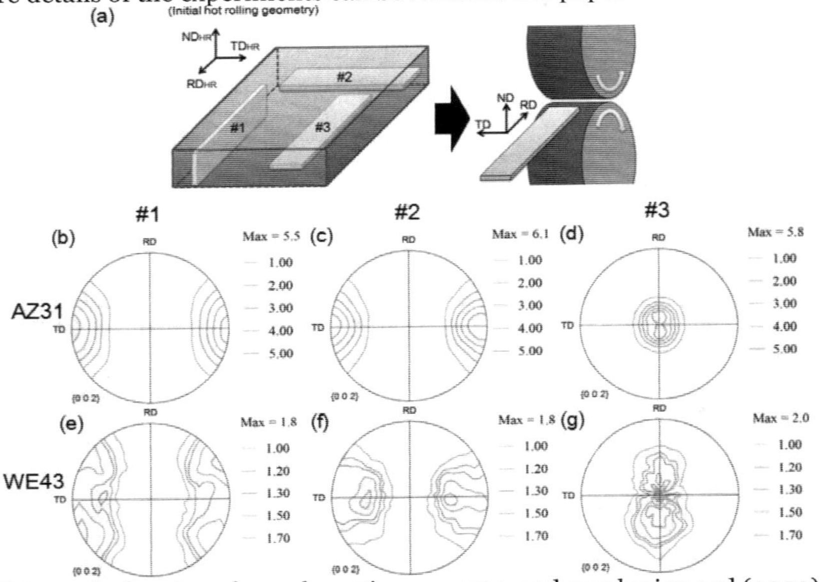

Figure 1 (a) Schematic drawing of sample cutting geometry and numbering and (0002) pole figures showing the initial textures of (b) ~ (d) AZ31 and (e) ~ (g) WE43.

3. Results and Discussion

3.1. High speed rolling of AZ31 with preheating at 573K

Figure 2 shows (0002) pole figures for AZ31 samples rolled with preheating at 573K. All the samples could be rolled without failure up to 40% thickness reduction. The features of the pole figures are similar to what we observed on the samples rolled with preheating at 473K [2]. Samples #1 and #2 have the pole distributions elongated toward TD as well as RD-split component while Sample #3 has only RD-split components. The split angle of the main components is ±20° in all the pole figures, which is the same as the samples preheated at 473K. The formation of RD-split basal components in Samples #1 and #2 suggests that the extension twinning is an active shear system even at 573K. However, in comparison to the case with preheating at 473K, higher densities are observed along the ND-TD lines and the maxima at RD-split components are lower, which suggests the activation of the slip systems is more enhanced and the twinning is suppressed by increasing the deformation temperature. It should be noted that the extension twinning is not a geometrically preferred shear system in all the samples when compression strain along ND and no strain along TD are assumed.

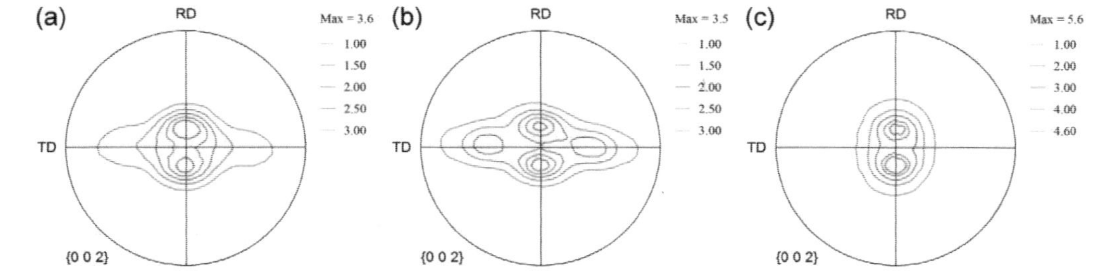

Figure 2 (0002) pole figures of the AZ31 samples after the high speed rolling test with preheating at 573K. (a) Sample #1, (b) Sample #2 and (c) Sample #3.

3.2. WE43

All the samples of WE43 were severely cracked even with preheating at 473K while the AZ31 samples could be rolled without cracking above 473K. Figure 3 is the macroscopic appearance of Sample #1 of WE43 rolled with preheating at 473K. All the samples showed equally spaced cracks parallel to the TD as seen in Figure 3, regardless of the initial texture or preheating temperature. The textures after the deformation with preheating are shown in Figure 4. The distributions of poles are somewhat similar to those in AZ31 (see Figure 2) but the pole densities are lower.

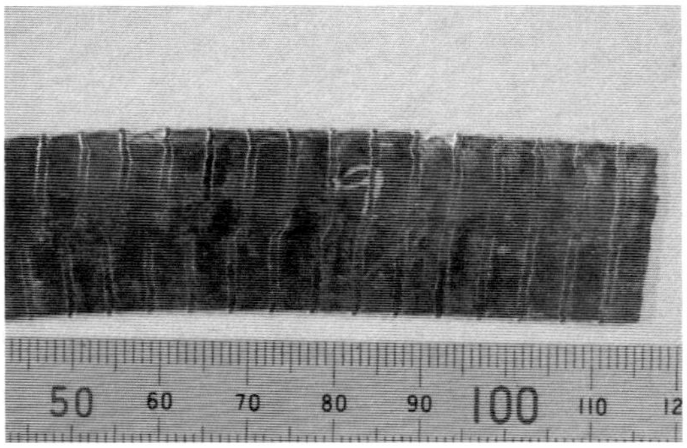

Figure 3 Photograph of Sample #1 of WE43 after high speed rolling with preheating at 473K.

There are a number of studies reporting the formations of weak deformation textures in rare-earth added magnesium alloys [6, 7]. This is usually explained by enhanced activities of non-basal slip systems and dynamic recrystallization [7]. However, we numerically showed the trend of the slip rotation that [0001] moves toward ND and subsequently rotates toward RD-split positions with the activation of <c + a> slip system [2]. Therefore, it is thought that the formation of the weak texture does not owe to the activation of non-basal slip systems.

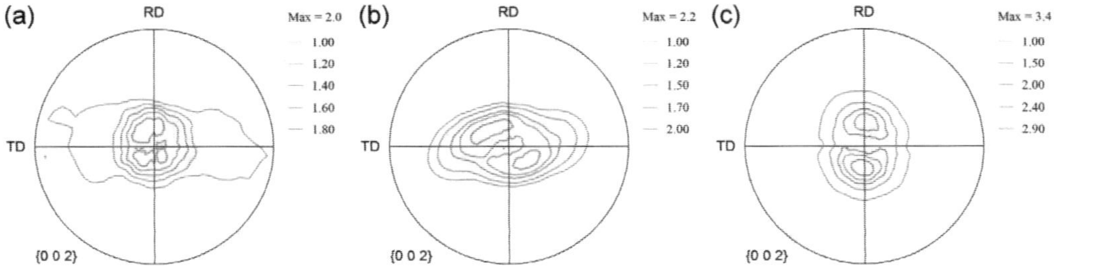

Figure 4 (0002) pole figures of the WE43 samples after the high speed rolling test with preheating at 473K. (a) Sample #1, (b) Sample #2 and (c) Sample #3.

The similar cracking was observed in the case of AZ31 samples at room temperature [2]. It was seen that the crack was introduced along the shear band, which was formed by laminated twins. Since the textures shown in Figure 4 have near-basal component, it can be considered that the extension twinning is active also in WE43. This can be confirmed in the result of EBSD measurement shown in Figure 5. In this band contrast map, extension twin boundaries are shown as red lines. In addition, the contraction and double twin boundaries are shown as yellow and blue lines, respectively. The contraction and double twin boundaries were limited in AZ31 but exist considerably in WE43. These twin systems can generate larger shear strain

IOP Conf. Series: Materials Science and Engineering **82** (2015) 012028 doi:10.1088/1757-899X/82/1/012028

than the extension twin [8], resulting in larger local stress concentration. Since the volume fractions of these thin twins are small and the initial texture is nearly random, specific alignment of crystal orientation cannot be achieved by the twinning. In addition, it is thought that the activities of slip systems in the region of stress concentration no longer agree with the macroscopic strain requirement. Therefore, there is no overall trend of crystal rotation to form any deformation texture.

It is of interest that the origin of the cracking during high speed rolling is the twinning in both alloys. Regardless of the alloy element or initial texture, twinning plays more important role than non-basal slip systems on the deformation behaviors. It is suggested that if one could develop the alloy which merely activates the twinning systems, it would have excellent ductility.

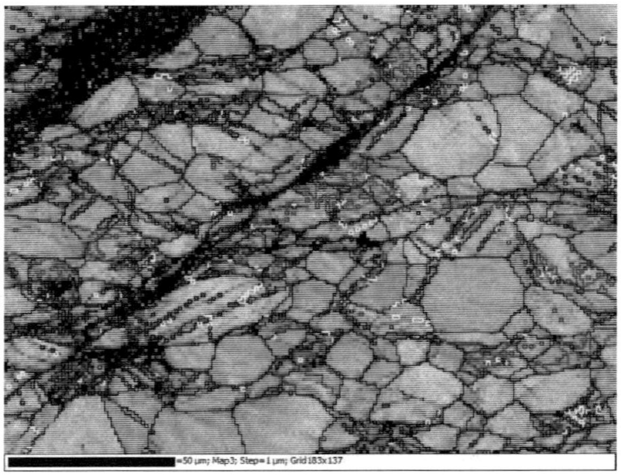

Figure 5 Band contrast map taken on the ND-RD cross section of Sample #3 rolled with preheating at 473K. Horizontal direction is the RD. Red, yellow and blue lines are the extension, contraction and double twin boundaries. Tolerance angle of 7° is applied to the boundary coloring.

4. Conclusions

High speed rolling tests of AZ31 and WE43 magnesium alloys were conducted in order to clarify the texture formation and deformation mechanisms. The deformation of AZ31 with preheating at 573K was conducted without cracking with all types of initial textures. The deformation textures showed higher ND-TD pole distribution and lower RD-split peak densities than deformation at lower temperatures. This suggests that the activation of the <c + a> slip system is enhanced and the extension twinning is suppressed with increasing temperature.

WE43 showed severe cracks after the rolling with preheating at 473K. The deformed microstructure showed high fractions of extension, contraction and double twin boundaries. It is considered that such inhomogeneous twin deformations generate the shear bands and cracks and result in weak texture, rather than the activation of non-basal slip systems.

References

[1] Li H, Hsu E, Szpunar J, Utsunomiya H and Sakai T 2008 *J. Mater. Sci.* **43** 7148

[2] Onuki Y, Hara K, Utsunomiya H and Szpunar J *2014 J. Mater. Eng. Perform. DOI: 10.1007/s11665-014-1318-8*

[3] Koh H, Sakai T, Utsunomiya H and Minamiguchi S 2007 *Mater. Trans.* **48** 2023

[4] Helming K, Lyubchenko M, He B and Preckwinkel U 2003 *Powder Diffr* **18** 99

[5] Pawlik K, Pospiech J and Lücke K 1991 *Texture Micrstruct.* **14-18** 25

[6] Barnett M 2004 *Mater. Sci. Eng.* A **386** 205

[7] Sandlöbes S, Zaefferer S, Schestakow I, Yi S and Gonzalez-Martinez R 2011 *Acta Mater.* **59** 429

[8] Christian J and Mahajan S 1995 *Prog. Mater. Sci.* **39** 1

IOP Conf. Series: Materials Science and Engineering **82** (2015) 012029 doi:10.1088/1757-899X/82/1/012029

Relationship between strain stored by compressive deformation and crystallographic orientation in a pure aluminum

Y Takayama[1], T Yoshimura[2] and H Watanabe[1]
[1] Department of Mechanical and Intelligent Engineering, Graduate School of Engineering, Utsunomiya University, Utsunomiya 321-8585, Japan
[2] Graduate Student, Utsunomiya University, Utsunomiya 321-8585, Japan
E-mail: takayama@cc.utsunomiya-u.ac.jp

Abstract. In order to investigate relationship between stored strain and crystallographic orientation, 99.99% purity aluminum cubes were compressed with uniaxial or with plane strain state up to a nominal strain of 30%. The aluminum cubes were examined on the same surface before and after compression by SEM/EBSD technique. Stored strain was estimated by Kernel Average Misorientation (KAM) derived from the EBSD analysis, and Taylor factor (TF) was measured before the compressive deformation. The analysis revealed that KAM value or the stored strain decreases until a certain value of TF and then increases with increment of TF.

1. Introduction

Recrystallization textures developed during annealing are governed by the energy stored in deformation state. It is well known that the stored energy depends on crystallographic orientation [1]. Alternatively, this is the reason why the recrystallization texture forms. Several attempts have been made to evaluate the stored energy by calorimetric [2, 3] and diffraction techniques [4-12]. The latter technique is divided into two types. One is the basic method [4-8] in which X-ray line broadening is measured to obtain the stored energy in grains having specific orientations, for example (100) plane parallel to the rolling plane in the sample. Another is the extended method [9-12] which was used to determine the stored energy for a much wider space described by the orientation distribution function. Using the extended method, Kallend and Huang, Gerber *et.al* [9, 12], Rajmohan *et.al* [10] and Rajmohan and Szpunar [11] carried out the measurement of the stored energy in copper, interstitial free steel and can body aluminum alloy after rolling. They also investigated the relation between the stored energy and the Taylor factor (TF) [13] which shows the total slip system activity expected during the deformation of a grain of specific orientation. Although the stored energy increases with an increase in Taylor factor, the obtained results indicated some exceptions and a lot of scatter. It was understood that these exceptions and the scatter were attributed to the rotation of the grain orientation to the final one and the inhomogeneity of the deformation. Takayama and Szpunar [14] performed the evaluation of the stored energy from the kernel average misorientation (KAM) on electron backscatter diffraction analysis in scanning electron microscopy (SEM/EBSD) to investigate the relation between the stored energy and Taylor factor in an Al-Mg-Mn alloy sheet worked by the continuous cyclic bending (CCB). Their analysis revealed that the stored energy is high for the high TF region, whereas a significant increase of the stored energy with the decrease of the Taylor factor appears in the vicinity of the minimum TF value of 2.

This paper presents the relationship between the stored energy and Taylor factor in aluminum subjected to compression with uniaxial or with plane strain state by using SEM/EBSD analysis.

2. Experimental procedure

A hot-rolled plate of 4N purity aluminum (99.99% Al) with a 12mm thickness was prepared through a process of DC casting, forging, annealing and hot-rolling at 673K. Cubic specimens with all edges of 10mm were made from the plate as each edge is respectively parallel to rolling direction (RD), transverse direction (TD) or normal direction (ND).

Specimens were compressed in uniaxial or plane strain compression (PSC) until a strain of about 20% or 30% at ambient temperature with measurement of load and displacement. The two kinds of PSCs, ND-RD and ND-TD were conducted as compression in ND and elongation in RD and TD, respectively. The Teflon sheet was inserted for lubrication between the specimen and the upper and lower dies. Analyzed surface was mechanically and electrochemically polished before compression.

SEM/EBSD analysis was carried out on the same area of 1000μm×1000μm with a step size of 5 or 10 μm before and after compression.

3. Results and discussion

3.1. Load-displacement curves

Figure 1 displays load-displacement curves during compression at ambient temperature. The compressive load becomes higher in order of uniaxial, ND-RD and ND-TD as shown in the figure. PSC shows the higher flow stress and is subjected to larger strain energy because it is accompanied with higher constraint of deformation compared with uniaxial compression. Thus, ND-TD PSC is larger in strain energy than ND-RD.

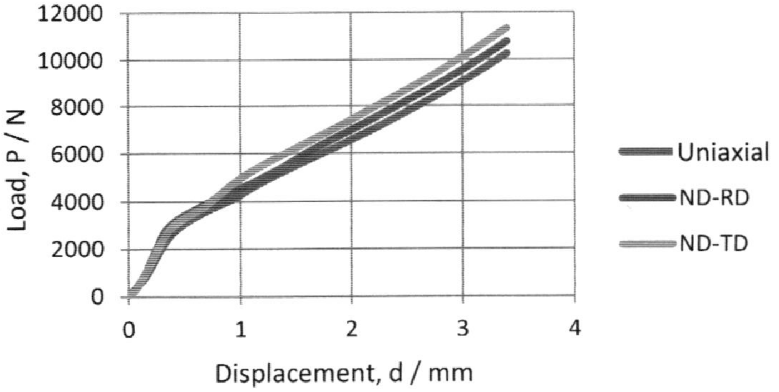

Fig. 1 Load-displacement curves during uniaxial and plane strain compression test.

3.2 Change in microstructure after compression

Changes in microstructure before and after uniaxial compression in ND until a strain of 20% and 30% are shown by inverse pole figure (IPF) maps in Fig. 2. For both specimens black-rectangular area in the map before compression corresponds to the analyzed area after compression. Comparison between the maps before and after compression reveals change in grain orientation and local deformation in grain interior. Grain rotation angles with compression were measured as 5.5 and 10.7 degrees in average for 20% and 30% compressed specimens, respectively. Moreover, a part of grains are found to approach a stable orientation ND//<101> in compression. An obvious grain subdivision in which a grain is divided into plural ones is observed in 30% compressed specimen.

Figure 3 shows IPF maps before and after PSC until a strain of 20% and 30% and kernel average misorientation (KAM) maps corresponding to stored strain of deformed materials. Remarkable changes in grain orientation or color over the IPF maps are observed in the PSCed specimens compared with uniaxially compressed ones. In the figures <111>//ND oriented (blue) grains change into <101>//ND oriented (green) while <001>//ND oriented (red) grains approach to <101>//ND via <103>//ND (orange) and <102>//ND (yellow). In ND-RD30 specimen a drastic grain subdivision, in which a <111>//ND grain was divided into three differently oriented grains with large change in orientation. As a result of orientation analysis, the grain subdivision likely occurred with active slips on different slip planes $(1\bar{1}\bar{1})$, $(\bar{1}\bar{1}1)$ and $(\bar{1}1\bar{1})$ for the [111]//ND grain.

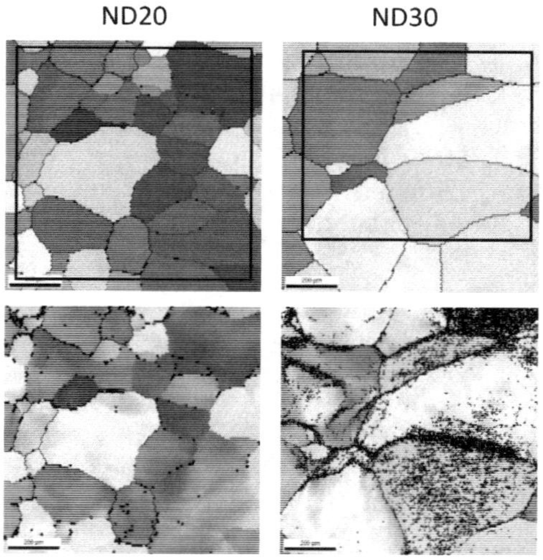

Fig. 2 Microstructures before and after uniaxial compression up to 20 and 30%. Top and bottom maps are before and after compression, respectively.

The KAM maps at the bottom of the figure indicate that stored strain concentrates near the grain boundaries in all of compressed specimens. It should be noted that grains with less strain exist. Near <001>//ND grains tend to gain less strain. This is probably related to orientations and misorientations of individual grain and neighboring ones. Mean KAMs were calculated to be 1.14, 1.33, 1.13 and 1.78 for ND30, ND-RD30, ND-TD20 and ND-TD30, respectively. This result is consistent with order of the applied load during compression test.

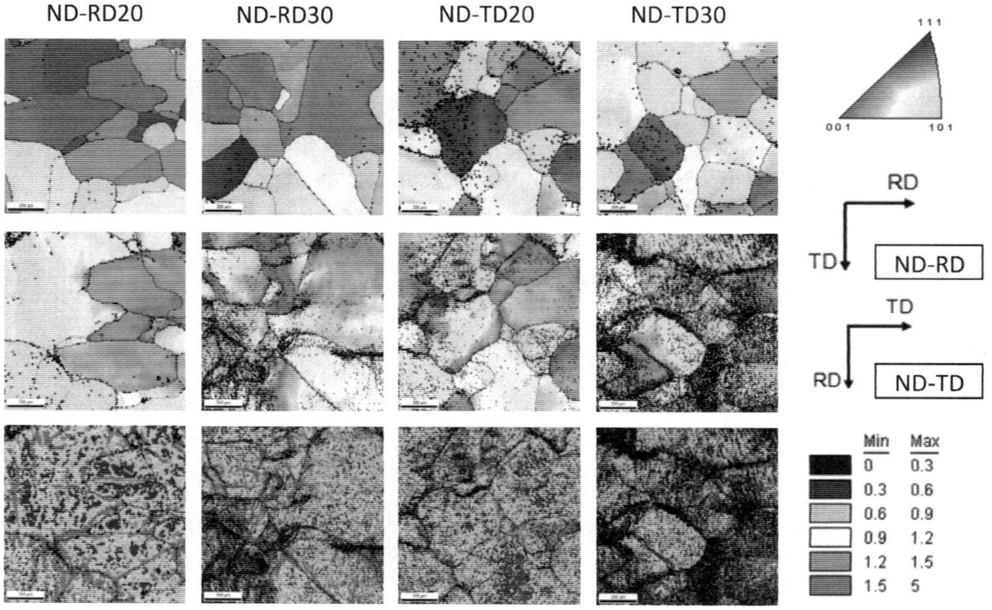

Fig. 3 Microstructures before and after PSC up to 20 and 30%. Top and middle maps are before and after compression, respectively. Bottom maps are kernel average misorientation (KAM) ones after PSC.

3.3 Relationship between KAM and Taylor factor

According to Taylor model [13], orientation before deformation should govern summation of shear strain. KAM after compression was investigated relating to TF before compression.

Figure 4 shows relationship between mean KAM after compression and Taylor factor before compression for the compressed specimens. Mean KAM was calculated for the specimens subjected to uniaxial compression and PSC up to the same strain level of 20% and 30%. TOTAL refers to mean KAM value for both strain level in each range of TF. As shown in the figure, KAM value decreases until a certain value of TF and then increases with increment of TF. TF for minimum KAM is a lower value of 2.75 at a strain level of 30% while it is 3.75 at 20%. TF dependence of KAM obtained here is good agreement with the previous result in Al-Mg-Mn alloy subjected to continuous cyclic bending [14].

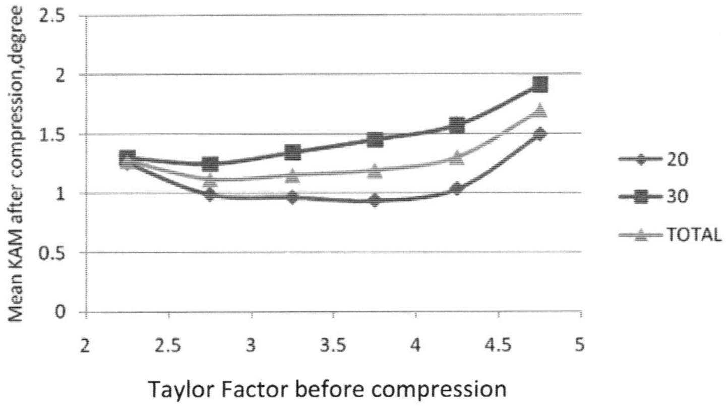

Fig. 4 Relationship between mean KAM after compression and Taylor factor before compression for specimens with the same strain level.

Summary

In order to investigate relationship between stored strain and crystallographic orientation, 99.99% purity aluminum cubes were compressed with uniaxial or with plane strain state up to a nominal strain of 30%. The aluminum cubes were examined on the same surface before and after compression by SEM/EBSD technique. After uniaxial compression a part of grains approached a stable orientation ND//<101>. Remarkable changes in grain orientation from <111>//ND or <001>//ND into ND//<101> were observed in the PSCed specimens compared with uniaxially compressed ones. Stored strain was estimated by Kernel Average Misorientation (KAM) derived from the EBSD data, and Taylor factor (TF) was measured before the compressive deformation. The analysis revealed that KAM value or the stored strain decreases until a certain value of TF and then increases with increment of TF.

References

[1] W.B. Hutchinson: Int. Metals Rev., 29(1984), 25-42.
[2] B.L. Averbach, M.B. Bever, M.F. Comerford & J.S. Leach: Acta Metall., 4 (1956), 477-484.
[3] H.U. Astron: Acta Metall., 3 (1955), 508-509.
[4] M. Matsuo, S. Hayami and S. Nagashima: Adv. X-ray Anal., 14 (1971), 214-230.
[5] H. Takechi, H. Katoh and S. Nagashima: Trans. Metall. AIME, 242 (1968), 56-65.
[6] R.L. Every and M. Hartherly: Texture, 1 (1974), 183-194.
[7] G. Mohamed and B. Bacroix: Acta Mater. 48 (2000) 3295-3302.
[8] G. Guiglionda, A. Borbely, J.H. Driver: Acta Mater., 52 (2004) 3413–3423.
[9] J.S. Kalland and Y.C. Huang: Metal Sci., 18 (1984), 381-385.
[10] N. Rajmohan, Y. Hayakawa, J.A. Szpunar & J.H. Root: Acta Mater., 45(1997), 2485-2494.
[11] N. Rajmohan and J.A. Szpunar: Mater. Sci. Tech., 15 (1999), 1259-1265.
[12] Ph. Gerber, J. Tarasiuk, Th. Chauveau and B. Bacroix: Acta Mater., 51 (2003), 6359-6371.
[13] G.I. Taylor: J. Inst. Met., 62 (1938), 307-324.
[14] Y. Takayama and J.A. Szpunar: Mater. Trans., 45 (2004), 2316-2325.

IOP Conf. Series: Materials Science and Engineering **82** (2015) 012030 doi:10.1088/1757-899X/82/1/012030

Study of deformation texture in an AZ31 magnesium alloy rolled at wide range of rolling speed and reductions

M. Sanjari[1,*], S. Tamimi[2], J. Su[1], A.S. Kabir[1], , K. Hara[3], H. Utsunomiya[3], R. Petrov[4], S.Yue[1] ,L. Kestens[4]

[1] Department of Materials Engineering, McGill University, Canada
[2] Center for Mechanical Technology and Automation, Department of Mechanical Engineering, University of Aveiro, Portugal
[3]Division of Materials and Manufacturing Science, Graduate School of Engineering, Japan
[4]Department of Materials Science and Engineering, Ghent University, Belgium

E-mail: mehdi.sanjari@mail.mcgill.ca

Abstract. The plasticity of Mg is restricted at low temperatures because: (a) only a small number of deformation mechanisms can be activated, and (b) a preferred crystallographic orientation (texture) develops in wrought alloys, especially in flat-rolled sheets. This causes problems in thin sheet processing as well as component manufacturing from the sheet. In this study, different rolling speeds from 15 to 1000 m/min were employed to warm-roll AZ31B magnesium alloy to different reductions. The results show that AZ31B sheets rolled at 15 m/min and 100 °C has fractured for reductions of more than 30% per pass. However, by increasing the rolling speed to 1000 m/min the rollability was improved significantly and the material can be rolled to reductions of more than 70% per pass. The results show that with increasing strain rate at 100°C, the splitting of basal poles was observed, indicating the activation of more contraction twins and secondary twins.

***Keywords*:** Magnesium alloys, Texture evolution, High speed rolling, VPSC

1. Introduction

Having the lowest density among all structural metals, magnesium has opened new horizons for developing commercial alloys with successful use in a wide variety of applications [1-2].
However, the plasticity of Mg is restricted at low temperatures because: (a) only a small number of deformation mechanisms can be activated [3-4], and (b) a preferred crystallographic orientation (texture) develops in wrought alloys, especially in flat-rolled sheets [5-7]. Therefore, manufacturing processes such as rolling and stamping should be performed at elevated temperatures [1, 8]. These barriers to the manufacturing process increase the price of magnesium wrought alloy products and limits the use of Mg to castings [9-10]. As a result, many studies have been conducted to improve formability by investigating the effect of manufacturing process.
Therefore the current sheet production techniques, based on DC casting and hot rolling, are basically slow because the demand is easily met [11]. Twin roll casting followed by hot rolling appears to be processing route which can fulfill high volumes and reduced costs.
The present authors succeeded in single-pass large draught rolling of various magnesium alloy sheets at low temperature(<473K) by high speed rolling [12]. Based on the data available in those works [13-17], the sheet obtained by high-speed rolling exhibited a fine-grained microstructure (mean grain size of 2-3 μm), with good mechanical properties. For these advantages, the high speed rolling is a promising process to produce high-quality rolled magnesium alloy sheets at a low cost.
For these advantages, the HSR is a promising process to produce high-quality rolled magnesium alloy sheets at a low cost. The goal of this research is thus to investigate the mechanisms responsible for the much higher rollability and the grain refinement after HSR. to do that, in this study, different rolling speeds from 15 to 1000 m/min were employed to twin rolled cast AZ31B magnesium alloy and different reductions.

2. Experimental procedure

The material used in the present study was commercial AZ31B (Mg-3.2%Al-0.77%Zn-0.34%Mn) rolled sheets with the thickness of 3.0mm. The rolling experiments were conducted through two different rolling mills. The high speed rolling of 1000m/min was performed in a high-speed, two-high mill with ϕ 530 mm rolls. A specimen was rolled at different reduction of 20-70% in a single pass and rolling temperature was varied from room temperature to 100 °C. The specimen was water-quenched immediately after rolling. The rolls were not heated. The low speed rolling of 15m/min was performed in a low-speed, two-high mill with ϕ 380 mm rolls. The specimens were processed in a single pass at the reduction of 30-60% and constant temperature of 100 °C. The rolls were neither heated nor lubricated.

The characterization has been done by optical microscopy , SEM with EBSD and XRD (Brucker X8) to observe the microstructures and analyze micro and macro texture of the sheets during rolling and annealing. The samples used for EBSD analysis were prepared by electro-polishing with a 10% Nital solution with a voltage of 20V at -20°C.

3. Results and discussion

3.1. Appearance of edge cracks

Fig. 1 shows the effect of rolling speed on the rollability of AZ31 alloys at different reductions and rolling temperature of 100 °C. As can be seen in this figure, the edge cracks can be classified into three main types:(a) minor edge cracks, (b) regular edge cracks, and (c) zigzag edge cracks. As can be seen in Fig.1, a heavy reduction of 70% with small zigzag edge cracking was attained for AZ31 in single pass by high-speed rolling at 100°C; however, at the lower speed of 15 m/min, cracks initiated at edges propagated to longitudinal fracture at the centre to form so-called "scissors cracks". The maximum reduction can be attained at low speed rolling is 30% although with "scissors cracks". Similar scissors cracks, although less intense, were observed in high speed rolling of 1000 m/min. By increasing the rolling speed to 1000 m/min and reduction of 70%, the sheets were produced without any edge cracks.

R %	15 m/min	100 m/min	1000 m/min
8%	◎	◎	◎
15%	◎	NA	◎
23%	○	○	○
30%	△	○	○
37%	X	○	○
51%		△	△
62%		△	△
71%		▲	▲

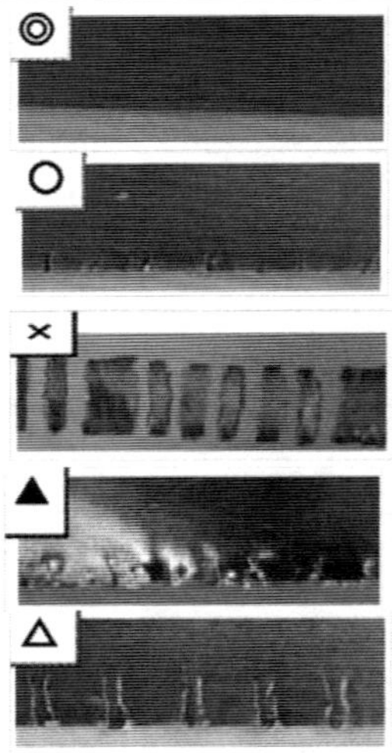

Figure 1 –Appearance of AZ31B specimen rolled at 100 °C at different rolling speed and reduction.

3.2. Texture evolution

To study the effect of rolling speed and reduction on the texture evolution, the pole figures of the samples deformed at 100°C and different rolling speed were analyzed (Fig. 3). According to these results, at all strains, the bulk texture of the deformed samples showed the typical basal texture of compressed Mg alloys, with the majority of c-axes of the grains aligned with the compression direction. At the early stage of deformation (reduction of 8% , ε=0.08), the rate of texture evolution is higher in the samples deformed with lower strain rate and the maximum intensity are 4.9 for the samples deformed at 15 m/min, in compare with the samples deformed at 1000 m/min with the maximum intensity of 4.2. By increasing the reduction to 16%, the maximum intensity increased to 4.8 and 4.5 for rolling speed of 15 and 1000 m/min; respectively. At rolling speed of 1000 m/min, by increasing the reduction from 23% to 30%, the maximum intensity does not change significantly and is slightly increased from 4.9 to 5.0. On the other hand in the case of 15 m/min, the maximum intensity increased from 5.8 to 6.5.

IOP Conf. Series: Materials Science and Engineering **82** (2015) 012030 doi:10.1088/1757-899X/82/1/012030

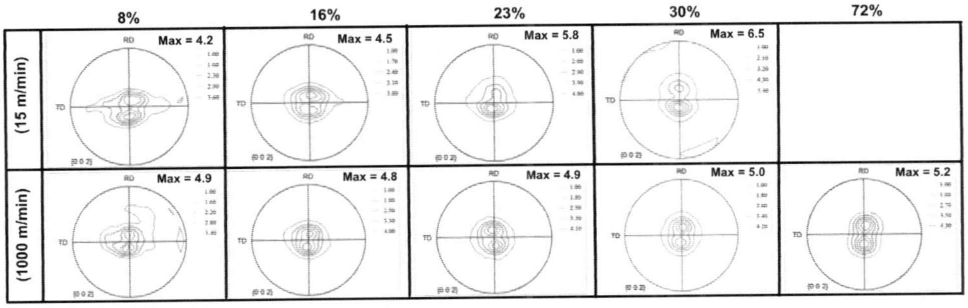

Figure 3 – {0002} pole figure of the finish rolled samples AZ31B specimen rolled at 100 °C at different rolling speed and reduction

It is worth noting that by forming C-twins the basal planes are reoriented by 56° around an <$2\overline{11}0$ > axis. This orientation change facilitates glide on the basal planes [18]. However; the activity of T-twinning is much higher than the C-twining particularly at the early stage of deformation [19]. It was reported that the T-twining would be activated in the grains tilted 45-90° away from the compression direction [20]. On the other hand, Yi et. al [20] reported that C-twins can be activated just in grains with c-axes is tilted less than 30° away from the normal direction (ND) of the sheet in their rolling experiments. However as shown in Figure5 (microstructure of the partially rolled and EBSD was performed of the entry to the exit of the deformed zone) by increasing the strain, the volume fraction of both type of C-twins increases especially in the samples deformed at high speed rolling of 1000 m/min. It has been hypothesized that potential mechanism for basal poles splitting toward the RD during the rolling can be promotion of S-twining [21] and/or <c+a> slip system [3]. The Critical Resolved Shear Stress (CRSS) of <c+a> slip system is very high at 100°C, leading to very low contribution of this type of slip system [22]. It was also reported that under dynamic conditions, the CRSS of non-basal slip systems remains higher than that of twinning [23].

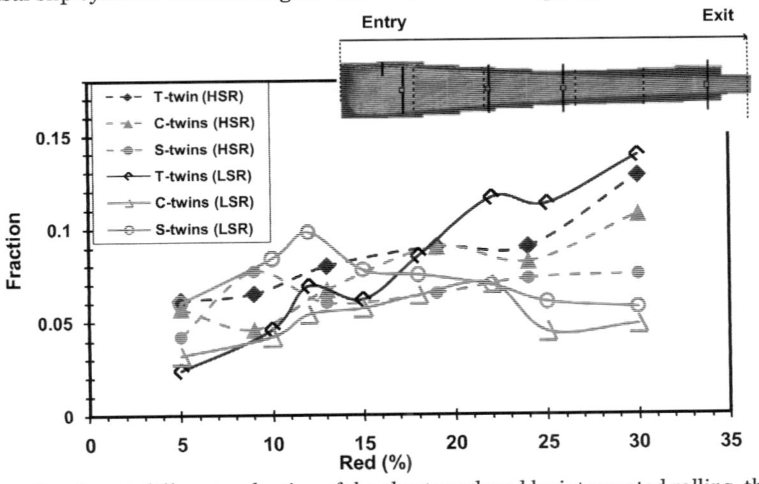

Figure 4 – The twin fraction at different reduction of the sheet produced by interrupted rolling, the microstructure of the partially rolled and EBSD was performed of the entry to the exit of the deformed zone.

3.3. Effect of rolling speed, annealing time and temperature on macro-texture

Texture evolution (the figure shown maximum intensity of basal texture at 200°C, 350°C and 500°C with increasing annealing time and (0002) plane pole figures at lowest value of basal texture and the longest annealing time at each temperature) of high speed rolled specimens during annealing are shown in Fig. 5. the basal texture of as rolled specimen at high speed is lower than that of low speed rolled one. And this tendency kept all the way along the annealing time at both 200°C and 350°C. Therefore the texture is more sensitive to annealing temperature, but less sensitive to rolling speed prior to annealing.

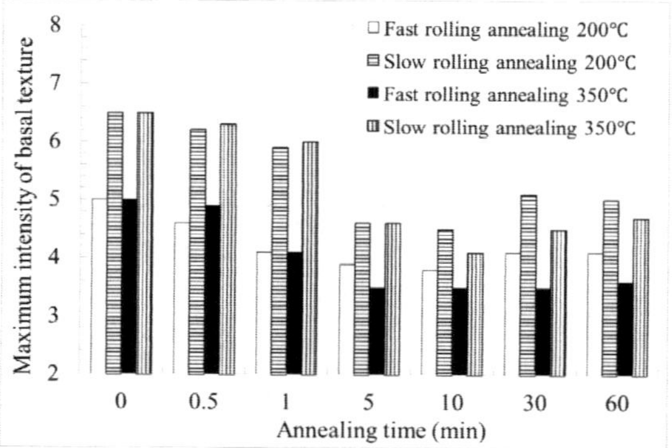

Fig. 5 The comparison of the maximum intensity of basal texture of high and low speed rolled specimens during annealing at temperatures of 200°C and 350°C and time range from 30s to 1h

4. Conclusion

AZ31B sheets are rolled in a wide range of rolling speed and reduction. The remarks obtained are listed as follows:

1. By increasing the rolling speed rollability of the samples are increasing.
2. The morphologies of edge cracks are classified into (a) minor edge cracks, (b) regular edge cracks and (c) zigzag edge cracks.
3. At the rolling speed of 1000 m/min and 1000 m/min, by increasing the rolling reduction tor 60%and higher, recrystallized grains can be observed in the microstructure.
4. Texture weakening occurred during annealing and reached a steady state at 200°C and 350°C while abnormal grain growth happened at 500°C which sharply strengthened the texture. With increasing annealing temperature, the maximum intensity of basal texture decreased.

5. References

[1] H.B. M. Avedesian, ASM specialty handbook. Magnesium and magnesium alloys. , ASM International, Materials Park, Ohio, 2000.
[2] K.U. Kainer, F. Kaiser, Magnesium alloys and technology, Wiley Online Library, 2003.
[3] S.R. Agnew, M.H. Yoo, C.N. Tomé, Acta Materialia, 49 (2001) 4277-4289.
[4] M.R. Barnett, Metallurgical and Materials Transactions A: Physical Metallurgy and Materials Science, 34 A (2003) 1799-1806.
[5] M.R. Barnett, M.D. Nave, C.J. Bettles, Materials Science and Engineering A, 386 (2004) 205-211.
[6] A. Styczynski, C. Hartig, J. Bohlen, D. Letzig, Scripta Materialia, 50 (2004) 943-947.
[7] L.W.F. Mackenzie, M. Pekguleryuz, Materials Science and Engineering: A, 480 (2008) 189-197.
[8] E. Doege, K. Dröder, Journal of Materials Processing Technology, 115 (2001) 14-19.
[9] Y. Nakaura, A. Watanabe, K. Ohori, Materials Transactions, 47 (2006) 1031-1039.
[10] A. Jäger, P. Lukáč, V. Gärtnerová, J. Haloda, M. Dopita, Materials Science and Engineering: A, 432 (2006) 20-25.
[11] M.M. Avedesian, H. Baker, ASM international, 274 (1999).
[12] T. Sakai, Y. Watanabe, H. Utsunomiya, in, 2009, pp. 483-486.
[13] G. Hamada, T. Sakai, H. Utsunomiya, in, 2010, pp. 227-231.
[14] T. Sakai, A. Hashimoto, G. Hamada, H. Utsunomiya, in, 2011, pp. 369-372.
[15] T. Sakai, Y. Watanabe, H. Utsunomiya, Materials Science Forum 618-619(PART 4), (2009) 483-486.
[16] M. Sanjari, S.A. Farzadfar, H. Utsunomiya, T. Sakai, E. Essadiqi, S. Yue, Materials Science and Technology (United Kingdom), 28 (2012) 928-933.
[17] H. Utsunomiya, T. Sakai, S. Minamiguchi, H. Koh, in, San Antonio, TX, 2006, pp. 201-204.
[18] L. Jiang, J.J. Jonas, A.A. Luo, A.K. Sachdev, S. Godet, Scripta Materialia, 54 (2006) 771-775.
[19] M. Sanjari, S.A. Farzadfar, I.H. Jung, E. Essadiqi, S. Yue, Materials Science and Technology, (2011).
[20] S. Yi, I. Schestakow, S. Zaefferer, Materials Science and Engineering: A, 516 (2009) 58-64.
[21] S.L. Couling, F. Pashak, L. Sturkey, Trans. ASM, 51 (1959) 94-107.
[22] M. Shamsi, M. Sanjari, A.Z. Hanzaki, Materials Science and Technology, 25 (2009) 1039-1045.
[23] I. Ulacia, N.V. Dudamell, F. Gálvez, S. Yi, M.T. Pérez-Prado, I. Hurtado, Acta Materialia, 58 (2010) 2988-2998.

IOP Conf. Series: Materials Science and Engineering **82** (2015) 012031 doi:10.1088/1757-899X/82/1/012031

Role of stacking fault energy on texture evolution revisited

R Madhavan[1], R Kalsar[1], RK Ray[2] and S Suwas[1]
[1] Department of Materials Engineering, Indian Institute of Science, Bangalore 560012, India
[2] Formerly Professor, Department of Metallurgical and Materials Engineering, Indian Institute of Technology, Kanpur 208016, India
E-mail: satyamsuwas@materials.iisc.ernet.in

Abstract. Three materials, pure aluminium, Al-4 wt.% Mg, α-brass have been chosen to understand the evolution of texture and microstructure during rolling. Pure Al develops a strong copper-type rolling texture and the deformation is entirely slip dominated. In Al-4Mg alloy, texture is copper-type throughout the deformation. The advent of Cu-type shear bands in the later stages of deformation has a negligible effect on the final texture. α-brass shows a characteristic brass-type texture from the early stages of rolling. Extensive twinning in the intermediate stages of deformation ($\varepsilon_t \sim 0.5$) causes significant texture reorientation towards α-fiber. Beyond 40% reduction, deformation is dominated by Bs-type shear bands, and the banding coincides with the evolution of <111>||ND components. The crystallites within the bands preferentially show <110>||ND components. The absence of the Cu component throughout the deformation process indicates that, for the evolution of brass-type texture, the presence of Cu component is not a necessary condition. The final rolling texture is a synergistic effect of deformation twinning and shear banding.

1. Introduction

The transition of deformation texture from copper-type to brass-type as a function of stacking fault energy (SFE) in fcc metals and alloys is a subject of investigation spanning over five decades [1]. Material with high/medium SFE develops a copper-type texture on rolling, while a brass-type texture is common in low-SFE material. Typically, a copper-type texture comprises equal fractions of copper (Cu) {112}<111>, brass (Bs) {110}<112> and S {123}<634> orientations, whereas a brass-type texture primarily consists of Bs and Goss {110}<001> components with negligible Cu fraction. It was earlier proposed that this texture change is attributed to the extent of cross slip, where easy cross slip (high SFE) leads to copper-type texture and otherwise to brass-type texture [2]. However, it was later suggested that the evolution of the brass-type texture in low SFE materials is related to deformation twinning [3]. Many experimental outcomes have indicated that the rolling texture in a low-SFE fcc material is similar to that of a high-SFE material up to certain deformation levels (~40%) and the deviation from copper-type to brass-type occurs only at the later stages of deformation [4,5]. It is argued that this transition is due to selective twinning in Cu-oriented grains. However, Leffers and co-workers [6,7] have noted clear differences between the two types of rolling textures from the early deformation stages. Further, Leffers [8] argued that twinning does not play a direct role in the evolution of the brass-type texture, but has an indirect effect. The dislocation slip subsequent to twinning is greatly hindered since the twin boundaries could act as obstacle for dislocation motion and hence slip would be confined to planes parallel to the twin habit plane.

It has been observed experimentally that the propensity for twinning ceases at intermediate levels of deformation and is followed by subsequent formation of shear bands at higher reductions. Shear bands can be either Cu-type (which are common in high- and medium-SFE materials and form as a result of increased obstruction to dislocation motion due to preexisting dislocation cell walls and micro-bands) [9] or Bs-type (common in low-SFE materials and are observed to form when the preceding obstacles are twin lamellae) [10]. In high- and medium-SFE materials, activity of shear bands has been observed only at very large deformation, when the texture is predominantly Cu-type. In Cu-Mn alloys, Engler [11] found that the volume of shear-banded regions in the microstructure increased with deformation and these microscopic shear bands were selectively found in grains having Cu orientation. In Al-Cu and Al-Mg alloys [12,13], where the formation of Cu-type shear bands is extensive at the later stages of deformation, it can be seen that the orientations of sub-grains within shear bands have hardly any effect on the final Cu-type rolling texture. On the other hand, in low-SFE material, subsequent to twinning, a continuous slip preferentially on twin habit planes rotates the {111} crystal plane parallel to the rolling plane [14]. This makes homogeneous deformation difficult and results in shear banding. The concurrence of the evolution of <111>||ND components and the origination of shear bands has been observed in many low-SFE materials [15,16]. These Bs-type shear bands are observed to form preferentially in grains with <111>||ND orientation [17]. In Cu-30Zn alloy, Duggan et al. [16] observed that the volume fraction of shear bands is large and has a significant role in the final texture. It is to be noted that in all the above studies, irrespective of the type of shear bands and SFE values, it has been reported that the crystallites within the bands have orientations close to <110>||ND.

In a recent work by the authors, the transition of texture from copper-type to brass-type has been observed at very large strain in a medium-SFE Ni-40Co alloy [18]. This texture deviation is attributed to the extensive activity of Cu-type shear bands preferentially in Cu-oriented grains. However, in low-SFE Ni-60Co alloy, the formation of brass-type texture initiates at the early stages of deformation, and it is observed that deformation twinning and Bs-type shear banding have a synergistic effect on the final texture [19]. The present study is aimed to understand the general trend in the evolution of microstructure and crystallographic texture as function of SFE in a range of fcc materials exhibiting different SFE values. A representative material from each of high-SFE (pure aluminum), medium-SFE (Al–4Mg) and low-SFE (Cu–30Zn) regimes has been chosen and its texture-microstructure correlation has been presented.

2. Experimental

The cast ingots of ultra pure Al (99.9999%), Al-4 wt.% Mg and Cu-30 wt.% Zn (α-brass) were obtained by vacuum melting. In order to weaken the initial texture, the ingots were subjected to 50% cross rolling followed by annealing at $0.5T_m$ for 6 h. The average grain size of the starting material is about 150 μm. The initial material has been cold rolled to 98% thickness reduction (equivalent true strain – 3.99). Textures for the initial and the cold rolled samples have been measured using Bruker D8 Discover texture goniometer with Cu K_α radiation. Four incomplete pole figures ($\alpha = 0 - 75°$, $\Delta\alpha = 5°$) of 111, 200, 220, 311 planes have been measured on the normal plane mid-thickness section. The Orientation Distribution Function (ODF) has been calculated from the measured pole figures using MTex software [20], which is available as MATLAB toolbox. Orthotropic sample symmetry has been imposed for the texture of rolled samples. The microstructural characterization for the rolled samples was done using a scanning electron microscope attached with an electron back-scattered diffraction (EBSD) camera. All the EBSD scans were taken on the transverse plane of the rolled sample.

3. Results

3.1 Evolution of texture

The $\varphi_2 = 0°, 45°, 65°$ sections of the ODF are shown for the 70%, 95% and 98% cold rolled conditions in Fig. 1. The presence of a strong <110>||ND texture is evident in the $\varphi_2 = 0°$ section in all materials at higher strain levels. For pure Al and Al-4Mg, the maximum intensity (in terms of f(g)) is seen near to Bs ($\varphi_1 = 35°$) locations (Figs. 1a, b). On the other hand, in α-brass, no unique intensity maximum is seen (Fig. 1c). Apart from the <110>||ND components, the simultaneous evolution of Cu and S components can be observed in the φ_2 = 45° and 65° sections, respectively, in pure Al and Al-Mg alloy. On the contrary, in α-brass, the evolution of brass-type texture is associated with a complete absence of Cu-component in the φ_2 = 45° section.

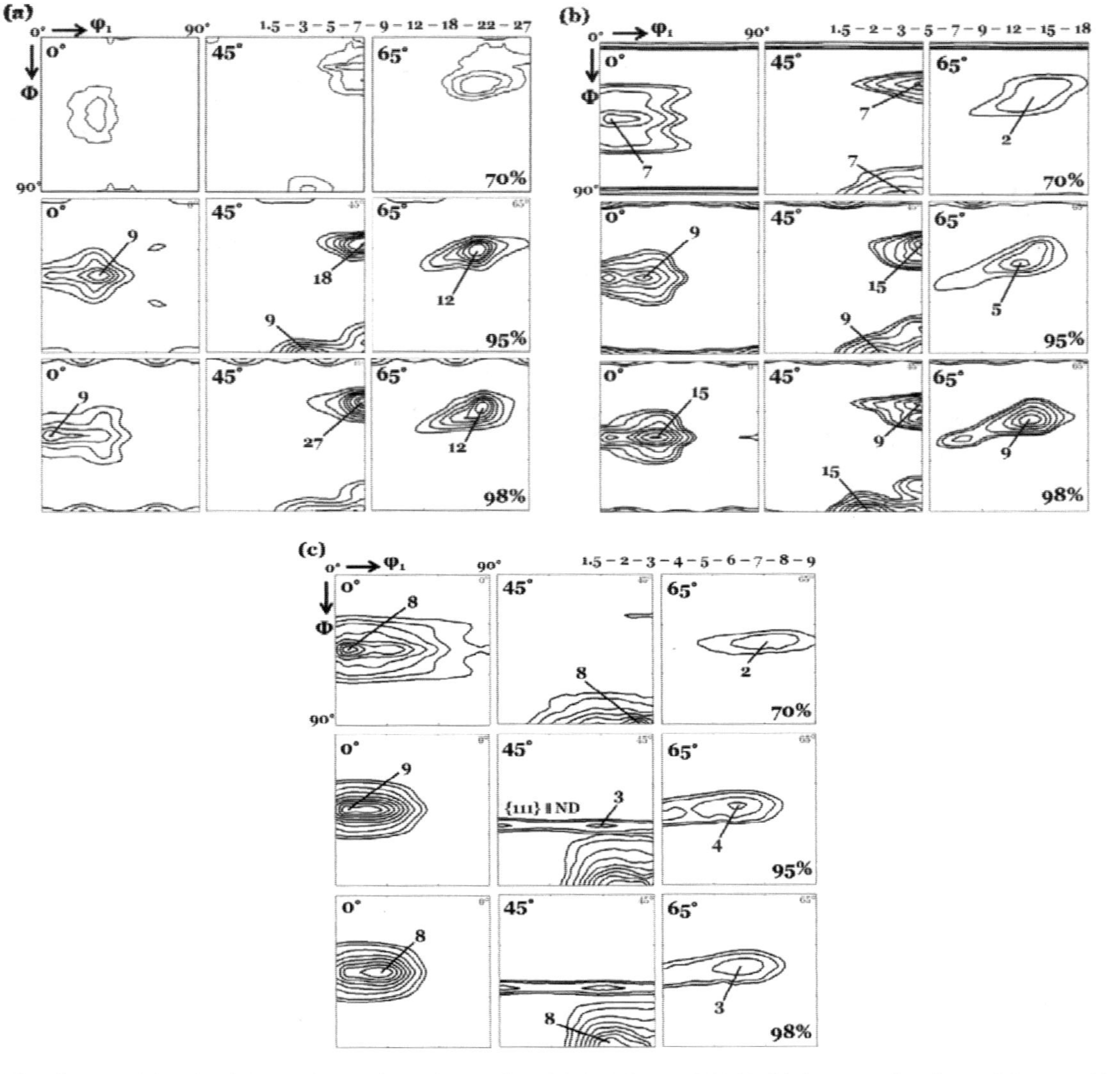

Fig. 1: φ_2 = 0°, 45°, 65° sections of the ODF after 70%, 95% and 98% thickness reductions: (a) pure Al, (b) Al-4Mg, (c) Cu-30Zn.

Unlike in Al and Al-4Mg alloy, the φ_2 = 45° section of the ODF corresponding to 70% reduction in α-brass shows a sudden emergence of <111>||ND components. However, the intensity of these components decreases with increasing deformation levels. It should be noted that in Al-4Mg alloy the intensity of Cu-component decreases beyond 95% reduction, which coincides with the corresponding rise in Bs-component.

3.2 Evolution of deformation microstructures

Fig. 2a shows the EBSD generated micrograph of pure Al subjected to 95% thickness reduction. The deformed microstructure is homogeneous and shows a combination of elongated and equiaxed grains. Within the elongated grains, prominent orientation gradients are seen which indicates the presence of intra-granular misorientation. In orientation imaging maps (OIM), these gradients are manifested as color variation within grains. The equiaxed grains are predominantly located adjacent to the prior deformed grains. The presence of a high fraction of high angle grain boundaries (HAGB) indicates the occurrence of restoration mechanisms at high

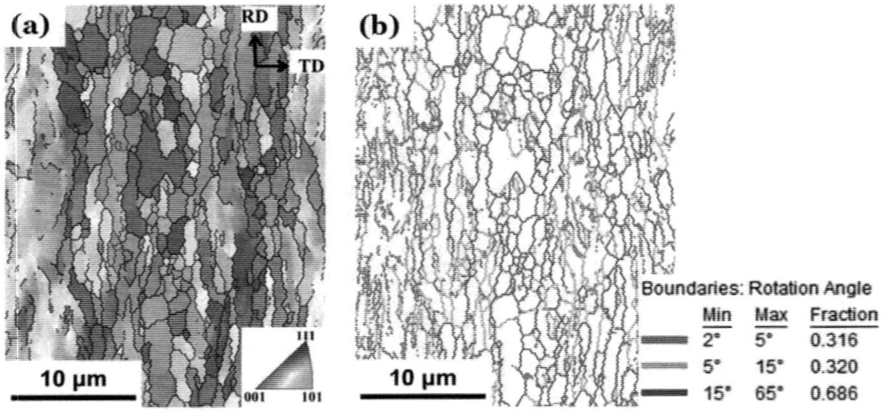

Fig. 2: EBSD generated deformed microstructure of pure Al after 95% thickness reduction: (a) OIM map, (b) grain boundary map.

reductions (Fig. 2b). The deformed micrographs of Al-4Mg alloy subjected to 70% and 90% rolling reductions are given in Fig. 3. The microstructure is homogeneous showing an extensive orientation gradient at 70% reduction (Fig. 3a), whereas after 90% rolling, the onset of shear banding can be observed. The bands are visible as dark patches in the image quality map and are found preferentially in Cu oriented grains (Fig. 3b). These parallel and widely spaced banded features are similar to the ones observed earlier in an Al-Mg alloy by Hurley and Humphreys [21]. They are micro-shear bands (also referred to as Cu-type shear bands) since they are mostly restricted within a grain, and have originated as a result of intense dislocation activity. The dislocation boundaries on either side of the bands show a slight curvature near the interface indicating that the bands cut through the prior sub-grain cells. The area fraction of these bands increases with deformation. In the case of α-brass, the microstructure of 40% rolled condition shows extensive deformation twinning (Fig. 4a). Grains show multiple layers of thin twin lamellae that extend across the grain interior. However, beyond 40% reduction the propensity of twinning almost ceases and band-like feature start to emerge. Similar to Cu-type shear bands, these bands are microscopic, and cut through the pre-existing twin lamellae. However, unlike Cu-type shear bands, these Bs-type shear bands in α-brass are found preferentially within <111>||ND oriented grains (Fig. 4b). The area fraction of Bs-type bands increases with deformation.

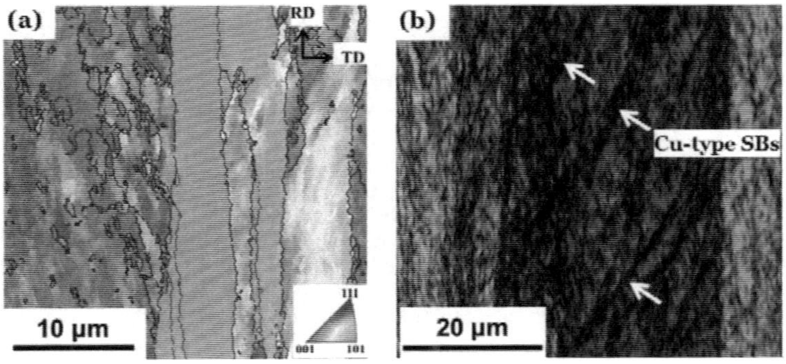

Fig. 3: EBSD generated OIM maps of deformed Al-4Mg alloy after (a) 70% rolled, (b) 90% rolled. Arrows indicate Cu-type shear bands.

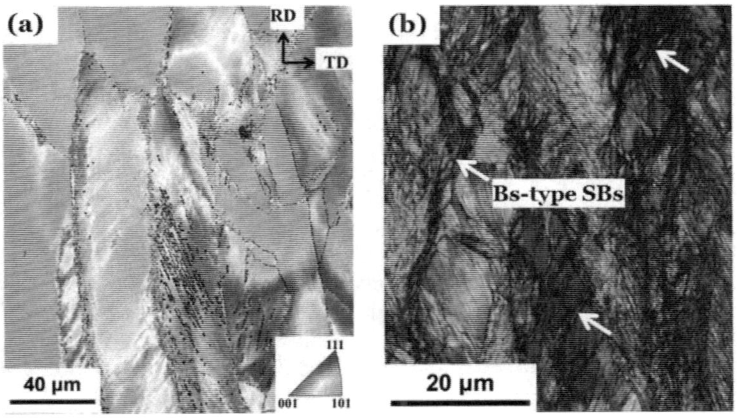

Fig. 4: EBSD generated micrographs of deformed Cu-30Zn alloy. (a) OIM map for 40% rolled, (b) superimposed OIM and image quality maps for 70% rolled. Arrows indicate Bs-type shear bands.

4. Discussion

The clear transition of rolling texture from copper-type to brass-type is evident with decreasing SFE from pure Al to α-brass. The pure Al shows distinct intensity peaks corresponding to Cu, S, and Bs components in the ODF sections. On the other hand, in α-brass, the initial weak texture completely gets transformed to a weak α-fiber texture. It is, therefore, evident that the characteristic evolution of Bs-type texture gets initiated at an early stage of deformation [6]. The Bs-component, which generally constitutes a major fraction of the final texture of low SFE materials, is not seen here as the stable end orientation at large deformation and in turn the intensity maximum is spread throughout the φ_1 values in the $\varphi_2 = 0°$ section (Fig. 1c). This observation is quite contrary to the early reports on the deformation texture of α-brass [5], where a distinct intensity maximum ($f(g)_{max}$) has been observed at Bs location. Nevertheless, this trend is in accordance with our earlier work on Ni-60Co alloy [19], where the $f(g)_{max}$ is not found close to Bs position. Earlier studies [4,5] have indicated that the formation of the Cu-component in the intermediate stages of deformation is crucial in the formation of the final brass-type texture, since this orientation would undergo preferential twinning and subsequent slip to reorient towards the Bs-component. However, in the present investigation, the fraction of Cu-component is insignificant in the initial material, and no sign of evolution of the Cu-component is evident at any stage of deformation (Fig. 1c). On the other hand, in the medium-SFE Al-4Mg alloy, the texture at a later stage of deformation shows a subtle variation in the intensities of ideal rolling texture components, viz. a decrease in the f(g) of Cu component and a significant rise in the fraction of Bs component.

4.1 Role of SFE in the evolution of microstructure

4.1.1 Pure aluminium

The development of the Cu-type texture in FCC metals is well understood and is generally attributed to cross slip [2]. The microstructures of deformed pure Al indicate the presence of orientation gradients within grains (Fig. 2). These orientation gradients are attributed to predominant dislocation activity as no other deformation mechanism is evident from the microstructures of the rolled materials. The intra-grain misorientation is a result of variation in the slip activity in different parts of the grain, resulting in localized plastic rotation. The sub-division of a grain into smaller domains (sub-grains) subsequent to local rotations is generally accompanied by the formation of low angle boundaries. The orientation spread from the mean orientation of the grain is very high near the triple junctions to account for grain boundary compatibility. However, the condition of boundary compatibilities decreases with increasing deformation due to thinning and elongation of grains. The increase in deformation causes the grains with unstable orientation to converge towards the orientations (like the Cu, S, Bs) which possesses a minimal divergence from its mean position [22].

4.1.2 Al-4Mg alloy

The microstructure and texture development in this alloy is similar to that of other medium SFE materials, viz. pure copper [23]. The deformed microstructure is homogeneous and shows grain fragmentation up to 70% thickness reduction indicating a predominantly dislocation slip based deformation process. The emergence of shear bands in already existing elongated sub-grains is an important observation in the microstructures of large deformation levels. In the present case, these shear bands are confined to single grains, due to the orientation dependence of Cu-type shear bands. Morii et al. [24] have proposed that layered obstacles like lamellar dislocation walls (or microbands) are essential for the initiation of Cu-type shear bands. Under these circumstances, it is relevant to examine the effect of solute atoms on the propensity of shear banding. Earlier studies on Al-Mg and Al-Cu alloys have shown the presence of shear banding at large strains, and the occurrence of shear bands has been attributed to the hindrance of dynamic recovery processes due to the presence of solute atoms (having large atomic misfit) resulting in severe work hardening [12,13]. In a previous work [18], the effect of the evolution of <110>||ND sub-grains within Cu-type shear bands on the transition of texture has been highlighted. However, in the present case, the influence of Cu-type shear banding and sub-grain evolution within the bands on the copper-type texture is minimal and is least visible in the bulk texture results.

4.1.3 α-brass

(i) Role of twinning

The deformation microstructures indicate a gradual transition in the active modes of deformation i.e., twinning during the initial stages of deformation to shear banding in the later stages. Since the strain accommodation by FCC deformation twins is large (0.707), twin thicknesses are small (20 – 30 nm). Hence, single twin lamellae in EBSD generated OIM micrographs may consist of a bundle of much finer twins within itself (Fig. 4a). The density of twins within the grains increases with deformation. It can be observed that twins within a grain are parallel to each other, which implies that only one twin variant is active. The formation of twins sub-divides the grains into many fragments which, in effect, would reduce the grain size. The grain size is an important parameter in the nucleation of deformation twins, since it influences the nucleation stress. In the present case, the tendency for twinning got saturated beyond ~40% reduction because of reduced grain size (due to grain fragmentation) and hence higher stress required for further nucleation. In a previous work, Madhavan et al. [19] observed, by a pseudo-in-situ method, a gradual reorientation of twin volumes towards <110>||ND orientation with increasing deformation, indicating subsequent slip within twin regions.

(ii) Role of shear bands

In the present case, the shear bands are first observed after 50% reduction (ε_t ~ 0.65). The fraction of matrix volume that undergoes banding increases with deformation. The appearance of <111>||ND components during the later stages is a crucial factor for the origin of shear bands in the microstructure. The formation of these {111} components can be explained as follows [25]: subsequent to deformation twinning, dislocation slip is largely confined to slip planes that are parallel to the twin habit planes. Slip will remain active till the twin plane (slip plane) rotates towards the rolling plane of the sample, beyond which the resolved stress acting on the slip plane is very low. The rotation of twin lamellae parallel to the rolling plane suppresses homogeneous crystallographic slip making the microstructure plastically unstable. The subsequent evolution of shear bands is the outcome of this plastic instability [16]. Fig. 4b depicts the situation where banding is extensive in one of the <111>||ND oriented grains and is cutting through pre-existing twins. The orientation of sub-grains within the shear band is of particular interest since it could play an influential role in deciding the final rolling texture. A careful examination of regions within the shear bands indicates that the distribution of crystallite orientations is not random, but preferentially towards <110>||ND.

5. Conclusions

An understanding of the evolution of texture and microstructure in pure Al, Al-4Mg and α-brass with stacking fault energies ranging from high to low has been developed. The key observations are as follows:

1. Pure aluminium develops a characteristic copper-type rolling texture at large rolling reductions comprising high fractions of S, Bs and Cu components. The evolution of texture in this material is primarily attributed to dislocation slip as evidenced by intense dislocation activity throughout the deformation.

2. In Al-4 wt.% Mg alloy, the overall texture is copper-type up to 98% reduction (ε_t = 3.9). The deformation mechanism up to a thickness reduction of ~70% is entirely slip-dominated. Beyond this deformation level, the initiation of Cu-type shear bands is observed. These shear bands are mostly confined to {112}<111> oriented grains. The effect of shear bands and the subsequent sub-grain evolution within the bands are least visible in the macro-texture of this alloy.

3. α-brass develops a characteristic brass-type rolling texture and the development of this texture starts in the early stages of deformation (ε_t ~ 0.5). The volume fraction of the Cu {112}<111> component is insignificant during the entire process of deformation. A clear transition in the deformation mechanisms from extensive deformation twinning in the intermediate stages (ε_t < 0.5) to shear banding in the later stages is observed. The major orientation changes towards the α-fiber, in the intermediate stages of deformation, take place by extensive twinning and slip within the twin volumes. In the later stages of deformation, shear bands originate due to plastic instability caused by the emergence of <111>||ND components in the microstructure. A non-random distribution of orientations close to <110>||ND orientation is found inside the shear bands.

References

[1] T. Leffers and R.K. Ray 2009 *Prog. Mater. Sci.* **54** 351-396.
[2] R.E. Smallman and D. Green 1964 *Acta Metall.* **12** 145-154.
[3] W. Heye and G. Wassermann 1966 *phys. stat. solidi (b)* **18** 107-111.
[4] J.S. Kallend and G.J. Davies 1972 *Texture* **1** 51-69.
[5] J. Hirsch and K. Lücke 1988 *Acta Metall.* **36** 2863-2882.
[6] T. Leffers and D.J. Jensen 1988 *Text. Microstruct.* **8** 467-480.
[7] T. Leffers 1993 *Text. Microstruct.* **22** 53-58.
[8] T. Leffers 1996 *Proc. ICOTOM 11* **1** 299-306.
[9] P. Wagner, O. Engler and K. Lücke 1995 *Acta Metall. Mater.* **43** 3799-3812.
[10] K. Morii, H. Mecking and Y. Nakayama 1985 *Acta Metall.* **33** 379-386.
[11] O. Engler 2000 *Acta Mater.* **48** 4827-4840.
[12] O. Engler, J. Hirsch and K. Lücke 1989 *Acta Metall.* **37** 2743-2753.
[13] P. Wagner, O. Engler and K. Lücke 1991 *Text. Microstruct.* **14-18** 927-932.
[14] T.C. Kamijo and K. Sekine 1970 *Metall. Mater. Trans. B* **1** 1287-1292.
[15] P.T. Wakefield, A.S. Malin and M. Hatherly 1977 *J. Aust. Inst. Met.* **22** 143-151.
[16] B.J. Duggan, M. Hatherly, W.B. Hutchinson and P.T. Wakefield 1978 *Met. Sci.* **12** 343-351.
[17] W.B. Hutchinson, B.J. Duggan and M. Hatherly 1979 *Met. Technol.* **6** 398-403.
[18] R. Madhavan, R.K. Ray and S. Suwas 2014 *Acta Mater.* **74** 151-164.
[19] R. Madhavan, R.K. Ray and S. Suwas 2014 *Acta Mater.* **78** 222-235.
[20] H. Schaeben, R. Hielscher and F. Bachmann 2010 *Solid State Phenom.* **160** 63-68.
[21] P.J. Hurley and F.J. Humphreys 2003 *Acta Mater.* **51** 1087-1102.
[22] Y. Zhou, L.S. Tóth and K.W. Neale 1992 *Acta Metall. Mater.* **40** 3179-3193.
[23] J. Hirsch, K. Lücke and M. Hatherly 1988 *Acta Metall.* **36** 2905-2927.
[24] K. Morii and Y. Nakayama 1985 *Scr. Metall.***19** 185-188.
[25] T. Leffers 1978 *ICOTOM 5* **1** 277-287.

IOP Conf. Series: Materials Science and Engineering **82** (2015) 012032 doi:10.1088/1757-899X/82/1/012032

Texture studies of hot compressed near alpha titanium alloy (IMI 834) at 1000°C with different strain rates

BK Kodli[1], KK Saxena[2], SR Dey[1], V Pancholi[2] and A Bhattacharjee[3]
[1]Department of Materials Science and Metallurgical Engineering, Indian Institute of Technology Hyderabad, Yeddumailaram-502205, Telangana, India
[2]Department of Metallurgical and Materials Engineering, Indian Institute of Technology Roorkee, Roorkee-247667, Uttarakhand, India
[3]Titanium Alloy Group, Defence Metallurgical Research Laboratory (DMRL), Kanchanbagh-500058, Telangana, India
E-mail: suhash@iith.ac.in

Abstract. IMI 834 Titanium alloy is a near alpha (hcp) titanium alloy used for high temperature applications with the service temperature up to 600^{0}C. Generally, this alloy is widely used in gas turbine engine applications such as low pressure compressor discs. For these applications, good fatigue and creep properties are required, which have been noticed better in a bimodal microstructure, containing 15-20% volume fraction of primary alpha grains (α_p) and remaining bcc beta (β) grains transformed secondary alpha laths (α_s). The bimodal microstructure is achieved during processing of IMI 834 in the high temperature α+β region. The major issue of bimodal IMI 834 during utilization is its poor dwell fatigue life time caused by textured macrozones. Textured macrozone is the spatial accumulation of similar oriented grains in the microstructure generated during hot processing in the high temperature α+β region. Textured macrozone can be mitigated by controlling the hot deformation with certain strain rate under stable plastic conditions having β grains undergoing dynamic recrystallization. Hence, a comprehensive study is required to understand the deformation behavior of α and β grains at different strain rates in that region. Hot compression tests up to 50% strain of the samples are performed with five different strain rates i.e. 10^{-3} s^{-1}, 10^{-2} s^{-1}, 10^{-1} s^{-1}, 1 s^{-1} and 10 s^{-1} at 1000^{0}C using Gleeble 3800. The resultant bimodal microstructure and the texture studies of primary alpha grains (α_p) and secondary alpha laths (α_s) are carried out using scanning electron microscopy (SEM)-electron back scattered diffraction (EBSD) method.

1. Introduction

IMI 834 Ti alloy with bimodal microstructure is suggested having better creep and fatigue properties [1] for service temperature up to 600^{0}C. Bimodal microstructure consists of equiaxed primary alpha (α_p) grains and secondary alpha laths (α_s) which are transformed from high temperature β grains upon cooling. Such a microstructure is only possible with an appropriate thermomechanical treatment in the α+β field followed by cooling. The final properties of the products are mainly dependent on the microstructure and texture evolution during plastic deformation at higher temperature. But there is limited literature available on the crystallographic texture evolution during hot deformation.

Therefore the present study concentrates on the evolution of microstructure and texture during hot compression tests of IMI 834 near alpha titanium alloy at 1000^{0}C at five different strain rates such as 10^{-3} s^{-1}, 10^{-2} s^{-1}, 10^{-1} s^{-1}, 1 s^{-1} and 10 s^{-1}. The microtexture have been investigated by using electron backscattered diffraction (EBSD) in a field emission scanning electron microscope. EBSD micro texture analysis through studies of orientations and misorientations of grain boundaries is a more powerful method to investigate the microstructure evolution of deformed materials [2] and to understand the deformation mechanisms [3-4]. The aim of the present study is to separate the primary alpha grains (α_p) from secondary alpha (α_s) laths in the orientation map for analyzing the deformation texture in primary alpha grains by using TSL software.

2. Experimental details

IMI 834 Ti alloy, was received as pancake (250 mm diameter, 12 mm height) from DMRL, Hyderabad. The pancake was cut into cylinders of 10 mm diameter and 12 mm height to maintain an aspect ratio of 1.2 by electrodischarge machining. These cylinders are chamfered to avoid folding during initial stage of testing. Hot compression tests were performed to in a Gleeble 3800 testing maschine. Thermo mechanical simulator at IIT Roorkee at a deformation temperature of 1000^{0}C with different strain rates such as 10^{-3} s^{-1}, 10^{-2} s^{-1}, 10^{-1} s^{-1}, 1 s^{-1} and 10 s^{-1} with their flow stress-strain captured during the test. Heating rate is maintained at 5^{0}C/s with further holding time of 5 min at the deformation temperature of 1000^{0}C during hot compression tests. After reaching 50% reduction in height of the specimens, the deformed samples are rapidly quenched by water spraying facility in Gleeble 3800. For microstructure characterization, the hot compressed specimens are cut parallel to the compression axis. In order to achieve the surface quality required for EBSD examinations, electro polishing is performed at 18 V and 5^{0}C in an electrolytic solution consisting of 600 ml methanol, 300 ml butyl alcohol and 100 ml perchloric acid. The prepared samples are probed with a FEG-SEM (Carl-Zeiss, Model number: Supra 40) operating at 20 kV and equipped with an EBSD orientation imaging system. Microstructures are taken at the center of the compression specimen with step size of 0.4 μm. The collected data are analyzed by using TSL (Tex Sem Laboratory) –OIM software.

3. Results and discussion

The orientation (EBSD) map of the separated primary alpha grains in the as-received material IMI 834 Ti alloy is shown in Fig. 1 (a). It should be noted that the as-received material consists of equiaxed primary alpha grains and secondary alpha laths. Both primary alpha and secondary alpha colonies having same crystal structure (hcp) but the morphology of primary alpha grains is equiaxed and secondary alpha grains is lath type. Primary alpha grains and secondary alpha grains were distinguished by using grain mode option in OIM analysis software. The largest grains observed in the inverse pole figure map (IPF) are primary alpha (α_p) and all small grains are associated with secondary alpha (α_s) lamellae.

The volume fraction of primary alpha grains observed in as-received sample is 45%. The average grain size of the equiaxed primary alpha grain size (α_p) is 20 μm, calculated by using grain size intercept length method. After 50% deformation at 1000°c with strain rates of 10^{-3} s^{-1}, 10^{-2} s^{-1}, 10^{-1} s^{-1}, 1 s^{-1} and 10 s^{-1} the observed orientation maps displayed deformed and fragmented primary alpha grains; as shown in Fig. 1. (b-f). The volume fraction and grain size of primary alpha grains in Fig. 1 (b-f) are 6.5% with 6.6 μm, 17.4 % with 8 μm, 26.2% with 10.7 μm, 19.5% with 8.4 μm and 27.6% with 8.3 μm respectively.

The pole figures and inverse pole figures of primary alpha grains in as received IMI 834 Ti alloy and hot deformed specimens are given in Fig 2. (a-f). A strong basal texture in the as-received material perpendicular to the CD-RD plane is noticed. In case of deformation conditions texture varies significantly with strain rate.

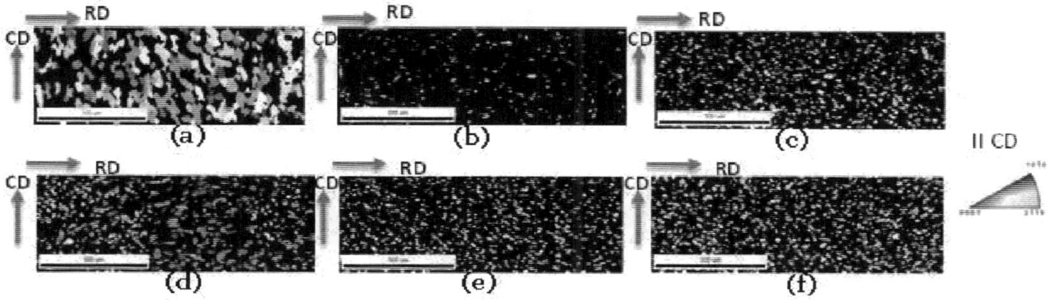

Figure 1 EBSD Orientation maps of primary alpha grains: (a) as-received IMI 834 Ti alloy hot compressed at 1000^{0}c with (b) 10^{-3} s^{-1} strain rate (c) 10^{-2} s^{-1} strain rate (d) 10^{-1} s^{-1} strain rate (e) 1 s^{-1} strain rate and (f) 10 s^{-1} strain rate.

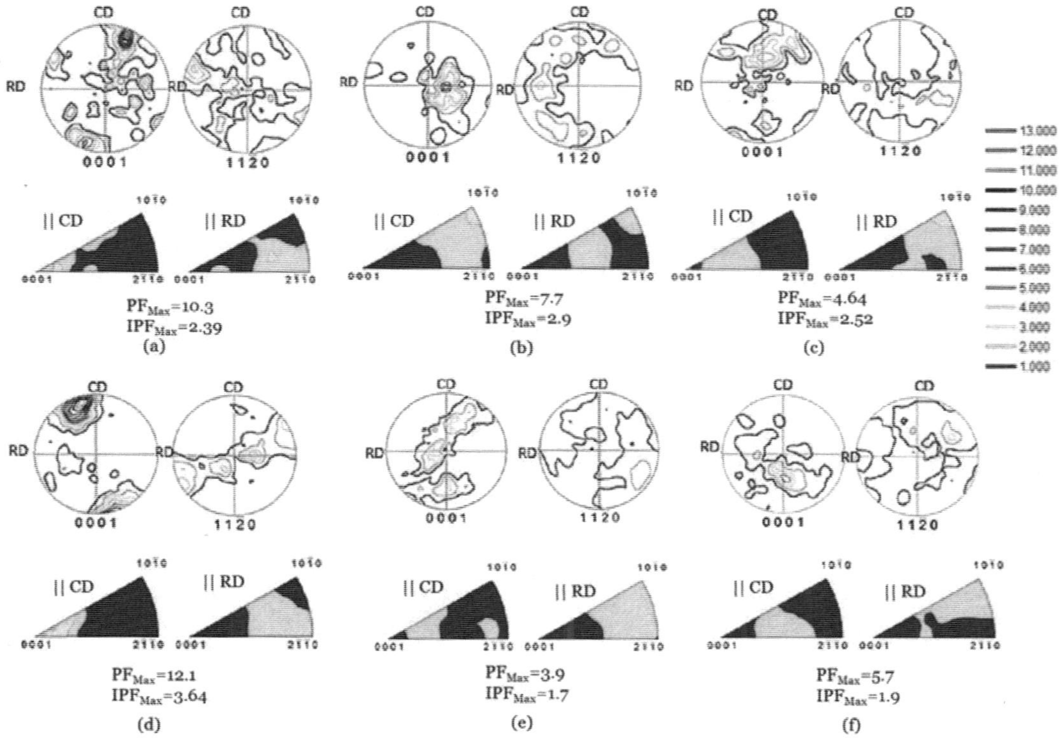

Figure 2 Pole figures and Inverse Pole figures of as-received and deformed samples: (a) as-received IMI 834 Ti alloy (b) 10^{-3} s^{-1} strain rate (c) $10^{-2}s^{-1}$ strain rate (d) 10^{-1} s^{-1} strain rate (e) 1 s^{-1} strain rate (f) 10 s^{-1} strain rate.

Figure 3 shows the Kernel average misorientation profile of the as-received and after hot deformation of IMI 834 titanium alloy. It clearly reveals that with increased strain rate the dislocations generate more misorientation inside the grains which ultimately aid in fragmentation of the α_p grains. Fig 4. shows number fraction of Schmid factor of various possible hcp slip systems in as-received and deformed specimens. The basal slip remains nearly similar in all strain rates. In as-received material, pyramidal and first order pyramidal slip systems are more favorable. At low strain rate (10^{-3} s^{-1}), prismatic and pyramidal slip systems are more active. Strain rate 10^{-1} s^{-1} showed higher first and second order pyramidal slip systems. Different strain rates showed different slip systems which ultimately resulted into various α_p textures.

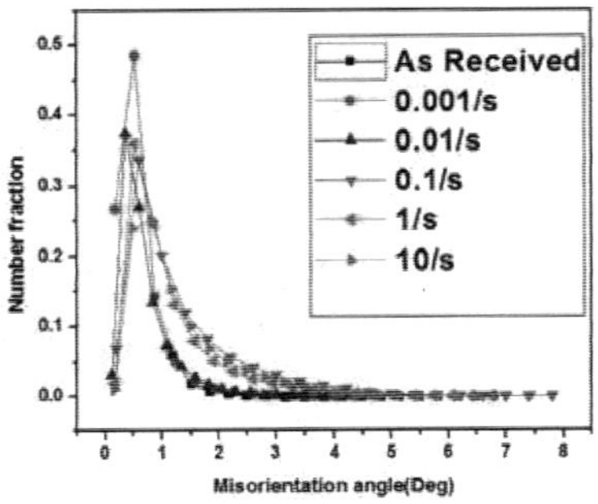

Figure 3 Kernel average misorientation profile of as-received and deformed IMI 834 Ti alloy.

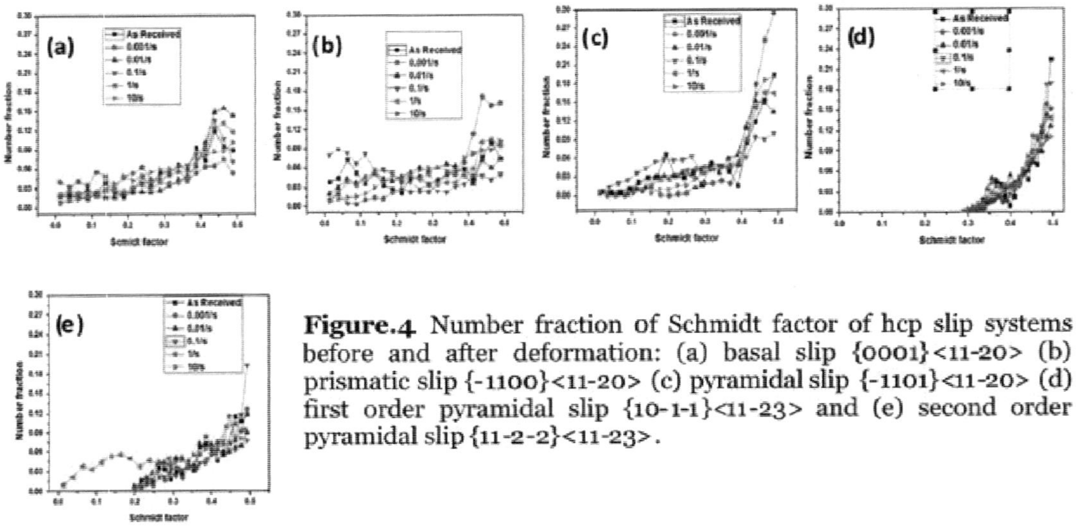

Figure.4 Number fraction of Schmidt factor of hcp slip systems before and after deformation: (a) basal slip {0001}<11-20> (b) prismatic slip {-1100}<11-20> (c) pyramidal slip {-1101}<11-20> (d) first order pyramidal slip {10-1-1}<11-23> and (e) second order pyramidal slip {11-2-2}<11-23>.

4. Conclusion

Volume fraction of primary alpha grains increases with increasing strain rate during hot deformation followed by water quenching. No twinning was observed in primary alpha grains after deformation but more fragmentation of primary alpha grains. The variety of textures during hot deformation at different strain rates suggests the activation of different slip systems which is confirmed from Schmid factor profiles of hcp slip systems.

References

[1] Neal DF 1988 *Proc. 6th World Conf. on Titanium* (France) p 253
[2] Randle V and Engler O 2000 *Introduction to Texture Analysis: Macrotexture, Microtexture and Orientation Mapping* (Gordon and Breach, Amsterdam, Holland) Chapter 6 pp 125-151
[3] Al-Samman T and Gottstein G 2008 *Mater. Sci. Eng.* A **488** 406
[4] Del Valle J A and Ruano OA 2008 *Mater. Sci. Eng.* A **487** 473

IOP Conf. Series: Materials Science and Engineering **82** (2015) 012033 doi:10.1088/1757-899X/82/1/012033

The deformation behaviour of hard and soft grains in RR1000 nickel-based superalloy

S Birosca[1]
[1] Materials Research Centre, College of Engineering, Swansea University, Swansea SA2 8PP, UK.
E-mail: s.birosca@swansea.ac.uk

Abstract: In the current study, the Electron Backscatter Diffraction (EBSD) plastic strain measurement is assessed to ascertain the deformation behaviour within individual oriented grains in polycrystalline RR1000 Nickel-based superalloy. Here, the crystallographic orientation correlations with Geometrically Necessary Dislocation (GND) contents, Schmid and Taylor factors as well as strain hardening were investigated. It is demonstrated that grain boundary is the major factor influencing GND accumulation in soft grains, while the orientation is the critical factor in hard grains' fragmentation. Furthermore, it is concluded that the total GND value for individual grains varies depending on the deformation stage, whereas the locations of high GND density within the grains indicates on strain hardening state. Keywords: GND, Strain Hardening, FCC, Deformation.

1. Introduction

It is argued that the classical plasticity theory cannot describe crystalline material behaviour at micron level. Subsequently the strain gradient theory was developed based on the framework of Geometrically Necessary Dislocations (GND) [1-4]. Since then, strain gradient plasticity has become an increasingly important topic in applied mechanics as it is widely reported that the yield strength of a material depends not only on the strain, but also on the strain gradient [2,4,5]. In general, during material plastic deformation the dislocations are generated, moved and their storage causes work hardening of the materials. Dislocations are stored because they accumulate by trapping each other in a random way (termed Statistically Stored Dislocations (SSD)s) or they are required for compatible deformation of various parts of the material (termed Geometrically Necessary Dislocations (GND)s) [1-6]. In other words, during plastic deformation, GNDs accommodate the orientation gradients within individual grains and they are required to maintain lattice continuity [7]. Ashby [4] proposed that work hardening reflects the way in which the arrays of stored dislocations obstruct the motion of other moving dislocations. He also suggested that in polycrystals the plastic strain gradient is produced by the mismatch of slip directions at grain boundaries. Furthermore, S. Sun, *et al.*, observed that during deformation, the constituent grains of polycrystals developed complex heterogeneous deformation patterns with variations in local strain and crystal orientation occurring over a wide range of length scales [2]. They also claimed that interfaces can resist the motion of lattice dislocations, or absorb or emit them depending upon the structure of the interface and local mechanical conditions. Their GND calculation also showed that the lattice curvature observed in the vicinity of interfaces and the GND accumulations at grain boundary was dependent on the strain level imposed on the material. Moreover, Raabe, *et al.*, reported that the orientation gradient may proceed from a variation in slip system activation throughout a grain leading to local domains with different orientation [8].

2. Materials and Methods

In the current study Electron Backscatter Diffraction (EBSD) is applied to measure the plastic strain within the individual orientation. The alloy used in this study was nickel based superalloy RR1000 with 18.5% Co and 15% Cr. The average grain size was in the range of 30 ± 5 μm diameter. The received samples from Rolls-Royce plc were subjected to thermo-mechanical fatigue cycles until failure. Standard EBSD working conditions was used and the orientation data was collected with 100 nm step size in order to determine the orientation spread within individual grains at small length scales. The EBSD obtained data was then analysed using the HKL-Channel 5 and TSL-OIM data analysis packages. The GND map was generated from EBSD data and the principle of GND calculations is reported in [9]. The microtexture was examined in the vertical section of the failed specimen parallel to the wider face of the gauge length, LD-TD plane (LD: Loading Direction, TD: Transverse Direction). Large area of EBSD scan was performed in the vicinity of the crack front. The experimental data collected from 10 (2 × 2 mm) EBSD maps for over 2000 grains.

3. Results and Discussion

The grains with specific texture component were identified and analysed individually as shown in Figure 1 and Table 1. It is evident from Figure 1 that Cube grains, identified in 1a, showed high Schmid Factor (0.45-0.48); 1b, low Taylor Factor (1.5-2.1); 1c, low overall GND content (102.4/μm2); 1d, and low Elastic stiffness (125 GPa); 1f. Whilst the Brass grains, identified in 1a, showed low Schmid Factor (0.26-0.31); 1b, high Taylor Factor (4.1-4.3); 1c, high overall GND content (104.3/μm2); 1d, with high Elastic stiffness (240 GPa); 1f. An example of a Brass grain containing high GND (104.5/μm2) is indicated by a dashed rectangle box in Figure 1c-f. In this study, the focus was on Cube and Brass orientations as they showed contrary characteristics and opposite response to the applied load. It was clear from the overall statistical analyses carried out for over 2000 grains, see Table 1, that Cube grains acted as "soft" grains and showed relatively low indication of strain hardening, (low GND density, especially in the centre of the grains), see grains 1 and 2 in Figure 1d. Whereas, the Brass grains showed to be more resistant to deform plastically, Figure 1b-d, and elastically, Figure 1f. Therefore, the Brass grains were behaved as "hard" grains and showed high strain hardening indicated in the GND map in Figure 1d. The ESBD elastic data obtained for RR1000 here, see Figure 1f, was comparable with the elastic stiffness data for pure nickel where the elastic stiffness in <111> direction was found to be higher than in <100> direction [10-12].

Table 1: Statistics for over 2000 grains showing Schmid factor and Taylor factor values as well as the calculated GND values for individual texture component.

Texture Co.	Symbol	Schmid Factor (0.1 – 0.5)	Taylor Factor (1 - 5)	GND (ρ_{tot})
Cube	□	High (0.42 – 0.5)	Low (1 - 2)	Low ($10^{2.4}/\mu m^2$)
Goss	△	Medium to High (0.35 – 0.45)	Low to Medium (2 - 4)	Low to Medium
Brass	⬡	Low (0.26 – 0.35)	High (4)	High ($10^{4.4}/\mu m^2$)
Copper	◇	Low to High (0.28 – 0.46)	Medium to High (3 -4)	Low to Medium
S	◺	Low to Medium (0.35 – 0.41)	Medium to High (3 - 4)	Medium to High

In literature, various crystallographic orientation correlations with Schmid factor, Taylor factor, and dislocation density are suggested [13-16]. However, there are ambiguities between the reported correlations and it appeared that some irregularities existed in the present study. For instance, although GND does not represent the total amount of dislocation density, it still is rather surprising from conventional metallurgical interpretations to have a grain with high Schmid factor, but contain low GND density and vice versa. Yet, the experimental studies clearly showed that such a relationship exists. Therefore, the formation and development of GND accumulation as well as the definition of hard and soft grains are reassessed here aiming to establish direct orientation relationship with deformation parameters.

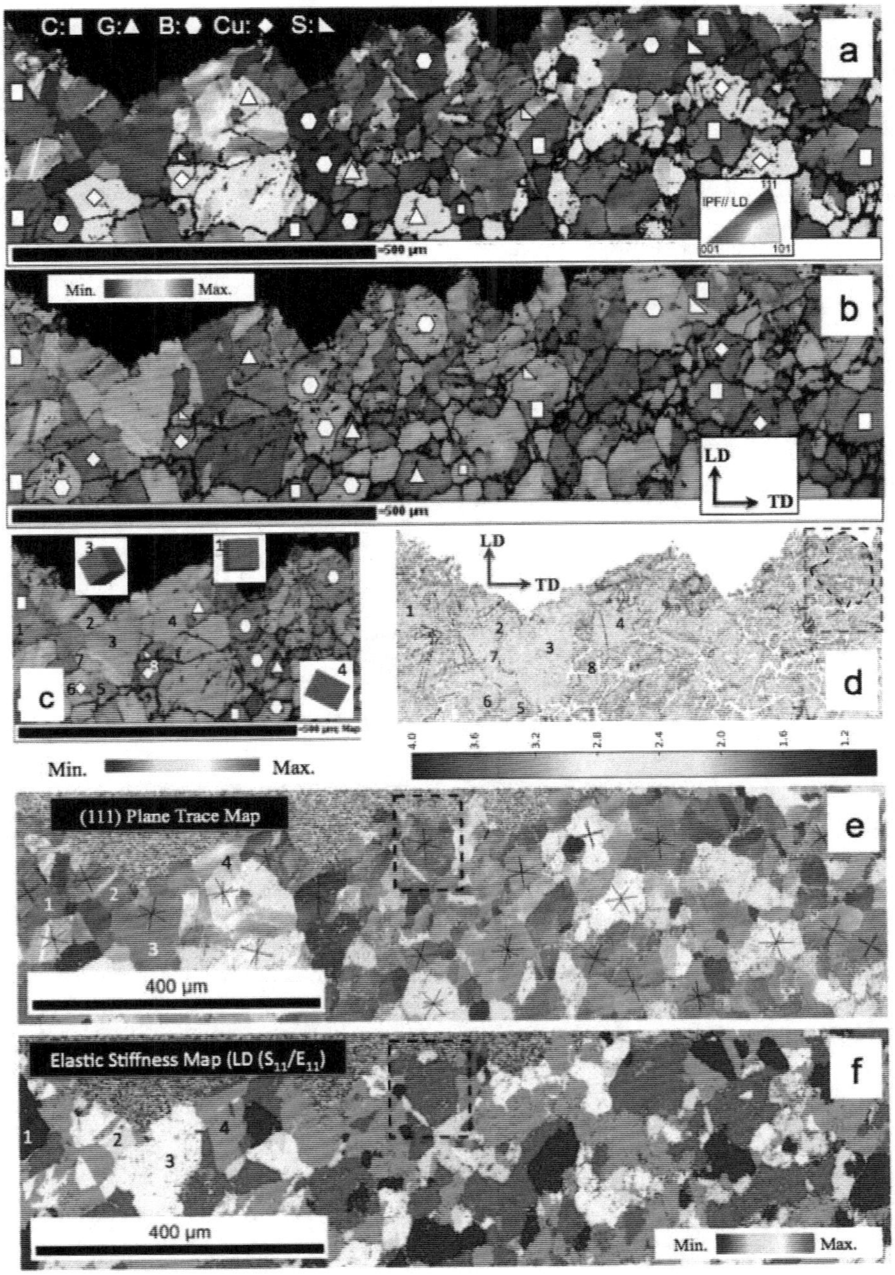

Figure 1: HKL-EBSD derived (a) IPF//LD map with individual texture components are indicated, C: Cube (100)<001>, G: Goss (110)<001>, B: Brass (110)<112>, Cu: Copper (112)<111>, S: (123)<634>, (b) (111)<1-10> Schmid factor map (Min.: 0.1, Max.: 0.5), (c) Taylor factor map (Min. = 1, Max. = 5), the 3D crystal viewer are shown for some grains, the calculated GND map is shown in (d), TSL-EBSD (e) calculated (111) slip trace map, (f) Elastic stiffness map (LD (S11/E11), where S_{ij}=stress and E_{ij}= strain component (Min. = 100, Max. = 250 GPa).

As proposed in [8] two main considerations are essential in order to understand the deformation mechanism in the individual grains (i) orientation (ii) neighbouring grain orientations aspects. From Figure 2a, for a soft grain, i.e., Cube grain, to deform plastically the dislocations including SSD and GND would generate on (111) plane, as this plane is perfectly aligned for slip if loaded in the direction indicated in the same figure, see also Figures 1b and 1e. Theoretically, for the ideal Cube ((001) // TD-

ND plane and <100> // TD) all four (111) planes have 54.74° angle with the LD. From slip trace map in Figure 1e, the angle between LD and two of the (111) trace are close to 45° in grains 1 and 2. See also the highlighted Cube grain in Figure 2b, showing (111) slip plane traces observed by SEM and fitted with highlighted EBSD-TSL (111) trace map in Figure 2c. It appeared that a slight deviation from ideal Cube orientation (typically ± 9°) makes the geometrical condition of <001> in the Cube oriented grains to be closely aligned with LD for relatively easy (111)<110>-slip system activation. This condition would facilitates uniform shear, consequently these high Schmid factor Cube grains yield easily and tend to plastically deform without reorientation, hence contain high strain concentration while stress concentration decreases within such grains. The generated and existed dislocations continue to move freely until they reach grain boundaries, see Figure 2b, as there is no obstacle to impede their movements. Thus, there is little strain hardening is expected upon this stage of deformation in soft grains.

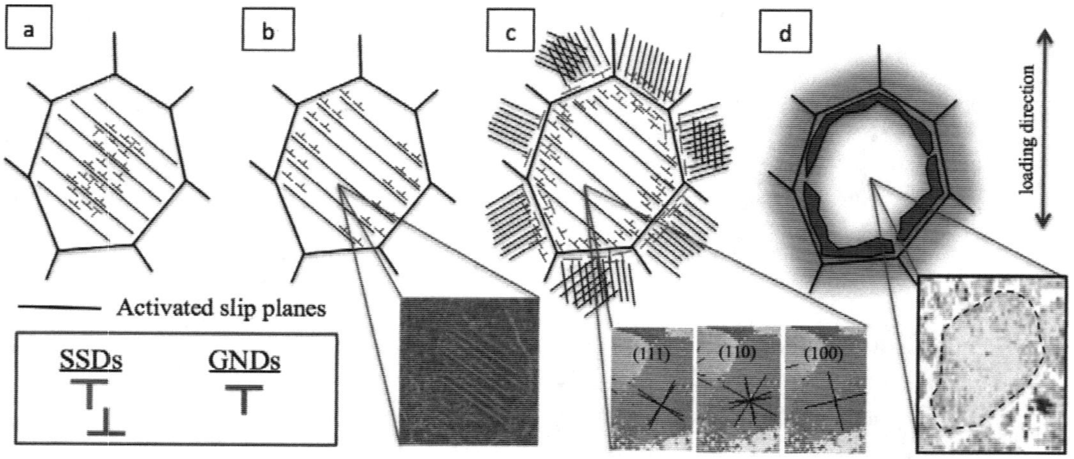

Figure 2: The development of GND accumulation model in soft grains.

Considering the orientation factor only, these grains expected to shear promptly on (111) plane under such a condition. However, from neighbouring grain orientations aspect, these dislocations would slowdown when reaching the grain boundaries as they face pronounced obstacles in movement from surrounding hard grains see Figure 2c. In other words, slip transfer is extremely difficult if a soft grain surrounded by hard grains. This is generally explained by the strain and stress incompatibilities between differently oriented grains [17]. At this stage of deformation the GND will accumulate near the grain boundaries and cause intense strain hardening, see Figure 2d. An experimentally observed GND accumulation on the grain boundary in a Cube is highlighted in the same figure. This explains minimal GND density in the center of most of the Cube grains observed in this study. Meanwhile the generated SSDs during deformation might remain in the center of the grain but this cannot be verified by EBSD. It should be emphasised here that because of the actual 3D dislocation projections on 2D EBSD GND map, the GND accumulations appeared all around the grain boundary of that particular grain, see Figure 2d. This soft grain strain-hardening model can explain Raabe, et al., findings as they showed that the Cube grain deformation is strongly dependent on the neighboring grains' orientation [8]. Also this verifies the work in [7,18] where they found that the hard grains did not deform significantly until the soft grains were hardened. Subsequently the grains with low Taylor factor developed less GNDs than the grains with high Taylor factor. The experimental evidence in support of this model is also in an agreement with the definition of GND in [19-21] as they stated that GNDs act as obstacles to the motion of SSDs and hence contribute to work hardening of the materials. It is also important to realise that this model only valid if the soft grain surrounded by hard grains. From overall GND and EBSD maps, it seemed that the Cube grain family might accommodate most of the imposed strain by SSD rather than GND when it surrounded by other soft grains.

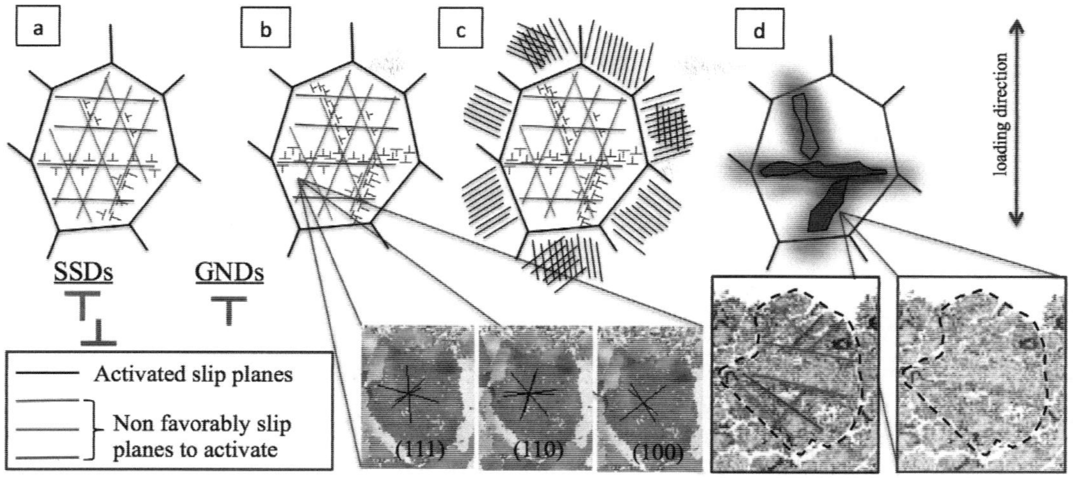

Figure 3: The development of GND accumulation model in hard grains.

As indicated in Figure 1b, visualised in Figure 1e and schematically drawn in 3a, in a hard Brass grain there was no particular slip system perfectly aligned for easy activation. From orientation aspects, these grains are not favourably oriented for octahedral (111)<110> slip system activation, therefore, the whole crystal tends to rotate in order to activate this or any other available slip systems including Cubic (100)<110> system. The crystal rotation (reorientation) during loading in order to accommodate the imposed strain is further explored in [22,23]. The crystal rotation will lead to high stress concentration, thus high strain hardening is expected at the very early stage of deformation in the hard grains. This will generate further dislocations including SSD and GND on different slip planes within the same grain, see Figure 3a, and the dislocation density would intensify during loading, see Figure in 3b-c. It seems that Brass grains resist the crystal rotation through accommodating further GND, i.e., strain hardening. The dislocations continue to accumulate on different primary and secondary slip planes, see high GND density locations in Figure 3d. In the same figure an experimentally measured GND density and the slip plane traces observed by EBSD measurements in a Brass grain is highlighted. Clearly, the locations of high GND in the highlighted Brass grain can be aligned with the certain (111), (110) and (100) primary and secondary slip traces in Figure 3c. This hard grain eventually fragmented after creating high orientation gradient. It was evident that the grain boundary played no major role in hard grain plastic deformation and strain hardening. However, considering the neighbouring grain interaction, when a hard grain finally deformed, the surrounding soft grains would deform accordingly. This explains the original high Taylor factor values in the hard grains as found in this and pervious studies.

4. Conclusions

The soft and hard grain GND accumulation models, described in this study, demonstrate that the grain boundary is the major factor influencing GND accumulation in soft grains, whilst the orientation is the critical parameter in hard grains' fragmentation. This is supported by experimental evidence where high GND content was found in the vicinity of the grain boundary in the Cube (soft) grains and on slip plane traces within the Brass (hard) grain. The GND absolute value varied as it is interrelated with deformation stages, but their location within the grains gives exact description of strain hardening state and deformation micromechanism in hard and soft grains.

Acknowledgements
The author would like to acknowledge the financial support of EPSRC Rolls-Royce Strategic Partnership in Structural Metallic Systems for Gas Turbines (EP/H500383/1). Also special thanks to

Dr. Mark Hardy in Rolls-Royce plc and Dr. Fabio Di Gioacchino at the University of Cambridge for developing the GND code.

References

[1] Arsenlis, A., Parks, D. M., *Acta Materialia* **47** (5), 1597 (1999).
[2] Sun, S., Adams, B. L., King W. E., *Philosophical Magazine A* **80** (1), 9 (2000).
[3] Nye, J. F., *Acta Metallurgica* **1**, 153 (1953).
[4] Ashby, M. F., *Philosophical Magazine* **21** (170), 399 (1970).
[5] Kysar, J. W., Gan Y. X., Morse T. L., Chen X., Jones M. E., *Journal of the Mechanics and Physics of Solids* **55**, 1554 (2007).
[6] Gao, H., Huang, Y., Nix, W. D, Hutchinson, J. W., *Journal of the Mechanics and Physics of Solids* **47**, 1239 (1999).
[7] Field, D. P., Trivedia, P. B., Wright, S. I., Kumar, M., *Ultramicroscopy* **103**, 33 (2005).
[8] Raabe, D., Zhao, Z., Park, S.-J., Roters, F., *Acta Materialia* **50**, 421 (2002).
[9] Pantleon, W., *Scripta Materialia* **58**, 994 (2008).
[10] Reed, R. C., The Superalloys: Fundamentals and Applications, Cambridge University Press, Cambridge, UK, 2006.
[11] Peralta, P., Laird, C., *Acta Materialia* **46** (6), 2001 (1998).
[12] Blochwitz, C., Brechbiihl, J., Tirschler, W., *Materials Science and Engineering A* **210**, 42 (1996).
[13] Lin, B., Zhao, L. G., Tong, J., *Engineering Fracture Mechanics* **78**, 2174 (2011).
[14] Peralta P., Dickerson, R., Dellan, N., Komandur, K., Jameel, M. A., *Journal of Engineering Materials and Technology* **127**, 23 (2005).
[15] Takayama, Y., Szpunar, J. A., *Materials Transactions* **45** (7), 2316 (2004).
[16] Choi, S., Jin, Y., *Materials Science and Engineering A* **371**, 149 (2004).
[17] Gourgues, A. F., *Materials Science and technology* **18**, 119 (2002).
[18] Kocks, U. F., Tome, C. N., Wenk, H. R., Texture and Anisotropy: Preferred Orientations in Polycrystals and their Effect on Materials Properties, Cambridge University Press, Cambridge, UK, 1998.
[19] Gao, H., Huang, Y., *Scripta Materialia* **48**, 113 (2003).
[20] Needleman, A., Gil Sevillano, J., *Scripta Materialia* **48**, 109 (2003).
[21] Kubin, L. P., Mortensen, A., *Scripta Materialia* **48**, 119 (2003).
[22] Hurley, P. J., Bate, P. S., Humphreys, F. J., *Acta Materialia* **51**, 4737 (2003).
[23] Clair, A., Foucault, M., Calonne, O., Lacroute, Y., Markey, L., Salazar, M., Vignal, V., Finot, E., Acta Materialia 59, 3116 (2011).

Effects of annealing on texture evolution of cross shear rolled high-purity Al foils

L Wang*, Y Liu, X Song, J He and L Zuo
Key Laboratory for Anisotropy and Texture of Materials, Northeastern University, China
*wanglei@mail.neu.edu.cn

Abstract: The effects of annealing on recrystallization texture of cross shear rolled high-purity Al foil were investigated by orientation distribution functions (ODFs) and electron backscattered diffraction (EBSD). The results show that the intermediate annealing is beneficial to the development of the cube texture. The cube texture can be promoted by annealing, and the critical annealing temperature is about 280 °C. The cubic orientation grains firstly nucleate, and then expand into other grains with a high growth speed, and large angle grain boundary ratio increases, finally can swallow up most of the original grains, which results in the cube texture.
Keywords: high-purity Al foils; cross shear rolling; cube texture; annealing

1. Introduction

High purity Al foil is mainly used as anode materials for high voltage capacitor, whose electricity capacitance is dependent on effective surface area ratio [1-2]. The special electrochemical corrosion technology has been developed, to produce a large number of tunnels and pore pits, thus the surface area of Al foil has successfully increased [3-4]. It has been reported that Al foil erosion performance is affected by the cubic texture. The more cubic texture, the higher specific capacitance is. Several studies have focused on the cubic texture controlling by means of rolling and heat treatment [5], but few studies are about texture evolution of Al foil during the cross shear rolling and the annealing process. Therefore, on the basis of the previous research [6], cross shear rolling was chosen in the present study, to investigate the cubic texture content during the following heat treatments, and the effect of annealing on texture evolution of Al foils.

2. Experimental procedures

The chemical compositions of 2mm thickness Al foils with 99.99% purity are Fe 0.002, Si 0.002, Cu 0.007 and balance Al(mass %). The foils with or without of intermediate annealing were prepared by cross shear rolling with the speed ratios of 1.06, respectively. Then the specimens were treated by final annealing. The textures were measured at 1/4 layer of the specimen thickness with incomplete {111}, {200} and {220} pole figures using the X'pert MRD type X-ray diffractometer (XRD) with Co-Kα radiation. The corresponding ODFs were calculated by Labtex 3.0 software. The textures were also characterized by EBSD. The specimens for EBSD were electrolytic polished, using a solution(CH_3OH： HNO_3=2:1) at 25V.

3. Results and Discussion

3.1 Effect of intermediate annealing on the cube texture

The texture fractions the samples with or without intermediate annealing are shown in Fig. 1. The final annealing are at 400℃ for 5 min. The mark "1.06t" means the rolling speed ratio is 1.06; t means the sample with intermediate annealing, while w means those without intermediate annealing. It is clearly shown that the volume fraction of cubic texture with intermediate annealing is higher than that without intermediate annealing. In addition, the volume fraction of cold rolled textures, such as S, copper in intermediate annealed foil are lower than those without intermediate annealing, but the volume fraction of R texture in intermediate annealed samples is higher than those without intermediate annealing. The cube and R texture are both recrystallization textures, transformed from the cold rolled textures. That means intermediate annealing can promote the changes from the cold rolling texture to the cube texture.

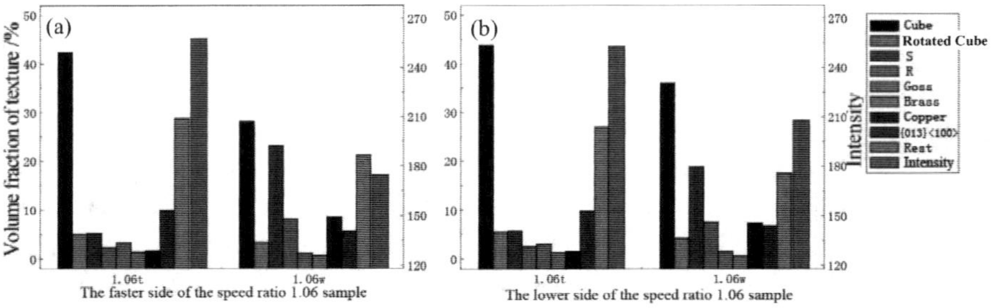

Fig. 1 Texture volume fraction in the samples with or without intermediate annealing: (a) the faster side; (b) the lower side

3.2 Effect of final annealing on the cube texture

Fig. 2 illustrates the texture volume fraction of the samples annealed at different temperatures. It shows that the increase of cubic texture is quite limited, if the samples are annealed at 250 °C for 1 h and 12h. If the annealing temperature increases to 280 °C, most of the cold rolled texture has completely transformed to the recrystallized cubic texture after 12h. It is obvious that annealing temperature dominates the recrystallization process, and the annealing times is still not enough to stimulate the development of the cubic texture if the annealing temperature is low. The critical annealing temperature is about 280 °C.

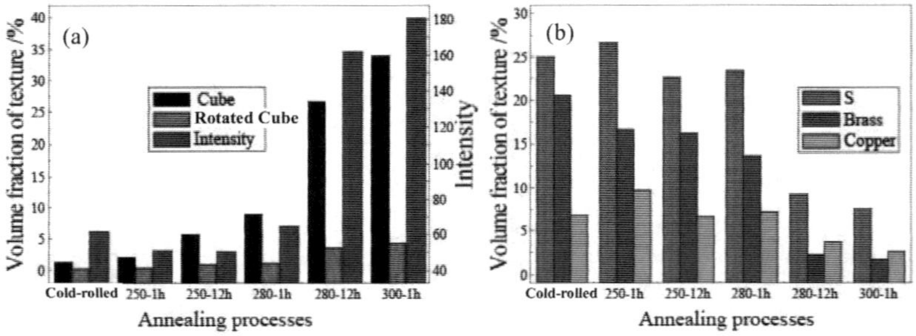

Fig. 2 Texture volume fraction of samples at low-temperature annealing: (a) cube, rotated cube texture; (b) cold-rolled texture

The effects of annealing time on the texture of Al foils with the final annealing of 300 °C are shown in Fig. 3. It is obvious that the volume fractions of recrystallization texture are still at a lower level, and the cubic texture does not illustrate the relative advantages compared with other textures, if the annealing time is within 10 min. When the annealing time prolongs to 20 min, the cubic texture gradually form. With the increasing of annealing time, the cubic texture and most other texture are kept constant, which means the effect of annealing time on the texture can be weaken, if the annealing time is higher than 20 min.

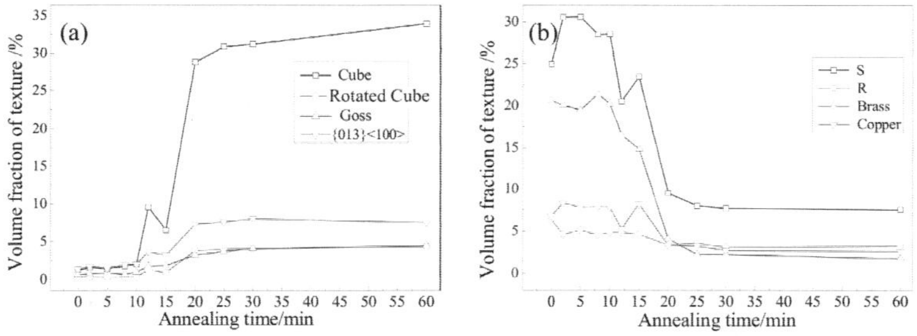

Fig. 3 Texture evolution of Al foils annealed at 300 °C: (a) cube, rotated cube texture; (b) cold-rolled texture

3.3 Texture evolution mechanism

In order to explore the texture evolution mechanism, the annealing time interval is enlarged to 2 min, and both ODF and EBSD are carried out. Fig. 4 shows the evolution of the ODF with series of annealing times at 300 °C. When the annealing time is 10 min, Brass texture is still the main part of the texture, as shown in Fig. 5 (a). If the annealing time extends to 12 min, the cubic texture starts to form and gradually develops with increasing of the annealing time. And until 20 min, they are almost cube textures.

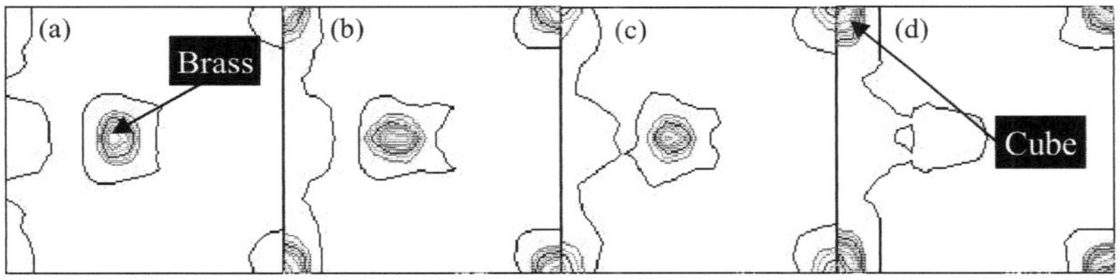

Fig. 4 Evolution of the ODF annealed at 300 °C for: (a) 10 min; (b) 12 min; (c) 15 min; (d) 20 min

Fig. 5 shows the orientation and texture map annealed at 300 °C. When the annealing time is 14min, the grain is priority to {101} orientation. When the time increases to 16 min, the red parts with cube texture increases. With the time increases, the cold rolled texture gradually disappears. It has been reported [7] that the growth of the cubic orientation ascribes to the 38°- 40° <111> axis rotation relationships between the cube orientation {001}<100> and the copper texture {112}<111>, or the 38° - 40°<111>axis rotation relationships between the cube orientation {001}<100> and S texture {123}<634>. That is why the cube texture preferentially grows, and there are almost cube textures when the annealing time is about 24 min.

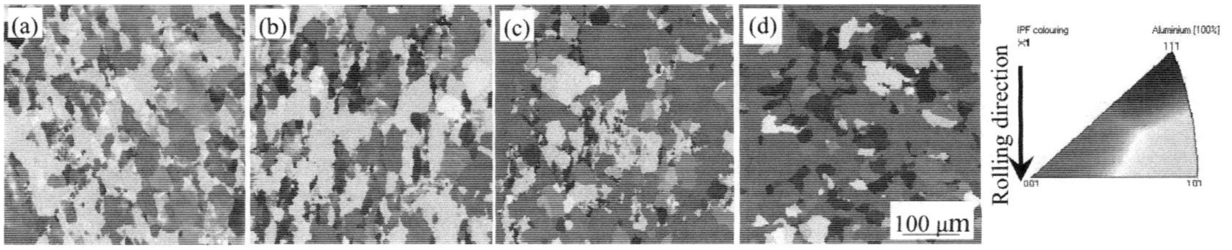

Fig. 5 Orientation and texture map annealed at 300 °C for: (a) 14 min; (b) 16 min; (c) 22 min; and (d) 24 min

In fact, the recrystallization attributes to grain boundary migration, so the grain boundary type changes with annealing can also reflect the texture transformation. The grain boundary types during annealing at 300 °C are shown in Fig. 6. It shows that when the annealing time is 14 min, the amount of small angle grain boundary is very high and almost near 6-7°, illustrating that a cold rolled texture component dominate the structure. While when the cold rolled texture starts to change to the cube texture, the large angle boundaries increase especially near 55°. Furthermore, the proportion of grain boundary orientation angle at 57° between adjacent boundaries is especially higher than that between non-adjacent ones, illustrating that part of the orientation of cold rolled grain has changes towards other orientation which is quite different and can also be confirmed to be cube orientation by ODFs and EBSD. So the grains with cube orientation gradually grow during the recrystallization, and finally, the most cube texture forms during the annealing.

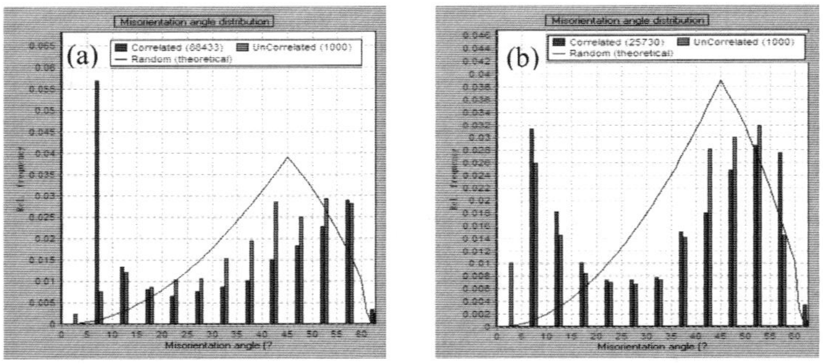

Fig. 6 Misorientations annealed at 300 °C for: (a) 14 min; (b) 24 min

Conclusions:

1. The cubic texture fraction of Al foil with the intermediate annealing is higher than that without intermediate annealing, which proves the intermediate annealing much beneficial to the development of the cube texture.
2. The critical annealing temperature of the cross shear rolled foil with the intermediate annealing is about 280 °C. Below 280 °C, the annealing temperature plays the leading role on the recrystallization process.
3. The texture transformation mechanism is as follows: Firstly, the grains with cubic orientation start to depressively nucleates, and then expand into other grains with a high growth speed, and large angle grain boundary ratio increases, finally can swallow up most of the original grains, which makes the cube texture becoming the main texture of the high-purity Al foil.

Reference:

1. Pan F S et al 2005 *Mate. Sci. and Tech.* **21** 1432
2. Wang H et al 2010 *J. of Rare Earths* **28** 134
3. Xiao R et al 2008 *Corrosion Sci.* **50** 1576
4. Tamara N et al 2004 *Mater. Trans.* **45** 1687
5. Du Y et al. 2006. *Trans. of Nonferrous Metals Soc. of China* **16** 1307
6. Wang L et al. 2011 *J. of Northeastern Univ. (Natural Science)* **32** 245
7. Liu C et al. 2001 *J. of Central South Univ. Tech. (Natural Science)* **32** 85

IOP Conf. Series: Materials Science and Engineering **82** (2015) 012035 doi:10.1088/1757-899X/82/1/012035

The influence of microchemistry on the recrystallization texture of cold-rolled Al-Mn-Fe-Si alloys

K Huang[1*], YJ Li[1] and K Marthinsen[1]

[1] Department of Materials Science and Engineering, Norwegian University of Science and Technology, Trondheim, N-7491 Trondheim, Norway

* E-mail: ke.huang@ntnu.no, huangke0729@hotmail.com

Abstract. The recrystallization textures of a cold-rolled Al-Mn-Fe-Si model alloy with three different microchemistry states after non-isothermal annealing were studied. The microstructure and texture evolution have been characterized by EBSD. It is clearly demonstrated that the actual microchemistry state as determined by the homogenization procedure strongly influence the recrystallized grain size and recrystallization texture after non-isothermal annealing. High Mn content in solid solution promotes stronger concurrent precipitation and retards recrystallization, which finally leads to a coarse grain structure, accompanied by strong P {011}<566> and/or M {113}<110> texture components and a ND-rotated cube {001}<310> component. A refined grain structure with Cube {001}<100> and/or a weak P component as the main texture components were obtained when the pre-existing dispersoids are coarser and fewer, and concurrent precipitation is limited. The different recrystallization textures are discussed with respect to the effect of second-phase particles using two different heating rates.

1. Introduction

The recrystallization texture of cold rolled aluminium alloys after annealing has been the subject of numerous investigations as part of a general effort to optimize their mechanical properties. Different from high purity aluminium alloys, almost all commercial aluminium alloys are multiphase, containing second-phase particles formed either before or during annealing. Considering Al-Mn-Fe-Si (AA3xxx) alloys, the supersaturated Mn in solid solution in the as-cast state will precipitate as Mn-bearing dispersoids during subsequent thermo-mechanical processing steps. The particles may affect the recrystallization texture in different ways. Coarse pre-existing constituent particles favour Particle Stimulated Nucleation (PSN) associated with a wide range of nucleus orientations, and thus yields an overall random recrystallization texture [1,2]. On the other hand, fine dispersoids especially those precipitated during annealing may pin boundaries leading to an elongated coarse grain structure [3] together with the P and ND-rotated cube texture components [4]. The microchemistry state of the alloys, i.e. amount of solutes and second-phase particle structures, determined by the chemical composition and homogenization procedures of the alloys is thus an important aspect in studying the recrystallization textures of cold-rolled Al-Mn-Fe-Si alloys. The recrystallization texture after isothermal annealing of various deformed aluminium alloys has been extensively investigated [3,4]. Much less studies on recrystallization texture after non-isothermal heat treatments exist, which are frequently carried out during industrial thermo-mechanical processing where extended recovery and precipitation of dispersoids are likely to occur, producing complex interactions with recrystallization.

In this paper, the effect of microchemistry on the recrystallization texture of cold-rolled Al-Mn-Fe-Si alloys is analysed after non-isothermal annealing. The different recrystallization textures obtained are discussed with respect to the second-phase particle structures, including pre-existing dispersoids and concurrently precipitated dispersoids.

2. Experimental

The starting material was a DC-cast AA3xxx extrusion billet, supplied by Hydro Aluminium. The chemical composition (wt.%) is 0.152% Si, 0.530% Fe, 0.390% Mn with the balance Al. The detailed description of the as-cast material can be found in Ref [5,6]. The as-cast material (C1-0) was subsequently homogenized at two different conditions to get different amounts of Mn in solid solution and dispersoid densities. One set of samples were heated with a heating rate of 50 °C /h to 450 °C and kept for 4 hours, referred to as C1-2, to be consistent with previous work on the same set of alloys [6-8]. Another set of samples were subjected to a two-stage homogenization treatment. The samples were first heated at 50°C/h to 600 °C for 4 hours, and then cooled at 25 °C /h to 500°C where they were kept for another 4 hours, which gave the C1-3 condition. Materials were water quenched to room temperature at the end of the homogenization procedure. Characteristics of the three microchemistry variants are listed in Table 1, while example images of their typical size and distribution can be found in Ref [6]. From Table 1, it is clear that the as-cast variant C1-0 has the highest potential for concurrent precipitation, while the C1-2 variant possesses a high density of fine pre-existing dispersoids. The homogenized samples were cold rolled at room temperature in multiple passes to a deformation strain of ε = 3.0. The rolled sheets were then non-isothermally annealed in an air circulation furnace to a target temperature, and then held for 10^5s at that temperature before water quench. The heating rate during non-isothermal annealing was 100°C/h and 200°C/h, respectively. The microstructure characterization details have been presented elsewhere [6].

Table 1 Electrical conductivity, solute (Mn) concentrations, equivalent area diameter and number density of particles in the alloys studied [6]

	Electrical conductivity (m/Ωmm^2)	Concentration of Mn (wt%)	Constituent particles		Dispersoids	
			Diameter (μm)	Number density (mm^{-2})	Diameter (μm)	Number density (mm^{-2})
C1-0	23.9	0.35	0.88	2.8e4	-	-
C1-2	27.5	0.16	0.96	2.9e4	0.054	1.3e6
C1-3	29.0	0.11	1.10	2.1e4	0.127	5.5e4

3. Results and discussion

The first set of results refers to the heating rate of 100°C/h to a target temperature of 400°C. The recrystallization textures of all three variants were characterized by EBSD and are presented in Fig.1. For the as-cast variant C1-0, which has the highest potential for concurrent precipitation (see Table 1), strong P and M {113}<110> texture components, together with a medium strength 35° ND-rotated cube component are obtained. For C1-2, the recrystallization texture is characterized by a very sharp P texture component and a 22° ND-rotated cube component. A much weaker P texture is obtained for C1-3, while a strong cube texture component is observed together with a medium strength S {123}<634> texture component, which is very weak for the other two variants. The weak P texture for C1-3 suggests that the strength of the P component is related to concurrent precipitation and/or pre-existing disperoids. The intensity of the P texture component was found to increase with the initial amount of Mn content during isothermal annealing [4]. The reason why C1-2 (which has less Mn in solid solution than C1-0) has the strongest P texture in the present case is related to the temperature at which it was recrystallized, an aspect which will be discussed in detail in a separate paper. When decreasing the amount of Mn in solid solution, a transition from 35° ND-rotated cube to 22° ND-rotated cube and finally to pure cube was found. In terms of recrystallized grain size, a coarse recrystallized grain structure was obtained for C1-0 and C1-2, while C1-3 exhibits a more refined grain structure, further confirming the latter being less affected by concurrent precipitation and pre-existing dispersoids, as shown in Table 2. The microstructure of the three variants, in terms of typical EBSD micrographs can be found in Ref [5].

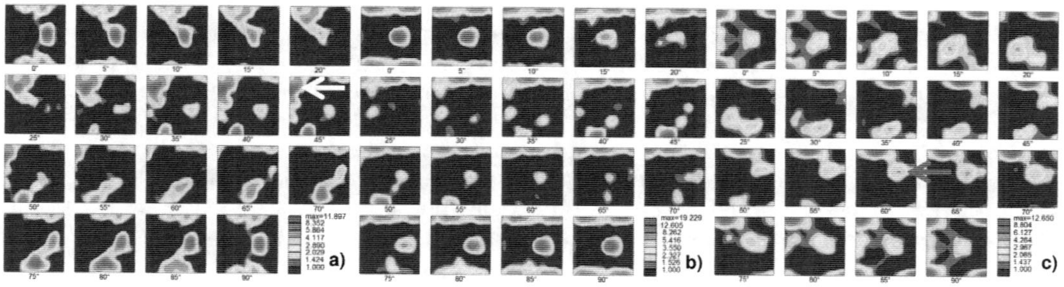

Fig. 1 ODF maps showing the textures of the three variants after heating at 100°C/h to 400 °C and kept for 10^5s. The M and S texture components are pointed out by white and red arrow, respectively
a) C1-0, b) C1-2, C) C1-3

Table 2 Effect of microchemistry on the recrystallized grain size after annealing at different heating rates to 400°C and kept for 10^5s

	Grain size (μm)		
	C1-0	C1-2	C1-3
100°C/h	157	157	49
200°C/h	117	131	34

In general, increasing the heating rate to 200°C/h tends to weaken the recrystallization textures, as shown in Fig.2. The variation of texture components with microchemistry, however, exhibits the same tendency as the cases annealed with the slower heating rate. With higher heating rate, there is less interaction time between recrystallization and concurrent precipitation, which is the main reason a weaker P texture were obtained with the higher heating rate. For C1-3, when the heating rate was increased from 100°C/h to 200°C/h, the intensity of cube texture decreased slightly, from 12x random to 8x random. This is related to the fact that recrystallization started at a higher temperature for the former case, due to stronger recovery and concurrent precipitation. It is reported that increasing annealing temperature enhances the strength of the cube texture for this material [6]. On the other side, the intensity of the P texture weakens at higher annealing temperatures due to fast recrystallization kinetics and less interaction with concurrent precipitation, the activation of other possible nucleation sites (grain boundaries, other deformation heterogeneities, etc.) might also have played a role [6,9]. As shown in Table 2, the recrystallized grain sizes for the three variants heated at 200°C/h are obviously smaller than their counterparts annealed at the lower heating rate of 100°C/h, again with the most refined grain structure obtained by the C1-3 variant where both concurrent precipitation and pre-existing particles are limited.

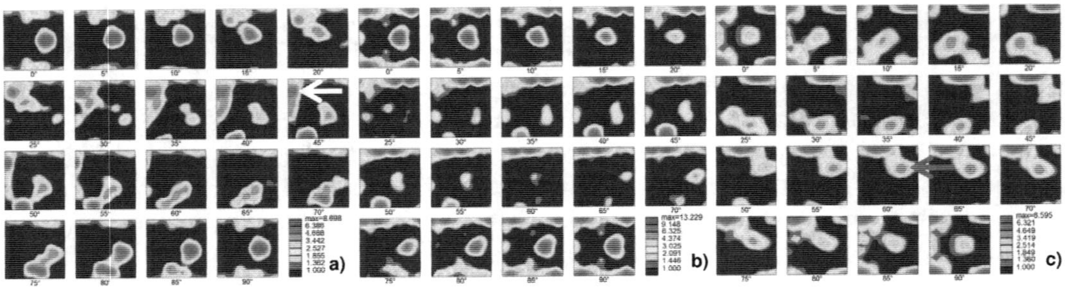

Fig. 2 ODF maps showing the textures of the three variants after heating at 200°C/h to 400°C and kept for 10^5s, where the M and S texture components are pointed out by white and red arrow, respectively.
a) C1-0, b) C1-2, C) C1-3

From the above analyses, different microchemistry states, as obtained after different homogenizations of the same alloy, have proved to significantly affect the recrystallization textures. It is hard, however, to distinguish and quantify how pre-existing dispersoids and concurrent precipitation affect the nucleation and growth behaviour. Further investigations focusing on the origin and evolution of different texture components need to be done to clarify this aspect [10]. Physically based numerical models [e.g.11,12] which take into account the interaction between recrystallization and second-phase particles may also be helpful to obtain a better understanding of this complicated interaction.

4. Conclusions

The microstructural evolution of an Al-Mn-Fe-Si alloy during annealing after cold rolling has been investigated. Specifically, the effect of microchemistry state, in terms of amount of Mn in solid solution, constituents, size and number density of dispersoids, introduced prior to cold rolling, through different homogenization treatments, on recrystallization texture has been analysed after non-isothermal heating experiments. It is clearly demonstrated that the microchemistry state strongly affects both the type of recrystallization texture components obtained and their strength. In general, high Mn content in solid solution promotes stronger concurrent precipitation and retards recrystallization, which finally leads to a coarse grain structure, accompanied by strong P and/or M texture components and a ND-rotated cube component. A refined grain structure with Cube and/or a weak P component as the main texture components were obtained when the pre-existing dispersoids are coarser and fewer, and concurrent precipitation is limited.

Acknowledgements

This research work has been supported by the KMB project (193179/I40) in Norway. The financial support by the Research Council of Norway and the industrial partners, Hydro Aluminium and Sapa Technology is gratefully acknowledged. KH acknowledges the financial support from NTNU through the Strategic Area Materials.

References

[1] Lücke K and Engler O 1990 Mater. Sci. Technol. 6 1113
[2] Vatne HE, Engler O and Nes E 1997 Mater. Sci. Technol. 13 93
[3] Somerday M and Humphreys FJ 2003 Mater.Sci.Technol. 19 20
[4] Tangen S, Sjølstad K, Furu T and Nes E 2010 Metall. Mater.Trans. A 41A 2970
[5] Huang K, Li YJ and Marthinsen K 2014 Mater. Sci. Forum 783-786 174
[6] Huang K, Wang N, Li YJ and Marthinsen K 2014 Mater.Sci.Eng. A 601 86
[7] Huang K, Li YJ and Marthinsen K 2014 Mater. Sci. Forum 793-796 1163
[8] Huang K, Li YJ and Marthinsen K 2014 Trans. Nonferrous Met. Soc. China 24 3840
[8] Liu WC, Li Z and Man C-S 2008 Mater.Sci.Eng. A 478 173
[9] Schäfer C and Gottstein G 2011 Int. J. Mater. Res. 102 1106
[10] Hersent E, Huang K, Friis J and Marthinsen K 2013 Mater.Sci.Forum.753 143
[11] Schäfer C, Song J and Gottstein G 2011Acta Mater. 57 1026

IOP Conf. Series: Materials Science and Engineering **82** (2015) 012036 doi:10.1088/1757-899X/82/1/012036

Effect of microalloying elements on the Cube texture formation of Fe48%Ni alloy tapes

Y. Ateba Betanda [a], A-L. Helbert [a], F. Brisset [a], M-H. Mathon [b], T. Waeckerlé [c] and T. Baudin [a]

[a] ICMMO UMR CNRS-8182, Université Paris-Sud, 91405 Orsay Cedex, France
[b] Laboratoire Léon Brillouin, CEA/ Saclay, 91191 Gif-sur-Yvette, France
[c] Aperam alloys Imphy, 58160 Imphy, France
Email: yanick.ateba-betanda@u-psud.fr

Abstract. The effect of microalloying elements sulfur and niobium was studied in Fe48%Ni sheets on the first stages of nucleation and on recrystallization. It was shown that the addition of sulfur promotes the formation of Cube grains while the addition of niobium prevents the Cube grains formation. Regarding sulfur, it combines with manganese which is an inhibitor of recrystallization to form the MnS precipitates. The stored energy difference between Cube component and others components $\Delta E_{Cube/other}$ increases when sulfur is added. This stored energy difference which is the main driving energy for Cube grains formation during recrystallization explains the sharpness of the Cube texture when sulfur is added. On the contrary the niobium microalloying element addition prevents the formation of Cube grains. This could be explained by the fact that stored energy of cold-rolled components decreases with the addition of niobium and thus decreases Cube grains fraction when niobium is added. In order to explain these results, the development of Cube texture during recrystallization has been investigated in detail by EBSD, furthermore, the effect of stored energy has been studied by carrying out neutron diffraction measurements on the deformed states.

1. Introduction

One way to develop the thin-film silicon photovoltaic cells with a high conversion efficiency is to make oriented silicon grains on a thin substrate. This substrate must have the Cube texture {100}<001> which promotes the silicon growing by epitaxy. Furthermore, the substrate must be non-fragile, flexible and must have a high melting temperature. The Fe48%Ni which develops the Cube texture after recrystallization [1-4] is suitable for this application [5]. In the present study, the effect of microalloying elements, sulfur and niobium on the development of the cube texture was studied.

The development of Cube texture in the FeNi alloys have been studied by some authors [1-2, 6-10]. But the effect of microalloying elements on the recrystallization of these alloys has not been yet studied. Nevertheless, this effect was studied in others alloys. Thus, Eickemeyer et al. [11] showed that sulfur (7 wt ppm content) can interfere in the formation of Cube grains during the recrystallization of Ni5%W alloys. In their study, Jakani et al. [12] demonstrated that sulfur can decrease the dislocation cell size after cold wire drawing and then decreases the kinetic of recrystallization in the copper alloys. For niobium microalloying element, most of the previous works showed that, this element in solid solution or as precipitates, delays the recrystallization process and increases the recrystallization temperature [13-15].

To obtain a highly textured Cube substrate, the material must be strongly cold-rolled and annealed at high temperatures [1-2]. After strong deformation by cold-rolling, the main components are formed: the aluminum component S {123}<634> , the copper component C {112}<111> and the brass component B {110}<112>. At the deformed state, the Cube orientation fraction is very low (about 1%). During recrystallization, it increases at the expense of cold-rolled orientations. Then the Cube orientation becomes the main orientation after recrystallization.

2. Experimental procedure

After bar casting, the alloys were hot-rolled to 5mm thick tapes at Aperam alloys Imphy. The tapes were subsequently cold-rolled to 50μm (99% thickness reduction). The samples where then annealed in vacuum atmosphere at 1000°C. To obtain flat surface, the specimens were polished. The microstructure and texture were analyzed by EBSD and OIM™ software (Orientation Imaging Microscopy). The results were collected using a scanning electron microscope FEG-SEM SUPRA 55VP operating at 25kV. The texture and microstructure measurements were done in the rolling plane (RD, TD) where RD is the rolling direction and

TD the transverse direction. In order to obtain statistical results, the scans explored an area of 1500µm*800µm with 0.1µm step size in the deformed samples and 1µm step size in the recrystallized samples. The stored energy of the deformed components in the cold-rolled samples were estimated from neutron diffraction measurement at the Laboratoire Léon Brillouin (CEA Saclay/France) on the four circles diffractometer 6T1. A pile up of samples was realized to obtain 1cm^3 cubic specimen. Moreover, the kinetic of recrystallization were followed by Vickers Micro-hardness measurement along the rolling plane. In the table 1, the chemical composition of the samples is shown.

Table 1: Chemical composition of the studied samples (% wt).

Samples	S-0	S-20	S-40	S-60	Nb-0	Nb-350	Nb-500
Ni	48	48	48	48	48	48	48
Mn	0.3	0.3	0.3	0.3	0.3	0.3	0.3
S	0	0.002	0.004	0.006	0.004	0.004	0.004
Nb	0	0	0	0	0	0.035	0.050

3. Results

3.1. Texture of the deformed state

After deformation, the microstructure and texture of sample were characterized by EBSD. It was shown the similar texture and microstructure for "S" and "Nb" samples. The texture was mainly composed of B, S and C components, the Cube component fraction is very low in these cold-rolled states.

3.2. Texture and microstructure after recrystallization

The corresponding volume fraction of the Cube orientation after recrystallization was calculated by OIM™ as well as misorientation of Cube grains around the ideal Cube orientation. When sulfur is added in the sample, the fraction of Cube grain increases and the misorientation around the ideal Cube orientation decreases (Fig.1). This result shows that the sulfur element favors the formation of Cube orientation. On the contrary, the fraction of Cube grains decreases and the misorientation increases with the niobium addition. Thus, contrary to the sulfur element, the niobium element prevents the Cube texture formation.

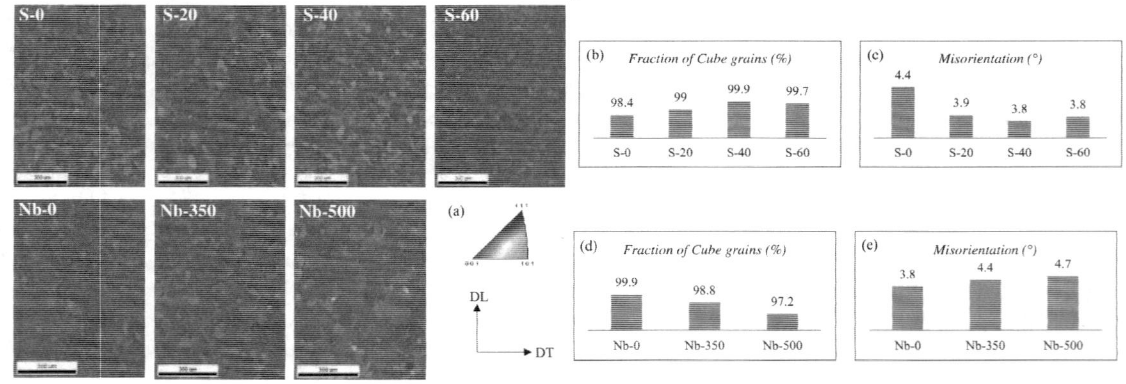

Figure 1: Orientation maps of samples after full recrystallization (a), fraction of Cube grains (b, d), average misorientation of Cube grains with 10° dispersion around the ideal Cube orientation (c, e). All samples were annealed at 1000°C during 60 minutes.

3.3. Kinetic of recrystallization and nucleation of cube grains

Fig.2 shows the effect of sulfur (a) and niobium (b) on the kinetic of recrystallization of Fe48%Ni alloy tapes. The extent of recrystallization was calculated from micro-hardness measurements, by using the following equation:

$$X_V = \frac{HV(cold-rolled\ state) - HV(t)}{HV(cold-rolled\ state) - HV(fully\ annealed)} \quad (1)$$

X_V is the extent of recrystallization and HV is the Vickers Hardness.

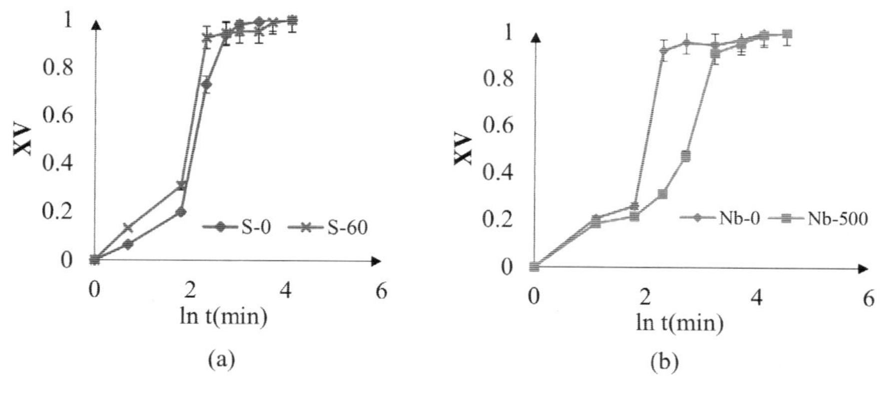

Figure 2: Evolution of the calculated recrystallized fraction determined from Vickers microhardness measurements as a function of annealing time at 600°C: (a) for S-0 and S-60 and (b) for Nb-0 and Nb-500 samples.

It is shown that recrystallization rate of S-0 samples is lower than the recrystallization rate of S-60 sample (Fig.2(a)) and that the recrystallization rate of Nb-0 is greater than the recrystallization rate of Nb-500 (Fig.2(b)). Thus, sulfur accelerates the recrystallization while niobium inhibits the recrystallization.

To complete the above results, an annealing treatment has been stopped at the beginning of recrystallization in order to study the first steps of nucleation in the extreme samples (S-0, S-60, Nb-0, and Nb-500). It was shown that the fraction of Cube nuclei increases with the sulfur content and decreases with the niobium content. The sulfur element favors the Cube grain nucleation, the main component after recrystallization, while the niobium element prevents this nucleation and these results are consistent with the recrystallization kinetic.

4. Discussion

The stored energy of cold-rolled samples was estimated from neutron diffraction measurement and Kernel Average Misorientation measurement (Fig. 3a and 3c). In the "S" samples, the stored energy of B, C and S components increases when sulfur content increases, the stored energy of the Cube component remains quasi-constant. In the "Nb" samples, the stored energy of B and S increases with the niobium content and that of C decreases with the niobium content. The stored energy of the Cube orientation is the lowest. By the SIBM (Strain Induced Boundary Migration) mechanism [10], the Cube orientation develops during the recrystallization to the detriment of B, C and S cold-rolling orientations. The increase of stored energy in the "S" sample can be attributed to the MnS precipitates formed in the Fe48%Ni in the presence of sulfur [1]. Furthermore, as reported in the literature, Mn element can hinder the recrystallization process by decreasing the grain boundary mobility [16]. Thus, in the present study, the precipitation of Mn into MnS reduces the detriment effect of Mn on the mobility of grain boundaries and could explain the speed up of the recrystallization kinetic (Fig.2). The variation of stored energy with niobium content can also be explained by the interaction between niobium and other elements in the microstructure and/or its presence in the grain boundary leading to decrease of grain boundary mobility. Consequently the nucleation is prevent by the niobium.

The stored energy difference ($\Delta E_{Cube/other}$) between the Cube orientation and the main B, C and S cold-rolling components (other) was calculated (Fig. 3b and 3d) to explain the fraction of Cube orientation evolution with sulfur and niobium content after recrystallization. This energy difference is the driving energy for the development of Cube grains. It is shown that the $\Delta E_{Cube/other}$ increases in the "S" samples when sulfur content increases and decreases in the "Nb" samples when niobium content increases. The increase of $\Delta E_{Cube/other}$ in the "S" samples promotes the Cube grain formation and growth during the recrystallization, while the diminution of $\Delta E_{Cube/other}$ in the "Nb" samples decreases the fraction of Cube grains with the niobium content.

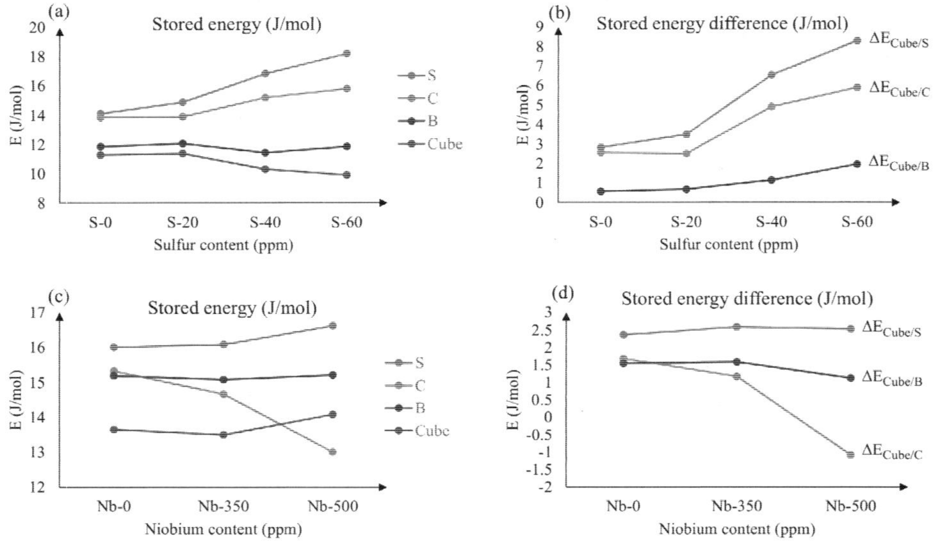

Figure3: (a), (c) Stored energy (J/mol) associated to B, C, S and Cube orientations as a function of sulfur and niobium content, respectively. (b), (d) stored energy difference between Cube orientation and B, C and S orientations and Cube orientation as a function of sulfur and niobium content, respectively.

5. Conclusions

The effect of microalloying elements sulfur and niobium on the formation of the Cube texture in the Fe-48%Ni has been investigated. It was shown that sulfur and niobium microalloying elements favors and inhibits the Cube orientation formation, respectively. It appears that the stored energy difference between the Cube orientation and the B, C and S cold-rolling orientations is the main factor for the development of the Cube orientation. In addition, sulfur accelerates the nucleation and the development of Cube grains while niobium inhibits theses process.

References

[1] Ateba Betanda Y A, Helbert A L, Brisset F, Wehbi M, Mathon M H, Waeckerle T, Baudin T 2014, *Adv. Eng. Mater.*
[2] Etter A L, Mathon M H, Branger V, Baudin T, Penelle R 2002 *Scripta Mater.*, **46**, 311.
[3] Branger V, Mathon M H, Baudin T, Penelle R 2000 *Physica B*, **276**, 914.
[4] Branger V, Mathon M H, Baudin T, Penelle R 2000 *Scripta Mater.*, **43**, 325.
[5] Reyal J P, Reydey P L, *European Patent* (2009) No. EP 2031082. *Munich, Germany: European Patent Office*.
[6] Baudin T, Julliard F, Paillard P, Penelle R 2000 *Scripta Mater.*, **43**, 63.
[7] Zaefferer S, Baudin T, Penelle R 2001 *Acta mater.*, **49**, 1105.
[8] Li X L, Liu WGodfrey, A, Juul Jensen D, Liu Q 2007 *Acta Mater.*, **55**, 3531.
[9] Caleyo F, Baudin T, Penelle R 2002 *Scripta Mater.*, **46**, 829.
[10] Penelle R, Baudin T 2010 *Adv. Eng. Mater.*, **12**, 1047.
[11] Eickemeyer J, Selbmann D, Opitz R, Wendrock H, Baunack S, Subramanya Sarma V, Holzapfel B, Schultz L, Maher E 2005 *Physica C*, **418**, 9.
[12] Jakani S, Baudin T, De Novion C H, Mathon M H 2007 *Mater. Sci. Eng. A*, **456**, 261.
[13] Wen Bo Don et al. 2007 *Material Sciences Forum*, **77**, 561.
[14] Fabregue P 2000 *Technique de l'ingénieur : Metallurgie du laminage à chaud*
[15] Storozheva L M, Escher K, Bode R, Hulka K, Burko D A 2002 *Metal Sciences and Heat treatment*, **44**, 102.
[16] Jeniski R A, Thanaboonsombut B, Sanders T H 1996 *Metall. Mater. Trans. A*, **27**, 19.

Misorientation characteristics of rapid moving boundaries during grain growth in Fe-3%Si alloys

W Mao and P Yang
Department of Materials, State Key Laboratory of advance Metals and Materials, University of Science and Technology Beijing, Beijing, China
E-mail: wmmao@ustb.edu.cn

Abstract. Based on a large amount of experimental observations on highly mobile Σ17, Σ19, and Σ9 coincidence site lattice boundaries (CSLB), the average jump distances of atoms in connection with misorientation were calculated during boundary migration, which indicated that Σ9 CSLB demonstrate not only a stable structure, but also higher mobility than other boundaries because of the actual lowest jump distance if the boundaries move. The higher mobility could even keep more or less if the misorientation of boundaries deviates from exact Σ9 in a limited extent. Therefore, Goss grains could grow rapidly in the matrix with many {111}<112> grains since the misorientation between Goss and {111}<112> is close to Σ9.

1. Introduction

The mobility of grain boundaries in Fe-3% Si alloys has long been a topic of concern and is important in clarifying the mechanism of Goss texture formation in grain-oriented electrical steels. It has been observed that both of the activation enthalpy and the corresponding pre-exponential factor for grain boundary migration in polycrystalline Fe-3%Si alloys are much lower than those observed in bicrystals [1], and the activation enthalpy should be also related with the atom jump distances during boundary migration. Ibe and Lücke conducted the first systematic investigation on boundary mobility in Fe-3% Si alloys, while growth competition of recrystallization grains was observed statistically [2]. The orientation relationship between the crystal matrix and the recrystallization grains that had survived growth competition was determined. The orientation relationships were characterized mainly around 26.7°<110> and 84°<110>, which are very close to the coincidence site lattices (CSL) Σ19 (26.5°<110>) and Σ17 (86.6°<110>), respectively [2], i.e., the corresponding Σ19 and Σ17 boundaries are more active than other boundaries. At the time the study was conducted, however, the reasons behind the phenomena observed were unclear.

Studies have confirmed that Σ9 boundaries in Fe-3% Si alloys are very mobile [3,4]. Goss grains bounded by high frequency Σ9 boundaries [2,5] have been found to grow abnormally with the help of inhibitor particles during secondary recrystallization. High-angle Σ9 boundaries are believed to have lower energies and are consequently less pinned by second phase particles, which induce their higher mobility [4,6]. High frequencies of Σ1 and Σ3 boundaries have been observed in Fe-3% Si alloys, and both boundaries are considered to have significantly low mobility, since Σ1 has low angle misorientation and Σ3 has a twin orientation [3,5]. High frequencies of Σ11 and Σ5 boundaries have also been observed around the periphery of recrystallized grains in Fe-3% Si alloys [3,5]. Given that Σ11 and Σ5 are high-angle and low-energy boundaries, why these boundaries (i.e., Σ11 and Σ5) are not as mobile as that of Σ9 [3] is not yet fully clear.

2. Atom jumps during boundary motion

Grain boundary migration during annealing is a thermal activation process where atoms of one grain jump over the boundary to the atom positions in the lattice of a neighboring grain. The process is rather complicated, and the jump distance plays an important role in the velocity of boundary migration. The activation energy of atomic jumps increases with increasing jump distance. A simplified impression of boundary mobility could be obtained if jump distance, which is related to the difficulty of the atom jump and the magnitude of the corresponding boundary mobility, is calculated and analyzed in a purely geometrical point of view.

Supposing that atoms of a grain rotate around the normal direction [hkl] of an (hkl) plane about an angle and reach the atom positions in the lattice of a neighboring grain, the atoms can jump for different distances on the common (hkl) plane of the two grains regularly when the boundary is in between moves. If the two grains have a common {110} plane, such as in bcc metals, or a common {111} plane, as in fcc metals, the jump distance should be, in general, much lower that those with other common {hkl} planes different from {110} or {111}, since these two planes are the most closely packed planes in bcc or fcc metals. Therefore, the most mobile boundaries often concern a common {111} or <111> in fcc metals [7,8], as well as a common {110} or <110> in bcc metals [2]. The atomic jump distance becomes generally larger than that on a common {110} plane if the atoms jump on another {hkl} plane or if the two grains do not have a common {110} plane, in which case high boundary mobility is not very favorable. Thus, a focused discussion on the misorientation relationships of <110> rotations is necessary when highly mobile boundaries in bcc metals are a concern.

If a (110) plane rotates around its normal direction [110] about a small angle (Fig. 1a), the removed atoms (open symbols) should jump for different distances to reach their original positions (gray symbols). If the plane rotates about such an angle that a CSL is formed between the rotated and original lattices (e.g., Σ9 in Fig. 1b), the atoms at the lattice points of CSL (black symbols) do not need to move in order to recover the original lattice. Figure 1c shows a Σ9 CSL boundary, in which the possible jumps between dashed circles in gray and white are indicated as the CSL boundary between two grains moves.

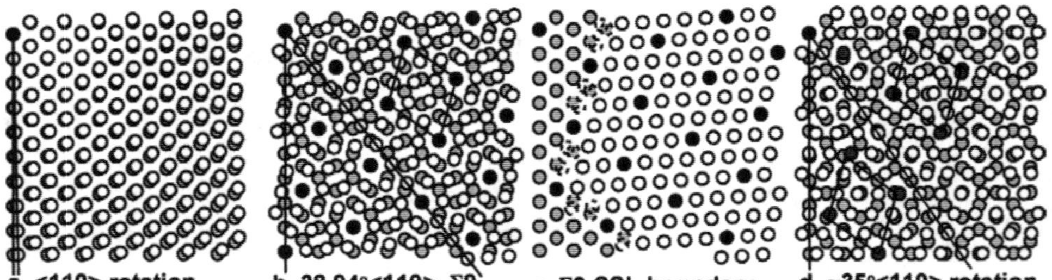

Figure 1. <110> rotation and Σ9 boundary around 38.94° <110> (in a {110} plane)

The average jump distance $\bar{d}$ of n atoms can be calculated according to the individual jump d_k as $\bar{d} = \frac{1}{n}\sum_{k=1}^{n} d_k$, in which the two possible atomic arrangements, i.e., the stacking sequence ...ABABAB... of (110) planes, should be considered.

The average jump distance of a (110) plane with about 8 million atoms (2000 × 2000 × 2) was calculated at different rotation angles ranging from 0° to 90°. The average distance, in general, is about 0.369b (in Fig. 2a, b is the length of Burgers vector) for the [110] rotation, except for some special misorientations with low Σ values.

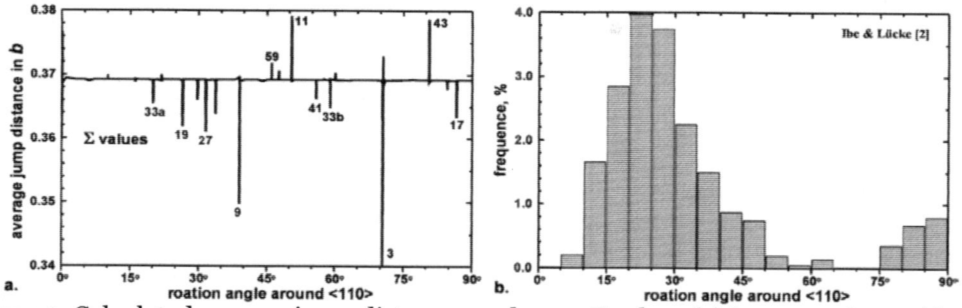

Figure 2. Calculated average jump distances and some Σ values (a) in comparison with the observed frequency of rapid moving boundaries under <110> rotation relations [2] (b)

The calculation implies that the lowest average jump distance is 0.215b at 70.53°<110> (Fig. 2a). The corresponding boundary under this misorientation is Σ3; however, the mobility of the boundary in Fe-3% Si alloys is very low because it has a twin orientation [3,5]. The second lowest average jump distance is about 0.3497b at 38.94°<110> (Fig. 2a). The corresponding boundary under misorientation Σ9 is very mobile in Fe-3% Si alloys and can migrate easily as previously reported [3–6]. The average jump distance at many positions with low Σ values is much lower than the general value, which may allow the higher mobility of corresponding CSL boundaries. On the other hand, the average jump distance at some low Σ positions, e.g., Σ11, is higher than the general value, indicating that not all CSL boundaries can offer higher mobility.

3. Discussion

The misorientation of <110> rotations leads to relatively lower average jump distances and higher mobility of corresponding CSL boundaries. Thus, the misorientations of <100>, <111>, or other <uvw> rotations are not discussed firstly. Table 1 shows possible reciprocal densities (Σ⩽19) of some CSLs of cubic metals, as well as their corresponding rotation axes and angles. The table shows that Σ5, Σ7, and Σ13 are not described by the misorientation of <110> rotations. As such, Σ5, Σ7, and Σ13 boundaries in bcc metals should not be as mobile as Σ9 boundary. Σ3 represents a twin orientation [3,5] and Σ11 shows a very high average jump distance (Fig. 2a). The only mobile boundaries left at Σ⩽19 are Σ9, Σ17b, andΣ19a [9] (Table 1), of which Σ9 may be expected to be the most mobile because it has the lowest average jump distance. The calculations (Fig. 2a), comparison (Table 1), and analysis above agree with observations from Fe-3% Si single crystal experiments for Σ17 and Σ 19 boundaries [2], as well as those of Fe-3%Si polycrystalline alloys for Σ9 boundaries [3–6].

Table 1. Possible rotation axis, angles, and reciprocal density Σ of CSL in cubic metals

Rotation	Reciprocal density Σ							
axis	3	5	7	9	11	13	17	19
<100>		36.87°				22.62°(a)	28.07°(a)	
<110>	70.53°			38.94°	50.48°		86.63°(b)	26.53°(a)
<111>	60.00°		38.21°			27.80°(b)		46.83°(b)

Figure 2b summarizes the misorientation distributions of rapidly moving boundaries in Fe-3% Si alloys, as observed by Ibe and Lücke [2], in which the misorientation distribution around 26.7°<110> also includes the areas of Σ33a, Σ27, Σ9, and other CSL boundaries with low average jump distances aside from Σ19 boundaries (Fig. 2b versus Fig. 2a). Therefore, other CSL boundaries aside from Σ19 are also responsible for the rapid movement of boundaries observed. The atom jumps between two grains without common {110} should not occur on single common plane if a boundary in between moves, and a quite mobile boundary could

then be obtained if the jump distance is short enough, in which the second fastest Σ5 boundary [3] might be the case. The detailed mechanism needs to be investigated.
Boundary mobility is misorientation-dependent and determined by the orientations of the growing grains and the matrix in which they grow. The texture of Fe-3% Si alloys plays an important role in boundary mobility. Inducing the nuclei of all kinds of orientations during the annealing of 20% rolled single crystals is not possible [2], and the nuclei do not always achieve the frequency necessary to be surrounded by Σ9 boundaries. Therefore, while they should show maximum mobility because they have minimum average jump distances (Fig. 2a), Σ9 boundaries do not appear with the highest frequency after the growth competition of recrystallized grains (Fig. 2b) [2]. A rather strong {111}<112> texture is often observed in rolled and recrystallized Fe-3% Si alloys sheets. The misorientation between the {111}<112> texture and the expected Goss {110}<001> texture is about 35°<110>, which is very close to 38.94°<110> (Table 1). This observation indicates that Goss grains are commonly bounded by high-frequency Σ9 boundaries [2,5], and that the high mobility of Σ9 boundaries [3,4] promotes the abnormal growth of Goss grains.
In fact, exact Σ9 boundaries are hardly obtained in real Fe-3% Si alloys. The reciprocal density of the CSL becomes extremely large and the jump distance approaches the general value 0.369b (Fig. 2a) by very small deviation from the ideal Σ9 boundary (Fig. 1b). The quadrangle in Fig. 1b indicates a unit cell in the Σ9 CSL. Similar unit cells can be found when the misorientation is very close to 38.94°<110>, e.g., at 35°<110> (Fig.1d), which indicates that a misorientation occurs between {111}<112> and {110}<001>. The ideal Σ9 CSL (Fig. 1b) is more likely to appear with higher stability than with random arrangements (Fig. 1d) if the misorientation of two neighboring grains is very close to 38.94°<110>. The tendency to form a Σ9 CSL in the boundary area of an unideal Σ9 CSL (Fig. 1d) increases as the boundary moves closer, and the atomic arrangement tends to change into ideal Σ9 CSL cells with some lattice defects. The boundary can then move over rapidly while the average atomic jump distance is shortened, which could explain the high mobility of 35°<110> boundaries [3,4].

4. Conclusion
Boundaries between two grains in Fe-3% Si alloys migrate generally faster when a common {110} plane exists between them, since the distance an atom jumps in the most closely packed plane becomes generally shorter as the boundaries move. The average jump distances calculated indicate that Σ9, Σ17, and Σ19 boundaries ($\Sigma \leqslant 19$) are highly mobile, of which Σ9 should be the most mobile. The observed phenomena agree with experimental observations and helps explain Goss texture formation if a certain deviation from Σ9 is considered. The mechanism of high mobile Σ5 boundary needs to be investigated in this aspect.

Acknowledgment
The authors would like to acknowledge the financial support given by the National Natural Science Foundation of China (Grant No. 51171019) and State Key Laboratory of advance Metals and Materials (Grant No. 2014Z-14).

References
[1] Mao W, Zhang M and Yang P 2014 *Steel research inter.* **85** 1215
[2] Ibe G and Lücke K. 1968 *Archiv für das Eisenhüttenwesen* **39** 693
[3] Lin P, Palumbo G, Harase J and Aust KT 1996 *Acta Mater.* **44** 4677
[4] Kumano T, Hatatani T and Ushigami Y 2004 *ISIJ International* **44** 1888
[5] Morishige N, Murakami K and Homma H 2007 *Mat. Sci. Forum* **558-559** 771
[6] Homma H and Hutchinson B 2003 *Acta Mater.* **51** 3795
[7] Ibe G, Dietz W, Fraher AC and Lücke K. 1970 *Z. Metallkd.* **61** 598
[8] Mao W. 2002 *Mat. Sci. Forum* **408-412** 269
[9] Brandon DG, Ralph B, Ranganathan S and Wald MS 1964 *Acta Met.* **12** 813

IOP Conf. Series: Materials Science and Engineering **82** (2015) 012038 doi:10.1088/1757-899X/82/1/012038

Regularities of recrystallization in rolled Zr single crystals

M. Isaenkova, Yu. Perlovich[1], V. Fesenko, O. Krymskaya, N. Krapivka and S. S. Thu
National Research Nuclear University MEPhI (Moscow Engineering Physics Institute), Moscow, Russia
E-mail: [1]yuperl@mail.ru

Abstract. Experiments by rolled single crystals give a more visible conception of the operating mechanisms of plastic deformation and the following recrystallization, than experiments by polycrystals. Studies by usage of X-ray diffraction methods were conducted by Zr single crystals. It was revealed, that regions of the α-Zr matrix, deformed mainly by twinning, are characterized with decreased tendency to recrystallization. Orientations of recrystallized α-Zr grains correspond to "slopes" of maxima in the rolling texture, where the level of crystalline lattice distortion is maximal and the number of recrystallization nuclei is most of all.

1. Introduction

The use of single crystals (SC) in experiments, investigating mechanisms of plastic deformation and following recrystallization, allows to reveal preferential modes of texture formation in polycrystalline zirconium and its alloys under main technological processes. Earlier studies of the plastic deformation mechanisms were carried out by coarse-grained polycrystalline zirconium [1-2]. A series of papers [3-4], performed on SCs, were devoted to estimation of critical shear stresses for different slip and twinning systems in zirconium, deformed by uniaxial tension or compression. In this paper, plastic deformation mechanisms by cold rolling and subsequent recrystallization of deformed Zr SCs were studied by use of X-ray diffraction methods.

2. Samples and methods of investigation

The studied plates 7x10x3 mm^3 in size were cut out by the electro-erosion method from cylindrical SC of pure Zr, obtained by solid-phase over-crystallization. SCs with different initial orientations were rolled on a laboratory mill up to various deformation degrees from 5 to 98% with ~5% reduction per pass. X-ray measurements of SCs were conducted after each stage of its plastic deformation. SC reorientation under rolling was analyzed by direct pole figures (PF). These PFs were constructed on the basis of data, obtained by the X-ray diffractometer DRON, equipped with an automatic texture attachment and utilizing the filtered radiation Cr. Distributions of strain hardening in rolled SCs were studied by X-ray method of generalized pole figures (GPF) [5], measured by diffractometer D8 Discover with the filtered Cu radiation and self-developed software for a description of the line profile at each point of the stereographic projection. All rolled samples were annealed in a vacuum furnace at 580°C for 3 h, which is a regime, used usually for recrystallization of products from Zr-based alloys.

3. Regularities of rolling texture formation

Development of texture in polycrystalline zirconium and its alloys during deformation and heat treatment is described in detail in [2]. The same steps of rolling texture formation as in polycrystalline zirconium-based alloys are observed in rolled SCs. Repeatedly described formation in α-Zr of the final stable rolling texture (0001) ± (30-40 °) ND-TD <$10\bar{1}0$> (texture T2), where ND and TD - normal and transverse directions in the rolled plate (Fig. 1-d), is preceded by the texture T1, stable at intermediate deformation degrees (50-70%) (0001) ±15-20°ND-RD <$11\bar{2}L$> (Fig. of 1-c), characterized by the deviation of the basal axes by 15-20 ° from ND to RD. Variation of the initial SC orientation with respect to the RD is conditioned by mechanisms, underlying the formation of rolling texture at successive stages

of its development. Incomplete PF {11$\bar{2}$0} with an angular radius of 80° for the investigated SC is shown in Fig. 1-a. The axis of the initial cylindrical SC, which coincides with the ND of a plate, cut of it, is in the center of the PF. The output of basal axis [0001] is shown in the same PF near the circumference of the stereographic projection with an angular radius of 80° . It is represented by circular contours and spaced at 90° from outputs of prismatic axes <11$\bar{2}$0>.
This investigation was performed for SC plates of three initial orientations, differing only in the RD. These orientations correspond to the following angular coordinates of basal axes: N 1 - (70°, 33°); N 2 - (70°, 152°); N 3 - (70°, 87°). Three columns of PFs in Fig. 1 demonstrate successive stages of rolling texture development in SCs of indicated orientations: N 1 – (*a-d*), N 2 – (*e-h*), N 3 – (*i-l*). First stages of all three successions in Fig. 1 correspond to deformation degree ε≈5-7% in order to create mutually differing distributions of SC fragments due to various positions of twinning systems relative to the RD.

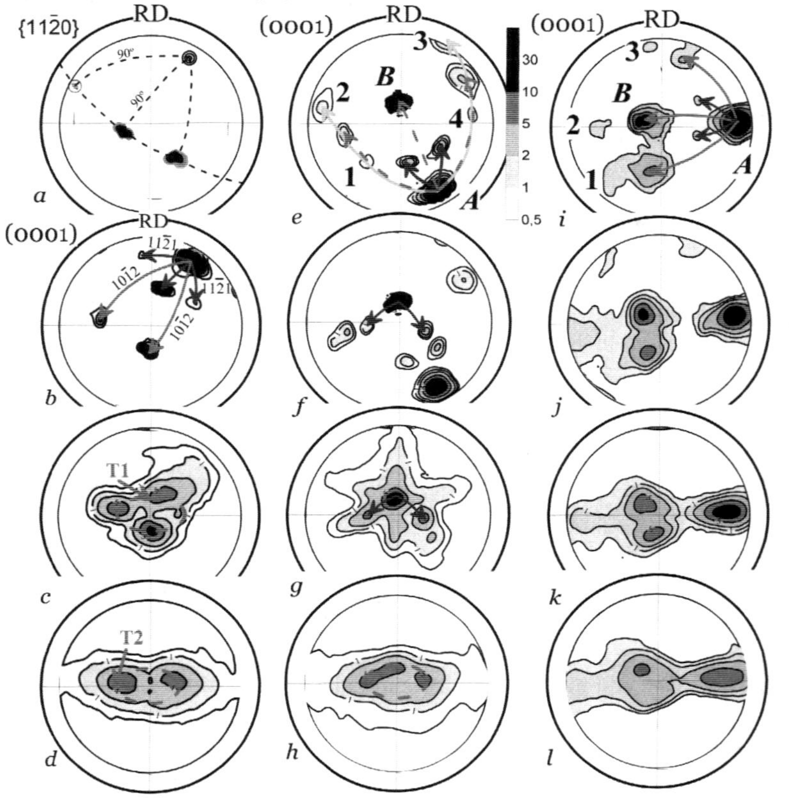

Fig. 1. PF {11$\bar{2}$0} of initial SC (a) and variation of PF(0001) under cold rolling of Zr SC with different orientations: SC N 1: *b* – ε≈5 %; c – ε≈53 %; *d* – ε≈80 %; SC N 2: *e* – ε≈5 %; *f* – ε≈13 %; *g* – ε≈51 %; *h* – ε≈71 %; SC N 3: *i* – ε≈7 %; *j* – ε≈30 %; *k* – ε≈50 %; *l*– ε≈70 %

4. Twinning as the main mechanism of deformation in SCs of α-Zr

When the strain is only 5-7%, the texture maximum of PF (0001), representing the initial orientation of the SC, is divided into 5 – 9 new maxima, corresponding to reorientation of the initial SC by twinning in planes {10$\bar{1}$2}, {10$\bar{1}$1} and {11$\bar{2}$1} (red, green and blue arrows in Fig. 1-b, *e, i,* respectively). These planes are identified by the characteristic angles of reorientation: 85, 57, 123 or 35°. At that, maxima, situated near ND, have the greatest intensity. Additional weak texture maxima, marked with figures 1-4 in Fig. 1-*e, i*, form due to twinning of the initial single crystal by planes {10 1} (Fig. 1-*e*). Twinning by planes {10$\bar{1}$1} previously did not considered by analysis of rolling texture formation in α-Zr [2-6]. Described above variations of basal axis orientations under cold rolling of SCs N 1 and N 2 are quite consistent with the diagram of Hobson [6], indicating mechanisms of plastic deformation, which are most active in α-Zr grains with different orientations of basal axis relative to ND, RD and TD. Twinning process continues as long as there are SC areas, having its initial orientation. At subsequent stages of deformation of SC N 2 it become possible to detect activation of twinning in planes $\{11\bar{2}1\}$. In Fig. 1-*f*, along with the initially formed texture maxima, new maxima appear. They are conditioned by secondary twinning on planes $\{11\bar{2}1\}$ in grains with orientation B (Fig. 1-*f*, *g*, blue arrows). Such twinning apparently also continues until there are grains with orientation corresponding to texture T1. Twinning is responsible for transition from texture T1 to texture T2, which is manifests in a shift of texture maxima from quasi-stable positions near the diameter of the stereographic projection ND-RD to its final stable positions at the diameter ND-TD. Such shift of texture maxima has an evident jump-like character along with slower flow of pole density in the same direction. This indicates, that along with pyramidal and basal slip, causing gradual flow of texture maxima, twinning by planes $\{11\bar{2}1\}$ is an active mechanism of α-Zr deformation. which is not suppressed by lattice distortions in separate blocks or coherent domains of the rolled SC.

Twinning in planes $\{10\bar{1}2\}$ and $\{11\bar{2}1\}$ operates also in SC N 3, where the basal axis is oriented along TD. The effect of twinning in plains $\{10\bar{1}2\}$ and $\{11\bar{2}1\}$ is seen especially at first stages of rolling. According to [6], in SC of this orientation prismatic slip systems work, but, due to equilibrium of these systems, the orientation proves to be stable and the basal axis does not deviate from TD (Fig. 1-i, j).

5. Recrystallization of rolled α-Zr single crystals

Analysis of texture changes in rolled Zr single crystals by recrystallization was carried out in detail for SC N 1, deformed by 53%. PF (0001) of rolled SC is shown in Fig. 1-c, and PF (0001) after its recrystallization - in Fig. 2-a. For rolled single crystals X-ray method of generalized pole figures (GPF) was used. This method consists in measuring of X-ray line profiles in each point of texture PF, so that the true width at half-height of X-ray line profile (the line broadening) characterizes strain hardening of corresponding grains [5].

The line broadening of X-ray line (0002) is denoted as β_{0002}. It is used as a measure of grain distortion, increasing by strain hardening and removing by recrystallization. At that, separate fragments of rolled SC respond to a wide spectrum of substructure conditions, i.e. values of β_{0002}, distribution of which in the stereographic projection of sample is depicted by GPF β_{0002}. For the studied rolled SC GPF β_{0002} is presented in Fig. 2-b.

Fig. 2-c represents the subtraction diagram $[PF(0001)_{recr} - PF(0001)_{def}]$, where differences of texture pole densities, calculated for each point (ψ, ϕ) of stereographic projection were drawn; pole densities were measured for basal axes distributions both in recrystallized and rolled samples. The subtraction diagram is presented within the region, limited by a dotted line and corresponding to pole density values above 0,5 in the PF (0001) for the rolled sample. This subtraction diagram visually demonstrates the location of maximal changes in PF (0001) as a result of recrystallization. Gradations of the red color in the difference diagram $[PF(0001)_{recr} - PF(0001)_{def}]$ indicate its positive values and correspond to an increase in the pole density due to recrystallization, whereas the lack of color agrees with a decrease in the pole density upon recrystallization and characterizes grains, deformed with predominant participation of twinning.

Due to the use of rolled α-Zr SCs, rolling textures, formed mainly by twinning, were obtained. Namely twinning is responsible for formation of the rolling texture with maxima, which, in spite of relatively high deformation degrees, take up transition positions between maxima of stable textures T1 and T2. The difference diagram shows, that by recrystallization of the deformed SC in PF regions, formed by twinning, the pole density does not grow (Fig. 2-c, e).

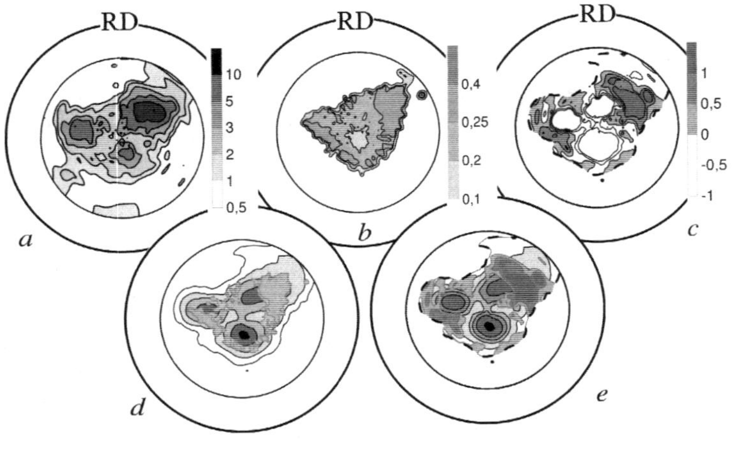

Fig. 2. Recrystallization of the rolled Zr SC N 1, deformed up to 53%:
a –PF (0001) of recrystallized sample;
b – GPF β_{0002} cold rolled SC N 1;
c – [PF(0001)$_{rec}$ - PF(0001)$_{def}$] (red color indicates regions of higher values of this difference);
d – superposition of PF(0001) and GPF β_{0002};
e – superposition of PF(0001) and regions of maximal growth of pole density upon recrystallization.

In grains, corresponding by their orientation of basal axes near normal direction (ND), only pyramidal slip can be activated under rolling, since values of shear stress for pyramidal slip in these grains are high. Other slip systems in these grains are suppressed because of low values of Schmidt factor. Therefore, in α-Zr grains, reoriented by predominant participation of twinning, strain hardening, i.e. distortion of crystalline lattice and fragmentation of substructure elements, is relatively small, so that in the course of subsequent annealing recrystallization nuclei do not form in these grains. As a result, they are swallowed up by grains, which are growing more actively and correspond by their orientations to regions of PF with increased strain hardening. At the same time difference diagrams, constructed for the rolled SC, show, that its texture maxima are bordered with zones of intensive recrystallization (Fig. 2-e). Namely in these regions of stereographic projection the increased distortion of grains is observed also (Fig. 2-b, d) due to above-shown simultaneous operation of pyramidal and basal slip. According to GPF β_{0002}, slopes of texture maxima respond to regions of the greatest fragmentation and distortion of crystalline structure [5]. Thus, in the considered case the main mechanism of α-Zr recrystallization consists in the growth of regions within the deformed matrix, characterized by increased energy of residual distortions.

6. Conclusions

1. The main plastic deformation mechanism in α-Zr SCs under rolling is twinning in planes $\{10\bar{1}2\}$, $\{10\bar{1}1\}$ and $\{11\bar{2}1\}$, but calculated diagrams, predicting active twinning systems by values of Shmid factor, are correct not for all grain orientations, - in particular, within the region of prismatic slip the twinning by planes $\{10\bar{1}2\}$ and $\{11\bar{2}1\}$ operates.
2. Regions of α-Zr matrix, deformed by predominant participation of twinning, show a low strain hardening and a decreased tendency to recrystallization.
3. Recrystallization of α-Zr SCs, rolled up to $\varepsilon \approx 50\%$, develops by growth of new grains at slopes of texture maxima, showing that slip participated in deformation along with twinning.

References

[1] Tenckhoff E 1988 *Deformation mechanisms, texture and anisotropy in Zirconium and Zircaloy* (ASTM, Special technical publication (STP 966), Philadelphia)
[2] Isaenkova M G, Perlovich Yu A 1987 *Fizika metallov i metallovedenie* **64** (1) p 107
[3] Akhtar A 1975 *Met. Trans.* A **6** p 1217
[4] Akhtar A 1973 *Acta Met.* **21** (1) p 1
[5] Perlovich Yu, Bunge H-J, Isaenkova M 2001 *Materials Science Forum* **378-381** p 180
[6] Hobson D O 1968 *Trans. Met. Soc. AIME* **242** p 1105

IOP Conf. Series: Materials Science and Engineering **82** (2015) 012039 doi:10.1088/1757-899X/82/1/012039

Modeling of recrystallization texture of aluminium: symmetric and asymmetric rolling

K Wierzbanowski[1], M Kotra[1], M Wronski[1], K Sztwiertnia[2], S Wronski[1] and A Lodini[3]

[1] AGH University of Science and Technology, Faculty of Physics and Applied Computer Science, al. Mickiewicza 30, 30-059 Kraków, Poland

[2] Institute of Metallurgy and Materials Science, Polish Academy of Sciences, ul. Reymonta 25, 30-059 Kraków, Poland

[3] LISM, UFR SEN, Université de Reims Champagne Ardenne, Moulin de la Housse BP1039, 51687 Reims, France

E-mail: wierzbanowski@fis.agh.edu.pl

Abstract. In some metallic materials the dominating recrystallization mechanism can be described by the oriented growth behaviour. Phenomenological laws state that in selected materials only these nuclei grow intensively which have a given misorientation relation with the deformed matrix. This description is frequently verified in f.c.c. metals and generally reported misorientations correspond approximately to 40^0 rotation around the <111> axis. Basing on the above ideas the recrystallization model, including the compromise condition, was formulated and applied to the study of recrystallization textures of rolled polycrystalline aluminium.

1. Introduction

Texture development in f.c.c. metals was studied and described in the literature using various kinds of models (e,g., [1-5]. In some of these models the stored elastic energy (and residual stresses) play the essential role (e.g., [5-9]). However, a relatively simple orientation relation between growing nuclei and deformed matrix was found in aluminium, which can explain the formation of its recrystallization texture. Preferred misorientations, Δg, between nuclei and deformed grains, correspond to high mobility of grain boundaries. They are traditionally described as a rotation $\omega_0 \cong 40^o$ around one of <111> axes. Therefore, the migration rate of grain boundary, u, between a nucleus and the deformed matrix depends very strongly on misorientation Δg. This relation was reported in [10] and [11]. Based on the above assumptions, a successful recrystallization model was formulated [5, 12]. It was also shown that predicted results are much improved if the so called *compromise condition* and, consequently, the *compromise function* are used. The proposed model was used to predict recrystalization textures of symmetrically and asymmetrically rolled aluminum alloys: Al-0.007% Fe and Al 6061. Let us describe now in more details the recrystallization model used in the present calculations.

2. Recrystallization model

2.1. Boundary migration rate

Each deformed grain is statistically consumed by many nuclei but with very different rates. The crystal misorientation Δg_{ij} between the nucleus i and the deformed grain j is expressed by the rotation axis $\mathbf{r}_{ij}$ and the misorientation angle ω_{ij}. Obviously, in general a determined rotation axis $\mathbf{r}_{ij}$ for a given misorientation Δg_{ij} does not exactly coincides with one of <111> axes; let us assume that it makes the angle α_{ij} with the nearest <111> axis of. Similarly, the ω_{ij} misorientation angle does not equal exactly to ω_0 (where ω_0 corresponds to the maximum migration rate). Therefore, basing on the results presented in [10] and [11] the boundary migration rate, $u_{ij}(\omega_{ij}, \alpha_{ij})$ between a nucleus i and a deformed grain j, was approximated as:

$$u_{ij}(\omega_{ij}, \alpha_{ij}) = u_{max}\, e^{\frac{-(\omega_{ij}-\omega_0)^2}{\Delta\omega^2}} \cdot e^{\frac{-\alpha_{ij}^2}{\Delta\alpha^2}} \tag{1}$$

The first exponential function describes the dependence of migration rate on the misorientation angle ω_{ij}. In the second exponential function the effect of deviation of a current rotation axis $\mathbf{r}_{ij}$ from the nearest <111> axis (α_{ij} angle) is taken into account. In the above formula the following constants were adjusted from experimental data [12]: $\omega_0=40^0$, $\Delta\omega=13.7^0$, $\Delta\alpha=7.5^0$ and $u_{max}=4.56$ mm/min.

2.2 Compromise function

The compromise function is defined as the fraction: $S(g_i)=V_{cons(i)}/V_0$, where V_0 is the total sample volume and $V_{cons(i)}$ is a volume fraction of all deformed grains (numbered by j) into which a nucleus i (with orientation g_i) can penetrate with some minimum migration rate: $u_{ij} \geq u_{min}$. Taking into account that $V_{cons(i)} = \sum_j V_j$, the compromise function is finally expressed as:

$$S(g_i) = \frac{\sum_j V_j}{V_0} \tag{2}$$

The value u_{min}=0.001 mm/min was used in the calculations.

2.3 Calculation of recrystallization texture

The deformed sample was presented as N=5000 grains of the same initial volume with crystal orientation distribution corresponding to the deformation texture. An array of nuclei was generated next with crystal orientations contained in the considered range of the Euler angles space. In the case of sample orthorhombic symmetry (and cubic crystal symmetry) 6859 nuclei were used and they had orientations in the $0 \leq \varphi_1, \phi, \varphi_2 \leq \pi/2$ range. Next, basing on the compromise function, a step function $c(g)$ is defined, which decides whether a given nucleus with orientation g_i will grow or not:

$$c(g_i) = \begin{cases} 1, & \text{if } S(g_i) \geq s_0 \\ 0, & \text{if } S(g_i) < s_0 \end{cases} \tag{3}$$

where s_0 is the assumed level of compromise [12].

It is considered next that a given deformed grain j can be consumed by many nuclei simultaneously. The following assumption was done: the volume ΔV_{ij} of the jth deformed grain 'consumed' by the ith nucleus is:

$$\Delta V_{ij} = \frac{u_{ij}}{\sum_{k=1}^{M} u_{kj}} \cdot V_j \tag{4}$$

where V_j is the initial volume of the jth grain and the summation on k is done on M 'active' nuclei, which fulfil the compromise condition (i.e., for them: $c(g_k)=1$). And finally, a total volume of the deformed matrix, 'consumed' by the ith nucleus is:

$$\Delta V(g_i) = \sum_{j=1}^{N} \Delta V_{ij} \tag{5}$$

where N is the number of grains in the deformed sample. Obviously, it is assumed for non-growing nuclei ($c(g_i)=0$) that: $\Delta V(g_i)=0$. Finally, the recrystallization texture is calculated as:

$$f(g_i) = a \cdot n(g_i) \Delta V(g_i) \tag{6}$$

where: $n(g_i)$ is the volume fraction of nuclei with g_i orientations and a is a normalization constant.

3. Results

3.1. Recrystallization texture of symmetrically rolled material

The ODF of starting rolling sample of aluminium was represented by 5000 grains of the same volume and corresponding to 50% rolling reduction. It is shown in Fig. 1a. This texture contains *copper* (φ_1=90°, ϕ=35°, φ_2=45°), *brass* (φ_1=35°, ϕ=45°, φ_2=0°) and *S* (φ_1=59°, ϕ=37°, φ_2=63°) ideal orientations. The compromise function, *S(g)*, calculated on the basis of this texture, is presented in Fig. 1b. The recrystallization texture (Fig. 1c) was calculated next assuming the compromise threshold s_0=0.45 and it is compared with experimental recrystallization texture of aluminium containing 0.007 % Fe (Fig. 1d) [2]. We observe a very good qualitative agreement between the calculated and experimental recrystallization textures - the same texture maxima appear in the two ODFs being compared. Three ideal orientations are marked in the predicted recrystallization texture (Fig. 1c): the dominating *cubic* component (0^0, 0^0, 0^0), *S* component (φ_1=59°, ϕ=37°, φ_2=63°) and also *rotated Goss* component (0^0, 90^0, 45^0). It should be noted that already the shape of the compromise function shows a very strong similarity with that of the recrystallization texture. Therefore, we conclude that the compromise condition is a very important factor in modelling of recrystallization textures.

3.2. Recrystallization texture of asymmetrically rolled material

The second part of the study was devoted to Al 6061 samples, which were rolled symmetrically and asymmetrically and then annealed at 500 ^{0}C during 1 h. The experimentally determined texture of asymmetrically rolled aluminium 6061, where the ratio of angular velocities of the top and bottom identical rolls was equal to 1.1, is shown in Fig. 2a. This texture was presented in φ_2=0^0 ODF section in the full φ_1 range (cf., [13]). This texture was discretized into 5000 grains of the same volume and used next for the calculation of recrystallization texture, which is shown in Fig. 2b. It contains the *cubic* orientation, which is the sole orientation of the experimental texture of Al 6061 (Fig. 2c). Moreover, the predicted recrystallization texture contains some weaker maxima around the *rotated Goss* component (this component appears in the recrystallization texture of the Al-0.007 % Fe sample but is absent in that of Al 6061).

It should be added that the experimental recrystallization textures of symmetric and asymmetric rolling of Al 6061 samples are qualitatively the same and they contain only the *cubic* component (the respective predicted recrystallization textures contain this component as the main one).

IOP Conf. Series: Materials Science and Engineering **82** (2015) 012039 doi:10.1088/1757-899X/82/1/012039

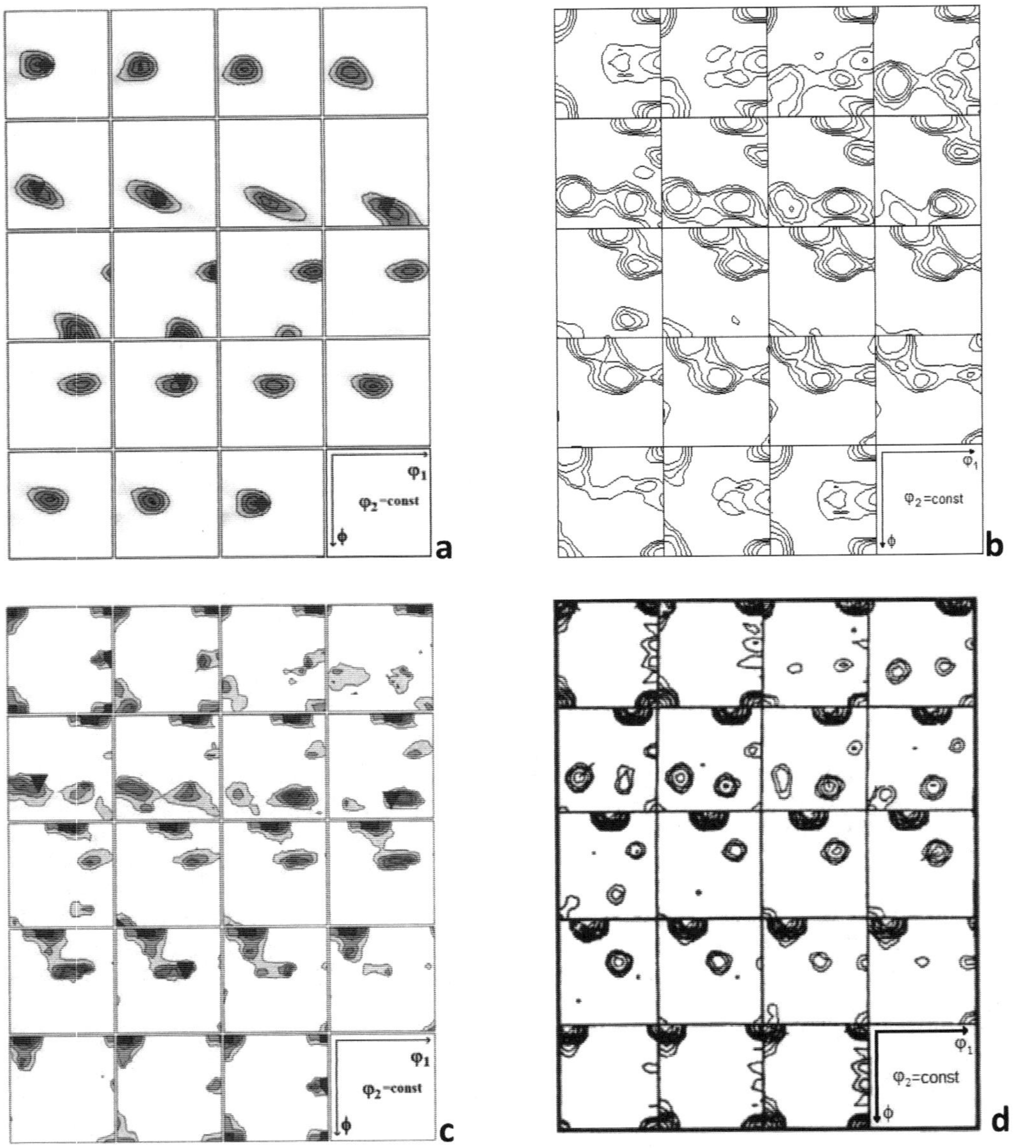

Fig. 1. *a) Model texture of symmetrically rolled aluminium (reduction 50%, max. intensity: 12), b) calculated compromise function S(g); max. intensity: 0.67, c) predicted recrystallization texture; max. intensity: 11, d) experimental recrystallization texture of rolled Al-0.007 % Fe alloy, annealed at 520⁰ C (after [2]).The following texture components are marked:* ● copper, ◆ brass, *cubic (■), S (▼), rotated Goss (◣).* φ_2*=const sections are shown.*

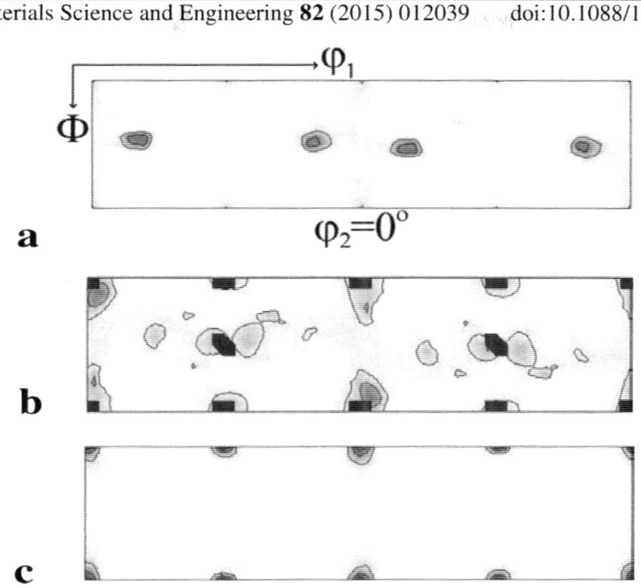

Fig.2. *a) Experimental texture of aluminium 6061 rolled asymmetrically (A=1.1.) to 40 % reduction (max. intensity: 20), b) predicted recrystallization texture (max. intensity: 28), c) experimental recrystallization texture of asymmetrically rolled aluminium 6061, annealed at 500°C during 1 h (max. intensity: 25). Cubic (■) and rotated Goss (◆) orientations are shown.*

4. Conclusions

The concept of preferred growth, defined by specific misorientations between nuclei and matrix, leads to a correct prediction of the main component of aluminium recrystallization textures (i.e., the *cubic* orientation) for symmetrically and asymmetrically rolled material. Some differences appear between experimental recrystallization textures of Al - 0.007 % Fe and Al 6061 (the first one contains also *S* and *rotated Goss* components). The compromise condition was included in our calculations in order to obtain a better prediction of recrystallization textures. The simple approach, used in the presented calculations, leads to relatively good results. This can be explained by the fact that there is principally one preferred misorientation relation in f.c.c. metals (40°<111>), while more complex characteristics of the oriented growth relations are reported for other metals (e.g., for b.c.c. metals).

Acknowledgment

This study was financed by the Polish National Centre for Science (NCN), grant number: DEC-2011/01/B/ST8/07394 and DEC-2013/11/B/ST3/03787.

References

[1] Humphreys J and Hatherly M 2004 *Recrystallization and Related Annealing Phenomena* (Elsevier)
[2] Lücke K 1984 *Proc. 7th Int. Conf. on Textures of Materials, Eds. Brakman et al.* (Noordwijkerhout, The Netherlands) p 195
[3] Piekos K, Tarasiuk J, Wierzbanowski K, Bacroix B 2008, *Comp. Mater. Sci.* **42** 584
[4] Piekos K, Tarasiuk J, Wierzbanowski K, Bacroix B 2008, *Comp. Mater. Sci.* **42** 36
[5] Wierzbanowski K, Tarasiuk J, Bacroix B, Sztwiertnia K, Gerber P 2003, *Met. Mater. Int.* **9** 9
[6] Baczmanski A, Wierzbanowski K, Tarsiuk J, Ceretti M, Lodini A 1997, *Revue de* Metallurgie-Cahiers d'Informations Techn. **94** 1467
[7] Wierzbanowski K, Baczmanski A, Lipinski P, Lodini A 2007, *Arch. Metall. Mater.* **52** 77

[8] Baczmanski A, Tidu A, Lipinski P, Humbert M, Wierzbanowski K 2006, *Mater. Sci. Forum* **524-525** 235
[9] Baczmanski A, Wierzbanowski K, Tarasiuk J 1995, *Z. Metallkd.* **86**, 507
[10] Liebmann B, Lücke K, Masing G 1956, *Z. Metallkd.* **47** 57
[11] Huang Y and Humphreys F J 1999, *Acta Mater.* **47** 2259
[12] Sztwiertnia K 1996, *Arch. Metall.* **41** 101
[13] Tarasiuk J, Wierzbanowski K, Baczmański A 1998, *Cryst. Res. Technol.* **33** 101

IOP Conf. Series: Materials Science and Engineering **82** (2015) 012040 doi:10.1088/1757-899X/82/1/012040

Influence of chemical composition on the evolution of texture in 441 ferritic stainless steels

CW Siyasiya [1], MG Maruma [1, 2] and WE Stumpf [1]
[1] Department of Materials Science and Metallurgical Engineering, University of Pretoria
[2] Mintek, South Africa

Abstract. The influence of addition of Nb+Ti (0.26 to 0.7wt%Nb and 0.15 to 0.3wt%Ti) to 441 on the development of annealing crystallographic textures was investigated by XRD and EBSD ODFs. It was found that addition of (Nb+Ti) to 441 improved the {111} texture to a limit while the resistance to ridging continued improving with more additions of Ti and Nb. It was found that the TiN nucleates heterogeneously first on the $MgO.Al_2O_3$ spinel and followed by Nb(C,N) on the TiN particle. The improvement in ridging was attributed to particle stimulated nucleation (PSN) of randomly oriented grains caused by these coarse compound particles.

1. Introduction

The 441 ferritic stainless steel (FSS) is (Nb+Ti) dual-stabilised steel. It is well known that Nb and Ti tie up both C and N, both of which retard nucleation of {111} texture grains [1, 2]. Furthermore, Lewis et al [3] investigated a wide range of Ti and Nb stabilised FSS and it was found that the (Nb,Ti)(C,N) particles promoted the formation of {111} texture through particle stimulation nucleation (PSN). The areas adjacent to the large and undeformable particles had high stored energy and, therefore, the effect was similar to that of higher levels of cold reduction. On the contrary, in Nb stabilized 430 FSS, De Sequeira et al [4] found that grains formed by PSN were generally randomly oriented and this minimized the clustering of the undesirable rotated cube ({001}<110>) grains which are difficult to recrystallize. Consequently roping and ridging are minimized. However, the volume fraction of {111} texture tends to be less.

In a related study which is currently in progress [5] as well as this study, it has been found that the TiN precipitates heterogeneously on the $MgO.Al_2O_3$ spinel and, thereafter, at lower temperatures, the NbC precipitates heterogeneously on the TiN. These compound inclusions have been found to increase with the increase in Ti content. These inclusions are not only known for causing Submerged Entry Nozzle (SEN) clogging problems during continuous casting but adversely affect the drawability behaviour leading to cracking and poor surface quality of the finished products. Hence, the need to modify the nature of these inclusions through the adjustment of the steel or slag chemistry in the process making them less favourable for the heterogeneous nucleation of TiN and NbC particles which form clusters of particles. This may have consequences on the evolution of texture because it should be these heterogeneously nucleated coarse particles which are responsible for the PSN of randomly oriented grains and not necessarily the finely distributed NbC in the matrix. Therefore, the influence of addition of (Nb+Ti) to 441 without $MgO.Al_2O_3$ spinel modification is investigated in this work. A comparative study will follow thereafter on 441 with modified spinel.

2. Experimental procedure

The chemical compositions of the steels used in this study are given in Table1. The reference steel B was industrially produced by Columbus Stainless (SA) and was supplied in 1.5mm thickness cold rolled and annealed strips. The steels A and C were cast under vacuum and hot forged into 25-30mm square bars. In order to simulate the industrial hot rolling process in which the bars were reheated to 1200°C, soaked for 1 hr and hot rolled to 4.5mm strips. The hot strips were immediately annealed at

1200°C for 300 s, followed by water quenching. These strips were thereafter cold rolled 62% and annealed at 1025°C for 300 s before quenching to room temperature in water to avoid formation of the Laves phase. Standard metallographic analysis was carried out by both optical and SEM microstructural analysis. The EBSD scans were performed in a FEI NOVA NanoSEM 230 FEG equipped with EDAX Digiview EBSD Camera operated at 20kV. Grain boundary analyses of the annealed samples were carried out using EBSD at mid thickness of the specimens in the RD-TD plane and were evaluated using the EDAX-TSL software package. EBSD data was calculated using the series expansion method with L_{max} = 22. Orthotropic sample symmetry was applied and the orientations were expressed in Euler angles such that $0^{\circ} \leq (\varphi_1,\theta,\varphi_2) \leq 90^{\circ}$ with 10° Gaussian spread around the ideal orientations. The surface ridging behaviour was investigated in the cold rolled and annealed (CRA) condition using a Taylor Hobson Talysurf stylus profilometer. The machined specimen dimensions complied with ASTM E517. The surface ridging heights were measured in the transverse direction after 10% tensile strain in the rolling direction. Seven measurements, 10 mm apart, were taken from each specimen (see insert in Figure 4) and the arithmetic mean deviation was plotted against distance. All specimens were thoroughly polished before tensile straining to ensure that prior surface roughness did not interfere with the post-deformation roughness measurements.

Table1: Chemical compositions of the steels in mass %

Steel	C	Cr	Mn	Al	V	Si	**Nb**	**Ti**	Ni	N	**(Nb+Ti)**
A	0.015	17	0.4	0.04	0.12	0.35	**0.26**	**0.20**	0.10	0.04	**0.46**
B	0.012	18	0.5	0.03	0.12	0.50	**0.44**	**0.15**	0.19	0.04	**0.60**
C	0.015	17	0.4	0.02	0.12	0.50	**0.70**	**0.32**	0.12	0.01	**1.02**

3. Results

3.1. Microstructures and mechanical properties

Figure 1 shows the optical microstructures of the three steels after 62% cold reduction and isothermal annealing at 1025°C for 300 s. The grain size for steel A and B was found to be 100±13μm while that of steel C with the highest (Nb+Ti) content, was found to be 34±10μm. Figure 2 shows clusters of a complex compound of $MgO.Al_2O_3$ spinel, TiN and Nb(C,N) in the form of stringers which are aligned in the rolling direction. Figure 2(b) shows heterogeneous nucleation of TiN on the $MgO.Al_2O_3$ spinel and the NbC on the TiN particle in that order. Apart from these coarse carbonitrides, fine NbC can also be seen in the same micrograph.

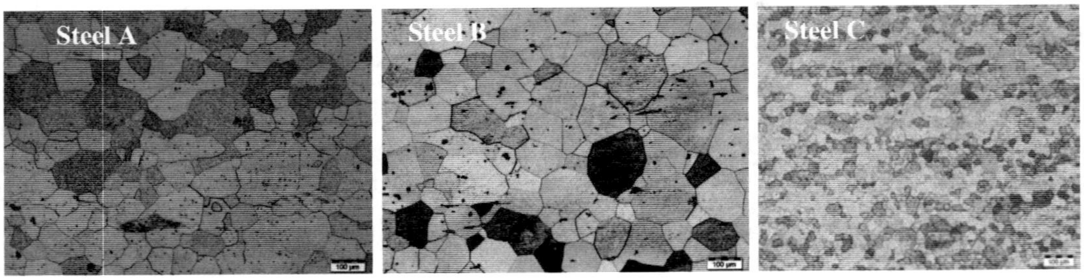

Figure 1: Micrographs after 62% cold reduction and isothermal annealing at 1025°C for 300 s.

IOP Conf. Series: Materials Science and Engineering **82** (2015) 012040 doi:10.1088/1757-899X/82/1/012040

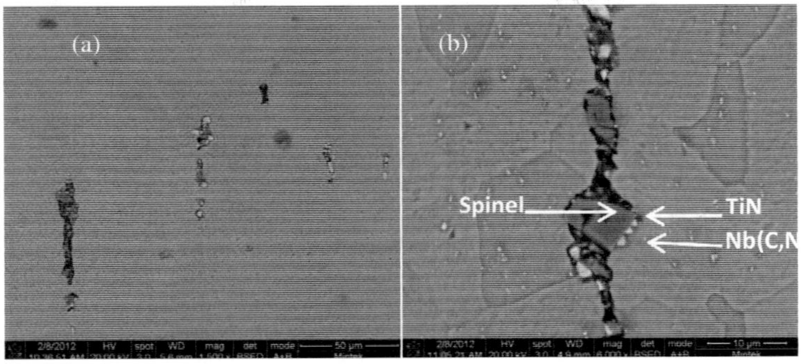

Figure 2: (a) Clusters of Spinel plus (Nb,Ti)(C,N) in the form of stringers in the rolling direction after cold rolling and annealing of steel B, (b) at higher magnification showing heterogeneous nucleation of firstly TiN on the $MgO.Al_2O_3$ spinel and followed by nucleation of NbC on the TiN particle.

3.2. Texture evolution after cold rolling and annealing

Figure 3 shows the $\varphi_2 = 45^o$ sections of the ODFs obtained from EBSD scans measured in steels A, B and C respectively after 62% cold work and annealing at 1025°C for 300 s. The rotated cube texture disappears as the (Nb+Ti) content increases while the γ-fibre texture does not seem to change much. Figure 4 shows the plot of the arithmetic mean deviation Ra versus distance in the transverse direction. Despite the lack of further improvement in texture as the (Nb+Ti) content is increased in steel C, the ridging resistance seems to increase with addition of (Nb+Ti). Steels A and C which were partially annealed for only 180 s exhibited more surface roughness due to material inhomogeneity.

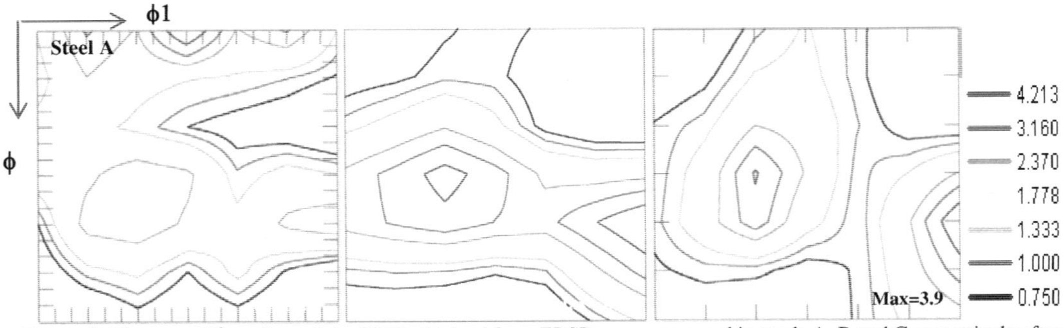

Figure 3: The $\varphi_2 = 45^o$ sections of the ODFs obtained from EBSD scans measured in steels A, B and C respectively after 62% cold work and annealing at 1025°C for 300 s.

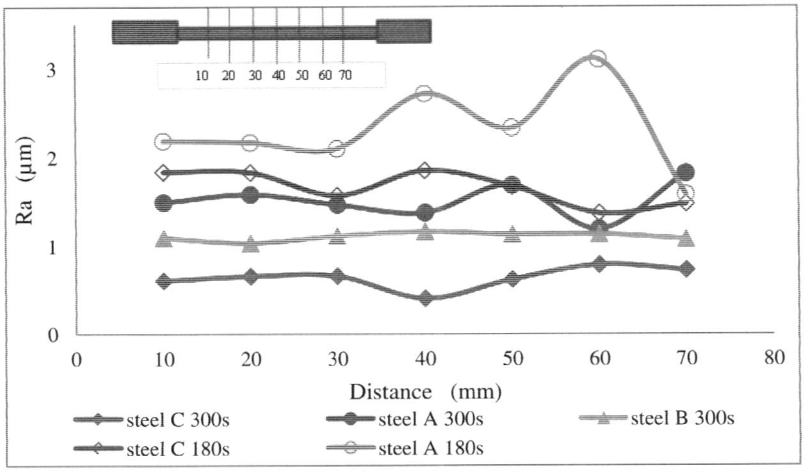

Figure 4: Plot of the average roughness Ra (arithmetic mean deviation) versus distance for the seven measurements corresponding to the insert

4. Discussion and conclusion

As may be seen from Figure 2, an increase in (Nb+Ti) leads to grain refinement caused by the pinning of grain boundaries by the fine Nb(C,N) particles [6]. The pinning effect comes from the finely dispersed Nb(C,N) particles in the matrix and not the clusters of the compound particles of spinel-TiN-Nb(C,N), see Figure 3(b). The fine grains may have contributed to improved ridging resistance possibly due to the breaking down of the as cast cube texture.

The texture of steels B and C after cold work and annealing was dominantly ND//<111> while steel A was a combination of RD//<110> and ND//<111>, see Figure 3. Similar results were also obtained in XRD texture measurements. It is evident that the addition of (Nb+Ti) in this 441 steel led to an improvement in the γ-fibre texture (notably the {111}<112> and {554}<225>) and a decline in the undesirable cube texture ({001}<110>) which has a negative effect on ridging resistance. This is not surprising because the amount of free C and N is reduced as more carbonitrides form in higher (Nb+Ti) content steels B and C, hence, the improvement in the {111} texture [1,2]. Furthermore, Sawatani et al [7] found that the fine Ti(C,N) precipitate suppresses recrystallisation of {110}<001> during an annealing heat treatment and {554}<225>, which is the shifted {111}<112> recrystallisation texture, is preferentially formed. In this study, the Nb(C,N) played a similar role in improving the γ-fibre texture. However, the improvement was only found up to steel B (0.44%Nb+0.15%Ti) and any further addition of both Nb and Ti did not seem to improve the γ-fibre texture in steel C. This could be attributed to the fact that there were no more substantial amounts of free C and N left in the steel and that any increase in the spinel-TiN-Nb(C,N) compound particles did not have significant influence on the development of the γ-fibre texture. On the contrary, the average surface roughness continued decreasing with an increase in (Nb+Ti) content, *i.e.* steel C exhibited the highest ridging resistance, see Figure 4. This agrees with the observations made by De Sequeira et al [4] who found that grains formed by PSN were generally randomly oriented and this minimized the clustering of the undesirable rotated cube ({001}<110>) grains which are difficult to recrystallize. Consequently roping and ridging are minimized but the {111} texture tends to be less, *i.e.* the addition of (Nb+Ti) led to more clusters of particles as seen in Figure 3 which resulted in more randomly oriented grains.

Although no microstructural evidence of PSN has been provided in this work (which is still in progress) it can be concluded that the coarse $MgO.Al_2O_3$-TiN-Nb(C,N) compound particles may have played a role in PSN of randomly oriented grains which resulted in the breakdown of the as-cast cube texture grains as evidenced by the grain refinement in steel C and its superior ridging resistance properties.

Acknowledgement

The authors would like to acknowledge the financial support and supply of materials from Columbus Stainless Steel (South Africa). Permission from the University of Pretoria to publish this work is also kindly acknowledged.

References

[1] Fukuda M, Journal of Iron and Steel Institute, Japan, 1967. 53: p. 559-561.
[2] Takahashi M and Okamoto A, Trans. of Iron and Steel Institute, Japan, 1979. 19: p. 391-400.
[3] Lewis DB and Pickering FB, *Development of Recrystallisation Textures in Ferritc Stainless Steels and their Relationship to Formability*, in *"Adcances in the Physical Metallurgy and Applications of Steels" Proc. Int. Conf. Metals Soc*1982, The Metals Society, London: University of Liverpool.
[4] De Siqueira R P, Sandim H RZ and Raabe D, Met. and Mat. Trans. A, Vol. 44A, (2013), pp 469 – 477
[5] Kruger D, Internal Communication from Columbus Stainless Steel, Middleburg, SA, (2014)
[6] Xiao F, Cao Y, Qiao G, Zhan G X and Liao B, Journal of Iron and Steel Research, International, 2012. Vol.19(No.11): p. 52-56.
[7] Sawatani T, Shimizu K, Nakayama T, and Miyoshi M, "The R-values and recrystallized textures of Ti stabilized low C,N-17%Cr stainless steel sheets", Tetsu-to-Hagane, vol. 63, 1977, p.843-854.

IOP Conf. Series: Materials Science and Engineering **82** (2015) 012041 doi:10.1088/1757-899X/82/1/012041

Improving magnetic properties by optimization of textures in non-oriented electrical steel with initial columnar grains

N Zhang, P Yang and W M Mao
University of Science and Technology Beijing, School of Materials Science and Engineering, Beijing, 100083, China
Email: yangp@mater.ustb.edu.cn

Abstract. This study investigates the processing route to optimize magnetic properties along both rolling and transverse directions, and the evolution of texture during the process is revealed by EBSD technique. The results show that, thinner hot-rolled bands accompanied with coarser structures after normalization are beneficial for promoting the magnetic properties of final sheets. Compared with the 35W300 high-grade NGO steel with a similar composition exhibiting B_{50} = 1.71T (along RD)/1.67 T (along TD), the B_{50} values of samples obtained by hot rolling to 1.5mm and subsequent processes are equal to or higher than 1.75T (along RD)/1.69T (along TD). Moreover, a greater quantity of {hk0}<001> oriented nuclei result in stronger {hk0}<001> recrystallization texture in recrystallized warm rolled samples heated at 300°C in advance, and stronger {100}<0vw> texture is achieved in the samples prepared by two-stage annealing method. In addition, the distinct deformation and recrystallization behaviors of {100}<001> and {100}<110> columnar grains are discussed.

1. Introduction

Improving magnetic properties of non-oriented electrical steel is a priority for electrical steel manufacturers. The processing technology mainly focuses on the decrease of iron loss in the past years, while higher magnetic induction is essential for improving the product efficiency, and it could be realized by texture control. In general, deleterious γ-fiber texture could be reduced with the increase of hot band grain size, which could conversely promotes {hk0}<001> nucleation inside shear bands and lead to stronger {hk0}<001> texture[1]. The texture evolution could also be affected by cold rolling and annealing methods, namely different rolling reduction, rolling and annealing temperatures [2-3].

Our previous study shows that retaining {100}<0vw> texture from initial columnar grains provides a promising way for preparation of non-oriented electrical steel[4]. This study further investigates the texture evolution of columnar grains. In addition, the development of γ-fiber and {hk0}<001> texture is investigated as well, thus optimizing the magnetic properties by texture control.

2. Experimental

The starting material is a columnar-grained electrical steel cast slab containing 2.5%Si, 0.52%Al and 0.0023%C, with the thickness of 30mm. After heating at 1200℃ for 30min, the slab was hot rolled to 2mm or 1.5mm thick in five rolling passes. For the bands hot rolled to 1.5mm, reheating at 1200℃ for 10min was carried out after three rolling passes. The hot rolled bands were heated up to 1000℃ with a rate of 5℃/min, and the annealed samples were named N1 and N2. N2 with the thickness of 1.5mm is thinner than N1, which is 2mm thick, and the thinner hot band leads to lower cold rolling reduction. Then N1 and N2 were further cold rolled and annealed with different methods displayed in table 1, in which quick, slow and two-stage annealing meant annealing at 1000°C for 10min with heating rates of 300℃/min, 5℃/min and firstly annealing at 700°C for 10min, followed by slowly heating up to 1000°C and remained for 10min respectively; and warm rolling meant heating at 300°C for 30min prior to rolling. In addition, the rolled samples from 1.5mm hot band were heated to 700°C to investigate the nucleation behaviors. Textures and microstructures in the cross sections of N1 and N2 as well final sheets were measured by EBSD in Zeiss Ultra 55 FESEM and HKL software. Magnetic inductions at 5000A/m (B_{50}) were measured by a single sheet tester in rolling and transverse directions of samples.

Table 1. Methods of cold rolling and final annealing

Annealed hot bands	Rolling	Annealing		
		Quick annealing	Slow annealing	Two-stage annealing
N1	Cold rolling	sample 1		
N2	Cold rolling	sample 2	sample 3	sample 4
	Warm rolling	sample 5	sample 6	sample 7

3. Results and discussion

As shown in figure 1, both N1 and N2 show recrystallized microstructure, and fewer small grains are observed in N2. {100}<0vw> texture is markedly retained in N2, and {hk0}<001> texture is obtained in both samples. Higher intensity of {100}<0vw> texture in N2 accompanied with weaker α and γ-fiber is assumed to be related with the recrystallization caused by reheating during hot rolling, showing the similar effect of reducing hot rolling reduction. When N1 and N2 are rolled and annealed, the magnetic inductions B_{50} values of sample1 are 1.72T(along RD)/1.67T(along TD), reaching the level of 35W300 NGO steel exhibiting B_{50}= 1.71T /1.67 T. Moreover, the B_{50} values of sample2 are higher as 1.75T/1.69T, and stronger beneficial {100}<0vw> and {hk0}<001> textures depicted in figure 2(b) result from their advantage in N2 and stronger heredity after cold rolling to moderate strain and final annealing. Meanwhile, the increasing grain size prior to cold rolling and lower rolling reduction lead to weaker γ-fiber texture in Sample2.

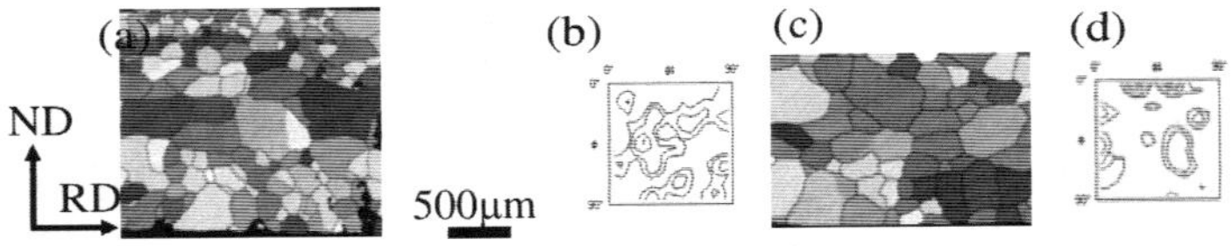

Figure 1. Orientation maps and ODFs at φ2=45° section of N1 (a-b) and N2 (c-d).

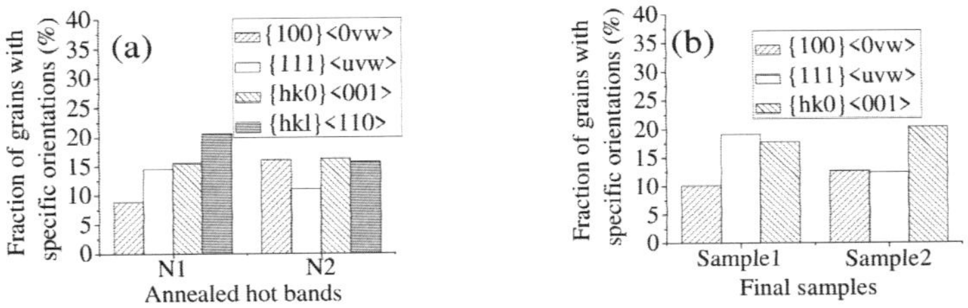

Figure 2. Fractions of grains with specific orientations in annealed hot rolled bands (a) and corresponding final samples (b).

As discussed above, thinner hot-rolled bands accompanied with coarser structures after normalization are beneficial for the magnetic inductions of final samples, and the magnetic induction could be further improved with modified rolling and annealing methods. Figure 3(a) illustrates higher B_{50} in recrystallized warm rolled samples, related with stronger {hk0}<001> texture in figure 3(b). Better magnetic inductions along TD are shown in the samples by slow annealing and two-stage annealing than quick annealing, and the smallest differences between B_{50} values along RD and TD are obtained in the samples processed by two-stage annealing, which is in accordance with the highest fraction of {100}<0vw> grains.

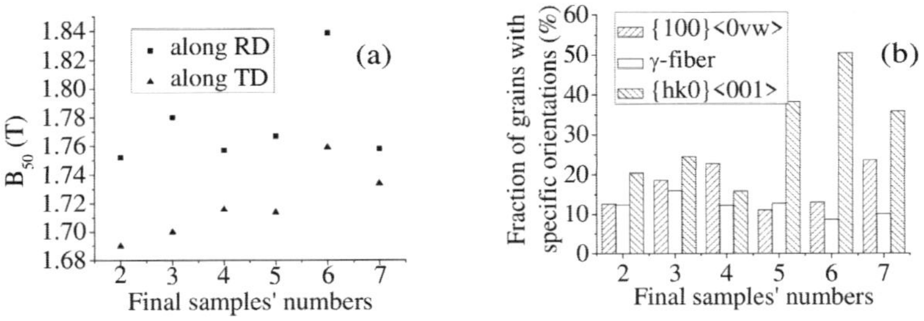

Figure 3. The Magnetic inductions (a) and fractions of specific oriented grains (b) of final sheets prepared with 1.5mm annealed hot band N2.

To clarify the texture evolution under different processing routes, partially recrystallized samples are measured to investigate nucleation behaviors. {100}<0vw>, γ-fiber and {hk0}<001> new grains are all observed in figure 4, while {hk0}<001> new grains show superiority in numbers than other grains in warm rolled sample, as shown in figure 4(b). The promoted {hk0}<001> nucleation contributes to stronger {hk0}<001> texture, and the effect of rolling methods on {hk0}<001> texture will be discussed in our other paper. The effect of annealing method on {100}<0vw> texture is attributed to varied nucleation behaviors at different temperatures as well as competition between {100}<0vw> and other nucleation. {100}<0vw> nucleation mainly occurs in {100}<uvw>-{113}<uvw> deformed regions, and it is due to stored energy difference. Compared with {113}<uvw> new grains, it is suggested that {100}<0vw> new grains show higher nucleation ability at 700°C [3]. Therefore stronger {100}<0vw> texture is achieved in samples by two-stage annealing, during which nucleation takes place at 700°C. In contrast, weaker {100}<0vw> textures are displayed in the samples prepared by slow annealing, and {hk0}<001> texture is even stronger, especially in warm rolled samples, which is in correspondence with the quantity difference between {100}<0vw> and {hk0}<001> new grains in figure 4(b). {hk0}<001> nucleation occurring at shear bands takes place at the earliest stage of recrystallization[5], and slow annealing could provide longer period for its nucleation and growth, reducing other texture consequently. In addition, there may exists effect of recovery on the stored energy and nucleation ability of {100}<0vw> and other grains, while more deep research is necessary to verify the effect and explain it.

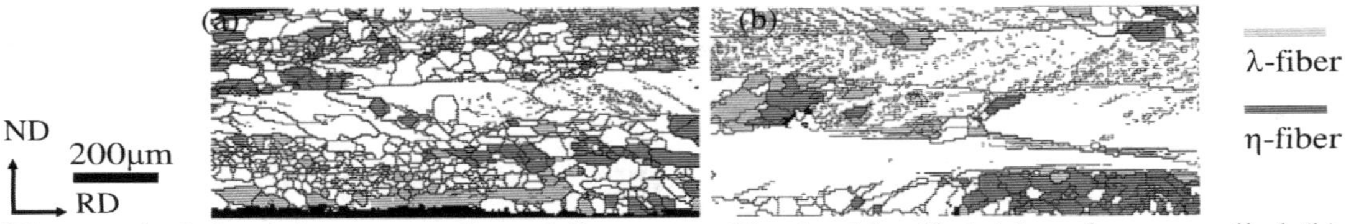

Figure 4. Orientation maps of partially recrystallized cold rolled (a) and warm rolled (b) samples.

It is worth to note that specific deformation and recrystallization behaviors of {100}<0vw> columnar grains, especially Cube{100}<001> and Rotate-Cube{100}<011> grains, exert extreme influence on the texture evolution along the processing routes [4]. Shear regions are obtained near the surface, and the effect of columnar grains mainly exists in the center layer of hot rolled bands. Initial {100}<011> grains transform to α-fiber deformed areas after hot rolling. These α-fiber areas would show strong heredity during following processes, and no nucleation sites for beneficial λ-fiber or η-fiber grains could be provided by them. However, fully recrystallization microstructure is obtained after long period normalization annealing in this study, which means that large α-fiber deformed grains after hot rolling are consumed, implying that the unbeneficial effect of initial {100}<011> grains is largely restrained. In contrast, initial Cube grains contribute to {100}<001>-<012> texture in annealed hot bands, and then Cube or {100}<012> texture can be achieved by the retaining after deformation at medium strain and subsequent recrystallization[3]. In particular, the intensity of {100}<001>-<012> texture is higher in annealed hot band N2 prepared with hot rolling and additional reheating processes, and α-fiber texture is weaker in the same sample, both leading to optimized texture in final samples

4. Conclusion

1) The magnetic properties are optimized by improving {100}<0vw> and {hk0}<001> textures in 2.5wt.%Si-0.52wt.%Al columnar-grained electrical steel. Thinner hot-rolled bands accompanied with coarser structures after normalization are beneficial for the magnetic inductions), the B_{50} values of samples obtained by hot rolling to 1.5mm and subsequent processes are equal to or higher than 1.75T (along RD)/1.69T (along TD).
2) A greater quantity of {hk0}<001> oriented nuclei result in stronger {hk0}<001> recrystallization texture in recrystallized warm rolled samples, and stronger {100}<0vw> texture is achieved in the samples prepared by two-stage annealing methods.
3) The unbeneficial effect of initial {100}<011> grains is restrained by long period normalization annealing, and the beneficial {100}<001>-<012> texture related with initial Cube grains is stronger in annealed hot band prepared with hot rolling and additional reheating.

Acknowledgments

This work was supported by the Natural Science Foundation of China (Grant No. 51071024).

References

[1] Dafe S S F, Paolinelli S C and Cota A B 2011 *J. Magn Magn Mater.* **323** pp 3234-3238
[2] Lee S and COOMAN B C 2011 *ISIJ Inter.* **51** pp 1545-1552
[3] Zhang N, Yang P and Mao W M 2012 *Acta Metall. Sin.* **48** pp 307-314
[4] Cheng L, Zhang N, Yang P and Mao W M 2012 *Scrip. Mater.* **67** pp 4068-4072
[5] Park J T and Szpunar J A 2003 *Acta Mater.* **51** pp 3037-3051

Influence of carbon content on cold rolling and recrystallization texture in polycrystalline 3% Si-Fe

M Takenaka, Y Shingaki, T Imamura and Y Hayakawa
Steel Research Laboratory, JFE Steel Corporation, JP
E-mail: m-takenaka@jfe-steel.co.jp

Abstract. The influence of carbon content on cold rolling and recrystallization texture in polycrystalline 3%Si-Fe under the relatively high rolling reduction condition has been investigated. The main component of recrystallization texture was {554}<225> orientation in ultra low carbon (ULC) 3%Si-Fe and {411}<148> orientation in low carbon (LC) 3%Si-Fe. The origin of {411}<148> recrystallization texture development in LC 3%Si-Fe is discussed in terms of the rotation of deformation twin from {100}<011> to {411}<148> orientation with the generation of the slip bands inside the neighboring matrix grain {111}<011>. The rotation axis of this crystal rotation was estimated <112> axis. Assuming the single slip system activation in BCC metal, crystal rotation around <112> axis indicates an activation of {110}<111> slip system. In terms of Schmid factor, {112}<111> slip system must be activated in {100}<011> matrix. This is not in agreement with the estimation of {110}<111> slip system activation. Detailed observation on the cold rolled sample revealed that common slip plane passed through the deformation twin and surrounding deformed matrix grains. It is considered that slip plane matching (SPM) with neighboring grains activates the lower Schmid factor slip system in deformation twin. These results suggest that not only Schmid factor but also SPM with neighboring grains should be considered to decide the active slip systems in polycrystalline metals.

Introduction

Recrystallization textures of cold rolled and annealed steels strongly influence their mechanical and magnetic properties. Development of <111>//ND fiber texture contributes the superior deep-drawability, which is preferred for automobile and white goods steels [1]-[3]. Development of near <100>//ND texture, for example, {100}<001> or {411}<148> improves the magnetic properties, which is preferred for electrical steels [4]-[5].

It is well known that the main component of recrystallization texture changes from <111>//ND fiber to {110}<001> orientation with increasing the carbon content in deep drawing steel (Si is not added intentionally) [6]-[8]. In the previous studies using single crystal, {110}<001> recrystallized grains nucleated from the shear bands in cold rolled specimen [9][10]. Ordinary steel sheets have polycrystalline microstructure, so it is important to investigate the deformation and recrystallization behavior in polycrystalline specimen. Furthermore, most previous studies were conducted under relatively low cold rolling reduction. Reducing the thickness effectively improves the iron loss of electrical steel products [11][12], so it is important to investigate the deformation and recrystallization behavior under high cold rolling reduction.

In our previous investigation, the influence of carbon content on recrystallization texture in polycrystalline 3%Si-Fe under relatively high cold rolling reduction has been investigated [13]. In this work, to further clarify the mechanism of {411}<148> recrystallization texture evolution under the above condition, more detailed investigation has been conducted by applying SEM-electron backscattered diffraction pattern (EBSP) analysis.

Experimental procedure

The experiments were carried out using two kinds of materials, Ti bearing ultra low carbon (ULC) 3%Si-Fe (Fe-2.96%Si-0.07%Mn-0.001%C-0.075%Ti (mass%)) and low carbon (LC) 3%Si-Fe (Fe-2.99%Si-0.07%Mn-0.018%C (mass%)). Two cast ingots were hot rolled to 3.0mm thickness after soaking at 1200°C. After the hot rolling, specimens were pre-cold rolled to 2.0mm to randomize the initial texture. These specimens were normalized at 1150°C for 60s in dry nitrogen atmosphere and cooled to room temperature by 50°C/s so that ample amount of carbon was kept solute. Grain sizes of the normalized specimens were about

1.2mm and both specimens have almost random texture. Normalized specimens were cold rolled to 0.2mm thickness (cold rolling reduction: 90%), followed by two types of annealing in dry nitrogen atmosphere, partial recrystallization annealing (annealed up to 625°C and N_2 gas cooled) and complete recrystallization annealing (annealed at 840°C for 120s and N_2 gas cooled). Textural characterization was performed using X-ray diffraction (XRD) analysis, and its orientation distribution function (ODF) was calculated using the discrete method [14]. SEM- EBSP analysis was used for microstructural study.

Results and discussion

Fig. 1 shows the orientation image maps of cold rolled specimens measured by SEM-EBSP analysis. In ULC 3%Si-Fe, uniform deformation occurred and slip bands propagated with increasing cold rolling reduction. On the other hand, deformation twins were generated in the early stage of cold rolling in LC 3%Si-Fe. Many of those twins had {100}<011> orientation. Increasing the cold rolling reduction, slip bands passed over those twins. Fig.2 shows the cold rolled and complete recrystallization textures [13]. The main component of recrystallization texture of ULC 3%Si-Fe was {554}<225> orientation. These results are almost the same as those obtained using deep drawing steel [8]. On the other hand, the main component of recrystallization texture of LC 3%Si-Fe was {411}<148> orientation.

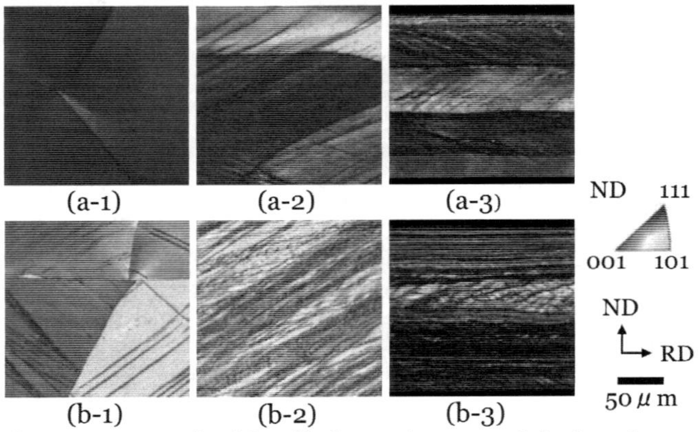

Fig. 1 Orientation image maps of cold rolled specimens. : (a) ultra low carbon 3%Si-Fe, (b) low carbon 3%Si-Fe, (-1) cold rolling reduction 20%, (-2) cold rolling reduction 60%, (-3) cold rolling reduction 90%.

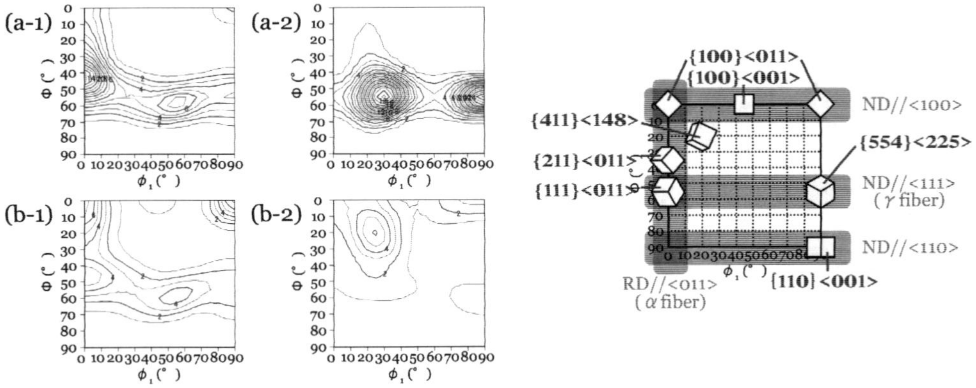

Fig. 2 Cold rolled and recrystallization textures of a) ultra low carbon 3%Si-Fe, (b) low carbon 3%Si-Fe, (-1) cold rolled specimens (cold rolling reduction 90%) (-2) completely recrystallized specimens [13] (Center layer; Bunge's ODF (φ2=45° cross-section)).

Some {411}<148> micro regions are observed around the crossing areas of the {100}<011> deformation twins and the slip bands in cold rolled LC 3%Si-Fe. Fig. 3(a) shows the orientation image map of cold rolled LC 3%Si-Fe measured by SEM-EBSP analysis. The orientations of deformation twin ({100}<011>) and {411}<148> micro region are shown by (211) pole figure of Fig. 3(b). It is found that there is a common <112> axis between deformation twin and {411}<148> micro region. In general, rotation axis is perpendicular to both slip plane normal and slip direction assuming the single slip system. Considering this relationship and the fact that main slip systems are {110}<111> and {112}<111> in BCC metal, crystal rotation around <112> axis indicates an activation of {110}<111> slip system. The orientations of deformation twin ({100}<011>) and neighboring grain ({111}<011>) are shown by (110) pole figure of Fig. 3(c). It is found that there is a common {110} plane between deformation twin and neighboring grain. This result confirms the activity of {110}<111> slip system in both deformation twin and neighboring grain. In terms of Schmid factor, {112}<111> slip systems must be activated in {100}<011> grains [15]. This is not in agreement with the result of {110}<111> slip system activation. It is considered that slip plane matching (SPM) with neighboring grains activates the relatively lower Schmid factor slip system [16]. Honma et al. also suggests that grain boundary plays an important role to nucleate the {411}<148> recrystallized grains [17]. These results suggest that not only Schmid factor but also SPM with neighboring grains should be considered to decide the active slip system in polycrystalline metals.

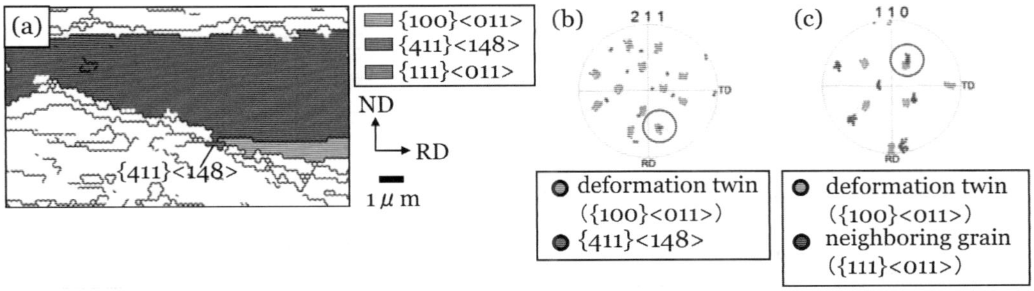

Fig. 3 EBSP analysis of cold rolled (rolling reduction 90%) specimen of low carbon 3%Si-Fe, (a) highlight only main orientations, (b) (211) pole figure, (c) (110) pole figure.

Fig. 4 shows the orientation image map of partially recrystallized LC 3%Si-Fe measured by SEM-EBSP analysis. {411}<148> recrystallized grain nucleated from {100}<011> deformation twin in the vicinity of slip bands inside the neighboring matrix grain. This result is strongly in agreement with SPM mechanism of {411}<148> recrystallization texture development in LC 3%Si-Fe.

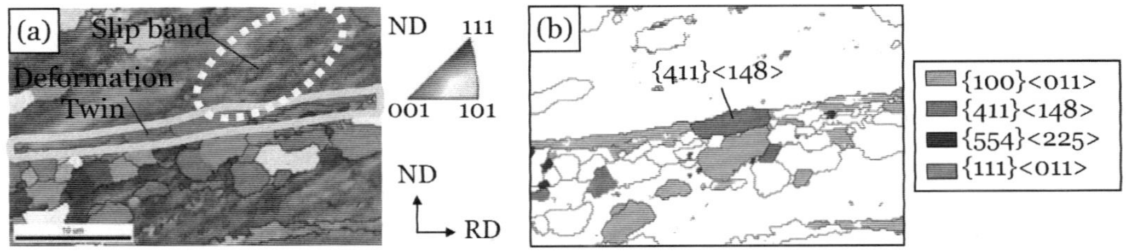

Fig.4 EBSP analysis of partially recrystallized (annealed up to 625°C and N_2 gas cooled) low carbon 3%Si-Fe, (a) orientation image map, (b) highlight only main orientations.

The mechanism of {411}<148> recrystallization texture development in 3%Si-Fe with increasing the carbon content is shown in Fig. 5. Increasing the carbon content induces the deformation twins in the early stage of cold rolling. SPM arises between deformation twin and neighboring grains with increasing cold rolling reduction. As a result, {110}<111> slip system is activated and {411}<148> micro regions are formed in the {100}<011> deformation twins. These results suggest that not only Schmid factor but also slip plane matching with neighboring grain should be considered to decide the active slip system in polycrystalline metals.

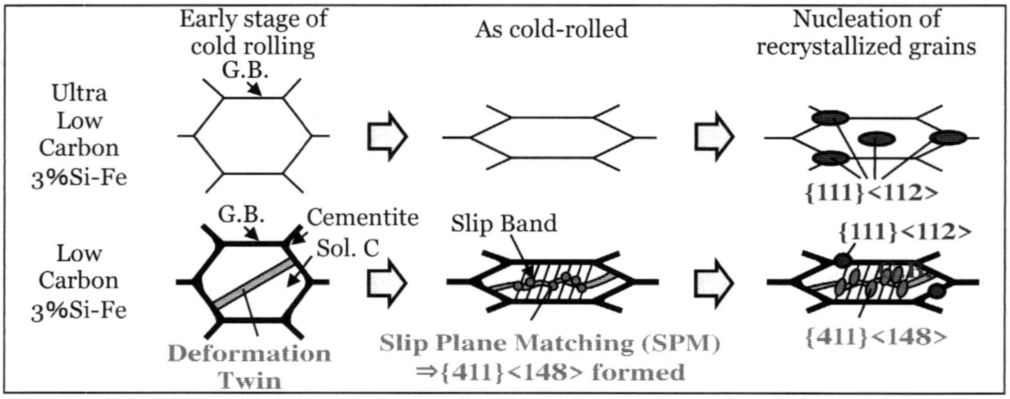

Fig. 5 Schematic drawing showing the mechanism of texture development in 3%Si-Fe with increasing the carbon content.

Conclusions

The mechanism of {411}<148> recrystallization texture development in 3%Si-Fe with increasing the carbon content has been investigated. Increasing the carbon content induces the deformation twins in the early stage of cold rolling. Slip plane matching arises between deformation twins and neighboring grains with increasing cold rolling reduction. As a result, {110}<111> slip systems are activated and {411}<148> micro regions are formed in the {100}<011> deformation twins.

References

[1] Goodenow R H and Held J F 1970 *Metall. Trans.* **1** 2507
[2] Hook R E, Heckler A J and Elias J A 1975 *Metall. Trans.* **6A** 1683
[3] Hutchinson W B 1984 *Int. Metall. Rev.* **29** 25
[4] Honda K and Kaya S 1926 *Sci. Rep. Tohoku Imp. Univ.* **15** 721
[5] Toda H, Oda Y, Kohno M, Ishida M and Zaizen Y 2012 *IEEE Trans. Magn.* **48** 3060
[6] Schlippenbach U V and Lüke K 1988 *Proc. of 8th Int. Conf. on Textures of Materials-ICOTOM 8* ed. by Kallend J S and Gottstein G 861
[7] Inagaki H 1994 *ISIJ Int.* **34** 313
[8] Hosoya Y and Nagataki Y 1996 *37th MWSP Conf. Proc. ISS* **XXXIII** 915
[9] Furubayashi E 1969 *Trans. ISIJ* **9** 222
[10] Haratani T, Hutchinson W B, Dillamore and bate P 1984 *Met. Sci.* **18** 57
[11] Braiseford F 1948 *Inst. Elec. Engrs.* **95[Ⅱ]** 38
[12] Pry R H and Bean C P 1958 *J. Appl. Phys.* **29** 532
[13] Takenaka M, Shingaki Y, Imamura T and Hayakawa Y 2013 *CAMP-ISIJ* **26** 1006
[14] Pawlik K 1986 *Phys. Stat. Sol (b)* **134** 447
[15] Hashimoto O, Ohashi N and Fujimoto K 1971 *Tetsu-to-Hagane* **57** 1155
[16] Imamura T, Hayakawa Y and Kurosawa M 2003 *Metall. Trans. A* **34A** 403
[17] Homma H, Nakamura S and Yoshinaga N 2004 *Tetsu-to-Hagane* **90** 510

IOP Conf. Series: Materials Science and Engineering **82** (2015) 012043 doi:10.1088/1757-899X/82/1/012043

Simulation of growing grains under orientation relation - dependent quadruple point dragging

K Ito
Emeritus Professor, Department of Materials Engineering, The University of Tokyo, JAPAN
E-mail: ito_e-prof@material.t.u-tokyo.ac.jp

Abstract. The growth behaviour of a specified grain embedded in matrix grains, for which the migration mobility of the quadruple points depended on the relation between the orientations of the growing and shrinking grains, was studied using a modified Potts MC-type three-dimensional simulation. Large embedded grains continued to grow without being overcome by coarsening matrix grains, whereas small embedded grains disappeared, under the influence of the relative mobilities of the quadruple points, the composition of the matrix grain texture and the width of the grain size distribution of the matrix grains. These results indicate that orientation relation-dependent quadruple point dragging can affect the recrystallization texture during the grain coarsening stage.

1. Introduction
It is known [1,2] that the quadruple junction mobility can affect the grain growth of polycrystalline materials. In the present report, the growth behaviour of a specified grain embedded in matrix grains, for which the mobility of the quadruple points depended on the relation between the orientations of the growing and shrinking grains, was observed using a simulation. The evolution of the texture component to which the specified grains belonged was estimated from the mean of the behaviour of individual embedded grains.

2. Simulation method
2.1 Construction of the specimen
The specimen contained 256×320×256 cubic cells, and the texture was composed of three components: S (specified grain component), M (matrix component) and B (background component). The starting structures that contained approximately 7100 grains were the same as those used formerly [3], and grains belonged to either the M or B components. Their grain size distributions were the same (figure 1). The number of faces of a grain is a number of grains sharing boundary planes with the grain concerned. The abbreviations 'standard 21' and 'standard 12' indicate standard starting structures with volume ratios M/B = 2/1 and 1/2, respectively, and 'wide 21' corresponds to a wide starting structure with M/B = 2/1.

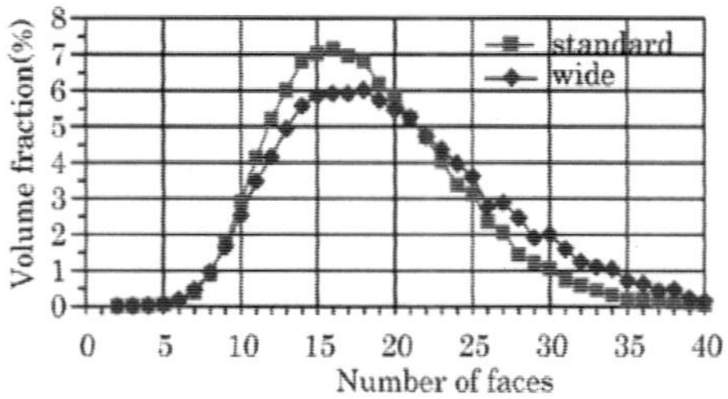

Figure 1. Grain size distributions in the standard and wide starting structures.

2.2 Coarsening algorithm

As reported previously [3,4], the unit process for boundary migration was presented as the occupation of passive cell j by active cell i and was permitted with the probability p. The value of ΔE was calculated with respect to the Moore neighbourhood of cell j :

$$p^{\Delta E} = \exp(-\Delta E/kT) \text{ when } \Delta E > 0 \quad \text{and } p^{\Delta E} = 1 \text{ when } \Delta E \leq 0 \qquad (1)$$

$$p = p^{\Delta E}/R \qquad (2)$$

The permission ratio R was specified by the category of the occupation event, while the multiplicity of event m was defined as the number of grains to which the cells in the Von Neumann neighbourhood of passive cell j belonged. For $m = 2$ (boundary migration) and $m = 3$ (triple line migration), R was equal to unity. On the other hand, for $m \geq 4$, R was equal to unity only when the event yielded the growth of the embedded grain with shrinkage of the grains in component M, and was larger than or equal to unity otherwise. The coarsening process was performed by repetition of a systematic MC step, as formerly described [3,4].

3. Simulation results

3.1 Relation between the permission ratio and the migration rate of a quadruple point

The migration of quadruple points was observed in a configuration composed of columnar hexagonal grains with edge length a after Barrales-Mora et al. [1, 5]. Figure 2 (a), where 'R1' and 'R10' respectively indicate $R = 1$ and $R = 10$, shows examples of the changes in the distance between the corresponding two points observed for $kT = 0.7$, 0.87, 1.1, and 1.3 (kT0.7, kT0.87, kT1.1, and kT1.3, respectively), and a=78/4 and 78 ($a78/4$ and $a78$, respectively). Runs for each set of the parameters were repeated 3 times. The straight parts of the curves represent the steady states. Figure 2 (b), where the gradients, being multiplied by a, relative to the value obtained for $a = 78$ and $R = 1$ are shown, demonstrates the effect of the permission ratio on the mobility. kT equal to 0.87 was used in this report as formerly [3].

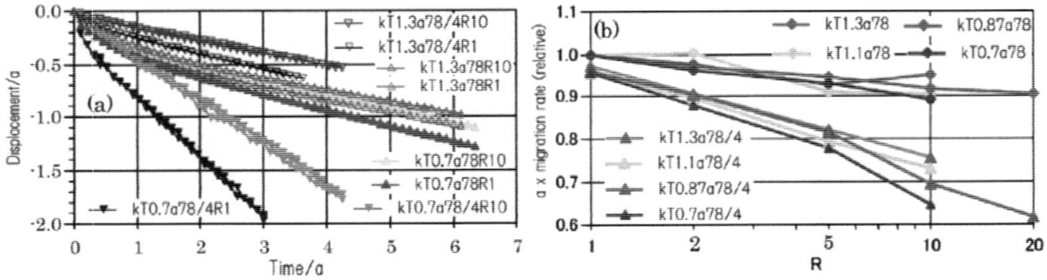

Figure 2. Migration behaviour of a quadruple point. $kT = 0.87$ was used in this report.

3.2 Coarsening behaviour of the embedded grains

A lenticular spheroid grain belonging to texture component S was embedded in a starting structure, as described previously [3]. Its size was controlled by the volume and the axial ratio. A reference structure was obtained using coarsening steps that decreased the number of the matrix grains by 10% (see figure 3 (a)). The volume of the embedded grain and the mean volume of the matrix grains in the reference structure are denoted by V_0 and $< V >_0$, respectively and those after coarsening are indicated by V and $< V >$, respectively. The following ratios are further defined:

$$referred\ coarsening\ ratio\ of\ the\ matrix\ grains = < V >/< V >_0, \qquad (3)$$

$$referred\ coarsening\ ratio\ of\ the\ specfied\ grain = V/V_0, \qquad (4)$$

$$relative\ growth\ ratio\ of\ the\ specfied\ grain = (V/V_0)/(< V >/< V >_0) \qquad (5)$$

Figure 3 shows 2D sections. The embedded grain is red; 'matrix' and 'relative' denote the *referred coarsening ratio of the matrix grains* and the *relative growth ratio of the specfied grain*, respectively; N denotes the initial number of faces of the embedded grain. Figure 4 shows a collection of evolution curves for the individual embedded grains whose number of faces in the reference structure N happened to be 21 (a) and 23 (b). As can be

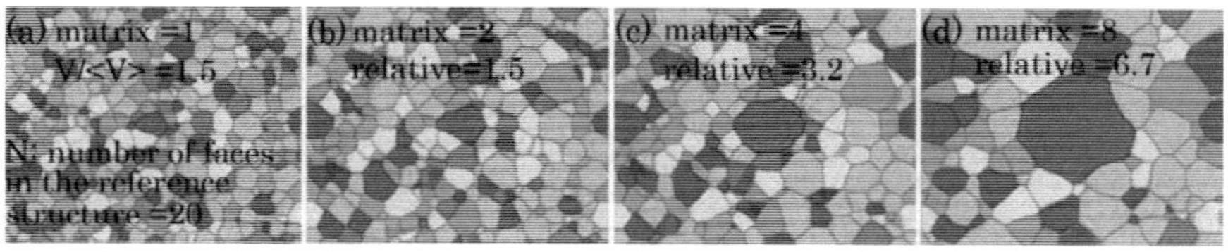

Figure 3. Structural development (a)->(b)->(c)->(d). Standard 21 R=10

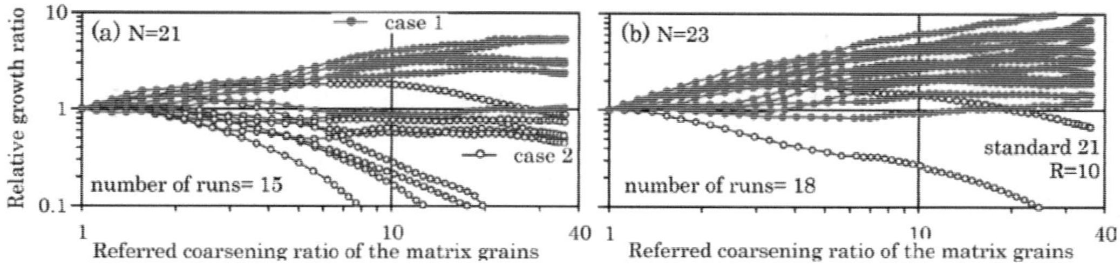

Figure 4. Behaviour of embedded grains shown by the relative growth ratio.

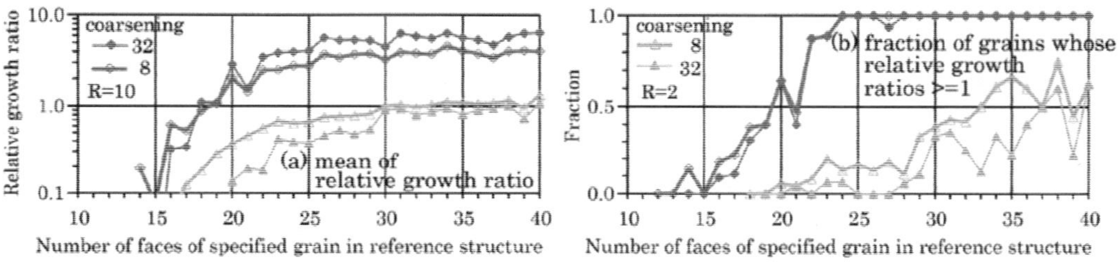

Figure 5. Growth behaviour dependence on the initial number of faces. Standard 21

observed in the figure, for case 1, the relative growth ratio at a referred coarsening ratio of 32 was greater than 1, whereas for case 2, it was less than 1. The mean values for the relative growth ratios at referred coarsening ratios of 8 and 32 shown in figures 4 (a) and (b) were plotted in figures 5 (a) at numbers of faces = 21 and 23, respectively. The fractions of grains whose relative growth ratios were greater than 1 were similarly plotted in figures 5 (b). The population parameter for the statistics was approximately 15 grains (figure 4). Figure 5 contains the values obtained for $R = 2$ as well. Note that throughout in figures 5 'coarsening' indicates $the\ referred\ coarsening\ ratio\ of\ the\ matrix\ grain$, and triangular and rhombic marks represent $R = 2$ and $R = 10$, respectively. When R was small ($R = 2$), the relative growth ratio decreased as coarsening progressed or as the coarsening ratio of the matrix grains increased (see $coarsening = 8$ and 32 in figure 5 (a)); namely, even larger embedded grains disappeared with time. Furthermore, the fraction of embedded grains that continued to grow generally increased with an increase in the initial size of the embedded grain (figure 5 (b)). This fraction never reached unity, however, even for the largest embedded grain or when the number of faces was 40 (see figure 1). On the other hand, when R was large ($R = 10$), the relative growth ratio increased as coarsening progressed, and the fraction reached unity.

4. Discussion

4.1 Migration rate of the quadruple point

The migration rate of the quadruple point decreased moderately with increasing R (figure 2 (b)). The dependence of the migration rate on a and R coincided roughly with the theoretical calculation [1] in a manner similar to that discussed for triple line migration in a 2D model [4]. Additionally, while the effect of the simulation parameter kT appeared more distinctly as the size of grain a decreased, it did not seem to be critical.

4.2 Estimation of changes in the texture

The ordinate of figure 6 (a) was calculated using equation (6) as a function of the referred coarsening ratio of the matrix grains j.

$$volume\ increasing\ ratio(j) = \sum_i jr(j,i)\, f(i) / \sum_i f(i) \qquad (6)$$

The numbers '8' and '32' in the legends of figure 6 (a) are the values for j.

In equation (6), $f(i)$ represents the volume fraction of the specified grains in the reference structure whose number of faces is given by i, and the size distribution shown in figure 1 was used for it. The parameter $r(j,i)$ represents the mean of the relative growth ratios exemplified in figure 5 (a). The summation in equation (6) was made for values of i ranging from 10 to 40, the value for $r(j,i)$ is practically zero when $i < 10$, whereas $f(i)$ is negligible when $i > 40$. Equation (6) is a good approximation for the volume change of the texture component S assuming that the initial grain size distributions are the same for all of the texture components, and the grains belonging to component S do not collide with one another.

In figure 6 (b), the critical values for the number of faces of the embedded grain in the reference structure are shown. These values were obtained from the graphs exemplified in figure 5 (b). In the legends of figure 6 (b), '100%' corresponds to the smallest number of faces for which the fraction shown in figure 5 (b) reached 1, whereas '50%' corresponds to the smallest number for which the fraction reached 0.5. It can be clearly observed from this figure that when the width of the grain size distribution of the matrix grains was large, the effects were reduced (cf. wide 21 vs. standard 21).

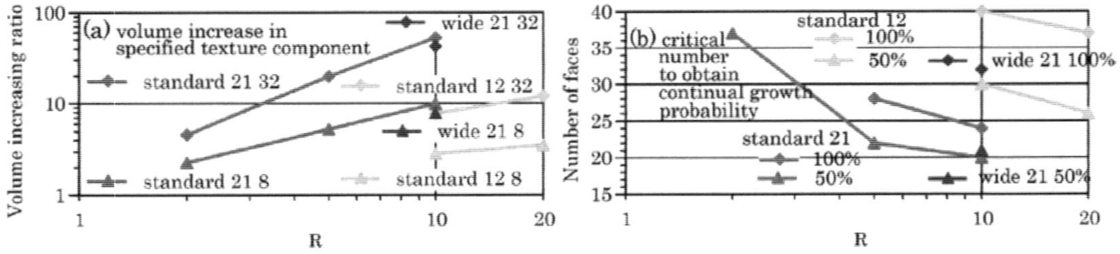

Figure 6. Summary of the orientation relation - dependent quadruple point dragging grain growth.

5. Conclusions

Under the conditions where the migration mobility of quadruple points depends on the relation of the orientations of the growing and shrinking grains, the behaviour of a specified grain embedded in matrix grains has been observed using a modified Potts MC-type 3D simulation.

Critical sizes were found for the embedded grain. Notably, smaller embedded grains shrink absolutely, whereas larger grains continue to grow continually without being overcome by the coarsening matrix grains. Furthermore, the growth of the grains depends on the relative mobilities of the quadruple points, on the composition of the matrix grain texture and the width of the grain size distribution of the matrix grains.

It was also demonstrated that the volume fractions of texture components may vary during grain coarsening when the mobility of the quadruple points is affected by combinations of the orientations of the growing and shrinking grains in correlation with their texture components.

References

[1] Barrales-Mora L A, Mohles V, Shvindlerman L S and Gottstein G 2008 Acta Materialia 56 pp 1151-1164
[2] Gottstein G and Shvindlerman L S 2010 Grain Boundary Migration in Metals (CRC Press second edition) chapter 4.6.6.7 p. 407
[3] Ito K 2013 Materials Science Forum 753 pp 121-124
[4] Ito K 2012 Computational Materials Science 62 pp 117-125.
[5] Barrales-Mora L A, Shvindlerman L S and Gottstein G 2013 Materials Science Forum 753 pp 373-376

IOP Conf. Series: Materials Science and Engineering **82** (2015) 012044 doi:10.1088/1757-899X/82/1/012044

Orientations of recrystallization nuclei developed in columnar-grained Ni at triple junctions

CL Xu[1, 2], S Huang[1], YB Zhang[2], GL Wu[1], Q Liu[1], D Juul Jensen[2]
[1] College of Material Science and Engineering, Chongqing University, Chongqing 400044, China
[2] Danish-Chinese Center for Nanometals, Section for Materials Science and Advanced Characterization, Department of Wind Energy, Technical University of Denmark, Risø Campus, DK-4000 Roskilde, Denmark
E-mail: chaoxu@dtu.dk

Abstract. A high purity columnar grained nickel sample with a strong <001> fiber texture was cold rolled to 50% reduction in thickness, followed by annealing at different temperatures. Optical microscopy was used to depict the grain boundaries prior to annealing and to detect nuclei formed on grain boundaries after annealing. Electron backscatter diffraction was performed to characterize the orientations of the nuclei and the deformed grains. Hardness tests were conducted at deformed grains. The potentials of triple junctions as preferential nucleation sites, the influence of deformation differences between adjacent grains on nucleation and orientation relationships between nuclei and parent matrix are analyzed.

1. Introduction

The nucleation process of recrystallization strongly affects the evolution of microstructure and texture and is thus of central importance for the properties of recrystallized materials [1, 2]. However, nucleation is difficult to study because of the small size of nuclei and the uncertainty about where the nuclei will form so that the investigation of nucleation is like a "needle in hay-stack" problem. Therefore, it is advantageous to focus on samples for which preferential nucleation sites may be active, such as materials with large particles [3, 4], with hardness indentations [5], scratches [6] and near triple junctions [7-9]. In the present study, we focus on nucleation at triple junctions and in particular on orientation relationships between the nuclei and their surrounding deformed matrix, where the nuclei are formed. The sample chosen for the investigation is high purity columnar grained nickel cold rolled to 50% reduction in thickness, thereby, nucleation from "below-the-surface deformed grains" is avoided.

2. Experimental

A high-purity columnar grained nickel sample with a strong <001> fiber texture was prepared by directional solidification. Viewed from the direction perpendicular to the grain growth direction, the material has a coarse grain structure with average grain diameter of 600μm, which is the same as the sample used in a previous study [9], and the grains have a perfect columnar shape. A sample was cold rolled to 50% reduction in thickness. During rolling, the grain growth direction is parallel to the normal direction (ND) of the sample. Three samples (named NO.1, NO.2 and NO.3) were cut from the rolled sheet. Each sample was cut perpendicular to the thickness direction into two identical pieces, the upper pieces for annealing experiments and the lower pieces for hardness tests. The samples were ground and electropolished. Extreme care was taken to get rid of surface imperfections, especially scratches, to avoid nucleation from such sites.

The upper pieces of the samples were first annealed at 350°C for one hour to somewhat recover the deformed microstructure, and then at 420°C, 400°C, and 395°C for one hour for sample No. 1, No.2, and No.3, respectively. After annealing, optical microscopy was used to reveal nucleation sites over the whole sample section. Electron backscatter diffraction (EBSD) mapping was then performed at the triple junctions where nuclei were found (using a Zeiss Auriga Dualbeam Station), to determine the orientations of the nuclei and the microstructures around the nuclei. Finally, microhardness tests were conducted using a force of 10g around triple junctions on the identical deformed lower pieces.

3. Results and discussion

3.1. Nucleation sites

A montage of optical micrographs of the deformed upper piece is used to show the grain boundaries for each sample, as demonstrated in Fig. 1 for sample NO.1. After annealing, nuclei are too small to be observed in a montage, so the nucleation sites are marked by red circles and numbered in the montage of deformed state. In total, 47 nucleation sites are observed in three samples: 35 right at triple junctions (e.g. nuclei 1-6 in Fig. 1), 12 at grain boundaries (e.g. nucleation site 7 in Fig. 1) and none within grain interiors. This result clearly shows the prevalence of triple junctions as the nucleation sites, which is in general agreement with previous investigations [7, 10]. It should, furthermore, be noted that all the nuclei formed at grain boundaries are very close to the triple junctions, within 50μm, and may well be formed close to the triple junctions at other section (although the grain boundaries are almost perpendicular to the RD/TD plane, there are of course some small variations).

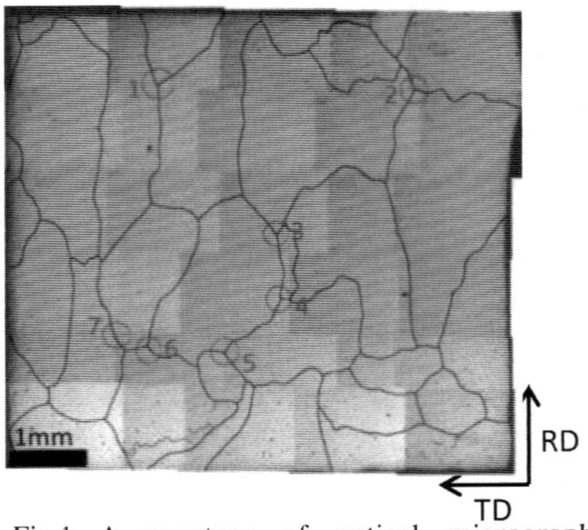

Fig.1 A montage of optical micrographs showing the grain boundary structure of the deformed upper piece of sample No.1. Numbers and red circles are used to identify nucleation sites of this sample after annealing.

3.2. Orientation relationship

The microstructure around 33 nucleation sites out of in total 47 were characterized using EBSD, including 22 at triple junctions and 11 at grain boundaries. Two examples are shown in Fig. 2, where nuclei and deformed grains are marked and numbered as Ni (i = 1-2.) and Gj (j = 1-6), respectively. Based on the misorientations between the nuclei and their surrounding deformed grains, the nuclei are divided into two groups. As seen in Fig. 2a, the nucleus N1 forms low angle boundaries to the deformed grain G1 at the triple junction, and its orientation lies at the outskirt of orientation cloud for grain G1 (see the pole figure). The nucleus N1 is therefore considered to originate from the deformed grain G1, and the nucleation is defined as parent type, Type I. A small grain, named T1, is also seen in Fig. 2a, which has a twin relationship with the nucleus N1. The grain T1 is therefore considered to develop by twinning during the growth of N1, and will not be considered as a new nucleus. In total Type I nucleation is found at 10 nucleation sites, including 8 at triple junctions and 2 at grain boundaries, where a deformed parent grain is found to be the same orientation with the nuclei. For these nuclei, subgrain growth or strain induced boundary migration could be the mechanisms for their formation.

It is found that there are 19 cases where the nuclei form high angle boundaries to the surrounding deformed matrix. But Σ3 boundaries (following the Brandon criteria, i.e. 8.66° deviation from ideal 60°<111> relationship) can be found between the nuclei and one of their surrounding deformed grains. For example, in Fig. 2b the misorientation between the nucleus N2 and the deformed grain G6 is around 56°<443>, which is about 8.1° away from the ideal 60°<111>. The nucleus N2 is therefore considered to originate from grain G6 through twinning. All these type of nuclei are defined as twin type, Type II. Twinning is a well-accepted mechanism for the nucleation of grains with new orientations. But it cannot be concluded from the present experiment whether the nucleus is formed by twinning directly from G6 or a nucleus Nx of parent type first forms below the surface and the nucleus N2 then forms by twinning during growth of the nucleus Nx. To answer this question, a full three dimensional characterization of the microstructure is necessary, which is planned for in the future.

At a few triple junctions, a series of nuclei are observed. These nuclei are often twin or double twin related to each other. For such cases the nuclei are classified as Type I and II if they are parent or twin related to any of the deformed grains respectively.

Finally, there are 4 cases where the nucleation does not belong to the two types. The nuclei do not form any Σ1, Σ3 or Σ9 boundaries to any of the surrounding deformed grains. This result is in agreement with reference [9, 11], in which some of the recrystallized grains have new orientations and are not twin related to the deformation matrix. Compared with the results in Ref. [9], the present investigation shows a few nuclei with new orientations. This might be related to the recovery process occurring during the initial low temperature annealing in the present experiment. As the deformation structure coarsening and the stored energy is reduced during the recovery, one may speculate that the formation of nuclei with new orientation may be less likely because this formation will require energy to form the new high angle boundaries. At present, the mechanism of their formation is unclear.

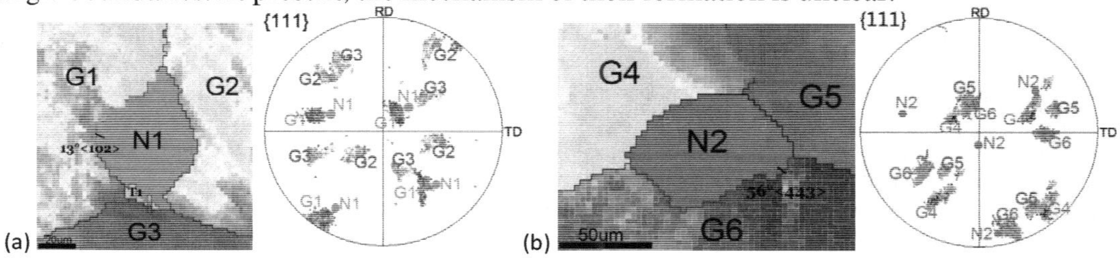

Fig. 2 EBSD orientation maps and {111} pole figures to show the orientations of deformed grains and nuclei; Fine white lines show low angle boundaries (>= 1.5° and <15°), coarse black lines high angle boundaries (>=15°), and red lines and purple lines 60°/[111] twin boundaries and 39.8°[110] second-order-twin boundaries, respectively.

3. 3. Microhardness tests

An example of microhardness measurements at a triple junction is shown in Fig. 3a. In total microhardness around 21 triple junctions is measured, including 13 triple junctions where nuclei are found at the same triple junctions in the identical upper pieces, defined as Group A triple junctions, and 8 triple junctions where no nucleus is found in the identical annealed pieces, defined as Group B triple junctions. The microhardness distributions around triple junctions within both groups are shown in Fig. 3b; all three microhardness values of the grains at each triple junction are included. It is seen that the microhardness distribution of Group A triple junctions spans a wider range than that of Group B triple junctions. On average, however, the hardness values for Group A and B triple junctions are almost similar, namely 2.065GPa and 2.072GPa, respectively. Combining the orientation analysis in section 3.2, where nucleation sites are determined to be specific grains at the triple junctions, and the hardness for all three grains around these triple junctions, it is found that the nuclei can originate from crystals with either lowest or highest microhardness values at Group A triple junctions. The hardness value of the grain at a triple junction can therefore not be used to indicate if nucleation will occur at that triple junction or not.

That the distribution of microhardness of Group A triple junctions is clearly wider than that of Group B triple junctions, may however imply that larger differences in hardness between the grains at a triple junction may lead to nucleation. The difference between the maximum and minimum microhardness, ΔHv, for each triple junction is therefore calculated and shown in Fig. 3c. It can be seen that the Group A triple junctions have a larger microhardness difference than the Group B triple junctions.

Nucleation by subgrain growth or strain induced boundary migration would be facilitated by a difference in stored energy (here taken proportional to the microhardness), which will drive the boundary towards the higher stored energy. It thus makes sense that more nucleation is seen at Group A than Group B triple junctions. It has, however, to be noticed that there are also Group A junctions with low ΔHv values and that the statistics of the present investigation is poor. Also it is not clear if the initial recovery in the present work may influence the result. However it would be of interest to investigate a possible effect of ΔHv on the nucleation potentials of triple junctions.

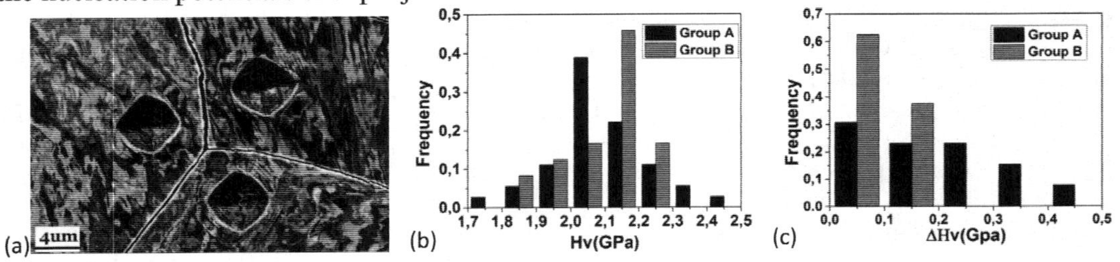

Fig. 3 (a) Example showing the hardness tests around a triple junction, and (b) the distribution of hardness and (c) ΔHv values of the maximum and minimum hardness at Group A and Group B triple junctions.

4. Conclusions

In a cold rolled and recovered columnar grained pure Ni sample it is found that all nuclei form at triple junctions (74.5%) and at original grain boundaries (25.5%). The vast majority of the nuclei have parent orientations or are twin related to a parent deformed grains. Only at 4 nucleation sites, (12.1%) nuclei are formed with new orientations. This is lower than in a previous investigation and may be an effect of the initial recovery of the sample in the present investigation.

Microhardness measurements at triple junctions in the initial deformed sample indicate a weak correlation between the difference in hardness of the grains at the triple junctions, ΔHv, and the potentials of the junction to form nuclei; the higher the difference, the more likely is nucleation. Further studies are needed to test this hypothesis.

Acknowledgements

CLX, GLW, SH and QL wish to thank the financial support from NSFC of China (Grant No. 51327805) and Fundamental Research Fund of Central Universities (Grant No. CDJZR14135502). All authors gratefully acknowledge the support from the Danish National Research Foundation (Grant No.DNRF86-5) and the NSFC of China (Grant No. 51261130091) to the Danish-Chinese Center for Nanometals. Helpful discussions with Drs. Xiaoxu Huang and Fengxiang Lin are grateful acknowledged.

References

[1] Haessner F 1978 *Recrystallization of Metallic materials*
[2] Doherty R D, Hughes D A, Humphreys F J, Jonas J J, Juul Jensen D, Kassner M E, King W E, McNelley T R, McQueen H J, Rollett A D 1997 *Mater. Sci. Eng. A* **238** 219
[3] Leslie W C, Michalak J T and Aul F W 1963 *Iron and its dilute solid solutions* pp.119-216
[4] Humphreys F J 1977 *Acta Metall.* **25** 1323
[5] Zambaldi C, Roters F, Raabe D, Glatzel U 2007 *Mater. Sci. Eng. A* **454** 433
[6] Fan G H, Zhang Y B, Driver J H, Juul Jensen D 2014 *Scripta Mater.* **72** 9
[7] Gordon P, Vandermeer R A 1962 *Trans. Metall. Soc. AIME* **224** 917
[8] Sabin T J, Winther G, Juul Jensen D 2003 *Acta Mater.* **51** 3999
[9] Wu G L, Juul Jensen D 2007 *Acta Mater.* **55** 4955
[10] Clemm P J, Fisher J C 1955 *Acta Metall.* **3** 70
[11] Skjervold S R, Ryum N 1996 *Acta Metall.* **44** 2395

Evolution of microstructure and texture during annealing of Al-2.5%Mg-0.2%Sc severely deformed by a combination of accumulative roll bonding (ARB) and conventional rolling

JR Gatti and PP Bhattacharjee
Department of Materials Science and Engineering, Indian Institute of Technology Hyderabad, Hyderabad, India
E-mail: ms10p003@iith.ac.in, pinakib@iith.ac.in

Abstract. Evolution of microstructure and texture during heavy cold-rolling and annealing of Al-2.5%Mg-0.2%Sc alloy was investigated. For this purpose recrystallized sheets of 1mm thickness having finely dispersed precipitates were processed to 3 cycles of ARB (equivalent strain, ε_{eq}=2.4) followed by conventional rolling to a final thickness of 200μm resulting in total equivalent strain of 4.0. Evolution of ultrafine microstructure and strong copper or pure metal type texture were observed during deformation. During annealing very stable microstructure was observed up to 400°C but further annealing resulted in formation of a layered microstructure with deformed layer sandwiched between recrystallized layers. Formation of strong cube texture is not observed in the recrystallized layers. Isothermal annealing for longer time at 500°C leads to abnormal growth of Q orientation ({013}<213>) within the deformed layer.

1. Introduction

Automobile industries are in real need of lightweight materials with superior mechanical properties. Aluminum alloys are widely used for this purpose. The development of annealing texture in aluminum alloys for automobile applications remains a subject of great importance as texture can greatly affect the mechanical properties of the processed materials [1, 2].
In the present work we investigate the development of deformation and annealing textures Al-2.5%Mg-0.2%Sc severely deformed by Accumulative roll bonding (ARB) [3]. Deformation of high stacking fault energy face centered cubic (FCC) materials leads to a strong copper or pure metal type texture consisting of copper (Cu) ((112)<111>), S ((123)<634>) and brass (Bs) ((110)<112>) orientations [4]. Annealing of this material leads to strong cube ((001)<100>) texture but the presence of second phase as fine precipitates changes the texture due to strong pinning effect as is observed in this work.

2. Materials and methods

The chemical composition of the alloy is given in Table 1. The as received blocks having dimensions of 150 mm^l x 60 mm^w x 10 mm^t were processed through various thermomechanical processing routes resulting in a final thickness of 1 mm. In Al-Mg-Sc alloy, very fine Al_3Sc precipitates are evolved during thermomechanical processing. These sheets are used as the starting materials for further processing. These 1 mm thick sheets are first processed through Accumulative Roll Bonding (ARB) to 3 cycles (equivalent strain of 2.4). A typical ARB processing cycle involves cutting of sheet to two halves, surface treatment including degreasing and wire brushing, stacking and roll-bonding of the sheets to 50% reduction in thickness in a single pass. The ARB was followed by conventional rolling to a final thickness of 200 μm (equivalent strain of ~4 corresponding to 97% reduction in thickness). The 97% deformed materials were isochronally annealed for 1hr at different temperatures to study the recrystallization behavior.

Table 1: Chemical composition of the Al-Mg-Sc alloy

Fe	Si	Ca	Cu	Mg	Sc	Al
20 ppm	10 ppm	5 ppm	22 ppm	2.5 wt%	0.2 wt%	Bal

The microstructures were observed using electron back scatter diffraction (EBSD) attached to scanning electron microscope (SEM) in order to characterize microstructural changes. The post processing of the data was done using TSL-OIM^{TM} software. The specimen along the longitudinal section including the normal direction (ND) and rolling direction (RD) were mechanically grinded and electropolished using 10% perchloric acid + 90% methanol at -30^{0}C and at a voltage of 20V. The volume fractions of texture components were calculated by taking a cut-off angle of 15° and assuming orthotropic sample symmetry.

3. Results and discussion

Figure 1 shows the orientation maps of starting and 97% cold rolled material. The merged orientation map of starting material shows recrystallized grains (average grain size ~ 115 μm) (Fig. 1(a)). The texture of the starting material is found to be weak (consisting mainly of cube and rotated cube grains denoted by Q as seen from Fig.1(a)) which is evident from the corresponding (111) pole figure (PF) (Fig. 1(b)). Deformation to 97% cold rolling (Fig. 1(c)) leads to the development of ultrafine lamellar structure extended parallel to the rolling direction (RD) with an average high angle grain boundary (HAGB) spacing of 980nm. The development of strong pure metal (copper) type texture characterized by the presence of Cu (red), S (yellow) and Bs (green) is observed in the orientation map in Fig. 1(c) which is also corroborated by the corresponding (111) PF (Fig. 1(d)).
When the heavily deformed material is isochronally annealed, the recrystallization behavior is found to be quite interesting. The material after annealing at 200°C (Fig.2(a)) and 350°C (Fig. 2(b)) shows the presence of deformation texture components. The stabilization of microstructure against discontinuous recrystallization is achieved by the addition of Sc through grain boundary pinning [5].
Discontinuous recrystallization is observed upon annealing at 400°C (Fig. 3(c)). When the material is annealed at higher temperature at 500°C (Fig. 3(d)), interestingly, a layered microstructure with two recrystallized regions separated by a deformed region is evolved. There is no grain growth into the deformed layer from either side and the grain growth is inhibited at the ARB interfaces on either sides.

IOP Conf. Series: Materials Science and Engineering **82** (2015) 012045 doi:10.1088/1757-899X/82/1/012045

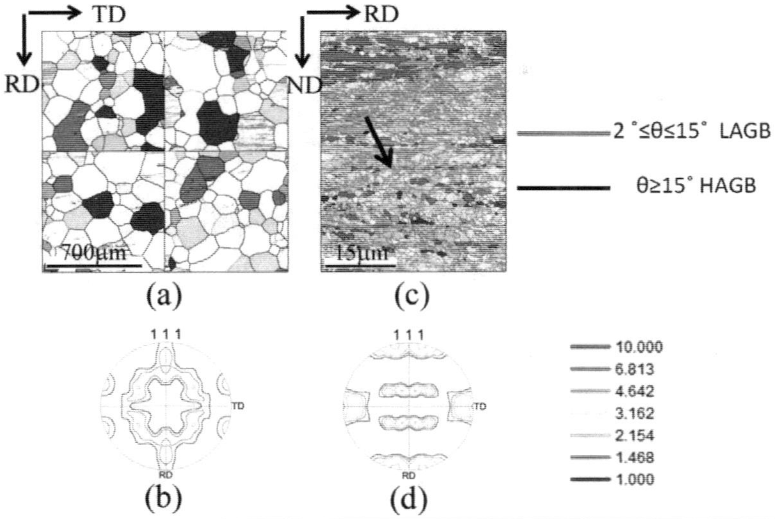

Cube	Cu	S	Bs	Goss	C_{RD}	Q	R
(001)<100>	(112)<111>	(123)<634>	(110)<112>	(110)<001>	(013)<100>	(013)<213>	(214)<112>

Fig. 1: Orientation maps along with corresponding (111) pole figures of (a) starting and (b) 97% ARB+CR processed AlMgSc alloy.

When the two layers are analyzed separately, the deformed layer has very fine grain structure and the deformation texture components are strongly present. In the recrystallized regions, strong cube texture is not observed. This is in contrast with the single phase Al-Mg alloy processed by ARB+CR where the dominating recrystallization texture component is cube [6].

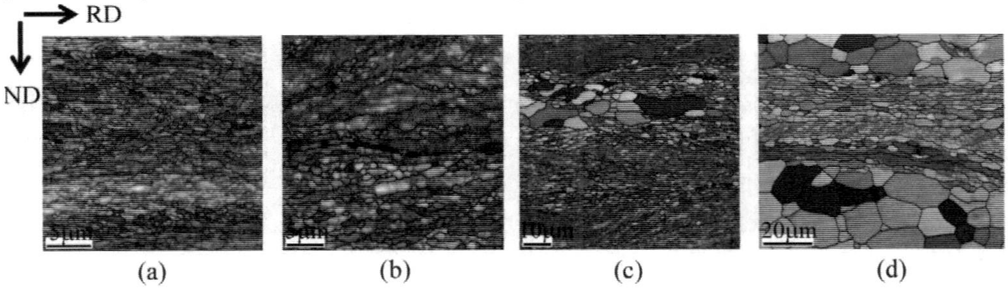

Fig. 2: Orientation maps of isochronally annealed 97% deformed AlMgSc alloy at (a) 200°C, (b) 350°C, (c) 400°C and (d) 500°C for one hour

In Al-Mg-Sc alloy the presence of precipitates plays an important role. Strong pinning due to the fine precipitates hinders strong recovery. The cube regions which undergoes preferential recovery during deformation results in early nucleation during recrystallization. However, absence of preferential recovery results in absence of preferential nucleation of cube grains and consequent absence of strong cube texture. Also the reason for the layered microstructure can be obtained from the processing route. Since in the ARB+CR processing route conventional rolling replaces the ARB, the contamination at the bonding interfaces are not distributed but concentrated at the center of the sheet which impart strong pinning leading to delayed recrystallization.

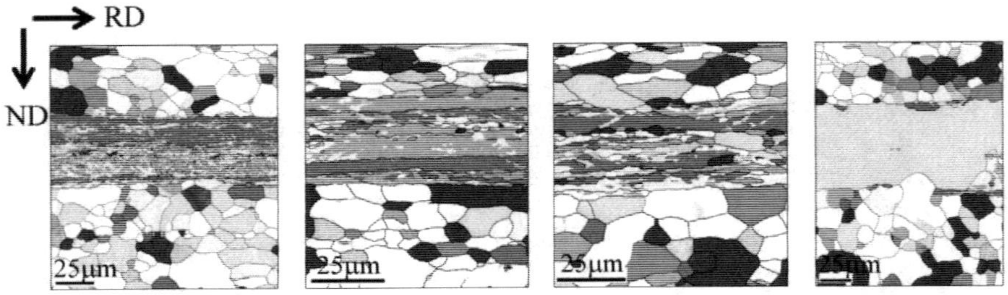

Fig. 3: Orientation maps of isothermal annealing at 500°C for (a) 2 seconds, (b) 30 min, (c) 90 min and (d) 180 min

To understand the stability of the layered microstructure isothermal annealing is carried out at 500°C for varying times. Figure 3 shows the orientation maps of material isothermally annealed at 500°C. It is very interesting to see that the layered structure developed at very short annealing time of 2 seconds (Fig. 3(a)). The recrystallization on either side of the deformed layer is very rapid. The orientations present in the recrystallized regions in all the cases seem to be similar. But within the deformed layer nucleation is observed after annealing for 90 minutes. It appears that preferential nucleation of ND rotated cube (designated as Q, (013)<213>)) grains (Fig. 3(c)). After 180 minutes the Q oriented grain consumes the entire deformed layer due to abnormal growth (Fig. 3(d)).

4. Conclusions

The severe deformation of Al-2.5%Mg-0.2%Sc alloy leads to very fine lamellar microstructure with pure metal type texture. Annealing at higher temperature results in an unusual layered microstructure with deformed layer sandwiched by recrystallized layers. Isothermal annealing for longer time leads to abnormal growth of Q orientation within the deformed layer.

Acknowledgement

The authors would like to acknowledge the financial support of DST, India (Grant No. SR/FTP/ETA-126/2010) for the research work.

References

[1] Doherty RD et al. 1997 *Mat. Sci. Eng.* A **238** 219-274
[2] Doherty RD 1997 *Prog. Mater. Sci.* **42** 39-58
[3] Saito Y et al. 1998 *Scripta Mater.* **39** 1221
[4] Leffers T and Ray RK 2009 *Prog. Mater. Sci.* **54** 351-396
[5] Miura Y, Shioyama T and Hara D 1996 *Mater. Sci. Forum* **217-222** 505
[6] Gatti JR and Bhattacharjee PP 2014 *J. Alloys Comp.* **615** 950-961

IOP Conf. Series: Materials Science and Engineering **82** (2015) 012046 doi:10.1088/1757-899X/82/1/012046

Microtexture of constituent phases in a heavily warm-rolled and annealed duplex stainless steel

M Zaid and PP Bhattacharjee
Department of Materials Science and Engineering
Indian Institute of Technology Hyderabad
Ordnance Factory Estate, Yeddumailaram 502205 AP, India

Abstract. Evolution of microtexture during isothermal annealing of a heavily warm-rolled Fe-0.08%C-24.18%Cr-10.5%Ni duplex stainless steel (DSS) having approximately equal volume fraction of ferrite and austenite was investigated in the present work. The DSS was warm-rolled to ~90% reduction in thickness at three different temperatures, namely, 225°C, 425°C and 625°C followed by isothermal annealing at 1175°C for different length of time. Austenite showed pure metal or copper type texture at different warm-rolling temperatures. In contrast, the texture of ferrite in different warm-rolled DSS revealed the presence of RD (RD//<110>) and ND (ND//<111>) fibers. The annealing texture of austenite showed retention of the deformation texture components while ferrite revealed strong RD-fiber.

1. Introduction

The evolution of microstructure and texture during recrystallization of heavily deformed two phase alloys has been reported extensively. However, amongst the two-phase alloys the main focus has been on those materials having the second phase present in the form of dispersoids or precipitates. In contrast, the recrystallization behavior of duplex alloys in which both the phases have grain structure, such as (α+β) brass or duplex stainless steels (DSS), have been studied to a much lesser extent. The focus of the present work is to study the recrystallization behavior of heavily warm-rolled DSS having approximately equal volume fraction of ferrite and austenite. Thermo-mechanical processing involving cold-rolling and annealing of DSS has been reported extensively [1-4], however, the evolution of microstructure and texture during warm-rolling has been clarified only recently by the present authors [5]. The present investigation is aimed to clarify the evolution of microstructure and texture during subsequent annealing which should be of interest for designing novel thermo-mechanical processing routes for DSS alloys.

2. Experimental

The DSS alloys were obtained in the form of an as-cast ingot having a chemical composition shown in Table 1. The as-cast ingot was initially processed through a series of thermo-mechanical treatments to obtain a hot-rolled slab which was homogenization annealed at 1175°C to obtain nearly equal volume fraction of the two phases [5]. The homogenization annealed slab was subjected to multipass warm-rolled up to ~90% reduction in thickness at three different temperatures, namely 225°C, 425°C and 625°C using a laboratory scale rolling equipment (SPX Precision Equipment, USA). In order to minimize the sudden quenching effect the rolls were pre-heated to 250°C during warm-rolling. The warm-rolled samples were immediately water quenched after every warm-rolling pass. The 90% warm-rolled samples were subsequently isothermally annealed in a conventional horizontal cylindrical furnace at 1175°C for different time intervals varying from 2 minutes to 120 minutes and immediately water-quenched. The microstructure and texture of the deformed materials were characterized using electron back scatter diffraction (EBSD) system (Oxford Instruments, UK) attached to a FEG-SEM (Make: Carl Zeiss, Germany, Model: SUPRA-40).The acquired EBSD dataset were analyzed using TSL-OIM™ software.

Table 1: Chemical composition of the experimental DSS

Element	**C**	**S**	**Cr**	**Mn**	**Ni**	**P**	**Si**	**Mo**	**Fe**
Weight%	0.08	0.001	24.18	0.14	10.5	0.017	0.45	3.11	balance

3. Results and discussion

Evolution of texture during warm-rolling
The texture of the two phases in the starting homogenization annealed material is shown in Fig.1. Figure 1(a) shows the φ_2=45° section of the ODF of ferrite indicating the presence of ND (ND//<111>) along with a rotated cube component shifted along the φ_1 by ∓ 15°. The φ_2=45° section of the ODF of ferrite in DSS warm-rolled at 225°C (Fig.1(b)) and 625°C (Fig.1(d)) shows the presence of both RD and ND fibers. However, φ_2=45° section of the ODF of ferrite in DSS warm-rolled at 225°C (Figure 1(c)) shows remarkably stronger RD-fiber as compared to the ND-fiber. The strongest component along the RD-fiber is {001}<110>. The evolution of strong RD-fiber texture in ferrite in DSS warm-rolled at 425°C is explained on the basis on dynamic strain aging effect due to which carbon atoms preferentially lock dislocations and restricts slip on the {110}<111> system [5]. Further deformation by slip occurs preferentially on the {112}<111> system. This leads to a stronger RD fiber than ND fiber.
(111) pole figure (PF) of the austenite in the starting homogenized DSS (Fig.1(e)) shows the presence of weak pure metal (or copper type) texture characterized by S ({123}<634>), Cu ({110}<112>) and the brass ({110}<112>) component. The (111) PFs of austenite in DSS warm-rolled at different temperatures (Fig.1(f)-(h)) shows the development of predominantly pure metal type texture. The mechanism for the above observation has been recently reported by the present authors [5].

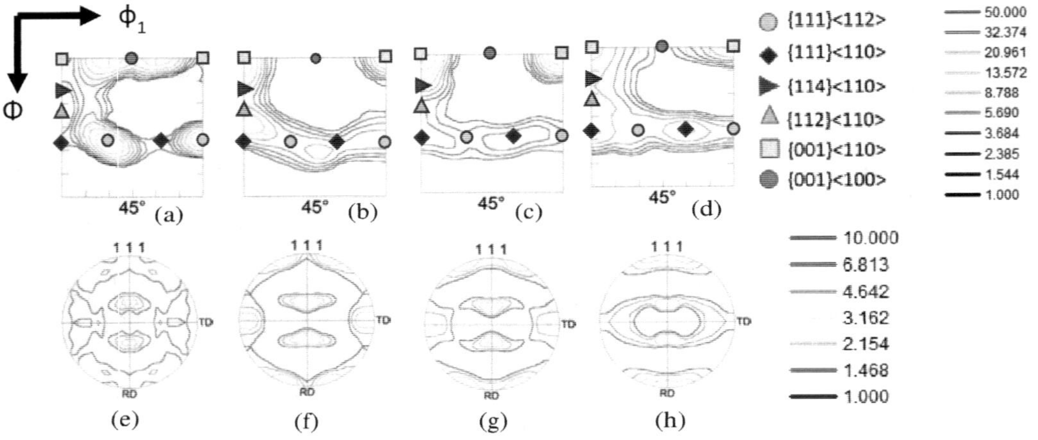

Fig. 1: φ_2=45° section of ODFs of ferrite in DSS (a) homogenized and warm-rolled at (b) 225°C, (c) 425°C and (d) 625°C. (111) PFs of austenite in DSS (e) homogenized and warm-rolled at (f) 225 °C, (g) 425°C and (h) 625°C.

3.2 Evolution of recrystallization texture during annealing

Isothermal annealing of the different 90% warm-rolled DSS is carried at 1175°C out for 2 minutes to 120 minutes. However, specimens annealed for 2, 30 and 120 minutes are only discussed here as major microstructural transformations happen in these temperatures. These microstructural transformations are clearly shown in Fig.2 for the DSS warm-rolled at 625°C but also holds good for DSS warm-rolled at other temperatures. The ultrafine microstructure obtained after warm-rolling (Fig.2(a)) is transformed to a bamboo type morphology for short isothermal holding time (Fig.2(b)). A globular morphology is gradually evolved due to the breakdown of the lamellar structure. Any significant changes could not be observed during prolonged annealing (Fig.2(c)).

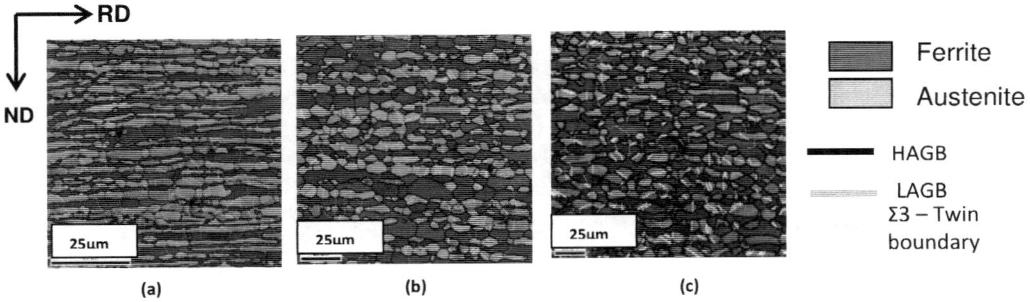

Fig. 2: Phase maps of DSS 90% warm-rolled at 625°c and annealed at 1175°C for (a) 2 minutes, (b) 30 minutes and (c) 120 minutes.

Despite the apparent differences in the texture of ferrite in as warm-rolled conditions as discussed above, the texture of ferrite in the three warm-rolled and annealed DSS (Fig.3(a)) shows much stronger RD-fiber than ND-fiber. The development of stronger RD-fiber components is attributed to recovery behavior of these components [1] as opposed to the typical recrystallization type behavior of the ND-fiber components. Warm-rolling of DSS results in diminished driving force for recrystallization and results in the formation of much stronger RD-fiber.

The recrystallization texture of austenite in different warm-rolled and isothermally annealed DSS is represented the by (111) pole figures (PFs) in Fig.3(b). The (111) PFs of austenite in annealed materials appear to be very similar to those of the as warm-rolled DSS (Fig.1(f)-(h)) indicating the retention of deformation texture even after prolonged annealing. Thus, annealing textures for the three warm-rolling conditions show retention of the deformation texture components. The retention of deformation texture during recrystallization indicates discontinuous recrystallization without preferential orientation selection. Interestingly brass recrystallization component ({236}<385>) is not found strong after annealing. Since Brass is not the dominant component of the warm-rolled texture, the preferential growth advantage of {236}<385> grains would be greatly reduced, preventing it from emerging a strong texture component after annealing.

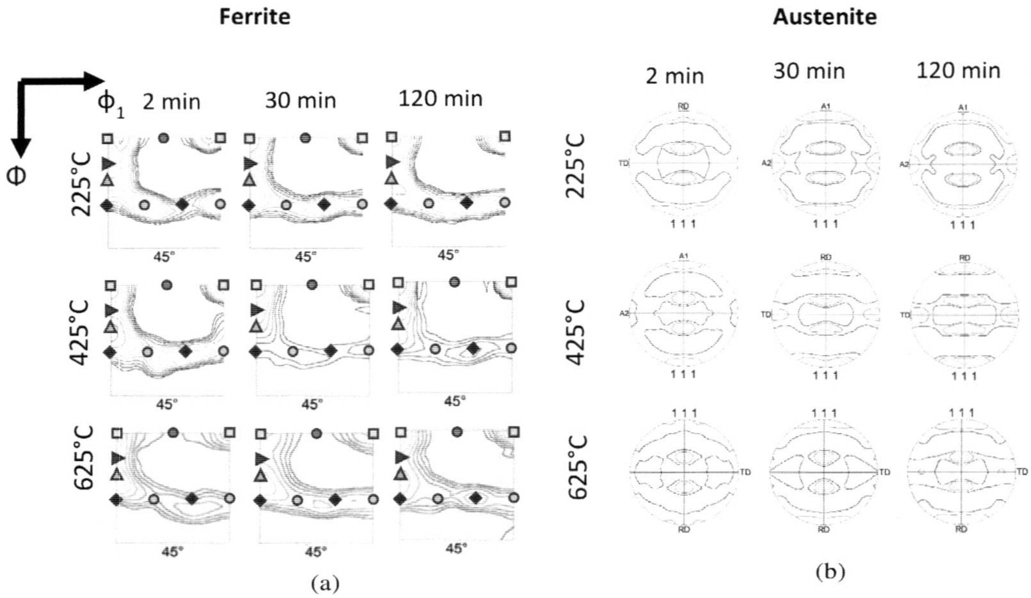

Fig. 3: (a) φ2=45° section of ODFs of ferrite and (b) (111) PFs of austenite in DSS 90% warm-rolled at different temperatures and then isothermally annealed at 1175°C for different time intervals. The intensities of the contour lines are same as those shown in Fig.1.

Conclusions

- The deformation texture of the two phases in DSS is significantly affected by the warm-rolling temperature, although, the recrystallization texture remains unaffected.
- RD-fiber strengthens due to strong recovery behavior of ferrite in DSS during annealing.
- Austenite in DSS shows retention of deformation texture components after annealing which is attributed to discontinuous recrystallization without preferential orientation selection.

References

[1] Keichel.J, Foct.J, and Gottstein.G, 2003, *ISIJ Int.,* **43;** 1788.
[2] Maki.T, Furuhara.T and Tsuzaki.K, 2001, *ISIJ Int.,* **41**, 571.
[3] Keichel.J, Foct.J, and Gottstein.G, 2003, *ISIJ Int.,* **43**, 1781.
[4] Keichel.J, Gottstein.G, and Foct.J, 1999, *Mater. Sci. Forum,* **318**, 785.
[5] Bhattacharjee.P.P, Zaid.M, Satiaraj.G.D and Bhadak.B, 2014, *Metall. Mater. Trans. A,* **45,** 2180.

IOP Conf. Series: Materials Science and Engineering **82** (2015) 012047 doi:10.1088/1757-899X/82/1/012047

Crystallographic aspects of nucleation and grain growth during recrystallization of high stacking fault energy metals as characterized on model Al-1%wt.Mn alloy crystals

M. Miszczyk[1,a], H. Paul[1], J.H. Driver[2] and C. Maurice[2]

[1] Institute of Metallurgy and Materials Science, PAS, Krakow, Poland
[2] Ecole des Mines de Saint Etienne, Centre SMS, Saint Etienne, France
[a] E-mail: m.miszczyk@imim.pl

Abstract. The objective of this paper is to identify the predominant crystallographic relations between deformed state and recrystallized grains during the early stages of recrystallization of the Goss{110}<001> and brass{110}<112> oriented single crystals of Al-1%Mn. The analysis was based on high resolution local orientation measurements in scanning electron microscopy. After annealing the disorientation across the recrystallization front 'defines' the final rotation by angles in the ranges of 35°-50° around axes located near the normals of all four {111} slip planes. Although the rotation axes approach the normal vector of the active slip planes during deformation, they only rarely coincide with the exact location of the <111> direction. For both initial orientations, preferred grain growth occurred along the {111} planes, the most active during strain.

1. Introduction

The literature on recrystallization pays a lot of attention to the role of high-angle boundaries and the conditions of their migration, e.g. [1 - 6]. Nevertheless, based on such an approach, it is unable to provide a comprehensive answer to the question of the mechanisms controlling the formation of a new grain inside the deformed structure and, in particular, its different orientation with respect to the deformed volume from which it grows. In the authors' view this results from neglecting the role of low-angle boundaries. Their thermally activated migration combined with the movement of dislocations stored inside (sub)grains and/or the diffusion of vacancies may play a key role in: (i) the 'transformation' of low-angle boundaries into high-angle ones and/or the formation of boundaries capable of fast migration [7 - 9], and (ii) the orientation change of a 'chosen area' with respect to the deformed matrix [10 - 13]. Still, the nature of these processes is not very clear at the moment. The purpose of this work is to identify and describe the preferences in the disorientation relationship appearing between a new grain and the deformed region during the early stages of recrystallization of high stacking fault energy (SFE) face centred cubic (fcc) metals. The current experimental investigations have focused on an analysis of transformations which occur in model single crystals od Al-1%Mn alloy, of brass {110}<112> and Goss {110}<001> orientations deformed in terms of plane strain compression (PSC). The deformation of stable orientation crystallites (up to large deformations) practically does not lead to the appearance of high-angle boundaries [14 - 17]. This is of paramount importance for the discussion of mechanisms controlling textural transformations during recrystallization. The basic research technique employed local orientation measurements based on scanning electron microscopy (SEM) equipped with electron backscattered diffraction (EBSD) facility.

2. Experimental

The material was Al-1%wt.Mn alloy. Single crystal bars were grown by directional solidification and the (110)[1-1-2] and (110)[00-1] oriented samples were carefully cut from the bars using a wire saw. The deformation was carried out in channel-die to a logarithmic strain of 0.51 at 293 K. In order to minimize friction between the sample, the punch and the walls of the channel-die, each sample was wrapped in 0.2 mm thick Teflon™ tape. Annealing was carried out in air furnace to the first stages of primary recrystallization. The microstructures were mostly analysed on the ND-ED plane (where: ND and ED are normal and extension directions, respectively) in the middle of the sample thickness by SEM using a JEOL 6500F. The SEM was equipped with electron backscattered diffraction (EBSD) facility. Post-processing analysis of the orientation maps was performed using HKL Technology Channel 5 software. The applied step size was ranged between 0.1 μm and 2 μm.

3. Results and discussion

3.1. Early stages of recrystallization in high SFE fcc crystals

After very short annealing the recrystallized phase is 'composed' of single grains of uniform orientation completely surrounded by deformed areas and the grains that typically form compact chains. In the following the orientation relationships between the adjacent regions of recrystallized grains have been ignored. The particular role of the orientations identified in the as-deformed state has been analysed by examining the orientation distribution of the new grains. A comparison of the pole figures only for the deformed and for recrystallized state shows that the <111> poles only rarely coincides. Most of new grains orientations form groups, which are symmetrically situated with respect to selected (external) axes. This symmetry results from positive and negative rotation of new grains orientations around axes grouped near the all the <111> poles of the deformed state.

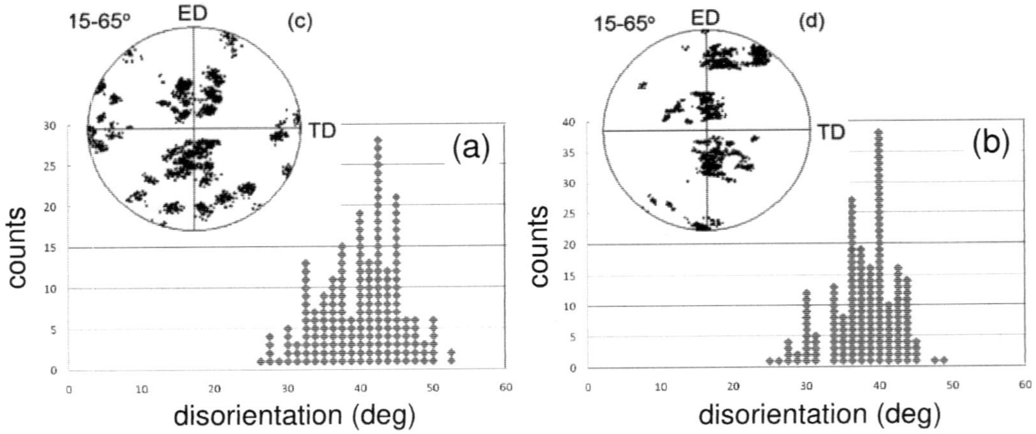

Fig. 1. Disorientation relation between recrystallized area and nearest as-deformed neighbourhood for (a) and (c) Goss (110)[00-1], (b) and (d) brass (110)[1-1-2] orientations. Distribution of disorientation angles - (a) and (b) and axes - (c) and (d).

3.2. The disorientation (angles/axes) distribution at early stages of recrystallization

The 'frequency' distributions of the disorientation angles between recrystallized and deformed/recovered, regions are presented in Fig. 1a and b. The range of potential disorientation angles was binned into intervals of 1.25° into which the number of disorientations across the recrystallization front corresponding to a given angular interval was counted. The prevalent range of angles across the migrating recrystallization front (between recrystallized and deformed grains and for the entire set of orientation maps) was within a wide range between 25° and 50°. The <111> poles of the deformed/recovered grains, do not coincide with the <111> poles of the recrystallized grains or more precisely overlapping of the <111> rotation axes was observed only rarely (Fig. 1c and d). Basically, the majority of the disorientation axes across the recrystallization front are concentrated around (but not coincide with) the <111> poles of the two slip planes (highly privileged) during deformation. Since the disorientation axes are slightly deviated from the normals of the {111} planes they 'occupy' a wide range near the <112>, <012>, <123> and <122> directions. A 'grouping' of rotation axes near the two remaining poles could, however, be detected but was clearly weaker.

3.3. The directionality of grain growth

The orientation maps (on the ND-ED plane) for both initial orientations show that a considerable number of recrystallized grains are of elongated shape. These grains exhibit a well-defined lattice rotation with respect to the deformed crystal orientations. An example of the crystal lattice rotation, related to the appearance of heavily elongated grains, was carried out on Al-1%Mn single crystal of Goss (110)[00-1] orientation, as presented in Fig. 2. To a first approximation, this aspect of nucleation and grain growth leads to the following conclusions: (i) the longest grain direction runs along the traces of {111} planes, the most active planes during strain, and which are symmetrically located to the ND-TD plane. To a lesser degree some grains are elongated along the traces of two other planes. (ii) The orientations of the elongated grains are related to those of the group of the deformed crystals. The crystal lattice of each of the grain groups is rotated (positively or negatively) around diverse groups of disorientation axes. The rotation axes are preferentially situated near the poles of {111} planes, the most active planes during strain. (iii) This may suggest a mechanism for the formation of a new grain by thermally activation movement of dislocation families, 'travelling' in pairs on one of the above-mentioned planes. Although the rotation axes approach the normal vector of the preferred slip planes they only rarely coincide with the exact location of the <111> directions.

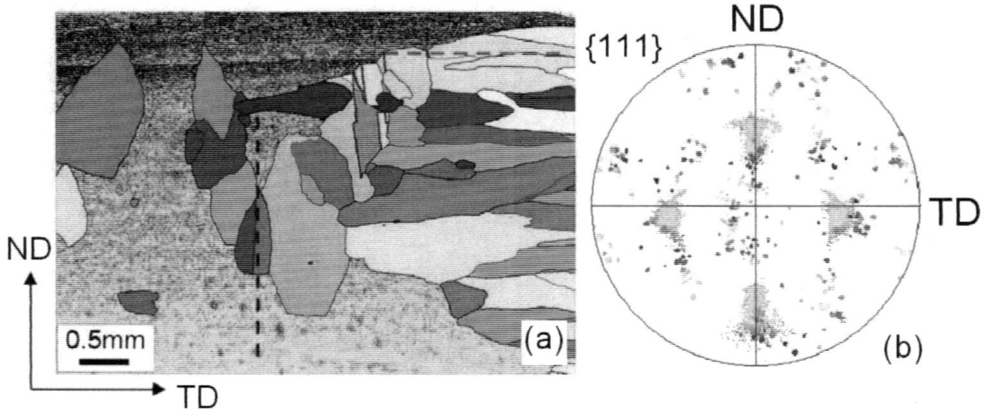

Fig. 2. Directionality observed at early stages of recrystallization in sample of Goss (110)[00-1] initial orientation. (a) Large area SEM/EBSD orientation map on the ND-TD plane and (b) corresponding the {111} pole figure. Step size of 2.8 μm. Sample deformed up to strains of 0.51 and then recrystallized at 680 K for 70 s.

An explanation of these results can be based on the thermally activated, coordinated movement of dislocations. But, many situations are possible due to screw or edge character of dislocations stored along the growth plane, as discussed in [18]. The spread of the lattice rotation angles of individual grains for a given category may result from the spread of orientations of the deformed state and/or an 'unequal' contribution of the two dislocation families (systems) grouped on one plane.

4. Conclusions

Based on local orientation measurement in SEM the orientation relations which appear at the initial stages of recrystallization between a new grain and the as-deformed areas of stable single crystals have been statistically analyzed. The results allow us to draw the following conclusions:

- For Al-1%wt.Mn alloy of both initial orientations at the initial stages of recrystallization, the appearance of a specific number of new orientation groups of new grains was demonstrated. The orientation relation across the recrystallization front 'defines' the final rotation by angles in the range 25°-50° around axes mostly grouped about the <122>, <012>, <112>, <123> and <111> directions.
- In the case of a single isolated nucleus of uniform orientation, these rotation axes are usually grouped around, but not at, the normals of all four {111} planes.
- A large fraction of new grains is characterised by a strongly elongated shape. For both initial orientations, preferred grain growth occurred mostly along the most active planes during strain. The orientations of the elongated grains were directly related to those of the group of as-deformed orientations.
- The variations of disorientations across recrystallization front tend to follow the orientation variations in the deformed state. This facilitates 'the orientation stability' of a new grain growing into the as-deformed structure.

Acknowledgments: This work was supported in part by the Polish National Centre of Science (NCN) and the Polish Ministry of Science and Higher Education, projects no: 3010/B/T02/2011/40 and 'Iuventus Plus' IP2011 036471, respectively. One of the author M. M. acknowledges the Region Alpes for a graduate student scholarship.

References

[1] Doherty RD et al 1997 *Mater. Sci. Engn.* A238 219-274
[2] Humphreys FJ and Hatherly M 2004 *Recrystallization and related annealing phenomena*, Pergamon, 2nd Edition
[3] Gottstein G and Shvindlerman LS 1992 *Scr. Metall. Mater.* 27 1515-1520
[4] Berger A et al 1988 *Prog. Mater. Sci.* 32 1-95
[5] Driver JH et al 2000 in *Proc. of the 21st RISØ Inter. Symp. on Mater. Sci.* ed Hansen N et al (Roskilde: Denmark) 35-48
[6] Humphreys F 2004 in *Proc. of 2nd Joint Int. Conf. on Rex&GGII*, ed Bacroix B et al, (Annecy: France) 107-116
[7] Rios PR and Sandim HRZ 2005 *Mater. Res.* 8/3 225-238
[8] Paul H et al 2012 *Metall. Mater. Trans.* A43/12 4777-4793
[9] Poplewska J et al 2013 *Mat. Sci. Forum* 753 267-270
[10] Inoko F and Kashihara K 1993 *Mat. Sci. Forum* 113-115 139-144
[11] Paul H, et al 2005 *Mat Sci Forum* 495-497 1249-1254
[12] Paul H and Driver JH 2006 *Microchim Acta* 155 235-242
[13] Inoko F et al 2010 *Mater. Trans.* 51/4 597-606
[14] Godfrey A et al 1998 *Acta Mater.* 46 823-833
[15] Godfrey A et al 2001 *Acta Mater.* 49 2429-2440
[16] Borbély A et al 2007 *Acta Mater.* 55 487-496
[17] Paul H et al 2010 *Acta Mater.* 58 2799-2813
[18] Miszczyk M et al 2015 *Acta Mater.* 83 120-136

IOP Conf. Series: Materials Science and Engineering **82** (2015) 012048 doi:10.1088/1757-899X/82/1/012048

Kinetics of individual grains during recrystallization of cold-rolled copper

F X Lin[1], Y B Zhang[1], S O Poulsen[2], N Schell[3], W Pantleon[4] and D Juul Jensen[1]
[1]Danish-Chinese Center for Nanometals, Section for Materials Science and Advanced Characterization, Department of Wind Energy, Technical University of Denmark, Risø Campus, 4000 Roskilde, Denmark
[2]Department of Materials Science and Engineering, Northwestern University, Evanston, IL 60208-3108, USA
[3]Helmholtz-Zentrum Geesthacht, Institute of Materials Research, Max-Planck-Straße 1, 21502 Geesthacht, Germany
[4]Section for Materials and Surface Engineering, Department of Mechanical Engineering, Technical University of Denmark, 2800 Kongens Lyngby, Denmark
E-mail: lnfe@dtu.dk

Abstract. The formation of a recrystallization texture is closely related to the nucleation and growth of recrystallizing grains, which may vary from grain to grain. Cube texture is a commonly observed recrystallization texture in face centered cubic metals of medium to high stacking fault energy after heavy cold-rolling and annealing. In this work, recrystallization of pure copper cold-rolled to a von Mises strain of 2.7 was investigated *in situ* using three-dimensional X-ray diffraction. Growth curves of 835 grains were determined, and the curves of cube and noncube grains were compared. It was found that the nucleation times of cube grains and non-cube grains were similar, whereas the growth rates of a few but not all cube grains were high. Effects hereof for the development of the cube texture were discussed.

1. Introduction

Recrystallization kinetics and formation of a strong cube texture in heavily cold-rolled face centered cubic metals have been extensively studied both experimentally and theoretically (e.g. [1-4]). Most experimental studies of recrystallization kinetics are based on a series of partially recrystallized samples characterized by optical or electron microscopy (e.g. [5]). More recently, 3D X-ray diffraction (3DXRD) microscopy was developed and used to study recrystallization kinetics of cold-rolled aluminium [6], which showed that each recrystallizing grain followed its own kinetics. The objective of the present work is to study recrystallization kinetics of individual grains for cold-rolled copper using 3DXRD. The kinetics of cube grains and noncube grains are compared to discuss the cube texture development.

2. Experimental

The initial material was oxygen free high conductivity copper (purity 99.95%), with a grain size of 22 μm (intercept length, ignoring twin boundaries). It was cold-rolled from 5 mm to 0.5 mm, corresponding to a von Mises strain of 2.7. Recrystallization in this material has been studied by electron microscopy, and a strong cube texture is known to develop during recrystallization [7,8]. A specimen was cut from the rolled sample of ~2 mm along the rolling direction (RD) and 0.75 mm along the transverse direction (TD), and was electropolished to avoid nucleation from surface imperfections during the subsequent annealing.

The 3DXRD experiment was conducted at beamline P07 at PETRA III, Deutsches Elektronen-Synchrotron (DESY) using a monochromatic X-ray beam of energy 50 keV and a square beam profile of size 500 × 500 μm². The specimen was mounted in a hot finger furnace with an inert gas atmosphere, with RD parallel to the vertical direction and TD parallel to the beam direction. The specimen was heated *in situ* at 130 °C, and the specimen was rotated around the vertical axis from 0° to 30° (the rotation angle was noted as ω) by sweeping a small 0.5° interval while acquiring diffraction images. Each 30° rotation took 5.7 min, and the entire measurement took about 250 min. According to hardness measurements, the sample was ~25% recrystallized at the end of the measurement.

Diffraction peaks from the {200} and {400} Debye-Scherrer rings were analyzed. The positions of diffraction peaks were identified from images collected during the last rotation. It should be noted that only peaks with relatively high peak intensities were considered. These peaks were then traced back to earlier times. Fig. 1 illustrates this process using a {200} peak as an example. Peaks were divided into two groups: originating from cube or noncube grains. A peak was considered as originating from a cube oriented grain, if the corresponding scattering vector deviated less than 10° from that of the ideal cube orientation. Since this treatment used only the partial orientation information from a single diffraction peak, cube grains in this work are actually grains with either RD or ND close to a <100> direction. However, according to EBSD data of the same sample after recrystallization [8], only very few grains (~2%) would be mislabeled; this treatment seems thus acceptable. The peak intensity was calculated as the integrated intensity over a peak subtracting its background. The intensities were converted to grain volumes as in [9], and equivalent sphere diameters (ESDs) were calculated from the obtained grain volumes. The growth curve of each grain was plotted as ESDs increasing as a function of annealing time. All growth curves were manually checked to remove artifacts.

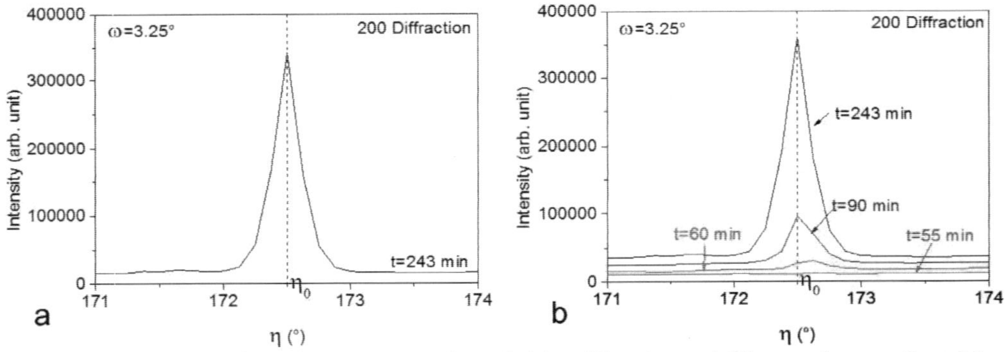

Fig. 1 Example showing the process of peak identification. a) The peak was first identified from the diffraction image collected during the last 30° rotation. The peak position (ω,η_0) were noted. b) The same peak at earlier times was found according to the peak position. This peak first appeared at t=60 min, and its intensity increased with time. For clarity, the peak is only plotted for a few time steps.

3. Results and discussion

In total, 101 cube grains and 724 noncube grains were identified in the last time step. The grain size (ESD) distributions are plotted in Fig. 2. Compared with noncube grains, cube grains have a higher fraction of grains smaller than 10 μm. All these small cube grains have peak positions near the azimuthal angles 0° or 180°. These peak intensities are thus enhanced by the Lorentz factor. It means that the detection limit (minimum grain size) is smaller for cube grains. To avoid this bias, all grains smaller than 10 μm are excluded in the following analysis. As shown by Fig. 2, the size distribution of noncube grains is observed to follow a lognormal distribution, but that of cube grains is not. There are a certain number of cube grains larger than 40 μm, and the largest cube grain is 88 μm. The size distribution of

IOP Conf. Series: Materials Science and Engineering **82** (2015) 012048 doi:10.1088/1757-899X/82/1/012048

cube grains is consistent with previous observations using electron backscatter diffraction in the same material after annealing at 150 °C [8]. The large cube grains have been termed supercube grains [8], and although few in number, these supercube grains make a significant contribution to the cube texture.

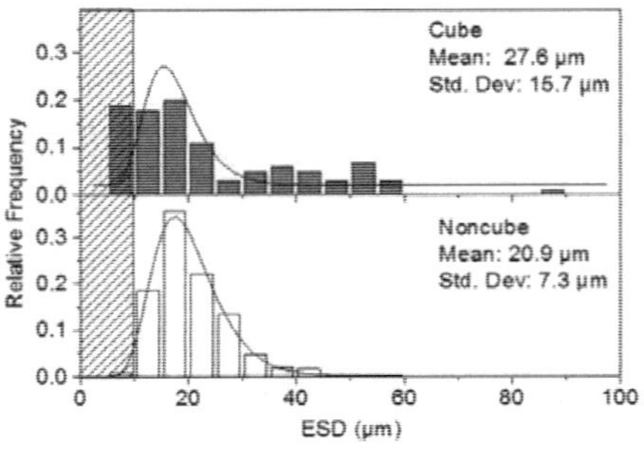

Fig. 2 Size distributions of cube and noncube grains at the end of the measurement (~25% recrystallized). There are more cube grains smaller than 10 μm, because of the different detection limits for cube and noncube grains (see text in Section 3 for details). To avoid this bias, a cut-off size of 10 μm was used, i.e. grains in the shaded part are not included in the following analysis. The size distributions of grains larger than 10 μm were fitted to lognormal functions.

A few typical growth curves are plotted in Fig. 3. Each grain follows its own growth curve, as has been observed for aluminium [6]. The growth rates vary with time and from grain to grain. Most grains tend to have a large growth rate right after nucleation, followed by a period during which the growth rate decreases. A few grains, such as the one marked by an arrow in Fig. 3a, exhibit a stagnation period for some time, after which they continue to grow. The nucleation times vary significantly from grain to grain, which is different from aluminium [6]. To compare the nucleation and growth behavior of cube and noncube grains, we obtained two parameters from each growth curve: the nucleation time t_n and the maximal growth rate g. Fig. 3b illustrates how the two parameters were obtained.

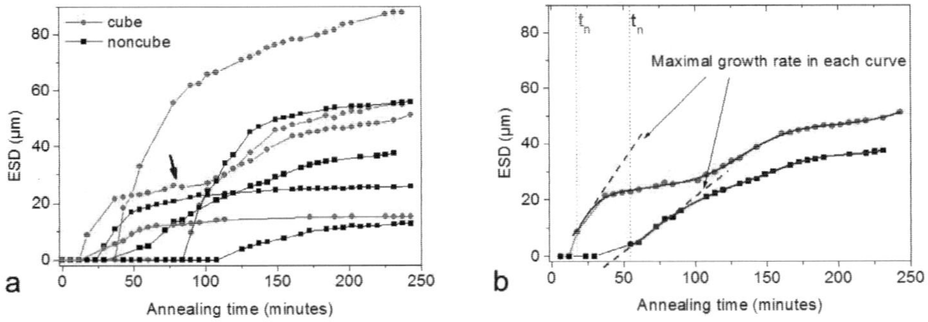

Fig. 3 Selection of measured growth curves. a) Growth curves of 4 cube grains and 4 noncube grains. The arrow marks a stagnation period. b) Examples explaining how the nucleation time t_n and maximal growth rates g were determined. t_n was the time when the grain was first observed. Each growth curve was fitted using a spline function (the blue curve), and the largest slope of the fitted curve was used as g.

The distributions of the nucleation times are plotted in Fig. 4a. Nucleation takes place during the entire measurement period for both cube and noncube grains. There is a slightly higher fraction of cube grains that nucleate during the first 30 min (see the first bin of the histograms in Fig. 4a), but according to the cumulative relative frequency, the nucleation behavior of cube and noncube grains is considered quite similar. Therefore, the different size distributions seem not to be associated with differences in the nucleation times.

The distributions of the maximal growth rates are plotted in Fig. 4b. For most cube and noncube grains, the maximal growth rates are less than 0.4 μm/min, but cube grains have more grains with higher growth rates. This tendency is clearer from the cumulative relative frequencies: the 70% slowest growing grains have similar maximal growth rates for both cube and noncube grains, whereas the 30% fastest growing cube grains show a significantly enhanced growth rate than the 30% fastest growing noncube grains. The few cube grains with larger growth rates may grow to larger grain sizes, which explains the longer tail in the size distribution of cube grains, i.e. the formation of supercube grains. However, as the growth rate of one grain varies with time, it is insufficient to use only the maximal growth rate in a growth curve to represent the growth behavior of a grain. We are planning to include more parameters to compare the growth behavior of cube and noncube grains in the future.

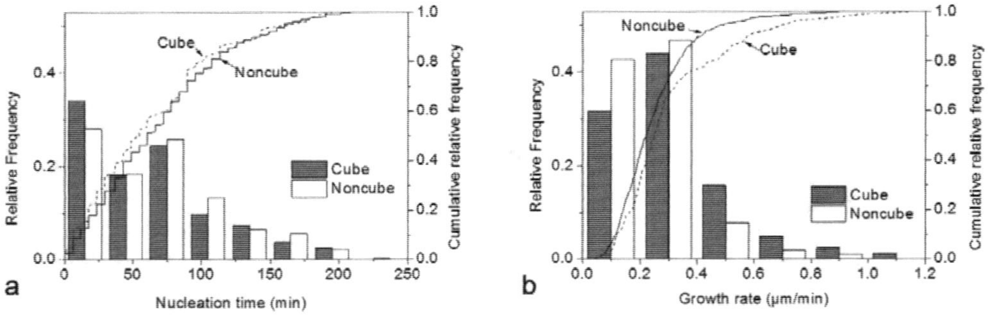

Fig. 4 Distributions of nucleation times t_n (a) and maximal growth rates g (b) for cube and noncube grains that are larger than 10 μm at the end of the measurement. The histograms show the relative frequencies, and the curves show the cumulative relative frequencies. The bin widths are 30 min and 0.2 μm/min for a) and b) respectively.

4. Summary

Nucleation and growth of individual grains in cold-rolled copper have been followed *in situ* during recrystallization at 130 °C by 3DXRD. A few cube grains having large sizes after recrystallization, named supercube grains, are important for the cube texture development. Cube and noncube grains seem to have similar nucleation behavior but different growth rates. More cube grains have larger growth rates, and this is considered to be related to the formation of the supercube grains.

Acknowledgements

The authors gratefully acknowledge the support from the Danish National Research Foundation (Grant No. DNRF86-5) and the National Natural Science Foundation of China (Grant No. 51261130091) for the Danish-Chinese Center for Nanometals, within which this work was performed. The authors also acknowledge beamtime by DESY under proposal number I-20110560 EC. We would like to thank Torben Fischer and Lars Lottermoser for assistance in using beamline P07.

References

[1] Vandermeer R A and Juul Jensen D 1995 *Metall. Mater. Trans. A* **26** 2227
[2] Juul Jensen D 1997 *Metall. Mater. Trans. A* **28** 15
[3] Wassermann G and Grewen J 1962 *Texturen metallischer Werkstoffe* Springer-Verlag Berlin pp 307-326
[4] Necker C T et al. 1991 *Texture. Microstruct.* **14-18** 635
[5] Vandermeer R A and Gordon P 1989 *Metal. Trans. A* **20** 391
[6] Lauridsen E M et al. 2003 *Acta Mater.* **51** 4423
[7] Lin F X et al. 2012 *Mater. Sci. Forum* **702-703** 398
[8] Lin F X et al. *To be submitted*
[9] Poulsen S O et al. 2011 *Scr. Mater.* **64** 1003

IOP Conf. Series: Materials Science and Engineering **82** (2015) 012049 doi:10.1088/1757-899X/82/1/012049

Formation behavior of basal texture under the high temperature plane strain compression deformation in AZ80 magnesium alloy

K. Kim[1,a], K. Okayasu[2,b] and H. Fukutomi[2,c]

[1] Department of Metallurgical Engineering, Pukyong National University, Busan 608-739, Korea
[2] Department of Solid State Materials and Engineering, Yokohama National University, Yokohama 240-8501, Japan

E-mail: [a]mrppeng@pknu.ac.kr, [b]kokayasu@ynu.ac.jp, [c]fukutomi@ynu.ac.jp

Abstract. The formation behavior of basal texture during high temperature deformation of AZ80 magnesium alloys in single phase was investigated by plane strain compression deformation. Three kinds of specimens with different initial textures were machined out from an extruded bar having a <10$\bar{1}$0> texture. Plane strain compression tests were conducted at temperatures of 623K and 723K and a strain rate of $5.0\times10^{-2}s^{-1}$, with a strain range of between -0.4 and -1.0. After deformation, the specimens were immediately quenched in oil. Texture measurement was carried out on the compression planes by the Schulz reflection method using nickel filtered Cu Kα radiation. Electron backscatter diffraction (EBSD) measurements were also conducted in order to examine the spatial distribution of orientations. Three kinds of specimens named A, B and C were prepared from the same extruded bar. In the specimens A, B and C, {0001} was distributed preferentially parallel to ND, TD, and RD, respectively. After deformation, texture evaluation was conducted on the mid-plane section. At the plane strain compression deformation, peaks appeared in the true stress-true strain curves irrespective of the kinds of specimen used. It was found that the main components and the pole densities of the textures vary depending on deformation condition and initial texture. Six kinds of texture components were observed after deformation. The (0001)<10$\bar{1}$0> has formed regardless of the initial texture. There are two types of texture components; one exists before the deformation, and the other does not. Either types are considered to have stable orientations for plane strain compression. Also, the basal texture is composed of two crystal orientation components – (0001)<10$\bar{1}$0> and (0001)<11$\bar{2}$0>. When (0001) existed before deformation, an extremely sharp (0001) (compression plane) texture is formed.

Introduction

Due to their low density and high specific strength, magnesium alloys have grown interests in various industrial fields such as electronics and vehicle parts. However, magnesium, having a hexagonal close-packed structure, has a poor formability at room temperature due to the limited number of slips and the twinning systems [1]. Texture control is one of the attractive approaches to improve the formability of magnesium alloys [2]. Many researches have conducted to understand the effect of texture on the deformation behavior in magnesium alloys [3].

The authors have previously investigated the deformation behavior and the texture development in AZ80 magnesium alloy [4]. The deformation was conducted at 723K with a strain rate of $5.0 \times 10^{-2}s^{-1}$. It was found that texture has formed along with the occurrence of DRX (Dynamic recrystallization), and the main texture component and its sharpness varied depending on the deformation conditions. It was reported that the deformation temperature had a strong effect on the texture [5]. However, the types of specimens and deformation conditions were limited in the previous studies [4]. Hence, in this study, the formation behavior of the AZ80 magnesium alloy was investigated on the specimens, which have different initial textures from the previous study, at 623K and 723K with a strain rate of $5.0\times10^{-2}s^{-1}$.

Experimental Procedure

AZ80 magnesium alloy bars with a strong <10$\bar{1}$0>(extrusion direction) texture were used. In order to investigate the characteristics of texture formation under plane strain compression deformation, three kinds of specimens; A, B and C; with different initial texture were machined out in the rectangular shape with a dimension of 10×10×6.7mm from the extruded bars. Figure 1 shows the geometry of the three specimens: A, B and C.

RD in the figure represents the elongation direction for plane strain compression deformation, and the hatched areas show the compression planes. The compression direction (ND) of specimen A, the constraint direction (TD) of specimen B, and the elongation direction (RD) of specimen C are the extrusion direction of the base material, respectively. All specimens were annealed at 723K for 1h before compression deformation.

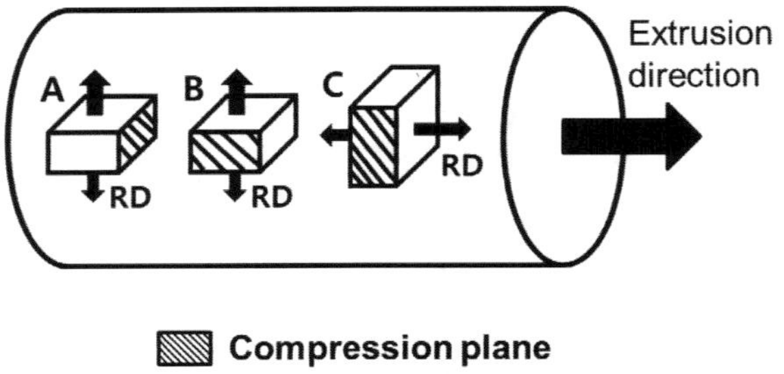

Fig. 1 Geometrical arrangement of specimens for the plane strain compression deformation.

Plane strain compression tests were conducted at 623K and 723K with a strain rate of $5.0\times10^{-2}s^{-1}$. The specimens were quenched in oil immediately after compression deformation in order to prevent a change in the microstructure. The mid-plane section of the specimens was taken out by the mechanical polishing for texture measurements. Texture measurements were carried out by the Schulz reflection method using Cu Kα radiation. Based on five pole figures, the crystal orientation distribution function (ODF) was determined using the Dahms and Bunge method [6]. Electron Backscatter Diffraction (EBSD) measurements were performed after the electrolytic polishing. The region of 0.45mm × 1.33 mm was scanned using the beam control option with a step size of 4μm.

Results and Discussion

The true stress – true strain curves for the three kinds of specimens are shown in Fig. 2. Deformation temperatures of (1) - (2) and (3) - (4) are 723K and 623K respectively. True strain of Fig. 2 is given by an absolute value. The flow curves of the specimen A indicates that the flow stress decreased with an increase in deformation temperature. At the beginning of the flow curve, work softening was observed in all kinds of specimens.

Figure 3 shows the (0001) pole figures for three kinds of specimens before and after deformation. The deformation was conducted at 623K, up to a strain of -1.0. Pole densities are projected onto the compression plane. Mean pole density is used as a unit to draw the contour lines. In fig. 3, (1) and (4) show the (0001) pole figures for specimen A before and after deformation. The (0001) pole densities are distributed at the periphery of the pole figure before deformation. The deformation changes the pole densities at the center, left and right ends, and top and bottom ends of the pole figure after deformation. The change of the pole density is similar to that at 723K [4].

Figure 3 (2) and (5) show the (0001) pole figures for specimen B before and after deformation. The (0001) pole densities are distributed on the great circle 90° away from the TD before deformation. The pole densities at the center, and the top and bottom ends of the pole figure changed with a deformation up to the true strain of -1.0. However, in case of the deformation at 723K, (0001) pole was distributed near TD at a strain of -1.0 similar to specimen A. As shown in Fig. 3 (4) and (5), the formation of basal texture is observed in specimens A and B. It seems that basal texture in (4) develops discontinuously, while it develops continuously in (5). In case of specimen C (Figure 3 (3) and (6)), the formation of basal texture was continuous. However, the maximum pole densities at 623K are different from the deformation at 723K. The maximum pole densities at 723K are 26 times the average level [4], while those of 623K are 7 times. It was indicated that basal texture of specimen C is affected by deformation temperature.

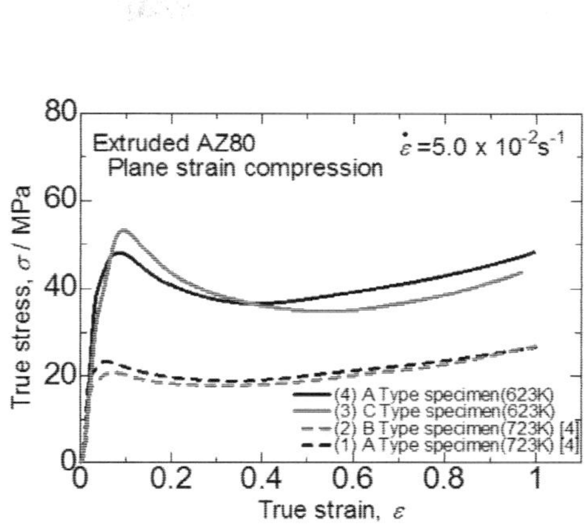

Fig. 2 True stress-true strain curves for the three kinds of specimens.

Figure 4 shows the φ2 =0° section for specimens A and B, which deformed at 623K up to -1.0. The preferred orientations are shown below the φ2 sections. It should be noted that the

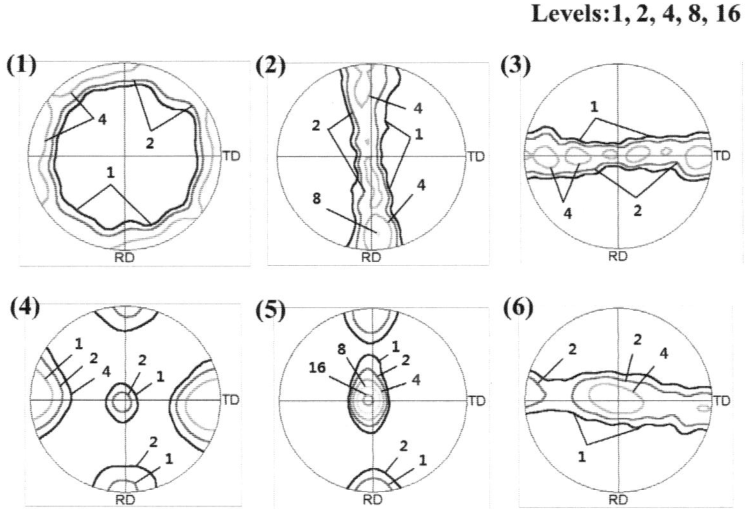

Fig. 3 (0001) Pole figure showing the crystallographic characteristics of the specimen before (upper) [4] and after (bottom) deformation - 623K with a strain rate of 5.0×10^{-2} s^{-1}, up to a true strain of -1.0. (1, 4), (2, 5) and (3, 6) show the specimens of A, B and C, respectively. Pole densities are projected onto the compression. Mean pole density is used as a unit to draw the contour lines.

present study uses a coordinate system wherein if ($\varphi 1$,Φ, $\varphi 2$)=(0°,0°,30°), then the coordinate system become $(0001)\langle 10\bar{1}0\rangle$ [4]. It is observed that the basal texture is composed of $(0001)\langle 10\bar{1}0\rangle$ and $(0001)\langle 11\bar{2}0\rangle$ texture components. The $(0001)\langle 10\bar{1}0\rangle$ component was found in all specimens, and had higher orientation density than other orientation component at 723K. In the specimen A, this texture component was formed by discontinuously, suggesting that the development of basal texture can be attributed to either DRX or twinning [4]. It was also confirmed that a sharp basal texture is formed when (0001) exists before deformation. The texture formation behaviors at 623K and 723K were observed to be similar.

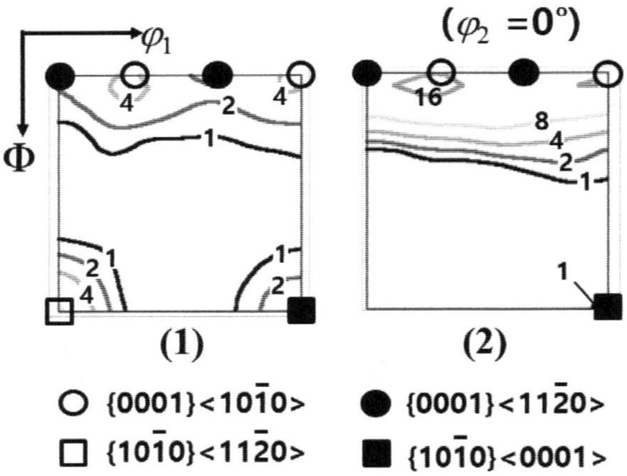

Fig. 4 φ_2 =0°section for specimens (1) A and (2) B, deformed at 623K with a strain rate of $5.0\times10^{-2}s^{-1}$ up to a true strain of -1.0.

Summary

In order to investigate the formation behaviors of basal texture during high-temperature deformation in AZ80 magnesium alloy, the texture formation by plane strain compression is investigated on the three kinds of specimens with different initial textures. It is found that the texture components vary depending on the initial texture and deformation conditions. Two basal texture components are formed by the continuous or discontinuous process. $(0001)\langle 10\bar{1}0\rangle$ texture component shows higher orientation density than other orientation densities in all specimens.

Acknowledgement

This work was supported by Japan Light metal foundation and the Research Grant of Pukyong National University(C-D-2013-0869).

References

[1] Barnett M R, Nave M D and Bettles C J 2004 Mater. Sci. Eng. A **386** 205

[2] Agnew S R and Duygulu 2003 Mat. Sci. Forum **419-422** 177

[3] Agnew S R, Tome C N, Brown D W, Holden T M and Vogel S C 2003 Scripta. Mater. **48** 1003

[4] Kim J, Okayasu K and Fukutomi H 2012 Mater. Trans. **63** 1870

[5] Kim J, Okayasu K and Fukutomi H 2013 Mater. Trans. **64** 192

[6] Dahms M and Bunge H J 1989 J. Appl. Cryst. **22** 439

IOP Conf. Series: Materials Science and Engineering **82** (2015) 012050 doi:10.1088/1757-899X/82/1/012050

Development of a strong Goss texture during annealing of a heavily rolled Al–0.3% Cu alloy

A Wakeel[1], T L Huang[1], G L Wu[1], O V Mishin[2], X Huang[1, 2]
[1] College of Materials Science and Engineering, Chongqing University, Chongqing 400045, China
[2] Danish-Chinese Center for Nanometals, Section for Materials Science and Advanced Characterization, Department of Wind Energy, Technical University of Denmark, Risø Campus, 4000 Roskilde, Denmark
E-mail: wugl@cqu.edu.cn

Abstract. The evolution of microstructure and texture during isochronal annealing of a heavily cold rolled Al–0.3% Cu alloy has been characterized using electron backscatter diffraction. It is found that the rolling texture of this alloy is dominated by the Brass component and that recrystallization during annealing leads to the formation of a pronounced Goss texture. It is suggested that the development of the strong Goss texture in Al–0.3% Cu is caused by preferred growth of Goss-oriented grains into the Brass-oriented matrix.

1. Introduction

The rolling texture in aluminum alloys typically contains three prominent components, i.e. Copper {112}⟨111⟩, S {123}⟨634⟩ and Brass {110}⟨112⟩ orientations. During annealing this texture is often replaced by a recrystallization texture with a pronounced Cube {001}⟨100⟩ component, although the deformation texture can also be retained in recrystallized samples [1]. Another frequently reported recrystallization texture of commercial aluminum alloys is a combination of P ~{110}⟨111⟩ and $Cube_{ND}$ {001}⟨310⟩ components [2–4]. A less frequently observed component in the annealing texture of Al alloys is the Goss {110}⟨001⟩ orientation. A strong Goss texture was observed after annealing of heavily rolled pure aluminum [5] and Al–1.8% Cu [6,7]. It was suggested that in these materials, Goss-oriented grains nucleated either in shear bands or in transition bands [5–7]. These suggestions were, however, based on a few orientation measurements made using the electron backscatter diffraction (EBSD) technique. Here we report a strong Goss texture observed after annealing of a heavily cold rolled Al–0.3%Cu alloy. To provide more detailed information on nucleation sites and to study the evolution of Goss-oriented grains, high-resolution EBSD analysis of large areas in rolled and annealed samples is conducted in the present work.

2. Experiment

The material used for the present study was Al–0.3%Cu (wt%) with a very coarse (1.5–2 mm) grain size in the as-cast condition. To produce a plate suitable for rolling, the as-cast material was first forged at 200 °C. An ODF obtained using X-ray diffraction data from the forged plate demonstrated a scatter of orientations near the Brass component with $f(g)_{max} = 7.2$. The plate was then cold rolled by multiple passes to a total thickness reduction of 98%, which corresponds to a true strain of 4. To ensure homogeneous deformation for each rolling pass, the l/d ratio (where l is the contact length between the roller and the sheet, and d is the average thickness of the sheet [8]) was controlled to be ~2.5. After each pass the material was immediately quenched in water. The cold-rolled material was then annealed at different temperatures for 1 h.

The microstructure and texture of the cold-rolled and annealed samples were characterized by EBSD in a Zeiss Auriga dual beam station. Specimens from the center layer of the longitudinal section containing the rolling direction (RD) and the normal direction (ND) were prepared for the EBSD analysis. Step sizes for EBSD scanning were 50–100 nm for the rolled and recovered conditions, and 0.5–5 μm for partially and fully recrystallized conditions. Low angle boundaries (LABs) and high angle boundaries (HABs) were defined in the EBSD maps as those with either 2–15° or >15° misorientations, respectively. Texture components were defined within 15° from their ideal orientations.

3. Results

Heavy rolling results in a lamellar-type microstructure with an average boundary spacing of 0.2 μm along the ND, where lamellae of similar orientations compose either broad or narrow texture bands (Fig. 1a). Similar to previous observations of the microstructure in heavily rolled aluminum [4], the band thickness is found to depend on the crystallographic orientation. As is seen in Fig. 1a, bands of the dominant Brass texture (~60% of the area sampled) are broader than individual bands of the other rolling texture components and contain predominantly LABs. The narrower S and Copper texture bands contain lamellae delineated by both HABs and LABs (see Fig. 1a). Localized shear banding at ~35° to the RD is also observed in the rolled microstructure. Goss-oriented crystallites are seen both within lamellar structures and in the vicinity of shear bands. Their area fraction in the deformed microstructure was significant, ~5%. In contrast, the area fraction of crystallites having orientations of the Cube texture was extremely small, 0.01%.

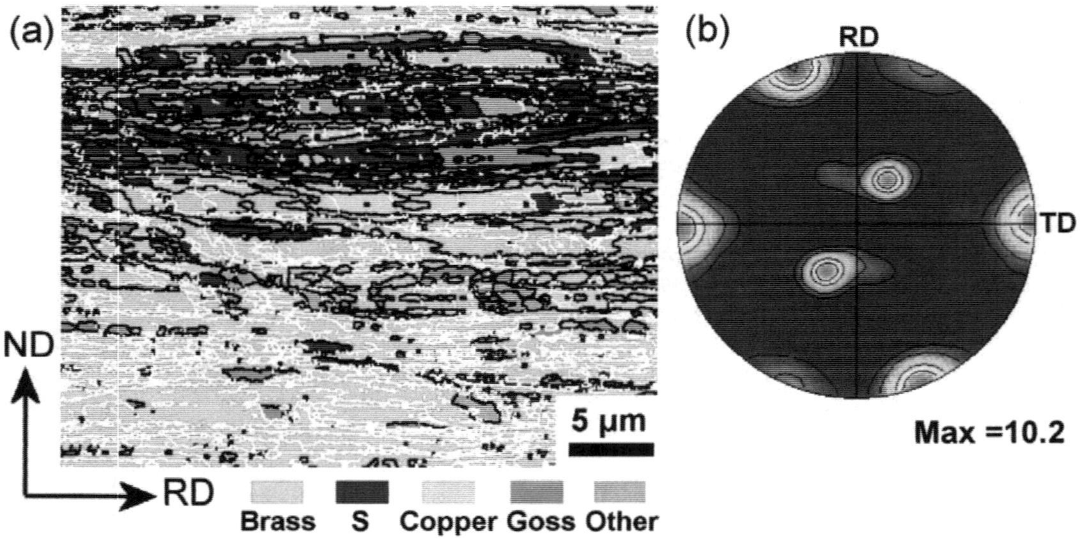

Figure 1. EBSD data for Al–0.3% Cu cold-rolled to a strain of 4: (a) orientation map with LABs and HABs shown by white and black lines, respectively; (b) {111} pole figure for the area sampled in (a).

After annealing at 190 °C for 1 h a number of recrystallized grains are seen in the microstructure (Fig. 2a), including a few Goss grains growing into the Brass-oriented matrix. Annealing at 250 °C for 1 h results in complete recrystallization with a very large fraction of the Goss texture, 54% (Fig. 3). Fig.3a demonstrates that in many cases recrystallized Goss-oriented grains are separated by LABs. Whereas a small number of S- and Copper-oriented grains are also present in the recrystallized microstructure, there are no grains of the Brass components in Fig. 3a.

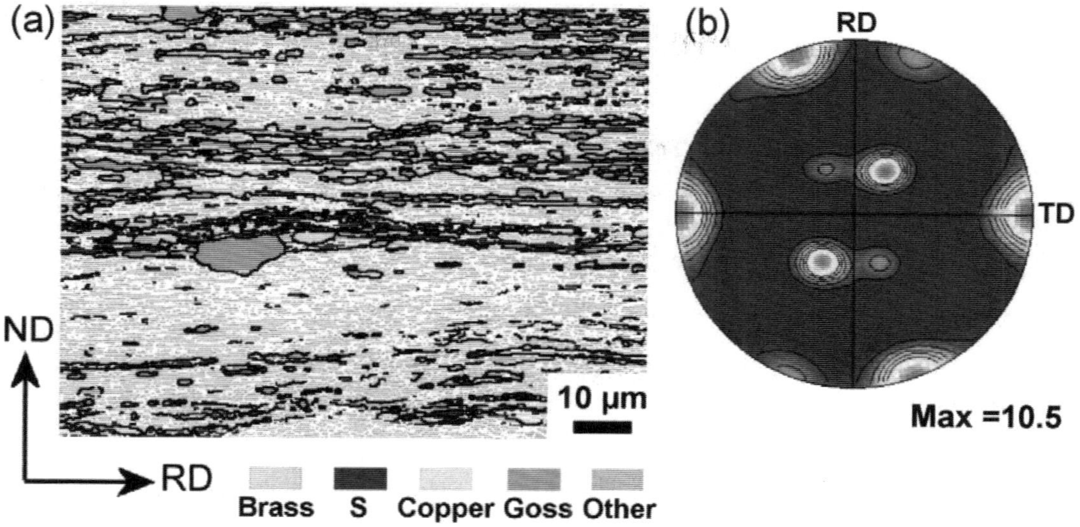

Figure 2. EBSD data for Al–0.3% Cu cold-rolled to a strain of 4 and subsequently annealed at 190 °C for 1 h: (a) orientation map with LABs and HABs shown by white and black lines, respectively; (b) {111} pole figure for the area sampled in (a).

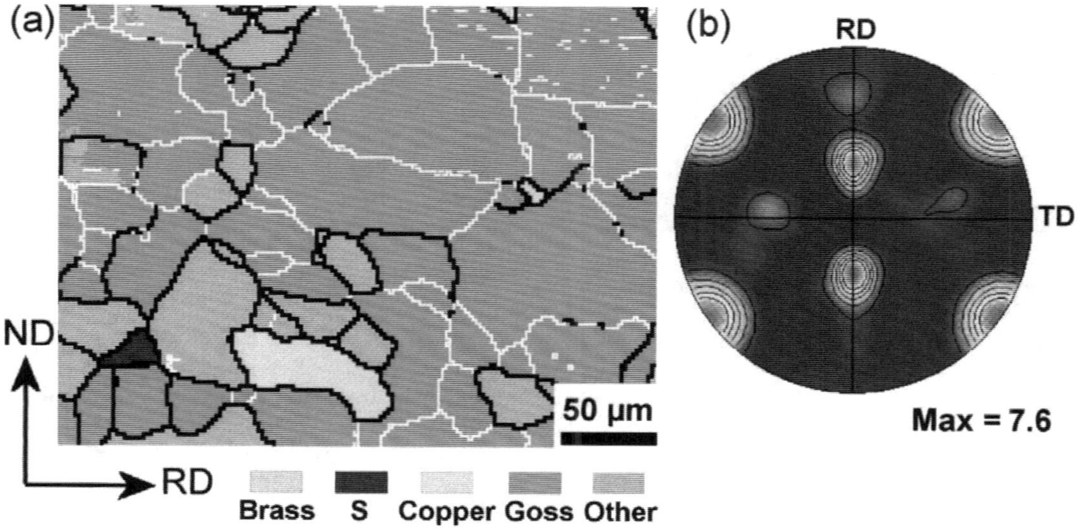

Figure 3. EBSD data for Al–0.3% Cu cold-rolled to a strain of 4 and subsequently annealed at 250 °C for 1 h: (a) orientation map with LABs and HABs shown by white and black lines, respectively; (b) {111} pole figure representing the texture in a large area.

4. Discussion

In contrast to the typical deformation texture developing after high-strain rolling of aluminum, in which the Brass orientation is usually much weaker than the S and Copper components [4], the present Al–0.3% Cu alloy, cold rolled to a strain of 4, contains a very strong Brass component. One possible explanation is that this strong component was inherited from the forged material where orientations near the ideal Brass component dominated the texture (see Section 2). For the given material, the presence of Cu in solid solution could slow down the development of the other rolling texture components during cold deformation. Furthermore, the effect of Cu in solid solution could be similar to the effect of solute Mg in Al–Mg alloys [6,9], i.e. increasing the tendency for shear banding during heavy rolling. According to Lücke and Engler [6] and Engler et al. [10], intense TD rotation of the Copper orientation takes place within such shear bands, first towards the Goss components and then towards the Brass component. The result of this rotation is a large area fraction of Brass-oriented deformation structures. Subgrains of the Goss component are also present in the deformed microstructure (Fig. 1a). In most cases, these Goss subgrains have at least one common boundary with the Brass-oriented volumes, which is similar to the observations of Hjelen et al. [5]. Such boundaries between the Goss subgrains and the Brass-oriented matrix are typically characterized by large misorientation angles across them (see Fig. 1a).

In early stages of the recrystallization process, nuclei with orientations of the rolling texture components are predominantly observed within coarsened lamellar structures, while nuclei having "random" orientations appear mainly within shear bands. Grains with the Goss orientation are also seen in the partially recrystallized microstructure, both near shear bands and within lamellar structures. These grains and grains with random orientations are expected to have growth advantage due to the presence of highly mobile HABs between such grains and the Brass-dominated deformation matrix (see Fig. 2a). For example, misorientations between the exact Goss orientation and the exact orientations of the Brass, S and Copper components are 35°⟨110⟩, 43°⟨112⟩ and 55°⟨110⟩, respectively. When recrystallization is almost complete, further grain growth takes place in the presence of a very strong Goss texture, which results in many low-mobility LABs formed between recrystallized Goss-oriented grains (Fig. 3a). Thus, at this stage further growth of the Goss-oriented grains is possible mostly at the expense of grains having other orientations, which is expected to further strengthen the Goss texture.

5. Summary

Microstructures and textures have been studied by EBSD in an Al–0.3% Cu alloy cold-rolled to a strain of 4 and subsequently annealed at different temperatures for 1 h. The cold-rolled material is characterized by a rolling texture with a strong Brass component and by a small spacing between lamellar boundaries, ~0.2 μm. Upon annealing recrystallization nuclei are developed both within lamellar structures and near shear bands. Nuclei with the Goss orientation appear to have growth advantage when they grow into Brass-oriented matrix, thus leading to a strong Goss texture in the fully recrystallized condition.

Acknowledgements

The authors wish to thank the financial support of the Natural Science Foundation of China (grants No. 51327805 and 51171214) and Fundamental Research Fund of Central Universities (grant No. CDJZR14135502 and CDJZR11130007). We also gratefully acknowledge the support from the Danish National Research Foundation (Grant No.DNRF86-5) and the National Natural Science Foundation of China (Grant No. 51261130091) to the Danish–Chinese Center for Nanometals.

References

[1] Humphreys F J and Hatherly M 2004 *Recrystallization and Related Annealing Phenomena,* Second Edition (Elsevier Ltd)
[2] Liu J, Banovic S W, Fields R J and Morris J G 2006 *Metall. Mater. Trans.* A **37** 1887
[3] Schäfer C and Gottstein G 2011 *Int. J. Mater. Res.* **102** 1106
[4] Mishin O V, Godfrey A, Juul Jensen D and Hansen N 2013 *Acta Mater.* **61** 5354
[5] Hejlen J, Ørsund R and Nes E 1991 *Acta Metall. Mater.* **39** 1377
[6] Lücke K and Engler O 1990 *Mater. Sci. Techn.* **6** 1113
[7] Engler O, Hirsch J and Lücke L 1995 *Acta Metall. Mater.* **43** 121
[8] Mishin O V, Bay B and Juul Jensen D 2000 *Metall. Mater. Trans.* A **31** 1653
[9] Ryen Ø, Laukli H I, Holmedal B and Nes E 2006 *Metall. Mater. Trans.* A**37** 2007
[10] Engler O, Hirsch J and Lücke L 1989 *Acta Metall. Mater.* **37** 2743

IOP Conf. Series: Materials Science and Engineering **82** (2015) 012051 doi:10.1088/1757-899X/82/1/012051

Grain growth simulation in an IF steel - Effect of grain boundary mobility

A Ayad[1,2] , N Rouag[2] and F. Wagner[3]
[1] Département de Pharmacie, Faculté de Médecine, Université Constantine 3, Algeria
[2]Laboratoire de Microstructures et Défauts dans les Matériaux, Université Constantine 1, Algeria
[3]LEM3, (CNRS-UMR 7239), Université de Lorraine, Ile du Saulcy, 57045 Metz, France
E-mail: abdelhak_ayad@yahoo.com

Abstract. In a recent paper, we have proposed a modified Potts model for the simulation of the normal grain growth stage in an IF steel. An original concept is introduced by a modular consideration of the grain size effect. It is based on experimental observations of partial textures of certain (sub) populations of grains. This allowed both better reflect the pressure at the grain boundaries and significantly speed up the calculations. This model was used to reproduce the main features of the microstructure and texture. Improvement of texture behavior, especially of the component {111}<112>, which differs slightly from the experiment, requires the introduction of strong assumptions on energy and mobility of grain boundaries. The objective of this study is to verify the ability of the modified Monte Carlo model to give satisfactory results for grain growth, by considering various assumptions on energy and mobility of grain boundaries. The validity of such assumptions and their impact on the simulation results are analyzed and discussed.

1. Introduction

Due to their excellent deep drawability, a large amount of research work has been carried out to study the development of recrystallization texture and microstructure in IF steels [1-6]. Therefore, many attempts have been made to model the microstructure evolution during heat treatments [7-12]. In a recently published work by the present authors, a modified Potts model for the simulation of the normal grain growth stage in an IF steel has been proposed [13]. It allows both better reflect the pressure at the grain boundaries (GBs) and significantly speed up the calculations. This model, which uses isotropic GB mobility and Read-Shockley(R-S) GB energy, reproduces globally the main features of the microstructure and texture. However, the texture shows a small increase of the component C-{111}<112> which differs somewhat from the experimental evolution. Enhancement of texture behavior requires the introduction of realistic energy and mobility of GBs. Some of the drawbacks are that it requires material data which are often not available and must then be replaced by assumptions. The first attempts to include anisotropic grain properties into Potts MC model simulations dates back to the eighties [14,15]. Holm et al. [16] were the first to introduce an anisotropic energy and mobility for GBs, with a switching probability.

It has often been suggested [17,18] that during normal grain growth, selective growth of grains of some particular orientations is controlled by specific types of coincidence site lattice (CSL) boundaries. Recently the Potts model was further generalized to also include the GB characteristics of certain CSLs [19,20].

In this work, a modified 2D Potts Monte Carlo model including anisotropy of GB mobility (dependent on misorientation) including the CSL GBs has been used, and a systematic study of their effect on grain growth texture of IF steel has been carried out. Some attention is also paid to the possible role of certain Σ <111> GBs during grain growth.

IOP Conf. Series: Materials Science and Engineering **82** (2015) 012051 doi:10.1088/1757-899X/82/1/012051

2. Simulation procedure

We briefly recall here the main points of the Monte Carlo grain growth method. The simulation method uses an interaction energy E_i associated to each site i [16]:

$$E_i = \sum_{j=1}^{n} \gamma(\theta_{ij}), \tag{1}$$

where γ, θ_{ij} and n are the energy per unit area of GB, the misorientation angle and number of neighbours of the selected site i, respectively. The total system energy is given by the sum over all lattice sites N.

The GB energy is defined by a Read-Shockley function with a cutoff angle $\theta^*=15°$:

$$\gamma_{RS}(\theta_{ij}) = \frac{\theta}{\theta^*}\left[1-\ln\left(\frac{\theta}{\theta^*}\right)\right] \quad \text{if } \theta \le \theta^* \quad \text{and} \quad \gamma_{RS}(\theta_{ij})=1 \quad \text{if } \theta > \theta^* \tag{2}$$

Simulation of grain growth involves the random selection of a site i and its neighbouring site j. The reorientation attempt is accepted with a probability P :

$$P = \frac{M(\theta_{ij})}{M_{max}}\frac{\gamma(\theta_{ij})}{\gamma_{max}} \quad \text{if } \Delta E \le 0 \quad \text{and } P=\frac{M(\theta_{ij})}{M_{max}}\frac{\gamma(\theta_{ij})}{\gamma_{max}}\exp\left(\frac{-\Delta E}{\beta\gamma(\theta_{ij})}\right) \quad \text{if } \Delta E > 0 \tag{3}$$

ΔE is the energy difference between altered and original state. The constants γ_{max} and M_{max} are the maximum allowed grain boundary energy and mobility, respectively. Both are equal to unity in these simulations. The constant β is an effective "lattice temperature" often chosen on an empirical basis. The mobility $M(\theta_{ij})$ is sometimes considered as being constant or can also follow an "S-curve" dependence on misorientation to give a sharp transition from low to high in the vicinity of the transition from low to high angle boundary type [21]. Both energy and mobility functions may include particular values for the CSL grain boundaries if data are available showing specific situations [10,22].

It is possible to increase the calculation speed by increasing the driving force in some specific situations. We have introduced such an effect in the following way. Instead of using the energy E_i associated to a site (see Eq. 1) we define a modified site energy E_i*:

$$E_i^* = \frac{E_i}{(D_i/D_m)^p} \tag{4}$$

where E_i is the standard energy associated to a site number i (Eq. 1), D_i is the diameter of the grain to which the site i belongs and D_m is the mean diameter of the considered set of grains. This definition for E_i* leads to favour the growth of large grains and, above all, to penalize the smallest grains. p is an empirical parameter which makes more or less effective this 'size effect' (p=0 suppresses it totally and corresponds then to the classical case whereas p=1 gives it the maximal influence). The motivation for such a modification as well as its consequences has been described in a previous work [13].

3. Experimental procedure

In this investigation, texture data were analyzed for an IF steels that had previously been studied [13, 23]. This IF steel was cold rolled up to 75% thickness reduction (final thickness 0.7 mm) and then submitted to an annealing treatment. The chemical composition of this steel is (% wt) :0.008 C, 0.2 Mn, 0.004 P, 0.01 S, 0.003N, 0.004 Si, 0.007 Cu, 0.02 Ni, 0.014 Cr, 0.04 Al, 0.1 Ti.

The data used here to start the simulation consist of an EBSD map made of such an IF sample heat treated at 630°C for 3 hours. A detailed study has shown that primary recrystallization is nearly finished for this heat treatment condition [23]. The map corresponds to a surface of 0.270 mm² which was obtained by using a JEOL 6490 FEG-SEM and an HKL EBSD system. The initial data set for the grain growth simulation is a set of 1 083 065 pixels taken as the lattice sites. This map corresponds to a set of 3104 grains, using a criterion of 5° misorientation to detect grains. When considering only grains with at least 4 pixels, 2982 grains with a size larger than 3 pixels are left. The mean grain size is 10.8 µm.

To compare the results obtained from simulations with experimental data, a second IF sample was heat treated at 800°C for 5 hours. The EBSD map contains 1 923 201 pixels, with a total analyzed area of 0.480 mm². The map contains 1635 grains with an area larger than 3 pixels. This gives a mean grain size of 19.4 µm. It represents twice the initial value (initial means at the end of primary recrystallization).

4. Results and discussion

4.1. Experimental texture

The ODFs of the two samples, at the end of primary recrystallization and after grain growth, have been calculated from the individual orientations of the EBSD maps. ODF calculations have been made assuming cubic-orthorhombic symmetries. The main features appear in the $\varphi_2=45°$ sections which are shown in Fig. 1. The texture mainly consists of a γ fiber with a relatively large spread around it, especially in the α fiber direction (i.e. for $\varphi_1=0°$). Inside this γ fiber the density is approximately evenly distributed with a slight peak around the orientation {111}<110> in the first case (Fig. 1a). In the second case (after grain growth) the ODF density decreases around the orientation {111}<112> and increases around {111}<110> (Fig. 1b). It could be simplified using the concept of texture components. A component is defined as an ideal orientation g_i and its associated weight, w_i, calculated as the integral of the ODF around the orientation g_i. In our case the surrounding of an ideal orientation consists of a sphere centered on g_i and with a radius of 15°. With such a definition the components C1 ({111}<110>) and C2 ({111}<112>) have a weight 24.1% and 17.7%, respectively, in the initial state (end of primary recrystallization) and a weight 26.3% and 14.3% after grain growth.

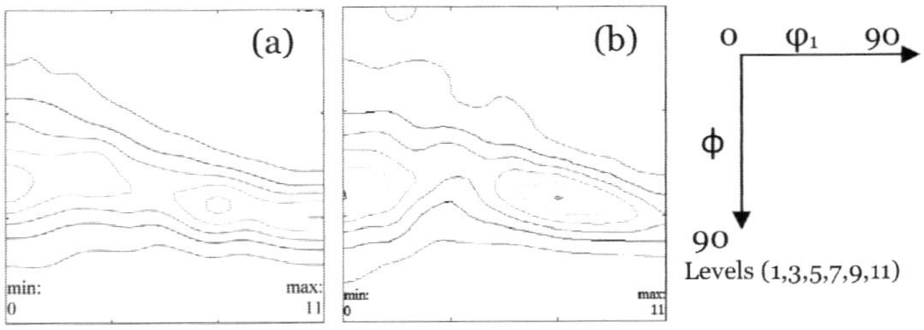

Fig. 1. φ_2 = 45° sections of the ODF for: (a) experimental initial state (630°C, 3 h); (b) experimental finale state (800°C, 5 h).

4.2. Simulated grain growth texture

To study the effect of GB mobility on grain growth, many types of mobility were proposed. Table 1 summarizes some key parameters for the different simulation cases, such as the ODF maximum, the intensities of {111}<110> and {111}<112> components and their weights.

Table 1: Texture features of the cases studied_(the maximum of ODF, ODF densities and weights of the two components {111}<110> and {111}<112>)

	Exp. states		Simulation : 1st case				Simulation : 2nd case			
			M=1		M=R-S		M=0		M=0.5	
	Ini.	Fin.	p=0	p=0 .5	p=0	p=0 .5	CSL	Σ3	CSL	Σ3
Fmax	10.4	11.2	12.5	13.4	12.1	13.1	12.0	11.9	12.5	12.5
F-C1	10.4	11.2	12.5	13.4	12.1	13.1	12.0	11.9	12.5	12.5
F-C2	9.0	6.0	9.9	11.1	9.8	11.0	10.1	9.4	9.2	9.8
w(%)C1	24.1	26.3	26.6	28.0	25.9	27.4	26.1	25.9	26.4	25.8
w(%)C2	17.7	14.3	19	19.7	19.1	19.6	19.0	18.7	18.4	19.6

First case: isotropic GB mobility (M=constant) and a R-S mobility has been introduced in this grain growth simulation through the reorientation probability (Eq. 3). The ODFs have been calculated using the same conditions as previously, where the simulations were stopped when the mean grain size reached twice the initial value. Figure 2 shows the $\varphi_2=45°$ sections of grain growth simulations, with isotropic and R-S GB mobilities, for p=0 and p=0.5 respectively (p is the grain size factor). All these ODFs are very similar regardless of the value of p used for the simulation. They resemble rather well the experimental ODF section of the

IF steel after grain growth shown in Fig. 1b. Nevertheless the decrease of the ODF density around {111}<112> is less marked than in the experimental case. In these simulations the weight of the component C1 increases slightly and reaches a level which corresponds to the reality (see Tab. 1). The simulation with p = 0 gives the lowest increase, in particular for R-S GBs mobility. The weight of the component C2 also slightly increases which is opposite to the experimental trend (see Tab. 1). Simulation with R-S GB mobilities presents also the lowest increase of the weight of component C2. It seems to be the closest to the experimental evolution. When the grain size factor is activated (p=0.5), The ODFs obtained from the isotropic and R-S mobilities are indistinguishable (Figs. 2b and 2c). Therefore, the mobility type has a very limited effect on the texture evolution when considering the grain size parameter.

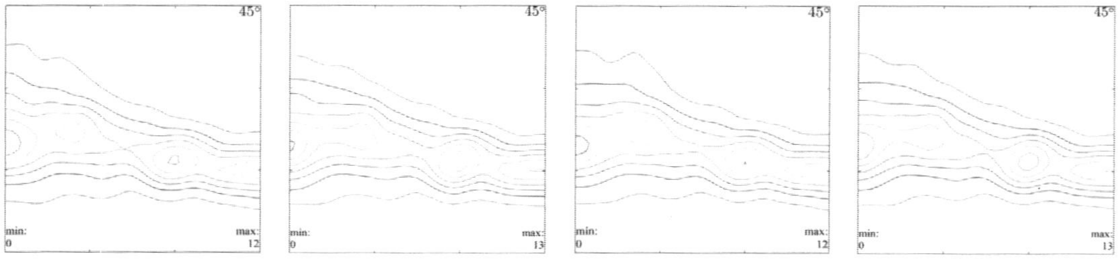

Levels (1,3,5,7,9,11,13)

(a) p=0, M=1 (b) p=0.5, M=1 (c) p=0, M(R-S) (d) p=0.5, M(R-S)

Fig. 2: φ_2 = 45° sections of the ODFs obtained from simulations with constant and R-S GB mobility for two values of p (0 and 0.5).

Second case: discrete mobilities for all CSL grains boundaries (all CSL have a unique value of mobility), and separately for only the Σ3 (60°<111>) CSL GB, were incorporated in the R-S function. This calculation assumed that all high angle GBs are equally mobile, except for certain "special" boundaries corresponding to particular misorientation angles (CSL GBs). The mobility values are M=0 and M=0.5. The case M=1 of CSL GB implicitly belongs to the R-S mobility previously studied in the first case. The deviation from the exact CSL misorientation was calculated using Brandon's criterion ($\omega = \pm 15° / \sqrt{\Sigma}$) [24]. The Σ3 CSL GB fraction represents the highest fraction of CSL GBs.
During experimental grain growth, the Σ3 CSL GB frequency increases from 4.30 to 6.32 % while the global CSL GB fraction increases, from 12.88 to 14.71 %. Figure 3 shows the ODF sections of the resulting texture.
These ODFs are quite similar and give the same previous trend (increase of both components in the γ fiber), which may be linked to the weak fraction of the CSL GBs. However, the case M=0 slightly enhances the texture evolution compared to M=0.5. When M=0, the maximum ODF values are 12.0 and 11.9 for all CSL GB and Σ3 GB simulations, respectively.

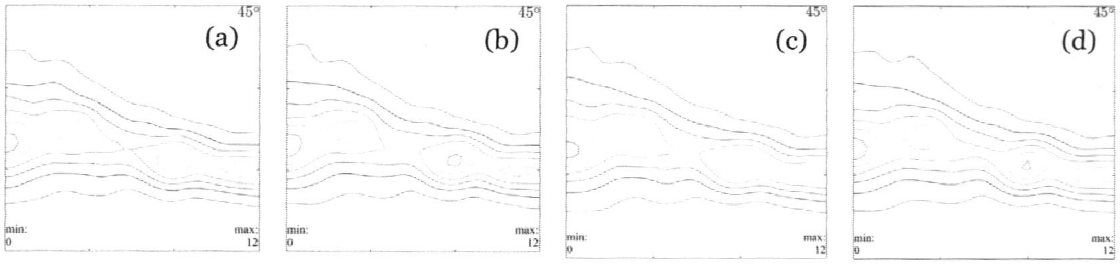

Levels (1,3,5,7,9,11)

Fig. 3: φ_2 = 45° sections of the ODF's obtained from simulations for: (a) all CSL GBs with M=0; (b) all CSL GBs with M=0.5 ; (c) Σ3 GBs with M=0.5 ; (d) Σ3 GBs with M=0.5.

5. Conclusions

The capability of a modified Potts model to reproduce the experimental grain growth texture of an IF steel has been checked. The following conclusions are obtained from the proposed model which incorporates various GB mobilities:

1) simulations with p=0 for constant and R-S GB mobilities globally give the same features of the experimental texture. Texture obtained with R-S GB mobilities is closest to the experimental evolution. It presents the lowest increase of the weight of the component {111}<112>;
2) the mobility type has a very limited effect on texture evolution when the grain size factor is activated ;
3) texture is slightly enhanced when considering a very low mobility for CSL GBs, in particular for Σ3 GBs.

6. References

[1] Juntunen P, Raabe D, Karjalainen P., Koio T and Bolle G 2001 *Met. Mater. Trans.* A **32** 1989–1995
[2] Magnusson H, Jensen DJ and Hutchinson B 2001 *Scripta Mater.* **44** 435-441
[3] Nakamichi H, Humphreys FJ, Brough I 2008 *J. Microscopy* **230** (3) 464-471
[4] Samet-Meziou A, Etter AL, Baudin T, Penelle R 2008 *Mater. Sci. Eng.* A **473** 342-354
[5] Réglé H, Grandemange D, Miroux A, Bacroix B 2002 *Mater. Sci. Forum* **408-412** 451-456
[6] Samajdar I, Verlinden B, Van Houtte P, Vanderschueren D 1997 *Mater. Sci. Eng.* A **238** 343-350
[7] Miodownik MA 2002 *J. Light Met.* **2** 125-135
[8] Humphreys FJ 2000 *Model. Simul. Mater. Sci. Eng.* **8** 893-910
[9] Ayad A, Rouag N 2012 *Mater. Sci. Forum* **715-716** 872-876
[10] Fjeldberg E, Marthinsen K 2010 *Comp. Mater. Sci.* **48** 267-281
[11] Ivasishin M, Shevchenko V, Vasiliev L, Semiatin L 2006 *Mater. Sci. Eng* .A **433** 216-232
[12] Francis AJ, Roberts G, Cao Y, Rollett AD, Salvador PA 2007 *Acta Mater.* **55** 6159-6169
[13] Ayad A, Wagner F, Rouag N, Rollett AD 2013 *Comp. Mater. Sci.* **68** 189-197
[14] Grest GS, Srlolovitz DJ, Anderson MP 1985 *Acta Metall.* **33** 509
[15] Rollett AD, Srlolovitz DJ, Anderson MP 1989 *Acta Metall.* **37** 1227
[16] Holm EA, Hassold GN, Miodownik MA 2001 *Acta Mater.* **49** 2981
[17] Hayakawa Y, Muraki M, Szpunar JA 1998 *Acta Metall.* **46** 1063
[18]Humphreys FJ, Hatherley M 2002 Recrystallization and Related Annealing Phenomena, Elsevier Publishing Company
[19] Brahme AP 2005 Modelling Microstructure Evolution during Recrystallization *Ph.D. Thesis* Carnegie Mellon University
[20] Brahme A, Fridy J, Weiland H, Rollett AD 2009 *Model. Sim. Mater. Sci. Eng.* **17** 015005
[21] Winning M, Rollett AD, Gottstein G, Srolovitz DJ, Lim A, Shvindlerman LS 2010 *Philos. Mag.* **90** 3107–3128
[22] Ono N, Kimura K, Watanabe T 1999 Acta Mater. **47** 1007–1017
[23] Ayad A, Allain-Bonasso N, Rouag N, Wagner F 2012 *Mater. Sci. Forum* **702-703** 269-272
[24] Brandon DG 1966 *Acta Metall.* **14** 1479-84

Modelling the stored energy of plastic deformation for individual crystal orientations

D Shore[1], A Van Bael[1], J Sidor[2], D Roose[3], P Van Houtte[1] and L Kestens[2]

[1] KU Leuven, Department of Materials Engineering, Kasteelpark Arenberg 44, 3001 Leuven, Belgium

[2] Ghent University, Materials Science and Engineering, Technologiepark 903, 9052 Gent, Belgium

[3] KU Leuven, Department of Computing Science, Celestijnenlaan 200a - box 2402, 3001 Leuven, Belgium
E-mail: diarmuid.shore@mtm.kuleuven.be

Abstract Recovery and recrystallisation processes in polycrystalline metals are driven by the release of energy stored in defect structures chiefly resulting from dislocation creation, motion and interaction during plastic deformation. Some statistical models of texture change during recrystallisation employ the Taylor factor to quantify the distribution of this stored energy amongst orientations. While the Taylor factor is an instantaneous measure of the plastic power dissipation per orientation for a given strain mode, it is technically only valid as an estimate of stored energy if strain path and texture can be assumed constant. This motivates the search for alternatives to the Taylor factor which do not neglect the effects of changing strain path and evolving texture. In this paper a first step is made toward this goal by comparing the Taylor factor with a possible alternative, the accumulated slip per orientation, for a plane strain compression deformation of a bcc material. It is discovered that even for this idealised deformation there are specific orientations for which there is a consistent difference between the two parameters at various magnitudes of strain. It is concluded that these results lend support to the case for replacing the Taylor factor with a history based parameter in recrystallisation texture models.

Introduction

Recovery and recrystallisation processes in polycrystalline metals are driven by the release of energy stored in defect structures chiefly resulting from dislocation creation, motion and interaction during plastic deformation. One approach to modelling the change in texture occurring during discontinuous recrystallisation is to consider the classical nucleation and growth processes as independent orientation selection mechanisms which each permit certain orientations to be present in the final recrystallised texture. In a recent example of such a model [1] nucleation is implemented as a process which selects orientations from the deformation texture on the basis that only those orientations which have both a large volume fraction in the deformation texture and a particular level of stored energy (relative to other orientations) may persist in the recrystallised texture. To date the Taylor factor M of grains in the final discrete texture has been used in the nucleation criterion of the above model as the estimate of the distribution of stored energy amongst the orientations. On its own M is an estimate of the dissipated plastic power per orientation at some moment in time, but can be

used as an estimate of the total dissipated plastic work per orientation on the assumption that (i) the deformation is one of constant strain mode, and that (ii) early in the forming process the orientation of the majority of grains have approached stable orientations. In other words due to assumption (i) the Taylor factor of a grain at any moment during the deformation would be a function only of its orientation at that moment, and due to assumption (ii) the orientation would be only slowly changing for most of the deformation process, so that the value of the Taylor factors for discrete grains in the final state would be approximately proportional to the total plastic work dissipated in those grains. Successful results have been obtained with the recrystallisation model for conventional rolling using the Taylor factor, which supports these assumptions at least for this classic process.
At the same time it is not clear what the limitations of the above assumptions really are, and to which extent they would be valid in a more general forming process, e.g. one with a changing strain path. Thus it is desirable to consider an alternative to the Taylor factor which can account in some way for the full history of deformation, without greatly increasing the complexity of the model.
The accumulated slip Γ is a scalar summation of the absolute value of slip activity on all slip systems of a single grain over some given interval of time. If the critical resolved shear stress τ_c is assumed equal on all slip systems, and strain hardening is assumed not to occur, then Γ can be considered an estimate of the distribution of plastic work amongst orientations and is a possible substitute for M in the recrystallisation model. This paper presents a comparison of M and Γ as estimates of the distribution of stored energy amongst orientations for a monotonic plane strain compression applied to a bcc material.

Computational Method

For a given orientation g both M_g and Γ_g can be calculated for a given macroscopic deformation history, starting from an annealed condition (state of zero stored energy), using the classic full constraints Taylor theory. In this case it is possible to calculate both M_g and Γ_g independently for any orientation g of interest as no grain interaction effects are considered. Furthermore the absolute values of M_g and Γ_g are not of interest, as only the relative values amongst orientations are necessary for the recrystallisation model. This means that for the purpose of the recrystallisation model M_g and Γ_g can be normalised as $\tilde{M}_g$ and $\tilde{\Gamma}_g$ per eqn. 1 without concern for units. In eqn. 1 $\min(A)$ and $\max(A)$ stand respectively for the minimum and maximum of the values of the generic parameter A for all orientations g under consideration.

$$\tilde{A}_g = \frac{A_g - \min(A)}{\max(A_g - \min(A))} \quad (1)$$

In this analysis $\tilde{M}_g$ and $\tilde{\Gamma}_g$ are compared numerically in the range (0,1) by contour mapping their values on a $\varphi_2 = 45°$ section in Euler space (Bunge convention). The difference between the two parameters is additionally expressed as a percentage per eqn. 2.

$$\text{rel\%} = \frac{\tilde{\Gamma}_g \times 100}{\tilde{M}_g} \quad (2)$$

Mapping $\tilde{M}_g$values is straightforward as for a given strain mode they can be calculated directly for any orientation desired. However to obtain corresponding $\tilde{\Gamma}_g$ values the crystal plasticity calculations must be such that, after the desired strain history, a sufficient quantity of grains have a final orientation which happens to fall on the $\varphi_2 = 45°$ plane at the strain of interest. Moreover, there can be a limit to the magnitude of strain that can be applied as the discrete orientations tend to cluster around stable orientations.
Considering the above the following method was employed to calculate contour maps: a full constraints Taylor simulation of plane strain compression was conducted starting with an

initial grid of 891,000 orientations ($0° \le \varphi_1, \Phi \le 90°, -10° \le \varphi_2 \le 100°$, with 1° spacing); {110}<111> and {112}<111> slip systems were active, with equal critical resolved shear stresses and no strain hardening. A map of $\tilde{\Gamma}_g$ values on the $\varphi_2 = 45°$ plane was produced by selecting those grains whose orientations were found to be within in the range $44° \le \varphi_2 \le 46°$ at the macroscopic strain of interest. The values of $\tilde{\Gamma}_g$ for these grains were extracted and interpolated on to a regular grid to allow generation of contours.
In this analysis the strain mode is constant, thus $\tilde{M}_g$ is constant and the contour map of $\tilde{M}$ for the $\varphi_2 = 45°$ section of Euler space does not change, and need only be calculated once. However in practise the same process was used to calculate both $\tilde{\Gamma}_g$ and $\tilde{M}_g$ at every step of the simulation, as a control on the quality of the interpolation, i.e. changes observed in the Taylor factor maps with increasing strain would indicate some numerical error to be present.

Results

Figure 1 (i) and (ii) show the normalised Taylor factor map at a von Mises equivalent strain of 5.7% and 57% respectively: only very small variations are visible, suggesting interpolation error at the largest strain considered to be low. Figure 1 (iii) shows a contour plot of normalised accumulated slip values $\tilde{\Gamma}_g$ at a von Mises strain of 57%: the main differences with the corresponding Taylor factors $\tilde{M}_g$ are local to ($\varphi_1 = 90°, \Phi = 10°$) where $\tilde{\Gamma}_g \gg \tilde{M}_g$, and local to ($\varphi_1 = 90°, \Phi = 75°$), where $\tilde{\Gamma}_g \ll \tilde{M}_g$.

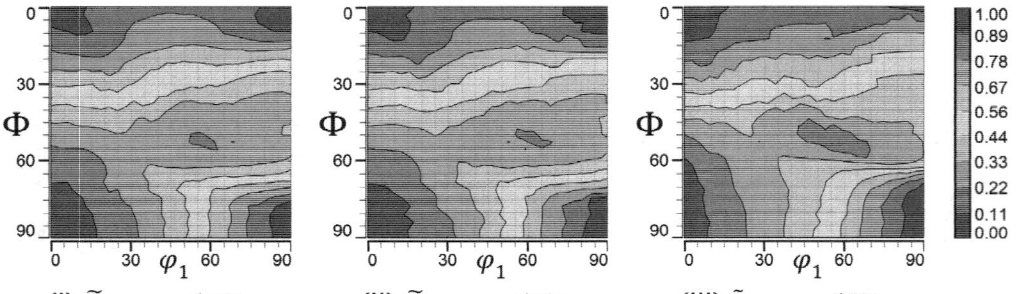

(i) $\tilde{M}, \varepsilon_{VM} = 5.7\%$ (ii) $\tilde{M}, \varepsilon_{VM} = 57\%$ (iii) $\tilde{\Gamma}, \varepsilon_{VM} = 57\%$

Figure 1. Normalised Taylor factor and normalised accumulated slip maps plotted for $\varphi_2 = 45°$ sections of Euler space for plane strain compression of bcc material: (i) and (ii) show normalised Taylor factor $\tilde{M}$ for von Mises equivalent strain ε_{VM} of 5.7% and 57% respectively; (iii) shows normalised accumulated slip $\tilde{\Gamma}$ at 57% ε_{VM}. All values are normalised per eqn. 1.

Figure 2 compares $\tilde{M}_g$ and $\tilde{\Gamma}_g$ using the relative values of eqn. 2. The red colour indicates regions where $\tilde{\Gamma}_g \gg \tilde{M}_g$, and blues indicates regions where $\tilde{\Gamma}_g \ll \tilde{M}_g$. At low strain ($\varepsilon_{VM} = 5.7\%$) Figure 2 (i) indicates no significant difference between $\tilde{M}_g$ and $\tilde{\Gamma}_g$, though two distinct regions of significant difference begin to appear at 11% ε_{VM}, as visible in Figure 2 (ii): one appears between the orientations $(001)[0\bar{1}0]$and $(112)[\bar{1}\bar{1}0]$, equivalently ($\varphi_1 = 90°, \phi = 0°$) and ($\varphi_1 = 90°, \phi = 17.6°$), where $\tilde{\Gamma}_g \gg \tilde{M}_g$, and another between $(332)[\bar{1}\bar{1}3]$ and the Goss orientation, equivalently ($\varphi_1 = 90°, \phi = 64.7°$) and ($\varphi_1 = 90°, \phi = 90°$),where $\tilde{\Gamma}_g \ll \tilde{M}_g$. These regions grow in size with increasing strain as visible in Figure (iii) at 57% ε_{VM}.

Discussion

Given the successful predictions of recrystallisation texture model for conventional rolling using the Taylor factor as the estimate of stored energy, it was expected no significant differences should be found between $\tilde{\Gamma}_g$ and $\tilde{M}_g$ for this idealised deformation. For most orientations in the $\varphi_2 = 45°$ section of Euler space the two parameters are within 25% of each other, which suggests Γ should at least provide comparably valid predictions when substituted for M in the recrystallisation model.

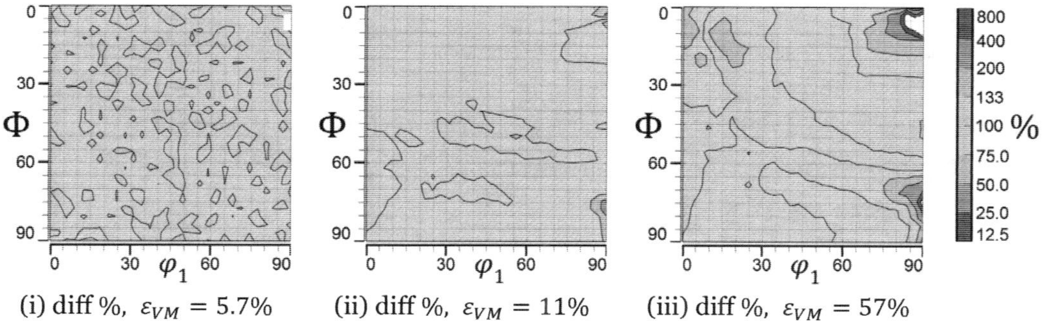

Figure 2. Ratio of accumulated slip divided by normalised Taylor factor per eqn. 2, plotted on $\varphi_2 = 45°$ sections in Euler space for three levels of strain: (i) 5.7%, (ii) 11% and (iii) 57% von Mises equivalent strain ε_{VM}. The white colour indicates regions where $\tilde{\Gamma}_g > 800\%$ of $\tilde{M}_g$.

It is somewhat surprising then to find some specific orientations where the two parameters differ significantly. Both of these regions of difference are close to important texture components in the deformation texture (γ-fibre and $(001)[0\bar{1}0]$ component), and notwithstanding other orientation effects (e.g. oriented growth) are thus likely to be significant for the recrystallisation texture.

It is interesting to note that the region where $\tilde{\Gamma}_g$ is much lower than $\tilde{M}_g$, near the $(331)[\bar{1}\bar{1}6]$ orientation, coincides with the tendency in some cold rolled steels for orientations in the gamma fibre to shift toward the neighbourhood of the $(331)[\bar{1}\bar{1}6]$ orientation [2]. On the other hand, if nucleation in such cold deformed steels is accepted to predominately occur in grains of high stored energy, a direct explanation for this shifting phenomenon would be forthcoming from the current analysis only if $\tilde{\Gamma}_g$ were to be higher than $\tilde{M}_g$, rather than lower as in fact found in these results.

Conclusions

- The normalised accumulated slip predicts a distribution of stored energy amongst orientations largely similar to the normalised Taylor factor for plane strain compression of bcc material.
- There are significant differences observed for the ranges of orientations between $(001)[0\bar{1}0]$ and $(112)[\bar{1}\bar{1}0]$, and between $(332)[\bar{1}\bar{1}3]$ and the Goss orientation .
- There may be some significance to the difference between the two parameters found near the $(331)[\bar{1}\bar{1}6]$ orientation: strengthening of the texture near this component has been observed in recrystallisation textures of cold rolled steels.

Acknowledgements

The authors acknowledge the financial support of KU Leuven project KP/12/007, and the Flemish Agency for Innovation by Science and Technology (IWT) for the PhD scholarship of D. Shore. Collaboration between KU Leuven and UGent is supported in part by the Belgian Science Policy Office Inter-University Attraction Poles Project P7/21.

References

[1] Sidor J, Petrov R and Kestens L 2011 *Acta Mater.* **59** (14) 5735–48
[2] Kestens L and Jonas J 1996 *Metall. Mater. Trans.* A **27** 155–64

IOP Conf. Series: Materials Science and Engineering **82** (2015) 012053 doi:10.1088/1757-899X/82/1/012053

Deformation behavior of electro-deposited pure Fe and its texture evolution during cold-rolling and subsequent annealing

N Sugiura[1] N Yoshinaga[1] and K Ushioda[1]
[1]Nippon Steel and Sumitomo Metal Corp, Futtsu, JAPAN
E-mail: ushioda.cd8.kohsaku@jp.nssmc.com

Abstract. Electro-deposited pure Fe has a characteristic of having very sharp isotropic ND//<111> fiber texture with a needle-like shaped fine grain elongated to ND. This Fe exhibits a surprisingly high r-value of over 7; however, such a high r-value might not be rationalized only from texture. Careful slip analyses reveal that restricted slips take place in the specific {110} slip planes, which are perpendicular to the sheet surface. Since grain boundaries with columnar structure are also perpendicular to the sheet surface, the slip plane in a certain grain may easily connect to the slip plane in adjacent grains having within ±30 degree rotation relationship around the common axis of ND//<111>. The operation of such a slip system is considered to cause the width strain much larger than the thickness strain. Furthermore, the texture evolution during cold-rolling and subsequent annealing was investigated using electro-deposited pure Fe as a starting material. Regardless of the amount of cold-rolling reduction, 65% to 90%, {111}<112> cold-rolling texture developed. After recrystallization, {111}<112> texture remained when material is cold-rolled by 65%, while {111}<011> texture developed when materials are cold-rolled by 80% and 90%. From the investigation into the mechanism on the development of recrystallization texture, the oriented nucleation and selective growth theories are concluded to contribute to the evolution of annealing texture.

1. Introduction

Steel products, which have been widely used for various applications, have greatly contributed to society. However, the potential capabilities of steels such as strength, formability and magnetic property, have not been fully exploited. In this sense, steels are very attractive materials having many possibilities.

In this paper, the studies of deep drawability [1] and tensile behavior [2, 3] of the electro-deposited pure Fe are reviewed because of its unique characteristic of having extremely sharp isotropic ND//<111> fiber textures with a fine columnar structure in the thickness direction. This Fe exhibits a surprisingly high r-value and local elongation, which indicates the remaining space for Fe to further improve its properties by controlling texture and structure. Moreover, the study on the texture evolution during cold-rolling and subsequent annealing using the electro-deposited pure Fe as a starting material [4] is reviewed so that we could have an insight into the nature of texture formation mechanism in Fe.

2. Electro-deposited pure Fe

The material used is the electro-deposited pure Fe produced by Toho Zink Co., Ltd., the chemical composition of which is listed in Table 1. The structure and texture close to

Table 1 Chemical composition of electro-deposited pure Fe used (mass %).

C	Si	Mn	P	S	Al	N	O
0.0006	0.0001	0.0001	0.0001	0.0002	0.0002	0.0009	0.0018

IOP Conf. Series: Materials Science and Engineering **82** (2015) 012053 doi:10.1088/1757-899X/82/1/012053

the electrode is different from that in the central region. Therefore, the original material with the thickness of about 2 mm is ground from both surfaces to 0.8 to 1.0 mm thickness. The electro-deposited pure Fe comprised a needle-like shaped fine grain elongated to ND (Fig. 1 (a)). The grain size observed from ND is about few μm to 30 μm. Moreover, the {100} pole figure, shown in Fig. 1 (b), shows that this Fe has extremely sharp isotropic ND//<111> fiber textures. The crystal growth in the electro-deposition process of bcc Fe implies that the Fe atoms are consistently deposited on {111} planes, despite this not being a close-packed or low energy surface in bcc metals.

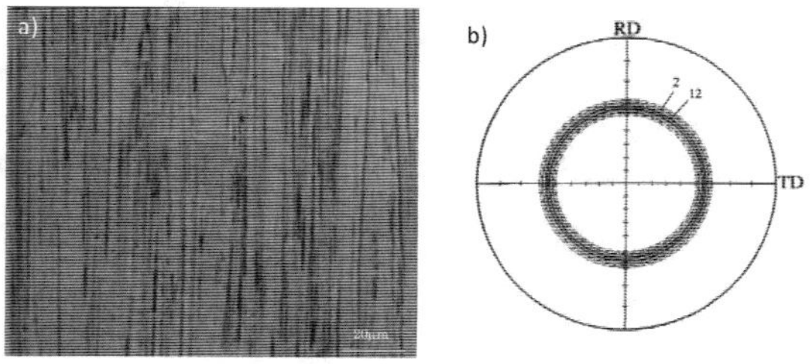

Fig. 1 a) Optical micrograph showing fine columnar structure and b) {100} pole figure of electro-deposited pure Fe.

3. Deformation behavior of electro-deposited pure Fe

3.1. Tensile property

The tensile properties of the electro-deposited pure Fe are shown in Table 2. The JIS 13B and JIS 5B (cf. Japanese Industrial Standard Z2241) tensile specimens were used to measure the r-values and the other tensile properties, respectively. The strength level of this Fe is similar to that of the conventional IF (interstitial atoms free) steel; however, the r-value is surprisingly as high as 7.1. Moreover, the total elongation is over 68%, in particular, the post uniform elongation is as large as 55%. The appearance of specimen after tensile test is shown in Fig. 2, which reveals that the specimen is necked significantly and fractured with large width reduction together with subtle thickness reduction.

The prediction of the r-value was carried out in order to check the possibility whether such a high r-value comes from the texture; software developed by Van Houtte [5] was employed for this. The calculation was based on both FC (Full Constraint) model and RC (Relaxed Constraint) model, where the condition of RC model is described in Table 3. The r-value calculated is 2.4 to 2.6 in FC model and 2.3 to 3.2 in RC model, which are significantly smaller than that of experimental values such as 4.8 to 7.1. Therefore, it is implied that there exist other reasons than texture for explaining the high r-value. However, it might be possible to predict r-value as high as 7 considering a needle-like shaped fine grain with sharp ND//<111> orientation assuming proper plasticity model and grain interaction [6-8].

Table 2 Tensile properties of electro-deposited pure Fe.

TS/ MPa	*YS*/ MPa	*T-El*(%)	*U-El*(%)	*L-El*(%)	r
288	215	68.5	13.0	55.5	7.1

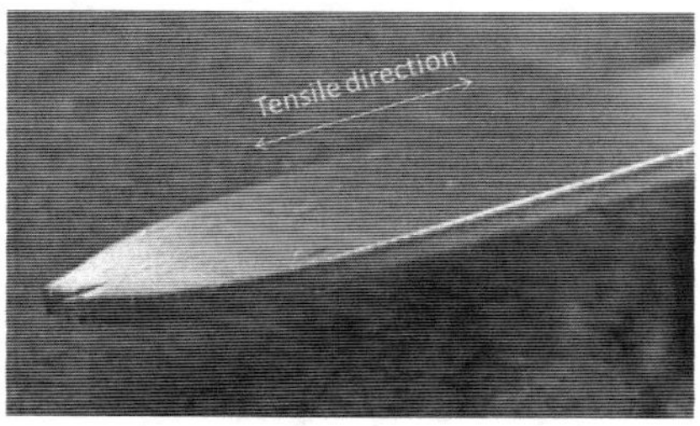

Fig. 2 Appearance of specimen after tensile test.

Table 3 Conditions for calculation of r-value and calculated results.

Relaxed condition	Slip system	r-value calculated
RC1	BCC2	3.16
RC2	BCC1	2.31
RC2	BCC2	2.53
RC3	BCC1	2.56
RC3	BCC2	2.98
RC4	BCC1	2.73
RC4	BCC2	3.17

Condition for Relaxed Constraint model
RC1: e_{13}, e_{23} RC2: e_{12}, e_{21}
RC3: e_{31}, e_{32} RC4: $e_{12}, e_{21}, e_{31}, e_{32}$
Slip system:
BCC1 {110}<111>+{112}<111>
BCC2 {110}<111>+{112}<111>+{123}<111>

3.2. Deep drawability

In order to verify such a high r-value obtained by the tensile test, a cup drawing test, *i.e.*, TZP test [1, 9], was carried out. The change in the relationship between the stroke and the force for both the first and second draw in the TZP test is described schematically in Fig. 3. Here, the maximum force for first draw and the fracture force in second draw with BHF (Blank Hold Force) are P_m and P_f, respectively. P_f is well known to be related to the r-value. Figure 4 clarifies that the fracture stress obtained experimentally is in good agreement with the one theoretically predicted using the r-value by tensile test [1]. Furthermore, T-value, defined as $(P_f - P_m) / P_f$, was confirmed to correspond to the r-value determined experimentally [1]. Therefore, the extremely high r-value of electro-deposited pure Fe obtained by tensile test may be considered trustable.

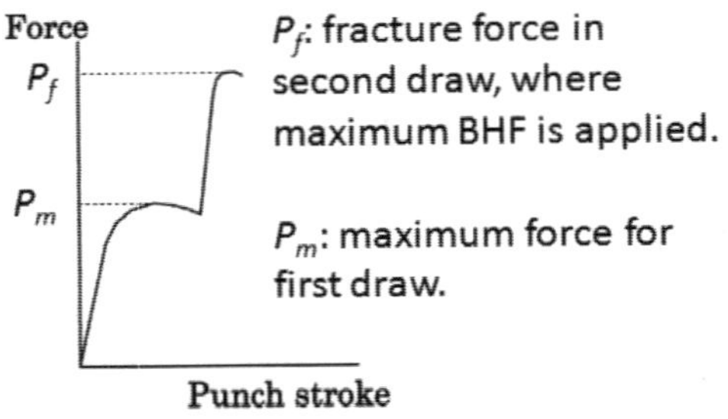

Fig. 3 Schematic illustration showing the change in stroke-force curve during TZP test.

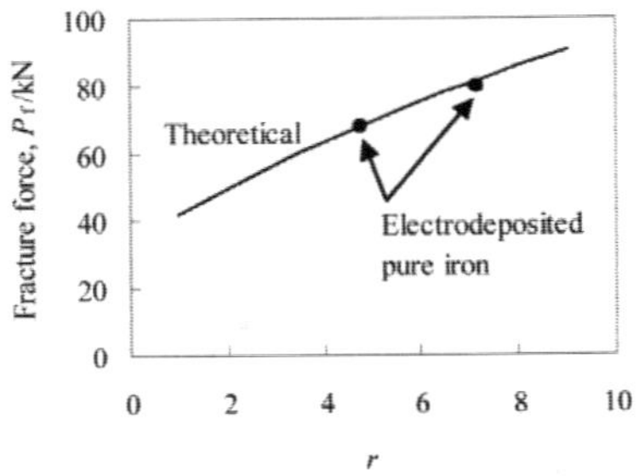

Fig. 4 Relationship between fracture force and r-value in TZP test.

3.3. Analysis of tensile deformation behavior

For the purpose of elucidating the nature of surprisingly high-r value of the electro-deposited pure Fe, the tensile behavior was analyzed in detail [2, 3].

Figure 5 shows the optical micrograph indicating the slip pattern observed from ND after 8% stretching. Many slip patterns aligning 30 to 60 degree from tensile direction were observed. EBSD-IQ map of the same stretched specimen is shown in Fig. 6, where the black contrasts reveal the boundaries of grains of the electro-deposited Fe with complicated shape, while the straight striations observed slightly within the grains marked as the red line correspond to the slip line. Crystal orientation analysis indicates that these striations are {110} slip planes perpendicular to the top surface. The red lines in the figure also show the three <112> directions parallel to the surface as schematically illustrated in Fig. 7. Moreover, the observed slip line is confirmed to be parallel to one of the three <112> directions.

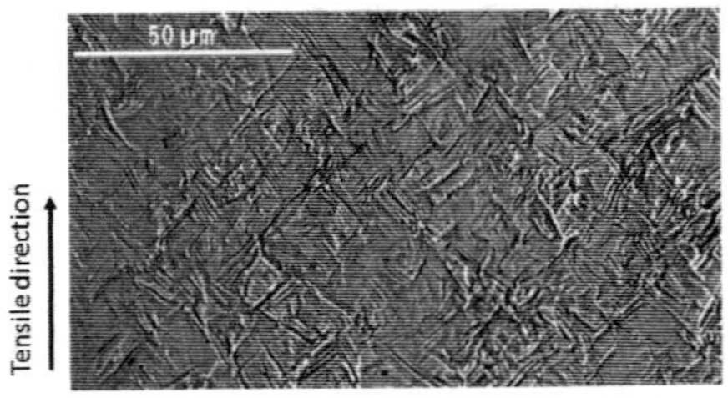

Fig. 5 Optical micrograph showing slip pattern observed from ND after 8% stretching.

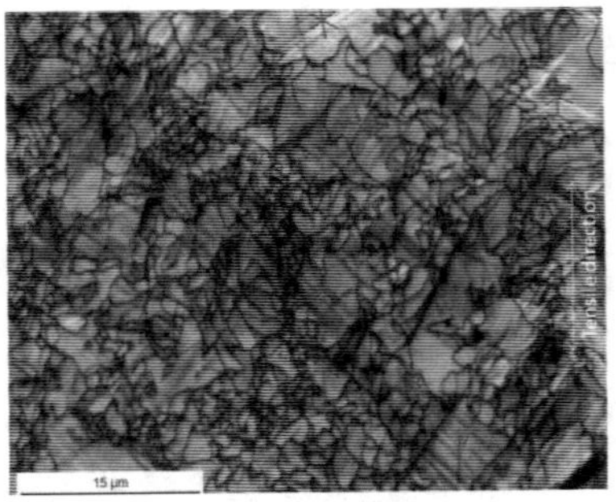

Fig. 6 EBSD-IQ map showing grain boundaries and slip lines observed from ND after 8% stretching.

Electro-deposited pure Fe is inferred to have an extremely high r-value owing to the operation of {110} slip system perpendicular to the surface. In the case of the slip system of {112} and {123}, there is only one slip direction of <111> which is parallel to the thickness direction. This implies that Schmid factors for these slip systems equal to zero. Therefore, the operation of {112} and {123} slip system becomes impossible. Meanwhile, there are six {110} slip planes with two different <111> slip directions, which leads to 12 slip systems. In the case of sharp ND//<111> texture, the three {110} planes are perpendicular to the surface (Type A), while the other three {110} planes incline 35 degree to the surface (Type B). Schmidt factors for these slip systems were calculated, and turned out that two slip systems belonging to Type A have the same maximum value as two slip systems in Type B. Despite the fact that these four slip systems are expected to be active, only the two slip systems belonging to Type A were confirmed to be active, the reason of which is discussed later.

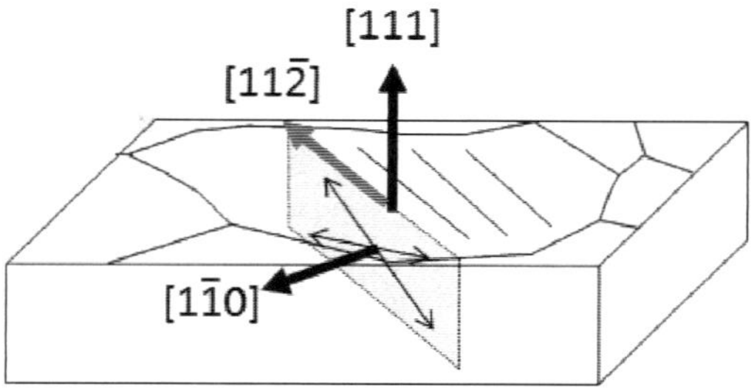

Fig. 7 Schematic illustration showing slip system and slip line at surface.

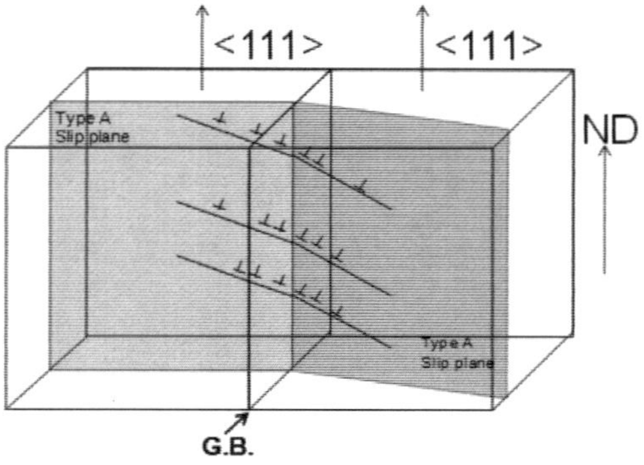

Fig. 8 Schematic illustration of Type A slip systems in neighboring two grains with common ND//<111> axis, suggesting easy transmission of internal stress through grain boundary.

In a grain with ND//<111> orientation like the electro-deposited pure Fe, there exist three slip planes belonging to Type A. Therefore, it is expected that a neighboring grain with ND//<111> orientation always has a slip system within ± 30 degree rotation relationship around the common ND//<111> axis with the original grain. Therefore, the crossing line of two Type A slip planes in the neighboring grains lie on the grain boundary schematically as shown in Fig. 8, because grains are elongated to the thickness direction. Consequently, it is inferred that the plastic deformation in the horizontal directions takes place smoothly because the slip planes are connected to every neighboring grain resulting in the easy transmission of internal stress through grain boundary (Fig. 8). Moreover, this kind of deformation suggests a large width strain compared to a very small thickness strain when the specimen is subjected to the tensile deformation. Meanwhile, the Type B slip systems have a twist relationship in each adjacent grain, which indicate the difficulty in internal stress transmission due to a lack of continuity in slip systems.

Furthermore, the electro-deposited pure Fe, having fine ND//<111> columnar structure in the thickness direction, is considered to have higher flow stress in the thickness direction as compared to the width direction. Therefore, a large difference in flow stress between thickness and width directions is expected, which also brings about a high r-value. The grain boundary sliding due to the fine columnar structure is reported to be responsible for the high r-value [10], but it is speculated difficult for grain boundary sliding to occur during tensile deformation at room temperature. The evidence of slip line within grain is another reason for the speculation.

4. Recrystallization and texture evolution of electro-deposited pure Fe

The recrystallization behavior and texture evolution in the process of cold-rolling and annealing were investigated using electro-deposited pure Fe as a starting material [4]. The focus was placed on the influence of cold-rolling reduction, which was changed from 65 and 80% to 90%. There are many previous studies using hot-rolled bands as starting materials; however, no study has been conducted using electro-deposited pure Fe with sharp isotropic ND//<111> texture as a starting material.

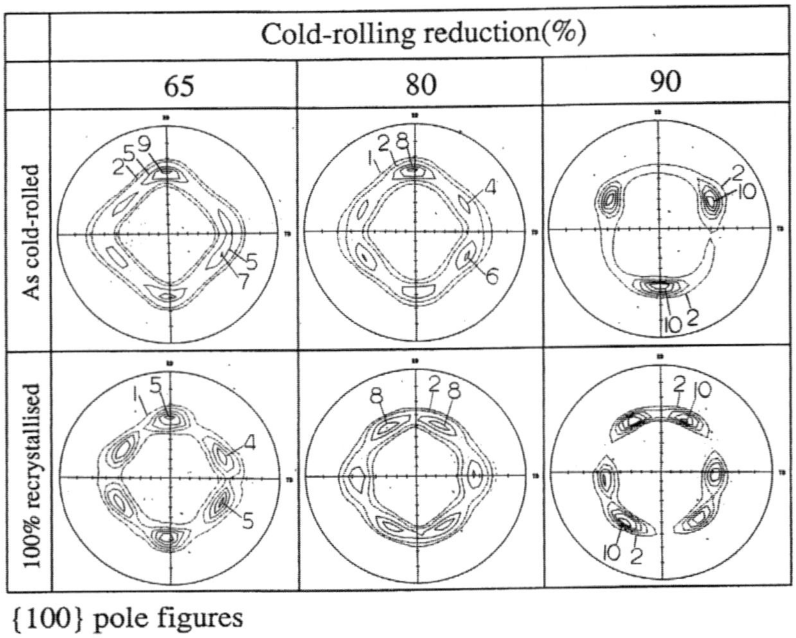

Fig. 9 {100} pole figures of cold-rolled and 100% recrystallized sheets as a function of cold-rolling reduction.

Recrystallization commences and completes during iso-thermal annealing at 400 °C of 65% and 80% cold-rolled electro-deposited pure Fe. This very low recrystallization temperature is another characteristic of electro-deposited pure Fe. Figure 9 shows {100} pole figures of cold-rolled and 100% recrystallized sheets as a function of cold-rolling reduction. It is generally known that {111}<011> texture rather than {111}<112> texture relatively increases with increasing cold-rolling reduction; however, it is worthy to note that {111}<112> orientation becomes major in this study. Furthermore, {111}<112> orientation remains after recrystallization when cold-rolling reduction is 65%; whereas, when the reduction increases to 80%, {111}<011> orientation becomes major and this tendency is strengthened by higher cold-rolling reduction by 90%.

In order to investigate the mechanism of the influence of cold-rolling reduction on texture evolution, the orientation of recrystallized nuclei was measured by EBSD, where the fraction recrystallized is about 5%. The ODFs determined by EBSD data for recrystallized grains are shown in Fig.10. The major orientation of recrystallized nuclei turned out

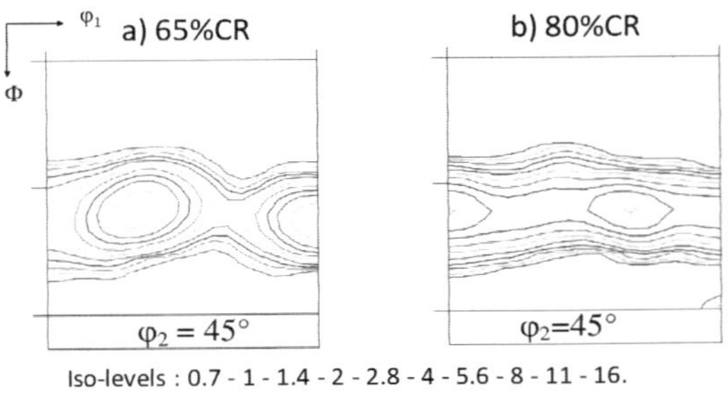

Fig. 10 ODFs of recrystallized grains determined by EBSD data. Fraction recrystallized is 5%, irrespective of cold-rolling reduction of a) 65% and b) 80%..

to be {111}<112> in the case of 65% reduction; whereas, in the case of 80% reduction the intensity of {111}<011> orientation is relatively high with the scattered γ-fiber textures. Therefore, it is obvious that the orientation of nuclei significantly influences the texture after the completion of recrystallization. However, the mechanism is not clearly rationalized only from the orientation nucleation theory.

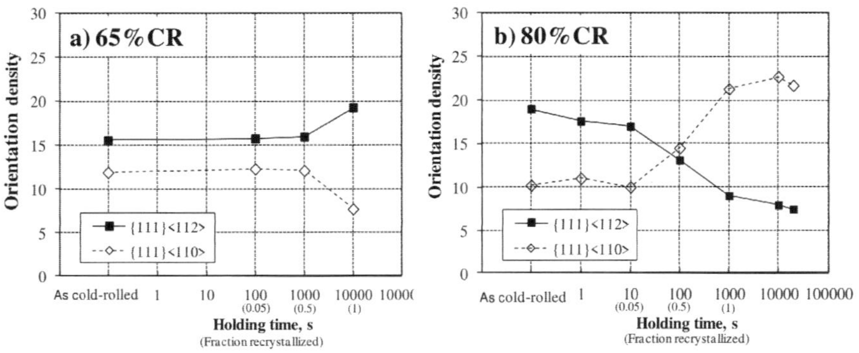

Fig.11 Changes in texture with evolution of recrystallization during annealing at 400 °C for a) 65% and b) 80% cold-rolled materials.

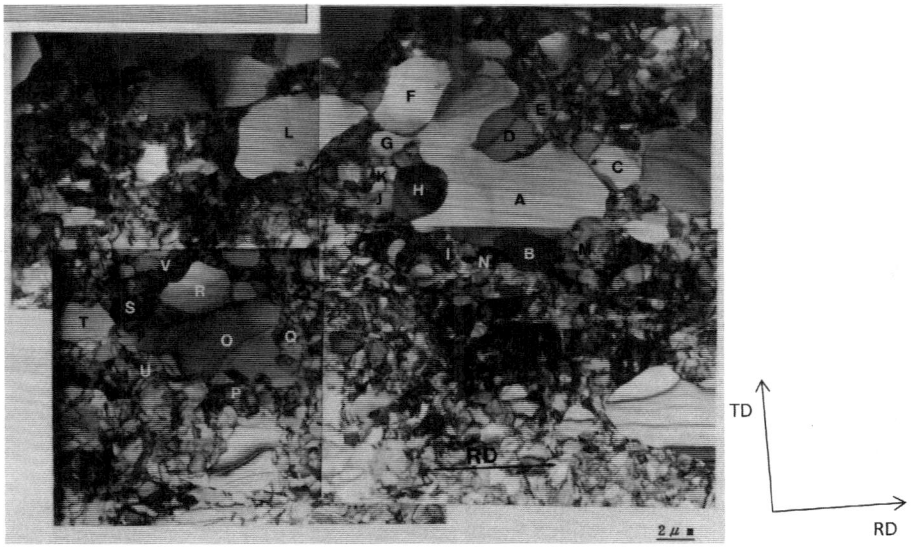

Fig. 12 TEM micrograph showing recrystallized grains with near {111}<011> and {111}<112> orientations. Cold-rolling reduction: 80%. Fraction recrystallized: 50%.

The texture change with the evolution of recrystallization is significantly different between the two materials cold-rolled by 65% and 80% reduction, as shown in Fig 11 a) and b). Namely, in the case of the material cold-rolled by 65% reduction the intensity of {111}<112> slightly increases while that of {111}<011> decreases. This indicates that the nuclei with {111}<112> orientation grows into the matrix with near {111}<112> orientation. Meanwhile, in the case of the material cold-rolled by 80% reduction the intensity of {111}<011> orientation increases intensively while that of {111}<112> decreases significantly. This indicates that the nuclei with {111}<011> orientation grows selectively into the matrix with {111}<112> orientation. In general, the recrystallization texture is determined by the product of nucleation and growth of each oriented grain. It was confirmed that the increase in cold-rolling reduction intensifies the rolling texture centering on {111}<112> orientation together with a rather homogenous deformation structure. Therefore, the selective interface migration for the nuclei with {111}<011> orientation into the {111}<112> matrix plays an important role when the cold-rolling reduction becomes over 80%. On the contrary, when the reduction is 65%, the development of stable cold-rolling texture, {111}<112> orientation, is not sufficient and the deformation structure is rather heterogeneous. Consequently, the nucleation and growth of recrystallized grains with {111}<112> orientation is inferred to take place. The TEM micrograph of the specimen cold-rolled by 80% followed by partial recrystallization annealing (50% fraction recrystallized), shown in Fig.12, supports the above speculation in the sense that the recrystallized grains with {111}<011> orientation are much larger than that with {111}<112> orientation.

5. Conclusion

The electro-deposited pure Fe has the fine columnar structure elongated in the thickness direction with isotropic sharp ND//<111> texture. It exhibits a surprisingly high r-value over 7 and extremely large local elongation. These properties are thought to be caused by the columnar structure with isotropic sharp ND//<111> texture, which leads to the large width strain as compared with the very small thickness strain because of the operation of {110} slip systems perpendicular to the surface during tensile test. Recrystallization proceeds at 400 °C in the process of cold-rolling and subsequent annealing, when the electro-deposited pure Fe is used as a starting material. With increase in cold-rolling reduction, {111}<112> cold-rolling texture develops, while annealing texture changes from {111}<112> to {111}<011>. The oriented nucleation and selective grwowth theory may explain the mechanism on these textures evolution. Thus, steels are considered to have potentials to further improve their properties by controlling microstructures and textures.

Acknowledgement

We are very grateful to Prof. A. Rollet, Carnegie Mellon University, for his valuable comments on the prediction of r-value.

References

[1] Yoshinaga N, Sugiura N, Hiwatashi S, Ushioda K and Kada O 2008 *ISIJ Int.* **48** 667
[2] Sugiura N and Yoshinaga N 2013 *J. Japan Inst. Metal Mater.* 77 361
[3] Sugiura N and Yoshinaga N 2014 *Advanced Materials Research* **922** *740*
[4] Yoshinaga N, Vanderschuren D, Kestens L, Ushioda K and Dilewijns J 1998 *ISIJ Int.* **38** 610
[5] Van Houtte P 1988 *Texture and Microstructure* **8-9** 313
[6] Daniel D and Jonas JJ 1990 *Metall. Trans. A* **21A** 331
[7] Delannay L, Van Houtte P, Van Bael A and Vanderschueren D2000 *Modelling and Simul. in Mater. Sci. Eng.* **8** 413
[8] Rollet A 2014 Private communication
[9] Siebel E and Beisswanger 1955 *Tiefziehen* (Carl Hanser Verlag)
[10] Motegi T, Masuyama S, Iida K, Suzuki T, Sato H and Tomota Y 2002 *Tetsu-to-Hagane* **88** 801

IOP Conf. Series: Materials Science and Engineering **82** (2015) 012054 doi:10.1088/1757-899X/82/1/012054

Deformation and recrystallization texture development in Fe-4%Si subjected to large shear deformation

AB Kustas[1,3,a], D Sagapuram[2,3], S Chandrasekar[1,3] and KP Trumble[1,3,b]
[1] School of Materials Engineering,
[2] School of Industrial Engineering,
[3] Center for Materials Processing and Tribology,
Purdue University, West Lafayette, IN 47906
Email: [a]akustas@purdue.edu, [b]driscol@purdue.edu

Abstract – Machining is used as a deformation technique to impose large shear strains ($\gamma \sim 2$) in a commercial Fe-4%Si alloy. The partial <111> and {110} – fiber texture components are generated throughout the as-deformed microstructure, which is expected of BCC metals deformed in simple shear. Using an annealing schedule similar to that in the commercial rolling process, samples retain the deformation texture, consistent with a continuous-type recrystallization mechanism. Fine-grained annealed samples reveal two different partial fiber orientations, one of which becomes the dominate texture, following the high-temperature growth treatment. The mechanisms of texture evolution and implications for texture control in the machining-based process are discussed.

Introduction

Silicon iron (Fe-Si) sheet is commonly used in transformer applications for its intrinsic magnetic properties, which are routinely achieved through multistep hot and cold rolling involving precise control of both deformation and annealing stages. Texture evolution of the thin Fe-Si sheet throughout the different stages in conventional processing has been well documented, with special interest regarding the development of the Goss texture [1]; however there has been limited evaluation of texture development in Fe-Si under deformation techniques other than rolling. The focus of this paper is to evaluate the texture evolution of Fe-4%Si following large shear deformation from machining.

Background

Machining is a technique that removes material through cutting, by using a sharp wedge tool set at a preset depth (t_o) into the bulk (Fig. 1). As a result of intense local shear (simple shear) deformation imparted by the cutting tool, material is removed from the surface of the rotating workpiece in the form a continuous chip/strip. This intense shear deformation occurs within a narrow zone, capable of creating a refined microstructure in the strip [2]. Furthermore, high cutting speeds can be used to produce large, local temperature rises from adiabatic heating.

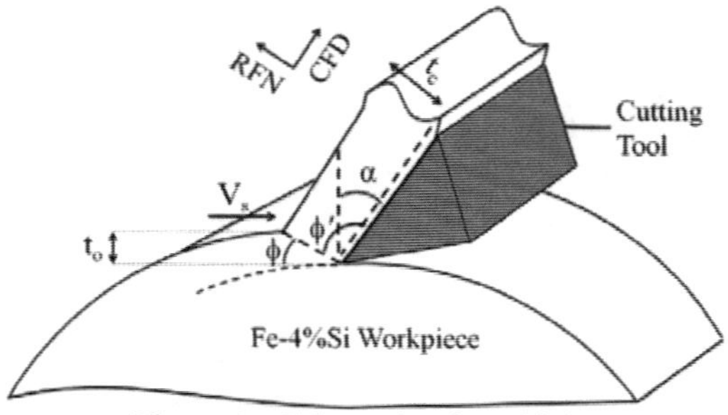

Figure 1 – Schematic of machining

To model the deformation zone, a single shear plane is often used, which allows for analysis of the imposed shear strain during cutting. Estimates of the shear strain (γ) can then be made from measurements of the chip thickness ratio ($\lambda = t_c/t_o$) and tool rake angle (α). In addition to control of the magnitude of the shear strain, α also controls the strain path as defined, for example, by the shear plane angle $\phi' = 90° + \alpha - \phi$ where $\phi = \tan^{-1}(\cos\alpha/(\lambda - \sin\alpha))$. Both strain path and slip system activity are known to control texture evolution during deformation [3]. In BCC systems, the {110} <111> slip system is most commonly activated; consequentially the slip plane and direction are expected to align with the macroscopic shear plane.

Experimental

A nominal Fe-4%Si plate (Scientific Alloys Inc, RI) was used as the bulk workpiece (actual analysis, wt%: Fe-3.83Si-0.32Mn-0.028C-0.018P-0.015S-0.006Al). The as-received hot-rolled plate possessed an initial coarse grain microstructure (d ~ 1 mm), which was refined through cold-rolling (r = 62%) and annealing to obtain a finer grain size of ~ 20 μm. A 10-cm diameter disk workpiece was removed from the plate with a final thickness of 0.32 cm, which corresponds to the width for the cut strip samples. Cuts were made using a cemented carbide cutting tool under a single rake angle ($\alpha = 20°$), at a constant surface velocity ($V_s = 2$ m/s) resulting in a shear strain of ~ 2 in strips ~ 250 μm thick (t_c). Strip samples were subjected to a low-temperature open-air anneal at 700 °C for 30 min followed by a high-temperature anneal at 1100 °C for 5h in a controlled atmosphere of Ar-5%H_2 flowing at ~ 1 cm^3/s at 1 atm. These conditions were selected to mimic commercial annealing. Texture measurements were conducted through electron backscatter diffraction in an FEI XL-40 field emission SEM. Data were analyzed using EDAX OIM software and processed through a confidence index (CI) filter ensuring a minimum CI value of 0.1 was achieved for each data point.

Results and discussion

As-deformed chip samples revealed a microstructure possessing grains aligned in the shear direction with a serrated top surface and a smooth bottom surface, which shows additional secondary shear deformation (Fig. 2). The axes used for pole figures were the rake face normal (RFN) and chip flow direction (CFD), which provided a convenient way to analyze texture components. Black regions in the inverse pole figure color map indicated areas that were severely deformed and consequentially produced low signal. Additionally, blurred grain boundaries were present, expected from microstructures containing high dislocation density [4]. Analysis of the as-deformed chips revealed a simple shear texture characterized by the partial {110} and <111> - fibers aligned with the shear plane and shear direction, respectively. These partial fibers have been observed by others using similar shear-based deformation techniques [5]. For the selected deformation condition, the measured textures were aligned at an angle $\phi' \sim 78°$, which is close to the theoretical value from the shear plane model ($\phi' \sim 81°$).

Figure 2 – As-deformed side view of the chip microstructure with dotted line illustrating the secondary shear zone. Inverse pole figure color map from the bulk showing inclined, elongated grain structure and corresponding (110) pole figure with ideal simple shear partial fibers (dotted lines) superimposed.

Following the low-temperature anneal, a fully recrystallized microstructure was achieved with grain size d ~ 20 μm (Fig. 3). Pole figures from the bulk indicated the deformation texture was retained following annealing. Additional scans near the chip rake face also showed the same fiber components but at significantly lower orientation ($\phi' \sim 39°$) than in the bulk of the chip, consistent with secondary shear effects at the material/tool interface during cutting.

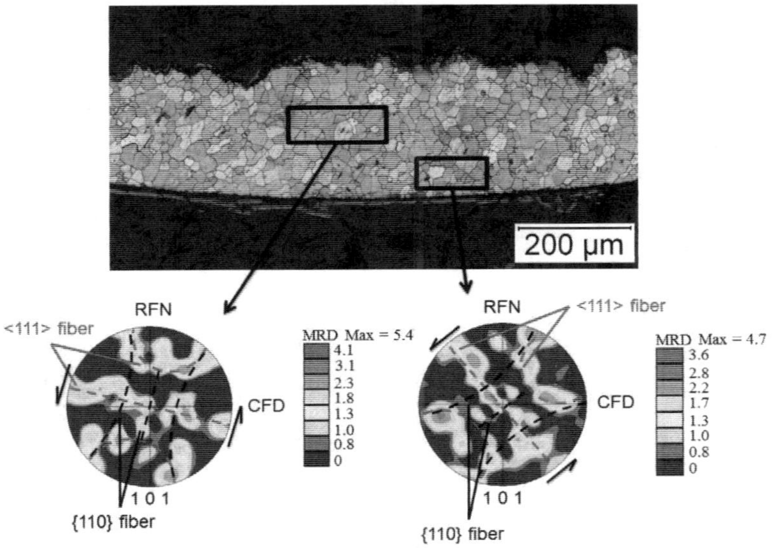

Figure 3 – Pole figures from the bulk (left) and secondary shear zone (right) for the low-temperature anneal condition. Textures indicate the simple shear fibers are retained following annealing.

The high-temperature (growth) annealing treatment resulted in significant grain growth (d ~ 100 - 200 μm) with 1 to 2 grains through the thickness of the strips (Fig. 4). Texture measurements indicated grains from the bulk of the strips consumed those within the secondary shear zone. This result is evident from the 45° slices of the ODFs for the low – and high-temperature annealed bulk (Fig. 4).

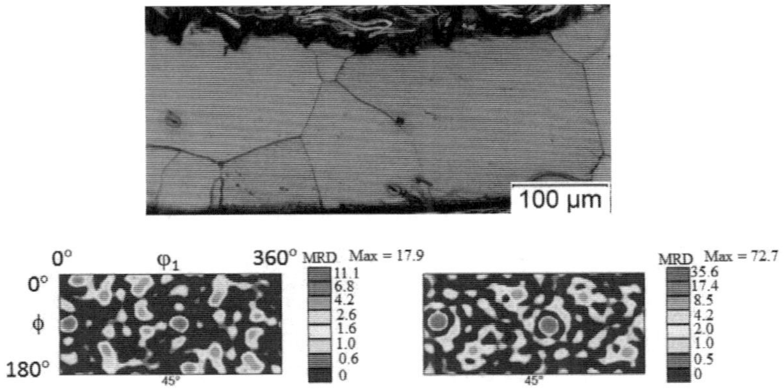

Figure 4 – Coarse grains that grew in the high-temperature annealing treatment. ODF sections for the bulk chip following low-temperature annealing (left) and the high-temperature growth treatment (right) showed strong intensity peaks at nearly identical Euler angles, indicating similar texture components were present. Note, ODF sections were generated from several grains.

The similarity between the deformation and recrystallization textures following large shear deformation has been observed in other systems in the literature [6, 7]. In general, it has been proposed that large strains imposed during deformation result in the formation of complex microstructures with non-equilibrium, high-angle sub-boundaries. Upon annealing, recovery initially transforms these boundaries into conventional grain boundaries, at which point homogeneous grain growth commences [8]. This process has been classified as continuous recrystallization (cRX). Since such a substructure was expected from the large strains experienced during machining, it is believed that the traditional inhomogeneous nucleation and growth of grains commonly observed in discontinuous recrystallization (dRX) was limited here, in favor of the cRX mechanism. Therefore, a recrystallization texture was not developed in a traditional sense and the same deformation texture was retained following annealing.

The homogeneous nature of cRX was confirmed experimentally from incremental annealing at 700 °C at times as low as 20 s (not shown).

After the high-temperature treatment, grains in the secondary shear zone were consumed by grains in the bulk due to the relative size advantage possessed by the bulk grains (d ~ 21 μm) over the grains within the secondary shear zone (d ~ 14 μm) following the low-temperature anneal. The finer size of the grains in the secondary shear zone is consistent with the additional deformation imposed from friction at the rake face, which provided a finer starting deformation sub-structure for grain coarsening. As a whole, machining results in a fundamentally different texture when compared with conventional rolling of Fe-Si, which is a product of the differences in the two deformation processes. For commercial sheet production, a hot rolled plate is subjected to multiple rounds of large cold rolling reductions with intermediate annealing. The first round of rolling leads to an inhomogeneous texture throughout the cold rolled sheet thickness, with texture components varying at the sheet surface, mid-plane and intermediate regions [1]. Machining brings about a nearly homogeneous texture regardless of the starting plate texture. Furthermore, the orientation, in contrast to rolling, can be varied as a function of α to engineer a wide variety of partial fiber orientations with respect to their inclination relative to the CFD. In rolling, a different recrystallization texture is produced when compared to machining. This annealed texture provides the essential environment for Goss grains to grow abnormally in the commercial grain-oriented sheet. As a consequence of the inclined orientations afforded by machining, a different microstructural environment may be induced around Goss grains in the machined strip, allowing for evaluation of the abnormal growth mechanisms in a new manner.

Summary

Machining, as a model simple-shear deformation technique, was used to produce chips from an Fe-4%Si alloy plate. Following machining, a classic simple shear type deformation texture developed defined by two partial fibers. This texture was retained following both a low-temperature and subsequent high-temperature annealing treatment. Texture conservation is believed to result from a continuous recrystallization mechanism for the low-temperature anneal while a size advantage offers an explanation for the mechanism behind the loss of the secondary shear zone following the high temperature growth treatment. Taken as a whole, shear deformation provides unconventional texture components and texture control when compared with current commercial processing for Fe-Si based on multistep rolling.

Acknowledgements

Support from NSF grants CMMI-1100712 and 1363524 and IIP-1237866 and US ARO grant W911NF-12-1-0012 is gratefully acknowledged.

References

[1] Mishra S, Därmann C, and Lücke K, 1984 *Acta Mater.* **32** 1984 2185-2201.
[2] Swaminathan et al., *J. Mater. Sci.* 22 (2007) 1529-1541.
[3] Sagapuram D, Efe M, Moscoso W, Chandrasekar S and Trumble KP, 2013 *Acta Mater.* **61** 6843.
[4] Humphreys FJ, 2001 *J. Mater. Sci.* **36** 3833-3854.
[5] Li S, Beyerlein IJ, and Bourke M AM, 2005 *Mater. Sci. Eng. A.* **394** 66-77.
[6] Jin YH, Huh MY and Chung YH, 2004 *J. Mater. Sci.* **29** 5311-5314.
[7] Yu CY, Sun PL, Kao PW and Chang CP, 2004 *Mater. Sci. Eng. A.* **366** 310-317.
[8] Belyakov A, Sakai T, Miura H, Kaibyshev R and Tsuzaki K, 2002 *Acta Mater.* **50** 1547-1557.

IOP Conf. Series: Materials Science and Engineering **82** (2015) 012055 doi:10.1088/1757-899X/82/1/012055

The evolution of recrystallized texture of cold-rolled pure copper annealed with a magnetic field in the transverse direction

T He[1] Y Wang[2] W Sun[1] and X Zhao[2]

[1]Research Institute, Northeastern University, Shenyang, 110189, China

[2]Key Laboratory for Anisotropy and Texture of Materials (Ministry of Education), Northeastern University, Shenyang, 110189, China

E-mail: zhaox@mail.neu.edu.cn

Abstract. In this paper, the effect of magnetic field annealing was investigated on the evolution of recrystallized texture of a cold-rolled pure copper sheet. During magnetic field annealing the transverse direction of the specimen was set to be parallel to the magnetic field. It was found, although, altering the orientation of the specimen to the magnetic field direction during annealing does not change the type of the recrystallized texture component, the intensity of cube texture in the specimen with field is higher than that without field.

1. Introduction

As an important external field, it has been verified through experiments that electromagnetic processing of materials can affect the evolution of recrystallized microstructure and texture in plastically deformed metallic materials [1-14], including ferromagnetic [1, 3] and non-ferromagnetic alloys. He and his research group [4-6] found that the application of electric field and magnetic field on the development of the microstructure and recrystallized texture of cold-rolled IF steels have a lot of significant results. Molodov and co-workers [8-10] found that depending on the specimen orientation to the applied magnetic field, the texture components can be strengthened, remain unchanged, or disappear entirely in zinc alloy and titanium sheet after the high magnetic field annealing. As reported in reference [15], the magnetic field annealing can promote the formation of the initial recrystallized *Cube* texture in cold-rolled pure copper sheets, when it was annealed at 240℃ for different times, with its rolling direction (RD) parallel to the magnetic field direction (MD). However, in most of the previous reports the field was applied in the rolling direction of the specimens and only limited study [16] was conducted with the magnetic field direction parallel to the transverse direction of the specimen. So in this research work, the effect of a magnetic field was particularly investigated on the recrystallized texture evolution of the cold-rolled pure copper sheet with the magnetic field direction parallel to the transverse direction (TD).

2. Experiments

A 78% cold-rolled pure copper sheet of 1.2mm thickness was used in the experiment, its chemical composition (mass %) of 0.01Fe, 0.08Pb and 99.9Cu. The specimens with the dimensions of 20mm×10mm×1.2mm were respectively subjected to isothermal annealing at 240℃ for different times (20, 60, 120min) in a furnace installed in a cryocooler-cooled super-conducting magnet at a heating rate of 5℃/min, with and without a magnetic field of 12-tesla, and then cooled in water. During annealing, a magnetic field was applied in the whole process of heating. The specimens were kept in the area with zero magnetic force, and with the transverse direction parallel to the magnetic field direction.

Textures were determined from the normal direction (ND) in the 1/4 layer of the sheet after annealing. Three incomplete {111}, {200} and {211} pole figures were measured using the X′ pert MRD type X-ray diffractometer of PANalytical B.V. company with $Co\text{-}K_{\alpha}$ radiation. The corresponding orientation distribution functions (ODFs) were computed with Labtex3.0 software [17]. Microstructures of longitudinal sections of the specimens were observed by orientation imaging microscopy (OIM). The scan was carried out over the area of 613×459 measuring points with a 0.5 μm step size, and the corresponding ODFs were computed using the Channel 5 software.

3. Results and discussion

Figure 1 gives the ODF Φ_2=45° sections of the specimens annealed at 240℃ for different times. As shown in the figure, the main recrystallized texture component is *Cube* in the annealing specimens with and without the field. The type of the recrystallized texture component in the annealing specimens with their transverse direction parallel to the field direction is still the *Cube* component which is same to the typical for the process of pure copper recrystallization. In the previous study [14], the recrystallized texture component is also the *Cube* when the specimens were placed with their rolling direction parallel to the field direction. The obtained results demonstrate that the magnetic field annealing of cold-rolled pure copper sheet with either their transverse direction or rolling direction being parallel to the magnetic field direction does not change the type of the final recrystallized texture.

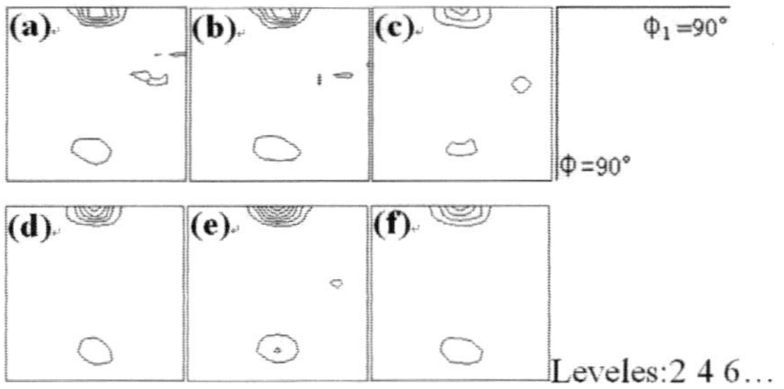

Fig. 1: ODF Φ_2=45° sections for the samples annealed at 240℃ for different times with and without magnetic field: (a) 0T×20min (b) 0T×60min (c) 0T×120min (d) 12T×20min (e) 12T×60min (f) 12T×120min

In Fig. 2, it shows the variation of *Cube* orientation intensity in the specimens annealing with and without the magnetic field for different holding times. The value of the *Cube* orientation intensity used in Fig. 2 is the direct intensity value obtained from ODF sections which come from the calculation results of three incomplete pole figures. It is obvious, from the figure, that the intensity of *Cube* component in the specimens with the magnetic field annealing is higher than that without the field at the same holding time. So the magnetic field annealing can promote the evolution of the initial recrystallized *Cube* texture when the magnetic field direction is parallel to the transverse direction of the cold-rolled pure copper sheet.

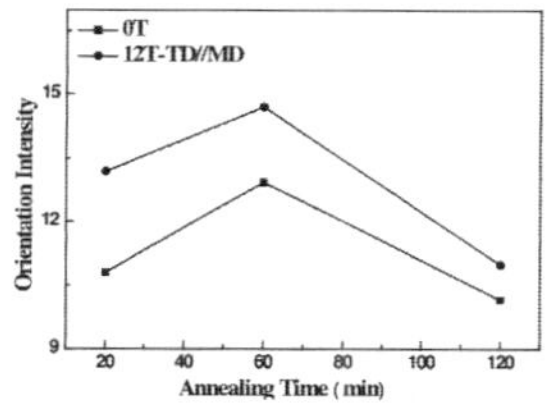

Fig. 2: Orientation intensity of cube texture in samples annealing for different times

In Fig. 3, it shows the orientation images of the specimens annealed for 20min, 60min and 120min, respectively. From Fig. 3, it can be seen that there is not obvious deformed microstructure in the specimens of the shortest holding time whether with the magnetic field annealing samples or not. There are many recrystallized grains with different sizes in the annealing microstructure, so the recrystallization nucleation process is nearly completed and grain growth is starting.

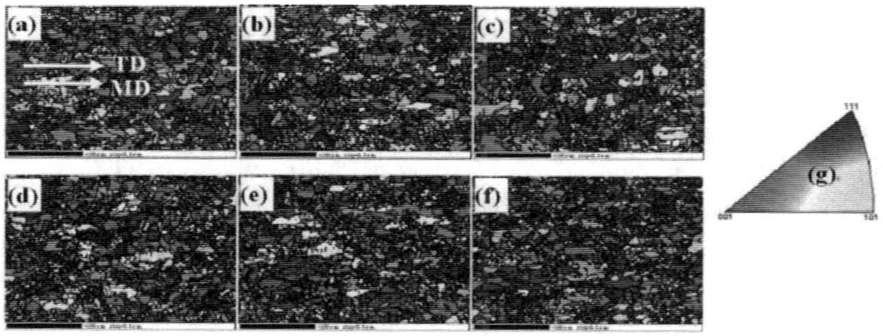

Fig. 3: OIM maps of samples annealed at 240℃ for different times: (a) 0T×20min (b) 0T×60min (c) 0T×120min (d) 12T×20min (e) 12T×60min (f) 12T×120min (g) Orientation identification

In this research work, it was further investigated the orientation features in the large, medium and small recrystallized grains of the specimens with and without the magnetic field annealing. Among the recrystallized grains, it was defined that an area of more than 200μm^2 is the large grain, more than 50μm^2 and less than 200μm^2 is the medium grain, and less than 50μm^2 is the small grain.The corresponding ODF Φ_2=45° sections of the recrystallized grains are shown in Fig. 4. It can be seen that the orientation of the small, medium and large recrystallized grains are mainly the *Cube* orientation for the specimens with and without the magnetic field annealing. It was supposed that the small grains are the nuclei and the large grains as the growth grains. The *Cube* orientation intensity of the small grains in the specimens with the magnetic field annealing is higher than that without the field. So the magnetic field annealing can promote the formation of the initial recrystallized *Cube* texture when the magnetic field direction is parallel to the transverse direction of the cold-rolled pure copper sheet.

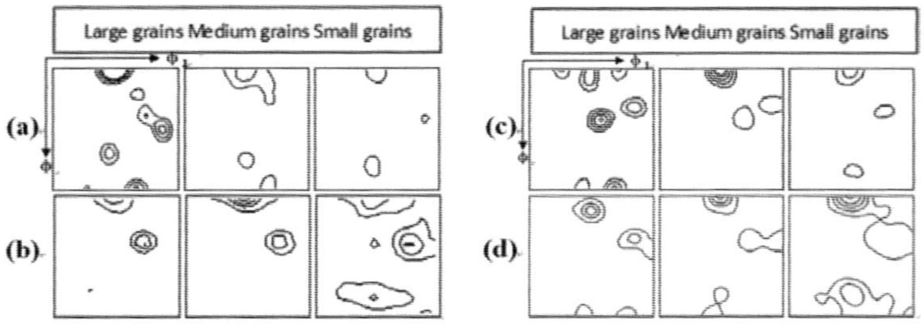

Fig. 4: ODF Φ_2=45° sections of recrystallized grains of different sizes in specimens annealed at 240ºC for different times (a) 0T×20min (b) 12T×20min (c) 0T×120min (d) 12T× 120min

In the experimental results of this research work, it shows that altering the specimen's orientation to the magnetic field direction during annealing does not change the type of the final recrystallized texture. The magnetic field annealing can promote the formation and evolution of the recrystallized *Cube* texture component when the transverse direction of the specimens is parallel to the field direction. As reported in literature [14], the magnetic field annealing can promote the evolution of the initial recrystallized *Cube* texture when the magnetic field direction is parallel to the rolling direction of the cold-rolled pure copper sheet. However, the high magnetic field can have different promoting effect on the intensity of recrystallized *Cube* texture components when the specimens are placed with the rolling direction and transverse direction parallel to the field direction respectively (as shown in Fig.5). The value of *Cube* orientation intensity used in Fig. 5 is also the direct intensity value obtained from ODF sections which come from the calculation results of three incomplete pole figures. The differences of the pole intensity may be attributed to the boundary selected migration under the high magnetic field induced by the difference of magnetic susceptibility of the grains with different orientation.

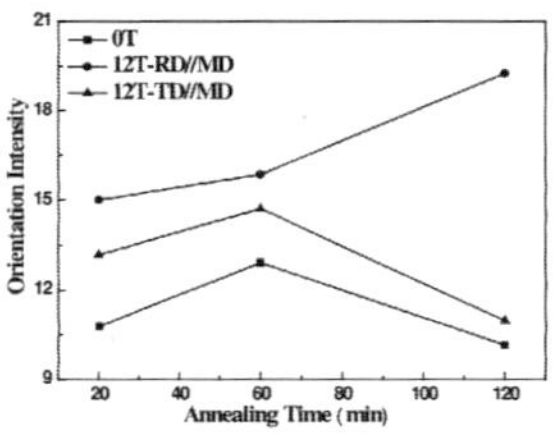

Fig. 5: Orientation intensity of cube texture of samples annealing for different times with the rolling direction and transverse direction parallel to the field direction

However, it is a very complicated process that includes magnetic field effects as well as the evolution of recrystallization textures. Further detailed research on this process will be done in the next step.

4. Summary

A 12-tesla high magnetic field was applied during the annealing of a cold-rolled pure copper sheet at 240℃ with the magnetic field direction parallel to the transverse direction of the specimens. Although altering the orientation of the specimen to the magnetic field direction during annealing does not change the type of the recrystallized texture component which is still the *Cube* component, the intensity of *Cube* texture in the specimen with field is higher than that without field. Moreover, the intensity of recrystallized *Cube* texture is higher in the specimens with the magnetic field direction parallel to the rolling direction than that parallel to the transverse direction.

Acknowledgements

This work was supported by the National Natural Science Foundation of China (Grants No. 51201028 and 51201027), the "111" Project (Grant No. B07015) and the Shenyang City Science and Technology Project Foundation (F12-277-1-48).

References

[1] Smoluchowski R and Turner R W 1949 *J. Appl. Phys.* **20** p 745.
[2] Mullins W M 1956 *Acta Metall.* **4** p 421.
[3] Bhandary V S and Cullity B D 1962 *Transactions of the Metallurgical Society of AIME* **224** p 1194.
[4] He C S, Zhang Y D, Wang Y N, Zhao X, Zuo L and Esling C 2003 *Scripta Mater*. **48** p737-742.
[5] Zhang Y D, He C S, Zhao X, Esling C and Zuo L 2004 *Adv. Eng. Mater*. **6** p310-313.
[6] He C S, Zhang Y D, Zhao X, Zuo L, He J C, Watanabe K, Zhang T, Nishijima G and Esling C 2003 *Adv. Eng. Mater.* **5** p579-583.
[7] Molodov D A, Gottstein G, Heringhaus F and Shvindlerman L S 1997 *Scripta Mater.* **37** p 1207.
[8] Molodov D A, Gottstein G, Heringhaus F and Shvindlerman L S 1998 *Acta Mater.* **46** p 5627.
[9] Sheikh-Ali A D, Molodov D A and Garmestani H 2002 *Scripta Mater.* **46** p 857.
[10] Sheikh-Ali A D, Molodov D A and Garmestani H 2002 *Mater. Sci. Forum* **408-412** p 955.
[11] Sheikh-Ali, A.D., and Garmestani, H 2008 *Ceramic Transactions.* **200** p 413.
[12] Molodov D A and Sheikh-Ali A D 2004 *Acta Mater.* **52** p 4377.
[13] Molodov D A and Konijnenberg P J 2006 *Scripta Mater.* **54** p 977.
[14] Bhaumik S, Molodova X, Molodov D A and Gottstein G 2006 *Scripta Mater.* **55** p 995.
[15] He T, Kang T Y, Wang Z J, Zhao X and Zuo L 2011 *Mater. Sci. Forum* **702-703** p 411.
[16] Molodov D A and Bozzolo N 2010 *Acta Mater.* **58** p 3568.
[17] Pawlik K, Ozga P: LaboTex: Texture Analysis Software, Göttinger Arbeiten zur Geologie und Paläontologie, SB4, 1999.

Orientation changes near the interface of explosively bonded (carbon steel)/Zr700 sheets

H Paul[1], T Baudin[2], F Brisset[2] and M Prażmowski[3]
[1]Institute of Metallurgy and Materials Science, Polish Academy of Sciences, Krakow, Poland
[2] Université Paris-Sud, ICMMO, CNRS UMR 8182, Laboratoire de Physico-Chimie de l'Etat Solide, Orsay, F-91405, France
[3]Opole University of Technology, Faculty of Mechanical Engineering, Opole, Poland
E-mail: h.paulg@imim.pl

Abstract. The microstructure and texture of explosively welded carbon steel (base) and Zr700 (flyer) plates were characterized by means of scanning electron microscopy equipped with a high resolution electron backscattered diffraction facility. The orientation maps demonstrate that the deformed zones near-the-interface are composed of several layers, the width of which depends on the applied bonding parameters. For both metals, the very thin layer of ultra-fine grains directly adheres to the interface. In the areas more distanced from the interface, the structure evolution depends on the plate material. In the case of a Zr 700 sheet the second layer is formed by highly dislocated (sub)grains, which progressively evolve, towards the structure composed of only lightly deformed grains. In the case of a carbon steel sheet, the second layer near the interface was composed of flattened grains.

1. Introduction

The mechanism responsible for the creation of a high strength bond during explosive bonding is a major unresolved problem. Explosive bonding is the process of metal joining as an effect of a collision of plates at high velocities by a controlled detonation of an explosive charge which propel a flyer plate towards a fixed metal base plate [1-3]. At the 'collision point', the perfectly clean surfaces are brought together under very high pressure. The high velocity oblique collision will 'produce' a *high temperature* and a *severe shear deformation* near the collision line in a very short time. This causes a local melting of the bonded metals, simultaneously with the severe plastic deformation of the interfacial layers, e.g. [1-7]. The plastic deformation leads to important structural and textural changes in the subsurface areas of the bonded plates and strongly affects the properties of the clad. The literature data on the cladding of steels with the use of zirconium and its alloys is very limited and only contributory works referring to selected aspects of the problem can be found, e.g. [7-9]. Therefore, the purpose of the present work is to identify the microstructure features and the corresponding texture changes occurring during dynamic loading in the clad (carbon steel)/(zirconium alloy)-type. We provide micro-scale analyses in the layers directly adhering to the interface. The basic research technique employed local orientation measurements based on scanning electron microscopy (SEM) equipped with the electron backscattered diffraction (EBSD) facility.

2. Experimental

The sheet of carbon steel (type P355NL2), 22 mm thick, as the base plate and the sheet of a Zr 700 alloy, 3.15mm thick, as a flyer plate were used for the clad preparation. A system of parallel arranged plates was applied to manufacture two-layered clads. The sizes of the plates were 300 x 500 mm^2. The clads were manufactured with two initial stand-off distances between the sheets (*h = 3 and 4.5 mm*) and at two detonation velocities (V_D = *2000 and 2200m/s*). All the samples were cut-off from the 'properly' bonded clad (in the 'as-bonded' state). The characteristic of the interfacial microstructure and the texture was performed by means of SEM, with the use of Zeiss Supra 55VP, equipped with a field emission gun and the EBSD facility. The orientation maps were created in the beam-scanning mode with the step size ranged between 50 nm and 100 nm in the ND-ED section (where: ND and ED are the normal and explosive directions, respectively).

3. Results and discussion

A detailed characteristic of microstructure in layers near-the-interface was performed against the background of the orientation maps displayed as a 'function' of the IPF colour code as well as the band contrast, as presented in Figs. 1a and b, respectively. The corresponding particular areas textures are presented in Fig. 1c and d, for the zirconium and carbon steel plates, respectively. The structure of the base plate (carbon steel) near-the-interface was composed of a very thin layer (1-2 μm in thickness) of fine equiaxed grains and a broader layer of flattened grains. The dimension of the flattened grains in the ND is at least one order smaller than that along the ED and TD. They were bent in the direction of the collision point (line) movement and their curvature reflects the operation of the shear stresses in near-the-interface layer. However, this directionality in the grain geometry was lost as the distance from the interface increased. In the areas distanced about 100 μm from the interface, nearly globular grains were observed, but still with an increased density of the dislocations (marked as low angle grain boundaries) inside the grains' interior. In the case of the flyer plate (Zr 700) the maps show a fine-grained layer, directly adhering to the melted zone. Those equiaxed grains (identified by the system as zirconium) form a layer of 5 – 20 μm in thickness (Fig. 2). However, a layer of flattened and/or bended primary grains was not observed in the zirconium plate.

The band contrast map revealed areas of a lower quality of the Kikuchi patterns (as darker areas) occupied by a melted zone and macroscopic shear bands. In the case of the melted zone, the low quality of the diffraction patterns results from the amorphous and/or ultra-fine-grained structure as well as the non-equilibrium chemical composition of the phases that are formed [4, 5, 7]. In the areas occupied by the melted zone, the layer of a very low band contrast is composed of a mixture of amorphous and (ultra-fine-grained) crystalline phases, as presented in [4 - 8].

The formation of the fine-grained structure near the interface is a common characteristic observed in both welded plates. The strongly refined structure is especially well-visible inside the wave crest. This area was always composed of very thin (with the thickness below 200 nm) and elongated grains, separated by large angle grain boundaries [6]. These thin grains were strongly curved and imitated well the rotational character of the material displacement during the wave formation, i.e. reflecting the non-homogeneous flow near the interface [6 - 8].

The flattened grains in the base plate form a layered structure which constitutes the background for the macro- shear bands (SB) formation. The shear banding process results from the operation of shear stresses (due to the oblique collision of plates) in the near-the-interface layers. The most important deformation mechanisms observed in the near-the-interface layers of the Zr 700 alloy plate are twinning and shear banding. Twins were observed also in the structure of the initial material and in the layers distanced ~ 1 mm from the interface. However, their density significantly increases as the distance from the interface decreases. With a contrast used, a rotation of the twins outside the SB and a displacement of their fragments within the band are clearly visible. The strong rotation within the shear bands is responsible for the fragmentation of twins. This leads to an ultra-fine-grained structure inside the well-developed SB. The internal structure of SB changes with the distance from the interface. The parts of the SB placed near the interface are composed of very fine (sub)grains elongated along the shear direction, whereas the parts of them more distanced from the interface are only bent. At the shear band boundary, the twin-matrix layers usually change the inclination, indicating a gradual incorporation of the layers into the SB area. Moreover, in the sheared layers near the interface, one of the twinning systems very often takes a position nearly parallel to the shear band plane. This can support the strain accommodation in the macro-scale.

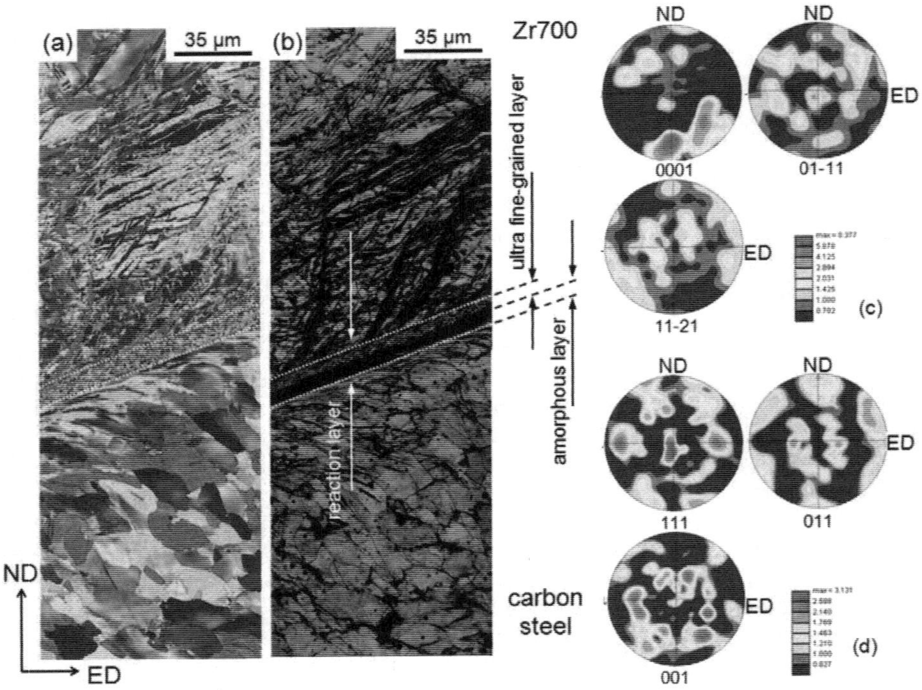

Fig. 1. Microstructure of near-the-interface layer. The orientation map displayed as a 'function' of the IPF colour code (a) and band contrast (b). (c) {0001}, {01-11} and {11-21} pole figures corresponding to the area of Zr 700 and (d) {111}, {011} and {001} pole figures corresponding to the area of carbon steel sheets. ND-ED section.

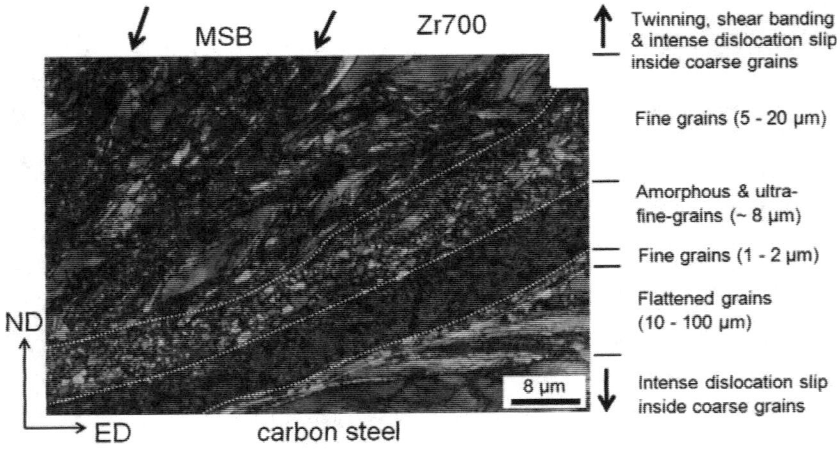

Fig. 2. Near-the-interface area showing formation of fine-grained layers in Zr 700 and carbon steel sheets. SEM/EBSD local orientation measurements with the step size of 100 nm. ND-ED section.

In the case of the carbon steel plate, the flattened grains also showed that a well-defined crystal lattice re-orientation occurred in some grains situated within the area of the broad macro-SB, although those grains initially had quite a different crystallographic orientation. Their crystal lattice rotated in such a way that one of the {110} slip planes became nearly parallel to the plane of the maximum shear. Moreover, one of the slip direction of the <111>-type, which is lying on this plane, is parallel to the shear direction. A natural consequence of this rotation is the formation of a specific macro-SB texture, which facilitates the slip propagation across the grain boundaries along the shear direction, without any visible variation in the slip direction. It was thereby established that the shear banding occurred across the grain boundaries by the continuity of the slip direction, although the slip plane did not coincide exactly in the adjacent grains.

4. Summary

This work describes the SEM/EBSD investigation of the microstructure and texture changes that occur in the interfacial layers of explosively welded sheets. Clads based on carbon steels as the base plate and Zr 700 as the flyer plate were analysed. The following detailed conclusions can be drawn.

- During the bond formation, the layers of the parent sheets near the interfaces underwent intense plastic (shear) deformation. The effect of the grain refinement was clearly observed in zirconium. The carbon steel deformed more uniformly in the significant part of the sheet, except for a very thin layer near the interface.
- In the case of base plate (carbon steel), the maps revealed a very thin layer (1 - 2 μm in thickness) of fine grains and a broader layer of flattened grains. In the case of the flyer plate (Zr 700), the maps show a thin fine-grained layer of 5 – 20 μm in thickness. However, a layer of flattened and/or bended primary grains was not observed. Both fine grained layers directly adhered to the melted zone.
- The most important deformation mechanisms observed in the near-the-interface layers of the Zr700 alloy plate are twinning and shear banding. In the case of the carbon steel plate, the flattened grains showed that a well-defined crystal lattice re-orientation occurred in some grains situated within the area of the broad macro-SB.
- 'Metallurgical bonding' was achieved by way of melting explosively welded sheets. The above proves that the proper explosive welding always incorporates the melting of thin surface layers of joined materials.

Acknowledgments

This work was partially supported by the Polish National Centre of Science (NCN), projects no.: UMO-2012/05/B/ST8/02522 and UMO-2012/04/M/ST8/00401.

References

[1] *Explosion Welding, Fundamentals of Process* 1992 Welding Handbook vol 3 pp 264-277
[2] Crossland B, Williams JD 1970 *Explosive Welding*, Metals Review vol 15 pp 79-100
[3] Akbari Mousavi SAA, Al-Hassani STS 2008 Materials and Design vol 29 pp 1-19
[4] Paul H, Lityńska-Dobrzyńska L, Prażmowski M 2013 Metall. Mater. Trans. 44A pp 3836-3851
[5] Song J, Kostka A, Veehmayer M, Raabe D 2011 Mat. Sci. Engn. A528 pp 2641-2647
[6] Paul H, Miszczyk M, Prażmowski M 2012 Mat. Sci. Forum, vol 702-703 pp 603-606
[7] Paul H, Morgiel J, Baudin T, Brisset F, Prażmowski M, Miszczyk M 2014 Arch. Metall. Mater. vol. 59 (2014), pp 1129-1136
[8] Paul H 2014 Advanced Materials Research vol 783-786 pp 1476-1481
[9] Prażmowski M 2014 Arch. Metall. Mater. vol 59, pp 1137-1142
[10] Paul H, Driver J, Maurice C, Jasieński Z 2003 Mat. Sci. Engn. A359 pp 178-191
[11] Paul H, Morawiec A, Driver JH, Bouzy E 2009 Int. J. Plast. 25 pp 1588-1608
[12] Paul H, Maurice C, Driver JH 2010 Acta Mater. 58 (2010) p. 2799-2813

IOP Conf. Series: Materials Science and Engineering **82** (2015) 012057 doi:10.1088/1757-899X/82/1/012057

Quantitative prediction of deformed austenite and transformed ferrite texture in hot-rolled steel sheet

Y. Tanaka[1], T. Tomida[2] and V. Mohles[3]
[1] Technival R&D Bureau,Nippon Steel and Sumitomo Metal Corp., Amagasaki, JP
[2] Amagasaki Unit, Nippon Steel and Sumikin Technology, Amagasaki, JP
[3] Institute of Physical Metallurgy and Metal Physics, RWTH Aachen, Germany
E-mail: yasuaki.km5.tanaka@jp.nssmc.com

Abstract. A model to quantitatively predict ferrite (α) textures in hot-rolled steel sheets has been developed. In this model, the crystal plasticity model, called "Grain Interaction model (GIA)", and the transformation texture model, called "Double K-S relation (DKS)", are linked together. The deformed austenite (γ) texture is predicted by GIA with taking not only the standard {111}<110> slip system but also non-octahedral slip systems into account. Then the transformed α texture is calculated by DKS, in which a nucleated α prefers to have orientation relationship near the Kurdjumov-Sachs relation with both of two neighboring γ grains. For validation, single pass hot-rolling tests on a C–Si-Mn steel were carried out. The comparison between the predicted and the experimental textures shows that the linked model (GIA & DKS) can lead to a remarkable reproduction of the texture of hot-rolled steel sheets.

1. Introduction

It is important to understand the texture development during hot-rolling processes because the texture strongly influences the material properties, and the texture of hot-rolled steel is known to largely influence (or inherited to) the texture even after cold-rolling and annealing. However, there is still much difficulty in predicting the texture development throughout the manufacturing process, since the influence of deformation and recrystallization of γ as well as phase transformation from γ to α on the texture evolution have not been thoroughly clarified. Among those phenomena, the texture development during deformation has been perhaps best understood. For instance it was reported that the Taylor-type crystal plasticity model, such as "Grain Interaction model (GIA) [1]" in which the interaction of the deformation of neighboring grains can be considered, could make a reasonably quantitative prediction for the deformation texture of cold-rolled fcc alloys [2]. Furthermore, recently, a new variant selection model called "the double K-S relationship (DKS)" has been proposed for the transformation texture simulation [3], which theoretically deals with an important issue in the phase transformation, which of the variants in the orientation relation (24 for K-S relation) are preferentially selected. DKS assumes that α nucleating on γ grain boundaries prefers to have orientation relationship near the Kurdjumov-Sachs relation with two neighboring γ grains at the same time. It was then reported that DKS could lead to an excellent reproduction of experimental transformation textures of low carbon steel [3]. In this research, GIA and DKS have been linked together to construct the through-process prediction model for hot-rolling processes in which deformed γ directly transforms to α without recrystallization, and the model has been experimentally examined.

2. Experiments

The hot-rolled steel sheets of 0.2%C-1.5%Si-1.45%Mn in chemistry (mass%) were prepared by single pass hot-rolling of 30 % or 50 % in reduction at 930°C as shown in Fig.1. The final thickness of hot-rolled samples was 3.4 mm. To avoid recrystallization of γ, the finish-rolled steel sheets were cooled immediately to 750°C at a rate of about 1000°C s^{-1} to promote precipitation of α. Then they were kept at 400°C for 10 minutes to promote bainite transformation and stabilize the retained γ. Fig.2 shows the OIM image of 50% hot-rolled steel sheet. The sample consisted of the retained γ of around 8% in volume fraction, polygonal ferrite, bainite and a trace amount of martensite. The textures for fcc and bcc phase in the thickness center were examined via XRD, and ODF was calculated using incomplete pole figures and the harmonics method. In this study, the model validation has been made on the assumption that the retained γ has the equivalent textures to that of the whole γ.

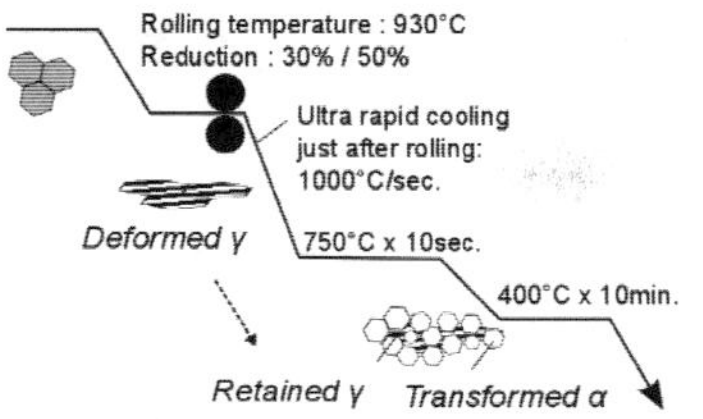

Fig. 1 Schematic representation of experimental hot-rolling.

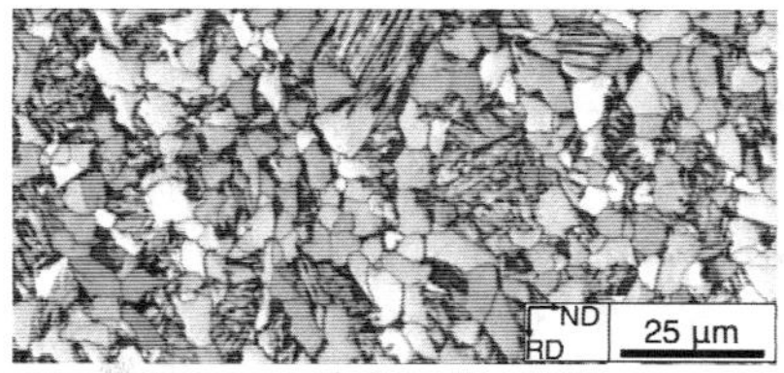

Fig. 2 OIM image of the steel sheet after 50% hot-rolling and transformation. The red filled area corresponds to the retained γ phase.

3. Simulations

3.1. The deformation texture model "GIA" [1]

GIA has been established as a crystal plasticity model in which the interaction of the deformation of neighboring grains is taken into account [1]. In this model, a material is deformed as a statistically suitable set of aggregates comprised of 8 grains as shown in Fig. 3. Although the aggregates follow the sample strain path, each grain within an aggregate is allowed to perform individual relaxations. Accumulating incompatibilities between the grains and their surroundings are compensated by geometrically necessary dislocations (GND), and the active slip systems are chosen such that the total consumed power for dislocation slips and GND generation is minimized. To improve the agreement with the measured rolled textures, a macroscopic plane strain deformation with random shear components were applied [4]. The proposed strain rate tensor takes the form as Eq. (1) [4].

$$\dot{\varepsilon}_{11} = 1 \pm 0.1, \dot{\varepsilon}_{11} = -1 \pm 0.1, \dot{\varepsilon}_{22} = -\dot{\varepsilon}_{11} - \dot{\varepsilon}_{33},$$
$$\dot{\varepsilon}_{13} = 0 \pm 0.6 \dots 0.75, \dot{\varepsilon}_{23} = 0 \pm 0.4, \dot{\varepsilon}_{12} = 0 \pm 0.3 \dots 0.15 \qquad (1)$$

The rolling simulation was conducted assuming the macroscopic plane strain deformation as above with standard {111}<110> as well as non-octahedral {110}<110> and {112}<110> slip systems activated. The non-octahedral slip systems have been reported to be activated due to deformation at elevated temperatures [5-7]. The preceding studies on Al alloys [5-7] have reported that the critical resolved shear stress (CRSS) of non-octahedral slip systems at the temperatures above 70% of their melting points are in the range from 0.75 to 1.4 as the ratio to that for standard {111} slips. Therefore, the CRSS's of {110}<110> and {112}<110> slips were assumed to be equal to that of {111}<110> slip system in this study to better reproduce the development of Cu-orientation {211}<111> during hot-rolling as will be described later.

The hot-rolling texture was thus calculated from the random texture, and its ODF was expanded applying the spherical harmonic method for the next transformation texture calculation.

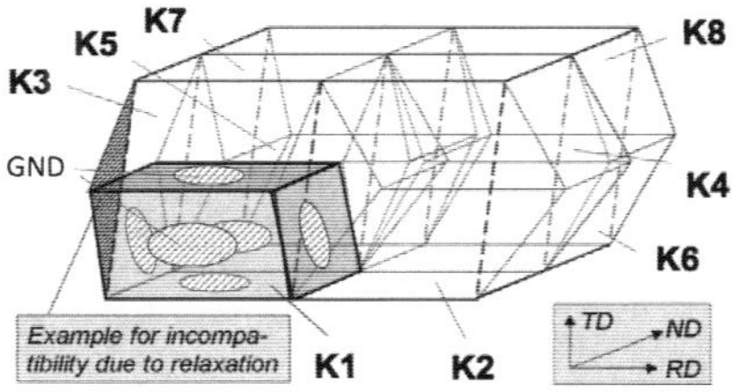

Fig. 3 8-grain aggregate described in the GIA Model.

3.2. The variant selection model "Double K-S relation" [3]

The transformation textures have been simulated assuming that the variants of α holding the K-S relation with two neighbor grains of γ are preferentially selected when α nucleates on the high angle grain boundaries of γ. This selection rule is called the double K-S relation: *DKS* [3]. Because the precise double K-S relation can be achieved only on the special grain boundaries, its probability in ordinary materials is negligibly small. In DKS, the tolerance angle of about 10° for the K-S relation on, at least, one side of the γ grain boundary is assumed, which would increase the probability of having double K-S relation on the general grain boundaries.

The variant selection function *ρ(g)* is expressed with the term due to DKS (the first term) and the one without selection (the second term) as given in Eq. (2). The value of the variant selection parameter ω was determined from the comparison of the measured texture of α and the simulated one based on the measured texture of γ.

$$\rho(g) = \frac{\omega}{N}\sum_{k} f(\Delta g^{-1} \cdot g_k^c \cdot \Delta g \cdot g) + \rho_c(g) \qquad (2)$$

Δg : crystal rotation of transformation. g_k^c : rotational operators for the cubic symmetry group. N : number of variants. ω : intensity of the variant selection.

The transformed texture of α can be related to the parent texture of γ by Eq. (3) with using spherical harmonic series expansion and the variant selection function of Eq. (2) [8]. ${}^{\alpha}C_{\lambda}^{\mu\nu}$ and ${}^{\gamma}C_{\lambda}^{\mu\nu}$ are the series expansion coefficients of the ODF's of α and γ respectively. $\dot{T}_{\lambda}^{\mu s}$ and $\dot{A}_{\lambda_1}^{m_1\mu}$ are the simmetrically invariant functions for cubic crystal symmetry and related coefficients [8]. $(\lambda_1\lambda_2 mr|\lambda s)$ and $\{\lambda_1\lambda_2\nu_1\nu_2|\lambda\nu\}$ are Clebsh-Gordan coefficients and related coefficients defined by Bunge et al. [8] for the symmetrically invariant functions.

$$ {}^{\alpha}C_{\lambda}^{\mu\nu} = \sum_{\lambda_1=0}^{\infty}\sum_{\mu=1}^{M(\lambda_1)}\sum_{\nu_1}^{N(\lambda_1)} {}^{\gamma}C_{\lambda_1}^{\mu\nu_1}\left[\sum_{\lambda_2=0}^{\infty}\sum_{\nu_1=1}^{N(\lambda_2)}\sum_{m,s=-\lambda}^{+\lambda}\rho_{\lambda_2}^{r\nu_2}\dot{A}_{\lambda_1}^{m_1\mu}(\lambda_1\lambda_2 mr|\lambda s)\{\lambda_1\lambda_2\nu_1\nu_2|\lambda\nu\}\,\dot{T}_{\lambda}^{\mu s *}(\Delta g)\right] \quad (3)$$

$$ s = m + r,\ |\lambda_2 - \lambda_1| \le \lambda \le |\lambda_1 + \lambda_2|,\ s \le \lambda \quad (4)$$

$\rho_{\lambda_2}^{r\nu_2}$ are the expansion coefficients of $\rho(g)$ by the invariant function for crystal symmetry $\dot{T}_{\lambda}^{\mu s}$, which can be given as Eq.(5). Thus, the α texture ${}^{\alpha}C_{\lambda}^{\mu\nu}$ can be calculated via Eq.(3) if ${}^{\gamma}C_{\lambda}^{\mu\nu}$, Δg and $\rho(g)$ are given. For the linked prediction of GIA and DKS, the ${}^{\gamma}C_{\lambda}^{\mu\nu}$'s were determined by GIA as aforementioned.

$$ \rho_{\lambda_2}^{r\nu_2} = \frac{\omega}{N}\sum_{\mu_2=1}^{M(\lambda_2)} {}^{\gamma}C_{\lambda_2}^{\mu_2\nu_2}\left[\sum_{k}\dot{T}_{\lambda_2}^{\mu_2 r}(\Delta g^{-1}\cdot g_k^C\cdot\Delta g) - \frac{1}{24}\sum_{i,k}\dot{T}_{\lambda_2}^{\mu_2 r}(\Delta g^{-1}\cdot g_k^C\cdot\Delta g\cdot g_i^C)\right] \quad (5)$$

4. Verification of the model

4.1. Deformation texture simulation

Figures 4 and 5 show the experimental ODF and β fiber intensity of retained γ and those predicted by GIA for hot-rolling. The measured texture of γ exhibits a rolling texture typical of cold-rolled pure Cu. It is also seen that the GIA has successfully reproduced the development of deformation textures in γ quantitatively. In particular, using both of the octahedral as well as non-octahedral slip systems, the agreement is excellent so that the predicted height of major texture components closely matches to the experimental ones. Without the non-octahedral slip being taken into account, a somewhat poorer agreement between the experimental and predicted textures has resulted as shown in Fig. 5.

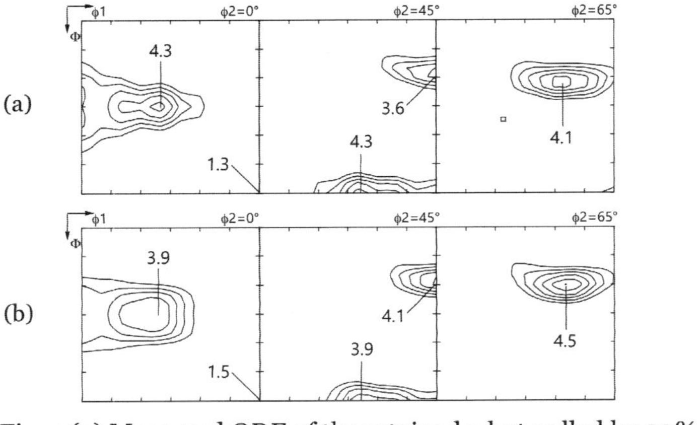

Fig. 4 (a) Measured ODF of the retained γ hot-rolled by 30% in reduction and (b) the predicted textures with octahedral and non-octahedral slip system. (Levels: 2.0,2.5,3.0...)

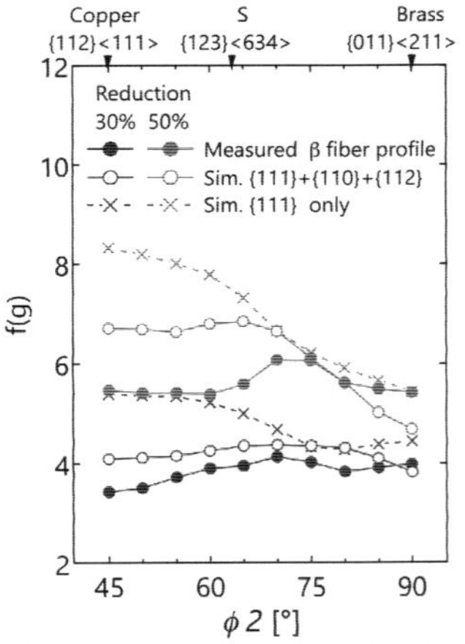

Fig.5 Comparison of the β fiber intensities between the measured and the simulated textures.

4.2. The Linked simulation results

Figures 6 and 7 present the comparison of the measured α textures and the simulated ones, which is calculated by the linked model of GIA and DKS. The measured texture has the typical transformation texture of α that is arisen from the unrecrystallized deformed γ. As

also seen in this figure, the calculated texture based on the deformation texture of γ by GIA is in a surprisingly good agreement with the measured α texture. The deviation from the experimental texture in ODF is as small as 0.5 or less, while the parameter of variant selection, ω, has been determined to be 0.8 to minimize the deviation. In Fig. 7 are shown the BCC α fiber plots, which compare the profiles of the major component in the experimental texture and the ones predicted with and without DKS. The simulation result without the variant selection rule is much weaker than the experimental one, while the adoption of DKS leads to a better reproduction of the experimental ones as abovementioned.
Furthermore, the comparative simulation in which non-octahedral slips were not taken into account, has led inadequate prediction results also for the transformation texture. A disagreement that much higher intensity of {211}<011> is predicted by the simulation without non-octahedral slip system seems to be caused by an improper prediction of γ texture with a too strong copper component. Introducing of non-octahedral slips could avoid such an excessive development of the copper component. However the CRSS of {112}<110> system has been reported to strongly affect the intensity of the copper component as well as that of the brass component [5]. For improving accuracy of prediction, the accurate values of CRSS for non-octahedral slips would be required. Since the strength of the variant selection according to DKS is determined by the parent γ texture, further improvement in the deformation texture simulation in hot-rolling would naturally lead to more accurate prediction of the transformation texture of hot-rolled steel sheets.

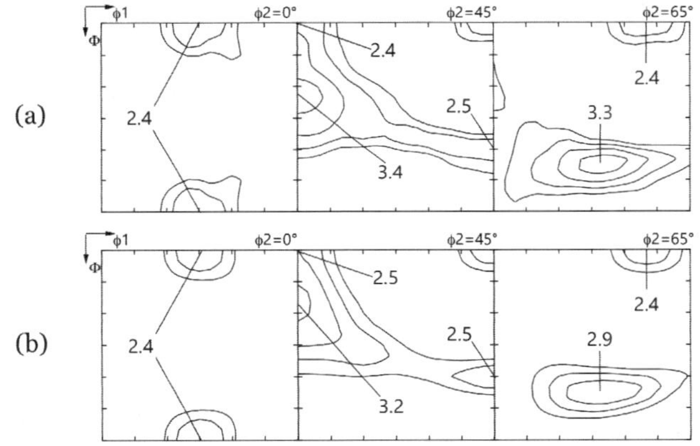

Fig.6 (a) Experimental textures of the transformed α in hot-rolling steel sheet of 30% reduction, and (b) the prediction by the linked simulation of GIA and DKS. (Levels: 1.5,2.0,2.5...)

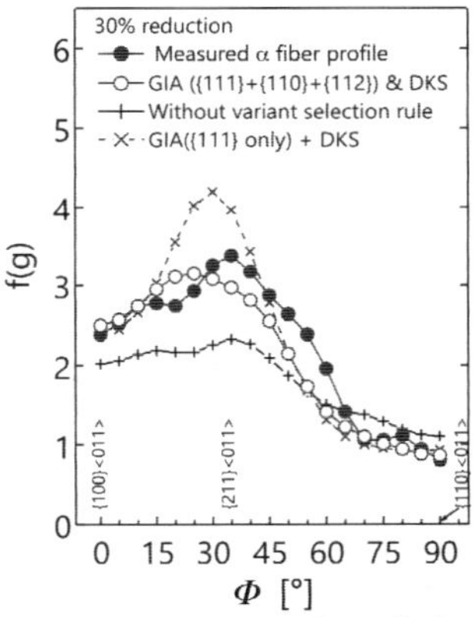

Fig.7 The measured and the simulated profiles of the BCC α fiber.

5. Conclusions

The linked model of GIA and DKS could provide us with a remarkable reproduction or prediction of the transformed α texture in hot-rolled steel sheets in the case where the deformed γ does not recrystallize before phase transformation. Incorporating non-octahedral slip system into the model of deformation texture development could lead to a better agreement between the measured and predicted textures in hot-rolled steel sheets. These results indicate the possibility of the through-process texture modeling of hot-rolling processes in steel, although further investigation is required, e.g., for the CRSS of non-octahedral slips as well as recrystallization texture development in γ.

References

[1] M. Crumbach, G. Pomana, P. Wagner, G. Gottstein 2001 *Proc. 1st Rex&GG* (Berlin), 1053
[2] P. Van Houtte, S. Li, M. Seefeldt and L. Delannay 2005 *Int. Journal of Plasticity* 21, 589
[3] T. Tomida, M. Wakita, M. Yoshida and N. Imai 2008 *Proc.15th ICOTOM* (Pittsburg),325
[4] O. Engler, 2002 *Advanced Engineering Materials* 4, 181
[5] B. BACROIX and J. J. JONAS 1988 *Textures and Microstructures*, Vols. 8 & 9, 267
[6] Cl. MAURICE and J. H. DRIVER 1997 *Acta mate.* , Vol. 45, No. 11, 4639
[7] F. Perocheau and J.H. Driver 2002 *Int. Journal of Plasticity*, 18, 185.
[8] H. J. Bunge, M. Humbert and P. I. Welch 1984 *Textures and Microstructures*, Vol. 6, 81.

IOP Conf. Series: Materials Science and Engineering **82** (2015) 012058 doi:10.1088/1757-899X/82/1/012058

Effect of re-austenitization on the transformation texture inheritance

A Kaijalainen[1], P Suikkanen[2] and D A Porter[1]

[1] Centre for Advanced Steels Research, Materials Engineering, University of Oulu, Finland

[2] Ruukki Metals Oy, Raahe, Finland
E-mail: antti.kaijalainen@oulu.fi

Abstract. Bainitic-martensitic microstructures produced by direct quenching austenite subjected to different degrees of pancaking have been re-austenitized and quenched to fully martensitic structures in order to investigate the effect of prior texture on the final martensite texture. Three different prior austenite pancaking states varying from convex-like to highly pancaked were investigated using an ultrahigh-strength strip steel hot rolled with various finish rolling temperatures followed by direct quenching. Microstructures were characterized using FESEM and transformation texture analysed using FESEM-EBSD at the strip surface, quarter-thickness and mid-thickness positions. The results show that an increase in rolling reduction below the non-recrystallization temperature increases the intensities of ~{554}<225>$_\alpha$ and ~{112}<110>$_\alpha$ texture components in the ferrite along the strip mid-thickness and of the ~{112}<111>$_\alpha$ component at the surface. The re-austenitization of the materials at 910°C for 30 min led to an inheritance of the same components from the parent specimens, but also increased the intensity of {001}<110>$_\alpha$, {110}<110>$_\alpha$ and {011}<100>$_\alpha$ components.

1. Introduction

Texture is known to influence the mechanical properties of steels [1-3]. It also well established that hot rolling parameters have a pronounced effect on the transformation textures of ultrahigh-strength steels [1-4]. Furthermore, in hot-rolled accelerated cooled or quenched steels, a memory effect has been observed by the variant selection of the Kurdjumov-Sachs (K-S) orientation relationship during the phase transformation, i.e. the original texture is partly retained after re-austenitization and cooling [4-6]. However, the effect of the degree of deformation in the non-recrystallization regime of austenite on the transformation texture inherence at different depths in the thickness direction is not widely covered. The aim of the present paper is to report the effects of total reduction in the non-recrystallization regime (R_{tot}) on the transformation texture of an ultrahigh-strength low-alloyed steel when quenched from as-rolled and re-austenitized material.

2. Experimental

In these experiments, 210 mm thick continuously cast slabs were thermomechanically rolled to a final strip thickness of 6 mm and direct quenched to room temperature at a rate of ~50-70 °C/s to produce three different austenite morphologies. The chemical composition of the steel was 0.1C-0.2Si-1.1Mn-1.2Cr-0.15Mo-0.03Ti-0.002B (wt.%). For comparison purposes, direct quenched material was also re-austenitized at 910 °C for 30 min and water quenched to room temperature at ~50-70 °C/s. The six investigated materials are given in Table 1.

Table 1. Investigated materials and austenite morphologies.

As-rolled	Re-austenitized
1Conv (Convex-like austenite)	1ReA (Equiaxed austenite)
2Panc (Pancaked austenite)	2ReA (Equiaxed austenite)
3HiPanc (Highly pancaked austenite)	3ReA (Equiaxed austenite)

General characterization of the transformation microstructures was performed with the aid of both a VK-X200, Keyence laser scanning microscope (LSCM) and a Σigma, Zeiss field emission scanning electron microscope (FESEM) using picric acid and nital etching. Electron backscatter diffraction (EBSD) measurements and analyses were performed using the EDAX-OIM acquisition and analysis software. The FESEM for the EBSD measurements were made at 15 kV with a step size of 0.10 μm. Prior austenite grain sizes (PAGS) were measured as described in Ref. [7].

3. Results and discussion

3.1. Austenite morphologies

The austenite morphologies of the specimens are shown in Figure 1. In the as-rolled specimens 1Conv, 2Panc and 3HiPanc prior austenite grain size (PAGS) were 9.6 μm, 8.1 μm and 8.0 μm and total reduction in the non-recrystallization regime (R_{tot}) were ~38%, ~52% and ~66%, respectively. In the re-austenitized specimens (1ReA, 2ReA and 3ReA) austenite morphologies were equiaxed and average PAGS were ~13.3 μm. The degree of R_{tot} did not affect the uniformity or magnitude of the PAGS. A typical re-austenized specimen is shown in Figure 1d.

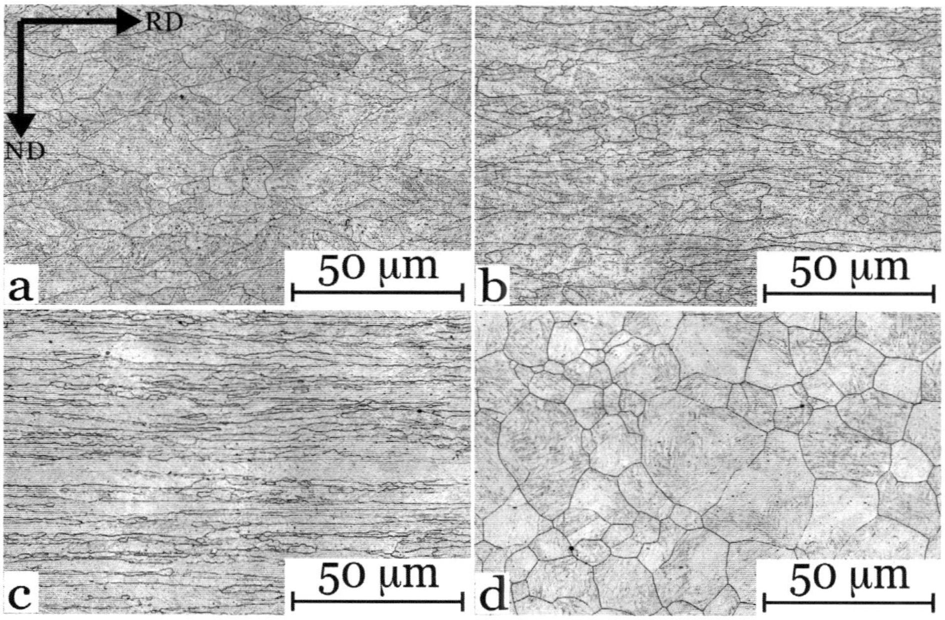

Figure 1. Prior austenite grain boundaries in 1Conv (a), 2Panc (b), 3HiPanc (c) and 1ReA (d).

3.2. Transformation texture

The orientations of importance in the present materials are found in the ϕ_2=45° section of Figure 2. Transformation textures were determined at the sheet subsurfaces, ¼-thicknesses and mid-thicknesses in as-rolled and re-austenitized specimens. Only surface and mid-thickness ODF sections are shown in Figure 3 (as-rolled) and Figure 4 (re-austenitized) as no difference was found between the ¼-thickness and mid-thickness specimens.

In the strip subsurface layer, increasing R_{tot} from 38% to 66% decreases the intensity of the {001}<110>$_\alpha$ component, indicating that austenite recrystallization was no longer taking place at this location during rolling due to surface cooling. Concurrently, the ~{112}<111>$_\alpha$ and ~{110}<111>$_\alpha$ - {110}<112>$_\alpha$ components increase in intensity quite significantly, Figures 3a, 3c, and 3e. Wittridge and Jonas [8] have reported, in the γ-to-α transformation, the shear components {111}<211>$_\gamma$ and {112}<110>$_\gamma$ lead to the formation of the shear components

{112}<111>$_\alpha$, {110}<112>$_\alpha$ and {110}<111>$_\alpha$, respectively. The high intensities of the bcc shear components indicate that the partially ferritic surface has been extensively deformed by shear.

In the specimen 1Conv, the strongest centerline texture component was the {001}<110>$_\alpha$, Figure 3b. This component is generally formed by the transformation of the fcc cube component ({001}<001>$_\gamma$), indicating that the centers of the plates had undergone recrystallization prior to transformation. In specimens 2Panc and 3HiPanc, the effect of increasing R_{tot} led to a decrease in the intensity of the {001}<110>$_\alpha$ component and to a significant increase in the intensities of the ~{554}<225>$_\alpha$, ~{112}<110>$_\alpha$ and ~{112}<131>$_\alpha$ components, Figures 3c and 3f. The latter are formed by transformation from the fcc Cu ({112}<111>$_\gamma$), Br ({110}<112>$_\gamma$) and Goss ({110}<001>$_\gamma$) rolling components, indicating that considerable "pancaking" had in fact taken place.

At the surface, the re-austenitized specimens contained the same texture components as the as-rolled starting materials, but with weaker intensities. Re-austenitizing 3HiPanc led to a decrease in f(g) from~15 to ~3 in the texture component close to {112}<111>$_\alpha$, Figures 3e and 4e. After Ray and Jonas [1], decreased texture caused from the multiplicity of variants on transformation. Also at the centerline {112}<131>$_\alpha$ and {554}<225>$_\alpha$ components are inherited reason of K-S orientation relation from the starting material, but specimen 2ReA contained significant intensities of {110}<110>$_\alpha$ and {011}<100>$_\alpha$ components and specimen 3ReA {001}<110>$_\alpha$ component, Figures 4d and 4f, respectively. Ray et al. [4] have been reported that the recrystallized austenite form a strong cube component {001}<001>$_\gamma$ and its, twin {122}<212>$_\gamma$, which can derive the {001}<110>$_\alpha$, {110}<110>$_\alpha$ and {011}<100>$_\alpha$ components in the phase transformation. These results indicate that intense texture in as-rolled material can intensify the texture in subsequently re-austenitized and quenched material.

In the future, re-austenitization followed by quenching can be used to study the effect of texture on strength and toughness without the uncertainties introduced by variations in chemistry, grain structure and inclusion content for example, because the chemical composition and prior austenite condition are the same in the each specimen (1ReA, 2ReA and 3ReA).

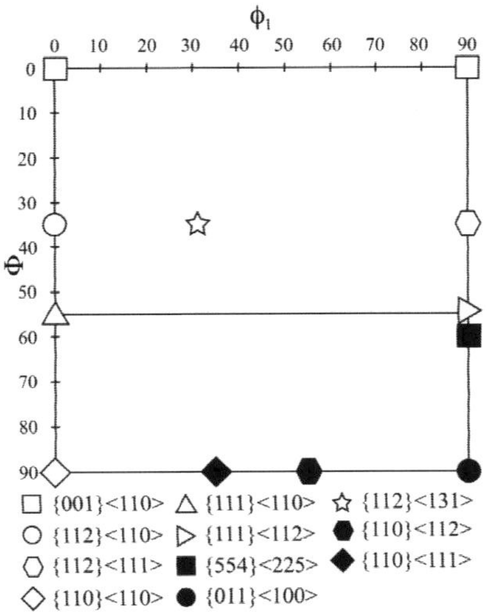

Figure 2. Important orientations of the bcc structure in the ϕ_2=45° section.

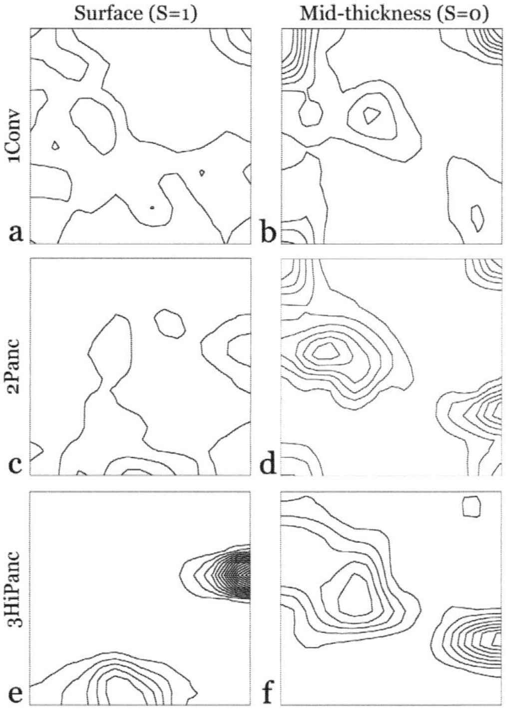

Figure 3. $\phi_2=45°$ sections of ODFs at the subsurface (S=1) and mid-thickness (S=0) positions in as-rolled specimens. 1Conv (a-b), 2Panc (c-d) and 3HiPanc (e-f) (Levels: 1, 2, 3...).

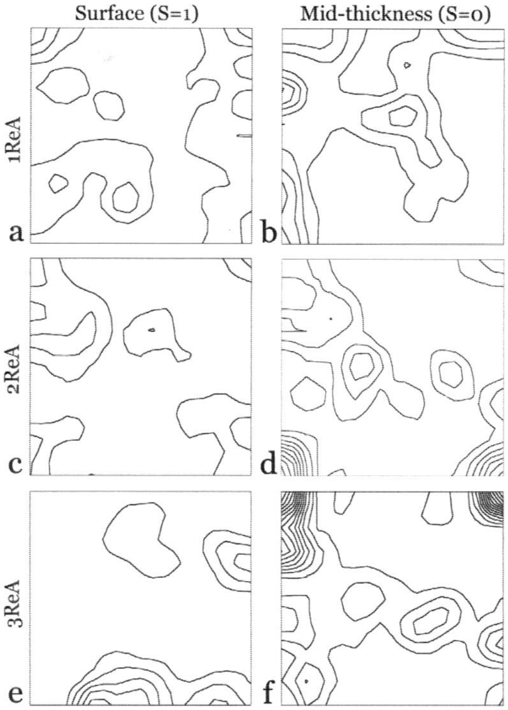

Figure 4. $\phi_2=45°$ sections of ODFs at the subsurface (S=1) and mid-thickness (S=0) positions in re-austenitized specimens. 1ReA (a-b), 2ReA (c-d) and 3ReA (e-f) (Levels: 1, 2, 3...).

4. Summary

The effects of total reduction the non-recrystallization regime (R_{tot}) on the transformation texture of an ultrahigh-strength low-alloyed steel has been investigated. The observations can be summarized as follows:

1. The as-rolled direct quenched condition consisted mainly of auto-tempered martensite and lower bainite while the re-austenitized and quenched material was fully martensitic.
2. An increase in R_{tot} strengthens the intensities of the ~{554}<225>$_\alpha$ and ~{112}<110>$_\alpha$ texture components at the ¼- and mid-thicknesses and the shear texture ~{112}<111>$_\alpha$ at the strip surface.
3. Re-austenitized specimens contained the same texture components as the as-rolled starting materials, but in weaker intensities. Texture components {001}<110>$_\alpha$, {110}<110>$_\alpha$ and {011}<100>$_\alpha$ were dominant at the mid-thickness.

Acknowledgments

The financial support of the Finnish Funding Agency for Technology and Innovation (Tekes) in the Breakthrough Steels and Applications Program of the Finnish Metals and Engineering Competence Cluster (FIMECC Ltd) is gratefully acknowledged.

References

[1] Ray R K and Jonas J J 1990 *Int. Mater. Rev.* **35** 1
[2] Ray R K et al 1994 *ISIJ Int.* **34** 927
[3] Joo M S et al 2012 *Mater. Sci. Eng.* A **556** 601
[4] Ray R K et al 1992 *ISIJ Int.* **32** 203
[5] Brückner G and Gottstein G 2001 *ISIJ Int.* **41** 468
[6] Lischewski I et al 2009 Microstructure and Texture in Steels eds. A Haldar, S Suwas and D. Bhattacharjee (London: Springer) chapter 24 pp 409-413
[7] Higginson R L and Sellars C M 2003 Worked Examples in Quantitative Metallography (London: Maney Publishing)
[8] Wittridge N J and Jonas J J 2000 *Acta Mater.* **48** 2737

IOP Conf. Series: Materials Science and Engineering **82** (2015) 012059 doi:10.1088/1757-899X/82/1/012059

Fully automated orientation relationship calculation and prior austenite reconstruction by random walk clustering

E Gomes[1] and L A I Kestens[1,2]

[1] Ghent University, Department of Materials Science and Engineering, Technologiepark 903, 9052 Gent, Belgium

[2] Delft University of Technology, Department of Materials Science and Engineering, Mekelweg 2, 2628 CD Delft, The Netherlands

E-mail: Edgar.Gomes@UGent.be

Abstract. Two new methods, one for determining the experimentally observed Orientation Relationship (OR) and another for reconstructing prior austenite phase, are proposed. Both methods are based on the angular deviation of the OR at the grain boundaries. The first algorithm identifies the optimum OR using the misorientation distribution of the entire scan i.e. without manual selection of parent grains. The second algorithm reconstructs the parent phase using a random walk clustering technique that identifies groups of closely related grains based on their angular deviation of the OR.

1. Introduction

Most of the steelmaking process occurs at elevated temperature which, for most steel grades, implies that part of the processing occurs in the austenitic phase. The characterization of the high-temperature austenitic phase is of great value to understand and control the microstructure evolution during the entire production chain. But the direct observation of the austenitic phase is not trivial. It can only be done at elevated temperatures, and therefore, it can only be observed within very specialized *in-situ* equipment like Electron Backscatter Diffraction (EBSD) with a hot stage[1]. Moreover, such sophisticated measurements are always done under simulating laboratory conditions, which are not necessarily representative of the real process.

Alternatively, indirect measurements are done at the room temperature. If, at this temperature, the phase transformation has already completed then the prior austenite microstructure can observed by: (a) optical microscopy with special etching techniques that revel the original grain boundaries[2], (b) reconstruction of Orientation Image Mapping (OIM) obtained with standard EBSD equipment[3, 4, 5, 6, 7]. The latter method identifies and groups product grains derived from a single parent (austenitic) grain. This technique depends on the presence of a specific OR between parent and product phases, and therefore, it only applies to diffusionless phase transformations. This work introduces two new methods aiming to improve the current-state-of-the-art in prior austenite reconstruction.

2. Orientation relationship calculation

One of the main characteristics of martensitic transformations is the presence of a specific OR between parent and product phases. Although there are some theoretical orientation relationships proposed such as the Kurdjumov-Sachs (K-S) or Nishiyama-Wassermann (N-W) correspondences for steels, a more accurate OR can only be derived from experimental data. The available methods[8, 9] for OR identification depend on the manual selection of parent grains, which is time consuming and in some cases, depending on the microstructure, impractical.

The method proposed here can derive the OR from an orientation scan in a fully automated manner and it is based on the misorientation between neighboring grains. There are two possible situations where the OR can be observed. The first one is between a parent grain and its product grain (when still a residual fraction of the parent phase is present in the product microstructure), cf. (1). The second case is between product grains sharing the same parent and involves a double OR transformation, cf. (2). In the ideal conditions that the proposed OR is valid, the angular deviation (ΔOR) is equal to 0.

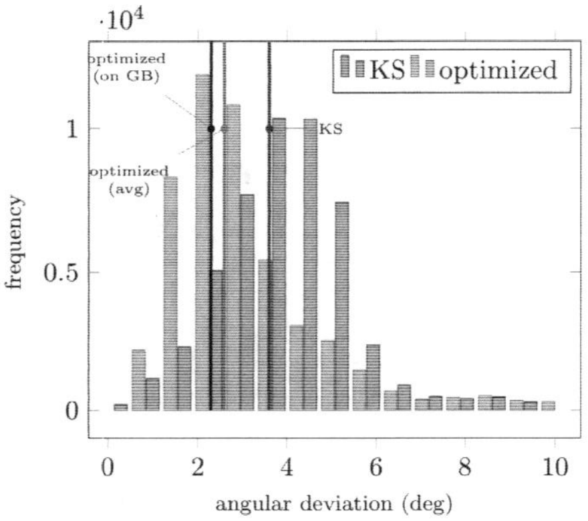

Figure 1. Histogram of angular deviation for KS OR and optimized OR.

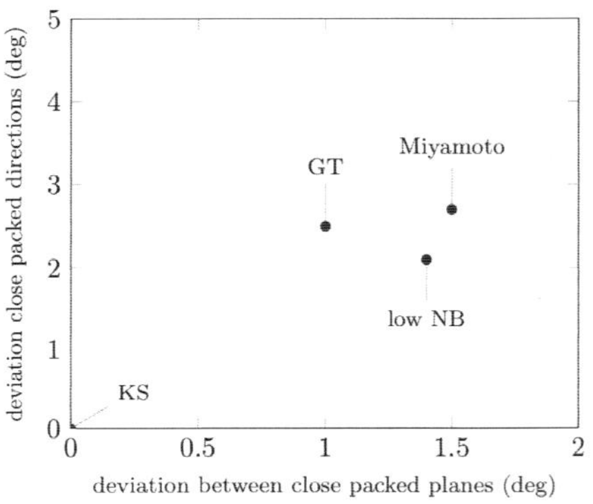

Figure 2. Comparison between optimized OR for low Nb sample and reference value (Miyamoto)[8].

The algorithm finds the optimized OR transformation ($T^{\alpha'\leftrightarrow\gamma}$) that minimizes the average angular deviation for all grain boundaries or for a selected subset of them. The number of grain boundaries exhibiting this OR should be vastly larger than the number of prior parent grain boundaries. This is a reasonable assumption since the martensitic phase transformation produces grain refinement and all the new grain boundaries have misorientations that obey the OR transformation.

$$T_m^{\alpha'\leftrightarrow\gamma}\left(Og^{\alpha'}\right)\left(Og^{\gamma}\right)^{-1} = \Delta OR \qquad T_m^{\alpha'\leftrightarrow\gamma}T_n^{\gamma\leftrightarrow\alpha'}\left(Og_a^{\alpha'}\right)\left(Og_b^{\alpha'}\right)^{-1} = \Delta OR \qquad (1, 2)$$

Two strategies were employed in order to calculate the grain boundary misorientations. One is based on the average orientation of both neighboring grains. The other one is based on the orientation in the immediate vicinity of the grain boundary, where only the first layer of orientations at each side of the grain boundary is selected. Slightly different values of OR are obtained by the two strategies due to plastic strain accommodation during phase transformation and associated orientation gradients in bulk of the grains[8].

This new method was validated with a low Nb steel alloy with carbon content around 0.05%. Figure 1 compares the distribution of angular deviation using the theoretical KS OR and the optimized OR, it can be noticed that the distribution shifts to the left, i.e. reducing the

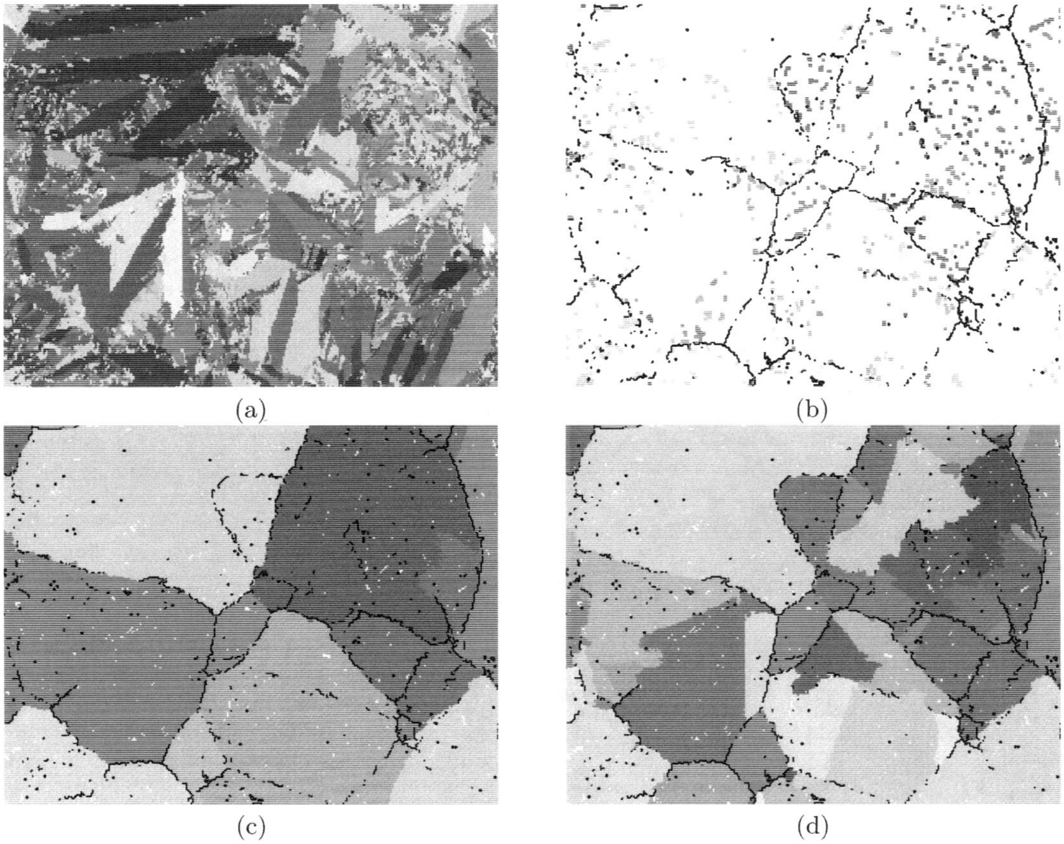

Figure 3. Validation of reconstruction algorithm on Fe-Ni alloy: (a) initial OIM before reconstruction, (b) retained austenite and expected prior austenite GB, (c) reconstruction after the first iteration and (d) second and final step of reconstruction.

overall error, when the optimized OR is used. Better results are obtained when considering orientations in the immediate vicinity of the grain boundary compared to considering grain averaged orientations. Figure 2 compares the optimized OR for the low Nb sample with two theoretical ORs and one reference value[8] that contains comparable amount of carbon. The optimized OR is in good agreement with the reference value.

3. Parent phase reconstruction

A new method, equally based on grain boundary misorientations, is proposed for the parent phase reconstruction. The key concept of this method is the use of a graph clustering algorithm to find groups of closely related grains that originate from a single parent grain. The Markov Clustering algorithm[10, 11] is used because of the following features:

- It is based on the probabilities of random walks through the graph.
- It identifies natural cuts and clusters in the graph without the need to specify threshold values of allowable angular deviations or the necessity to specify a predefined number of clusters

- There is only one parameter that controls the cluster coarsening.
- Excellent noise tolerance.
- Fast calculation time.

In order to use the clustering algorithm, a graph has to be derived from the OIM in the following way: (a) each grain, product or parent, becomes a node in the graph, (b) each grain boundary becomes an edge connecting two neighboring nodes (grains) and (c) each edge receives a value based on the equation (1) or (2), which determines the probability of the two neighboring grains being products of the same parent. Once the graph is constructed, the clusters are determined using the highest level of coarsening. Then each cluster forms a parent grain and its orientation is calculated by an optimization algorithm that minimizes the average angular deviation using the equation (1).

The first iteration described above may result in some of the clusters containing more than one parent grain because of the initial coarsening level. These grains can be identified by its higher average angular deviation, and therefore, they should be further refined by running an extra iteration step with reduced coarsening level. The refinement procedure should be repeated many times until all parent grains have low average error.

A Fe-Ni 29% sample was used to validate the proposed reconstruction method. The OIM after the martensitic phase transformation with still 10% retained austenite is shown in the figure 3a. The expected parent microstructure is visualized in figure 3b by the retained austenite and the traces of high angular deviation ($> 10°$). Figure 3c shows the reconstruction after the first iteration, showing that some reconstructed grains incorporate more than one parent grain. The final microstructure is shown in figure 3d after an extra step of refinement.

A careful look at the figure 3d shows that some unexpected features are visible at twin boundaries in parent grains and some grains are over refined. The current implementation of the reconstruction software is quite simple and does not include any special rule for treating this type of boundaries or for re-merging over-divided clusters.

4. Conclusion

Tolerance error reduction and noise immunity are two fundamental strategies for the success of prior austenite reconstruction. The use of optimized ORs reduces the overall error but the identification by previous techniques requires manual selection of parent grains. The method proposed here identifies an optimized OR, whereby the entire orientation scan is considered and the result is obtained without user intervention.

Markov clustering seems a promising technique for parent grain reconstruction due its noise tolerance and fast calculation time. The present results are satisfactory considering the initial stage of development and the simple implementation. Further work on the special treatment of twin boundaries and cluster re-merging will be done.

References

[1] Seward G G E, Prior D J, Wheeler J, Celotto S, Halliday D J M, Paden R S and Tye M R 2002 *Scanning* **24** 232–240 ISSN 1932-8745
[2] Bechet S and Beaujard L 1955 *Rev. Met* **52** 830–836
[3] Cayron C, Artaud B and Briottet L 2006 *Materials Characterization* **57** 386–401 ISSN 1044-5803
[4] Miyamoto G, Iwata N, Takayama N and Furuhara T 2010 *Acta Materialia* **58** 6393–6403 ISSN 1359-6454
[5] Abbasi M, Nelson T W, Sorensen C D and Wei L 2012 *Materials Characterization* **66** 1–8 ISSN 1044-5803
[6] Tari V, Rollett A D and Beladi H 2013 *Journal of Applied Crystallography* **46** 210–215 ISSN 0021-8898
[7] Bernier N, Bracke L, Malet L and Godet S 2014 *Materials Characterization* **89** 23–32 ISSN 1044-5803
[8] Miyamoto G, Takayama N and Furuhara T 2009 *Scripta Materialia* **60** 1113–1116 ISSN 1359-6462
[9] Humbert M, Blaineau P, Germain L and Gey N 2011 *Scripta Materialia* **64** 114–117 ISSN 1359-6462
[10] Dongen S M v 2001 *Graph clustering by flow simulation* Ph.D. thesis Universiteit Utrecht
[11] Schaeffer S E 2007 *Computer Science Review* **1** 27–64 ISSN 1574-0137

IOP Conf. Series: Materials Science and Engineering **82** (2015) 012060 doi:10.1088/1757-899X/82/1/012060

Quantitative prediction of transformation texture in steel by Double Kurdjumov-Sachs relation

T Tomida[1], M Wakita[2], S Vogel[3], HRZ Sandim[4]

[1] Technical R & D Bureau, Nippon Steel & Sumitomo Metal Corp., Amagasaki, Japan; now at Amagasaki Unit, Nippon Steel & Sumikin Technology, Amagasaki, Japan

[2] Technical R & D Bureau, Nippon Steel & Sumitomo Metal Corp., Amagasaki, Japan

[3] Los Alamos Neutron Science Center, Los Alamos National Laboratory, Los Alamos, NM, USA

[4] Escola de Engenharia de Lorena, Universidade de São Paulo, Lorena, Brazil
E-mail: tomida-toshiro@nsst.jp

Abstract. The transformation texture prediction by so-called double K-S relation is described. Various types of transformation textures such as the one in hot-rolled steel sheets and those of texture memory in hot-rolled steel and cold-rolled pure iron have been reported to be able to be quantitatively predicted by this variant selection rule. Recently such an experimental investigation has been extended to the texture memory in ECAP-processed pure iron. Although the type of texture as well as the symmetry of samples in ECAP iron is very different from the previous materials, the investigation has clearly indicated that the double K-S relation should indeed be the mechanism governing variant selection on the phase transformation in iron and steel.

1. Introduction

To develop new products in steel industry, it is desired to develop through-process texture simulation methods, since crystallographic texture largely influences the formability as well as elastic and magnetic properties of steels. However, the development of such a prediction method has been hampered by insufficient understanding of the mechanisms that govern the texture change during the manufacturing processes.

One of the phenomena on texture prediction that have long been investigated but not sufficiently understood is the transformation texture. During the phase transformation, the texture of parent phase is inherited to the texture of product phase due to the orientation relationship such as Kurdjumov-Sachs (K-S) relation [1] for the austenite (γ) to ferrite (α) transformation, $\{111\}_{\gamma}//\{011\}_{\alpha}$ and $<110>_{\gamma}//<111>_{\alpha}$. Then according to the orientation relation, the orientations in the parent crystals are rotated to develop the texture in product crystals. Therefore the prediction of transformation texture is seemingly rather straight forward. However, the prediction has been still only qualitative, since it is still unclear which of the variants in the orientation relation (24 for K-S relation) are preferentially selected. Understanding of such a variant selection is certainly indispensable to construct the models for through-process texture prediction. In this respect, we have proposed a variant selection rule that we call the "double K-S relation" [2, 3]. This mechanism relies on the nucleation of the product phase that preferentially occurs on grain boundaries of the parent phase, holding K-S relation or the one near it with adjoining two parent grains at the same time. In other words, the nucleation needs two parent grains having the K-S or near K-S relation. With this variant selection mechanism, the ferrite textures of hot-rolled steel [2] and the texture memory in the hot-rolled steel and cold-rolled pure iron [3] have been quantitatively predicted so far. In this article, the most recent experimental verification of the double K-S relation by observing in-situ the texture memory in the ECAP (equal channel angular pressing) -processed pure iron by pulsed neutron diffractometer HIPPO in Los Alamos Neutron Science Center [4] is described.

2. Texture prediction method and variant selection model

Although the calculation of change in texture during transformation becomes rather complicated if many variants operate at different probabilities, it can be calculated by the equations based on the spherical harmonic expansion of ODF (orientation distribution function) as shown below [5],

$$^{P}C_{\lambda}^{\mu\nu}=\sum_{\lambda_1=0}^{\infty}\sum_{\mu=1}^{M(\lambda_1)}\sum_{\nu 1=1}^{N(\lambda_1)}{}^{M}C_{\lambda_1}^{\ \mu\nu_1}\left[\sum_{\lambda_2=0}^{\infty}\sum_{\nu_2=1}^{N(\lambda_2)}\sum_{m=-\lambda}^{\lambda}\sum_{s=-\lambda}^{\lambda}\rho_{\lambda_2}^{r\nu_2}\dot{A}_{\lambda_1}^{\ m_1\mu}(\lambda_1\lambda_2 mr\,|\,\lambda s)\{\lambda_1\lambda_2\nu_1\nu_2\,|\,\lambda\nu\}\dot{T}_{\lambda}^{\ \mu s}*(\Delta g)\right] \quad (1)$$

$$s=m+r,\quad |\lambda_2-\lambda_1|\le\lambda\le|\lambda_2-\lambda_1|,\quad s\le\lambda, \quad (2)$$

where $^{P}C_{\lambda}^{\mu\nu}$ and $^{M}C_{\lambda}^{\mu\nu}$ are the series expansion coefficients of the ODF's of product and parent phases respectively. The Δg represents the crystal rotation due to the K-S relation, and $\rho_{\lambda 2}^{\mu,\nu 2}$'s are the expansion coefficients of the variant selection function ρ(g).
In this study, ρ(g) is derived from the double K-S relation [2, 3]. To satisfy the double K-S relation, an α particle, which nucleates on the grain boundary having the i-th K-S relation to γ_1 (with orientation g) as shown in Fig.1, has to hold the K-S relation to γ_2 on the opposite side as well. Therefore if such K-S related γ_2 abundantly exists in the parent phase, the possibility for the i-th K-S variant to operate would become large. Hence such probability should be in proportion to the probability to find such K-S related γ_2 in the parent texture. Then, as described elsewhere [3], ρ(g) is derived as follows,

$$\rho(g)=\frac{\omega}{N}\sum_{k}f(\Delta g^{-1}\cdot g_k^c\cdot\Delta g\cdot g)+\rho_C(g), \quad (3)$$

where f(g) is the ODF of the parent phase, and g_k^c's are the 24 rotational operators for the cubic symmetry group. Note that the variant selection is entirely determined by the texture of parent phase.

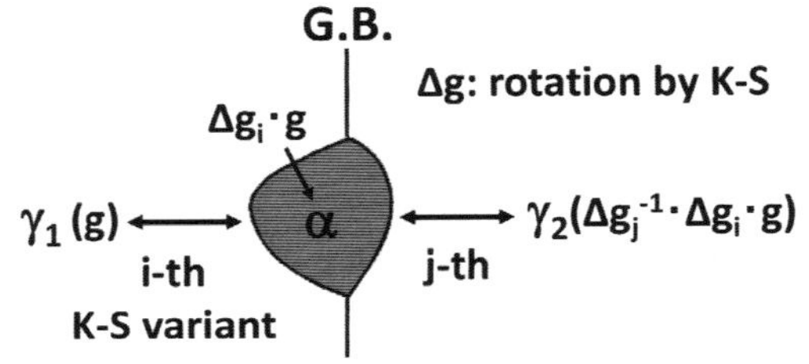

Fig.1 Schematic representation of double K-S relation

It should be also emphasized that we assume at least one of the K-S relations with the two neighboring parent grains to be allowed to deviate up to about 10° from the exact K-S relation. With this deviation allowed, the double K-S relation can occur on ordinary grain boundaries in ordinary parent structures. The proper value of the only parameter ω in Eq. (3) is known to be from 0.6 to unity for the transformation during hot-rolling processes in steel [2,3].

3. Verification using the texture memory in ECAP-processed pure iron

The texture memory effects, in which an initial texture is well retained after two transformations on heating and cooling (α→γ→α in iron and steel), are thought to be ascribed to variant selection. However the selection mechanism has not been clarified thoroughly. If the double K-S relation is a universal property for the phase transformation between α and γ, such a memory effect should be explained thereby. Therefore the texture memory effects in different types of material have been observed by pulsed neutron diffraction and examined based on the double K-S relation. Here, such investigation using ECAP-processed pure iron is described. A rather different texture as well as symmetry of samples in the ECAP iron than before [2,3] particularly makes this investigation significant. The experimental results on this ECAP-processed iron have been published elsewhere [6].

3.1 Experimental

A 99.98% pure iron ingot was hot-forged and machined into a 25 x 25 x 170 mm^3 square bar and processed by ECAP at room temperature using the route B_c. The ECAP die had an internal angle (Φ) of 90°. The pressing speed was 2.5 $mm\cdot s^{-1}$. The von Mises equivalent strain per pass is 1.15, corresponding to a total strain of 9.2 after 8 passes. Then in-situ pulsed neutron diffraction texture measurements were performed on HIPPO at LANSCE using a 8 mm-edge cube cut out from the central part of the ECAP-processed bar. Measurements were performed in vacuum at 50 and 800 °C on heating, at 950 °C after completion of $\alpha \rightarrow \gamma$ transformation and at 800 °C on cooling to measure the entire cycle of texture change during the $\alpha \rightarrow \gamma \rightarrow \alpha$ transformation. Heating and cooling rates were about 10 $°C \cdot min^{-1}$. The dwell time at each temperature to acquire the data for texture analysis was about 5 min. The neutron diffraction data at each temperature were analyzed by simultaneous Rietveld analysis and a harmonics expansion method to reconstruct pole figures and ODF. Note that ODF's were calculated using different coordinates than those in ref. 6.

3.2 Experimental results and comparison with prediction based on the double K-S relation

The observed (111) pole figures of the ECAP-processed iron are shown in Fig. 2. Since the deformed structure by ECAP recrystallizes on heating, a large change is seen between the pole figures at 50 and 800 °C (Figs. 2(a) and (b)). Then it changes again by phase transformations from α to γ and from γ to α as shown in Figs. 2 (b) to (d). Although somewhat weakened, texture memory is apparent in those pole figures with the almost identical peak positions seen in Figs. 2 (b) and (d).

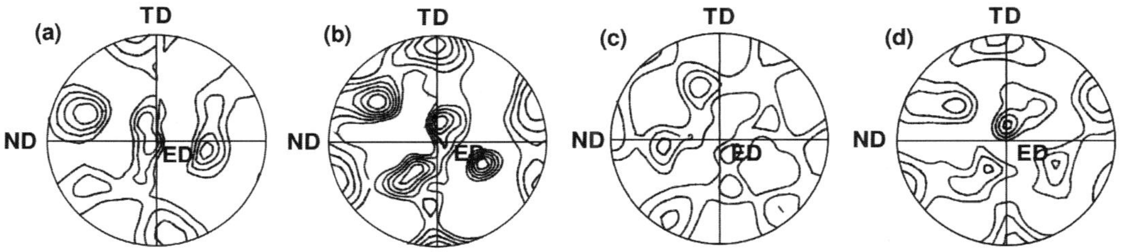

Fig. 2 Experimental (111) pole figures of ECAP-processed pure iron (a) at 50 °C (as deformed), after heating to (b) 800 °C and (c) 950 °C, and (d) after cooling to 800 °C; (a), (b) and (d) are for α, while (c) is for γ. Levels are 1.0, 1.2, 1.4 …...

The ϕ_2=45° sections of corresponding ODF except for as-deformed states are shown in Fig.3 along with the predicted ODF's by the double K-S relation using Eqs. (1) to (3). At 800 °C on heating, major components lie around {100}<510>, {110}<211>, {111}<211>. Then after the first transformation, they change to those around {100}<120>, {110}<110>, {221}<102>, etc., and finally a similar ODF to the one at 800 °C on heating reappears as seen in Figs. 3 (a) to (c), exhibiting that the texture memory certainly exists in the present experiment.

Even more interesting is that although the recrystallized texture is seemingly reproduced after two transformations, this reproduction or the phenomenon of texture memory is not what this word literally means but only a change into a similar (but different) state of texture. As indicated by arrows in Figs. 3 (a) and (c), the component around ϕ_1=150°, Φ=40° and ϕ_2=45°, which exists at 800 °C on heating, is seen to be missing in the final texture. Similar absence of texture components in the final texture has been seen in the other sections of ODF. The predicted textures for the ones after the first and second transformations are shown in Figs. 3(d) and (e) respectively; the values of ω for the first and second transformations are 0.9 and 2.5 respectively. They are calculated from the experimental recrystallized texture in Fig. 3(a); note that for the final texture in Fig. 3(e), transformation calculation was applied to the predicted texture in Fig. 3(d). By comparing with the experimental textures, it is evident that the prediction by the double K-S relation is capable of quantitatively reproducing the texture change during $\alpha \rightarrow \gamma \rightarrow \alpha$ transformation in the ECAP iron, as has been observed in hot-rolled steel and cold-rolled iron [3]. Almost every detail in the experimental texture is observed in the predicted ones including the absence of texture components around ϕ_1=150°, Φ=40° and ϕ_2=45° in the final texture, although slight difference is seen in peak intensity.

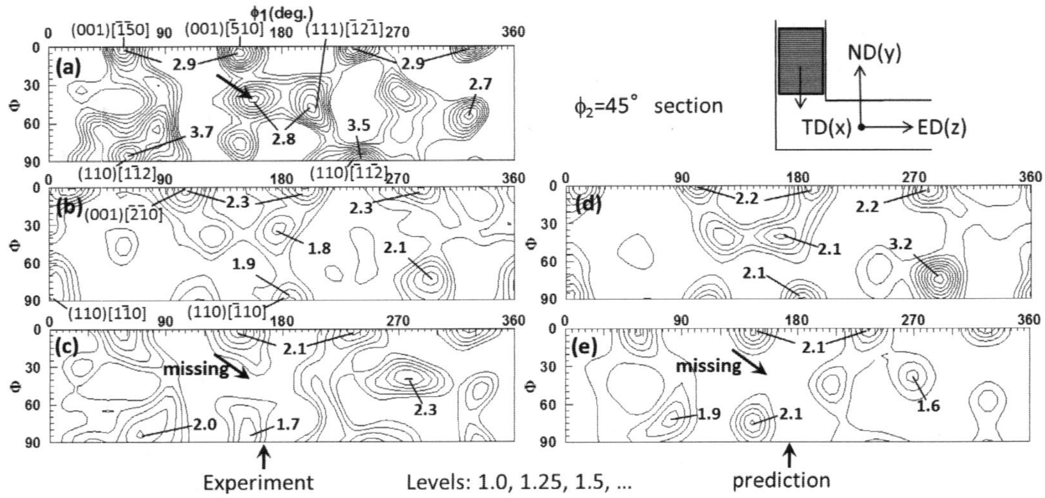

Fig.3 (a to c) Experimental and (d and e) predicted ϕ_2=45° sections of ODF (a) at 800 °C on heating for α, (b and d) at 950 °C for γ and (c and e) at 800 °C on cooling for α.

Therefore, the present analysis of the texture change during the successive transformations in ECAP-processed pure iron indicates that the double K-S relation should indeed be the governing variant selection rule on the phase transformation between α and γ in iron and steel. Finally in Fig. 4 are presented the predicted {111} pole figures corresponding to the Figs.3 (d) and (e), in which a good agreement with the experimental pole figures is seen.

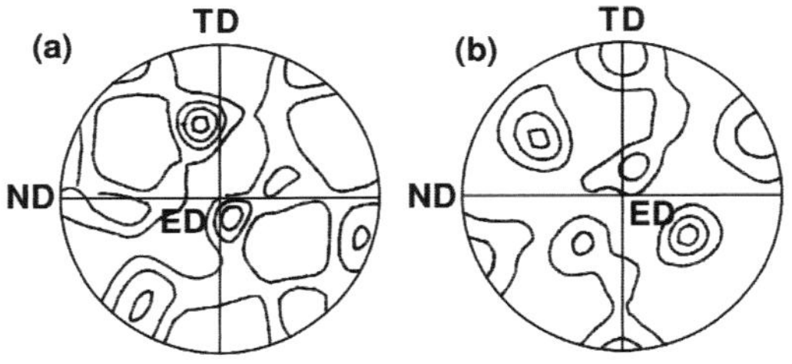

Fig. 4 Predicted (111) pole figures of ECAP-processed iron for (a) γ after heating to 950 °C and (b) α after cooling to 800 °C. Levels are 1.0, 1.2, 1.4 ·····.

4. Conclusions

As presented above, it is most likely that the mechanism governing the variant selection on phase transformation between α and γ in iron and steel should be the one described by the double K-S relation. Therefore, interphase boundary structures and interfacial energy seem to be the most influencing factor to determine the active variants on such transformation.

References

[1] Kurdjomov G and Sachs G 1930 *Z. Phys.* **64** 325
[2] Tomida T, Wakita M, Yoshida M and Imai N 2008 *Proc. 15th ICOTOM* (Pittsburg: Pennsylvania) p 325
[3] Tomida T, Wakita M, Yasuyama M, Sugaya S, Tomota Y and Vogel S C 2013 *Acta Mater.* **61** 2828
[4] Wenk H R, Lutterotti L, Grigull S and Vogel S 2003 *NIMA* **515** 575
[5] Bunge H J, Humbert M and Welch P I 1984 *Textures and Microstructures* **6** 81
[6] Sandim H R Z, Bolmaro R E, Renzetti R A, Sandim M J R, Hartwig K T, Vogel S C, Raabe D 2014 *Metall. Mater. Trans.* A **45A** 4235

Investigation of phases and textures of binary V-Si coating deposited on vanadium-based alloy (V-4Cr-4Ti) using electron backscatter diffraction

P Nowakowski[1], HS Ubhi[2] and S Mathieu[3]
[1]Oxford Instruments, Gometz La Ville, France,
[2]Oxford Instruments, High Wycombe, UK,
[3]Université de Lorraine, Institut Jean Lamour, Vandoeuvre-lès-Nancy, France

Abstract. Barrier coating consisting of binary silicide compounds Si_xV_y were deposited on a V-4Cr-4Ti vanadium alloy substrate. Samples were cycled in a furnace for 122h at 650°C and 1100°C. The electron backscattered (EBSD) combine with X-ray energy dispersive spectrometry (EDS) techniques were employed to identify the phases in the multi-layered coating and to determine growth texture for each phase. The microstructure evolutions occurringduring cycling at 1100°C in the protective coating and the crystal orientation relationships between Si_xV_y were determined.

1. Introduction

The low capture cross-section for thermal neutrons and the capability of vanadium to withstand high temperatures makes vanadium attractive as structural material for fuel cladding either in gas- or sodium-cooled fast reactors in generation IV nuclear systems. However, an oxygen concentration as low as 1 ppm in Na [1] at 550°C is required to avoid oxygen embrittlement of vanadium. Thus, a diffusion barrier for oxygen based on a multilayered silicide coating was developed [2]. It was demonstrated [3] by oxidation tests in air and low-O2 environments that this type of coating is very efficient in preventing embrittlement by oxygen in the expected operating temperature range of 550 to 750°C. Moreover, it was compatible with liquid Na (10 ppm O_2) during 1000h at 550°C. To further valid the capability of this system for cladding application, its behaviour was also evaluated under severe conditions that could arise in case of accidental overheating [4]. In these cases temperatures as high as 1000°C in liquid Na can be reached. The microstructure evolution and origin of failure in these conditions has to be fully analysed and understood.

The electron backscattered technique (EBSD) combined with electron dispersive spectroscopy (EDS) was employed to characterize the multilayered silicide coating , in terms of phase distribution, growth texture for each phase as well as orientation relationships between the silicide layers. The present communication reports this microstructure characterisation work on two coated specimens after cyclic oxidation tests: 1) After 122 1h-cycle at 650°C in air. This specimen did not undergo much change from the "as deposit" state, 2) After 430 1h-cycle at 1100°C which corresponds to severe accidental overheating conditions.

2. Material and methods

V-4Cr-4Ti substrate was provided by GfE Metalle und Materialien GmbH, Nuremberg, Germany; the manufacturing process is detailed elsewhere [5]. Coupons 10 mm × 10 mm × 1 mm were cut from a rolled plate and recrystallised at 1000°C. Then polished down to 1200 grit SiC paper and their corners rounded. The silicide coating was deposited onto the coupons by halide-activated pack cementaton [2].

Cyclic oxidation tests were performed on the coated samples in a tubular furnace in air for a 1-h cycle at 650 and at 1100°C. The specimens were then removed from the furnace, between

IOP Conf. Series: Materials Science and Engineering **82** (2015) 012061 doi:10.1088/1757-899X/82/1/012061

each cycle, cooled for 10 min at room temperature and weighed. Cyclic oxidation conditions allowed for the evaluation, in a single experiment, of both the oxidation resistance of the coating and the effect of thermo-mechanical stresses applied to the coating-substrate system. The 650°C sample was cycled 122 times, while the 1100°C sample was cycled 430 times. After the test, the oxidized samples were cut transversally, embedded in epoxy resin, manually ground using wet SiC paper down to 4000 grit. Then the samples were polished to 0.4 μm silica colloidal solution and cleaned with warm water. In the last step of polishing, H_2O_2 was added to the colloidal silica. The polished samples were cleaned with warm water, acetone, ethanol, dried in air and then coated with carbon.
The Oxford Instruments NordlysNano EBSD detector combined with an X-Max80 EDS detector coupled with Oxford Instruments AZtec software, attached to a FEG-SEM was employed for phase identification and microstructural characterization. The FEG-SEM was operated at 20kV acceleration voltage.

3. Results and discussion

3.1 Phase Identification
Figure 1 shows a calculated Si-V phase equilibrium diagram, using MTDATA [6] and the Scientific Group Thermodata Europe (SGTE) Solution Database. This indicates the existence of the following: Si_2V, Si_5V_6, Si_3V_5 and SiV_3 phases. A phase database was created for these phases using TWIST in CHANNEL5 and used for phase search and indexing the EBSD Kikuchi patterns obtained from the various phases found in the coating. *PhaseID* module from AZtec software was used for phases identification. Crystal structure details for these phases were taken from Pearson's Handbook [7]. The similar diagram was also proposed by C. Zhang et al. [8].

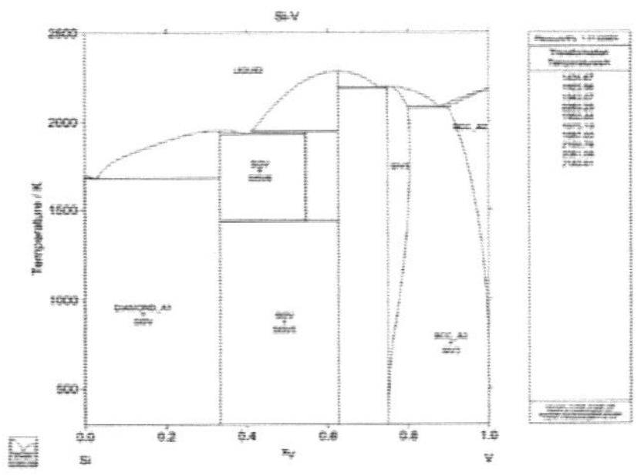

Figure 1: Si-V equilibrium phase diagram, calculated using MTDATA [6] and the Scientific Group Thermodata Europe (SGTE) Solution Database

Figure 2 shows a SEM image from a section of the coating. EDS quantitative results from the points shown in Figure 2 are summarized in Tables 1. It is clear from these results that all the four phases predicted in the phase diagram were found to exist in the coating. The EDS quantitative results show that low levels of Ti and Cr have partitioned into these phases and the atomic ratios of V to Si are close to the theoretical values, despite the data being acquired at 70 degree tilt angle used for EBSD. A summary of the phases identified is given in Table 1.

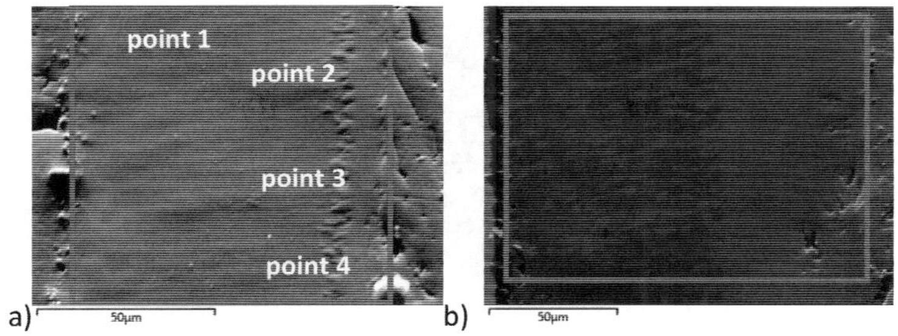

Figure 2: SEM images from analyzed samples: a) after 122 cycles (1h-cycle) at 650°C, b) after 430 cycles (1h-cycle) at 1100°C. Analysed area is defined by the red marked rectangle.

Table 1: EDS quantification results and phases identification results from points 1-4 (Figure2)

Element	Si at.%	Ti at.%	V at.%	Cr at.%	Phase	Space Group	Crystal System
Point 1	65.41	1.81	31.61	1.17	Si_2V	180	hexagonal
Point 2	41.88	3.89	52.32	1.91	Si_5V_6	72	orthorhombic
Point 3	35.89	3.62	58.61	1.88	Si_3V_5	140	tetragonal
Point 4	19.48	1.97	75.10	3.46	SiV_3	223	cubic

3.2 EBSD mapping

The EBSD maps were acquired following phase identification, using the phases information determined and listed in Table 1. The EBSD phases distribution and Inverse Pole Figure (IPF) color coded maps are shown in Figures 3 (a, b) and Figure 3 (c, d) respectively. These clearly show that all the phases in the coating have been successfully identified and mapped and that each layer has a distinctive grain shape and size. The average grain sizes, aspect ratio (GAR) and thickness for each of the phases are shown in Table 2. A significant change of microstructure and layers thickness is observed between the two annealing temperatures (Figure 3, Table 2). For both temperature of cycling the Si_2V grains are relatively coarse and elongated with the highest aspect ratio, while the Si_5V_6 phase is characterized by smaller and more regular shape grains. However, one can notice the increasing of grain aspect ratio in the case of SiV-1100 sample. The Si_3V_5 phase has more regular shape of grain for SiV-650 sample, but with the increased annealing temperature of 1100°C, two types of grain populations are observed: small and regular shape grains closed to Si_5V_6 layer and bigger and elongated shape (with higher GAR) grains closed to SiV_3 layer (Figure 3, Table 2). This microstructure and thickness changes can be explained by the high diffusion rate of vanadium and its alloying elements (Cr and Ti) at 1100°C (Figure 3). Figure 3 also shows the orientation inverse poles figure (IPF) for the color code EBSD map. It is clear that the Si_5V_6 and Si_3V_5 layer grains have the same [001] orientation.

Contoured EBSD pole figures from each of the phases in the coating are shown in Figures 4 for SiV-1100 samples.

It is clear from the Figure 4 a), b), c) that the external layers are strongly textured with a {0001} and {11-20} fiber in the SiV_2 and {001} fibers in the Si_5V_6 and Si_3V_5 phases. There also exists the following orientation relationship between the Si_3V_5 / Si_5V_6 and SiV_2 / Si_5V_6 phases: {001} Si_3V_5 // {001} Si_5V_6 and {11-20} SiV_2 // {001} Si_5V_6. The same fiber textures and crystallographic relationships were observed for the sample SiV-650.

Figure 4 d) shows the pole figures for the SiV_3 layer, close to the substrate. This layer has the lowest thickness, only few grains are present in the analyzed area (Table 2, Figure 3). For that reason one could notice that the statistics are poor for this data set, which also shows that the SiV_3 layer is not strongly textured. The random texture of SiV_3 layer could also be observed on IPF color maps for both samples (Figure 3 c, d).

Table 2: Average grain sizes for Si-V sample annealed at 1100°C

Phase	SiV-650			SiV-1100		
	Thickness	Grains Size	Aspect ratio	Thickness	Grains Size	Aspect ratio
Si_2V	80 μm	16 μm	4	80 μm	15 μm	3.5
Si_5V_6	6 μm	4 μm	2	20 μm	12 μm	2.5
Si_3V_5	4 μm	3 μm	2	35 μm	6 ; 17 μm	2 ; 3 μm
SiV_3	3 μm	2 μm	2	10 μm	7 μm	2.5

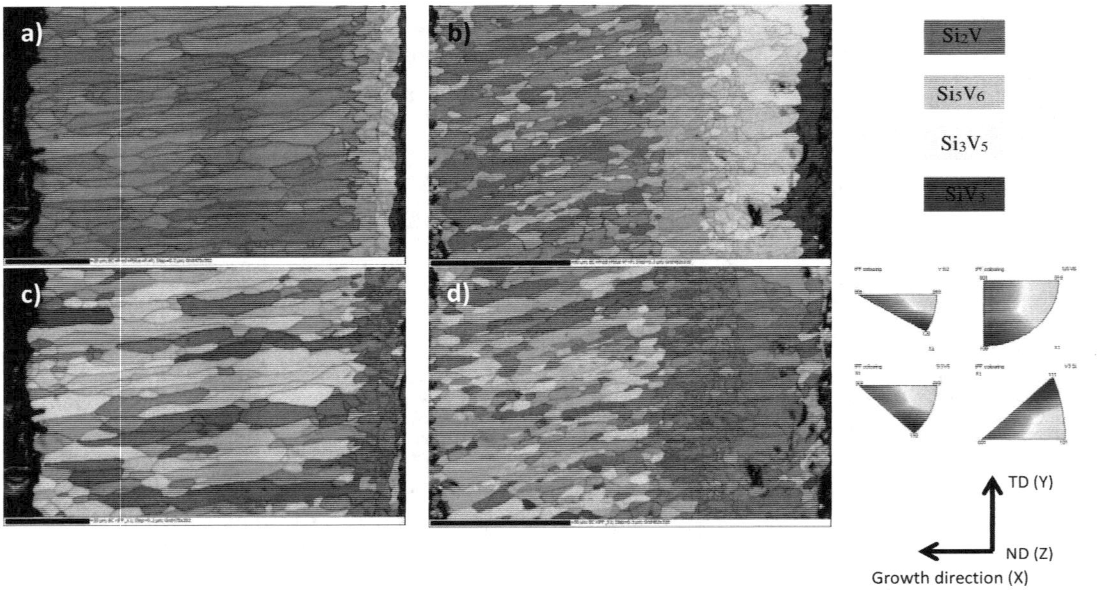

Figure 3: EBSD phases distribution maps: a) after 122 cycles (1h-cycle) at 650°C, b) after 430 cycles (1h-cycle) at 1100°C and IPF color coded map in layer growth direction: c) after 122 cycles (1h-cycle) at 650°C, d) after 430 cycles (1h-cycle) at 1100°C from Si-V samples.

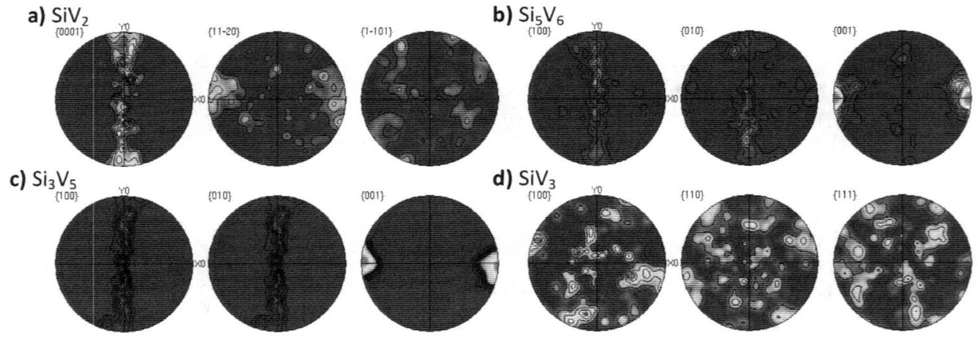

Figure 4: Set of pole figures from the data in Figure 3 for SiV-1100 sample : a) SiV_2 phase with {0001} and {11-20} fiber texture, b) Si_5V_6 phase with {001} fiber texture, c) Si_3V_5 phase with {001} fiber texture, d) SiV_3 with random texture.

4. Conclusions

With the combination of EDS/EBSD and phase diagrams for the Si-V system, all the phases present in the multilayered silicide coating have been accurately identified. Additionally the full microstructural characterization of the phases nature and distribution, of the grains sizes and textures was made using EBSD. It was shown that the fiber texture is generally the dominating texture for the Si-V multilayer. The crystallographic orientation relationships were found between the V-Si coating phases: {11-20} Si_2V // {001} Si_5V_6 and {001} Si_3V_5 // {001} Si_5V_6. Microstructural changes in the coating due to a severe accidental increase of temperature in Sodium Fast Reactor were clearly identified.

Acknowledgements

Sandrine Mathieu, Stéphane Mathieu and Nabil Chaia from University of Nancy are thanked for permitting the study.

References

[1] Klueh R L, Devan J H 1970Effect of oxygen in sodium on vanadium and vanadium alloys *J. Less Common Metals* **22** pp 389-398

[2] Mathieu S Chaia N, Le Flem M Vilasi M 2012 Multi-layered silicides coating for vanadium alloys for generation IV reactors, *Surf. Coat. Technol.* **206** pp 4594-4600

[3] Chaia N Mathieu S Cozzika T Rouillard F Desgranges C Courouau J -L Petitjean C David N Vilasi M 2013 An overview of the oxidation performances of silicide diffusion coatings for vanadium-based alloys for generation IV reactors, *Corros. Sci.* **66** pp 285-291

[4] Chaia N Mathieu S Rouillard F Vilasi M 2014 The ability of silicide coating to delay the catastrophic oxidation of vanadium under severe conditions, J. *Nucl. Mater.* **457** pp 124-129

[5] Duquesnes V Guilbert T Le Flem M 2012 French investigation of a new V–4Cr–4Ti grade: CEA-J57 – Fabrication and microstructure, *J. Nucl. Mater.* **426** pp 96-101

[6] Davies R H Dinsdale A T Gisby J A Robinson J A J Martin S M 2002 MTDATA - Thermodynamics and Phase Equilibrium Software from the National Physical Laboratory, *Calphad* **26** pp. 229-271

[7] Villars P 1997 Pearson's Handbook, *Crystallographic Data for Intermetallic Phases ASM international* **2**

[8] Zhang C Du Y Xiong W Xu H Nash P Ouyang Y and Hu R 2008 Thermodynamic Modeling of the V-Si System Supported by Key Experiments *Calphad* **32** pp 320-325

IOP Conf. Series: Materials Science and Engineering **82** (2015) 012062 doi:10.1088/1757-899X/82/1/012062

Crystallographic texture of light tinplate coatings made in various electrolytes

R Gburík[1], M Černík[2], R Leggat[3] and P Vranec[2]
[1] CTS manager, U. S. Steel Košice, s.r.o., Slovakia
[2] Specialist, U. S. Steel Košice, s.r.o., Slovakia
[3] Technical Manager, United States Steel Corporation, USA
E-mail: rgburik@sk.uss.com

Abstract. Two electrolytic tinplating processes are currently used in Europe: PSA (based on phenolsulfonic acid) and MSA (based on methanesulfonic acid). The Halogen Process is used in other parts of the world. The electrolyte composition and process parameters affect the electrodeposit and ultimately the tinplate appearance and performance. In order to better understand the impact of electrolyte composition on the crystallographic texture of tin coating tinplate, light tin coatings on single reduced, continuously annealed (CA) tinplate produced in three electrolytes: Halogen, PSA and MSA were analyzed. The crystallographic texture of thin tin coating ($<2.8 gm^{-2}$) was analyzed by X-ray Diffraction and Electron Backscatter Diffraction. The effect of reflow (melting of the tin followed by rapid solidification) and ironing during drawn and wall ironed (DWI) can forming on the tin crystallography were evaluated. Both texture analysis by XRD and EBSD confirmed that all un-melted tin coatings, made in three different electrolytes, contain texture fibers. The effect of steel sheet crystallographic texture was investigated by comparing the tin crystallographic orientation on continuously annealed steel substrate (with α and γ fiber texture) versus batch annealed (BA) steel with a strong γ fiber texture. The main electrolytic parameters, current density and line speed, did not affect the texture formation of tin coating produced in MSA-based electrolyte within the commercial ranges. Un-melted tin coatings produced in the MSA-based electrolyte showed sharper texture than those produced in PSA and Halogen electrolytes. The $FeSn_2$ alloy structure was not observed in un-melted tin coatings; however, it was detected after ironing in the DWI process.

1. Introduction

Many performance attributes of polycrystalline materials (e.g. drawability, magnetic polarization or corrosion resistance) can be optimized by control of the crystallographic texture. Electrochemical tin deposition has two main processing parameters: current density (CD) and bath chemical composition, both of which affect the tin deposition's morphology and crystallographic orientation. The un-melted tin coating texture of laboratory-produced PSA electrolyte consists of (321) texture, of which the fraction decreases with increasing CD [1]. Girin established two axial texture components (110) and (100) for laboratory produced un-melted tin coatings, where the texture (110) influences the protective ability of the tin coating [2]. Ichiba et al., studied the effect of MSA bath chemistry and CD on the tin coating appearance [3]. On the basis of X-Ray texture analysis these authors concluded that tin coatings with a (101) crystal orientation have a beneficial effect on glossy appearance, while the (200) crystal orientation was formed by dendritic tin deposits. Most of the abovementioned papers used X-ray diffraction (XRD) for tin coating crystallographic orientation analysis. In the last decade, texture measurement of polycrystalline materials has increasingly been performed using EBSD, which enables analysis of individual grain orientation, while conventional XRD is more useful for bulk texture measurement [4]. The un-melted tin coating texture was analyzed using both XRD and EBSD. The crystallographic textures of the tin deposit are presented in Orientation Distribution Function (ODF) sections. The volume fractions of main fiber textures from ODF were calculated.

2. Materials and Experimental Procedures

The texture of commercial un-melted tin coatings, produced in three electrolytes (Halogen, PSA and MSA), was analyzed by XRD and EBSD to establish the effect of electrochemical bath composition on light tin coating texture formation. In addition to the electrolyte effect, the effect of steel substrate texture produced by continuous or batch annealing on tin texture was analyzed. The tinplate samples included in the analysis are summarized in **Table 1**.

Table 1 Analyzed commercial tinplate BA and CA samples produced in USS Corporation

Electrolytic tinning line Type	Electrolyte	Sample Name	Tin Coating weight [g.m^{-2}]	Line Location
Horizontal	Halogen	H	2.3	USA
	MSA	MSA-H	2.3	
Vertical	PSA	PSA	2.8	Europe
	MSA	MSA-V	2.5	

The crystallographic texture of un-melted tin coating was measured on the surfaces of the samples as produced without additional preparation. Five incomplete pole figures of un-melted tin coating, (200), (101), (220), (301) and (112), were measured by XRD with filtered Co radiation using a texture goniometer and a position sensitive detector (PSD). The XRD crystallographic texture was measured using point focus, which expands profiles of diffraction lines much more than line focus. ODF was calculated from the measured pole figures data using the Arbitrarily Defined Cells (ADC) method [5].

The tin coatings were also analyzed by EBSD. The samples were analyzed as-produced because the thin and relatively soft tin coatings precluded the use of the grinding and polishing typically used for EBSD preparation. All samples were analyzed at 200-2000x magnification with 0.15-1.5 μm step intervals, depending on the surface quality. The textures of the un-melted tin coatings were evaluated from measured points with a confidence index (CI) >0.1 that resulted in volume fraction in the interval of 5–16%, [6, 7]. Those small volume fractions are hardly satisfactory values for reliable texture evaluation. However, the EBSD texture of the Halogen and PSA, which showed the highest CI values, were compared with the XRD results.

3. Results

None of the X-ray diffractograms of the four un-melted tin coatings revealed the presence of the $FeSn_2$ structure, which forms after the reflow process and can be detected in melted tin coating. However, the presence of $FeSn_2$ was confirmed after ironing during the DWI can production. The diffractograms were also evaluated by the Rietveld method, [8].

The ODF were calculated in orthorhombic sample symmetry because all the pole figures showed orthorhombic sample symmetry observed in the (200) and (101) pole figures. ODF section $\varphi_2=0°$ of all the samples are summarized in **Figure 1**. All the ODF sections confirmed an incomplete fiber texture. The same ODF section $\varphi_2=0°$ determined by EBSD analysis for the Halogen and PSA sample are shown in **Figure 2**, which correspond well to those calculated from the X-ray pole figures. In spite of the small texture fraction analyzed using EBSD, the fiber texture can be seen in both ODF.

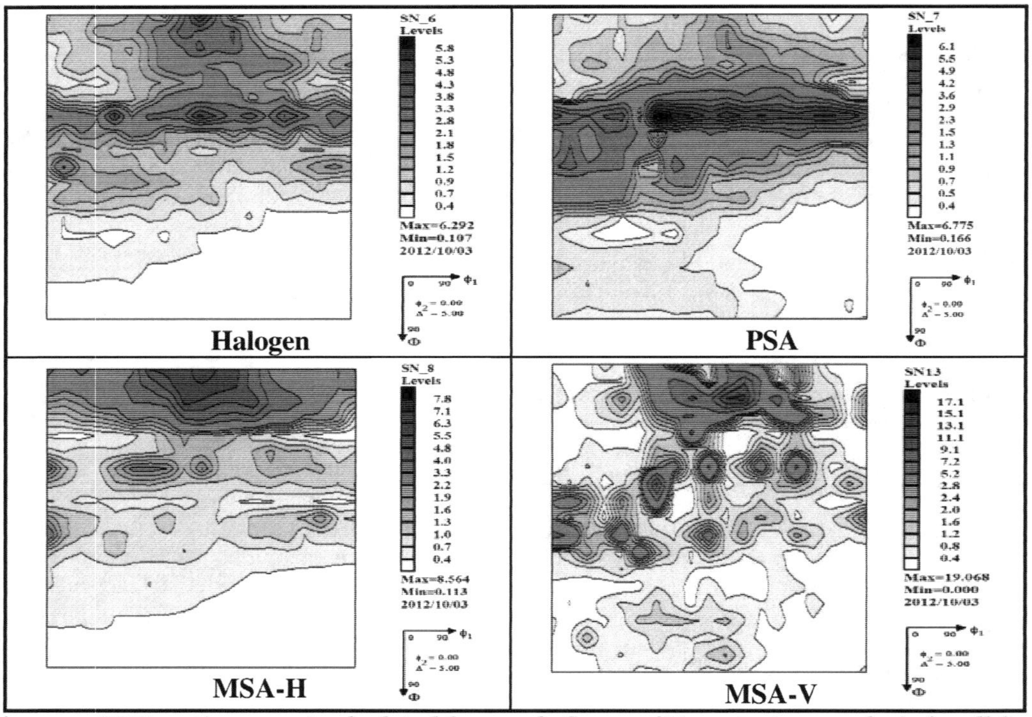

Figure 1 ODF section φ_2=0° calculated from pole figure of X-ray texture analysis for all four samples of CA tinplate sheets

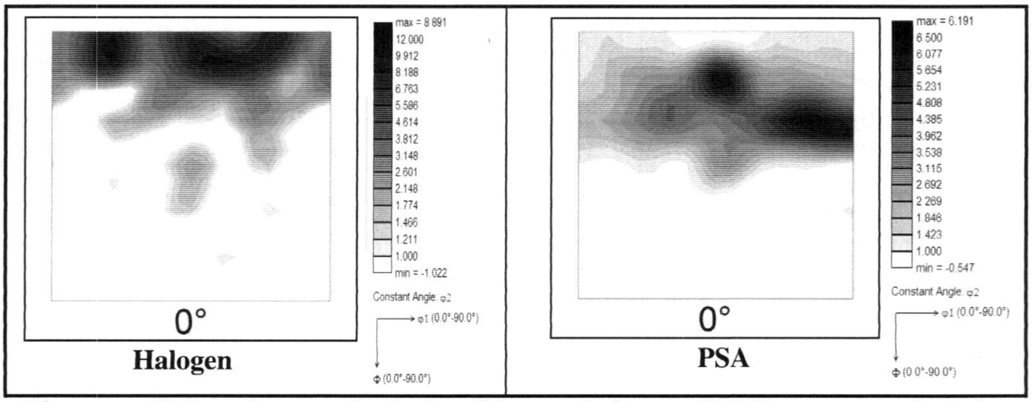

Figure 2 ODF section φ_2=0° calculated from pole figure of EBSD texture analysis for Halogen and PSA samples of CA tinplate sheets

The main texture fibers with a volume fraction greater than 4% from the XRD ODF were calculated using LaboTex software. The main texture fiber components of all four CA samples are summarized in **Figure 3**. Analysis of the texture fibers processed from ODF showed that approximately 70% of the total crystalline volume of the un-melted tin coatings is textured, while the remaining volume is randomly oriented.

In spite of being plated on lines with different configurations and with different processing parameters, in terms of line speed and CD value, both MSA tin coatings showed very similar

IOP Conf. Series: Materials Science and Engineering **82** (2015) 012062 doi:10.1088/1757-899X/82/1/012062

crystallographic texture. Those textures consisted of fibers: (321), (221), (213), (011) and (001) with similar volume fraction, see **Figure 3**, with the strongest fiber of (321) plane which is in agreement with the literature [1]. All five of the analyzed fibers of the MSA-V and PSA un-melted tin coatings have similar volume fractions with highest (321) and lowest (001) fibers. MSA-H and Halogen samples showed weaker (321) fibers than PSA and MSA samples. No significant difference was seen in the tin coating crystallographic texture between CA and BA tinplate steel strip coated in the MSA and PSA electrolyte bath as shown in **Figure 3**.

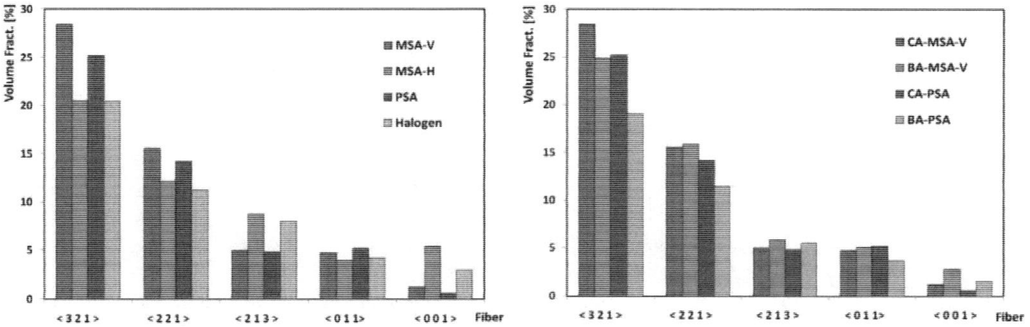

Figure 3 Volume fraction (>4%) of the main fiber texture of un-melted tin coatings. Four different electrolyte bath of CA tinplate (left). MSA vs PSA samples of CA and BA tinplate (right)

4. Conclusion

X-ray diffraction phase analysis of un-melted tin coatings did not reveal the presence of a $FeSn_2$ phase. However, such intermetallic phase was detected after ironing during DWI can production. All un-melted tin coatings, made in three different electrolytes, contain incomplete texture fibers, mainly (321)<uvw> fiber. Crystallographic textures of the both continuously and batch annealed tinplate strips have the same effect on the crystallographic texture component of the tin coating which is slightly influenced by the electrolyte.

References

[1] Lempereur J, Renard L, Weymeersch A and Cauwe, B 1988 Electrotinning at high current densities, *4th International Tinplate Conference, ITRI (*London*)* p 75
[2] Girin O B and Kolesnyk V 2011 Crystallographic Texture of Electrochemical Tin Coating on Non-Reflow Tinplate as Related to their Protective Ability, *Electroplating & Finishing,* **30** (11)
[3] Ichiba M, Kubo H and Yomura Y 1996 Characterization of Electrolytic Tin Deposits from Methane Sulphonic Acid Bath, *6th International Tinplate Conference, ITRI (*London*)* p 28
[4] Bunge H J 1991 The Basic Concepts of Texture Investigation in Polycrystalline Materials, *Steel Research* **62 (12)** p 530
[5] Pawlik K 1986 Determination of the Orientation Distribution Function from Pole Figures in Arbitrarily Defined Cells, *Physica Status Solidi (b)* **134** p 477
[6] Bunge H J 1982 Texture Analysis in Materials Science, Mathematical Methods, (Butterworths, London)
[7] Kocks UF, Tome C N and Wenk R 1998 Texture and Anisotropy (Cambridge University Press, Cambridge)
[8] Young R A 1982 Introduction to the Rietveld Method, (Oxford University Press, Oxford)

EBSD characterization of crystallographic orientations and twin interfaces of modulated martensite in epitaxial NiMnGa thin film

Bo Yang[1,2,3], Yudong Zhang[2,3]*, Zongbin Li[1], Gaowu Qin[1], Claude Esling[2,3], Xiang Zhao[1], Liang Zuo[1]*

[1] Key Laboratory for Anisotropy and Texture of Materials (Ministry of Education), Northeastern University, Shenyang 110819, China

[2] Laboratoire d'Étude des Microstructures et de Mécanique des Matériaux (LEM3), CNRS UMR 7239, Université de Lorraine, 57045 Metz, France

[3] Laboratory of Excellence on Design of Alloy Metals for low-mAss Structures (DAMAS), Université de Lorraine, 57045 Metz, France

*E-mail: yudong.zhang@univ-lorraine.fr; lzuo@mail.neu.edu.cn

Abstract. The crystal structure and the global texture of 7M modulated martensite in epitaxial $Ni_{50}Mn_{30}Ga_{20}$ thin films were investigated by X-ray diffraction (XRD), and the local crystallographic orientations correlated with microstructural features were revealed by electron backscatter diffraction (EBSD). The microstructure of 7M martensite can be classified into two distinct constituents. One refers to long straight strips with relatively homogeneous contrast, running parallel to one edge direction of the substrate. The other refers to shorter and bent plates with relatively high relative contrast, being oriented with the plate length direction roughly in 45°with respect to the substrate edges. Each 7M martensite plate is designated as a single orientation variant, and four orientation variants that are twin-related one another constitute one variant group. With the local crystallographic orientations of martensite plates and the orientations of inter-plate interface traces, the Type I, Type II and compound twin interfaces in the low and high relative contrast zones were determined.

1. Introduction

The discovery of ferromagnetic Ni-Mn based Heusler alloys has quickly promoted a breakthrough in the development of magnetically-controlled functional materials. To date, giant magnetic-field-induced strain up to 12% can be achieved in bulk Ni-Mn-Ga single crystals [1]. It is established that the type and the character of twin boundaries are dominant factors affecting the magnetic-field-induced strains in ferromagnetic Ni-Mn-Ga alloys, as the so-called Type II twin boundaries of modulated martensite possess higher mobility [2-4]. So far, for bulk materials, these factors have been well clarified through direct orientation determination on individual martensitic variants.

As for thin films, the multi-scale microstructures formed through epitaxial growth are quite different from those in polycrystalline bulk alloys [5-10]. Efforts have been constantly made to reveal the microstructural features and crystallographic organizations of martensitic variants in epitaxial Ni-Mn-Ga films [11-15]. However, the character of ultrafine microstructural constituents brings a major challenge to conventional crystallographic characterization in terms of feasible experimental techniques and correct result interpretations. For instance, only the so-called a_{7M}-c_{7M} twin interfaces could be deduced by examining X-ray pole figures and local surface morphologies [14, 16]. In this connection, the spatially-resolved electron backscatter diffraction (EBSD) technique may be exploited to obtain precise local crystallographic information of martensitic variants in epitaxial thin films having ultrafine microstructures.

In this work, the crystal structure and the global texture of 7M modulated martensite in epitaxial Ni-Mn-Ga thin films were investigated by X-ray diffraction (XRD), and the local crystallographic orientations correlated with microstructural features were revealed by EBSD. Based on the local crystallographic orientations of martensitic variants, the type and the character of twin boundaries were analyzed.

2. Experimental

The Ni-Mn-Ga thin films with nominal composition (at.%) of $Ni_{50}Mn_{30}Ga_{20}$ and nominal thickness of 1.5 μm were epitaxially grown on MgO monocrystalline substrate. Prior to the microstructural observations and orientation measurements, the thin film samples were subject to thickness-controlled electrolytic polishing with a solution of 20%HNO_3 in CH_3OH at 12 volts at room temperature. The microstructures and crystallographic orientations of the thin films were analyzed using a field emission gun scanning electron microscope (SEM, Jeol JSM 6500 F) equipped with an EBSD system. The EBSD patterns from martensitic variants were manually acquired using Channel 5 Flamenco's interactive option.

3. Results and discussion

3.1 Orientations and twin relationships of martensitic variants

Fig. 1 presents the SEM secondary electron images of 7M martensite plates after electrolytic polishing. According to the direction and the brightness of martensite plates (Fig. 1a), the self-accommodated groups of 7M martensite can be classified into low relative contrast zones (Group 1) or high relative contrast zones (Group 2). The low relative contrast zones contain long straight plates parallel to one of the substrate edges, whereas the high relative contrast zones contain shorter and bent plates oriented at specific angles (roughly 45°) with respect to the substrate edges. Detailed inspections show that the traces of the inter-plate interfaces have one orientation in the low relative contrast zones (marked by the white dotted line in Fig. 1a), but two distinct orientations in the high relative contrast zones (indicated by the yellow and blue dotted lines in Fig. 1b).

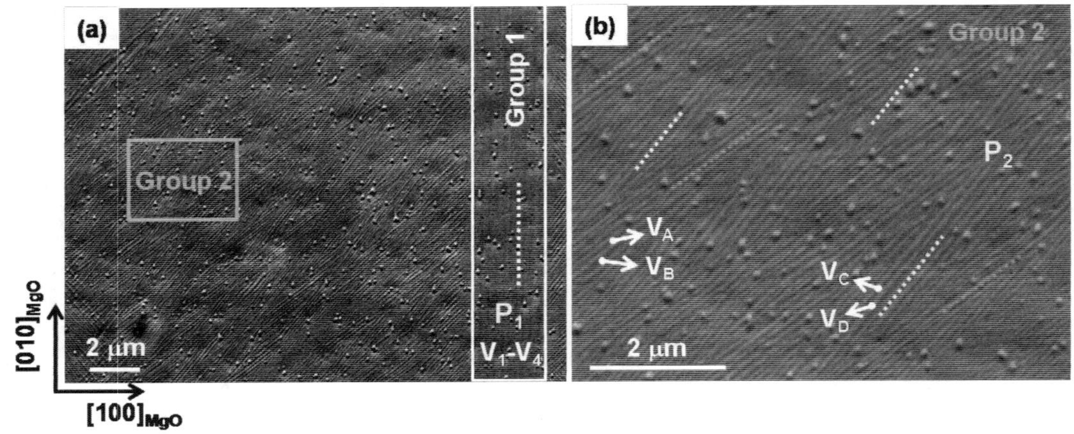

Fig. 1. (a) SEM secondary electron image of 7M martensite in thin films. (b) The magnified view of Group 2. The white dotted line in (a) and the blue and yellow dotted lines in (b) outline inter-plate interface traces in Group 1 and Group 2, respectively.

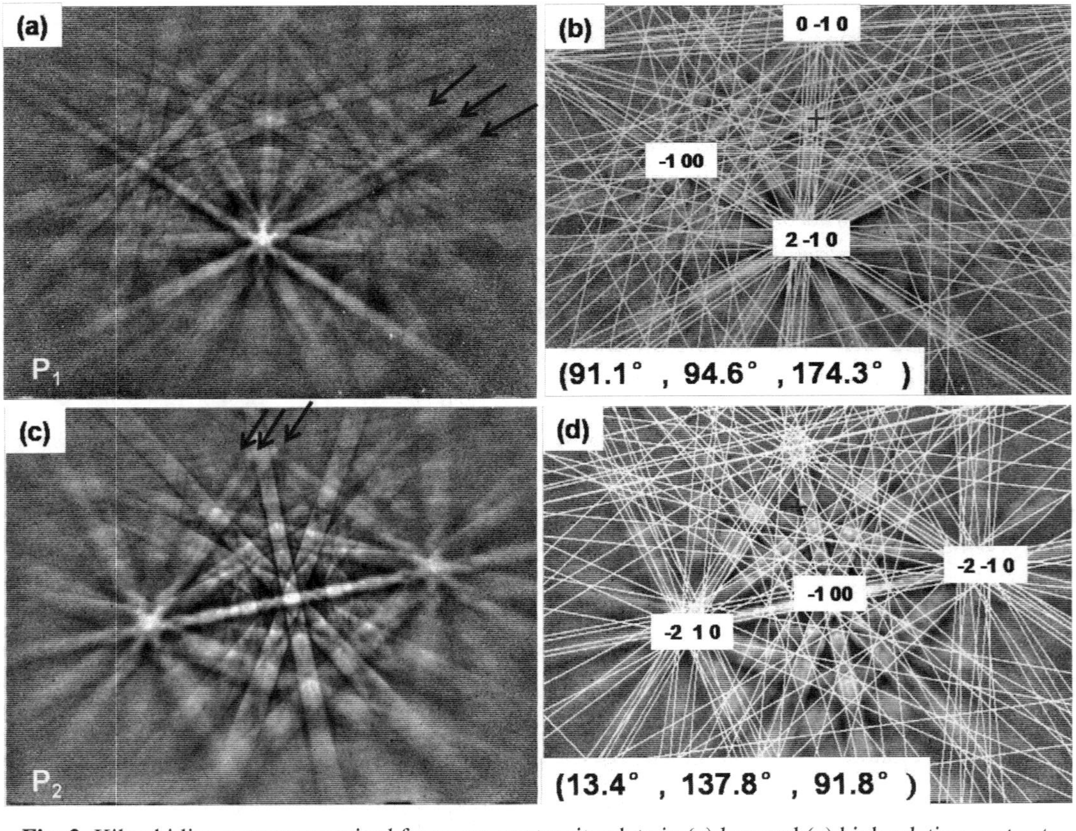

Fig. 2. Kikuchi line patterns acquired from one martensite plate in (a) low and (c) high relative contrast zones and (b and d) the corresponding indexed results.

Fig. 2 presents the Kikuchi line patterns acquired from one martensite plate in the low and high relative contrast zones. The measured patterns can be well indexed with the monoclinic superstructure of the 7M martensite. As compared to the case of two overlapping Kikuchi patterns for the NM martensite [17], there is only a single pattern with some satellite diffractions (marked with arrows in Figs. 2a and 2c). Thus, each 7M martensite plate is specified by a single crystallographic orientation, being designated as one orientation variant.

Further characterization demonstrates that there are in total four different orientation variants distributed in one plate group (Group 1 or Group 2). Here, the four orientation variants - representing one plate group with low or high relative contrast - are denoted by the symbols V_1, V_2, V_3, V_4 (Fig. 1a) and V_A, V_B, V_C, V_D (Fig. 1b), respectively. To specify the four orientation variants, the individually measured orientations of the 7M martensite plates in Group 1 and Group 2 are presented in the form of $\{2\ 0\ 20\}_{mono}$, $\{2\ 0\ \text{-}20\}_{mono}$ and $\{0\ 4\ 0\}_{mono}$ pole figures, as shown in Fig. 3. In the low relative contrast zone (Group 1), the four variants V_1, V_2, V_3, V_4 are all with their $(2\ 0\ 20)_{mono}$ plane nearly parallel to the substrate surface (Fig. 3, left). These variants correspond to the ***a*-*c*** twins with their common ***b*** axis perpendicular to the substrate surface, the so-called ***b***-variants in literature [12-14]. However, in the high relative contrast zone (Group 2), the variants V_A and V_D are with their $(2\ 0\ \text{-}20)_{mono}$ plane nearly parallel to the substrate surface (Fig. 3, middle), and the variants V_B and V_C with their $(0\ 4\ 0)_{mono}$ plane nearly parallel to the substrate surface (Fig. 3, right). Furthermore, the variants V_B and V_C and the variants V_A and V_D are of low brightness and high brightness, respectively. They correspond to the ***a*-*c*** twins with their common ***b*** axis parallel to the substrate surface, the so-called ***a***-variants for V_A and V_D and ***c***-variants for V_B and V_C [14].

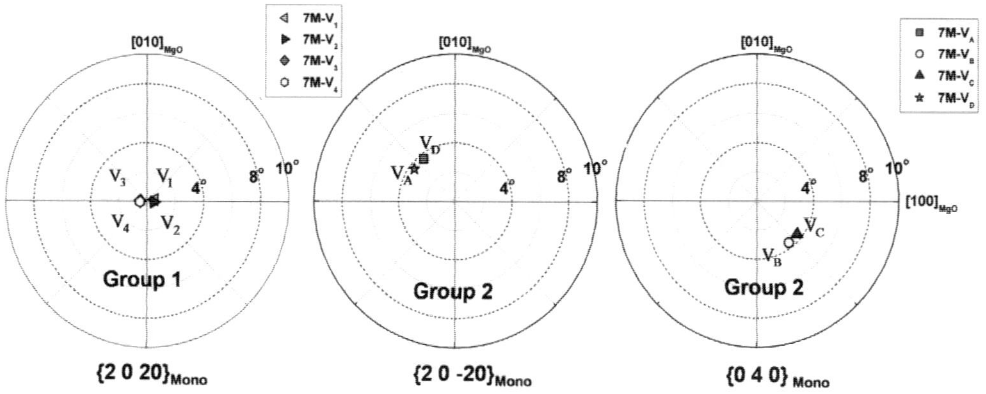

Fig. 3. Representation of individually measured orientations of 7M variants in Group 1 and Group 2 in the form of $\{2\ 0\ 20\}_{mono}$, $\{2\ 0\ \text{-}20\}_{mono}$ and $\{0\ 4\ 0\}_{mono}$ pole figures.

With the determined crystallographic orientations (represented with three Euler angles in Bunge's notation [18]) of the four 7M variants in each plate group, the orientation relationships between neighboring variants can be further evaluated. Detailed analysis has revealed that each pair of adjacent variants can be identified as either Type I or Type II or Compound twins. In the low relative contrast zone (Group 1), the variant pairs V_1 & V_3 and V_2 & V_4 belong to Type I twins, V_1 & V_2 and V_3 & V_4 to Type II twins, and V_1 & V_4 and V_2 & V_3 to Compound twins. The adjacent variants in the high relative contrast zone (Group 2) possess exactly the same twin relationships, where V_A & V_C and V_B & V_D belong to Type I twins, V_A & V_B and V_C & V_D to Type II twins, and V_A & V_D and V_B & V_C to Compound twins. The complete twinning elements - $\boldsymbol{K}_1$, $\boldsymbol{K}_2$, η_1, η_2, $\boldsymbol{P}$ and s - of the above three types of twins were derived using a general method [19, 20] and are shown in Table 1. The twinning modes are exactly the same as those in the bulk materials [6, 20].

Table 1. Twinning elements of 7M variants represented in the monoclinic crystal coordinate frame. $\boldsymbol{K}_1$ is the twinning plane, $\boldsymbol{K}_2$ the reciprocal or conjugate twinning plane, η_1 the twinning direction, η_2 the conjugate twinning direction, $\boldsymbol{P}$ the shear plane, and s the magnitude of shear.

Twin elements	Type I twin (V_A & V_C; V_B & V_D; V_1 & V_3; V_2 & V_4)	Type II twin (V_A & V_B; V_C & V_D; V_1 & V_2; V_3 & V_4)	Compound twin (V_A & V_D; V_B & V_C; V_1 & V_4; V_2 & V_3)
$\boldsymbol{K}_1$	$(1\ \bar{2}\ \overline{10})_{mono}$	$(1.1240\ \bar{2}\ \overline{8.7602})_{mono}$	$(1\ 0\ 10)_{mono}$
$\boldsymbol{K}_2$	$(\overline{1.1240}\ \bar{2}\ 8.7602)_{mono}$	$(\bar{1}\ \bar{2}\ 10)_{mono}$	$(\bar{1}\ 0\ 10)_{mono}$
η_1	$[11.0973\ 10\ \overline{0.8903}]_{mono}$	$[\overline{10}\ \overline{10}\ 1]_{mono}$	$[\overline{10}\ 0\ 1]_{mono}$
η_2	$[10\ \overline{10}\ \bar{1}]_{mono}$	$[11.0973\ \overline{10}\ \overline{0.8903}]_{mono}$	$[10\ 0\ 1]_{mono}$
$\boldsymbol{P}$	$(1\ \overline{0.1161}\ 11.1610)_{mono}$	$(1\ 0.1161\ 11.1610)_{mono}$	$(0\ 1\ 0)_{mono}$
s	0.2537	0.2537	0.0295

3.2 Orientations of twin interfaces

Using the indirect two-trace method [21, 22], the planes of twin interfaces in the low and high relative contrast zones were further calculated with the measured crystallographic orientations of individual variants and the measured orientations of their interface traces on the film surface. Results show that the twin interface planes are coincident with the respective twinning planes, i.e. $\{1\ -2\ -10\}_{mono}$ for Type I twins, $\{1.1240\ -2\ -8.7602\}_{mono}$ for Type II twins, and $\{1\ 0\ 10\}_{mono}$ for Compound twins, as in the case for bulk materials. For easy visualization of the through-film-thickness orientations of twin interfaces in two relative contrast zones, the $\{1\ -2\ -10\}_{mono}$, $\{1.1240\ -2$

-8.7602}$_{mono}$ and {1 0 10}$_{mono}$ planes of the four 7M variants in Group 1 and Group 2 are represented in the form of stereographic projections, as displayed in Fig. 4a and Fig. 4b. In both figures, the interface trace directions on the film surface are indicated by the arrowed lines.

For the low relative contrast zone (Group 1), it is seen from the {1 -2 -10}$_{mono}$ pole figure (Fig. 4a, left) that two different Type I twin interfaces are oriented at +89.9° (V_1 and V_3) and -89.9° (V_2 and V_4) with respect to the substrate surface, although their interface traces on the film surface are both parallel to the [0 1 0]$_{MgO}$ direction. Similar cases can be found for the Type II twin interfaces. As shown in the {1.1240 -2 -8.7602}$_{mono}$ pole figure (Fig. 4a, middle), the two Type II twin interfaces that have parallel traces on the film surface intersect the substrate surface at +85° (V_3 and V_4) and -85° (V_1 and V_2), respectively. Apparently, the four Type I and Type II twin interfaces are oriented differently through the film thickness, but they have the same interface trace orientation on the film surface. As for the two compound twin interfaces, they are parallel to the film surface, as shown in the {1 0 10}$_{mono}$ pole figure (Fig. 4a, right). That is why no compound twin interfaces could be observed by examining solely the film surface microstructure in the present work. Such compound twin interfaces have been detected by the cross section observation using SEM [14].

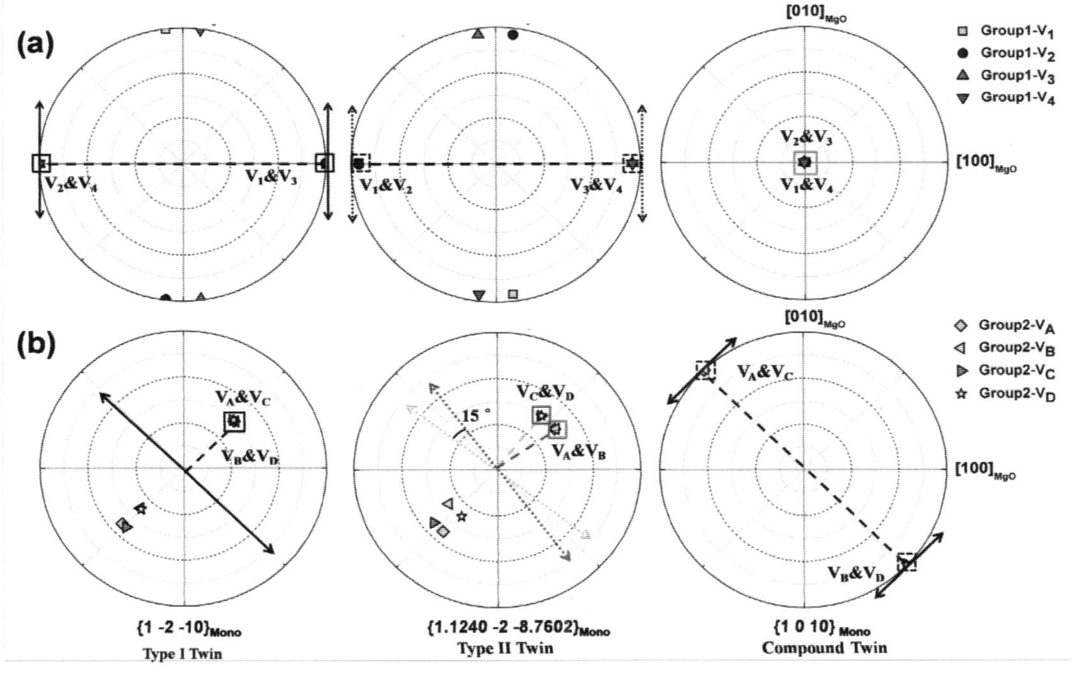

Fig. 4. Stereographic projections of {1 -2 -10}$_{mono}$, {1.1240 -2 -8.7602}$_{mono}$ and {1 0 10}$_{mono}$ planes of 7M variants in (a) low relative contrast zone (Group 1) and (b) high relative contrast zone (Group 2). The macroscopic sample coordinate frame is set to the crystal basis of the MgO substrate. The arrowed lines represent the trace directions of twin interfaces.

For the high relative contrast zone (Group 2), it is seen from the {1 -2 -10}$_{mono}$ pole figure (Fig. 4b, left) that the Type I twin interface between variants V_A and V_C and that between variants V_B and V_D possess roughly the same orientation in the film. They incline at about 45° to the substrate surface and their traces on the film surface are along the [-1 1 0]$_{MgO}$ direction. However, for the Type II twin interfaces shown in the {1.1240 -2 -8.7602}$_{mono}$ pole figure (Fig. 4b, middle), although the twin interface between variants V_A and V_B and that between variants V_C and V_D incline at about 45° to the substrate surface, their interface traces on the film surface do not possess the same orientation. They are 37.3° (V_A and V_B) and 52.5° away from the [1 0 0]$_{MgO}$ direction, respectively. The angle between the two differently oriented Type II twin interface traces on the film surface is about 15°, being in accordance with the results reported in the literature [23]. As for the compound twin interfaces, the twin interface between variants V_A and V_D and that between variants V_B and V_C possess roughly the same orientation (Fig. 4b, right). They are nearly perpendicular to the substrate surface and their traces on the film surface are nearly parallel to the [1 1 0]$_{MgO}$ direction. It is found that the bending of martensite plates in the high relative contrast zones is associated with the orientation change from V_A to V_D or from V_B to V_C, which results in the plate interface change from Type I twin interface to Type II twin interface or from Type II twin interface to Type I twin interface.

4. Summary

By means of the EBSD-based orientation characterization, a comprehensive analysis was made on the microstructural and crystallographic features of 7M modulated martensite in $Ni_{50}Mn_{30}Ga_{20}$ thin films epitaxially grown on MgO(100) substrate. The plate-like microstructure of the 7M martensite is composed of two distinct kinds of plate groups. One refers to long straight strips with relatively homogeneous contrast in the SE images, running parallel to one edge direction of the substrate. The other refers to shorter and bent plates with relatively high relative contrast, being oriented with the plate length direction roughly in 45° against the substrate edges. Each 7M martensite plate contains one orientation variant. There are four orientation variants in one plate group. The inter-plate interfaces are either Type-I or Type-II twin interfaces.

Acknowledgements

This work is supported by the National Natural Science Foundation of China (Grant No. 51431005), the 863 Program of China (Grant No. 2014AAQ00297), the 111 Program of China (Grant No. B07015), the Program for Liaoning Innovative Research Team in University (Grant No. LT2013007), the Fundamental Research Funds for the Central Universities of China (Grant Nos. N130110001 and N130523001), and the French State through the Investment in the Future Program operated by the National Research Agency and referenced by ANR-11-LABX-0008-01 (LabEx DAMAS), are highly appreciated.

References

[1] Sozinov A, Lanska N, Soroka A and Zou W 2013 *Appl. Phys. Lett.* **102** 021902

[2] Heczko O, Straka L and Seiner H 2013 *Acta Mater.* **61** 622

[3] Chulist R, Straka L, Lanska N, Soroka A, Sozinov A, and Skrotzki W 2013 *Acta Mater.* **61** 1913

[4] Neudert A, Lai Y W, Schäfer R, Kustov M, Schultz L and McCord J 2012 *Adv. Eng. Mater.* **14** 610

[5] Li Z B, Zhang Y D, Esling C, Zhao X and Zuo L 2012 *Acta Mater.* **60** 6982

[6] Li Z B, Zhang Y D, Esling C, Zhao X and Zuo L 2011 *Acta Mater.* **59** 2762

[7] Li Z B, Xu N, Zhang Y D, Esling C, Raulot J M, Zhao X and Zuo L 2013 *Acta Mater.* **61** 3858

[8] Cong D Y, Zhang Y D, Esling C, Wang Y D, Lecomte J S, Zhao X and Zuo L 2011 *Acta Mater.* **59** 7070

[9] Cong D Y, Zhang Y D, Wang Y D, Humbert M, Zhao X, Watanabe T, Zuo L and Esling C 2007 *Acta Mater.* **55** 4731

[10] Yang S, Wang C and Liu X 2012 *Intermetallics* **25** 101

[11] Niemann R, Rößler U K, Gruner M E, Heczko O, Schultz L and Fähler S 2012 *Adv. Eng. Mater.* **14** 562

[12] Jakob A M, Müller M, Rauschenbach B and Mayr S G 2012 *New J. Phys.* **14** 033029

[13] Tillier J, Bourgault D, Odier P, Ortega L, Pairis S, Fruchart O, Caillault N and Carbone L 2011 *Acta Mater.* **59** 75

[14] Kaufmann S, Niemann R, Thersleff T, Rößler U K, Heczko O, Buschbeck J, Holzapfel B, Schultz L and Fähler S 2011 *New J. Phys.* **13** 053029

[15] Kaufmann S, Rößler U K, Heczko O, Wuttig M, Buschbeck J, Schultz L and Fähler S 2010 *Phys. Rev. Lett.* **104** 145702

[16] Kauffmann-Weiss S, Gruner M E, Backen A, Schultz L, Entel P and Fähler S 2011 *Phys. Rev. Lett.* **107** 206105

[17] Yang B, Li Z B, Zhang Y D, Qin G W, Esling C, Perroud O, Zhao X and Zuo L 2013 *Acta Mater.* **61** 6809

[18] Bunge H J, Esling C, and Muller J 1980 *J. Appl. Crystallogr.* **13** 544

[19] Zhang Y D, Li Z B, Esling C, Muller J, Zhao X and Zuo L 2010 *J. Appl. Crystallogr.* **43** 1426

[20] Li Z B, Zhang Y D, Esling C, Zhao X, Wang Y D and Zuo L 2010 *J. Appl. Crystallogr.* **43** 617

[21] Zhang Y D, Esling C, Zhao X and Zuo L 2007 *J. Appl. Crystallogr.* **40** 436

[22] Cong D Y, Zhang Y D, Wang Y D, Esling C, Zhao X and Zuo L 2006 *J. Appl. Crystallogr.* **39** 723

[23] Backen A, Kauffmann-Weiss S, Behler C, Diestel A, Niemann R, Kauffmann A, Freudenberger J, Schultz L and Fähler S 2013 *arxiv.org* 1

IOP Conf. Series: Materials Science and Engineering **82** (2015) 012064 doi:10.1088/1757-899X/82/1/012064

Crystallographic texture and the preferential orientation of a martensite in the polycrystalline $Ni_{2.08}Mn_{0.96}Ga_{0.96}$ alloy

I.I. Musabirov[1], **R.R. Mulyukov**[2], **V.V. Koledov**[3]

[1] Institute for Metals Superplasticity Problems of RAS, Ufa, Russia

[2] Institute for Metals Superplasticity Problems of RAS, Ufa, Russia

[3] Institute of Radioengineering and Electronics of RAS, Moscow, Russia
E-mail: irekmusabirov@imsp.ru

Abstract. A study of the relationship of the crystallographic texture, the preferential orientation of martensite twins and the geometry changes during the martensite transformation of polycrystalline NiMnGa alloy is shown. Two states of alloy are investigated: as-cast state and the state after annealing at 923 K for 5 hours. It is shown that in the initial state the austenite phase has crystallographic axial texture <001>, while martensite has the two-component texture of the <001> and <110>. The preferential orientation of the martensitic twins in the martensitic structure is found. The crystallographic texture of the alloy after annealing is not changed, however, no preferential orientation of the martensitic twins is observed. In the martensitic phase structure martensitic twins are oriented randomly. Anisotropy of the thermal expansion during the martensitic transformation is shown, while in the annealed state such effect is not found.

1. Introduction

The Heusler NiMnGa alloys attract attention in consequence of the martensitic transformation, which occurs in the material at the room temperature. Depending on the composition, alloy can proceed to the ferromagnetic or paramagnetic state. In the transformation temperature range alloys have the magnetic shape memory effect (or the effect of magnetic field induced deformation) [1-7], the giant magnetocaloric effect [8-11] and giant magnetoresistance [12-13]. Control of the shape and dimensions of the material by applying an external magnetic field opens prospects of using this material as a functional component for various kinds of microelectronic devices. The greatest effect of irreversible magnetic field induced deformation achieving up to 6% is observed on single-crystal samples [14]. On the polycrystalline samples, the reversible magnetic field induced deformation had a value less than 1% [15]. However, it would be very promising to obtain the significant magnetic deformation on polycrystalline materials, as they are cheaper to manufacture in comparison with single crystals.

In the process of martensitic transformation, on the single-crystal samples of alloys the anisotropy of the thermal expansion was observed [16]. In some cases, it was also observed in polycrystalline materials [17]. In literature such effect is explained by the formation of preferential orientation of martensitic twins [18], however, there is rather little experimental data. To clarify this, deeper study of the structure of polycrystalline materials in different structural states is required. In this paper, experimental data confirming the dependence of the formation of the martensite twins preferential orientation and the anisotropy of the thermal expansion of the material in the process of transformation are presented.

2. Material and methods

The polycrystalline $Ni_{2.08}Mn_{0.96}Ga_{0.96}$ alloy was selected for the investigation. The alloy was prepared by arc melting of high purity Ni, Mn and Ga in Ar atmosphere. To homogenize the structure, samples are annealed for 9 days at 1100 K, which is followed by quenching in ice water. The phase transformation temperatures are: M_S=294 K (martensite start); M_F=267 K (martensite finish); A_S=287 K (austenite start); A_F=304 K (austenite finish); T_C=375 K (Curie temperature).

IOP Conf. Series: Materials Science and Engineering **82** (2015) 012064 doi:10.1088/1757-899X/82/1/012064

Crystallographic orientations are measured by diffraction of back-scattered electrons (Electron Back-Scattered Diffraction - EBSD) at an accelerating voltage of 20 kV. EBSD analysis of texture by scanning electron microscopy is widely used to study the materials texture [19].The mapping is carried out by the detector back-scattered electrons (Nordlys detector, Oxford Instruments) mounted on a high-resolution scanning electron microscope Mira-3LMH (Tescan) with a cold cathode field. Scanned area in the process of shooting typically is about 1.8 mm × 1.7 mm.
The alloy is investigated in two states: initial as-cast state and after annealing at 923 K (the temperature is below the “order-disorder” ($B2'$- $L2_1$) temperature) for 5 hours followed by cooling to the room temperature in the furnace. Annealing is carried out for the possible removal of the internal stresses which may appear during the melt crystallization.

3. Results and discussion

The directions in the ingot were chosen so that the axis N1 is parallel to the ingot axis. The axis N2 is perpendicular to the axis N1 and lies in a horizontal plane of the ingot. The axis N3 is perpendicular to the axes N1 and N2. The sample was is cut so that its plane of investigation is parallel to plane N1N2.
In the Fig. 1 (a) the pole figures (PF) for the {100} and {110} planes of the austenitic phase in the initial as-cast state are shown. . These two planes fully reflect the structure of the austenite because of the cubic symmetry of the phase. Planes {101} and {011} are equivalent to {110}. Analysis of the figure shows that the structure of the austenite phase of the alloy in the initial state has a highly axial texture <001> in direction N3 of the ingot.

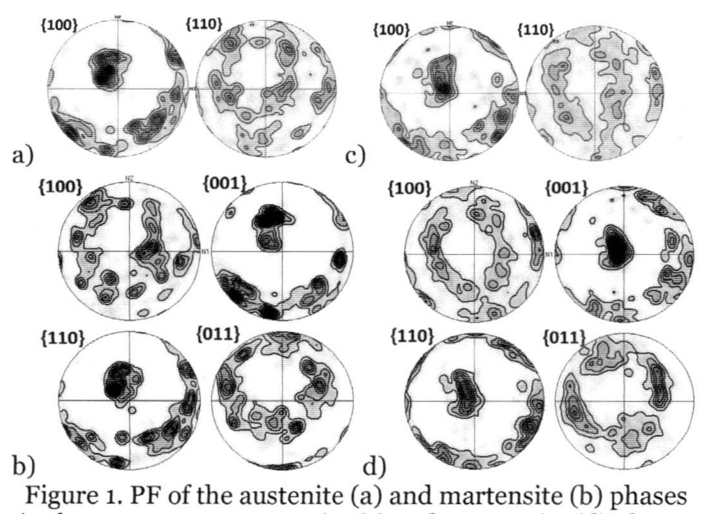

Figure 1. PF of the austenite (a) and martensite (b) phases in the as-cast state, austenite (c) and martensite (d) phases of annealed alloy.

In fig. 1 (b) the PF of the alloy martensitic phase in the initial state for the set of planes {100}, {001}, {110} and {011} are presented. The analysis of PF for the martensitic phase shows the two-component texture: <001> and <110> in the N3 direction.
The results of the structure investigation of austenite and martensite phases of the same sample after annealing at 923 K for 5 hours are presented on Fig. 1 (c, d). The PF for planes {100} and {110} of the austenitic phase of annealed alloy are presented in Fig. 1 (c). In Fig. 1 (d) the PF for planes {100}, {001}, {110} and {011} of the martensitic phase are presented. On the PF the equipotential contours of the alloy in the initial state and after annealing have the same the texture indexes. The comparison the PF of the austenite and martensitic phases of the alloy in initial and annealed states shows that after the heat treatment texture does not change. Only the slight diffusion of the intensity maxima is observed. Thus, annealing of the alloy at 923 K for 5 hours does not lead to the removal of the crystallographic texture.
The microstructure analysis of the martensite phase of both studied structural states is carried out by the optical methods. To study the microstructure from the ingot, the samples

with the size of 1 mm × 2 mm × 7 mm are cut in two mutually perpendicular directions. The long side of the first sample (the sample axis) coincides with the direction N1, in the second sample it coincides with the direction N2. The image of the martensite twin structure of both samples is shown in fig.2 (a) and 2 (b), respectively.

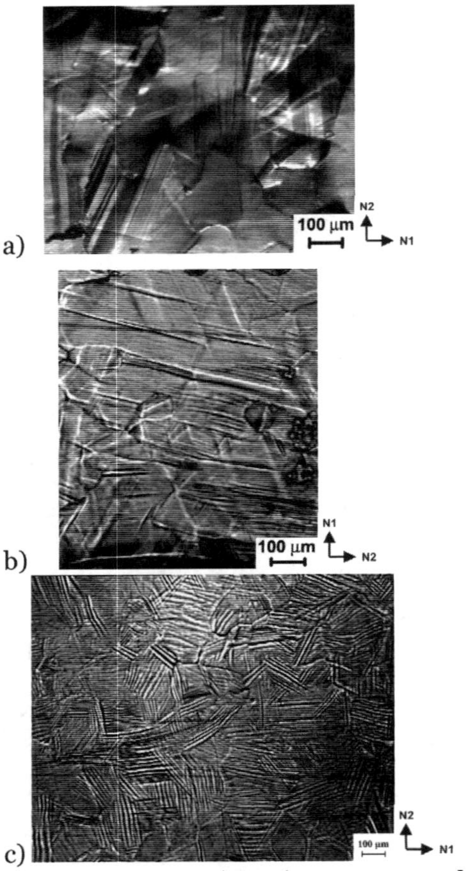

Figure 2. Martensite twin structure: a, b – initial state, c – after annealing.

It is shown that the structure of martensite has the martensite plates with width from units to several tens of micrometers. The image contrast is observed due to the formation of micro-relief on the sample surface. There is the preferential orientation of martensite plates in both cases. The majority of the martensite twins of both samples lie along the direction N2. It should be noted, that upon the repeated cycles of martensitic transformation the distribution and size of twins are not changed.

In Fig. 2 (c) the martensite structure of annealed alloy is presented. The martensite plates have the width about 15 microns. The martensite twins have a random distribution. The absence of a preferential orientation of the twins in the martensite structure can be explained by the removal of internal stress as a result of the annealing.

From the above mentioned data, it can be seen that the martensitic structures of the alloy in the initial state and after annealing are different. It is necessary to take into account the sample geometry changes during phase transformation. These processes should be correlated.

The thermal expansion was investigated on the samples, which was used for the structure investigations. In all cases the measurements was carried out along the long side of the samples. The results for both types of the samples in the initial state are shown in Figs. 3 (a) and (b). As it can be seen from the figures, the length of the samples in the austenitic and martensitic phases is linearly reduced. In the transformation temperature range curves are different. The sample which is cut in the N1 direction is abruptly shortened on 0.35% as a result of the martensitic transformation. The sample which is cut in the N2 direction is elongated on 0.35%. The martensitic structure of the first sample has a preferential orientation of the martensitic twins which are oriented perpendicularly to the axis of the sample. The structure of the second sample has a preferential orientation of the martensitic twins which are oriented parallel to the sample axis [see Figs. 2 (a) and (b)]. It can be conclude that the abrupt length change during the martensitic transformation is caused by the formation of the preferential orientation of the twins.

Figure 3 (c) shows the results for thermal expansion of the annealed sample. The thermal expansion of both types of the samples after annealing does not show the abrupt length change. Over the whole cooling range, only the monotonous reduction of sample's length is observed.

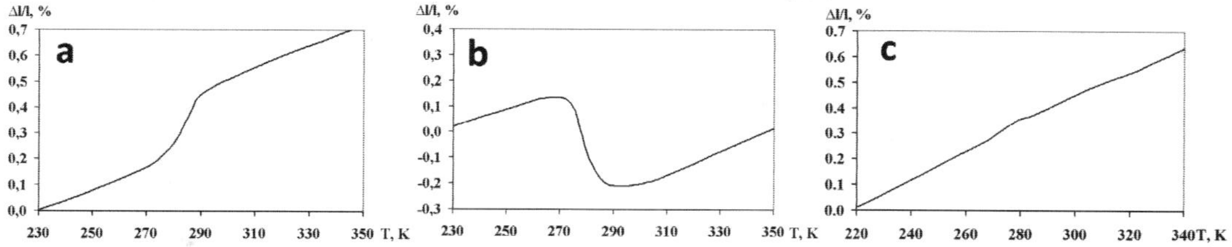

Figure 3. Thermal expansion curves of alloy in the initial state: a – the long side of the sample coincides with N1 direction, b – the long side of the sample coincides with N2 direction. c – alloy after annealing at 923 K (5 hr).

For the martensitic transformation, no abrupt change of the sample length is observed. It is shown that the preferential orientation of the martensitic twins of the annealed alloy is absent. Thus, if during the martensitic transformation the structure with a random distribution of twins is formed, then the abrupt length change of the sample is not occurred.

4. Conclusions

The polycrystalline $Ni_{2.08}Mn_{0.96}Ga_{0.96}$ alloy in the initial state has a crystallographic texture. The austenite phase has the crystallographic axial texture <001>, at the same time the martensitic phase has a two component crystallographic texture <001> and <110>.
Annealing of the alloy at 923 K (5 hr) does not lead to removal of the crystallographic texture, but affects the distribution of the martensitic twins.
The differences in the thermal expansion curves for the initial and annealed states can be explained by the formation of the preferential orientation of the martensitic twins.

This work was supported by Russian Foundation for Basic Research grant 14-02-31699-mol-a.

References

[1] Murray S J et al 2000 *Appl. Phys. Lett.* 77 **6** 886-888
[2] Sozinov A et al 2002 *Appl. Phys. Lett.* **80 10** 1746-1748
[3] Sutou Y et al 2004 *Appl. Phys. Lett.* **85 19** 4358-4360
[4] Oikawa K et al 2006 *Appl. Phys. Lett.* **88** 122507
[5] Mulyukov Kh Ya et al 2008 *Technical Physics* **53 6** 802-803
[6] Mulyukov Kh Ya et al 2011 *Phys. Met. and Metall.* **112 5** 488-494
[7] Musabirov I I 2014 *Basic Problems of Material Science* **11 2** 251-256
[8] Pramanick S et al 2013 *J. Alloys and Compounds* **578** 157-161
[9] Krenke T et al 2005 *Nat. Mater.* **4** 450
[10] Han Z D et al 2006 *Appl. Phys. Lett.* **89** 182507
[11] Mulyukov K Y et al 2012 *Letters on Materials* **2 4** 194-197
[12] Koyama K et al 2006 *Appl. Phys. Lett.* **89** 182510
[13] Yu S Y et al 2006 *Appl. Phys. Lett.* **89** 162503-1-162503-3
[14] Sozinov A et al 2002 *Appl. Phys. Lett.* **80 10** 1746-1748
[15] Albertini F et al 2001 *Journal Appl. Phys.* **89 10** 5614-5617
[16] Wang W H et al 2000 *Appl. Phys. Let.* 77 **20** 3245-3247
[17] Vasilev A N et al 2000 *Int. J. Appl. Electromagn. Mechan.* **12** 35-40
[18] W H Wang et al 2000 *J. Phys.: Condens. Mater.* **12 28** 6287
[19] Sergeev S N et al 2012 *Letters on Materials* **2 2** 74-77

IOP Conf. Series: Materials Science and Engineering **82** (2015) 012065 doi:10.1088/1757-899X/82/1/012065

Effects of thermomechanical processing on the recrystallization texture and grain size of Al-1%Si sputtering target material

XR Li[1], CL Xu[1], TL Huang[1], Y Luo[1], GL Wu[1], Q Liu[1] and X Huang[1, 2]

[1] College of Materials Science and Engineering, Chongqing University, Chongqing 400045, China

[2] Danish-Chinese Center for Nanometals, Section for Materials Science and Advanced Characterization, Department of Wind Energy, Technical University of Denmark, Risø Campus, 4000 Roskilde, Denmark

E-mail: wugl@cqu.edu.cn

Abstract. An Al-1%Si alloy was solution treated and deformed by conventional cold rolling to different strains, followed by annealing at various temperatures until complete recrystallization. The microstructures of annealed samples were characterized by electron backscatter diffraction. It is found that under optimal conditions of cold rolling and annealing, the microstructure desired for sputtering target materials with fine, uniformly sized and randomly textured grains can be obtained for the Al-1%Si alloy.

1. Introduction

Al-Si alloys are widely used as sputtering target materials in semiconductor industry for deposition of thin films [1, 2]. The key factors influencing the properties of sputtered films are the crystallographic texture and grain size. In general, a weak texture and a fine and uniform grain structure are desired for sputtering target materials. It has been well established that grain refinement in metals can be achieved by severe plastic deformation and such an approach is applicable for sputtering target materials. For example, severe plastic deformation by equal channel angular extrusion has been applied to produce sputtering targets [3]. However, this method does not align grains in a manner that facilitates uniform sputtering [4, 5]. Fine grain sizes were also obtained by deforming the materials at very low temperatures [6, 7]. It is more challenging to produce a weak or random texture in sputtering target materials. Plastic deformation in a complicated manner, i.e. clock rolling, has been developed to obtain weak textures for sputtering targets [8]. In this study, we report that Al-1%Si target with a random texture, and fine and uniform grain structure and can be produced by thermomechanical processing involving solid solution treatment, cold rolling to high strains and recrystallization annealing.

Experiment

An Al-1%Si alloy was casted from ultra-high purity (99.9996%) Al and high purity Si. The ingot was deformed at about 200°C into a thick slab, which had a mean grain size of about 150 μm. A plate with dimensions of 50 × 100 × 20 mm³ was cut from the slab, and solid solution treated at 540°C for 6 hours followed by water quenching. Then the plate was cold rolled at room temperature to 63, 86, 95 and 98% reductions in thickness, respectively. These samples were annealed at 250, 350 and 450°C for different times until complete recrystallization. The microstructures of the samples were observed in a Zeiss Auriga dual beam station, and local crystallographic orientations were characterized using an Oxford HKL Channel 5 electron backscatter diffraction (EBSD) detector attached to the microscope. For EBSD measurements, the specimen surface was mechanically polished followed by electrochemical polishing in the A2 solution. A step size of 2μm was used in EBSD scanning. The grain size was determined by the equal circular diameter method of the Channel software with a 5° critical angle in the data processing.

Results

The distribution of Si particles after solid solution treatment and recrystallization annealing at 350°C for 6 h are shown in Fig. 1(a) and Fig. 1 (b), respectively. It is seen that after solid solution treatment at 540°C for 6 hours, majority of Si particles have been dissolved. However, a small amount of coarse particles or particle clusters are retained, distributing along the rolling direction (RD) of the raw plate. After recrystallization annealing at 350°C for 6 h, fine nano-sized Si particles precipitate and fairly randomly distribute in the matrix. There are some large particles of a few micrometers. These large particles could be formed by ripening of the retained particles during annealing.

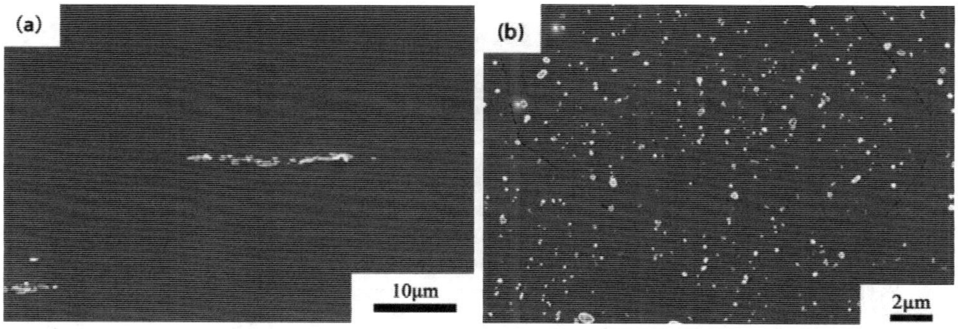

Fig. 1 SEM micrographs of Si particles. (a) after solid solution treatment and (b) after cold rolling to 86% and annealing at 350°C.

Figure 2a shows the average grain size of all treated samples after complete recrystallization. A general observation is that the samples annealed at 450°C show larger average grain sizes than the samples annealed at 250°C. For both annealing temperatures of 250 and 450 °C, the average grain size decreases with increasing deformation strain, except for the case of 98% deformed sample annealed at 450°C. However, for the annealing at 350°C, the evolution shows a more complicated pattern with the 86% deformed sample showing the finest grain size after annealing. Another general observation is that with increasing strain, the grain structure becomes more homogeneous, as shown by the standard deviation of grain size distribution in Fig. 2b. The 63% deformed sample has the largest spread of grain sizes (20-40μm), whereas the 98% deformed sample the smallest (15-20μm). As for the recrystallization texture, a rather weak (random) texture was developed in all recrystallized samples. The 250°C annealed samples have ~9-23% rolling texture components, while 350°C and 450°C annealed samples only have less than 10% rolling texture components. All samples have ~3-5% of cube texture.
It is found that the sample deformed to 86% and annealed at 350°C for 6h has the smallest grain size and a random texture. Figure 3 (a) shows the EBSD map of the sample. It is clear

that the material is completely recrystallized, and the grains are fine and equiaxed. The recrystallization texture of Al-1%Si is shown in Fig. 3 (b). It can be seen from the orientation distribution function (ODF) map that the recrystallized texture of the alloy is relatively random. Beside weak textures around cube {100}<001>, rotated cube {100}<011> (R-cube) and rotated Goss {110}<110> (R-Goss), other are random orientations.

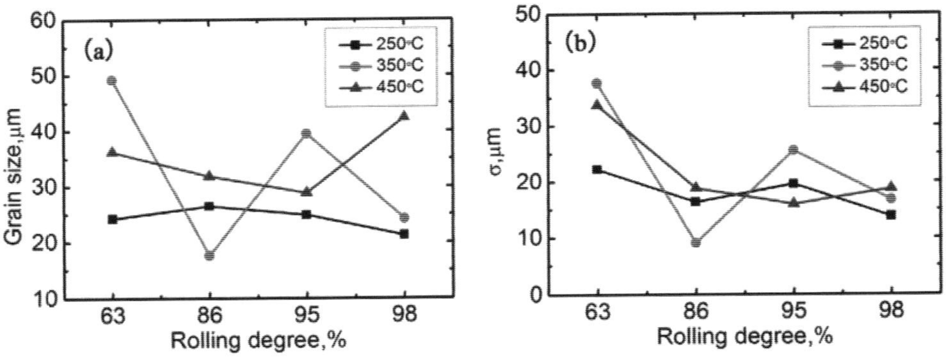

Fig. 2. (a) Average grain size and (b) standard deviation of grain size distribution of Al-1%Si after complete recrystallization.

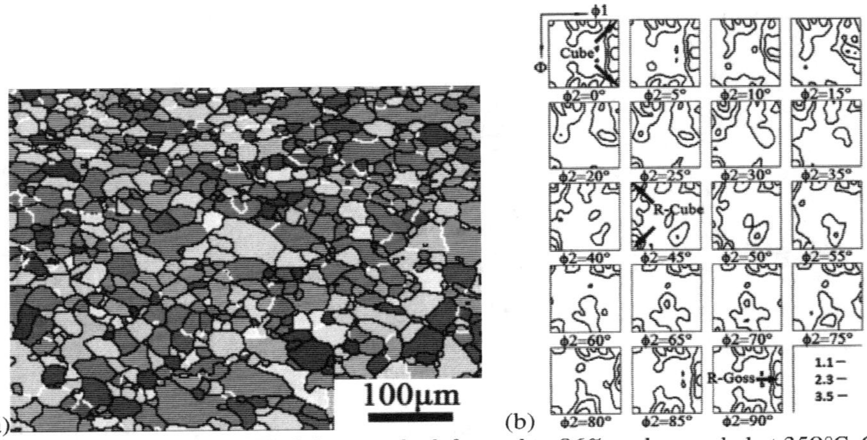

Fig. 3 (a)EBSD map and (b) ODF of the sample deformed to 86% and annealed at 350°C. On the EBSD map, black and white lines represent boundaries of misorientation≥ 15° and 2-15°, respectively.

Table 1 shows the orientation content ratio (represented by the percentage of grains with their normal parallel to a specific crystallographic direction [7]) of the sample deformed to 86% and annealed at 350°C for 6h. It is seen that the percentages of {200} and {220} orientations are about 20-25%, and the {311} orientations account for about 31%, while the {111} orientations are greatly suppressed. The combination of a weighted {200} orientation and balanced {111}, {220} and {311} orientations is expected to give uniform sputter properties from the sputter target [7] made from the present alloy.

Table 1 Orientation content ratio of the sample deformed to 86% annealed at 350°C

	{111}	{200}	{220}	{311}
Ratio, %	9	26	21	31

Discussion

Usually the average recrystallization grain size decreases with increasing deformation strain since the stored energy increases with strain and thus promotes recrystallization nucleation, which results in a reduced average grain size after recrystallization. For the present alloy, the Si element was present in the solid solution before deformation. The over-saturated Si atoms precipitate during annealing concurrent with the recrystallization of the Al matrix. Precipitated fine particles can pin the migration of grain boundaries during recrystallization due to the Zener-drag effect. Therefore the grain size is mainly controlled by the distribution and size of Si precipitates. For the samples annealed at 250°C, the Si precipitates are all fine and randomly distributed. So the average grain sizes are generally small. For the samples annealed at 450°C, precipitates are coarsened at the temperature, so the grain size are relative large. Especially for the sample deformed to 98% and annealed at 450°C, precipitations are ripped to large sizes so that the pinning effect of particles is rather weak. Very coarse grain sizes are obtained in the sample after recrystallization. For the samples annealed at 350°C, the large grain size may be related to the short annealing time so that the precipitations are not sufficient and extensive growth occurred. As the case of the sample deformed to 86% and annealed at 350°C, Si particles are dispersed and uniformly distributed (Fig 1b) and thus a small grain size is achieved (Fig. 3a).

A texture dominated by random orientations is developed in this alloy after recrystallization. The volume fractions of rolling texture components are very small. This manifests that nucleation from deformation microstructure with rolling texture components is very limited. The reason for the formation of the random texture could be the particle stimulated nucleation since there are large sized particles in the materials before deformation (Fig. 1a) [9, 10]. Another reason could be the confined growth of recrystallizing grains during annealing due to the presence of fine Si particles. The growth advantage of grains with specific orientations, i.e. cube, will be inhibited. Therefore all grains may grow at similar rates, leading to a random texture.

Conclusion

An Al-1%Si alloy was solid solution treated followed by cold rolling and annealing. The microstructural and textural evolutions were characterized by EBSD. Most of Si element has been dissolved after solution treatment and fine Si precipitates form during recrystallization annealing. Fine and uniform grains as well as a random recrystallization texture are developed in the alloy deformed to 86% followed by complete recrystallization at 350°C.

Acknowledgements

The authors wish to thank the financial support of the NSFC of China (51327805 and 51171214) and Fundamental Research Fund of Central Universities (CDJZR14135502, CDJZR11130007 and CDJZRPY14130001).

References

[1] Legresy J M, Marticou M H 1992 US Patent 5160388
[2] Fukuyo H, Nagasawa M 1995 US Patent 5456815
[3] Segal V, Willett W B, Ferrasse S 2004 US Patent 6723187B2
[4] Valiev R Z, Islamgaliev R K 1998 Phys Metal. Metallog. **85**, 367
[5] Nakashima K, Horita Z, Nemoto M 1998 Acta Mater. **46**, 1589
[6] Lo C F, Draper D 1999 US Patent 5993575
[7] Perry A C, Gilman P S, Van den Sype P 2005 US Patent 6942763B2
[8] Wickersham C E, Levit V, Alexander P T 2012 US Patent 8231745B2
[9] Humphreys F J 1977 *Acta Metall.* **25** 1323
[10] Chen YL, Huang TL, Gong X, Wu GL, Liu Q, Huang X 2013 Mater. Sci. Forum **753**, 243

IOP Conf. Series: Materials Science and Engineering **82** (2015) 012066 doi:10.1088/1757-899X/82/1/012066

Uncertainty in retained austenite measurements applied to individual crystallographic orientations

A Creuziger[1] **and T Gnäupel-Herold**[2]

[1] Materials Science & Engineering Division, National Institute of Standards and Technology (NIST), Gaithersburg, MD, USA 20899

[2] NIST Center for Neutron Research, National Institute of Standards and Technology, Gaithersburg, MD, USA 20899

E-mail: adam.creuziger@nist.gov

E-mail : tgh@nist.gov

Abstract. A technique to measure the phase volume fraction of an individual orientation and the uncertainty in the measurement is demonstrated in this paper. The technique of complete pole figure averaging using neutron diffraction was used to assess the phase fraction of retained austenite in transformation induced plasticity (TRIP) steels and quantify the uncertainty in the phase fraction. In parallel, an ensemble of orientation distribution functions was calculated to assess crystallographic volume fractions of particular orientations and the uncertainty of these volume fractions using Monte Carlo techniques. These methods were combined to measure the retained austenite phase volume fraction of an individual orientation.

1. Introduction

The accurate measurement of the phase fraction of retained austenite in manufactured steels remains a challenge. Protocols for measurement of retained austenite using x-ray diffraction [1], [2] are often employed, despite the caveat that these methods only apply to uniform (random) texture distributions. Due to the strong crystallographic texture caused by deformation during processing, these assumptions are typically not valid for rolled sheet steel. Current interests in transformation induced plasticity (TRIP) steels have brought a new application to retained austenite measurements that are outside the scope of uniformly distributed textures.

The overall phase fraction of austenite in TRIP steels is of interest as the rate of transformation is thought to strongly affect mechanical strength and ductility [3]. There is data to indicate that the transformation will not be distributed evenly as a function of orientation, but particular stress states will cause some orientations to transform at a higher rate than other orientations [4]. Therefore, the distribution of retained austenite as a function of crystallographic orientation and the rate of transformation of a particular orientation is also of interest.

The accurate measurement of retained austenite is difficult enough [5], but to compare between retained austenite values, not only are the values themselves required, but a metric on the uncertainty as well. Recently, a technique using complete pole figure averaging using neutron diffraction was developed to measure the retained austenite in textured TRIP steels as well as provide an estimate on the uncertainty in the phase fraction [6]. These techniques provide a method to assess the overall phase fraction of retained austenite.

To assess individual orientations, volume fractions from orientation distribution functions (ODFs) are required. Pole figures recorded as part of the phase fraction measurement were used to calculate these ODFs. As with the phase fraction of retained austenite, not only is a metric on the volume fraction required, but also the uncertainty in the volume fraction.

A technique has been demonstrated to measure the uncertainty at each point in the ODF using Monte Carlo methods on the diffracted neutron counts at each location in the pole figures [7]. By combining the values and uncertainties of phase fraction and crystallographic orientation volume fraction measurements, the retained austenite volume fraction at a particular orientation can be calculated.

2. Experimental methods

These methods were applied to an as-received TRIP 700 sample. The sample was measured using the BT8 beamline at the NIST Center for Neutron Research [8]. Four complete austenite pole figures and three complete ferrite pole figures were measured on the same sample with three different pole figure resolutions on a hexagonal mesh: 4° (highest resolution), 5° and 6° (lowest resolution). The sample was not removed or remounted between scans. The acquisition time was constant per point, so the total time for a complete pole figure is approximately 1.4 times more for the 5° mesh compared to the 6° mesh, and 2.2 times longer for the 4° mesh compared to the 6° mesh.

3. Results and discussion

Pole figure data were analyzed using the program PF [9] to measure the phase fraction of retained austenite. Each point in the pole figure was summed together to create a single peak for each reflection. The channel width for peak fitting (using Gaussian functions) was set to 4 standard deviations on either side of the peak to ensure an accurate background value, with the exception of the 111 austenite peak which was done manually due to interference from the 110 ferrite peak. These peak and background values were then used to calculate the phase fraction as described in [6]. From these data, phase fractions and uncertainties were calculated, shown in Table 1.

Table 1: Phase fraction of retained austenite, calculated by the procedure described in [6].

Mesh	Retained Austenite Phase Fraction	Absolute Uncertainty	Relative Uncertainty
6° (331 pts)	0.147	0.0142	9.66%
5° (469 pts)	0.146	0.0134	9.18%
4° (721 pts)	0.145	0.0131	9.03%

Table 1 shows that the phase fraction is the same within the uncertainty of the measurement. The relative uncertainty values are nearly the same as those observed in [6]. The relative uncertainty decreases slowly as the number of points increase, decreasing only by 0.5% when twice as many points were measured.

ODF calculations were performed using the texture analysis package mTex [10]. The rest of this paper will focus on the austenite ODF. Triclinic sample symmetry and cubic crystal symmetry were applied in the analysis. In order to estimate the uncertainty, a Monte Carlo approach was implemented in mTex. Individual points on the pole figure were perturbed using Poisson noise proportional to the peak and background intensities. These perturbed pole figures were then used to calculate an ODF. Crystallographic volume fractions (VFs) of selected crystallographic orientations were extracted over a series of misorientation angle thresholds.

For the examples shown here 5000 perturbed ODFs were calculated. From this ensemble of ODFs a mean and standard deviation were calculated for seven different orientations commonly investigated in face centered cubic materials: Cube, Goss, RGoss, RCube (Shear), Copper, Brass and S. Three different misorientation angle thresholds were investigated for each of the orientations: 30°, 10°, and 1°. The term error in this calculation is defined as (unperturbed – mean) / unperturbed. This paper does not address the absolute error quantified from a known ODF, nor is the difference between the input pole figures and the recalculated pole figures considered. Table 2 shows the ODF volume fractions for selected orientations (Cube and Brass) as a function of mesh density and misorienation angle.

Table 2: Volume fractions of Brass and Cube orientations as a function of misorientation for mesh resolutions of 6°, 5° and 4°. The terms 'Mis Angle' refers to the misorientation angle threshold, 'g' for the particular crystallographic orientation. 'Uniform VF' is the volume fraction of a uniform texture distribution for the orientation and misorientation angle indicated. 'ODF VF' refers to volume fraction extracted from the ODF calculated from the unperturbed pole figures. 'Mean VF' refers to mean volume fraction extracted from the ensemble of ODFs calculated from perturbed pole figures. 'St Dev VF' refers to standard deviation of volume fractions extracted from the ensemble of ODFs calculated from perturbed pole figures. 'Rel Uncert' refers to the standard deviation divided by the mean volume fraction. ×10

Mesh	Mis Angle	g	Uniform VF	ODF VF	Mean VF	Relative Error	St Dev VF	Rel Uncert
6°	30°	Cube	1.80E-01	1.55E-01	1.56E-01	0.81%	2.43E-03	1.56%
6°	10°	Cube	6.76E-03	6.00E-03	6.18E-03	3.01%	4.24E-04	6.86%
6°	1°	Cube	6.77E-06	6.13E-06	6.40E-06	4.26%	7.09E-07	11.08%
6°	30°	Brass	1.80E-01	2.46E-01	2.49E-01	1.11%	2.58E-03	1.04%
6°	10°	Brass	6.76E-03	2.02E-02	2.04E-02	1.10%	7.54E-04	3.70%
6°	1°	Brass	6.77E-06	2.34E-05	2.36E-05	0.75%	1.35E-06	5.72%
5°	30°	Cube	1.80E-01	1.54E-01	1.54E-01	0.18%	1.50E-03	0.97%
5°	10°	Cube	6.76E-03	6.18E-03	6.19E-03	0.18%	3.14E-04	5.07%
5°	1°	Cube	6.77E-06	6.43E-06	6.46E-06	0.55%	6.33E-07	9.79%
5°	30°	Brass	1.80E-01	2.49E-01	2.50E-01	0.34%	1.83E-03	0.73%
5°	10°	Brass	6.76E-03	2.05E-02	2.08E-02	1.41%	5.73E-04	2.75%
5°	1°	Brass	6.77E-06	2.51E-05	2.56E-05	1.73%	1.38E-06	5.39%
4°	30°	Cube	1.80E-01	1.56E-01	1.56E-01	0.08%	1.46E-03	0.93%
4°	10°	Cube	6.76E-03	7.05E-03	6.96E-03	1.26%	3.43E-04	4.93%
4°	1°	Cube	6.77E-06	7.92E-06	7.75E-06	2.16%	8.31E-07	10.72%
4°	30°	Brass	1.80E-01	2.47E-01	2.48E-01	0.34%	1.63E-03	0.66%
4°	10°	Brass	6.76E-03	2.15E-02	2.18E-02	1.37%	5.33E-04	2.45%
4°	1°	Brass	6.77E-06	2.52E-05	2.56E-05	1.54%	1.46E-06	5.72%

In all cases, uncertainty increases as the misorientation angle decreases. This behavior is expected as a small misorientation angle means the ODF is only being evaluated at a few points, which are varying. When summing many points together for a volume fraction, the effect of variation at each of those points often attenuates the uncertainty. Similarly, the error generally decreases as the mesh resolution is reduced, but increases as the misorientation angle is decreased. Error is less than uncertainty in most cases, but often of a similar magnitude.

The root sum of squares method of uncertainty propagation was used to assess the combined effects of phase fraction uncertainty and ODF volume fraction uncertainty, shown in Table 3. These values shown in Table 3 are then a measure on the phase volume fraction of an individual orientation and the uncertainty of that phase volume fraction.

Table 3: Phase volume fraction of an individual orientation and the uncertainty for 10° misorientation angle and 3 different mesh resolutions. 'V' represents the phase volume fraction, and 'sigma' the propagated standard deviations.

Mesh	Cube		Brass	
	V	Sigma	V	Sigma
6°	9.08E-04	1.08E-04	2.99E-03	3.10E-04
5°	9.04E-04	9.47E-05	3.04E-03	2.91E-04
4°	1.01E-03	1.04E-04	3.16E-03	2.96E-04

4. Summary

The combination of phase fraction and crystallographic volume fractions has been demonstrated in this paper. The uncertainties in the phase fraction measurement and crystallographic volume fractions have been quantified and used to provide a metric on the uncertainty in the phase volume fraction. By measuring the phase volume fraction as a function of applied strain, the effect of orientation on the overall deformation can be measured and compared to predictions for transformation rate [4].

Acknowledgements

The authors acknowledge the support of the National Institute of Standards and Technology, U.S. Department of Commerce, in providing the neutron research facilities used in this work, as well as the NIST Center for Automotive Lightweighting.

References

[1] Jatczak C F et al 1979 SAE SP-453
[2] ASTM International 2003 E975
[3] Lani F et al 2007 *Acta Mat* **55** 3695
[4] Creuziger A and Foecke T 2010 *Acta Mat* **58** 85
[5] Jacques P J et al 2009 *Mat Sci Tech* **25** 567
[6] Gnäupel-Herold T and Creuziger A 2011 *MSEA* **528** 3594
[7] Creuziger A et al 2014 *Scripta Mat* **72-73** 55
[8] Brand P C et al 1998 *Physica B* **241-243** 1244-1245
[9] Gnaeupel-Herold T (2012) BT8 Data analysis program PF
http://www.ncnr.nist.gov/instruments/bt8/BT8DataAnalysis.htm
[10] Hielscher R and Schaeben H 2008 *J of Appl Cryst* **41** 1024

IOP Conf. Series: Materials Science and Engineering **82** (2015) 012067 doi:10.1088/1757-899X/82/1/012067

Texture developed during deformation of Transformation Induced Plasticity (TRIP) steels

M Bhargava[1*], C Shanta[2], T Asim[1] and M Sushil[1]
[1]Department of Mechanical Engineering, IIT Bombay, India
[2]Department of Metallurgical Engineering & Materials Science, IIT Bombay, India
E-mail: mbhargava@iitb.ac.in

Abstract. Automotive industry is currently focusing on using advanced high strength steels (AHSS) due to its high strength and formability for closure applications. Transformation Induced Plasticity (TRIP) steel is promising material for this application among other AHSS. The present work is focused on the microstructure development during deformation of TRIP steel sheets. To mimic complex strain path condition during forming of automotive body, Limit Dome Height (LDH) tests were conducted and samples were deformed in servo hydraulic press to find the different strain path. FEM Simulations were done to predict different strain path diagrams and compared with experimental results. There is a significant difference between experimental and simulation results as the existing material models are not applicable for TRIP steels. Micro texture studies were performed on the samples using EBSD and X-RD techniques. It was observed that austenite is transformed to martensite and texture developed during deformation had strong impact on limit strain and strain path.

1. Introduction

Advance high strength steels (AHSS) are often the material of choice for light weight structures as they combine various attractive properties such as formability, strength and weld ability. It is predicted that the usage of AHSS will increase to 35% by 2015 whereas mild steel would decrease to 29% [1]. TRIP steels are first generation steels with low content in Mn. It has a microstructure consisting of ferrite and austenite. During plastic deformation the metastable retained austenite phase is transformed into martensite. This transformation helps in enhancing the strength and ductility of the steel. Mechanical and microstructure characterization were done for TRIP steel to study the forming behaviour [2-6]. However limited work is done on complete Strain path diagram (SPD). Due to texture development and phase transformation during deformation there is huge difference in experimental and simulation results. This is the motivation for the present work. The present work is focused on to study the complete SPD by experimental and simulated method. It is observed that there is a difference between experimental and simulated SPD as the existing material models are not applicable for TRIP steels and there is a strong impact on limit strain and strain path as during deformation austenite is transformed to martensite. A microstructure analysis was done to understand these behaviours.

2. Experimental procedure

The samples were cleaned with acetone and painted with black on one side. The samples were dried for few hours and white paint was sprinkled on it to produce a random speckle pattern. Lubrication system was applied on the other side of the sample to reduce friction. The samples were deformed in a hydraulic press with a hemispherical punch. The deformation process was captured with two high resolution high-speed cameras and post processing was done to calculate the surface strains using Digital Image Correlation. The simulation was done on PAMSTAMP 2G with the help of mechanical properties obtained from tensile tests. EBSD and X-RD techniques were used for micro structural characterization. Samples were cut near failure zone and followed by mechanical polishing, diamond polishing and finally with colloidal silica polishing. EBSD measurements were done on an area of 45 x45 micron with step size of 0.05 and confidence index of 0.1 is maintained for all the scans. Results obtained from EBSD measurements were used to calculate phase fraction and grain average misorientation (GAM) values. The orientation distribution function (ODF), maximum ODF, texture index and r value is computed with the help of MTM-FHM.

3. Results and discussion
Experimental and simulated SPD of TRIP steel was calculated and shown in Fig. 1. The experimental SPD was plotted by creating a point near the localized region. The simulated strain path was plotted by selecting points where the mesh elongation starts. Solid lines and dotted lines represent the experimental and simulation results respectively. Black solid curve and black dotted curve represents failure and green solid curve and green dotted curve represents localisation curves respectively. Simulation results are matching with the experimental data in drawing and plane strain region but not in stretching region. It is observed that strains are decreasing in stretching region. Microstructural studies are done for this peculiar behaviour. The difference in simulation results was due to the material models used in this study does not include microstructure evolutions.

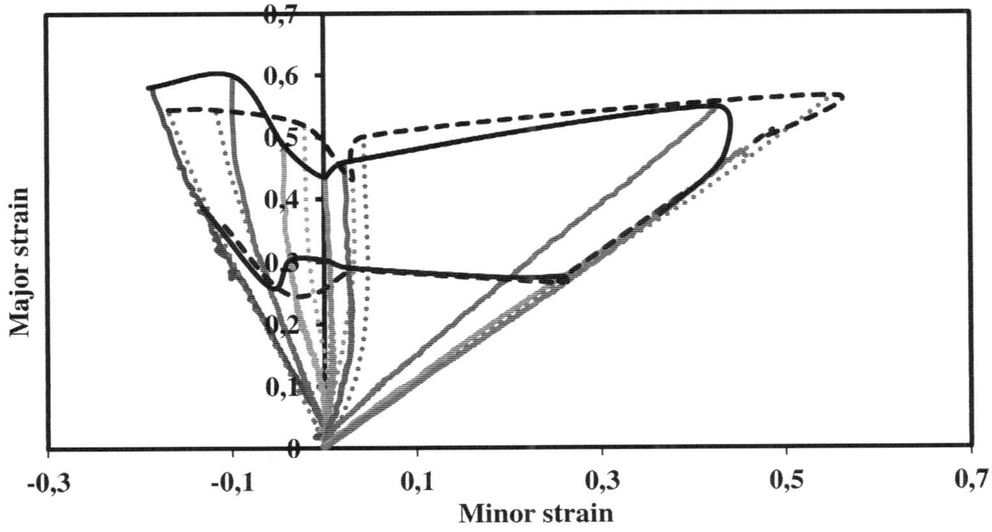

Fig.1: Experimental and simulated strain path diagram of TRIP steel

The variation of maximum ODF and texture index at failure zone for TRIP steel are shown in Fig. 2. It is observed that ODF and texture index decreases from uniaxial drawing to plane strain followed by stretching region. This gives some indication about the effect of strain path on the development of texture in TRIP material. Figure 3 shows the variation of normal anisotropy 'r' value for deformed specimens of TRIP steel and compared with the undeformed sample. It is observed that r value decreases in deformed samples as compared to undeformed but the rate of decrease of r value is more in uniaxial drawing than that of biaxial stretching.

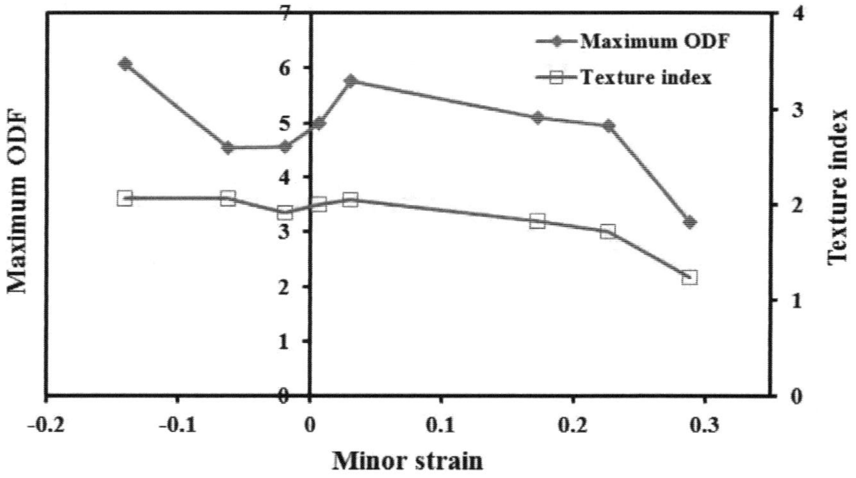

Fig. 2: Maximum ODF and Texture index of TRIP specimens for all strain paths with respect to minor strain.

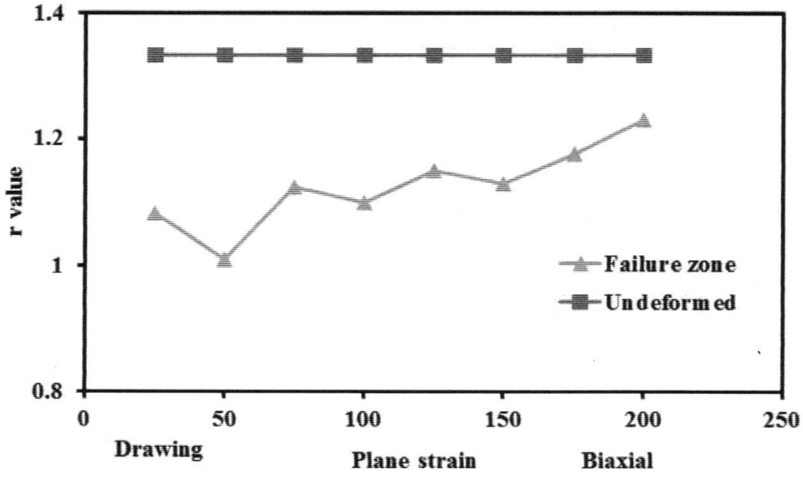

Fig. 3: Comparision of r value for undeformed and deformed specimens of TRIP steel for all strain paths.

Micro-texture and phases of undeformed and deformed specimens of TRIP steel for extreme strain paths, such as uniaxial drawing, plane strain and biaxial stretching at high strains were studied. It is observed that austenite was decreasing in deformed samples and transforming to martensite. This transformation helps in improving the strength and ductility of steel. Effect of strain paths on development of GAM and phase transformation are shown in Fig. 4. In biaxial stretching the development of GAM and rate of phase transformation is more compared to plane strain and uniaxial drawing.

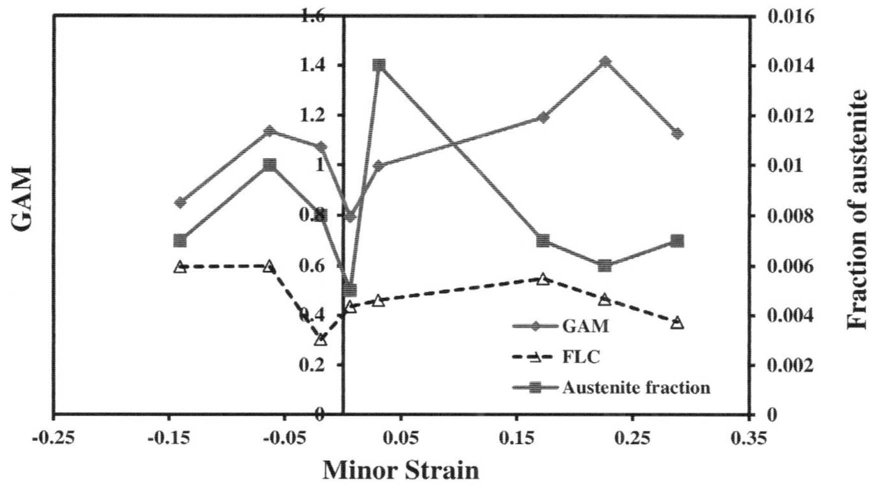

Fig. 4: Phase transformation and development of GAM in deformed specimens of TRIP steels for different strain paths.

The phase transformation is mostly taking place in stretching region. So after finite element (FE) simulation it was observed that there is a large difference between experimental and simulated strain paths in stretching region. To reveal this observation a material model for TRIP steel based on the evolution of microstructure has to be developed.

4. Conclusion

The phase transformation and development of texture during deformation had strong impact on limit strain and strain path. The strain decreases in stretching region. In stretching region the development of GAM and rate of phase transformation is more compared to plane strain and uniaxial drawing. Maximum ODF value, texture index decreases in stretching region. Deviation of experimental and simulated strain paths were observed in stretching region as the existing material models are not applicable for TRIP steels.

References

[1] Emmanuel M et al 2010 *J. Iron and Steel Technol.* **7** 1-13
[2] Zhao J et al 2012 *J. Iron and Steel Research, Int.* **19** 57-62
[3] Tian Y and Zhuang L 2012 *J. Iron and Steel Research, Int.* **19** 47-52
[4] Ding H et al 2011 *J. Iron and Steel Research, Int.* **18** 36-40
[5] Koh-Ichi S et al 2006 *J. Mat. Proc. Technol.* **177** 390–395
[6] Zhang M et al 2011 *J. Iron and Steel Research, Int.* **18** 73-78

IOP Conf. Series: Materials Science and Engineering **82** (2015) 012068 doi:10.1088/1757-899X/82/1/012068

Evolution of microstructure and crystallographic texture in severely cold rolled high entropy equiatomic CoCrFeMnNi alloy during annealing

GD Sathiaraj[1], C Lee[2], CW Tsai[2], JW Yeh[2], PP Bhattacharjee[1]
[1]Department of Materials Science and Engineering, IIT Hyderabad, AP, India
[2]Department of Materials Science and Engineering, NTHU, Hsinchu, Taiwan
E-mail: ms12p0003@iith.ac.in, pinakib@iith.ac.in

Abstract. An equiatomic FCC CoCrFeMnNi high entropy alloy (HEA) was heavily cold rolled up to 90% reduction in thickness followed by isochronal annealing for 1 hour at temperatures ranging between 700°C to 1100°C. A strong brass texture was observed in the cold-rolled condition indicating the low stacking fault energy of the material. A fine stable microstructure was observed during annealing at low temperatures. The recrystallization texture was characterized by the presence of deformation texture components, in particular, the α-fiber (ND//<110>), S ({123} <634>) and the typical brass recrystallization texture component ({236} <385>). Annealing twins were shown to have important effect on the formation of annealing texture.

1. Introduction

High entropy alloys (HEAs) are recently developed multicomponent alloys consisting of at least five or more elements in equimolar or near equimolar proportion but with simple crystal structures e.g. FCC, BCC or FCC+BCC [1]. The retention of simple crystal structure instead of formation of complex intermetallic phases is attributed to large configurational entropy due to mixing of a large number of elements [1]. Recently, many multicomponent alloy systems with simple crystal structures have been proposed. Amongst these FCC CoCrFeMnNi alloy has attracted much attention due to excellent thermal stability, good ductility and other attractive properties [2-3].

Recent studies show that the single phase structure is stable in this alloy even after annealing for 72h at 1000°C [4]. It is expected that appropriate thermo mechanical processing can greatly enhance the properties of these alloys. However, this necessitates appropriate understanding of microstructure and texture evolution during thermo-mechanical processing in these alloys. In the present work effect of heavy cold-rolling and annealing on microstructure and texture evolution in CoCrFeMnNi is investigated.

2. Experimental procedure

An equiatomic CoCrFeMnNi alloy was used in the present study. As received block having dimensions 50 mm (length) × 8 mm (width) × 8 mm (thickness) were produced by vacuum arc melting. To enhance the chemical homogeneity the blocks were homogenized for 6 h at 1100°C. In order to breakdown the coarse microstructure (average grain size ~200 μm), samples of ~25 mm (length) × 8 mm (width) × 5 mm (thickness) were cut from the blocks and then cold rolled to ~ 50% reduction in thickness using a laboratory scale rolling mill followed by annealing at 800°C for 1 h in a salt bath furnace. The fully recrystallized samples with thickness of 2.5 mm were used as starting material. The grain size of starting material was ~8 μm. The starting materials were cold-rolled to ~90% reduction in thickness to a final thickness of 0.25 mm.

The microstructure and microtexture of the deformed and annealed samples were investigated by Electron Back Scatter Diffraction (EBSD) unit attached to a FE-SEM (Supra 40; Carl-Zeiss, Germany) operated at 20kv. The EBSD measurements were taken from ND-RD plane for 90% cold rolled samples.

For annealed samples the scans were taken in both ND-RD and rolling planes (RD–TD plane). The volume fraction of individual texture components were calculated using a cut off angle of 15° around the ideal locations in the Euler space. TSL-OIM™ software was used for an analysis of the acquired EBSD data.The samples were prepared by mechanical polishing followed by electropolishing using a mixture of methanol and percholoric acid (9:1 by volume).

3. Results and discussion

Figure 1 (a) shows the grain boundary (GB) map of starting material. The GB map clearly shows the typical microstructure of fully recrystallized material. The average grain size of the starting material is ~8 μm (excluding the Σ3 boundaries). Figure 1 (b) shows the pole figure (PF) of the starting material which indicates the presence of rather weak texture.

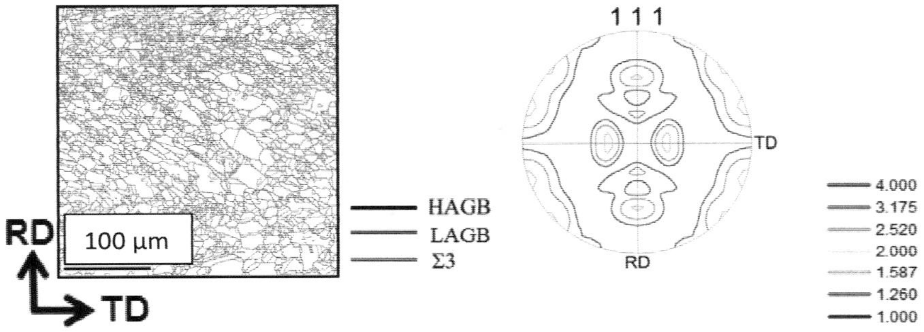

Fig. 1: (a) Grain boundary map and (b) (111) PF of the starting material.

Microstructure and texture of the severely deformed material

The image quality map clearly shows the heavily fragmented structure and thin shear band also observed in the microstructure (~25° to rolling direction indicated by arrow mark). Fig 2(b) and (c) show the grain orientation map (GO Map) and pole figure of the 90% cold-rolled sample. It shows the strong brass (green) texture indicating low stacking fault energy this alloy [5].

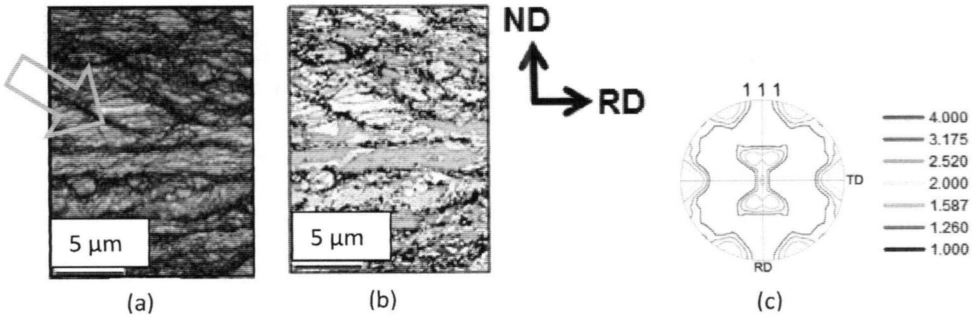

C	Cu	S	Bs	G	G/B	A	BR
{001}<100>	{112}<111>	{123}<634>	{110}<112>	{110}<001>	{110}<115>	{110}<111>	{236}<385>

Fig. 2: (a) IQ map, (b) GO map and (c) (111) PF of 90% cold rolled material.

Microstructure and texture of the annealed material

Figures 3(a) and (b) show the grain orientation (GO) maps obtained from the RD-TD and ND-RD planes of the 700°C material, respectively. The GO maps obtained from rolling and ND-RD planes show fully recrystallized microstructure with average grain size 1.4μm and 1.1μm, respectively. The appearance of the (111) pole figures (Figures 3(c) and 3(d)) suggests presence of α-fiber component, namely, Brass/Bs (green), G/B (deep rose; {011}<115>)) and (A ((110)[1$\bar{1}$1]). Presence of S (yellow) is also noticed. Significant presence of the brass recrystallization component (BR) {236} <3 8 -5> is observed. A very similar average grain size ~34 μm is obtained from the RD-TD (Fig.4 (a)) and ND-RD planes (Fig.4 (b)), respectively after annealing at 1100°C. The GO maps also show the presence of very similar texture components observed in case of the material annealed at 700°C. The PFs do not reveal any substantial change in texture (Fig. 4 (c) and (d)).

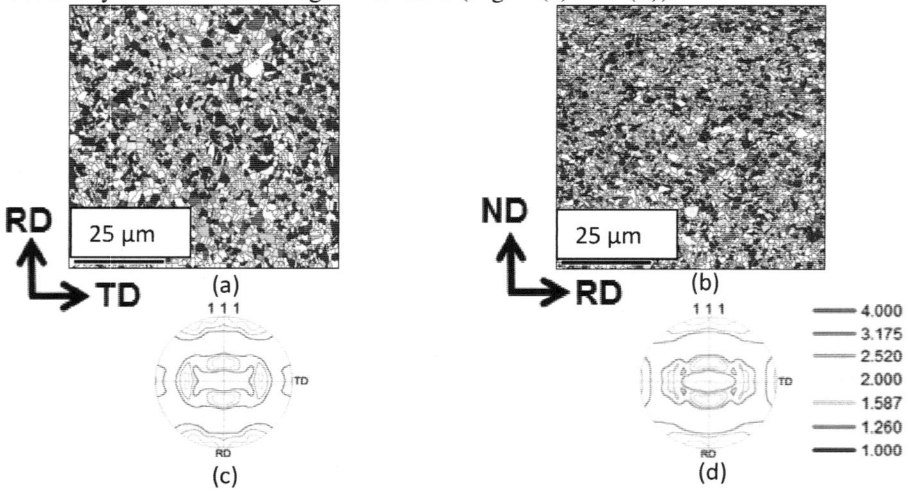

Fig. 3: GB maps (a, b) and (111) PFs (c, d) of the material annealed at 700°C obtained from the RD-TD (a,c) and ND-RD (b,d) planes. Color code is same as in Fig.2.

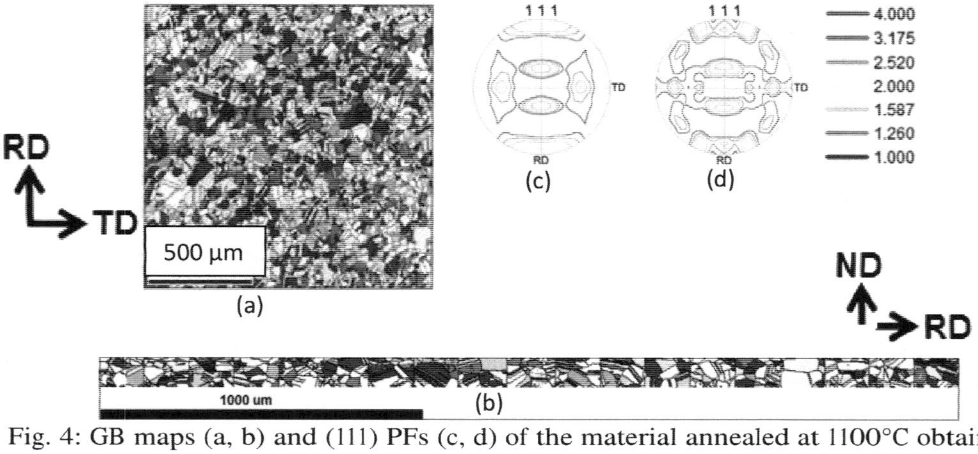

Fig. 4: GB maps (a, b) and (111) PFs (c, d) of the material annealed at 1100°C obtained from the RD-TD (a,c) and ND-RD (b,d) planes. Color code is same as in Fig.2.

The retention of deformation texture components is explained by the mechanism of discontinuous recrystallization without preferential orientation selection [6]. The development of α – fiber components, such as, A is explained by the formation of annealing twins [6]. The present analysis shows significant presence of the BR component in heavily deformed and annealed FCC HEAs.

The through-thickness homogeneity in texture of materials annealed at 700°C (Fig.5 (a)) and 1100°C (Fig.5 (b)) is shown quantitatively in Fig.5. Remarkable through-thickness homogeneity in texture is obtained after annealing at different temperatures.

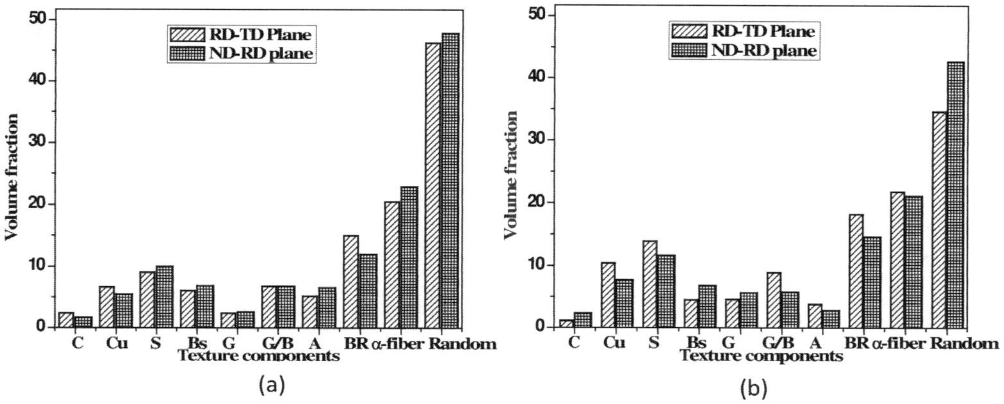

Fig. 5: (a) and (b) show the individual texture components in materials annealed at 700°C and 1100°C, respectively.

4. Conclusions

- Development of heavily fragmented microstructure after severe cold rolling is observed. The development of the strong brass type texture indicates the low stacking fault energy of this alloy.
- Ultrafine grain structure with mean grain size of 1.1 μm was obtained through conventional cold rolling and subsequent annealing at 700°C.
- The recrystallization texture has significant presence of the α – fiber components, S and BR components.
- Remarkable through-thickness homogeneity of microstructure and texture is obtained after annealing.

Acknowledgement

The authors acknowledge the financial support of DST-India through grant no. SB/S3/ME/0047/2013.

References

[1] Yeh JW, Chen SK, Lin SJ, Gan JY, Chin TS, Shun TT, Tsau CH and Chang SY 2004 *Adv. Eng. Mater.* **6** 299.
[2] Liu WH, Wu Y, He JY, Nieh TG and Lu ZP 2013 *Scripta Mater.* **68** 526.
[3] Otto F, Dlouhy A, Somsen Ch, Bei H, Eggeler G and George EP 2013 *Acta Mater.* **61** 5743
[4] Otto F, Yang Y, Bei H and George EP 2013 *Acta Mater.* **61** 2628.
[5] Bracke L, Verbeken K, Kestens L and Penning J 2009 *Acta Mater.* **57** 1512.
[6] Bhattacharjee PP, Sathiaraj GD, Zaid M, Gatti JR, Lee C, Tsai CW and Yeh JW 2014 *J Alloys Compd.* **587** 544.

IOP Conf. Series: Materials Science and Engineering **82** (2015) 012069 doi:10.1088/1757-899X/82/1/012069

Texture evolution of cold rolled and reversion annealed metastable austenitic CrMnNi steels

A Weidner[1], K Fischer[1], C Segel[1], G Schreiber[2] and H Biermann[1]
[1] Institute of Materials Engineering, Technische Universität Bergakademie Freiberg, Freiberg, Germany
[2] Institute of Materials Science, Technische Universität Bergakademie Freiberg, Freiberg, Germany
E-mail: weidner@ww.tu-freiberg.de

Abstract. A thermo-mechanical process consisting of cold rolling and subsequent reversion annealing was applied to high-alloy metastable austenitic CrMnNi steels with different nickel contents. As a result of the reversion annealing ultrafine grained material with a grain size in the range between 500 nm up to 4 µm were obtained improving the strength behavior of the material. The evolution of the texture of both the cold rolled states and the reversion-annealed states was studied either by X-ray diffraction or by EBSD measurements. The nickel content has a significant influence on the austenite stability and consequently also on the amount of the martensitic phase transformation. However, the developed textures in both steel variants with different austenite stability revealed the same behavior. In both investigated steels the texture of the reverted austenite is a pronounced Bs-type texture as developed also for the deformed austenite

1. Introduction

Low carbon, metastable austenitic stainless steels (CrMnNi-steels) exhibit excellent mechanical properties like concurrently high strength and high ductility as well as a high capability of deformation energy absorption [e.g. 1] due to the TRIP (TRansformation Induced Plasticity) and TWIP (TWinning Induced Plasticity) effects. However, produced as cast steels the yield strength of this class of materials is quite low (about 200 MPa) caused mainly by the large grain size (up to 2 mm). A well established method for enhancing the yield strength of the material is the thermo-mechanical controlled processing (TMCP) mainly applied on wrought steels [e.g.2]. However, even cast CrMnNi steels can be treated by TMCP as shown elsewhere [3]. A cold rolling process up to 90 % of thickness reduction followed by a conventional heat treatment at 700°C and different holding times lead to ultra-fine grained (ufg) reverted austenite. Mean grain sizes varying between 500 nm and 4 µm revealed an increase in the yield strength of the material up to 1 GPa. Several authors studied the evolution of the deformation and transformation textures already mainly in TWIP steels [e.g. 4], but to minor extend also in TRIP steels [e.g. 5]. The deformation texture of the deformed austenite at high cold rolling (CR) degrees is of brass (Bs) type, which is typical for materials with low stacking fault energy (SFE). The Bs-type texture is mainly described by two texture components – (i) a strong Brass (Bs) component ({110}<112>) and a well developed Goss (G) component ({110}<001>). The martensitic phase transformation during the plastic deformation of TRIP steels leads to the development of a transformation texture of deformation-induced α'-martensite. This texture can be characterized by strong components with <110> || rolling direction (RD) and <111> || normal direction, respectively. It was shown

that the variant selection between the austenite and the α'-martensite is related to Kurdjumov-Sachs relationship [6].
The aim of the present paper is to have closer look on the texture evolution in the low carbon high-alloy, metastable austenitic CrMnNi steels. It is focused on the influence of the austenite stability on the evolution of deformation as well as transformation texture.

2. Experimental details

The investigated materials were two high-alloy CrMnNi steels (16 wt% Cr, 6 wt% Mn, , 1 wt% Si, 0.05 wt% N, 0.08 wt% C) with different content of nickel (3.7 wt% and 7.1 wt%, respectively) in cast conditions [7] after solution heat treatment (0.5 h at 1050°C, N2 gas quenching). The initial cast state was characterized by a mean grain size up to 2mm (see ref. [3]). Due to the chemical composition and the dendritic solidification segregations some amount of ferromagnetic δ-ferrite phase (2 % for steel with 7.1 % Ni and up to 10 % for the steel with 3.7 % Ni) is apparent in the microstructure. Stripes of the cast material were cold rolled up to CRs of 50 %, 70 % and 90 %, respectively. The reversion annealing was performed at different temperatures (650°C up to 800°C) and different holding times (1 sec, 2.5 min, 5 min, 10 min, 20 min and 120 min). Details on the cold rolling procedure and the heat treatment are given in [3]. The volume fractions of the *b.c.c.* phase (deformation-induced α'-martensite and δ-ferrite) of the cold rolled plates and the reversion annealed states were investigated by X-ray diffraction (XRD) phase analysis, ferritescope measurements and measurements using magnetic balance. The texture measurements were performed using XRD (cold rolled states) and electron backscatter diffraction (EBSD) (reversion annealed states).

3. Results and discussion

3.1. Volume fraction of deformation-induced α'-martensite

The evolution of the volume fraction of deformation-induced α'-martensite was determined using XRD, ferritescope and magnetic balance. The average values of all three methods are summarized in Tab. 1. In addition, Tab. 1 contains the martensite start temperature M_s calculated according to [8] and the SFE [9]. It becomes evident that with decreasing nickel content the austenite stability as well as the SFE is reduced, which results in a higher tendency of martensitic phase transformation during plastic deformation. This is well expressed by the higher volume fraction of the ferromagnetic phase in the steel with 3.7 % nickel. Whereas the steel with 7.1 % nickel has about 70 % of ferromagnetic phase after a CR of 90% the steel with 3.7 % nickel is nearly completely martensitic already at CR of 50%.

Table 1: Volume fraction of ferromagnetic phase for the as-cast state and different cold rolled states, M_s, SFE.

	as-cast	**50 % CR**	**70 % CR**	**90 % CR**	**M_s [8] [°C]**	**SFE [9] [mJ/m²]**
7.1 % Ni	2	45	48	70	-30	17.5 ± 1.4
3.7 % Ni	25	90	100	100	60	8.1 ± 0.9

However, it should be noted that the steel with 3.7. % nickel contains already in the as-cast state about 25 % of ferromagnetic phase, which is related to the above mentioned δ-ferrite (10 %) and 15 % martensite coming from the cooling process.

3.2. Texture of cold rolled austenite

According to the different level of martensitic phase transformation differences in the deformation textures of both steel variants were expected. However, for the steel with 3.7. % nickel the texture evolution of the deformed austenite could not be measured since the volume fraction of the retained austenite (10 %) was too small for XRD texture measurements. Therefore, Figure 1 shows only the texture of the deformed austenite of the steel with 7.1 % nickel at CR = 90 %. The deformation texture of the austenite is clearly a Bs-type texture with a dominant Bs-component and a weaker G component. There is no indication for other components like Copper (C) {112}<111> or S-component {123}<624>, which would be typical for materials with higher SFE.

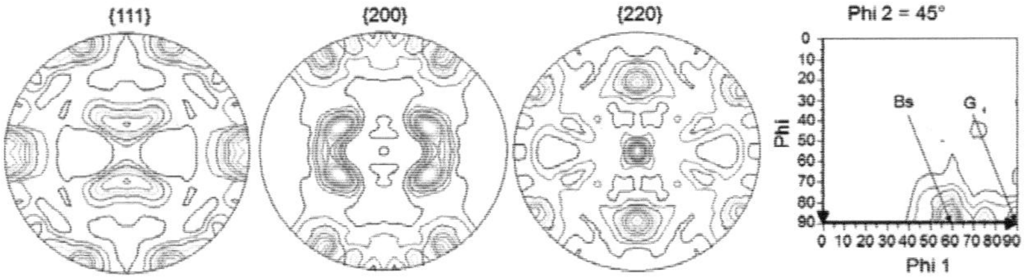

Figure 1: Texture of cold rolled austenite of the steel with 7.1 % nickel at cold rolling reduction of 90 %. {111}, {200} and {220} pole figures and the ODF section at constant Phi 2 = 45°.

3.3. Texture of deformation-induced α'-martensite

The texture evolution of the deformation-induced α'-martensite was followed for the steel with 3.7 % nickel for different CRs. Figure 2 illustrates the texture evolution using typical texture fibers known to develop in *b.c.c.* materials. At CR ≤ 70 % the formation of a strong {001}<110> component and a weaker {112}<110> component was observed. In addition, a well developed {111}<112> component is present, too. However, a significant change in the texture evolution occurred at CR > 70%. The {112}<110> component becomes more intense whereas the intensities of the other two components are decreased. A comparison with the texture in the steel with 7.1 % nickel at CR = 90% shows similar tendency. However, the texture components {112}<110> and {111}<112> are much more pronounced. Most probably, this related to the fact that in the steel with lower nickel content the martensitic phase transformation starts already at lower deformation degrees and no such strong Bs-type texture is developed during the lower CR levels.

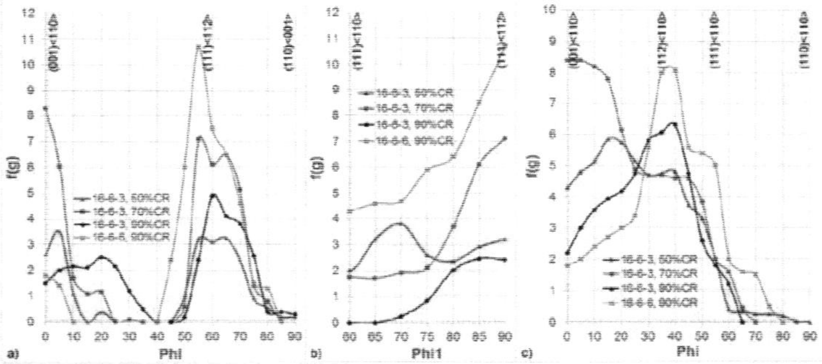

Figure 2: Texture evolution of the deformation-induced α'-martensite in the steel with 3.7 % nickel for CR of 50 %, 70 % and 90 %, respectively, in comparison to the steel with 7.1 % nickel at CR 90 %. (a) α-fiber (Phil = 0°, Phi2 = 45°). (b) γ-fiber (Phi2 = 45°, Phi = 55°). (c) ε-fiber (Phi 1 = 90°, Phi2 = 45°).

3.4. Textures of reverted states

Figure 3 shows the texture of the steel with 7.1 % nickel after CR = 90 % and different reversion annealing parameters. For all reversion annealing parameters a similar texture has developed, which is characterized by a pronounced Bs-component and a minor G-component as seen from the α-fiber shown in Fig. 3a. Thus, the transformation texture after the reversion annealing is comparable to the texture of the deformed austenite. However, as seen in Fig. 3b the τ-fiber reveals new components, which were not present in the cold rolled states. The main component along the τ-fiber is the C-component {112}<111> and its related components – the C_{twin} {552}<115> as well as the $C_{twin, rot}$ {332}<113>. However, it should be noted that about 30 % of deformed austenite remained in the microstructure at CR = 90 %. Therefore, there should be a similar amount of recrystallized and/or annealed austenite after the heat treatment. The components C, C_{twin} and $C_{twin, rot}$ are related according to [10] to the recrystallization process. However, the strong Bs-type texture after the reversion annealing can be taken according to

[10] as an indicator for a diffusionless shear transformation of the α'-martensite back into the austenite.

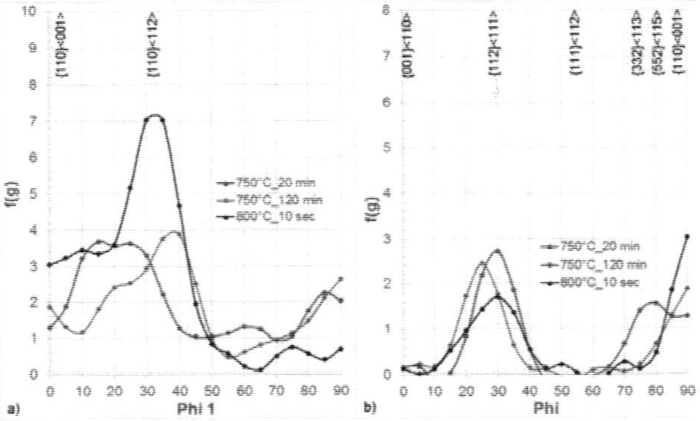

Figure 3: Texture evolution of the austenite in the steel with 7.1 % nickel at CR = 90% and different heat treatment parameters. (a) α-fiber (Phi = 45°, Phi2 = 0°). (b) τ-fiber (Phi1 = 90°, Phi2 = 45°).

The textures of the reverted states of the steel with 3.7 % nickel show similar behavior. However, in contrast to the steel with 7.1 % nickel recrystallized δ-ferrite is observed in the fully reverted austenitic microstructure.

4. Summary

The path of the texture transition in metastable CrMnNi steels during TMCP can be described by the following steps: (i) pronounced Bs-type texture with a strong Bs-component and weaker G-component, (ii) texture of α'-martensite with strong {112}<110> and {111}<112> components and a weaker {001}<110> component and (iii) pronounced Bs-type texture after reversion annealing again with strong Bs-component and weaker G-component. This transformation path is according to the literature related to the Kurdjumov-Sachs relations ship. In addition, the textures of the reverted states contain C, C_{twin} and $C_{twin, rot}$ components, which were not present in the cold rolled states. These components are related to the recrystallization process.

Acknowledgment

The authors acknowledge gratefully the German Research Foundation (DFG) for the financial support of the Collaborative Research Centre TRIP-Matrix-Composites (CRC 799). Furthermore, the authors thank Mr. U. Heinze (Institute of Metal Forming) and Dr.-Ing. T. Kreschel (Institute of Iron and Steel Technology) for the support performing the cold rolling and the heat treatment, respectively. Ms. K. Zuber is grateful acknowledge for the careful specimen preparation for EBSD measurements.

References

[1] S.Wolf, S. Martin, L. Krüger, U. Martin, *Mater. Sci. Eng. A* **594** (2014) 72–81.
[2] J. Mola, B.C. de Cooman, *Scripta Mater.* **65** (2011) 834–838.
[3] A. Weidner, A. Müller, A. Weiss, H. Biermann, *Mater. Sci. Eng. A* **573** (2013) 68–76.
[4] N.K. Tewary, S.K. Ghosh, S. Bera, D. Chakrabarti, S.Chatterjee, *Mater. Sci. Eng. A* **615** (2014) 400–415.
[5] S.G. Chowdhury, S. Datta, B.R. Kumar, P.K. Deb, R.N. Ghosh, *Mater. Sci. Eng. A* **443** (2007) 114–119.
[6] I.L. Dillamore and W.T. Roberts, *Acta Metall.*, **12**, (1964) 281–293.
[7] A. Weiß, H. Gutte, M. Radke, P.R. Scheller, *Patent specification* WO002008009722A1.
[8] A. Jahn, A. Kovalev, A. Weiss, P. R. Scheller, *Steel Res. Int.* **82** (2011) 1101–1108.
[9] D. Rafaja, C. Krbetschek, C. Ullrich, S. Martin, *J. Appl. Cryst.* **47** (2014) 936–946.
[10] B.R. Kumar, A.K. Singh, S. Das, D.K. Bhattacharya, *Mater. Sci. Eng. A* **364** (2004) 132–139.

IOP Conf. Series: Materials Science and Engineering **82** (2015) 012070 doi:10.1088/1757-899X/82/1/012070

Influence of cubic texture intensity of hot rolled ferritic non-oriented electrical steels on the microstructure and texture in the final processed material

A Stöcker[a], J Schneider[a], T Scholze[a], A Franke[b], H Hermann[a], R Kawalla[a]
[a] Institute of Metal Forming, TU Bergakademie Freiberg
[b] Stahlzentrum Freiberg e.V., Leipziger Straße 34, D-09599 Freiberg
E-mail: anett.stoecker@imf.tu-freiberg.de

Abstract. The magnetic properties of non-oriented electrical steels are determined by the microstructure and texture of the material. Besides optimum grain size (microstructure) for low values of specific magnetic losses, a high intensity of θ-fibre texture and low intensity of γ-fibre and α-fibre texture is desirable. Each of the processing steps influences the intensity of the θ-fibre in the final processed material. In this paper the interplay of the various processing steps on the intensity of the θ-fibre is regarded for ferritic Iron-Silicon steels with 2.4 wt.% Si and 3.0 wt.% Si.

1. Introduction

Ferritic Iron-Silicon (FeSi) steels form the basis for low loss grades of non-oriented electrical steels. As well known the magnetic properties of electrical steels generally depend on the crystallographic direction and structural heterogeneity. In bcc iron the <100> direction is the easiest direction of magnetization. For excellent magnetizing behaviour a high intensity of θ-fibre (<100>||ND) and a low intensity of α- and γ-fibre is preferable. FeSi steels with a silicon content higher than 2 wt.% and a carbon content of less than 0.002 wt.% exhibit no phase transformation [1]. Therefore, no homogenisation of the microstructure during hot rolling takes place. Besides the specific technological rolling parameters, the existing gradients of strain and temperature affects much more the resulting microstructure and texture at the production of the hot strip compared to steels with a phase transformation [2-4]. It was found that higher intensities of θ-fibre can be obtained in the hot strip. It was also observed that higher intensities of the preferable θ-fibre texture of the hot strip result in a higher intensity after cold rolling. The results of previous studies indicate that there is an interaction between the microstructure and texture of the various processing steps: hot rolling, optional hot strip annealing, cold rolling and final annealing [3].

2. Experimental procedure

The hot rolled material, see Table 1, was prepared on the four stand hot rolling pilot line at the Institute of Metal Forming (IMF), TU Bergakademie Freiberg. An additional hot strip annealing was performed with sample E and F (T_{max} = 900°C for 120 s, heating rate 4 K/s, cooling rate 20 K/s). Cold rolling was realised on a two-high cold rolling mill at the IMF within 5 passes with a total reduction of 70% to 75%. Final annealing was done in an annealing furnace at the IMF for the simulation of a continuous annealing line.

Table 1 : Parameters of hot rolling

Sample	*d mm*	*Si wt.%*	*RT °C*	*FT °C*	*Rapid cooling to °C*	*A-T °C*	*A-t min*	*Cooling to °C with K/h*	
A	1.1	3.0	1250	645	-	-	-	200°C, 270 K/h	
B	2.3	3.0	1150	690	-	-	-	200°C, 130 K/h	
C	1.3	2.4	1250	1000	925	750	20	200°C, 50 K/h	
D	1.3	2.4	1250	930	700	-	-	200°C, 50 K/h	
E	2.0	2.4	1150	1010	950	750	270	200°C, 50 K/h	*
F	2.0	2.4	1150	820	400	-	-	-	*

d – thickness after hot rolling; RT – Reheating temperature; FT – hot rolling finishing temperature; A-T – annealing temperature; A-t – annealing time; * additional hot strip annealing before cold rolling

Light microscopy, SEM and EBSD was used to investigate the microstructure. The IFW Dresden and the Stahlzentrum e.V. Freiberg did the EBSD measurements. Both use the EDAX system. The optical microscopy and EBSD measurements were carried out in the plane perpendicular to the sample transverse direction. Measurement settings were like in [6], the map size depends on the cross section of the sample, which was measured completely.
To analyse the texture data the program MTEX was used [5]. The orientation distribution functions ODFs were calculated by the harmonic series expansion method assuming orthorhombic sample symmetry, with a half width of 5° and a resolution of 5°. Results of the texture measurements are presented as constant $\varphi_2 = 45°$ section of the ODF. Of particular interest are the θ-, α- and γ-fibre [7].

3. Results and Discussions

3.1 Hot rolled material

Figure 1 shows the cross section of hot rolled material with a low finishing temperature without any further heat treatment (specimen A and B). The cross section reveals a heterogeneous character of the microstructure. The grains in the mid-thickness region are elongated bands and pancake shaped. Within the bands no grain boundaries appear and the orientation is identical. The region close to the surface has elongated grains but has also small equiaxed grains and many grain boundaries. The reason for the heterogeneity is the gradient of strain during the hot rolling and the temperature schedule after finishing hot rolling. The texture consists of rotated cube {100}<110> components and the thinner material has a lower intensity of the α-fibre and γ-fibre then the thicker material.

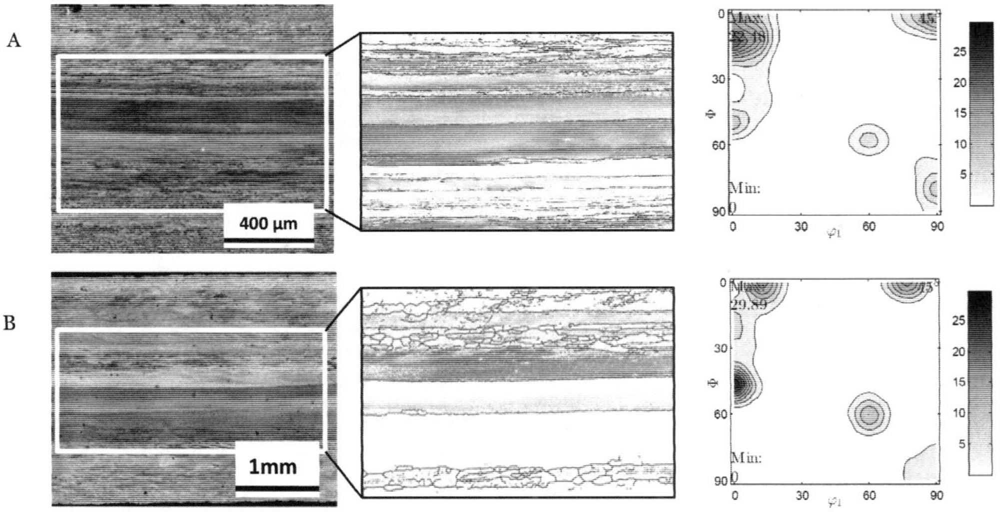

Figure 1 : Hot rolled sample A and B; microstructure (BSE); orientations {100}<110> (gray shading) deviation of 15° and grain boundaries calculated with misorientation > 10°; texture ODF φ_2=45° for total cross section

Figure 2, sample C and D with strip thickness 1.3 mm, illustrates the resulting texture depending on a thermal treatment immediately after the last pass of hot rolling. At appropriate finishing temperatures at hot rolling and annealing conditions immediately after the last hot rolling pass a higher intensity of the desired θ-fibre texture can be obtained. For different thicknesses of the hot rolled material, in the range of 1.3 mm to 2.3 mm, similar conditions of hot rolling finishing temperature and annealing conditions after the last hot rolling pass lead qualitatively to the same image of the texture.

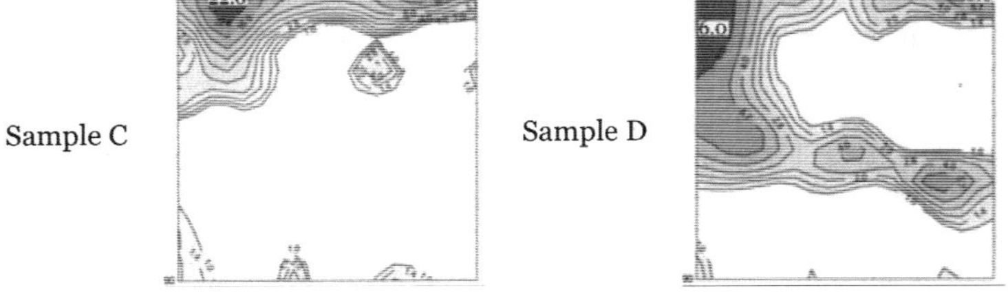

Figure 2: Comparison of texture of hot rolled material with annealing (sample C) and without annealing directly after finish rolling (sample D), ODF section φ_2=45° for mid-thickness layer

Hot strip annealing, as an additional processing step before cold rolling, can also cause a higher intensity of preferable magnetic texture, see Figure 3. The microstructure changes from a deformed to an equiaxed structure with few bands in the mid-thickness layer.

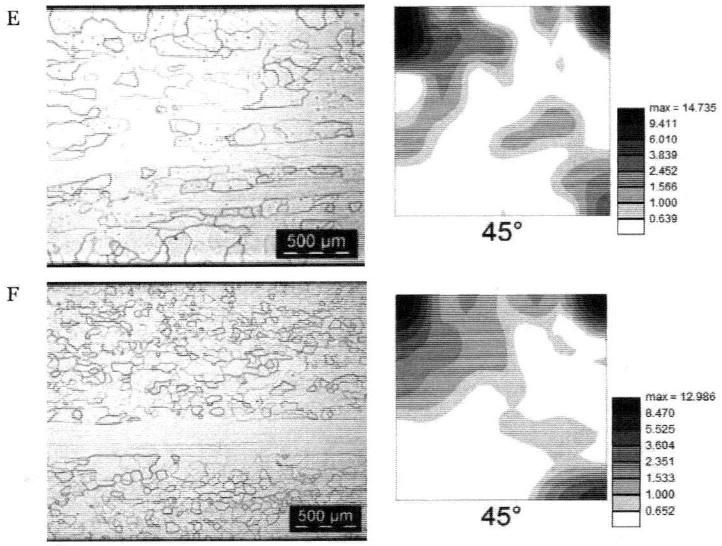

Figure 3: Microstructure (light microscopy) and ODF (total cross section) sample E and F after additional hot strip annealing before cold rolling

3.2 Cold rolling and annealing

The intensity of the rotated cube component remains after cold rolling, see Figure 4. A higher intensity of this components were found for the samples, which had already more pronounced rotated cube components in the hot rolled state. The appearance of the microstructure has not changed much in comparison with the hot rolled state. The surface layer has no equiaxed grains, almost no grains are visible and every part shows a deformed structure. The mid-thickness layer stills shows the elongated bands without grain boundaries and with a homogenous orientation.

Results obtained after final annealing of the cold rolled material are exemplary presented in Figure 5. The texture intensities of the final annealed samples depends sensitively on the annealing conditions. It was possible to keep the θ-fibre but the intensities and the recurrence of the γ-fibre cannot be prevented. It seems that a smaller heating rate is beneficial. Similar results were found for the other samples.

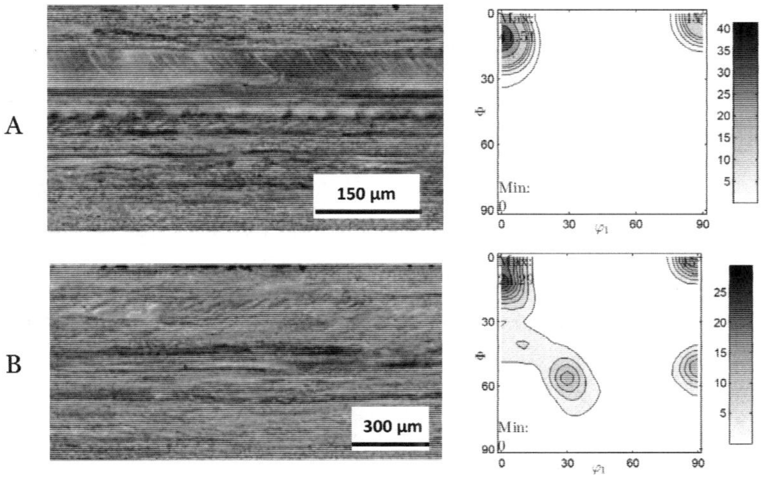

Figure 4 : Sample A and B, cold rolled with total reduction of 70%, microstructure (BSE) and ODF section $\varphi_2 = 45°$ (total cross section, EBSD-data)

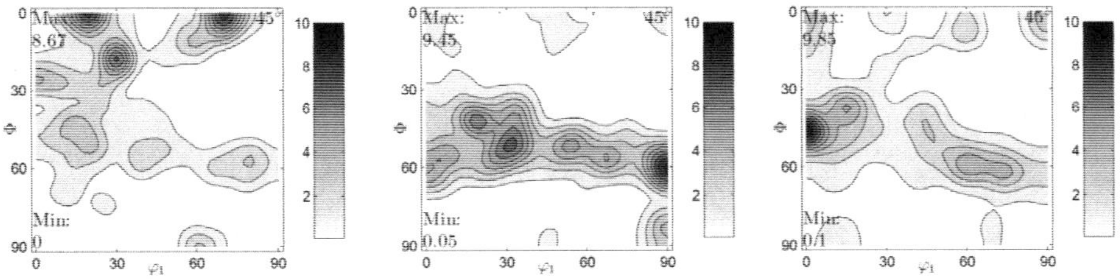

a) hr = 2.6 K/s, T_{max} = 900°C, t = 120s, quenching to RT
b) T_1 = 650°C, t_1 = 60s, T_2 = 950°C, t_2 = 60s
c) T_1 = 650°C, t_1 = 120 s, T_2 = 950°C, t_2 = 120s

Figure 5 : Sample A different final annealing treatments, hr – heating rate, cr – cooling rate, T – annealing temperature, t – holding time at T, RT– room temperature; b) and c) hr = 35 K/s; cr = 10 K/s

4. Conclusions

A high intensity of θ-texture in the hot strip can be obtained by different hot rolling parameters as described. Cold rolling reduces the preferable cube texture intensity. In the cold rolled state the typical rolling textures α- and γ-fibre appear. Nevertheless, the results show a higher intensity of θ-fibre texture after cold rolling for hot strip with higher intensity of this texture fibre component. The final annealing destroys the desired high intensity of θ-fibre, if the process is not controlled in an appropriate way (heating rate, maximum annealing temperature, two step annealing). More studies are necessary to collect experimental data for the interplay of the different steps and to realise a modelling and of the underlying processes.

Acknowledgments

The author would like to thank the DFG within contract KA 1591/23-1 for financial support.

References

[1] Rodríguez-Calvillo P et al 2012 *Materials Chemistry and Physics* **136** p 710
[2] Gomes E et al 2010 *Mater. Sci. Forum* **638-642** p 3561
[3] Verbeken K et al 2008 *IEEE Tras. Mag.* **44** p 3820
[4] Sidor J J et al 2012 *Materials Characterization* **71** p 49
[5] Bachmann F et al 2010 *Solid State Phenomena, Trans Tech Publ* **160** p 63
[6] Calvillo P et al 2008 *Defect and Diffusion Forum* **273-276** p 69
[7] Raabe D 2003 *Steel Research* **74** p 327

17th International Conference on Textures of Materials (ICOTOM 17) IOP Publishing
IOP Conf. Series: Materials Science and Engineering **82** (2015) 012071 doi:10.1088/1757-899X/82/1/012071

Applied, residual strain and texture investigations by means of neutron time-of-flight diffraction combined with acoustic emission detection: Application to a gneiss sample from Forsmark, Sweden

Ch Scheffzük[1,4,a], A Zang[2], O Stephansson [2], K Ullemeyer[3] and FR Schilling[1]

[1] Karlsruhe Institute of Technology, Institute of Applied Geosciences, Kaiserstr. 12, 76131 Karlsruhe, Germany

[2] Helmholtz Centre Potsdam, GFZ German Research Centre for Geosciences, Section 2.6 Seismic Hazard and Stress Field, Telegrafenberg, 14473 Potsdam, Germany

[3] Institute of Geosciences, University of Kiel, Otto-Hahn-Platz 1, 24098 Kiel, Germany

[4] Joint Institute for Nuclear Research Dubna, Frank Laboratory of Neutron Physics, ul. Joliot-Curie 6, 141980 Dubna, Russia

e-mail': [a] christian.scheffzuek@kit.edu

Abstract. The mechanical behaviour of a gneiss sample from the Forsmark drilling site (Sweden) consisting of quartz, plagioclase (oligoclase), microcline and biotite, is characterized by residual strain and crystallographic preferred orientation measurements. Cyclic *in situ* uniaxial deformation experiments were carried out on a cylindrical sample at four uniaxial stress levels (55, 84, 114 and 141 MPa) and examined using time-of-flight neutrons and simultaneous recording of acoustic emissions. At each stress level applied, a residual strain scan was performed around the sample axis and residual strains were determined. The residual compressional strain detected for quartz was -1.0×10^{-3}, the maximum value of residual strain approached 1.2×10^{-3}. The onset of acoustic emissions is an indicator of the previous *in situ* maximum stress. Comparison of acoustic emission data with applied and residual stress values at different load levels is the key to understand the freezing-in mechanism of residual stress and the Kaiser Effect in rock.

1. Introduction and sample characterization

The combination of neutron strain analysis and acoustic emission measurements aims to gain a better understanding of the physics behind the Kaiser effect [1]. 'Kaiser effect' describes the phenomenon that a material undergoing cyclic load/unloading with increasing stress level emits acoustic emissions only after the previous maximum stress (PMS) has been exceeded. If the recalled maximum stress under cyclic load equals the PMS, the Kaiser effect par excellence is confirmed. The experiments presented in this study are important to gain insight into the freezing-in mechanism of rock stress *in situ*. This process is closely related to the conservation of residual stress in rocks in general and the actual origin of acoustic emission during laboratory cyclic loading.

For the investigation of the physical properties of polycrystalline rocks the neutron diffraction has advantages due to the large penetration depth of neutrons [2, 3]. Time-of-flight neutrons allow the simultaneous registration of a large number of Bragg diffraction lines over a wide lattice plane *d*-spacing range. Hence, the texture diffractometer SKAT [4] and the strain diffractometer EPSILON-MDS [5] at the IBR-2M pulsed reactor at JINR (Dubna) were used for our investigations.

The gneiss sample investigated was taken from the Forsmark area (Sweden) and was selected because of high residual stresses to be expected. Sample KFM01-D1 shows a foliation in the **xy**-plane and a coarse-grained granoblastic fabric consisting of quartz, plagioclase (oligoclase), alkali feldspar and biotite. The phase content has been determined by X-ray phase analysis on powder sample. Large quartz grains (< 5 mm in diameter), which exhibit undulose extinction, are surrounded by smaller recrystallized neoblasts (< 200 μm). The neoblasts show homogenous extinction, indicating post-deformational growth at high temperatures. The plagioclase is affected by intense hydrothermal alteration and widely replaced by fine-grained sericite and/or saussurite (Fig. 1).

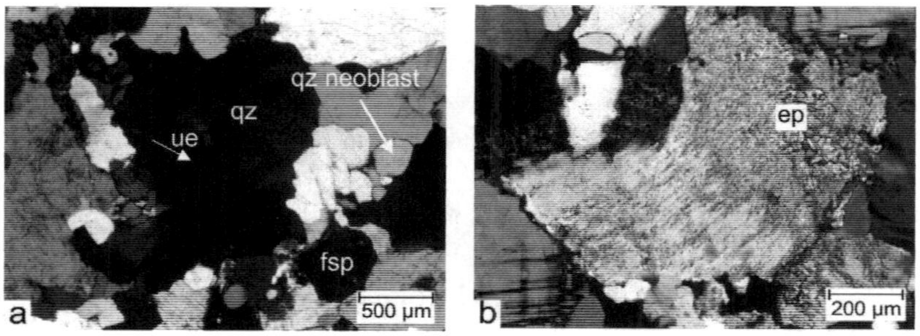

Fig. 1: *Microstructure of the gneiss sample KFM01-D1. Quartz-feldspar matrix: a) quartz (qz) grain showing undulose extinction (ue) and marginal recrystallization, b) saussuritized large plagioclase grain (ep - epidote),* ***xy****-plane.*

2. Textures

For the texture evaluation from the time-of-flight patterns gained at SKAT the 'Rietveld Texture Analysis' (RTA) [6] was applied using the MAUD software [7]. In contrast to X-ray phase analysis, RTA revealed potassic feldspar (microcline) and slightly different volume fractions of the rock constituents (quartz 39%, oligoclase 42%, microcline 17%, biotite 2%; Fig. 2).

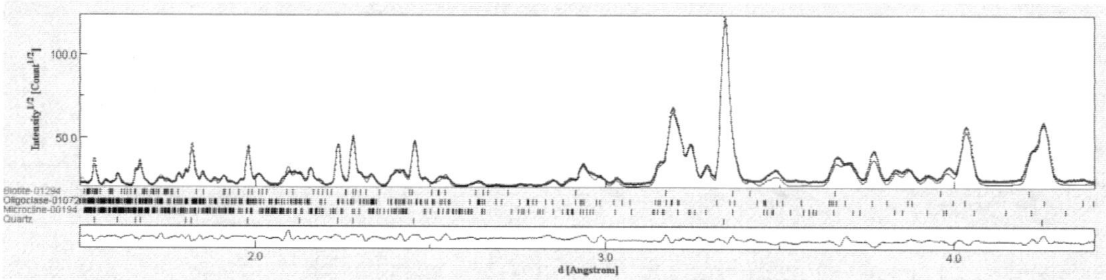

Fig. 2: *Normalized neutron TOF diffraction pattern of the sample measured at SKAT. Blue color: experimental data, black color: approximated data. The line patterns below indicate peak positions.*

The quartz texture is well-pronounced and characterized by a single maximum in the [0001] pole figure close to the **z**-axis and 120° pole density distributions of the other lattice directions around the [0001] cluster (Fig. 3). These are hints for postkinematic recrystallization. In contrast, the textures of the other phases are very weak, see example pole figures in Fig. 3.

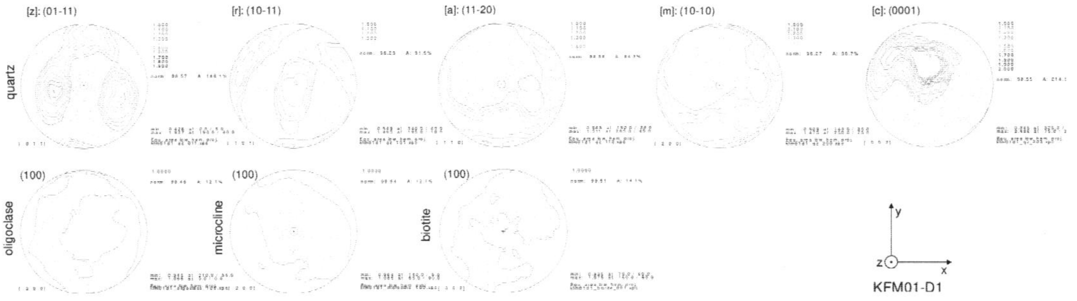

Fig. 3: *Textures of quartz (top), oligoclase, microcline and biotite (bottom). Lower hemisphere, equal area projection, linear scale contours.*

3. Residual and applied strain

The residual strain data gained at Epsilon-MDS are related to a ground (< 62 µm) and heated powder of the sample serving as reference for the strain-free state. The cylindrical sample (d= 30 mm, l= 60 mm) has been rotated around the cylinder axis **z** to detect the residual strain in steps of 15° in the **xy**-plane (Fig. 4a). The residual strains vary between -1 x 10^{-3} and 1.2 x 10^{-3}.

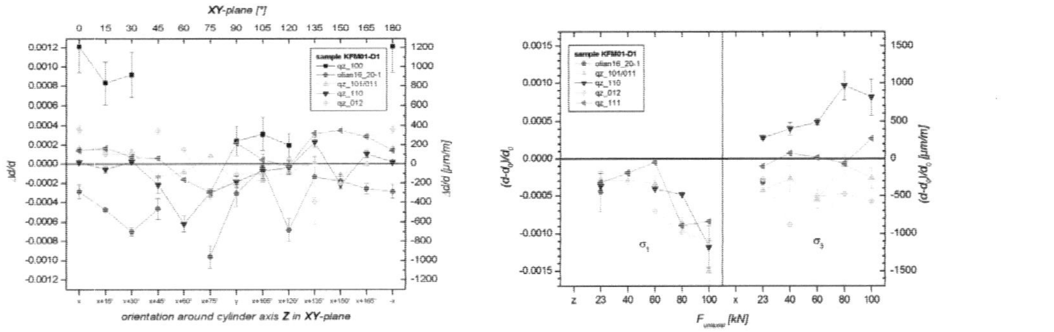

Fig. 4: *a) Residual strain of the quartz (10-10), (10-11)/(01-11), (11-20), (01-12), 11-21) and oligoclase (20-1) lattice planes. The sample was rotated around the* ***Z****-axis in steps of 15°. b) Applied strain of the quartz (10-11)/(01-11), (11-20), (01-12), (11-21) and oligoclase (20-1) lattice planes. The* ***z****-axis is the main stress direction (σ_1), the* ***x****-axis is oriented perpendicular to σ_3.*

For applied strain measurements, inclination of the **z**-axis with respect to the incident beam was 45°, i.e., σ_1 and σ_3 were measured simultaneously with two detectors in opposite position [8]. The experiments were carried out at loads of 40, 60, 80 and 100 kN (55, 84, 114 and 141 MPa). The negative strain in **z** direction corresponds to Hooke's law, the positive strain in **x** direction is lower and depends on Poisson's ratio (Fig. 4b). Simultaneously, grain-scale cracking processes have been detected by eight piezoceramic transducers positioned on two rings around the **z** axis (preamplifier: 40 dB, filter: 95-850 kHz, threshold: 31 dB, sensor frequency: 200 kHz). In each cycle, the onset of acoustic emission (cracking events) was close to the previous applied maxiumum stress. In Figure 5, the results for load step 40 kN (Fig. 5a) and 60 kN (Fig. 5b) are shown for reference.

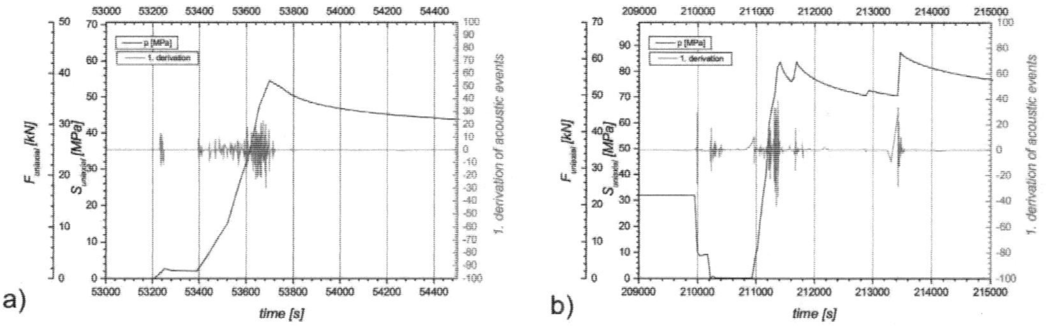

Fig. 5: *Force and rate of acoustic emissions versus time of uniaxial deformation experiment. Presented are only two of four loading steps: a) at 40 kN, and b) at 60 kN at previous maxiumum load.*

4. Conclusions

The Kaiser effect on a granite rock sample has been investigated using uni-axial *in situ* deformation of the sample in four steps up to 100 kN. The strain analysis has been carried out in two perpendicular directions parallel to σ_1 and parallel to σ_3 using neutron TOF diffraction. Texture and residual strain measurements indicate that the material is characterized by a quartz recrystallization texture, and preserved residual strains. In the next step, the ratio of acoustic emission crack strain and applied and residual strain from neutron diffraction will be used together with pre- and post-mortem microstructural analysis of rock cores to identify the freezing-in mechanism of residual stresses and the physics of the Kaiser effect in rock.

The work has been funded by the German Federal Ministry of Education and Research (contract 05K10VKF). We thank A. A. Kruglov for help in acoustic emission registration.

References:

[1] Kaiser J 1950 PhD Thesis, Munich.
[2] Frischbutter A (Ed.) 1998 *Schriftenr. f. Geowiss.* **6** 64 p
[3] Schäfer W 2002 *Eur. J. Mineral.* **14** 263-289
[4] Ullemeyer K et al 1998 *Nucl. Instr. & Meth. Phys. Res.* **A 412** (1) 80-88
[5] Walther K et al 2005 Trans Tech Pub, *Solid State Phenomena* **105**, 67-70.
[6] Matthies S et al 1997 *J. Appl. Cryst.* **30** 31-42
[7] Lutterotti L 1999 *IUCr: Newsletter of the CPD* **21**, 14-15
[8] Scheffzük C et al 2012 *J. Phys.: Conf. Series* **340** 012038

A numerical and experimental investigation on seismic anisotropy of Finero peridotite, Ivrea-Verbano Zone, Northern Italy

X Zhong[1], M Frehner[2], K Kunze[3] and AS Zappone[4-5]
[1] Institute for Geochemistry and Petrology, ETH Zurich, Switzerland
[2] Geological Institute, ETH Zurich, Switzerland
[3] Scientific Centre for Optical and Electron Microscopy, ETH Zurich, Switzerland
[4] Institute for Process Engineering, ETH Zurich, Switzerland
[5] Swiss Seismological Service, Zurich, Switzerland
E-mail: xin.zhong@erdw.ethz.ch; marcel.frehner@erdw.ethz.ch; karsten.kunze@scopem.ethz.ch; alba.zappone@sed.ethz.ch

Abstract. The Ivrea-Verbano Zone offers a unique opportunity to study lower crust/upper mantle rocks unaffected by serpentinization. Peridotites from Finero (North-East Ivrea-Verbano Zone) have been collected to investigate the influence of individual mineral phases on seismic anisotropy. The methods involve experimental ultrasonic P-wave velocity measurements using a hydrostatic pressure vessel, and electron backscatter diffraction (EBSD) analysis to investigate the crystallographic preferred orientation (CPO). CPO was analyzed by large area EBSD scans for major minerals including olivine, orthopyroxene, but also for metasomatic hornblende. The CPO of olivine is confirmed to be the major source of detected seismic anisotropy. Minor influence can arise from hornblende if it is sufficiently abundant.

1. Introduction

One of the causes of seismic anisotropy in the lower continental crust and in the upper mantle is the crystallographic preferred orientation (CPO) of volumetrically dominant rock forming minerals [1] or of textured minerals that are volumetrically less abundant, but with strong single crystal anisotropy [2]. Several velocity measurements have been made for the most representative lithologies of the crust and upper mantle [3–4], and compared to the results of numerical models of rock seismic properties. A common method for calculating the seismic bulk properties of a rock is based on measurements of CPO of the constituent minerals, obtained by electron backscatter diffraction (EBSD) techniques [5]. The Voigt-Reuss-Hill (VRH) estimates [6] are then used to calculate the bulk elastic tensor and various averages are applied to the elastic stiffness tensor. In this study, we accomplished EBSD measurements on two Finero peridotites and microtextural calculations based on these EBSD measurements. Meanwhile, laboratory measurements using a hydrostatic pressure vessel were performed for comparison with the calculated seismic velocity. A novel numerical method, which incorporates EBSD-data into a finite-element (FE) model [7, 8] to simulate elastic waves propagating through an anisotropic and heterogeneous rock model, is presented and discussed.

2. Material and methods

2.1. Sample description

We selected sample ZAP202 and ZAP207, representative of an upper mantle with crustal metasomatic alterations, which were drilled in the Finero ultramafic complex (Ivrea Zone, Northern Italy). The two samples are harzburgites (olivine and orthopyroxene) and contain phlogopite and hornblende in different modal proportions. Clinopyroxene (Cr-diopside) is in low modal proportion and spinel represents less than 1% of the whole rock. Secondary minerals such as oxides and serpentine occur only in negligible quantities. The samples are coarse grained (up to 3 mm diameter). The lineation and foliation were determined on hand specimen and in thin section; the foliation is defined by the planar disposition of phlogopite, and a lineation is defined by the elongation of olivine and/or amphibole.

IOP Conf. Series: Materials Science and Engineering **82** (2015) 012072 doi:10.1088/1757-899X/82/1/012072

2.2 Seismic velocity measurements in laboratory

Cores of 26 mm diameter and ca. 30–50 mm length were drilled along three mutually perpendicular directions aligned with lineation and foliation. A hydrostatic pressure vessel is utilized to measure the travel time of compressional waves through the sample up to 200 MPa. Due to the closure of microfractures, the seismic velocity measured during pressurization is lower than depressurization. We report the velocity measured during depressurization as it is reproducible. After systematic calibration, seismic velocities are calculated and extrapolated to 0 MPa from above 100 MPa downwards to reflect the intrinsic mechanical properties.

2.3 Determination of CPO

The core surfaces were ground, polished, and lapped with colloidal silica in order to obtain a smooth surface low in defect content and thus suitable for EBSD measurements. The sides of each core were covered with silver paint to reduce charging problems, but no surface coating was applied. EBSD data were acquired on a scanning electron microscope EOscan (Tescan, Brno CZ), equipped with a Pegasus OIM™ version 6.2 analytical system by Ametek-EDAX (Mahway, NJ, USA). EBSD patterns and energy dispersive X-ray spectroscopy (EDS) counts of specific elements were mapped simultaneously with a beam current of 3–5 nA and 20 kV acceleration voltage. Only one high-symmetry phase was indexed online to increase acquisition speed. The data were later reprocessed using the ChiScan routine of OIM™ Data Collection to re-index all relevant mineral phases filtered by user-defined windows for EDS counts. The MATLAB toolbox MTex [9] is utilized to perform textural analysis, i.e. calculating and plotting the stereograms of CPO and of seismic velocity. Hill averages are presented by averaging the Voigt and Reuss bounds [6]. Thus the effects from CPO are reflected, whereas the grain shape preferred orientations (SPO) are not considered. As the selected samples are massive and lack very strongly developed SPO, such approximation is reasonable. We have also investigated the influence of SPO on seismic anisotropy separately, confirming that assumption [7, 8]

3. Results

The laboratory measured Vp at around 8 km/s are typical for peridotite (Table 1). The seismic anisotropy calculated for room pressure as $(Vp_{max}-Vp_{min})/Vp_{average}$ for sample ZAP202 is significantly lower than for sample ZAP207 (Table 1). The velocities calculated from the EBSD dataset using MTex toolbox [9] are illustrated in Fig. 1. The CPO and seismic velocity stereograms (Hill average) are calculated for the three most important and abundant mineral phases, olivine, enstatite, and hornblende. The results yield that olivine is the most influential mineral phase, which dominates the velocity stereogram. Minor influence can arise from the rest mineral phases as the bulk velocity is generally reduced compared with the olivine velocity stereogram (see colorbars in Fig. 1). Hornblende in ZAP202 yields different maxima velocity direction compared to olivine but is not sufficient to cause a significant change in the bulk velocity distribution. Hornblende abundance (9%) and its sharp CPO are most likely the cause of the higher anisotropy in sample ZAP207.

Table 1 Experimental results on Vp measurements

Sample	Density [g cm^{-3}]	Vp at 0 MPa [km s^{-1}]	Vp at 250 MPa [km s^{-1}]	Vp pressure derivative [10^{-4} km s^{-1} MPa^{-1}]	Anisotropy
ZAP 202 X	3.29	8.318	8.491	6.938	2.4%
ZAP 202 Y	3.29	8.317	9.524	8.249	
ZAP 202 Z	3.29	8.122	8.231	4.343	
ZAP207 X	3.26	8.371	8.523	6.082	7.2%
ZAP 207 Y	3.26	8.084	8.199	4.599	
ZAP 207 Z	3.25	7.762	7.931	6.762	

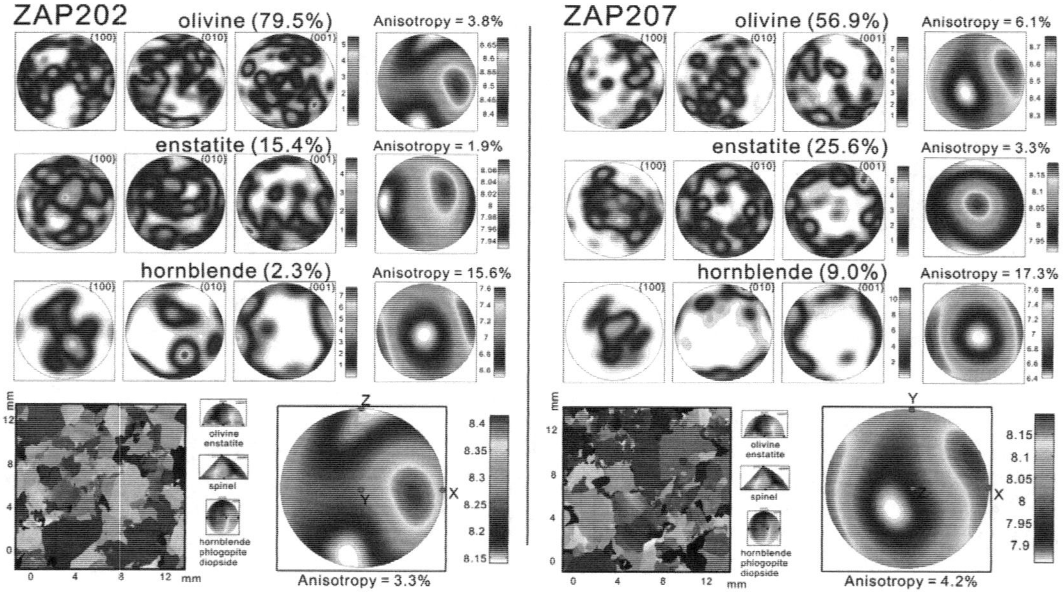

Figure 1 Pole figures (in m.u.d.) and stereograms of calculated Hill Vp velocities (in km/s) of samples ZAP202 and ZAP207. The modal compositions are given after the mineral names (in area %). Calculated anisotropies are given for individual mineral phases and the bulk rock. Hand specimen orientations equal to core directions are indicated on the bulk rock velocity stereograms. EBSD crystallographic orientation maps are provided on the bottom left. Minor amounts of spinel, phlogopite and clinopyroxene are not listed here.

4. Discussion

We performed laboratory measurements using a hydrostatic pressure vessel and classic VRH calculation on Finero peridotite samples. Moderate degree of seismic anisotropy has been observed in the reported peridotite samples ZAP202 and ZAP207. Olivine is confirmed to be the most dominant mineral phase on seismic anisotropy. Enstatite has lower degree of anisotropy in the abundant mineral phases. Although highly anisotropic, phlogopite has very low modal concentration and can be neglected. If its modal percentages would be higher, a significant influence on bulk velocity can be caused by such strongly anisotropic phase. SPO is observed but is not considered based on current calculation. Hornblende has higher modal composition and is sufficient to influence on the overall higher anisotropy of sample ZAP207. There is some extent of discrepancy between the seismic anisotropy of laboratory data and the calculated one, especially for ZAP207; the potential cause may be the presence of preferentially aligned fractures that were not recorded on EBSD data. Therefore, more EBSD sections, or even three-dimensional analysis of larger sample volumes are required to achieve closer correlation between laboratory measurements and numerical calculations.

The advantage of laboratory measurements is that the velocity is directly representative for the full sample core, but the involved experimental effort often hinders researchers from measuring seismic velocity in many directions. In many situations, the directions of maximum and minimum velocity do not coincide with the directions of the selected (typically three perpendicular) cores. The numerical calculation, on the other hand, provides complete information in the form of stereogram but may neglect important features, such as fractures and SPO. Therefore, comprehensive investigation is imperatively required to provide thorough understanding on seismic anisotropy. New approaches, such as EBSD-based FE model can provide a promising alternative. The main idea of such FE simulation we have developed is to use the measured CPO on the X-Y grid in the sample surface plane to construct a map of rotated elastic tensors on the same spatial coordinates, which in turn are the mesh to perform FE wave propagation simulations [7, 8].

5. Conclusions

Olivine dominates the seismic anisotropy in these reported Finero peridotite samples. Minor influence can arise from highly anisotropic hydrous mineral phases if they are textured and their modal percentage is high enough. Coherence is found between the laboratory measurements and the calculation but discrepancies still exist possibly due to fractures or SPO.

References

[1] Nicolas A and Christensen N I 1987 *American Geophysical Union Series* **16** 111-123
[2] Llana-Funez S and Brown D 2012 *Geological Society America Bulletin* **124** 1495-1513
[3] Birch F 1960 *Journal of Geophysical Research* **65** 1083–1102
[4] Ji S, Wang Q and Xia B 2002 Handbook of Seismic Properties of Minerals, Rocks and Ores *Polytechnic International Press* 630 pp
[5] Erdman M et al 2013 *Geophysical Journal International* **195** 1211-1229
[6] Hill R 1952 *Proceedings of Physical Society* **65** 349-354
[7] Zhong X 2014 Physical Properties of Amphibole and Phlogopite bearing Peridotites from Finero, Ivrea Zone (Italy): An Experimental and Numerical Investigation *unpublished master thesis. ETH Zurich*
[8] Zhong X, Frehner M, Kunze K and Zappone A 2014 *Geophysical Research Letters* **41** 7105-7114
[9] Mainprice D, Hielscher R and Schaeben H 2011 *The Geological Society Special Publications* **360** 175-192

IOP Conf. Series: Materials Science and Engineering **82** (2015) 012073 doi:10.1088/1757-899X/82/1/012073

Characterization of multiple twinned structural units in pulse-electrodeposited nickel

U Klement, A Kahrimanidis and Y Yao
Materials and Manufacturing Technology, Chalmers University of Technology, Gothenburg, SE
E-mail: uta.klement@chalmers.se

Abstract. The investigation was performed on pulse-electrodeposited Nickel with sub-microcrystalline microstructure containing slightly elongated grains having a <110> fibre texture in growth direction. Structural units in form of groups of elongated grains possessing a common <110>-zone axis in growth direction and CSL boundaries (in some cases twins) between them have been found in the microstructure by use of EBSD. Grain growth sets in above 325°C but the texture is conserved up to at least 600°C. This means that the arrangement of twins and other CSL boundaries stabilized the structural units; there is no orientation change (by further twinning) when grain growth occurs as seen in previous studies on Ni and Ni-Fe of different initial texture. The observed structural units were characterized in detail and the occurring grains and grain boundaries are described.

1. Introduction

Materials with nano- and sub-microcrystalline grain sizes have been intensively studied due to their potential in high strength applications and as wear resistant coatings. A common way to produce materials with such length scales is pulse-electrodeposition. Depending on the process parameters, the resulting grain size and texture can be tailored. For example, for Nickel electrodeposits produced from an additive-free sulfamate bath, <100>-, <110>- and <210>-textures were reported [1-3]. Texture formation is said to be governed by the adsorption of an inhibiting species, i.e. hydrogen [4]. The higher the adsorption on a specific crystallographic plane, the slower is its growth velocity. As a consequence, the growth in some crystallographic directions is faster than in others and a textured material is obtained.

For obtaining a nanocrystalline material the electrolyte often contains stress reliever and grain refining agents. Organic additives such as saccharin have been frequently used in electroplating operations to moderate deposit growth rates and to control film quality. According to literature [5-7], the addition of saccharin leads to the formation of fine crystallites that have a preferred <111> orientation with the growth direction (GD). But also double fibre textures have been observed in Ni and Ni-base alloys in the presence of additives: a <200> <111> double fibre texture was found for Ni with 10 and 20 nm grain size and a sulphur content of 1580 ppm and 1200 ppm, respectively [8], and a <411> <111> double fibre texture was observed in Ni and Ni-Fe electrodeposits by using X-ray diffraction and by performing measurements using electron backscatter diffraction (EBSD) technique [9]. Upon annealing, abnormal and normal grain growth sets in and the final <111> texture with respect to GD is obtained with one or two twinning operations from the initial texture of the as prepared material [8-12].

The presence of twins and other low-Σ boundaries (use of coincidence site lattice (CSL) description [13]) in the microstructure is known to significantly improve mechanical properties (grain boundary engineering). Related to a high twin density, 5-fold twins have been observed for different materials [14-16]. However, also structural units including other low-Σ boundaries than twins have been reported [17,18]. Common to the grains in 5-fold twins and low-Σ structural units is a <110>-orientation parallel to GD. Moreover, investigations perpendicular to GD showed clearly the 3-dimensional character of structural units [17,18]. In this paper, these structural units are investigated in more detail both parallel and perpendicular to GD.

2. Experimental

Two different pulse-electrodeposited Nickel materials were investigated. One of which was produced at TU Dresden, Germany, the other one at the Indian Institute of Science (IISc) in Bangalore, India. The experimental procedure will be described elsewhere but in both cases a saccharine free sulfamate bath was used and the materials had a <110> fibre texture in GD. The material from Germany had slightly elongated grains in GD (aspect ratio of 1.8) and a grain size of about 180 nm. The Ni electrodeposit from India had an average grain size of 190 ± 100 nm. Both materials did not show noticeable grain growth when heat treated at temperatures below 325°C. To investigate thermal stability, differential scanning calorimetry (DSC) measurements were performed up to 823 K in argon atmosphere at a heating rate of 10 K/min.

EBSD measurements were performed in a Leo 1550 Gemini field emission gun scanning electron microscope (SEM). The instrument was equipped with a Channel 5 EBSD system by Oxford Instruments and Nordlys II detector. Samples oriented parallel as well as perpendicular to GD of the electrodeposit were prepared by mechanical grinding with SiC paper down to grit size 4000, followed by electropolishing using a Struers Lectropol-5 and A2 electrolyte consisting of perchloric acid, ethanol, 2-butoxyethanol and distilled water. In addition, samples were prepared by mechanical grinding and polishing. Grinding was performed on P2400 and P4000 sandpaper to achieve a flat surface. Afterwards, the sample was polished on MD-Nap with 6 μm, 3 μm and 1 μm diamond paste. Final polishing was performed with OPS on MD-Chem. Between the individual steps, the samples were carefully cleaned with ethanol. After the final polishing step, the samples were cleaned for 3 minutes in an ultrasonic bath containing isopropanol. Furnace annealing treatments were performed for 20 min at 350, 400, 450, 500, 550, and 600°C followed by air cooling.

3. Results and discussion

The DSC curve showed the typical shape of a nanocrystalline electrodeposits consisting of a low-energy exotherm and a main heat release peak. However, it is important to notice that for the investigated materials both the onset of the low-exotherm and the main heat release peak have higher values than the ones recorded for Ni samples with 20 nm grain size and <100><111> double fibre texture. In fact, the onset of the low-energy exotherm was at 325°C (material from Germany) instead of 84°C, while the heat release peak was with 395°C more than 100 degrees higher than in the Nickel material with 20 nm grain size (289°C) [19]. Hence, the Nickel material with <110> fibre texture has a higher thermal stability.

Figure 1 shows the EBSD orientation map in inverse pole figure colouring (coloured parallel to GD) of a Ni sample investigated perpendicular to GD (side view). The orientation map is superimposed to the band contrast image and the sample was heat treated for 20 min at 350°C. As can be seen, the microstructure consists of elongated grains arranged in structural units which all have a <110>-axis parallel to GD (grains are green according to inverse pole figure colour code). When investigating the grain boundaries within the structural units, mainly twin boundaries are found. However, due to the arrangement of the grains, Σ9 grain boundaries are observed in the centre of the structural units. Two examples encircled in Fig. 1 are shown in higher magnification in Fig. 2. Notice, that the structural units shown in Figs. 2

(c) and (d) are not showing a <110> fibre texture as they are not coloured in green. Nevertheless, the example may be a 5-fold twin (the same goes for the encircled structural unit at the lower left corner in Fig. 1). The presence of the $\Sigma 9$ boundary in the centre (only 1 pixel between the blue grain at the top and the purple grain at the bottom; illustrated by an arrow in Fig. 2d) may be due to the resolution of the EBSD map. On the other hand, a 5-fold twin may be surprising since the twins are described by 60°<111> orientation relationships (6 grains with 60°<111> orientation relationship and common <111>-axis would be required for achieving a full circle/identity). However, the grains do not have a common <111>-axis as seen from the pole figure in Fig. 3. In fact, the grains have a common <110>-axis which is tilted out of the y-direction which is parallel to GD.

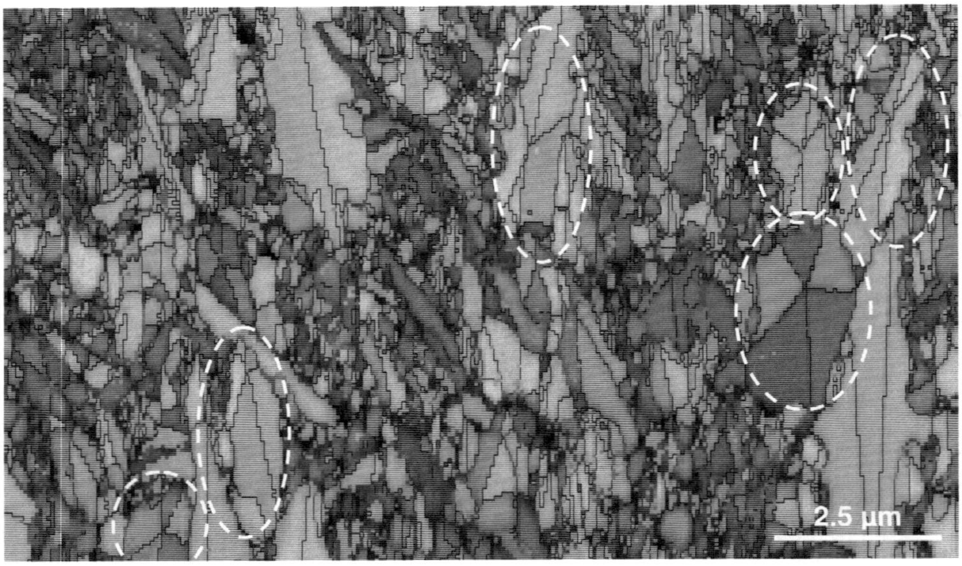

Fig. 1: Inverse pole figure superimposed on band contrast image of a Ni-sample investigated perpendicular to the GD (the colouring is chosen parallel to GD). Black lines illustrate large angle grain boundaries (rotation angle > 10°).

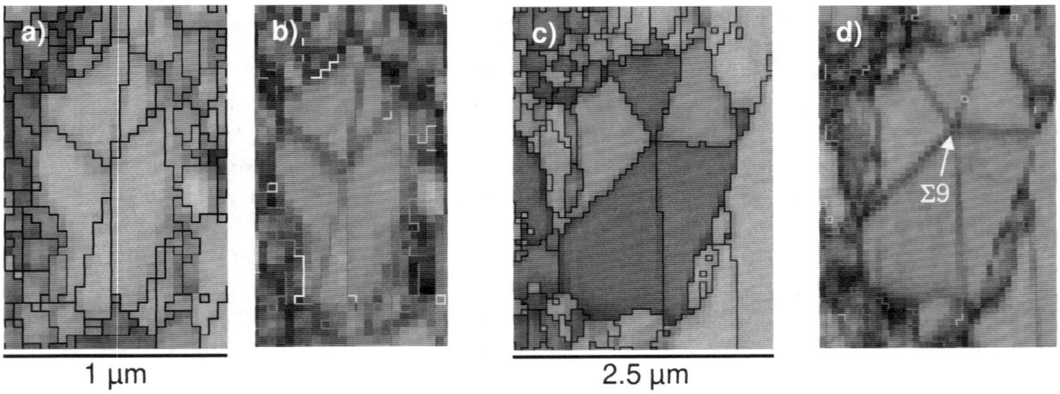

Fig 2: Magnified areas from Fig. 1 ((a) and (c)) and respective band contrast images ((b) and (d)) with Σ boundaries given in colour: Σ3: red, Σ9: pink, Σ11: yellow, Σ13a: light blue.

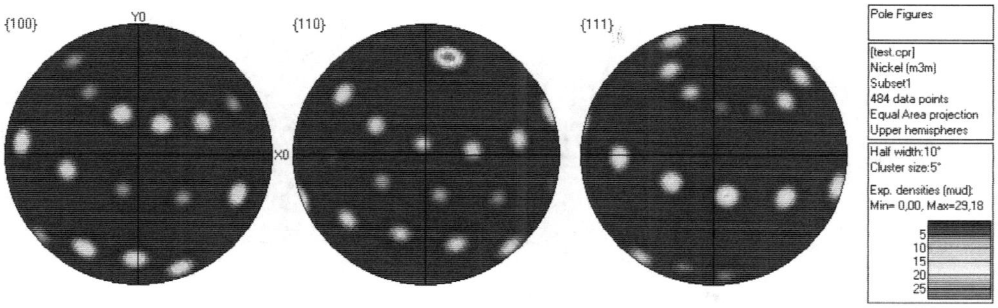

Fig. 3: Pole figures of the structural unit (subset of 5 grains) in Fig. 2 (c) (y-direction parallel GD).

As seen in the examples shown above (Fig. 1), the grains of structural units are usually not meeting in the centre, i.e. they are not forming a common junction. In Fig. 4, a structural unit obtained after heat treatment of 20 min at 500°C and measured perpendicular to GD (side view) is provided in higher magnification (the orientation map is coloured perpendicular to GD while a colouring parallel to GD is shown in the inset in the upper left corner). As can be seen, the hit rate at the centre of the structural unit is too low and no orientation information was obtained (only wild spikes are removed and no additional noise control in form of addressing orientations to zero solutions was performed). For identification of orientation relationships between the grains of the structural unit, the coincidence site lattice (CSL) concept and Brandon's criterion were used [13,20] and all grain boundaries could be identified as low-Σ boundaries, i.e. below Σ27 (Fig. 4 (b)). Solid lines represent grain boundaries that are clearly identified, while dotted lines indicate the position of assumed grain boundaries (not measured due to low local hit rate). Position and character of these grain boundaries were chosen in a way to not violate the combination rule [21]. Following this concept, the grains of the structural unit have twin boundaries between neighbouring grains while in the centre Σ9 and Σ27 grain boundaries (multiple twins) are assumed to be present.

When investigating structural units in top view, i.e. parallel to GD, 5 grains with twin orientation relationship to each other (70.53°<110>) make almost a full circle (352.64° instead of 360°). The missing 7.35° are supposed to be accommodated by internal elastic strain [22]. Figure 5 shows a structural unit in which one of the grains has an orientation gradient (up to 6°; best visible when coloured in y-direction) to accommodate for the low-Σ grain boundary arrangement. The orientation map is processed only slightly, i.e. wild spikes are removed and minor noise reduction is performed (6 nearest neighbours required). Consequently, there are still a large number of zero solutions (mainly at grain boundaries) where the measured patterns could not be identified unequivocally. In Fig. 6, position and character of the grain boundaries are provided which were obtained by taking into account Brandon's criterion and combination rule [20,21]. Again, solid lines are used to describe grain boundaries that are clearly identified by measurement, while dotted lines are used to indicate the position of assumed grain boundaries. As in the previous case, the structural unit can be described by low-Σ grain boundaries (Σ ≤ 33).

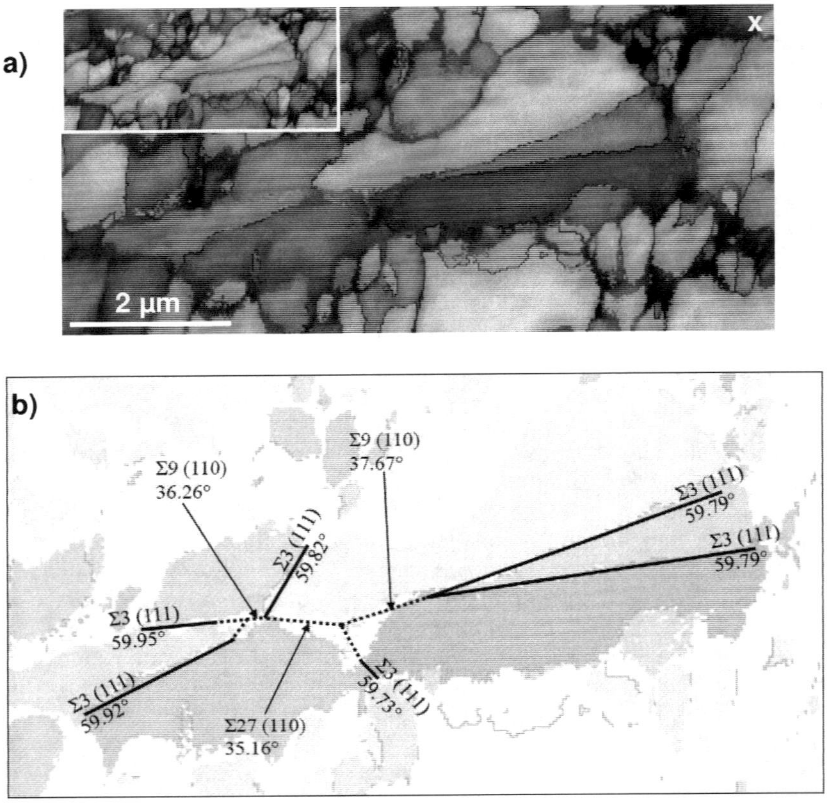

Fig. 4: (a) EBSD orientation map of structural unit in inverse pole figure colouring (chosen perpendicular to GD) superimposed on band contrast image after heat treatment for 20 min at 500℃. The inset (upper left corner) shows the orientation map coloured parallel to GD. (b) Illustration of the grain boundary arrangement in the structural unit seen in (a).

Fig. 5: EBSD orientation map in inverse pole figure colouring superimposed on band contrast image of a structural unit measured parallel to GD (z || GD).

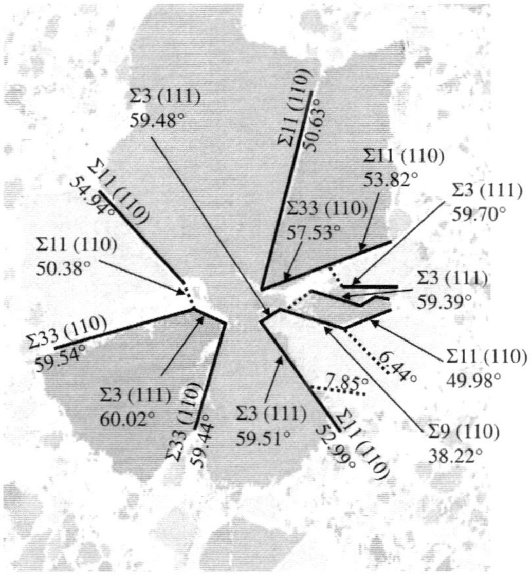

Fig. 6: Grain boundary arrangement of the structural unit in Fig. 5.

In the measurements parallel to GD, no 5-fold twin structure was observed. Instead, the structural units are built up by grains having low-Σ grain boundaries to each other and additional smaller grains in the centre and/or orientation gradients in some of the involved grains accommodate for internal stresses. Changing only one of orientation relationships between neighbouring grains would cause other (energetically favourable) low-Σ boundaries to disappear and be replaced by (energetically less favourable) random grain boundaries. Overall, the formation of these distinct structural units seems to be energetically favourable (low-Σ boundaries) as they are still present and dominate the microstructure after annealing (see Fig. 4). In fact, similar structural units were observed after annealing the material for 20 min at 600°C [16]. However, the performed annealing treatments were obviously not sufficient to create 5-fold twins, which are expected to be energetically the most favourable configuration. Overall, the microstructure development is different from what was observed in Nickel electrodeposits having another initial texture. In those materials, twinning is observed to form new, fast growing, low energy grain orientations dominating the final texture. In the <110> textured material, in contrast, the twins stabilize the grain configuration.

An attempt was made to investigate the structural units in three dimensions. This was done by localizing the structural unit of interest by use of EBSD, marking the position with help of an indent, and then electropolishing the material step-wise. For determining how much material was removed, a silver paint droplet was put on the surface before electropolishing. After the polishing step, the remaining droplet was removed and the height difference between the location of the droplet (initial surface) and the electropolished surface at the location of the structural unit was measured with help of a Wyko Optical Profiler. It was possible to perform three consecutive polishing steps of 1 μm, 1 μm and 1.5 μm, respectively. Hence, it is possible to perform 3-dimensional measurements in this way. However, the orientation maps were partly distorted due to drift and the grains of the structural unit were more or less the same over the measured depth of 4.5 μm, i.e. the area of interest in the centre of the structural unit was not captured. This means, that 3D EBSD measurements by employing focused ion beam and/or investigations by TEM are required for achieving better resolution and for being able to fully and reliable describe the (3-dimensional) grain arrangement in the structural units.

4. Summary and conclusion

Pulse-electrodeposited Nickel with sub-microcrystalline microstructure and a <110> fibre texture in growth direction was investigated. By use of EBSD, structural units consisting of grains possessing a common <110>-zone axis in growth direction and low-Σ grain boundaries (often twins) between each other have been found in the microstructure. The grain boundaries of the structural units have been found to be strongly connected, i.e. changing only one of the orientation relationships between neighbouring grains would cause other low-Σ boundaries to disappear and be replaced by energetically less favourable random grain boundaries. Even though grain growth sets in above 325°C, the structural units remain in the microstructure and the <110> fibre texture is conserved up to at least 600°C. This means that the grain boundary arrangement can be held responsible for the preservation of both the structural units and the <110> texture (low energy configuration even though some stress is involved). In this case there is no orientation change (e.g. by twinning) when grain growth occurs as seen in previous studies on Ni and Ni-Fe of different initial texture. The occurrence of structural units in form of 5-fold twins (parallel to growth direction) seems to be a special case and has not been observed, yet.

Acknowledgement
W. Skrotzki, TU Dresden, Germany, and A. Chokshi, IISc Bangalore, India, are acknowledged for providing the material.

References

[1] Lin C S et al 2001 *J. Appl. Electrochem.* **31** 925
[2] Kim I and Lee S G 2000 *Texture Microstruct.* **34** 159
[3] Nielsen C B and Horsewell A 1997 *J. Appl. Electrochem.* **27** 839
[4] Amblard J et al 1997 *J. Appl. Electrochem.* **9** 233
[5] Nakamura Y et al 1994 *J. Appl. Electrochem.* **24** 227
[6] Gao C-Z et al 1999 *Trans. IMF* **77** 75
[7] Kelly J J and Yang N Y C 2002 Effect of Substrate Configuration on the Grain Structure and Morphology of Electrodeposited Ni for Prototyping *LIGASandia report* SAND2001-8609
[8] Klement U et al 1995 *Mater. Sci. Eng.* **A203** 177
[9] Klement U et al 2008 J. *Microscopy* **230** 455
[10] Klement U et al 2012 *Mater. Sci. Forum* **706-709** *1618*
[11] Czerwinski F and Szpunar J A 1999 *NanoStrMater.* **11** 5669
[12] Rasmussen A A et al 2004 *Mater. Sci. Forum* **467–470** 1345
[13] Kronberg M L and Wilson F H 1949 *Trans. Am. Inst. Min. Metall. Eng.* **185** 501
[14] Lucadamo G et al 2005 *Phil. Mag.* **85** 2549
[15] Bringa E M et al 2008 *Scripta Mater.* **59** 1267
[16] Huang P et al 2009 *Appl. Phys. Lett.* **95** 203101
[17] Klement U et al 2012 *Mater. Sci. Forum* **702-703** 928
[18] Kahrimanidis A and Klement U 2014 *Mater. Sci. Forum* **783-786** *2573*
[19] Klement U et al 1995 *Mat. Sci. Eng.* **A203** 177
[20] Brandon D G 1966 *Acta Metall.* **14** 1479
[21] Gertsman V Y and Tangri K 1995 *Scripta Metall. Mater.* **32** 1649
[22] Zhu Y T 2005 *Appl. Phys. Letters* **86** 103112

Study of texture, microstructure and mechanical properties of asymmetrically rolled aluminium

M Wronski[1,2], K Wierzbanowski[1], S Wronski[1], B Bacroix[2], M Wróbel[3] and A Uniwersał[3]

[1] *AGH University of Science and Technology, Faculty of Physics and Applied Computer Science, al. Mickiewicza 30, 30-059 Kraków, Poland*

[2] *LSPM-CNRS.* Université Paris 13, 99, av. J.B. Clément, 93 430 Villetaneuse, France

[3] *AGH University of Science and Technology, Faculty of Metals Engineering and Industrial Computer Science, al. Mickiewicza 30, 30-059 Kraków, Poland*

Abstract. Asymmetric rolling is a promising forming technique offering numerous possibilities of material properties modification and the improvement of technological process parameters. This geometry of deformation is relatively easy to implement on existing industrial rolling mills. Moreover, it can provide large volume of a material with modified properties. The study of microstructure, crystallographic texture and mechanical properties of asymmetrically rolled aluminium is presented in this work. The above characteristics were examined using EBSD technique and X-ray diffraction. The rolling asymmetry was realized using two identical rolls, driven by independent motors, rotating with different angular velocities. It was found that asymmetric rolling leads to microstructure refinement, texture homogenization and decreasing of residual stress.

E-mails: mwronskii@gmail.com, wierzbanowski@fis.agh.edu.pl, wronski@fis.agh.edu.pl, brigitte.bacroix@univ-paris13.fr, mwrobel@agh.edu.pl, auniwers@agh.edu.pl

1. Introduction

Symmetric rolling (SR) is commonly used in industrial practice and a huge majority of materials is formed in such a way. Asymmetric rolling (AR) presents some alternative - in this deformation mode the peripheral velocities of the top and bottom rolls are different [1-5]. The velocity difference can result from different angular velocities of rolls, their diameters or friction coefficients between rolls and a material. In the present study two identical rolls, of 180 mm diameter, were turning with different angular velocities, ω_1 and ω_2 (the bottom roll turned with 10 rpm). The rolling process asymmetry was defined as: $A=\omega_1/\omega_2$. AR is of potential importance for industrial applications, because it reduces the applied rolling forces and torques and modifies the rolled plate shape and its microstructure. In particular, it can lead to grain refinement. The associated strong shear deformation offers also a possibility of tailoring deformation and recrystallization textures and resulting mechanical properties. In the present work we study microstructure, crystallographic texture and mechanical properties of asymmetrically rolled aluminium 6061.

2. Microstructure of asymmetrically rolled aluminium

The material microstructure was examined by EBSD technique using Cambridge S360 (W-GUN) electron microscope located in LSPM laboratory, University Paris 13. The presented EBSD maps were determined in the RD-TD plane and they show topology of the orientation distribution of normal direction, ND, in the crystal reference frame. They were determined in the central sample layers. The maps for the initial material and for samples rolled symmetrically ($A=\omega_1/\omega_2=1$) and asymmetrically (A=1.1 and A=1.5) are shown in Figs. 1a and 1b, respectively. After AR we observe elongated grains, which are strongly fragmented in their interiors (Fig. 1b). The enhanced grain fragmentation after AR is confirmed by a systematic increase of the *Kernel average misorientation* vs. degree of rolling asymmetry (Fig. 2a).
It is common industrial practice that an annealing is applied after rolling in order to recover material properties. Therefore, we examined the microstructure of the material rolled to 36% reduction and annealed during 1 h at the temperature 500 °C. The obtained EBSD maps are shown in Fig. 1c. We observe equiaxed grains with decreasing size vs. rolling asymmetry. This

result is quantitatively confirmed in Fig. 2b, where one observes a decrease of the average grain area vs. rolling asymmetry ratio (for its two measures :*'number'* and *'area'* - according to *OIM TSL* software). The grain boundary was defined under condition of misorientation higher than 15°.

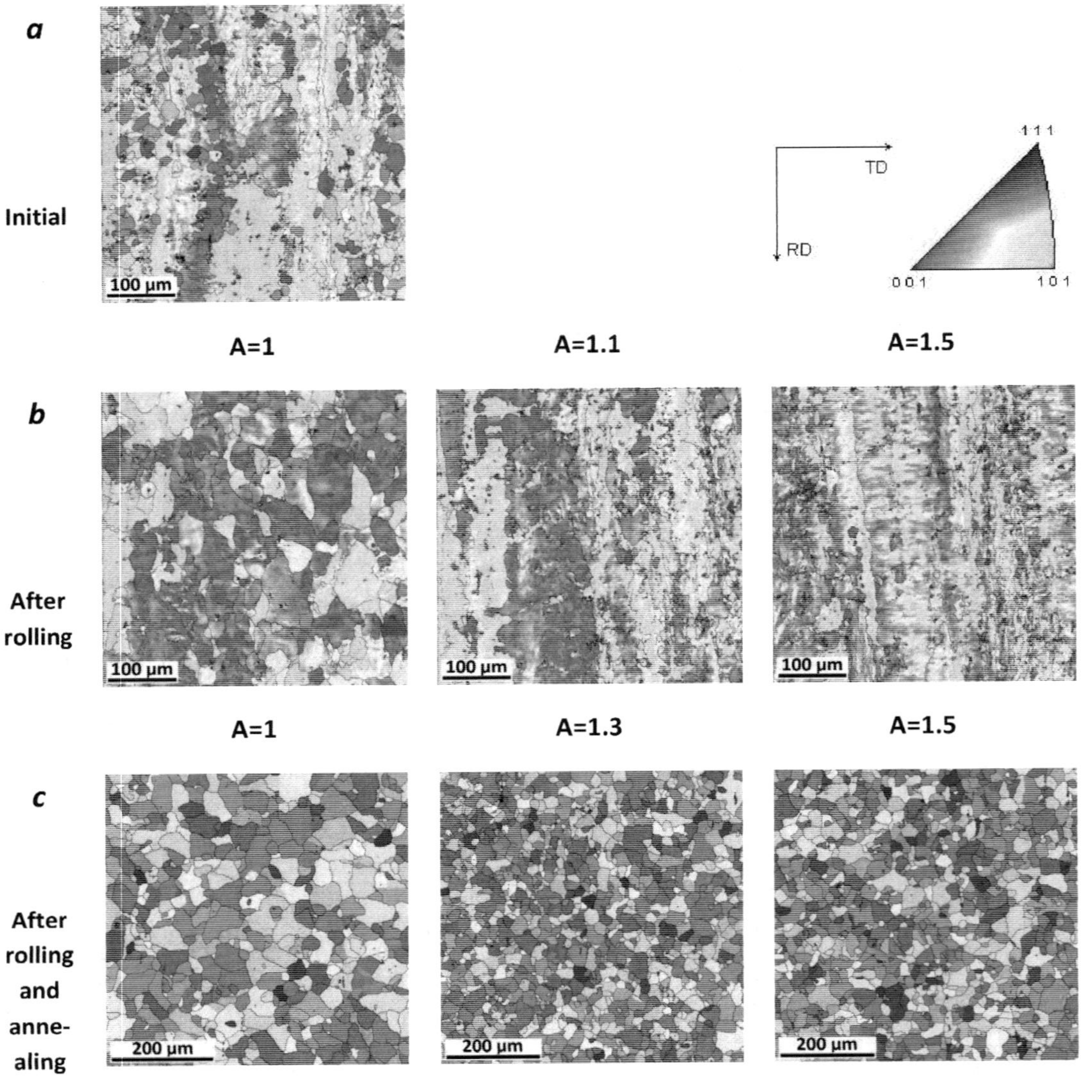

Fig. 1. *EBSD maps for the initial material (a), for the samples rolled symmetrically (A=1) and asymmetrically (A=1.1 and A=1.5) to the reduction of 36% (b), and for the samples annealed after rolling (with A=1, A=1.3 and A=1.5) at 500 °C for 1 h. Topology of ND orientation is shown.*

IOP Conf. Series: Materials Science and Engineering **82** (2015) 012074 doi:10.1088/1757-899X/82/1/012074

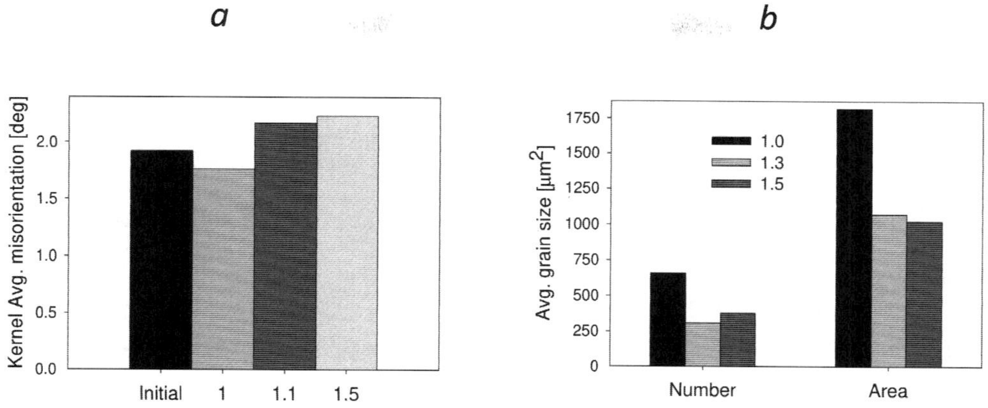

Fig.2. *Influence of rolling asymmetry (A=1, A=1.1, A=1.3 and A=1.5) on:*
a) Kernel average misorientation of samples rolled to 36% reduction, b) average grain area of samples rolled to 36% reduction and annealed at 500 °C for 1 h, calculated as 'number' and 'area' averages (μm²). Results for centre layers are shown.

3. Texture variation

Pole figures were determined in three regions of the rolled samples, i.e. in the top, bottom and central layers, using X-ray diffraction. ODFs were calculated with WIMV method using popLA calculation package [6] (cf. also [7]). The crystallographic texture of the initial material was very sharp and contained Brass (*B*), Cubic (*W*), *S*, and Copper (*C*) orientations. This material was rolled symmetrically (*A*=1) and asymmetrically (*A*=1.5). The ODF sections at $\varphi_2=0^o$, presenting the rolling textures, are shown in Fig. 3. A full analysis showed that the texture of SR sample contained the following components: *W* (dominating one), *B*, rather weak *C* orientation, and traces of *S* and *G* orientations. Characteristic effects appearing during SR and AR can be followed examining *B* component which is visible in $\varphi_2=0^o$ section. In the central layer of SR material all equivalent B component maxima lie on the horizontal red line drawn at $\Phi=45^o$ (Fig 3c-left). In the surface layers of SR material one observes slight shifts of these maxima. In the top layer the first two maxima of B component (in the [0°, 180°] range of φ_1) are shifted up while the next two maxima are shifted down (Fig. 3b - left). All discussed textures are expressed in the same coordinates system: RD, TD, ND. The opposite shifts appear in the bottom layer of the material (Fig.3d - left). Another behaviour is observed in AR material - the shifts of *B* maxima are the same in the three examined layers (the first two maxima are shifted up while the next two maxima are shifted down) as seen in Fig. 3b-d - right. It was checked that the discussed shifts of maxima are caused by corresponding texture rotations around TD. In the case of AR they are of the same sign in the three material layers. Therefore we conclude that the texture becomes more homogeneous after AR process as compared to the SR process.

The aluminium samples were recrystallized after rolling (at 500 °C for 1 h). The recrystallization textures from the centre layer which are corresponding to SR (*A*=1) and AR (*A*=1.5) are shown in Fig.4. They contain only Cube (*W*) component. We do not observe any qualitative influence of the rolling asymmetry ratio, *A*, on the annealing textures.

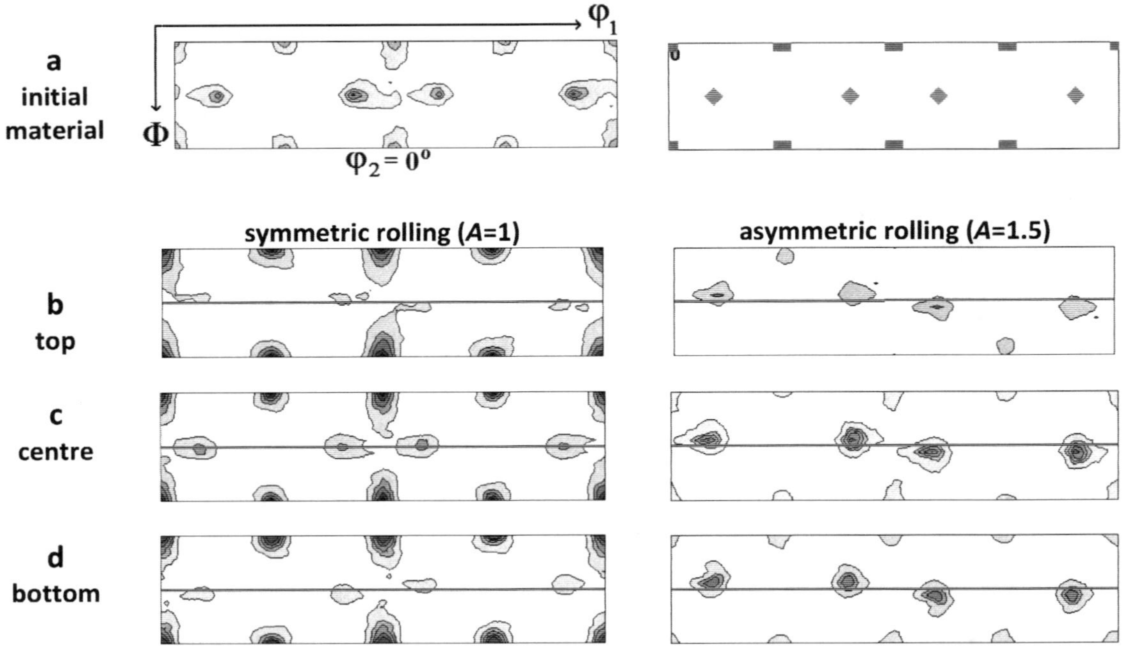

Fig. 3. *Textures of: homogeneous initial material (a) and of symmetrically and asymmetrically rolled samples determined in top (b), centre (c) and bottom (d) material layers. φ_2=0° sections are shown and red lines are located at Φ=45°. W (■) and B (◆) orientations are marked. Textures are presented using levels: 2.40, 8.00, 12.22, 17.61, 22.00, 35.00.*

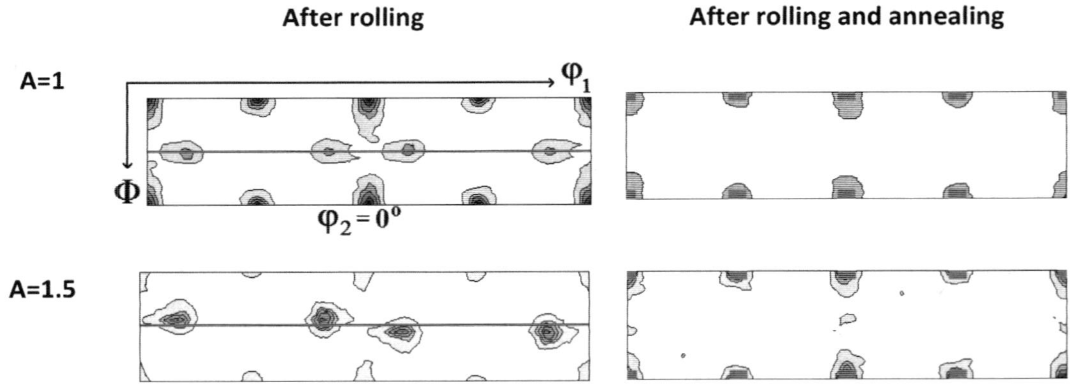

Fig.4. ODF sections (φ_2=0°) *for symmetrically and asymmetrically rolled and recrystallized aluminium 6061 samples (annealed at 500 °C for 1 h). Textures of centre layers are shown and W (■) component is marked. Textures were presented using levels: 2.40, 8.00, 12.22, 17.61, 22.00, 35.00.*

4. Selected mechanical characteristics

Asymmetric rolling has a noticeable influence on mechanical properties. The stress-strain curves from tensile test of samples with different ratios of rolling asymmetry are shown in Fig. 5. It can be noted that samples rolled asymmetrically have higher strength and are less ductile. This effect is caused by a stronger microstructure fragmentation.

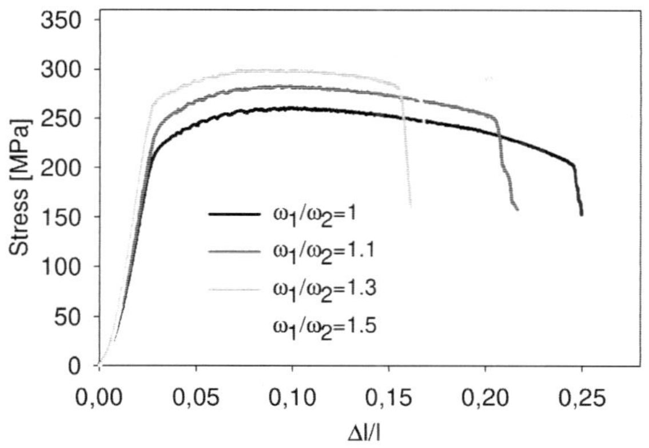

Fig.5. *Stress-strain curves from tensile test of samples rolled symmetrically (A=1) and asymmetrically (A equal to 1.1, 1.3 and 1.5).*

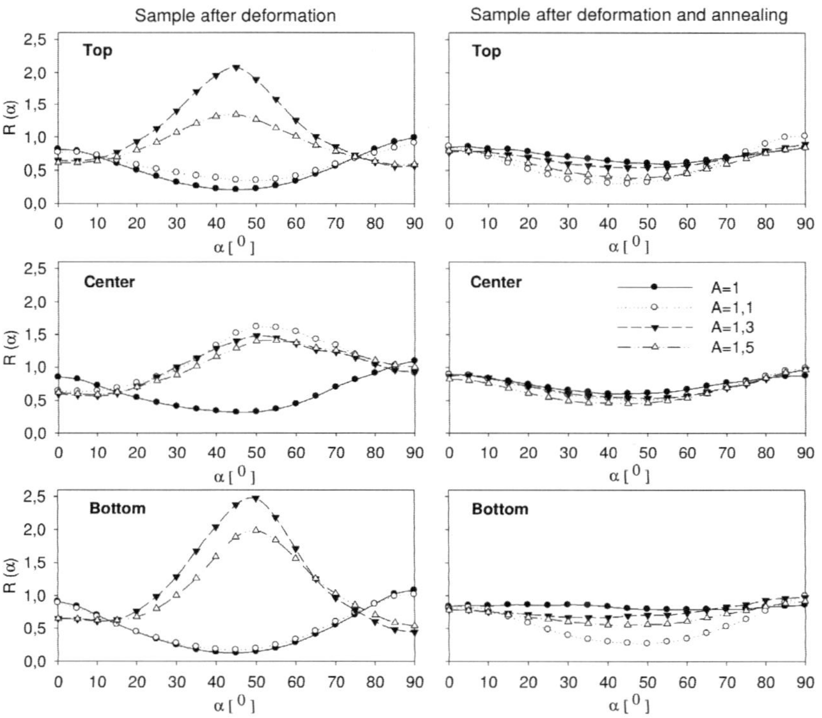

Fig.6. *Lankford coefficient, R(α), for rolled and annealed material in three sample layers (top, centre, bottom) for different ratios of rolling asymmetry, A. α is the angle between the tensile direction and rolling direction.*

The plastic anisotropy, defined by Lankford coefficient, was examined for the rolled and recrystallized samples (Fig. 6). This coefficient was calculated using deformation model and rolling textures of examined samples according to [8] (cf. also [9,10]). We observe an important increase of plastic anisotropy in the case of AR samples (compared to SR ones) in

the three considered material layers. In contrast, the influence of rolling asymmetry on plastic anisotropy is strongly reduced after recrystallization.
Determined by X-ray diffraction [11-13], residual stresses in the aluminum samples rolled to 70% reduction were considerable lowered after AR compared with SR. The σ_{11} residual stress component was reduced from -60 MPa to -23 MPa in the top surface layer, while for the bottom surface layer it was reduced from -90 MPa to -16 MPa (the experimental error was around 10 MPa), when passing from A=1 to A=1.5.

5. Conclusions

The obtained results show that asymmetric rolling of polycrystalline aluminium 6061 leads to the following changes in material properties:
- microstructure is more fragmented after AR than after SR,
- rolling textures are more homogeneous across the sample thickness after AR (they are homogeneously rotated around TD),
- recrystallization textures are similar after SR and AR and they contain only Cubic component,
- AR samples exhibit higher strength (visible in stress-strain curves from tensile tests) than SR,
- AR samples show much higher anisotropy of Lankford coefficient in the rolling plane (in three material layers) than the SR samples; this anisotropy is strongly reduced after recrystallization.

Acknowledgements

This study was financed by the Polish National Centre for Science (NCN) under decision numbers: DEC-2011/01/B/ST8/07394 and DEC-2011/01/D/ST8/07399.

References

[1] Gao H and Chen G 1998, *Iron and Steel* **33** 63
[2] Lee S H and Lee G N 2001, *Int. J. Mech. Sci.* **43** 1997
[3] Chhann S et al. 2007, *Mat. Sci. Forum* **550** 551
[4] Wronski S et al. 2011, *Mater. Character.* **62** 22
[5] Wronski S et al. 2009, *Arch. Metall. Mater.* **54** 89
[6] Kallend J S et al. 1990 *Operational texture analysis, Report LA-UR-90-2852* (Center for Materials Science, Los Alamos National Laboratory, USA)
[7] Tarasiuk J et al. 1998, *Cryst. Res. Technol.* **33** 101
[8] Arminjon M and Bacroix B 1991, *Acta Mech.* **88** 219
[9] Wierzbanowski K et al. 2007, *Arch. Metall. Mater.* **52** 77
[10] Baczmanski A et al. 1995, *Zeit. Metallkd.* **86** 507
[11] Baczmanski A et al. 1997, *Revue de Metallurgie - Cahiers d'Informations Techniques* **94** 1467
[12] Baczmański A et al. 2006, *Mat. Sci. Forum* **524-525** 235
[13] Baczmanski A et al. 1999, *Archives of Metallurgy* **44** 39

Practical applications of the method of generalized pole figures

Yu Perlovich[1], M Isaenkova, V Fesenko, O Krymskaya and P. Dobrokhotov
National Research Nuclear University MEPhI (Moscow Engineering Physics Institute), Moscow, Russia
E-mail: yuperl@mail.ru

Abstract. Several actual practical applications of the new X-ray method of generalized pole figures are considered. Among them there are determination of c- and a-dislocation densities in shell tubes from Zr-1%Nb alloy, analysis of strain hardening at opposite sides of shell tubes from ferrritic-martensitic steel with oxide disperse strengthening particles for high-temperature nuclear reactor and revealing of substructure non-uniformity in rods of Cu, subjected to equi-channel angular pressing.

1. Introduction.

The X-ray method of generalized pole figures (GPF), realizable by texture diffractometers of last generations, includes successive registration of profiles for the same X-ray line at each point of a texture pole figure in the course of its measurement [1]. The method of GPF gives the most complete information on substructure conditions of grains with different crystallographic orientations. By computer treatment of obtained data, distributions of various substructure parameters are calculated.

According to the theory of X-ray diffraction [2], a profile of X-ray line (hkl) characterizes the condition of crystalline lattice along the normal to reflecting planes {hkl}. The GPF method allowed to reveal the systematic difference in strain hardening of crystallites, corresponding to maxima and minima of the deformation texture, and to disclose experimentally the equilibrium of residual elastic microstress in the deformed metal matrix [3]. X-ray studies testify, that these rules are general lows for deformed textured metal materials. This paper contains some examples, where the GPF method was applied efficiently.

2. Dislocation density distributions in shell tubes from Zr-1%Nb alloy

Determinations of the dislocation densities ρ_c and ρ_a in α-Zr matrix by X-ray method [3] gives reasonable, statistically significant results, devoid of possible errors, connected with preparation of samples for direct local observation of dislocations. This approach was developed by use of GPF method, so that GPF ρ_c and GPF ρ_a were constructed for as-rolled shell tubes from Zr-1%Nb alloy. The recalculation of the coherent domain size and the lattice distortion into the dislocation density ρ bases on the supposition that the total energy of dislocations is equal to the energy of lattice distortions, whereas the distance between neighboring dislocations is assumed to be equal to the coherent domain size. But X-ray diffraction allows to detect only those dislocations, which cause some rearrangements of atoms along the normal to reflecting planes. When taking into account that the dislocation produces a displacement of neighboring atoms along its Burgers vector, presence in α-Zr crystallites of c-dislocations with the Burgers vector parallel to axis <0001> and a-dislocations with the Burgers vector parallel to axis $<10\bar{1}0>$ can be revealed by features of X-ray reflections from basal and prismatic planes, respectively. When calculating the dislocation density ρ for all orientations (ψ,φ) of basal and prismatic planes, presented in the texture of the sample, we obtain GPF $\rho_c(\psi,\varphi)$ and GPF $\rho_a(\psi,\varphi)$.

X-ray study of tubes included measurements of three sections, perpendicular to radial, tangential and axial directions, denoted as R-, T- and L-directions, respectively. These sections were prepared from segments, cut out the tube with the help of electro-erosion method. Three partial texture pole figures (PF) with an angular radius of 70° were measured and sewed into complete PF with angular radius 90°. The sewing procedure consists in mutual adjustment of corrected incomplete PFs by leveling of distributions within their overlapping regions (Fig. 1-a).

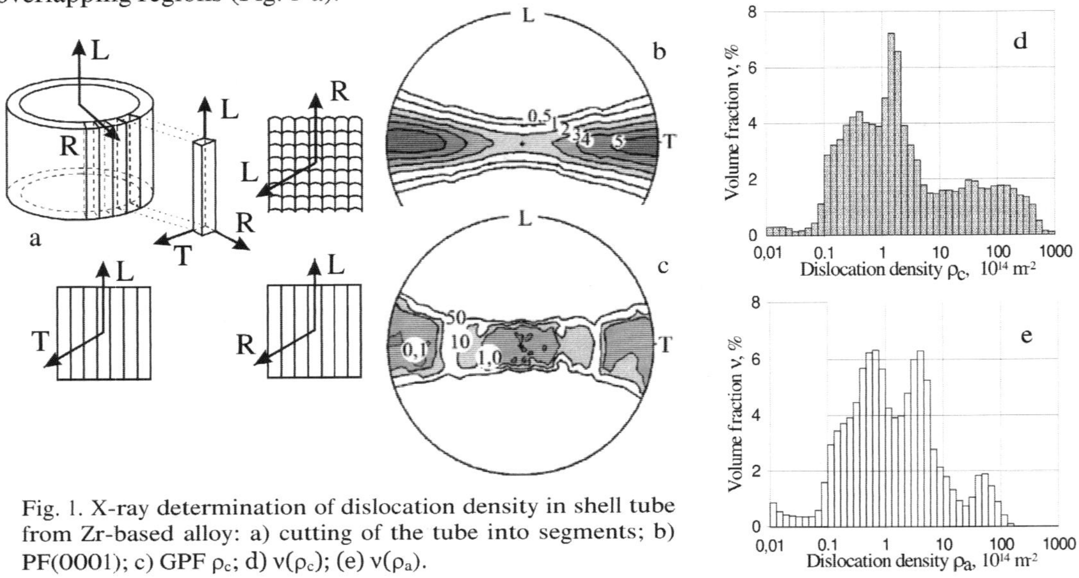

Fig. 1. X-ray determination of dislocation density in shell tube from Zr-based alloy: a) cutting of the tube into segments; b) PF(0001); c) GPF ρ_c; d) $\nu(\rho_c)$; (e) $\nu(\rho_a)$.

Since each crystallite has only one axis <0001>, GPF $\rho_c(\psi,\varphi)$ fully describes the correspondence between the density of c-dislocations and the orientation of α-Zr crystallites. A distribution of volume fractions of grains with different densities of c-dislocations $\nu(\rho_c)$ can be constructed, using values of pole density as weight coefficients by the statistical treatment of GPF ρ_c. In the case of a-dislocations the situation is more complicated. However in rolled products from Zr-based alloys there is a texture maximum, coinciding with rolling direction, to which each α-Zr grain gives one of its axis <10$\bar{1}$0>, and namely this maximum was used for measurement of X-ray line profiles, aimed to construction of GPF ρ_a and $\nu(\rho_a)$.

Complete PF(oo01) and GPF ρ_c for α-Zr matrix of the as-rolled Zr-1%Nb tube ø=20 mm are shown in Fig. 1-b, c. Distributions of volume fractions $\nu(\rho_c)$ and $\nu(\rho_a)$ for the same as-rolled tube are presented in Fig. 1-d, e. Obtained results testify, that estimation of the dislocation density in tubes by the standard method is unrepresentative, since it does not take into account the inhomogeneity of the dislocation distribution. According to data of the used GPF method, in studied tubes the dislocation density varies within 4 - 5 orders of magnitude depending on grain orientations and attain value of 10^{17} m^{-2}.

3. Shell tubes for high-temperature nuclear reactors

The GPF method was used for the study of residual deformation effects in tubes from ferritic-martensitic steel with oxide dispersed strengthening (ODS) particles. Such steels presently are most prospective materials for production of shells and covers for high-temperature nuclear reactors. This is conditioned by their small induced activity, low vacancy swelling, high stability relative to high-temperature embrittlement and stability to high-temperature creep [5]. By manufacture of concrete products from these steels a structure inhomogeneity inevitably arises, and its analysis promotes understanding of mechanisms, responsible for properties, which distinguish these steels from other reactor materials.

Studied shell tubes of 13,0 mm in diameter with wall thickness 1,0 mm were made from ferritic-martensitic ODS steel EP-450, containing 0,35 mass.% of yttrium oxide Y_2O_3.

Application of the GPF method allows to estimate the strengthening caused by disperse oxide particles, which were added in steel at the stage of powder mechanical alloying. Features of plastic deformation mechanisms in steel in the presence of strengthening disperse phase can be revealed as well. Changes of substructure condition of matrix in the presence of disperse strengthening particles helps to estimate indirectly the influence of additions on the steel structure, though direct observation of the oxide phase by diffraction pattern proves to be impossible.

Since manufacture of tubes is connected with large plastic deformations, tube material proves to be strongly textured and, as in any metal material with a developed deformation texture, significant substructure inhomogeneity arises in it. This inhomogeneity consists in sharp differences of strain hardening in different grains depending on their place in the tube texture. At that, additions of disperse strengthening particles into steel can result in redistribution of strain hardening, typical for the initial matrix, which did not contain these particles. As an illustration of possibilities of the used approach to study tubes from ODS steel, the following results are considered. Fig. 2 shows direct PF{001} for inward (a) and outward sides (d) of tube wall, GPF β_{002}, presenting distributions of true half-width for X-ray line for both sides of the same tube (b and e, respectively), and diagrams of correlation

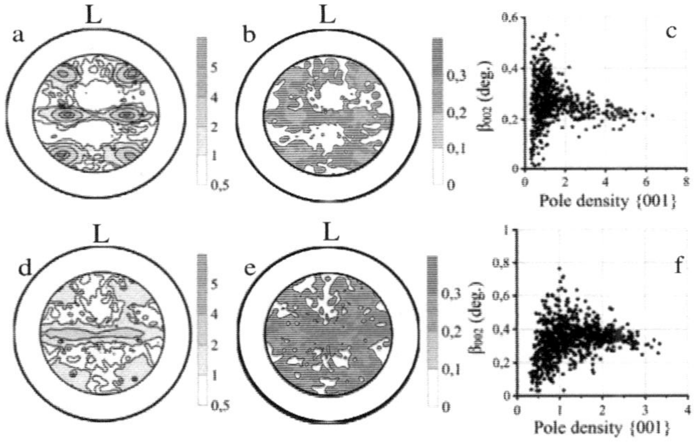

Fig. 2. Strain hardening of steel at opposite sides of tube wall from ODS ferritic-martensitic steel:
a, b, c – inward side;
d, f, g – outward side;
a, d – PF(001);
b, f – GPF β_{002};
c, g – correlation diagrams between PF(001) and GPF β_{002}.

between GPF β_{002} and PF{001} for inward and outward sides, where abscissa of each point (φ, ψ) in the correlation diagram is equal to pole density in this point of PF{001}, while ordinate is equal to angular half-width β_{002} in the same point of GPF β002 (c and f, respectively). PF and GPF, constructed for opposite sides of tube wall, are characterized by significantly different scattering. At that, the mean level of values β_{002} for outward side of the tube wall significantly exceeds that for its inward side.

Near outward surface of the rolled tube the texture is more scattered. This fact testifies that some factors prevent grains from attainment of texture maxima, make the crystallographic slip more difficult and less regular. An evident reason of observed differences between inward and outward tube surfaces consists in increased strain hardening of the latter one, connected with greater quantity of ODS particles due to their redistribution by extrusion and cold rolling. Thus, considerations of PF and GPF for opposite sides of steel ODS shell tubes lead to the mutually agreed conclusion concerning the non-uniform distribution of ODS particles within the tube.

4. Substructure non-uniformity in rods of Cu, subjected to ECAP

Though structure and texture formation under equal-channel angular pressing (ECAP) is now the main topic of numerous studies, still a number of questions remains to be considered more accurately on the basis of additional experimental data. A new experimental approach, based on X-ray GPF methods, makes clear crucial aspects of ECAP deformation process. Cu rods had a rectangular cross-section 8 x 8 mm in size and initially were cut out from rolled and annealed slabs. The ECAP axis coincided with the transversal direction of slabs. The temperature of ECAP was 20°C. For X-ray study of ECAP rods cubic samples 3x3x3 mm in size were cut out by the electro-sparkle method from different regions of the cross-section (Fig. 3-a). Complete texture PF were constructed by sewing of incomplete PF, measured for three facets of cubic samples (Fig. 3-b).

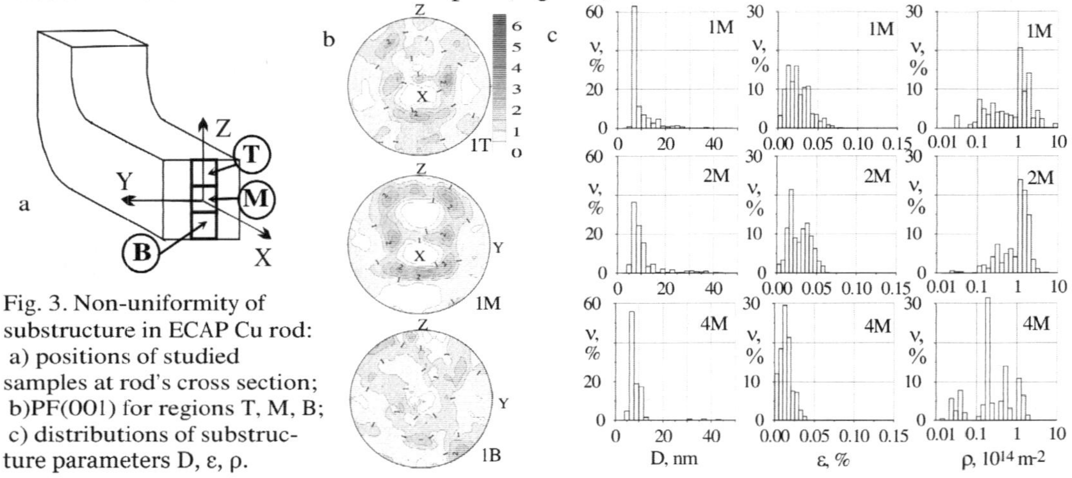

Fig. 3. Non-uniformity of substructure in ECAP Cu rod: a) positions of studied samples at rod's cross section; b)PF(001) for regions T, M, B; c) distributions of substructure parameters D, ε, ρ.

By X-ray study it was assumed, that any textured metal material can be adequately characterized only by distributions of substructure parameters instead of their single values. From this point of view the new X-ray GPF method was used. On the basis of obtained GPF, distributions of coherent domain size ν(D), lattice distortion ν(ε), dislocation density ν(ρ), characterizing the condition of crystalline lattice in reflecting grains along axes <hkl>, were constructed (Fig. 3-c).

In Cu rods the substructure anisotropy decreases with the number of ECAP passes and in most grains some perfection of the crystalline lattice takes place. In Cu rods within M-region most often the coherent domain size D_{111} varies in limits, typical for highly deformed metal materials – from 5 nm to 20 nm. At the same time there are minor fractions with the coherent domain size D up to 40 nm and more. After 4th ECAP pass the Cu rod becomes more isotropic and homogeneous, so that everywhere D is equal 7-8 nm with probability ~60%.

The substructure of ECAP Cu rods shows an evident correlation with the texture non-uniformity through the rod section: weak disordered textures in T- and B-regions after 1st ECAP pass are accompanied by increased intervals of lattice distortion. Hence, irregular texture and relatively increased lattice distortion are conditioned by the same mode of deformation process, when the material cannot form the specific ECAP texture due to turbulence of the plastic flow.

5. Summary

By use of X-ray GPF method grains of deformed metal polycrystals can be split into fractions, differing by substructure parameters and, in particular, by strain hardening.

References

[1] Bunge H J, Perlovich Yu, Isaenkova M 1997 *Textures and Microstructures* **29** *p 241*
[2] Warren B E 1969 *X-ray diffraction* (Addison-Wesley Publishing Company, Inc., Reading, Massachussetts)
[3] Griffiths M, Winegar J E, Mecke J F, Holt R A 1992 *Advances in X-Ray Analysis* **35** p 593

IOP Conf. Series: Materials Science and Engineering **82** (2015) 012076 doi:10.1088/1757-899X/82/1/012076

Grain boundary character distribution of CuNiSi and FeNi alloys processed by severe plastic deformation

H Azzeddine [1], T Baudin [2], AL Helbert [2], F Brisset [2], F Hadj Larbi [1], K Tirsatine [1], M Kawasaki [3,4], D Bradai [1] and TG Langdon [4,5]

[1] Faculté de Physique, USTHB, BP 32 El-Alia, Dar El Beida, Alger, Algérie
[2] Université de Paris-Sud, ICMMO, UMR CNRS 8182, Bâtiment 410, 91405 Orsay Cedex, France
[3] Division of Materials Science and Engineering, Hanyang University, Seoul 133-791, South Korea
[4] Departments of Aerospace & Mechanical Engineering and Materials Science, University of Southern California, Los Angeles, CA 90089-1453, USA
[5] Materials Research Group, Faculty of Engineering and the Environment, University of Southampton, Southampton SO17 1BJ, U.K.
E-mail: azehibou@yahoo.fr

Abstract. In this work the Grain Boundary Character Distribution (GBCD) in general and the relative proportion of low-Σ CSL (Coincidence Site Lattice) grain boundaries are determined through EBSD in Cu-2.5Ni-0.6Si (wt.%) and Fe-36Ni (wt.%) alloys after processing by high-pressure torsion, equal-channel angular pressing and accumulative roll bonding.

Introduction

The processing of ultra-fined grains and/or nanostructured materials by the top-down approach of severe plastic deformation (SPD) has developed significantly during the last two decades. A considerable volume of current information on Bulk Nanostructured Materials (BNMs) has been reviewed recently [1]. It includes all aspects of mechanical, magnetic and physical properties of BNMs after SPD processing. However, the nature and geometry of the grain boundaries evolution has been seldom studied whereas the grain boundary characteristic is critical to control different bulk properties of a polycrystalline material.

In the present work, the grain boundary character distribution (GBCD) in general and the relative proportion of special boundaries, defined within the Coincidence Site Lattice (CSL) classification, are determined through Electron Backscatter diffraction (EBSD) in Cu2.5-Ni0.6-Si (wt.%) and Fe-36Ni (wt.%) alloys after SPD by high-pressure torsion (HPT) [2], equal-channel angular pressing (ECAP) [3] and accumulative roll bonding (ARB) [4].

Experimental

The materials used in this investigation were Cu-2.5Ni-0.6Si (wt.%) and Fe-36Ni (wt.%) alloys. The billets of Cu-2.5%Ni-0.6%Si were processed by ECAP at 150 °C for N = 2, 8 and 12 passes via route A with a die-angle of 135°. Disks of the same alloy were processed via HPT under 6.0 GPa for 1/2, 5 and 10 revolutions at room temperature. The ARB process was carried out for Fe-36%Ni sheet alloy to 2, 4, 6, 8 and 10 cycles at 550 °C without lubricant. The detailed sample preparations and the processing histories of the SPD processing are reported elsewhere [5, 6].

EBSD was used to determine the GBCD and the microtexture via calculated orientation distribution functions (ODFs). The experimental data were obtained using a scanning electron microscope FEG-SEM SUPRA 55 VP operating at 25 kV with a TSL orientation imaging system, OIM™ software and a cut-off angle of 1° for the minimum boundary misorientation. The grain boundaries were classified into three categories based on the misorientation between the neighboring grains. They are low-angle boundaries (LAGBs) with misorientation between $5° \leq \theta \leq 15°$, special boundaries (CSL) $\Sigma 3$ to 25 and high-angle boundaries (HAGBs) with a misorientation angle of $\theta > 15°$. A boundary is usually classified as belonging to a certain Σ value, if the deviation from an exact CSL misorientation is specified by the Brandon criterion [7].

Results and Discussion

Fig.1 shows the number fractions of CSL boundaries of as-received samples for (a) Cu-2.5%Ni-0.6%Si alloy and (b) Fe-36%Ni alloy, respectively. It is clear that the fractions of $\Sigma3$ and $\Sigma9$ type twin boundaries are dominant for both as-received alloys. The $\Sigma3$ boundary is known as the coherent annealing twin boundary that is characterized by a 60° rotation around the <111> axis. It is generally understood that thermomechanical treatments on f.c.c metals and alloys with low to medium stacking fault energies (γ_{SFE}) leads to microstructures with high fractions of $\Sigma3^n$ type boundaries [8]. The texture intensity is low in Cu-2.5%Ni-0.6%Si alloy and there is a β type texture classically observed in f.c.c. alloys after rolling. The as-received Fe-36%Ni alloy shows the typical texture of annealed f.c.c material sheets. The dominant components are the Cube {001}<100> component.

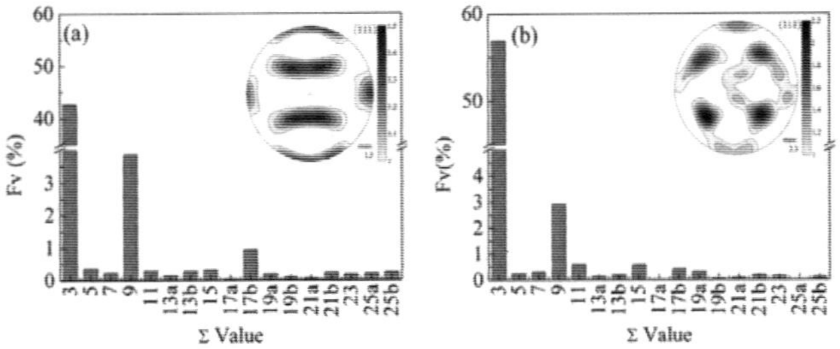

Fig.1: Number fraction of CSL boundaries of as received samples: a) Cu-2.5%Ni-0.6%Si alloy and b) Fe-36%Ni alloy.

Fig.2 shows the GBCD after (a) ECAP and (b) HPT on the Cu-2.5%Ni-0.6%Si alloy and (c) ARB on the Fe-36%Ni alloy. It is apparent that the fraction of HAGBs increase and LAGBs decrease with increasing processing numbers for all deformation processing. In the initial stage of deformation, a lamellar boundary structure is formed having a large fraction of LAGBs and with increasing strain the boundaries transform to HAGBs [9] with refining microstructure leading to an equi-axed grains structure [6,10].

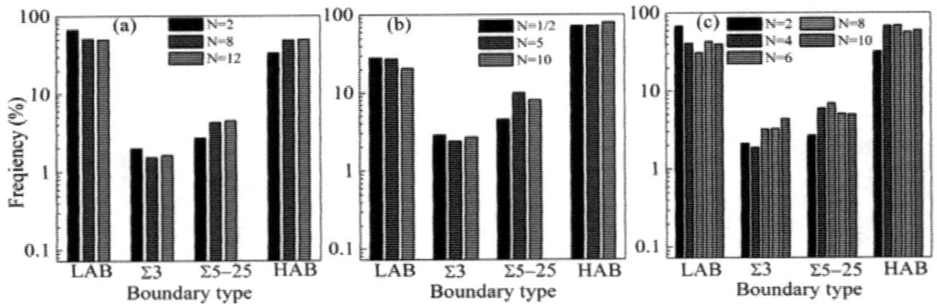

Fig.2: Grain boundary character distribution after a) ECAP, b) HPT (Cu-2.5%Ni-0.6%Si alloy) and c) ARB (Fe-36%Ni alloy) processing.

It is clearly seen that the HAGB frequency increases up to 4 cycles and then slightly decreases after 6 cycles of ARB processing. Similarly, a saturation of HAGBs is depicted after 0.5 turn of HPT, and after 8 passes of ECAP. The saturation in the microstructural refinement and the grain boundary evolution was reported earlier in metallic material after severe plastic deformation [11]. It is worth noting that ECAP generates a highest fraction of LABs while HPT process leads to the highest fraction of HAGBs, including the fraction of CSL boundaries. The $\Sigma 3$ boundaries fraction decreases initially and then increases after reasonably high processing numbers of ECAP and HPT processing, while after ARB processing it increases continuously with increasing numbers of cycle. Moreover, high fractions of $\Sigma 3$ are observed in all processing, thereby suggesting active formation of twinning during SPD even at room temperature [12]. Nevertheless, there is a reasonable difference in the GBCD between the SPD processes. These differences are attributed to the alloys composition and deformation conditions. The stacking fault energy (γ_{SFE}) is expected to have a considerable effect on the fraction of $\Sigma 3^n$ boundaries.

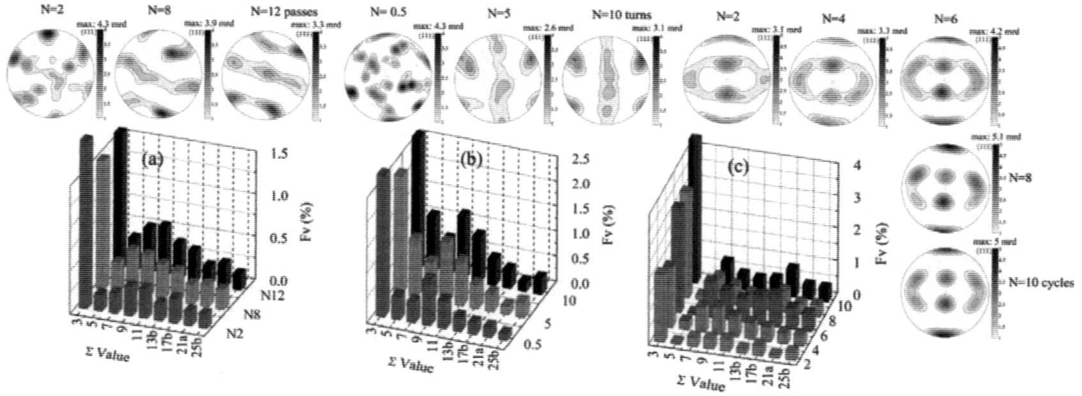

Fig.3: Number fraction of CSL boundaries after: a) ECAP, b) HPT and c) ARB processing.

Fig.3 shows the number fraction of CSL boundaries after (a) ECAP and (b) HPT on the Cu-2.5%Ni-0.6%Si alloy and (c) ARB on the Fe-36%Ni alloy. The sample deformed by ECAP shows an increase and saturation at Σ 7, Σ 11 Σ 13b and Σ 21a boundaries after only 2 passes. A continuous increase of Σ9 fraction is observed while that of Σ 17b decreases then saturates. HPT processing leads to an increase and stabilisation of Σ 7, Σ 17b and Σ 21a fractions. However, Σ 11, Σ 9 and Σ 13b show a continuous increasing as strain increases. The sample deformed by ARB displays a slight difference from ECAP and HPT. In practice, there is a transition in CSL boundary evolution after 6 cycles of ARB and only a net increase of Σ 3, 7, 9, 13b and 17b is observed after 6 cycles. The increase in the fractions of special boundaries may be caused by dynamic recovery.

The evolution of the texture after SPD is presented as {111} pole figures for each processing condition in Fig.3. The textures are generally not sharp in the initial stage of deformation whereas ARB leads to the formation of α-fiber. After 6 cycles of ARB in Fig. 3(c), the Goss component disappears and the S component is more visible. The fraction of Brass component decreases, while the intensity of Copper increases considerably leading to the formation of β-fiber. The texture after 6 cycles of ARB may be characterized by a transition of the formation of α-fiber to β-fiber.

All three components of shear texture revealed earlier [13], namely C component {001} <110>, A-fiber {111} <uvw > and B-fiber {hkl} <110>, are observed after ECAP and HPT processing of the Cu-2.5%Ni-0.6%Si alloy. There were no changes in the type of texture components after ECAP and HPT processing compared to ARB Fe-36%Ni alloy. The intensity of both fibers increases with increasing strain and the intensities of the B-fibre are higher than A-fibre for both deformations. It is worth noting that the HPT texture is sharper than after ECAP and this is probably because of the degree of imposed strain through processing. Moreover, it is suggested that the GBCD of several alloys depends strongly on grain size. Thus, the fraction of coincidence boundaries with Σ values lower than Σ 29 increases with decreasing average grain sizes. In practice, the Σ value is known to be closely related to the common rotation axis of adjacent grains crossing the boundary and it is reported that the Σ value of the coincidence boundary depends on the {100} or {111} fibre texture [14].

Conclusions

The fraction of HAGBs increases as further continuing plastic deformation by the three SPD processes of ECAP, HPT and ARB. A saturation of HAGBs is obtained after 0.5 turn of HPT, 8 passes of ECAP and 4 cycles of ARB. ECAP processing generates the highest fraction of LABGs while HPT processing leads to the highest fraction of HAGBs. High fractions of $\Sigma3$ are observed in all processing. Most of the special boundaries where the fraction increases upon increasing strain are related to rotations around <111> and <110> axes (Σ 3, Σ 7, Σ 11, Σ 13, Σ 17 and Σ 21).

Acknowledgements

The authors thank Pierre-Louis REYDET from APERAM-alloys Imphy Society and Mr. Hacen HADJ LARBI from the CLAL-FRANCE for kindly providing the Fe-36%Ni and Cu-2.5%Ni-0.6%Si alloys.

References

[1] Valiev R Z, Zhilyaev A P and Langdon T G 2014 Bulk Nanostructured Materials: Fundamentals and Applications, WILEY-VCH, Weinheim, Germany.
[2] Zhilyaev A P and Langdon T G 2008 *Prog. Mater. Sci.* **53** 893.
[3] Valiev R Z and Langdon T G 2006 *Prog. Mater. Sci.* **51** 881.
[4] Saito Y, Tsuji N, Utsunomiya H, Sakai T and Hong R G 1998 *Scripta Mater.* **39** 1221.
[5] Khereddine Y, Hadj Larbi F, Kawasaki M, Baudin T, Bradai D and Langdon T G 2013 *Mater. Sci. Eng.* **A576** 149.
[6] Tirsatine K, Azzeddine H, Baudin T, Helbert A L, Brisset F, Alili B and Bradai D 2014 *J. Alloys Compd.* **610** 352.
[7] Brandon D G 1966 *Acta Metall.* **14** 1479.
[8] Lee D S, Ryoo H S and Hwang S K 2003 *Mater. Sci. Eng.* **A 354** 106.
[9] Suresh K S, Sinha S, Chaudhary A and Suwas S 2012 *Mater. Charact.* **70** 74.
[10] Khereddine Y, Hadj Larbi F, Azzeddine H, Baudin T, Brisset F, Helbert A L, Mathon M, Kawasaki M, Bradai D and Langdon T G 2013 *J. Alloys Compd.* **574** 361.
[11] Loucif A, Figueiredo R B, Baudin T, Brisset F and Langdon T G 2010 *Mater. Sci. Eng.* **A527** 4864.
[12] Sitarama Raju K, Ghanashyam Krishna M, Padmanabhan K A, Muraleedharan K, Gurao N P and Wilde G 2008 *Mater. Sci. Eng.* **A491** 1.
[13] Montheillet F, Cohen M and Jonas J J 1984 *Acta Metall.* **32** 2077.
[14] Matsumoto K, Shibayanagi T and Umakoshi Y 1997 *Acta Mater.* **45** 439.

IOP Conf. Series: Materials Science and Engineering **82** (2015) 012077 doi:10.1088/1757-899X/82/1/012077

EBSD characterization of an IF steel processed by Accumulative Roll Bonding

F Cruz-Gandarilla[1], A M Salcedo-Garrido[1], M Avalos[2], R Bolmaro[2], T Baudin[3, 4], J G Cabañas-Moreno[5] and H J Dorantes-Rosales[6]
[1]Instituto Politécnico Nacional, ESFM. Ed 9, México, D.F., C.P. 07738.
[2]Instituto de Física Rosario. CONICET-UR. 27 de Febrero bis 2000, Argentina.
[3] CNRS, UMR8182, ICMMO, SP2M, Orsay, F-91405, France.
[4] University of Paris-Sud, Orsay, F-91405, France.
[5] CINVESTAV IPN, México D.F., 07360.
[6] Instituto Politécnico Nacional, ESIQIE-DIM, AP 118-430, México, D. F. 07051.
E-mail: fcruz@esfm.ipn.mx

Abstract. The objective of this work is to study the texture and microstructure evolution of an IF steel deformed by Accumulative Roll Bonding (ARB) using Electron Backscatter Diffraction. Texture changes occur with increasing number of ARB cycles. For the early cycles, the main components are the α and γ fiber components characteristic of steels. With increasing the number of ARB cycles a tendency towards a random texture is obtained. In the initial state, the mean grain size is 30 μm and after 5 cycles it decreases to 1.2 μm. For the first ARB cycles, the fraction of high angle grain boundary is low but it increases with the number of cycles to about 80% for 5 cycles. The Kernel Average Misorientation (KAM) has no appreciable changes with the number of ARB cycles for all the texture components.

1. Introduction.

Materials with ultra-fine grain size (UFG) have been produced by different severe plastic deformation (SPD) processes, including equal-channel angular pressing, high-pressure torsion, Accumulative Roll Bonding (ARB), etc. [1, 2].
The interest in these materials is related with two aspects. One is their attractive mechanical properties, in particular the increase in strength and ductility for industrial applications [3, 4]; the other aspect is related with the clarification of the microstructure evolution and deformation mechanisms occurring during the processing of these materials. Grain Boundary Sliding (GBS) takes place when the materials have a nanocrystalline structure, with preponderance of High Angle Grain Boundaries (HAGB), and very low strain inside the grains. However in UFG structure, despite a mixture of deformation mechanisms have been detected as active by many authors and that GBS could be active at least at the latest stages of deformation, dislocation glide could be still one of the most important deformation phenomena [5]. Texture, which is highly correlated with microstructure development, is another aspect that can influence the mechanical properties.
The Electron BackScatter Diffraction (EBSD) technique is ideal for characterizing these microstructural features, including texture components and their volume fraction, grain size (GS), boundary character (HAGB, and Low Angle Grain Boundary, LAGB), grain inner misorientation (Kernel Average Misorientation, KAM, which is an indicator of the dislocation density) [6].
In this work we study texture and microstructure evolution by EBSD of an interstitial free (IF) steel deformed by ARB after different number of strain cycles.

2. Material and Experimental Techniques.

The ARB process is described in references [1-3]. In our particular case as received samples were reduced until 1 mm thickness by several cold rolling passes. Then the samples were recrystallized in a high vacuum furnace 3 h at 750 ^{o}C. The mill cylinders for ARB have 20.4 cm in diameter and the rolling speed was 13.4 m/min. Between each ARB pass, the samples were brushed, degreased and heated in a cylindrical furnace for 10 min at 550 ^{0}C under Ar atmosphere. The accumulated equivalent plastic strains for each pass [5] are 0 (o cycle), 0.8 (1 cycle), 1.6 (2 cycles), 2.4 (3 cycles), 3.2 (4 cycles), 4 (cycles). The composition of the steel, provided by ArcelorMittal, France is shown in table 1.

Table 1 Chemical composition of the steel (ppm).

C	Mn	P	S	Si	Al	Ni	Cr	Cu	Mo	Sn	Nb	V	Ti	B	N
15	1022	131	107	58	600	158	225	138	7	29	2	20	532	1	26

For EBSD study, the samples were prepared in a conventional form (mechanical polishing with colloidal silica) on a plane formed by the Rolling Direction (RD) and the Normal Direction (ND). Kikuchi pattern acquisition was made using a TSL EDAX equipment mounted on a FEG-SEM (Quanta, FEI). Scans were run with a step of 0.04 μm. The data were processed with TSL OIM 7.1 software. For grain size calculation, a grain tolerance angle of 5^0 was used, and the data were cleaned up with grain dilation method with a minimum grain size of 4 points. Limits of HAGB (>15^0), LAGB (2^0-15^0) and KAM (<5^0) and third neighbors were used.

3. Results and Discussion.

3.1 Grain Size.

Figure 1 shows the Inverse Pole Figure (IPF) maps (Normal direction to the sheet) measured on the RD-ND plane.

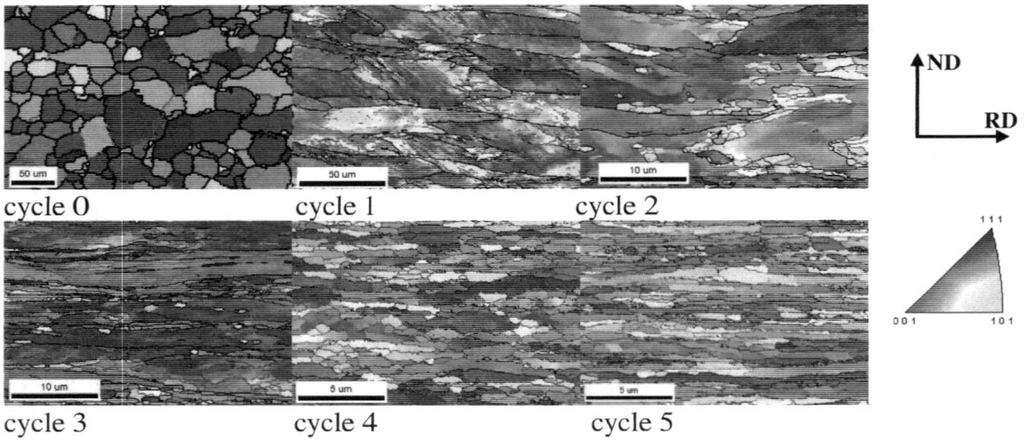

Fig 1 EBSD maps (IPF contrast) from the RD-ND plane for different numbers of ARB cycles (LAGB HAGB). The stereographic triangle shows the color code for ND.

It is clear from figure 1 that the grain morphology changes with increasing number of ARB cycles, going from equiaxed grains to elongated ones along RD. The grains in specimens subjected to 1, 2 and 3 cycles are grouped together forming bands (pancake-like grain structure) with grains in approximately the same orientation; within these bands, the (sub-) grains are separated by LAGB. The samples with

IOP Conf. Series: Materials Science and Engineering **82** (2015) 012077 doi:10.1088/1757-899X/82/1/012077

cycles 4 and 5 show a reduction of grain size and an increase of orientation randomness.

In figure 2, the evolution of the mean grain diameter with the number of cycles is shown. In the first cycles, the evolution is fast, but after 3 cycles there is a decrease in the rate of reduction of grain size, which tends to a size close to 1 μm.

This tendency to grain size stabilization for cycles 4 and 5 could be related with several factors, such as dynamical recrystallization.

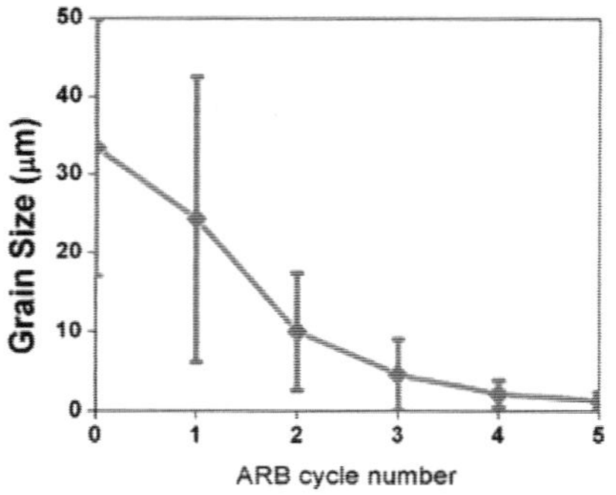

Fig 2 Evolution of the grain size vs. the number of ARB cycles.

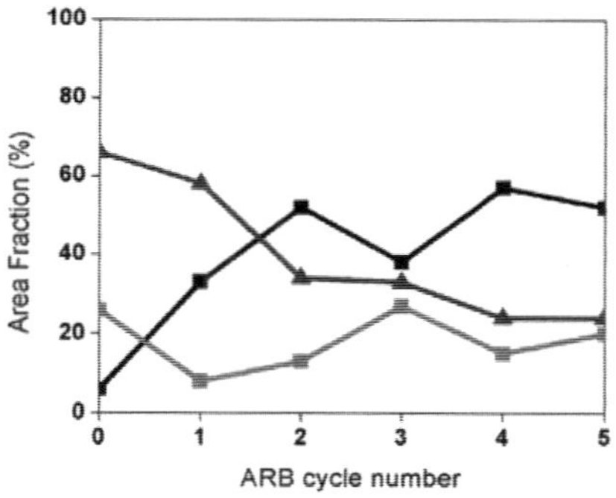

Fig 3 Evolution of the area fraction of α (red), γ (blue) fibers and of other components (black) vs. ARB cycles.

3.2 Texture components.

Figure 3 shows that area fraction of grains randomly oriented tends to increase and stabilizes after 2 ARB cycles. This figure shows also that the α fiber tends to remain approximately constant during the ARB cycles, while the γ fiber tends to decrease and to stabilize after 2 cycles.

Let us note that it was observed that both orientations {111}<110> and {111}<121> of the γ fiber have approximately the same acuity.

3.3 Grain boundaries (HAGB and LAGB).

Figure 4 shows the evolution of HAGB (>15°) and LAGB (2°-15°) fractions with the number of ARB cycles. After the first cycle the fraction of LAGB is greater than the one of HAGB, but after 3 cycles this proportion is reversed. After 5 cycles the HAGB percentage is about 80%.

3.4 Misorientation (KAM).

In figure 5, an example of the KAM value distribution is shown for the sample that underwent 5 ARB cycles. The great majority of points have KAM values between 0° and 1° (blue color). This means that very low misorientation values are present inside the grains. No appreciably differences in values are present between cycles. This implies that all the orientations have more or less the same dislocation density.

4. Conclusion

The results show that the grain size decreases very quickly for the first three cycles, and at a slower rate with additional cycles, until a mean size of about 1 μm is obtained after 5 cycles.

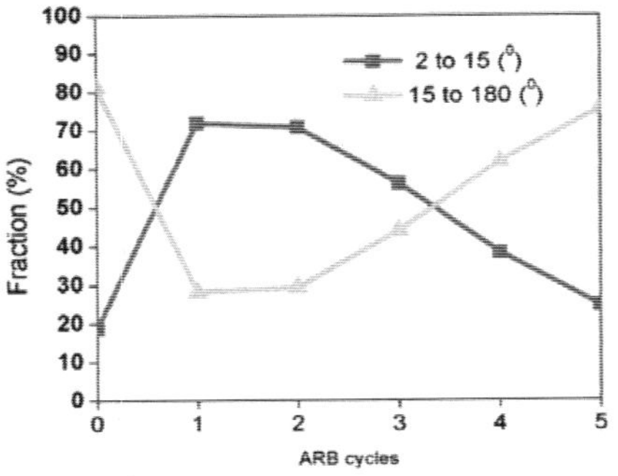

Fig 4 Evolution of HAGB and LAGB fractions with the ARB cycle.

Fig 5 EBSD map of KAM contrast for the sample subjected to 5 ARB cycles.

At the same time, changes in the grain boundary type happened with a tendency to form HAGB (≈80 %). The texture tends to become random with increasing number of ARB cycles.

Acknowledgments
FCG, AMSG, MA and RB acknowledge funding from CONACyT (Mexico) –MINCYT (Argentina) project 208438. FCG, AMSG and HJDR acknowledge support from COFAA-IPN in this research. FCG kindly thanks to S. Martinez, EDAX Mexico, for his support with EDAX software. FCG and TB express their recognition to H. Réglé from Arcelor Mittal Maizières Les Metz, France for the supplied material.

References

[1] Tsuji. N2009 *Bulk Nanostructure Materials* Ed. Zehetbauer M.J. and Zhu Y T. Ed. WILEY-VCH (Weinheim, Germany) pp 235-252.
[2] Valiev R Z, Zhilyaev A P and Langdon T G, 2014 Bulk Nanostructured Materials: Fundamentals and Applications, TMS-WILEY, Hoboken, New Jersey, USA.
[3] N. Tsuji 2002 Strength and ductility of ultrafine grained aluminum and iron produced by ARB and annealing. *Scripta Materialia* 47 pp 893–899.
[4] http://www.autospeed.com/cms/title_Steel - Identification - Using - Hardness - Testing/A_109717/article.html
[5] Toth L S, Gu C. 2014 Ultrafine-grain metals by severe plastic deformation. Materials Characterization **92**, pp 1–14.
[6] Wright S I, Nowell M M, Field D P. 2011 A review of strain analysis using electron backscatter diffraction. *Microscopy and Microanalysis* **17** pp 361-329.
[7] Grains in OIM. OIM ANALYSIS 7.0. TSL-EDAX 2013, TexSEM Laboratories, Incorporated. pp 440.
[8] Kolahi A, Akbarzadeh A, Barnett M R. 2009 Electron back scattered diffraction (EBSD) characterization of warm rolled and accumulative roll bonding (ARB) processed ferrite. *Journal of Materials Processing Technology* **209** pp 1436–1444.

IOP Conf. Series: Materials Science and Engineering **82** (2015) 012078 doi:10.1088/1757-899X/82/1/012078

A variety of microstructures in Mg/Cu super-laminate composites caused by competitive reactions during hydrogenation

K Tanaka[1], K Shibata[2], Y Nishida[2, 3], K Kurumatani[2, 4], R Kondo[2], S Kikuchi[5] and H T Takeshita[2]
[1] National Institute of Advanced Industrial Science and Technology (AIST), Osaka, Japan
[2] Kansai University, Osaka, Japan
[3] Present address: LIXIL Corporation, Tokyo, Japan
[4] Present address: Fukunishi Foundry Co., Ltd., Osaka, Japan
[5] The University of Shiga Prefecture, Shiga, Japan
E-mail: koji.tanaka@aist.go.jp

Abstract. A variety of microstructures in Mg/Cu super-laminate composites (SLCs) caused by competitive reactions during hydrogenation has been shown experimentally. Two types of $MgCu_2$ structures, three-dimensional (3-D) network and layer, were observed after initial hydrogenation of Mg/Cu SLCs under the conditions of 573K, 86.4ks in H2 of 3.3MPa. It was proposed that Mg/Cu SLCs could be hydrogenated by two kinds of processes. The one is alloying Mg with Cu to form Mg_2Cu followed by hydrogenation of Mg_2Cu, leading to the formation of 3-D network of $MgCu_2$. The other is hydrogenation of Mg followed by the reaction of MgH_2 to Cu, leading to the formation of layer $MgCu_2$. SEM observations revealed that there existed Mg_2Cu nano-crystals at the interface between Mg and Cu in as-rolled Mg/Cu SLCs, and layer $MgCu_2$ at the interface between MgH_2 and Cu in pellets of MgH_2 powder and Cu powder heated under the conditions of 673K, 86.4ks in H_2 of 8.0 MPa. The existence of Mg_2Cu nano-crystals enables alloying Mg with Cu at low temperatures (<473K).

1. Introduction

Mg is attractive as hydrogen storage materials owing to the advantages such as light weight, great abundance and high hydrogen storage capacity [1]. However, sluggish dehydrogenation rate and low equilibrium H_2 pressure at ambient temperatures is the issue to be solved for practical application of MgH_2. Thus, various Mg-based alloys and compounds have been investigated to improve the rate and to lower the temperature of hydrogen desorption [2-5].

The improvement of hydrogen absorption/desorption kinetics, its relations with microstructures, and the effect of initial structures of Mg/Cu super-laminate composites (SLCs) on hydrogen absorption/desorption properties have been reported in previous papers for Mg_2Cu-H_2 system [6-8]. Through these investigations, two types of processes have been proposed for hydrogenation of Mg/Cu SLCs: The one is alloying Mg with Cu to form Mg_2Cu followed by hydrogenation of Mg_2Cu, leading to the formation of three-dimensional (3-D) network of $MgCu_2$ coexisting with MgH_2. The other is hydrogenation of Mg followed by the reaction of MgH_2 to Cu, leading to the formation of layer $MgCu_2$ existing between MgH_2 layers and Cu layers. The dominant process depends on the initial structures of Mg/Cu SLCs [7]. However, alloying Mg with Cu at low temperatures (<473K) and the reaction of MgH_2 to Cu under H_2 atmosphere are not known in general. In this paper, we reveal these reactions are competitive in Mg/Cu SLCs during initial hydrogenation and therefore microstructures of Mg/Cu SLCs are diverse.

2. Experimental procedure

Commercial Mg (99.9 % purity, 40 μm in thickness) and Cu (99.99 % purity, 10 μm in thickness) metal foils were used as raw materials. The foils were cut into pieces of 20 mm x 30 mm in size. The Mg and the Cu foils were stacked in one alternately. By combining Mg foils with Cu foils in different thickness, a stack with molar ratio Mg/Cu of 2 was prepared.

Mg/Cu SLCs were prepared by a repetitive fold and roll method using a conventional two-high roll mill [9]. Two types of specimens were prepared by changing rolling reduction in cold-rolling. Each of them had a fine and a coarse microstructure, respectively.

Commercial MgH_2 (Alfa Aesar, 98 % purity, 25-150 μm in practical particle size, the main impurity of Mg) and Cu (Kojundo Chemical Laboratory Co., Ltd., 99.9 % purity, 45 μm in nominal particle size) powder were used as raw materials. MgH_2 powder and Cu powder were mixed in the molar ratio of 2 for 300 s by using agate mortar and pestle, and the mixed powder was cold-pressed at the pressure of 1.73 GPa to obtain pellets with 7 mm in diameter.

Mg/Cu SLCs were hydrogenated under the conditions of 573 K, 86.4 ks in H_2 of 3.3 MPa with a Sieverts' type instrument, whereas pellets of MgH_2 powder and Cu powder were heated under the conditions of 673 K, 86.4 ks in H_2 of 8.0 MPa with a Sieverts' type instrument for acceleration. Constituent phases were analyzed by powder X-ray diffraction (XRD). Microstructures of Mg/Cu SLCs, and pellets of MgH_2 powder and Cu powder were observed with a scanning electron microscope (SEM). SEM specimens were prepared by conventional mechanical grinding and polishing followed by ion milling.

3. Results and discussion

Fig. 1 shows backscattered electron (BSE) SEM images of the cross-section of two types of as-rolled Mg/Cu SLCs and those of Mg/Cu SLCs after hydrogenation at 573 K for 86.4 ks in H_2 of 3.3 MPa, "Fine" ((a), (b)) and "Coarse" ((c), (d)). The microstructure of as-rolled "Fine" (fig. 1(a)) was much finer than that of as-rolled "Coarse" (fig. 1(c)). As-rolled "Fine" had the uniform microstructure due to its thinness of Cu layers. In addition, cracks were often observed, and gaps between Mg layers and Cu layers were seldom observed in as-rolled "Fine", whereas cracks were seldom observed and gaps were often observed in as-rolled "Coarse". The cracks were formed because of the severe deformation of accumulative rolling and the gaps were probably formed because of the insufficient contact of foils. It can be thought that an effective cold-rolling was not achieved in as-rolled "Coarse" and therefore the mixing of Mg and Cu was not well done in it, different from as-rolled "Fine".

Microstructures of hydrogenated "Fine" (fig. 1(b)) and "Coarse" (fig. 1(d)) are different from each other. The domain size and the total area of MgH_2 layers are larger in "Coarse" than "Fine. It is clearly seen in fig. 1 (d) that there exist two types of $MgCu_2$ structures: the one is 3-D network of $MgCu_2$ coexisting with MgH_2 and the other is layer $MgCu_2$ existing between MgH_2 layers and Cu layers. It should be pointed out that pores often stay adjacent to the layer $MgCu_2$. The layer $MgCu_2$ was relatively popular and the 3-D network of $MgCu_2$ was rare in hydrogenated "Coarse". The situation was reverse in hydrogenated "Fine". These results indicate that there exist two different hydrogenation mechanisms in Mg/Cu SLCs and the dominant mechanism depends on their initial structures.

Since Mg/Cu SLCs were heated from RT to 573 K in H_2, hydrogenation of Mg and alloying Mg with Cu should be competitive. The hydrogenation mechanisms of Mg/Cu SLCs can be considered as follows. Mg metal can react with Cu metal and H_2 gas under the present conditions. When alloying Mg with Cu occurs prior to hydrogenation of Mg (A -> H process),

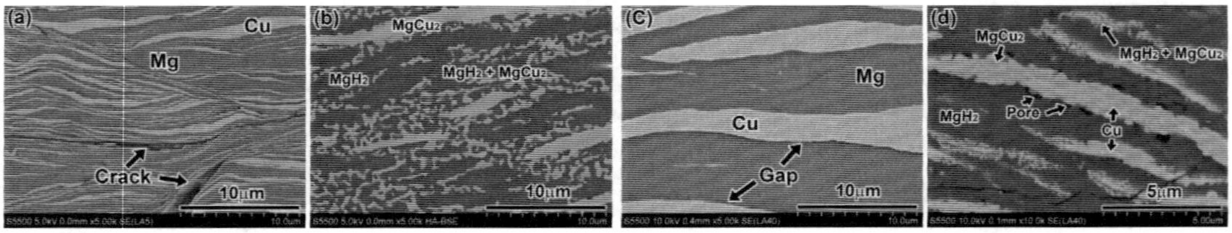

Fig. 1. BSE- SEM images of the cross-section of two types of as-rolled Mg/Cu SLCs ((a), (c)) and those of Mg/Cu SLCs after hydrogenation at 573 K for 86.4 ks in H_2 of 3.3 MPa ((b), (d)). "Fine" ((a), (b)) and "Coarse" ((c), (d)).

Mg/Cu SLCs can be finally hydrogenated by the following two steps,

$4Mg(s) + 2Cu(s) = 2Mg_2Cu(s)$ (1)

$2Mg_2Cu(s) + 3H_2(g) = 3MgH_2(s) + MgCu_2(s)$ (2)

On the other hand, when hydrogenation of Mg occurs prior to alloying Mg with Cu (H -> A process), Mg/Cu SLCs can be finally hydrogenated by the following two steps.

$4Mg(s) + 4H_2(g) = 4MgH_2(s)$ (3)

$4MgH_2(s) + 2Cu(s) = 3MgH_2(g) + MgCu_2(s) + H_2(g)$ (4)

These two hydrogenation processes are illustrated in fig. 2 simply.

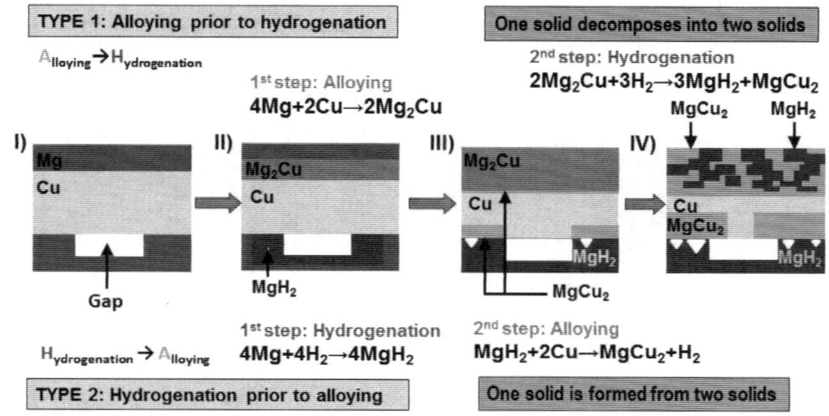

Fig. 2. Schematic models of the microstructure formation mechanism by competitive reactions in Mg/Cu SLCs.

As-rolled "Fine" had a large area of interface between Mg layers and Cu layers, and a short diffusion length for alloying Mg with Cu, due to its fine and uniform structure. Therefore, alloying Mg with Cu followed by hydrogenation of Mg_2Cu should have mainly occurred in as-rolled "Fine" according to the equations (1) and (2). On the other hand, since the mixing of Mg and Cu was not well done in as-rolled "Coarse", hydrogenation of Mg prior to alloying Mg with Cu significantly should have occurred in as-rolled "Coarse" according to the equations (3) and (4).

Fig. 3 shows SEM images of the cross-section of as-rolled "Fine". Fig. 3(a) is a secondary electron (SE) image and fig. 3(b) a BSE image of the same area. It is seen that there exist Mg_2Cu nano-crystals at the interface between Mg and Cu in as-rolled "Fine". Therefore, it is expected that alloying Mg with Cu during heating from RT to 573 K is possible.

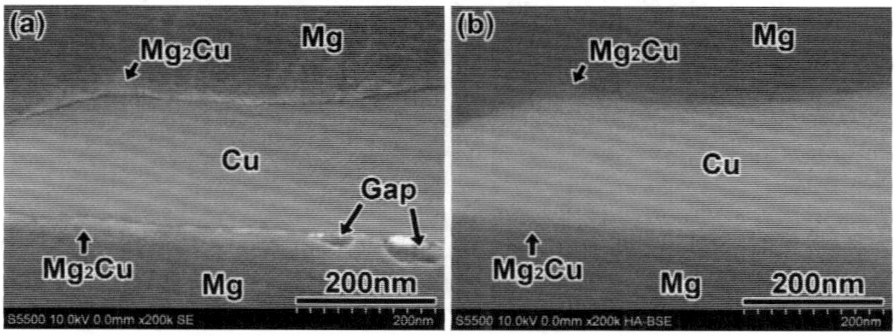

Fig. 3. SEM images of the cross-section of as-rolled "Fine": (a) SE image, (b) BSE image.

In order to confirm the reaction of MgH_2 to Cu to form $MgCu_2$ in H_2, microstructures of pellets of MgH_2 powder and Cu powder before and after heating in H_2 were examined with a

SEM. Fig. 4 shows BSE- SEM images of an as-pressed pellet of MgH_2 powder and Cu powder (a) and one heated under the conditions of 673 K, 86.4 ks in H_2 of 8.0 MPa (b). It is seen from fig. 4 (a) that MgH_2 particles and Cu particles contact tightly to each other, although gaps due to their original shapes were observed. On the other hand, it is seen from fig. 4 (b) that layer $MgCu_2$ was formed between MgH_2 particles and Cu particles, together with large gaps. These structural features are the same as those of the layer $MgCu_2$ observed in hydrogenated "Coarse", fig. 1 (d). It is proved that the not-well-known reaction (4) can occur under H_2 atmosphere. Fig. 3 and 4 support our hypothesis.

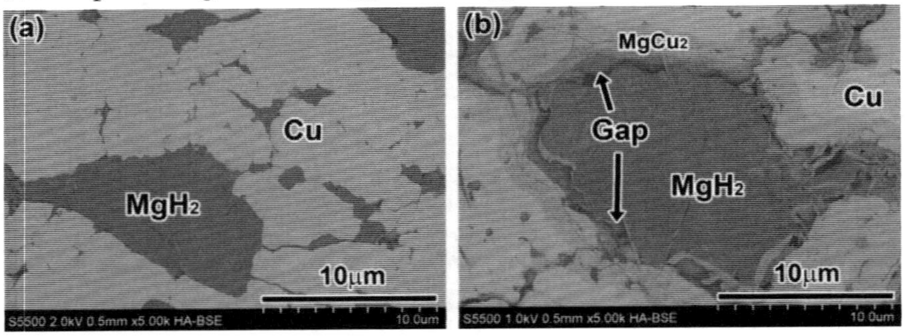

Fig. 4. BSE- SEM images of an as-pressed pellet of MgH_2 powder and Cu powder (a) and one heated under the conditions of 673 K, 86.4 ks in H_2 of 8.0 MPa (b).

4. Conclusions

The present study can be summarized as follows:

(1) Two types of process are competitive for hydrogenation of Mg/Cu SLCs: the one is alloying expressed by the equation of $4Mg(s) + 2Cu(s) = 2Mg_2Cu(s)$, followed by hydrogenation, $2Mg_2Cu(s) + 3H_2(g) = 3MgH_2(s) + MgCu_2(s)$ (A -> H process), and the other is hydrogenation, $4Mg(s) + 4H_2(g) = 4MgH_2(s)$, followed by alloying, $4MgH_2(s) + 2Cu(s) = 3MgH_2(g) + MgCu_2(s) + H_2(g)$ (H -> A process).

(2) The former process forms 3-D network of $MgCu_2$ coexisting with MgH_2, and the latter layer $MgCu_2$ existing between MgH_2 layers and Cu layers.

(3) The existence of Mg_2Cu nano-crystals formed during repetitive cold-rolling enables alloying Mg with Cu at low temperatures (<473K), therefore hydrogenation of Mg and alloying Mg with Cu can be competitive during initial hydrogenation.

Acknowledgement

This work was supported by JSPS KAKENHI Grant Number 23560794.

References

[1] Stampfer, Jr. J F, Holley, Jr. C E and Suttle F 1960 *J. Am. Chem Soc.* **82** 3504
[2] Orimo S, Fujii H and Ikeda K 1997 *Acta Mater.* **45** 331
[3] Liang G, Huot J, Boily S and Schulz R 2000 *J. Alloy. Compd.* **305** 239
[4] Reilly J J and Wiswall R H 1967 *Inorg. Chem.* **6** 2220
[5] Reilly J J and Wiswall R H 1968 *Inorg. Chem.* **7** 2254
[6] Tanaka K, Takeichi N, Tanaka H, Kuriyama N, Ueda T T, Tsukahara M, Miyamura H and Kikuchi S 2008 *J. Mater. Sci.* **43** 3812
[7] Tanaka K, Takeshita H T, Kurumatani K, Miyamura H and Kikuchi S 2013 *J. Alloy. Compd.* **580** S222
[8] Tanaka K, Takeshita H T, Shin H, Kurumatani K, Kiyobayashi T, Takeichi N, Miyamura H and Kikuchi S 2014 *Mter. Trans.* **55** 1122
[9] Ueda T T, Tsukahara M, Kamiya Y and Kikuchi S 2005 *J. Alloy. Compd.* **386** 253

17th International Conference on Textures of Materials (ICOTOM 17) IOP Publishing
IOP Conf. Series: Materials Science and Engineering **82** (2015) 012079 doi:10.1088/1757-899X/82/1/012079

Simulation of the effects of interstitial content and temperature on texture and substructure evolution of commercially pure titanium during ECAP

X Guo and **M Seefeldt**
KU Leuven, Department of Materials Engineering, Kasteelpark Arenberg 44, 3001 Leuven, Belgium
E-mail: xiaodong.guo@mtm.kuleuven.be

Abstract. The effects of interstitial content and temperature on texture and substructure evolution of commercially pure titanium during the first pass of equal channel angular pressing were investigated. Different values of critical resolved shear stresses were proposed for different interstitial contents and processing temperatures. Simulation results show that texture was affected by both interstitial element and temperature. Also, in substructure simulation, the cell size was affected by interstitial content, while the fragment size was more dependent on temperature change.

1. Introduction

Commercially pure (CP) titanium with hexagonal crystal structure develops strong textures during deformation due to the uneven activation of dislocation glide and twinning systems, in which five modes are generally considered, i.e. prismatic $\langle a \rangle$ slip, basal $\langle a \rangle$ slip, first order pyramidal $\langle c+a \rangle$ slip, $\{10\bar{1}2\}\langle 10\bar{1}\bar{1}\rangle$ tensile twin, and $\{11\bar{2}2\}\langle\bar{1}\bar{1}23\rangle$ compressive twin. The activation of different modes in deformation, and thus the texture and substructure evolution are quite sensitive to the detailed material and process parameters. Therefore, the present paper is discussing the response of the modelling framework proposed by Seefeldt and Guo in this volume to variations of the interstitial content and the processing temperature.

2. Simulation details

2.1. Simulation setup

Two temperatures, $298K$ and $473K$, and two interstitial contents, Grade 1 and Grade 4 with an oxygen equivalent content of $1800\,ppm$ and $4000\,ppm$ respectively, were studied. The virtual die of ECAP had an intersection angle of $\Phi = 90\,^\circ$. The imposed strain for one pass was 1.15 and the strain rate was $10^{-3}/s$. The initial texture was random. More details involved in simulation were referred to the model description in this volume.

2.2. Effect of interstitial content and temperature on simulation parameters

In the texture simulation, the interstitial content and the temperature enter only through τ_c. Several experimental and simulation works have used different selections of τ_c [1–6]. Besides, Conrad [4] also reported that, τ_c of dislocation slip increase with increasing interstitial content

Table 1. τ_c (in MPa) of slip and twinning modes used with respect to interstitial element content and temperature [4–6].

Interstitial Content	Temperature	Prismatic	Basal	Pyramidal	Tens. Twin	Comp. Twin
Grade 1	$298K$	75	100	250	150	300
Grade 4	$298K$	120	150	300	120	200
Grade 4	$473K$	35	60	200	120	200

or decreasing temperature, while for twinning, it has a different tendency that τ_c of twin systems increase with decreasing interstitial element content and keep unchanged with changing temperature. Based on these studies, in this paper, τ_c of slip and twinning modes were deduced for two interstitial contents and two temperatures, see table 1.

The interstitial content will also change the stacking fault energy (SFE) significantly [6], and thus cross-slip, dislocation annihilation and multiplication. However, by lack of experimental or theoretical estimations for this dependency, SFE data calculated in high purity titanium are used [7], and the effect of interstitial content on SFE was neglected.

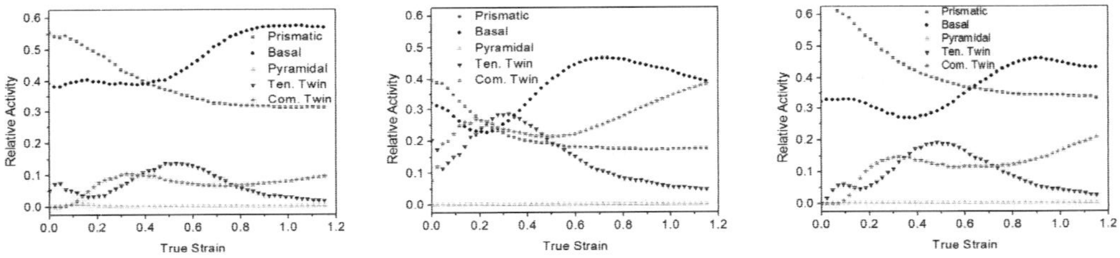

Figure 1. Slip activities of slip and twinning modes during the first pass for materials (left) Grade 1 at $298K$; (mid) Grade 4 at $298K$; (right) Grade 4 at $473K$.

3. Results and discussion

3.1. Texture evolution

Figure 1 shows the slip activities of all deformation modes for three simulations with different interstitial contents and process temperatures. Figure 2 and 3 present the texture evolution during the first pass with interstitial content change and temperature change, respectively. By comparing the left and middle images in Figure 1, prismatic $\langle a\rangle$ and basal $\langle a\rangle$ slip are suppressed with increasing of interstitial content, while two twin modes are promoted, thus the texture component along the normal shear direction which is mainly attributed to prismatic $\langle a\rangle$ and basal $\langle a\rangle$ slip becomes weaker with higher interstitial content. Moreover, by comparing the middle and right images in Figure 1, we can see that, at $473K$, the prismatic $\langle a\rangle$ slip is promoted and twins are suppressed. Basal $\langle a\rangle$ slip keeps almost unchanged. Therefore, as shown in Figure 3, higher temperature strengthens the texture component at normal shear direction. For all simulations, the pyramidal $\langle c+a\rangle$ slip has a negligible contribution to the deformation. The results are in line with previous findings and match well with experimental observations, e.g. by Saiyi Li's work [8].

IOP Conf. Series: Materials Science and Engineering **82** (2015) 012079 doi:10.1088/1757-899X/82/1/012079

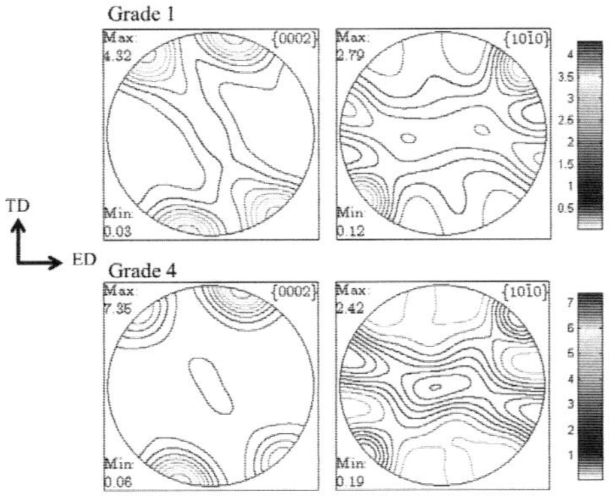

Figure 2. Simulated texture after one pass for Grade 1 and Grade 4 at $298K$.

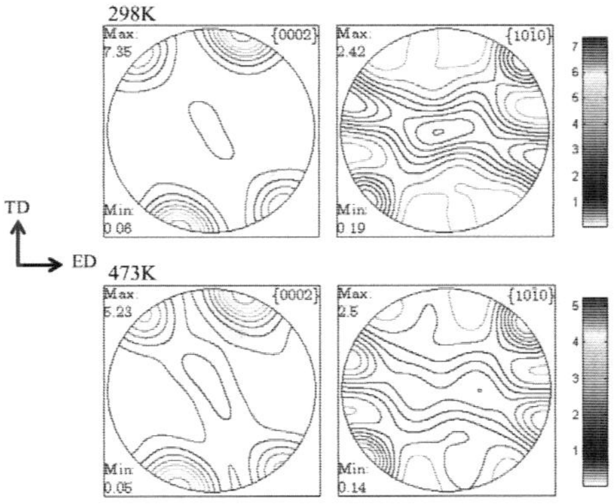

Figure 3. Simulated texture after one pass at $298K$ and $473K$.

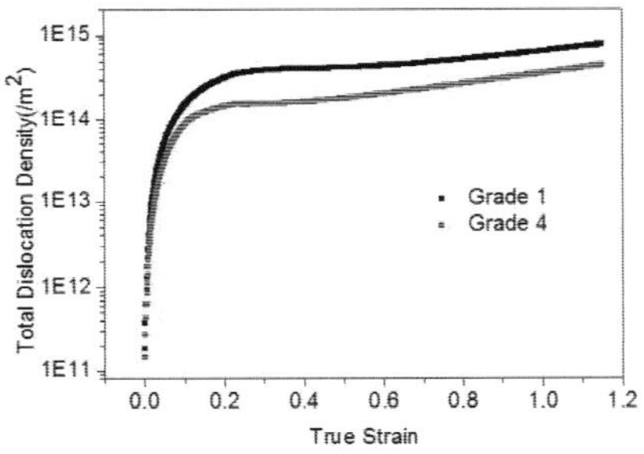

Figure 4. Total dislocation density of Grade 1 and Grade 4 CP titanium at $298K$.

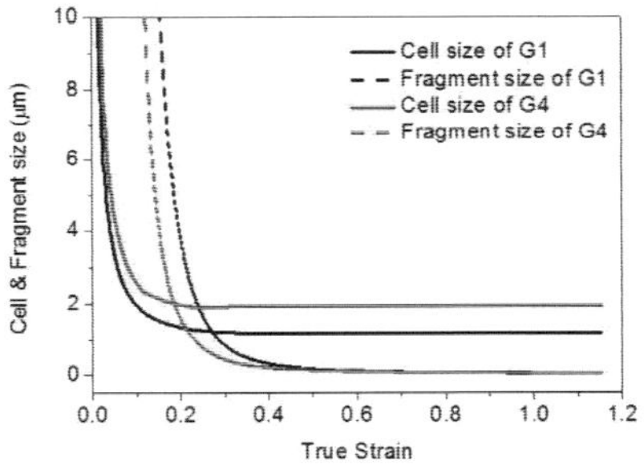

Figure 5. Cell and fragment size of Grade 1 and Grade 4 CP titanium at $298K$.

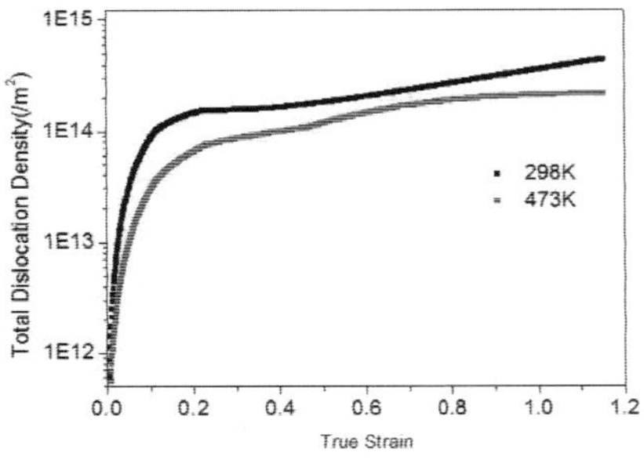

Figure 6. Total dislocation density of a Grade 4 CP titanium at $298K$ and $473K$.

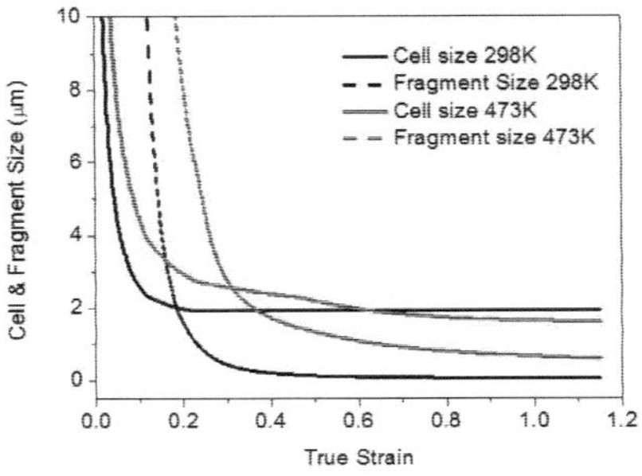

Figure 7. Cell and fragment size of a Grade 4 CP titanium at $298K$ and $473K$.

3.2. Substructure evolution

The effects of interstitial content and temperature on substructure evolution are discussed in terms of total dislocation densities and cell and fragment size evolution. The cell and fragment sizes are calculated via scaling laws from the total dislocation density and from the fragment boundary surface density. Figure 4 shows that, the total dislocation density was decreased with increasing interstitial content, indicating that the suppressed activities of dislocation slip with more interstitial content resulted in a decrease of total dislocation density. Therefore, as shown in Figure 5, the saturation value of cell size increased from about 1.2 μm in Grade 1 to about 2 μm in Grade 4 materials. The saturation value of fragment size were almost unchanged. On the effect of temperature, in Figure 6, a decreasing trend of total dislocation density was observed with increasing temperature. The cell size was almost the same, however, the saturation value of fragment size was higher at an elevated temperature. This simulation result indicates that interstitial element affects the evolution of cell structure while temperature affects the evolution of fragment structure. Experimental observations on substructure evolution of pure titanium during ECAP are referred to the works of Kim et al. [9], Gunderov et al. [10], Zhao et al. [11, 12]

4. Conclusion

The effects of interstitial content and temperature on texture and substructure evolution of CP titanium materials were investigated for the first pass of ECAP. Simulation results show that prismatic and basal slip were suppressed and twinning was promoted at lower temperature or with higher interstitial content. Also, the saturation value of cell size was mostly affected by interstitial element content, while the fragment size was more dependent on the temperature change.

Acknowledgment

The authors wish to acknowledge the financial support provided by the Belgian Science Policy BELSPO (IAP Projects P7/21) and the ViNaT Project (Contract No. 295322, FP7-NMP-2011-EU-Russia, NMP.2011.1.4-5 coordinated with State Contract No. 16.523.12.3002 of the Russian Ministry of Education and Science) is gratefully acknowledged.

Reference

[1] S. Zaefferer, *Materials Science and Engineering A*, Vol **344**, 1-2, 20-30, 2003.
[2] S. Suwas, B. Beausir, L.S. Toth, J.-J. Fundenberger, G. Gottstein, *Acta Materialia*, Vol **59**, 3, 1121-1133, 2011.
[3] M. J. Philippe, M. Serghat, P. Van houtte and C. Esling, *Acta Metallurgica et Materialia*, Vol **43**, 4, 1619-1630, 1995.
[4] H. Conrad, *Progress in Materials Science*, Vol **26**, 2-4, 123-403, 1981.
[5] M. A. Meyers, O. Vohringer and V. A. Lubarda, *Acta Materialia*, Vol **49**, 4025-4039, 2001.
[6] C. P. Biswas, *Doctoral Thesis*, 1973.
[7] M. Ghazisaeidi, D.R. Trinkle, *Acta Materialia*, Vol **60**, 1287-1292, 2012.
[8] S. Li, *Acta Materialia*, Vol **56**, 5, 1031-1043, 2008.
[9] I. Kim, J. Kim, D.H. Shin, C.S. Lee, S.K. Hwang, *Materials Science and Engineering A*, Vol **342**, 1-2, 302-310, 2003.
[10] D.V. Gunderov, A.V. Polyakov, I.P. Semenova, G.I. Raab, A.A. Churakova, E. I. Gimaltdinova, I. Sabirov, J. Segurado, V.D. Sitdikov, I.V. Alexandrov, N.A. Enikeev, R.Z. Valiev, *Material Science Engineering A*, Vol **562**, 128-136, 2013.
[11] X. Zhao, X. Yang, X. Liu, X. Wang, T. G. Langdon, *Materials Science and Engineering A*, Vol **527**, 23, 6335-6339, 2010.
[12] X. Zhao, X. Yang, X. Liu, X. Wang, T. G. Langdon, *Material Science Engineering A*, Vol **527**, 6335-6339, 2010.

IOP Conf. Series: Materials Science and Engineering **82** (2015) 012080 doi:10.1088/1757-899X/82/1/012080

Serial sectioning of grain microstructures under junction control: An old problem in a new guise

D Zöllner and P Streitenberger
Institute for Experimental Physics, Otto von Guericke University, Magdeburg, Germany
E-mail: dana.zoellner@ovgu.de

Abstract. In the present work the importance of 3D and 4D microstructure analyses are shown. To that aim, we study polycrystalline grain microstructures obtained by grain growth under grain boundary, triple line and quadruple point control. The microstructures themselves are obtained by mesoscopic computer simulations, which enjoy a far greater control over the kinetic and thermodynamic parameters affecting grain growth than can be realized experimentally. In extensive simulation studies we find by 3D respectively 4D microstructure analyses that metrical and topological properties of the microstructures depend strongly on the microstructural feature controlling the growth kinetics. However, the differences between the growth kinetics vanish when we look at classical 2D sections of the 3D ensembles making a differentiation of the controlling grain feature near impossible.

1. Introduction

Today there still exist many gaps in our fundamental understanding of polycrystalline microstructures and in predicting their behavioral changes due to recrystallisation and grain growth. Closing this gap is of utmost importance since the grain size of a material determines its properties. Common stereological 2D investigations [1], e.g., using the mean linear intercept method, yield information on the average grain size, but under certain assumptions regarding the morphology of grains. Hence, only the 3D data provide the necessary metrical and topological properties of polycrystals. To that aim, in recent years much effort has been put into the global task to develop new, advanced materials, which has led to new characterization techniques enabling materials scientists and engineers to analyze bulk materials and thin films in 3D and 4D by X-rays, electrons, and neutrons with high resolutions in real time [2].

An analysis of grain microstructures by means of focused ion beam (FIB) tomography allows the creation of serial sections in an automated manner. After reconstruction the shapes and sizes of individual grains in the network can be described (compare, e.g., [3]). The combination of a focused ion beam and scanning electron microscope with the aim of a tomographic orientation analysis is already developed for about 10 years, but the use in the nanocrystalline grain size range, where the grain boundary junction can control the microstructural evolution, is not trivial as it leads the 3D tomographic Electron Backscattering Diffraction method (3D EBSD) method in conjunction with FIB sectioning to their resolution limit concerning the determination of the local orientation [4]. Then again, while recently the spatial resolution in 3D for the tomographic EBSD was about 50×50×50 nm^3 [5], today resolutions well below 50 nm (see [3]) can be achieved using low-voltage methods. For example Balach et al. [6] succeeded in generating 10 nm thin slices in FIB–SEM tomography experiments.

Experiments in nanocrystalline iron [7] have shown that microstructures with an average grain radius of well below 100 nm show anomalous behavior (e.g., a linear average growth law). Hence, such microstructures are important to investigate more closely. Using

IOP Conf. Series: Materials Science and Engineering **82** (2015) 012080 doi:10.1088/1757-899X/82/1/012080

mesoscopic computer simulations [8] and analytic mean-field theories [9] it has been found that such behavior can be associated with triple junction drag leading among other things to a very strong increase in small grains. With an average grain size of less than 100 nm as in the experiments this yields a high number of small grains with radii as small as 10 nm. These can, however, not be portrayed with the resolution necessary to describe their morphology correctly. Hence, for such nanocrystalline grain microstructures we still might have to use stereological 2D investigations.

In the present work we study polycrystalline grain microstructures obtained by grain growth under grain boundary, triple line and quadruple point control. In extensive simulation studies we find by 3D respectively 4D microstructure analyses that metrical and topological properties of the microstructures depend strongly on the microstructural feature controlling the growth kinetics. Those differences between the growth kinetics become blurred when we look at classical 2D sections of the 3D ensembles making a differentiation of the controlling grain feature very difficult.

2. Grain growth under grain boundary junction control

Unlike conventional materials, metals and alloys of nanocrystalline size have quite different mechanical properties like high values of hardness, yield and fracture strength and superplastic behaviour at low temperatures implying a size-effect. They show stable grain sizes even up to relatively high temperatures and linear or even exponential growth kinetics in clear contradiction to parabolic normal grain growth [7, 10]. Such investigations of the stability of nanocrystalline materials during grain growth are, of course, of intense technological interest because an increase in grain size from nm to µm can result in a loss of important materials properties making them unusable in applications. Already in 1997 Malow and Koch [11] summarised significant works concerning the stabilization of nanocrystalline grain structures in many materials and the number of factors influencing the grain boundary mobility in nanocrystalline alloys, like grain boundary segregation, solute drag, pore drag, second phase (Zener) drag and chemical ordering. A universal explanation has not been found yet, but the discussion is still on.

In particular, Gottstein and Shvindlerman [12-14] proposed that grain growth can be controlled by grain boundary junction mobility. The established structures are rather stable, in particular, for ultra-fine grained and nanocrystalline materials. Streitenberger and Zöllner [9, 15] considered grain growth as a dissipative process driven by the reduction of Gibbs free interface and junction energy. A general grain evolution equation has been derived separating into nine types of growth kinetics. The corresponding self-similar grain size distributions were in agreement with first results from modified MC simulations considering size effects in triple and quadruple junction limited grain growth. Details on the simulation method as used in the present work can be found in [8, 16].

3. Evolution of the average grain size: 3D versus 2D measurements

If the grain microstructural evolution is controlled solely by the mobility of one particular structural feature (grain boundary, triple line or quadruple point) it has been shown [8, 9, 14] that the average growth law of the 3D structure follows in each case a unique time law. Under grain boundary (gb) control the classical parabolic growth law, $\langle R\rangle \propto \sqrt{t}$, holds (compare Fig. 1a). Under triple line (tl) control (Fig. 1b) a linear growth law results, while grain growth controlled by the quadruple points (qp) yields the exponential increase in annealing time as shown in Fig. 1c. In each case it can be seen that there is also an initial period of time that is not described by the respective laws depending on the initial microstructure. In addition, for each of the three cases the temporal development of the average grain size calculated from 2D sections, $\langle r\rangle(t)$, is also shown, from which it is evident that the growth laws follow the same general behavior as their 3D counterpart, while showing fluctuations that are due to the limited number of observed grains in the sections. Hence it seems that the microstructural feature controlling grain growth can be deducted from the observation of the growth law in its 3D or 2D version.

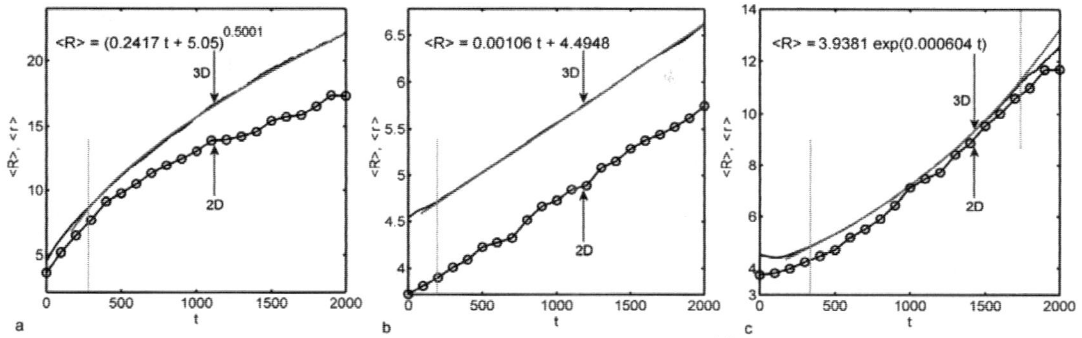

Figure 1. Temporal development of average 3D grain size $\langle R\rangle(t)$ and grain size in 2D sections $\langle r\rangle(t)$ for grain growth under: a – grain boundary control; b – triple line control; c – quadruple point control.

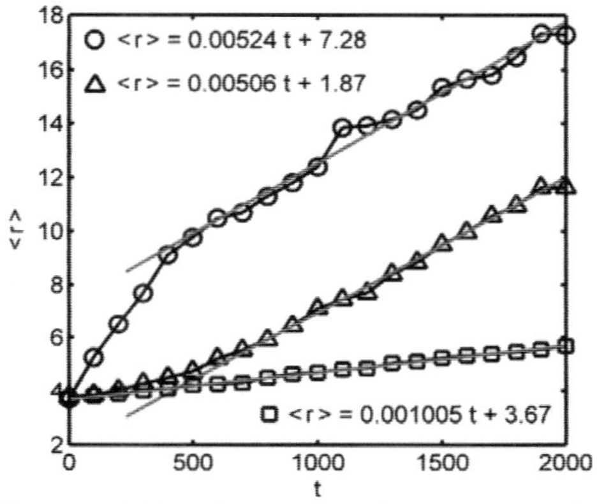

Figure 2. Development of average grain size in 2D sections for grain growth under: gb control (circles), tl control (squares), and qp control (triangles).

However, this is not entirely true as Figure 2 shows, where the three kinetics (gb control – circles; tl control – squares; qp control – triangles) are shown again as $\langle r\rangle(t)$, only now each together with a linear least-squares fit. It is evident that due to the limited number of measured data points a linear relation gives a good representation independent of the microstructural feature controlling the growth kinetics allowing no conclusions!

More frequent or longer observations can solve this problem, and although this is particularly important, it is not always possible to detect more observation times, since, e.g., for long time annealing the growth kinetics changes for example under triple line control from linear to parabolic (see [8]) restricting the observation range.

In order to allow nevertheless a reliable deduction of the microstructural feature that is in control of the growth kinetics, we show in the following that both, the metrical and topological properties of the grain network have to be investigated very thoroughly.

4. Grain size distribution in sections and real

Each of the above growth kinetics of a 3D polycrystal (as presented in Fig. 1) is associated with a distinct scaled grain size distribution $f(x)$ [8, 9], where x is defined as grain radius of a grain divided by the average radius of the ensemble. This is shown indeed in Figure 3a-c, where it can be seen that grain growth under limited triple line mobility with a linear increase in the average grain size (Fig. 3b) as well as under a limited quadruple point mobility with exponential growth law (Fig. 3c) are characterized by a remarkably high number of small grains compared to normal, parabolic grain growth (Fig. 3a). All three types of grain size distributions are found to be self-similar (i.e. time-independent) and in agreement with theoretical predictions (black lines in Fig. 3a-c; compare [9, 15]).

Then again, it is well known that sectioning of 3D polycrystals results in deviating grain size distributions compared to their 3D counterpart. For normal grain growth the scaled size distribution $f(y)$, where y is the scaled grain size of the section, is shifted to smaller grain sizes, while the distribution becomes broader and less peaked (Fig. 3d). Similar changes can also be observed for triple line controlled grain growth (Fig. 3e) and grain growth under quadruple point control (Fig. 3f). However, only the latter is distinctively different from normal grain growth allowing clear conclusions regarding the associated exponential growth regime. The differences between the distributions in Figs. 3d and 3e are basically negligible giving us nearly no indication, which distribution belongs to parabolic normal grain growth and which to linear triple junction controlled growth. The only real difference is the maximum grain size: Under grain boundary control we can observe grains as big as 2.5 times the average size, whereas under triple junction control grains growth as large as 3.0 times the average size giving the distribution (Fig. 3e) a slightly larger tail.

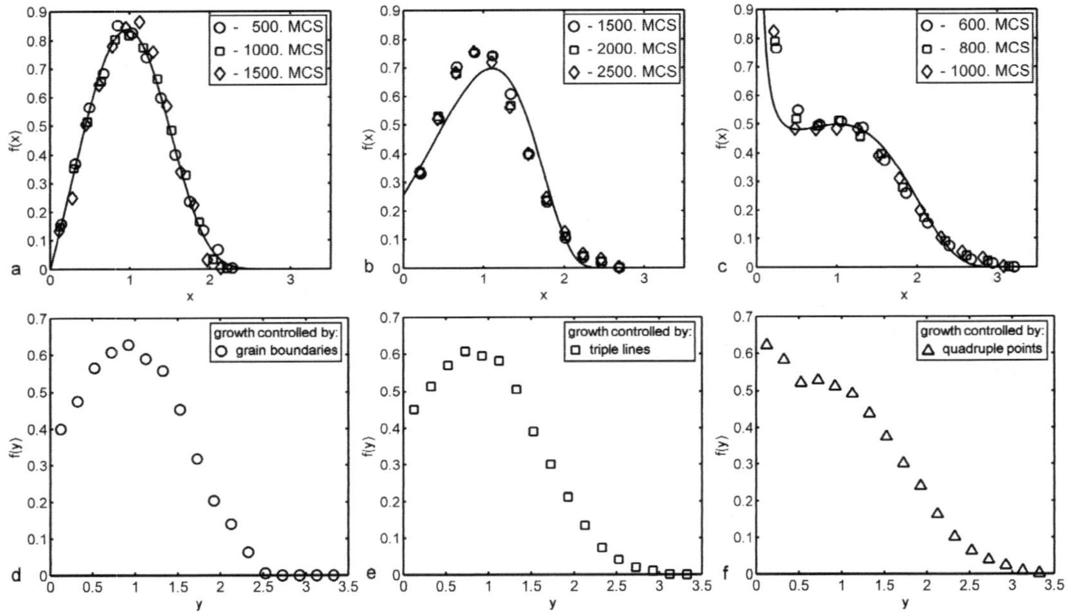

Figure 3. Relative grain size distributions: a – 3D normal grain growth; b – 3D triple line controlled growth; c – 3D quadruple point controlled growth; d – 2D sections from normal grain growth; b – 2D sections from triple line controlled growth; c – 2D sections from quadruple point controlled growth.

5. Topological investigations of 2D sections

In addition to metrical properties, polycrystalline grain boundary networks are also characterized by topological properties like the number of faces and edges of the grains. It is shown particularly in Figure 4a that the grain size y of the grains in the 2D sections is strongly correlated to the number of edges.

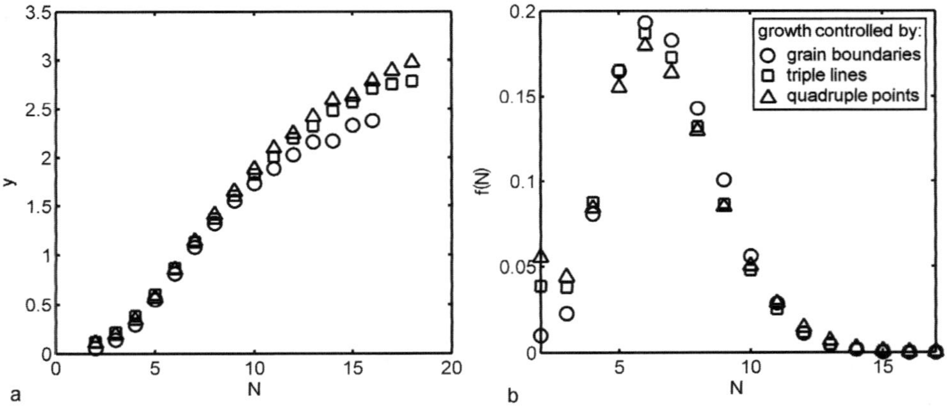

Figure 4. Topological properties of the grain networks: a – Relative grain size versus number of edges (averaged y-values for each edge class); b – Distribution of number of edges.

While it has been shown previously in [8] that the number of faces as a function of the grain size of the 3D microstructures allows a clear deduction of the microstructural feature that is in control of the growth kinetics, Fig. 4a shows that the 2D sections show virtually no differences for small and average grains. Only for large grains with many faces the triple and quadruple junction controlled kinetics differ significantly from normal grain growth. Then again there exist only very few of such large grains as the distributions of the number of edges show in Fig. 4b. In contrast the latter also illustrate the strong increase in the number of grains with few faces.

All in all, it can be concluded that while the metrical and topological properties of the 3D polycrystalline grain structures depend strongly on the microstructural feature controlling the growth kinetics (see, e.g., [8]), it is shown in the present work that those differences between the three growth kinetics become blurred when we look at classical 2D sections of the 3D ensembles. In particular, we have shown that due to a limited number of measured annealing times the average growth law can be described by a linear relation independent of the microstructural feature controlling the growth kinetics, which is in strong contrast to the 3D growth laws. Only from a combined observation of metrical and topological properties of the grain networks a reliable deduction of the microstructural feature that is in control of the growth kinetics can be drawn.

References

[1] DeHoff RT 1983 *J. Microsc.* **131** 259
[2] Hansen N, Juul Jensen D, Nielsen SF, Poulsen HF and Ralph B 2010 *Challenges in materials science and possibilities in 3D and 4D characterization techniques* (Risø National Laboratory for Sustainable Energy)
[3] Uchic MD, Holzer L, Inkson BJ, Principe EL and Munroe P 2007 *MRS Bulletin* **32** 408
[4] Bastos A, Zaefferer S and Raabe D 2008 *Journal of Microscopy* **230** 487
[5] Liu T, Raabe D and Zaefferer S 2008 *Science and Technology of Advanced Materials* **9** 035013
[6] Balach J, Soldera F, Acevedo DF, Mücklich F and Barbero CA 2013 *Microscopy and Microanalysis* **19** 745
[7] Krill CE, Helfen L, Michels D, Natter H, Fitch A, Masson O and Birringer R 2001 *Physical Review Letters* **86** 842
[8] Zöllner D 2011 *Computational Materials Science* **50** 2712
[9] Streitenberger P, Zöllner D 2011 *Acta Materialia* **59** 4235
[10] Ames M, Markmann J, Karos R, Michels A, Tschöpe A and Birringer R 2008 *Acta Materialia* **56** 425
[11] Malow TR, Koch CC 1997 *Acta Materialia* **45** 2177
[12] Gottstein G, Shvindlerman LS 1998 *Scripta Materialia* **38** 1541
[13] Gottstein G, Shvindlerman LS 2006 *Scripta Materialia* **54** 1065
[14] Gottstein G, Shvindlerman LS and Zhao B 2010 *Scripta Materialia* **62** 914
[15] Streitenberger P, Zöllner D 2012 *Materials Science Forum* **715-716** 806
[16] Zöllner D 2014 *Modelling and Simulation in Materials Science and Engineering* **22** 025028

IOP Conf. Series: Materials Science and Engineering **82** (2015) 012081 doi:10.1088/1757-899X/82/1/012081

Advanced characterization of microstructural changes during recrystallization in aluminum alloy 6013

M Bieda[1], J Kawalko[1], F Brisset[2] and K Sztwiertnia[1]

[1]Institute of Metallurgy and Materials Science of Polish Academy of Sciences Laboratory of Anisotropic Structures, Reymonta 25, 30-059 Krakow, Poland

[2]Universite Paris-Sud, CNRS, Institut de Chimie Moléculaire et des Matériaux d'Orsay, 91405 Orsay, France

E-mail: m.bieda@imim.pl

Abstract. Aluminum alloy 6013 was chosen as an example of a material that, after thermal treatment, possesses a relatively uniform and stable bimodal distribution of fine (<<1 μm) and coarse (>1 μm) particles. Samples of this alloy were subjected to plastic deformation by cold rolling. The presence of large and small particles has an influence on the behavior of this material during the recrystallization process. A complex investigation of the microstructural changes during annealing were carried out by means of advanced SEM and TEM techniques. Orientation mapping (OM), i.e., automatic determination of the topography of the crystallographic orientations, was performed using electron backscatter diffraction (EBSD) in the SEM and microdiffraction in the TEM experiments. These techniques were combined with in-situ heating experiments in the TEM and SEM experiments. The quantitative description of the microstructure at each step of recrystallization is presented. Changes in the microstructure of the investigated material during annealing reveal the role and impact of both types of particles on recrystallization and grain growth. The obtained results are in agreement with parallel calorimetric studies.

1. Introduction

Advanced microstructural characterization of changes in a material under rising temperature is still a challenge and many in-situ devices have been developed for annealing a material using TEM or SEM. However, in-situ experiments in electron microscopy were considered difficult and unreliable, mainly because of different experimental conditions, sample preparations and observations which all can influence the behaviour of materials. Existence of one (in SEM) or two (in TEM) free surfaces can introduce differences in processes occurring in microscopic samples with respect to bulk material. [1, 2]. Fortunately, by taking into account many factors influencing in situ measurements, proper conditions can be assured, making them source of scientific observation [3]. Moreover, in order to avoid artificial results and confirm observations of in situ annealing, supplementary calorimetric studies should be performed.

New benefits of in situ experiments come from development of orientation microscopy, both in TEM and SEM. To obtain the most comprehensive data, not only a heating stage but also the technique of acquisition of crystallographic data should be used.

With these techniques, one can compare the changes in texture or orientation between the grains at each annealing step.

In this paper, some examples of combined in situ heating and orientation mapping study applied to aluminum alloy 6013 is presented. These results focus mainly on in-situ SEM and EBSD results. In situ TEM and calorimetry results have been presented previously [4-6].

2. Material

Aluminum alloys 6xxx series are especially interesting because of their very good properties

[7]. Aluminum alloy 6013 [table 1] was chosen because of the characteristic bimodal distribution of the second phase particles, which plays a particular role during the recrystallization process.
Sample of the as-received alloy was supersaturated (1 h - 530°C) and then aged (5 days - 165°C). As a result, the alloy contains a characteristic structure with stable secondary phase particles, which are both small (<<1 μm) and large (>1 μm). After proper heat treatment, the alloy was deformed up to 90% by cold rolling.

Table 1. Chemical composition of 6013 aluminum alloy (% wt.).

Mg	Si	Cu	Mn	Fe	others	Al
1.15	1.0	1.1	0.3	0.5	0.15	remainder

3. Experimental methods

To determine the changes in the material under annealing, a parallel study by means of scanning electron microscopy (SEM), transmission electron microscopy (TEM) and calorimetry was performed.
Experiments using TEM [4-6] were conducted using a heated sample holder. While the thin foil was annealed inside the holder, changes in the microstructure were observed and orientation maps were recorded using microdiffraction. Microdiffraction was used because of the very small grain size in the vicinity of the large particles, which are called deformation zones (DZ). However, to ensure that the observed changes are at least close to those occurring in bulk samples, it is necessary to maintain proper conditions during the experiment [5].
For the scanning electron microscope measurements, the limited spatial resolution is offset by the larger observable area of the sample, which allows for the investigation of bulk materials [3]. In addition, by using a dedicated, special heating stage for simultaneous heating and EBSD mapping (pretilted or with the ability to tilt to 70 ° depending on the model), it is possible to observe the nucleation and grain growth, or phase transformations [2,8-12].
The investigations in the SEM were performed using two heating stages, the H1002D heating stage module for SEM for EBSD applications (Universite Paris-Sud) and the Gatan Murano 525 (Institute of Metallurgy and Materials Science PAS, Krakow). EBSD analysis was performed with the TSL EDAX OIM system. The results obtained by both heating stages were comparable and repeatable.
The results from the TEM and SEM were confirmed by a calorimetric study [4-6].

4. Results and discussion

The microstructure of the deformed aluminum alloy is composed of grains elongated in the rolling direction (RD). The grain size of the matrix in the normal direction (ND) does not exceed 0.5 μm (fig. 1 and 2 - temperature 23°), and the boundaries between them were characterized as HAGBs (high angle grain boundaries). In the areas near the large secondary phase particles, deformation zones (DZ) were identified. This area is composed of grains of approximately 50-200 nm, which could not be examined using the SEM EBSD system. In fig. 1 and 2, areas of large particles are marked by an Image Quality (IQ) map and are surrounded by points with low confidence index (CI), which means that diffraction images in this area were not indexed or were indexed incorrectly.
Global deformation textures are typical for FCC metals with high stacking fault energies and contain two fibers, α with components S {123}<634> and Cu {112}<111> and β with components S {123}<634> and Cu {112}<111>.
After acquiring the orientation maps in the deformed state, the samples were heated inside the SEM to the temperatures represented by peaks in the calorimetric diagram (fig. 1). This allowed for an observation of changes in the microstructure.

At the temperature of the first calorimetric peak (maps from temperatures 270°, 290° and 310° in fig. 1 and fig. 2), nuclei/new grains in the areas of the DZ (fig. 2) are found, and only small changes in the matrix areas can be observed as the arrangements of the low angle grain boundaries (LAGB) with no migration of the HAGBs. In addition, the phenomenon of particle stimulated nucleation PSN [2] was identified.

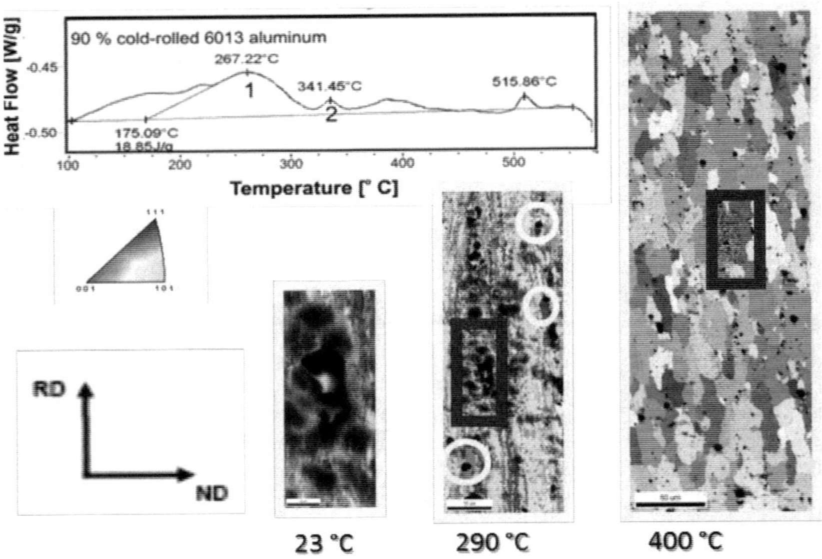

Fig. 1. Heat flow representing the release of stored energy from the 6013 aluminum alloy after 90% cold-rolling with two peaks marked, and EBSD maps with IQ+IPF from the TSL OIM 6.2 system presented before (23°C) and after annealing (290 and 400°C) at the temperatures indicated for each peak. The area of the PSN is marked near the deformation zones (DZ) (yellow circles).

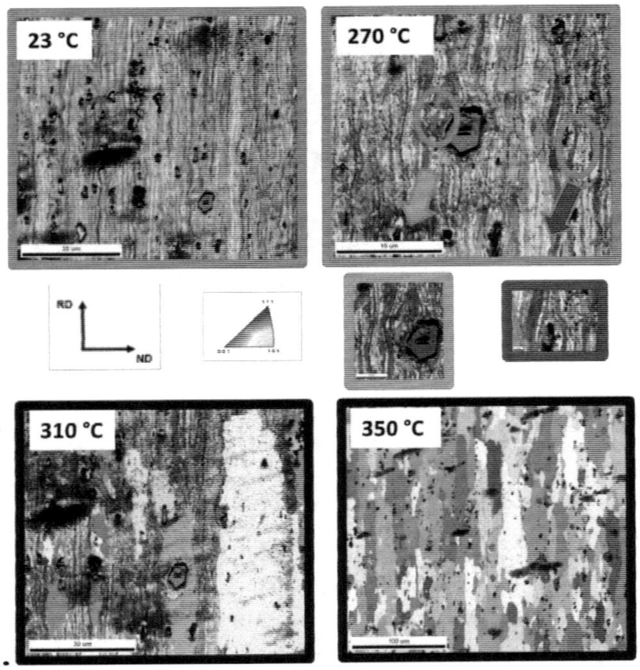

Fig. 2. EBSD maps (IQ+IPF) at each step of the recrystallization with the DZs marked.

At the temperature of the second peak, the migration of the HAGB in the matrix and further growth of the nuclei are observed, but the growth is limited in the direction parallel to the ND.
The relationship of the orientation between grains in the deformed state and between the new grains that appeared at the same location after annealing is characterized by a random distribution of misorientations that suggests that there was no favorable grain growth inside the DZs [5].
To confirm the results from the in-situ studies, parallel calorimetric research was also conducted. Samples were heated in a calorimeter to the temperature of each peak, and orientation mapping was performed [5]. The results obtained were identical to those determined from the in situ studies.

5. Conclusions

Complex investigations of the microstructural changes during annealing of the material after deformation were carried out by means of advanced scanning electron microscope (SEM) techniques such as EBSD and in situ annealing. The experiments were performed in parallel with a calorimetric study. The applied methods, such as combined in situ studies and orientation mapping in the SEM, are powerful tools in providing complex information about the material behavior during annealing.
These investigations confirm previous results from TEM analysis, which found that the recrystallization process in aluminum alloy 6013 is composed of several overlapping processes. The influence of the secondary phase particles is significant. The large particles act as sites for particle stimulated nucleation (PSN).

Acknowledgements

This work was supported by the National Science Center (Poland) UMO - 2011/03/D/ST8/04106.
Authors are grateful to Prof. M. Faryna, Dr. K. Berent and Dr. P. Bobrowski for their assistance in SEM in-situ part.

References

[1] Mullins W W 1958 Acta Metall. **6** pp.414-427
[2] Humphreys F J and Hatherly M 2002 Recrystallization and Related Annealing Phenomena (Oxford: Pergamon Press)
[3] Sztwiertnia K, Haessner F 1994 Mater. Sci. Forum **157-162** pp.1069-1075
[4] Sztwiertnia K, Bieda M and Korneva A 2012 Application of rientation Mapping in TEM and SEM for Study of Microstructural Evolution during Annealing (Recrystallization) ed K Sztwiertnia (Croatia: InTech) chapter 3 pp 43-58
[5] Bieda M, Sztwiertnia K, Korneva A, Czeppe T, Orlicki R 2010 Solid State Phenom. **16** pp 13-18
[6] Sztwiertnia K, Bieda M, Korneva A 2013 Mater. Sci. Forum **753** pp 221 -224.
[7] Hirsch J, Al-Samman T 2013 Acta Mater. **61** pp. 818–843
[8] Humphreys F J, Ferry M 1996 Mater. Sci. Forum **529** pp 217–222
[9] Hurley P J, Humphreys F J 2004 J. Microsc. **213** 225
[10] Lens A, Maurice C, Driver J H 2005 Mater. Sci. Eng. A **403** 144
[11] Wright S and Nowell M 2009 Review of In Situ EBSD Studies (Electron Backscatter Diffraction in Materials Science) eds A J Schwartz et al. (Springer) chapter 24 pp 329-337
[12] Brisset F, Helbert A-L, Baudin T 2013 Microsc. and Microanal. **19** pp 969-977

IOP Conf. Series: Materials Science and Engineering **82** (2015) 012082 doi:10.1088/1757-899X/82/1/012082

Texture related unusual phenomena in electro-deposition and vapor deposition

DN Lee[1] and HN Han[2]
[1] Professor Emeritus, Department of Materials Science and Engineering and Research Institute of Advanced Materials, Seoul National University, Seoul 151-744, Korea
[2] Professor, Department of Materials Science and Engineering and Research Institute of Advanced Materials (RIAM), Seoul National University, Seoul 151-744, Korea
E-mail: dnlee@snu.ac.kr

Abstract. The tensile strength of electrodeposits generally decreases with increasing bath temperature because the grain size increases and the dislocation density decreases with increasing bath temperature. Therefore, discontinuities observed in the tensile strength vs. bath temperature curves in electrodeposition of copper are unusual. The tensile strength of electrodeposits generally increases with increasing cathode current density because the rate of nucleation in electrodeposits increases with increasing current density, which in turn gives rise to a decrease in the grain size and in turn an increase in the strength. Therefore, a decrease in the tensile strength of copper electrodeposits at a high current density is unusual. The grain size of vapor deposits is expected to decrease with decreasing substrate temperature. However, rf sputtered Co-Cr deposits showed that deposits formed on water-cooled polyimide substrates had a larger grain size than deposits formed on polyimide substrates at 200°C. These unusual phenomena can be explained by the preferred growth model for deposition texture evolution.

1. Introduction

Thin films used in a wide variety of electronic, magnetic, and optical devices and systems are fabricated by electro-deposition, electroless-deposition, and vapor deposition. Thin films generally have textures, which influence their electronic, magnetic, optical, and other properties, and are closely linked to their microstructures and surface morphologies. Therefore, it is important to understand the texture evolution in films. The textures can develop in films during deposition and post-deposition annealing. The deposition texture can vary with deposition conditions and may change after annealing. The annealing texture may be classified into the recrystallization texture and the abnormal grain growth texture. Recrystallization occurs through nucleation and growth in a matrix in which the dislocation density is sufficiently high. On the other hand, the abnormal grain growth takes place in fully recrystallized matrices and several nm crystalline materials, in which energy due to dislocations is negligible compared with the grain boundary energy, the surface energy or the strain energy. The evolution of annealing textures of thin films has been reviewed [1].

2. Preferred growth model for deposition texture evolution

The evolution of deposition texture may be classified into the pre-coalescence stage and the coalescence and thickening stage. In the case that the film material does not fully wet the substrate, the film will grow initially as independently nucleated islands which grow into the vapor and along the vapor:substrate interface. The pre-coalescence stage films develop random to weak textures [2]. These islands will eventually coalesce to form a continuous film, which will subsequently thicken. The texture of the vapor deposits and electrodeposits varies with their deposition conditions. Lee [3,4] has proposed a model for the evolution of the vapor deposition texture. In the model a texture is developed by the selection for survival of grains which are favorably oriented. The model is a growth theory for the texture evolution unlike Walton's nucleation theory [5]. It is generally accepted that a random polycrystalline structure is obtained up to a critical film thickness unless epitaxial growth conditions are satisfied. The preferred growth model [3,4] suggests that the texture of deposits changes from the orientation that places the smaller surface energy crystal facet parallel to the deposit

IOP Conf. Series: Materials Science and Engineering **82** (2015) 012082 doi:10.1088/1757-899X/82/1/012082

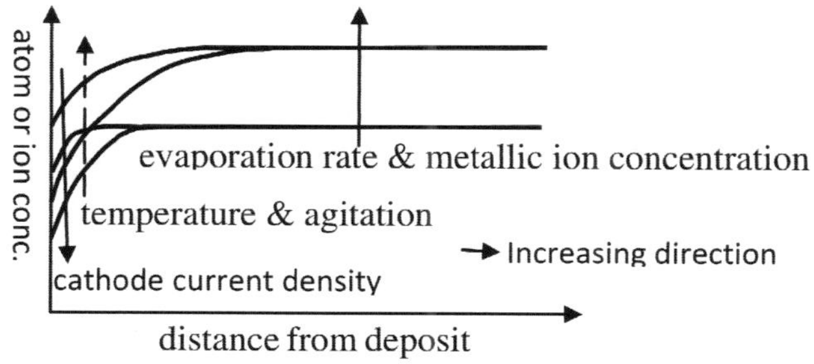

Figure 1. Schematic drawing showing effects of vapor deposition and elecrodeposition conditions on metallic atom or ion concentration adjacent to deposit.

surface under the condition of low atom or ion concentration adjacent to the deposit into the orientation that places the higher surface energy crystal planes parallel to the deposit surface as the atom or ion concentration adjacent to the deposit increases, unless the deposition temperature is high enough for active volume diffusion to take place. At such high temperatures, the minimum surface energy crystal plane tends to be placed parallel to the deposit surface to minimize the total energy of the system because the surface energy may be of the largest energy source in the growing system. Figure 1 shows effects of deposition conditions on metallic atom or ion concentration adjacent to deposit. The relationship between the deposition condition and texture is described in detail in [3,4].

3. Tensile strength of electrodeposited copper foils

Figure 2a shows the tensile strengths (TSs) and elongations of Cu foils obtained from a Cu plating bath at current densities (CDs) of 16-20 A/dm^2. The texture of the foils obtained below 30°C was approximated by <111>+<311>//ND, whereas those above 30°C had the <110>//ND texture. The texture results are compatible with the preferred growth model taking into account the surface energies of fcc crystals ({111}<{100}<{311}<{110}), where {hkl} indicates the surface energy of {hkl} planes [8]. The TS decreases and the elongation increases with increasing bath temperature, which is expected because the grain size generally increases with increasing bath temperature. Therefore, the discontinuity in the TS

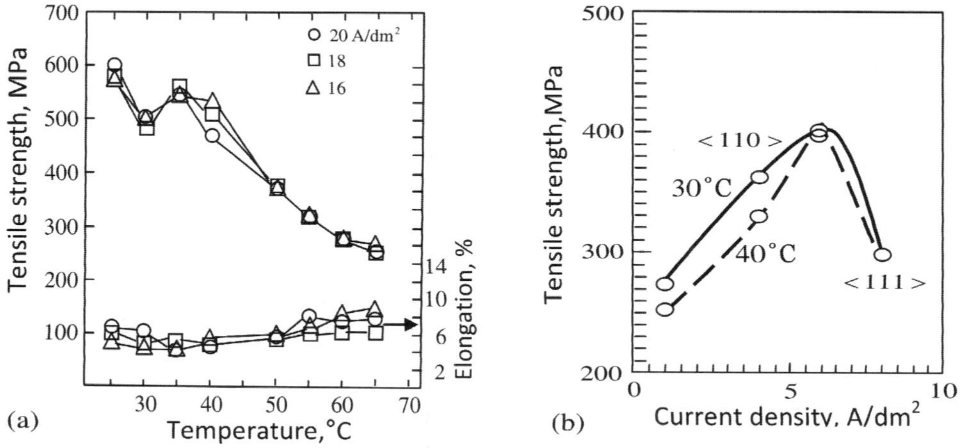

Figure 2. Tensile properties of 35-μm-thick Cu foils as functions of (a) bath (240 g/l $CuSO_4 \cdot 5H_2O$ + 80 g/l H_2SO_4) temperature [6] and (b) cathode current density in bath (280 g/l $CuSO_4 \cdot 5H_2O$ + 80 g/l H_2SO_4) at 30 and 40°C [7].

vs. bath curves are unusual. Two factors should be considered to explain the discontinuity. One is the texture itself. Different orientations may activate different slip systems and in turn give rise to different mechanical properties. Calculated yield loci [9] indicate that the TS along the film surface does not seem to be affected by the texture. Another is that the grain size may be influenced by the texture. Since the deposits have fiber textures, the grain boundaries can be approximated by the tilt boundaries. If electrolysis conditions are similar, the deposit whose grain boundary energy is smaller will have the smaller grain size than the deposit with the larger grain boundary energy and the higher tilt boundary mobility is likely to give rise to the higher grain growth rate. The average boundary energy decreases in order of the <100>, <111>, and <110> tilt boundaries [10], and the average tilt boundary mobility in Al decreases in order of <111>, <100> and <110> tilt boundaries [11]. Assuming that the average tilt boundary mobility is similar to that in Al, the <110>//ND deposits will have finer grains than the <111>//ND deposits, if they are obtained under similar electrolysis conditions. As the bath temperature increases, the ion concentration adjacent to deposits are likely to increase as can be seen in Figure 1 and the deposit texture tends to be <110>//ND according to the preferred growth model and so the grain size of the <110>//ND deposit is likely to decrease and in turn its TS is likely to increase. Therefore, the discontinuities in Figure 2a can be attributed to the grain size difference influenced by the texture.

Figure 2b shows another interesting example of the TS discontinuity. In general, the TS of electrodeposits increases with increasing CD. It is believed that the rate of nucleation in electrodeposits increases with increasing CD, which in turn gives rise to a decrease in the grain size and an increase in TS. Figure 2b shows a drop in the TS of the deposit with the <111>//ND texture, whereas the TS of the <110>//ND deposits increases with increasing CD. This result can be explained as in Figure 2a. As the CD increases, the ion concentration adjacent to the deposit decreases as shown in Figure 1, and the deposit texure is likely to be <111>//ND by the preferred growth model. Therefore, the <111>//ND textured deposit can have larger grain size and lower tensile strength. The TS of deposits also depends on the dislocation density in them. It is not known how much the dislocation density varies with texture. However, recrystallization texture of a Cu electrodeposit with a duplex texture of <111>+<110>//ND indicates that the <110>//ND grains have the higher defect density than the <111>//ND grains [7]. The defect can be grain boundaries and/or dislocations.

4. Microstructures of radio frequency (rf) sputter-deposited Co-17% Cr thin films

Jang [12], in his rf sputtered Co-Cr deposits, found an unusual phenomenon that the deposits formed on water-cooled polyimide substrates had a larger grain size than deposits formed on 200°C polyimide substrates. Figure 3 shows TEM images and SAD patterns of 50 nm-thick Co-17%Cr films formed on the water-cooled and 200°C substrates. The SAD pattern of the film deposited on the water-cooled substrate shows very strong (200), (110) and (100) rings indicating a good c-axis alignment to the growth direction, whereas (002), (101), (102), and (103) rings which should be excluded for a good c-axis alignment appear with relatively strong intensity in the film formed on the 200°C substrate. The results are compatible with the preferred growth model in which a good c-axis alignment to the growth direction is favored at lower deposition temperatures. Figure 3 also shows that the film deposited on the water-cooled substrate has a larger grain size than the film formed on the 200°C substrate. The result is at variance with the usual observation that the deposit grain size decreases with decreasing substrate temperature. The grain size of deposits is controlled by the nucleation rate and the lateral growth rate. The lateral growth rate influenced by the surface diffusion of depositing atoms is increasingly difficult with decreasing temperature, resulting in finer grains. This situation holds in the case of the same texture. For differently oriented deposits, complications can arise. A deposit with the texture such that lower surface energy facets are parallel to the substrate has a lower ledge density on its surface. The lower

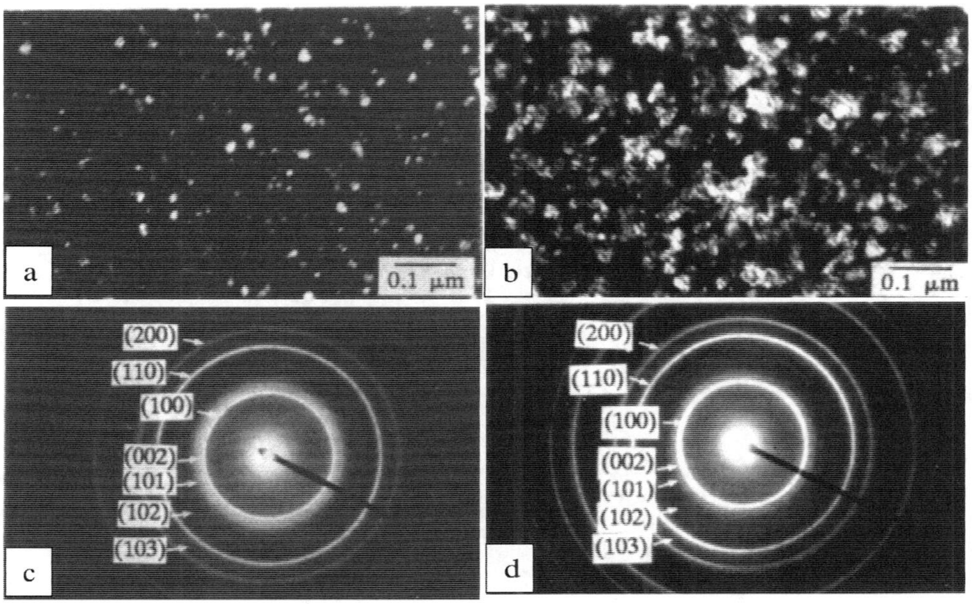

Figure 3. TEM microstructures and SAD patterns of 50 nm thick Co-17% Cr films rf sputtered on (a,c) 200°C and (b,d) water cooled polyimide sustrates [12]

ledge density is likely to make the lateral growth easier and the nucleation rate more difficult,
resulting in the coarser grains, and vice versa. Another explanation is suggested as follows: The grain boundaries of deposits can be approximated by the tilt boundaries. If boundaries of grains whose texture is such that lower surface energy planes are parallel to the deposit surface have higher boundary mobility, their grain size is likely to be larger, and vice versa, as in Section 3. Thus, the coarser grains in the deposit formed on the water-cooled substrate are attributed to the [002]//ND texture.

Acknowledgement

This work was supported by the Center for Iron and Steel Research at RIAM, Seoul National University.

References

[1] Lee DN 2005 *Z. Metallkde.* **96** 259
[2] Thompson CV and Carel R 1995 *Mater. Sci. Eng.* B **32** 211
[3] Lee DN 1989 *J. Mater. Sci.* **24** 4375
[4] Lee DN 1999 *J. Mater. Sci.* **34** 2575
[5] Walton D 1962 *Phil. Mag.* **7** 1671
[6] Lee DN, Kim YK 1985 *Proc. 2nd Asian Metal Finishing Forum* (The Metal Finishing Society, Tokyo: Japan) p 130
[7] Kang SY, Lee DN 2002 *Mater. Sci. Forum* **408-412** 895
[8] Sundquist BE 1964 *Acta Metall. Mater.* **12** 67
[9] Kwon JW 1995 *Ph.D. Thesis Seoul National University* (Seoul: Korea)
[10] Wolf D 1990 *Acta Met. Mater.* **38** 781
[11] Aleshin AN, Aristov VY, Bokstein BS, Shvindlerman LS 1978 *Phys. stat. sol.* **A45** 359
[12] Jang PW 1991 *Ph.D. Thesis Seoul National University* (Seoul: Korea)

Evolution of microstructure and texture during recovery and recrystallization in heavily rolled aluminum

O V Mishin[1], A Godfrey[2], T Yu[1], N Hansen[1] and D Juul Jensen[1]

[1] Danish-Chinese Center for Nanometals, Section for Materials Science and Advanced Characterization, Department of Wind Energy, Technical University of Denmark, Risø Campus, 4000 Roskilde, Denmark

[2] Key Laboratory of Advanced Materials (MOE), School of Materials Science and Engineering, Tsinghua University, Beijing 100084, China
E-mail: olmi@dtu.dk

Abstract. The annealing behavior of nanostructured aluminum AA1050 prepared by cold rolling to an ultrahigh strain (ε_{vM} = 6.4) has been investigated using both transmission electron microscopy and electron backscatter diffraction techniques, paying particular attention to changes in microstructure and texture during recovery and their influence on subsequent recrystallization. It is found that coarsening of lamellar structures during recovery can occur via triple junction motion, and that this process can modify the proportion of different boundary types and texture components compared to those in the cold rolled material. Additionally, the heavily deformed material is characterized by different textures and different spatial arrangements of rolling texture components in the center and subsurface. It is found that changes in the misorientation distribution and texture during coarsening are greatly affected by the initial spatial distribution of crystallographic orientations. In particular, the reduction in the fraction of high angle boundaries observed during recovery is much more pronounced in the subsurface layers than in the center layer. The initial through-thickness heterogeneity is thus greatly enhanced during recovery, which leads to significant differences in recrystallized microstructure and texture in the different layers.

1. Introduction

Materials deformed to high strains are characterized by very small boundary spacings and typically contain large fractions of high angle boundaries (HABs), which can result in a modified annealing behavior compared to materials deformed to low strains. For example, it has been recently reported that coarsening of deformation structures in heavily rolled aluminum may occur via triple junction (TJ) motion [1,2]. Depending on the spatial distribution of different texture components, such coarsening can lead to very different recovered microstructures in which subsequent recrystallization takes place [3]. This has been demonstrated in our previous electron backscatter diffraction (EBSD) study of an AA1050 sample cold rolled to a von Mises strain of 6.4, which had different proportions and different spatial distributions of rolling texture components in the center and subsurface layers [3]. In the present work, we complement the EBSD data with new transmission electron microscopy (TEM) observations of TJ motion, and discuss the evolution of texture and microstructural parameters during recovery and recrystallization in heavily rolled aluminum.

2. Experimental

A plate of aluminum AA1050 was cold rolled by multiple passes from 10 cm to ~0.4 mm. Cold rolling was conducted with lubrication, unidirectionally by alternating the top and bottom sides between passes, and applying predominantly intermediate draughts [4]. The cold-rolled material was then annealed in air at 300 °C for different periods of time. The microstructure and texture were characterized in the section containing the rolling direction (RD) and the normal direction (ND). For the deformed and recovered samples, EBSD maps were obtained in the center and at 30–130 μm from the surface. For texture analysis of all samples and for microstructural analysis of partially and fully recrystallized samples, the entire sample thickness was covered by EBSD. Each through-thickness data set was divided into three subsets of equal size to obtain information separately for the center and subsurface layers. The data from the two opposite subsurface layers were combined for determining boundary spacings, proportions of different boundary types and fractions of different texture components. In addition, thin foils were investigated in a JEM 2100 transmission electron microscope equipped with a double-tilt heating holder. The foils were heated to 270 °C within ~4 min, and then to 300 °C over a period of 2 minutes. TEM images were taken both in the cold-rolled condition and during in-situ annealing using a TVIPS FastScan camera.

3. Results and Discussion

Lamellar structures combined with microshear bands are seen in the heavily rolled microstructure (see Fig. 1). The EBSD data indicate that the dominant texture in the center is the S {123}⟨634⟩ component, whereas the texture in the subsurface layers is dominated by one of the symmetric variants of a component near the ideal Cu {112}⟨111⟩ orientation [3]. These dominant components compose broad texture bands subdivided both by HABs (between symmetric variants of one texture component) and by LABs. The broad bands are interspersed with narrower bands representing weaker components of the rolling texture, of which the Brass (Bs) {110}⟨112⟩ component was the weakest (Fig. 1). Interestingly, despite the significant differences in the spatial distribution of the rolling texture components, the average boundary spacing along the ND and the fraction of HABs are very similar in the different layers, being ~0.2 μm and 54–56%, respectively.

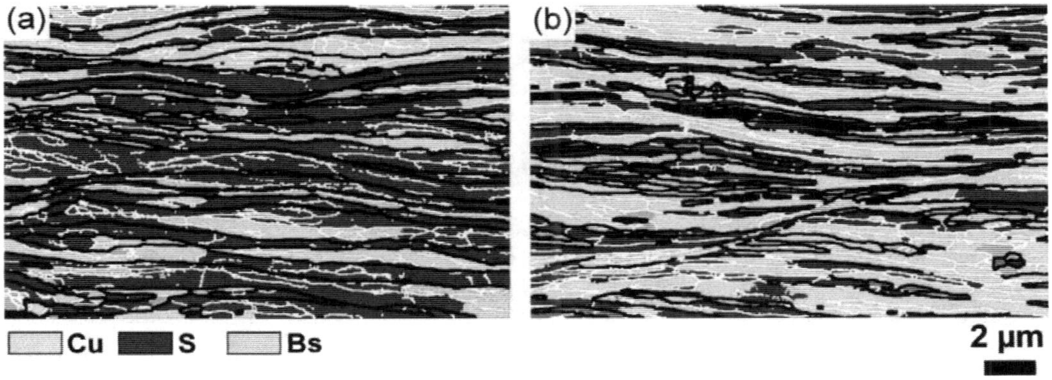

Figure 1. EBSD maps showing deformation structures in the center (a) and the subsurface (b) of the cold-rolled sample. LABs (θ =1.5–15°) and HABs (θ >15°) are shown by white lines and black lines, respectively. The RD is parallel to the scale bar. Reprinted from [3] with permission from Elsevier.

Upon annealing the microstructure coarsens during recovery, followed by pronounced discontinuous recrystallization after ~10 minutes at 300 °C. The coarsening occurs without significant changes in the lamellar morphology and is accompanied by reductions in the HAB fraction to ~40% in the center and 20% in the subsurface. The reduction in the fraction of HABs observed during coarsening in this work is consistent with several previous observations in heavily deformed and annealed materials [5–8].

In our previous publication [3], based both on EBSD maps collected after short-time annealing at 300 °C and on TEM findings in samples annealed at lower temperatures [1,2], the coarsening of the lamellar structures was attributed to TJ motion. In the present work, we reinforce this conclusion by direct TEM observations of TJ motion at temperatures approaching 300 °C. Although structural coarsening takes place already during the heating-up to 270 °C, significant changes continue to occur as the temperature approaches 300 °C. Figure 2 shows a sequence of TEM images taken during in-situ heating as the temperature reached ~290 °C. Initially the TJ marked by an arrow in Fig. 2a was pinned by an interconnecting boundary in a neighboring lamella on the left side (this interconnecting boundary is shown as a dashed line near the TJ marked in Fig. 2e). Within 0.37 s the TJ moved rapidly over a distance of ~2 μm (Fig. 2b), then slowed down as it migrated past two more interconnecting boundaries on the left side (Fig. 2c), after which the TJ moved rapidly again (Fig. 2d and Fig. 2f). The migration of this TJ therefore increased the boundary spacing in the inspected region (cf. Fig. 2e and Fig. 2f), while preserving the lamellar morphology.

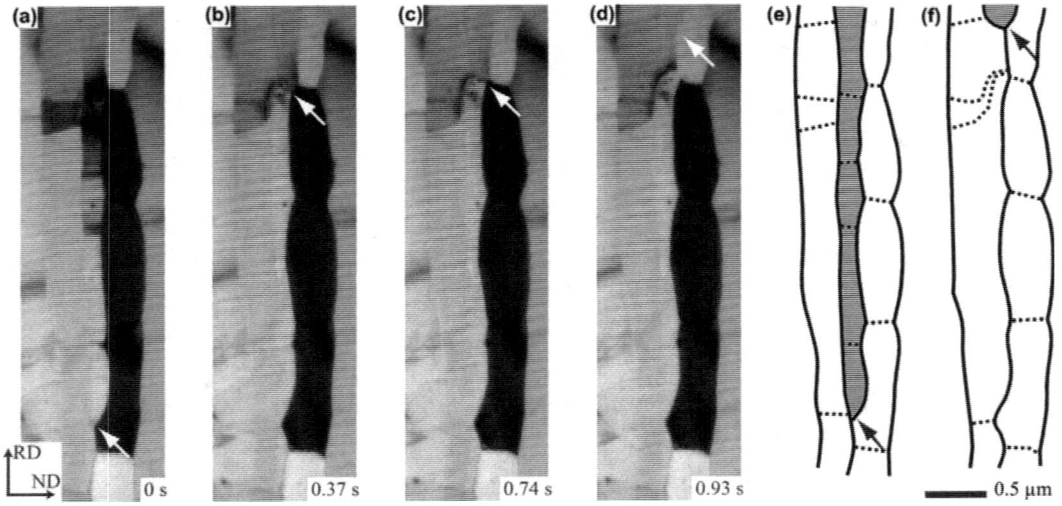

Figure 2. TJ motion observed in a TEM foil (center layer) at ~290 °C: (a-d) TEM images obtained after different periods of time (0 to 0.93 s), where a rapidly migrating TJ is marked by an arrow. The changes before and after the TJ motion are shown schematically in (e,f). Solid and dashed lines in (e,f) indicate lamellar and interconnecting boundaries, respectively.

Most subgrains and narrow lamellae of non-dominant texture components are located between coarser lamellae of the dominant texture components, and as such are typically surrounded by HABs. Therefore, their shrinkage by TJ motion leads to a reduction in the HAB fraction as newly formed boundaries between the broad lamellae of similar orientations are LABs (Fig. 3).

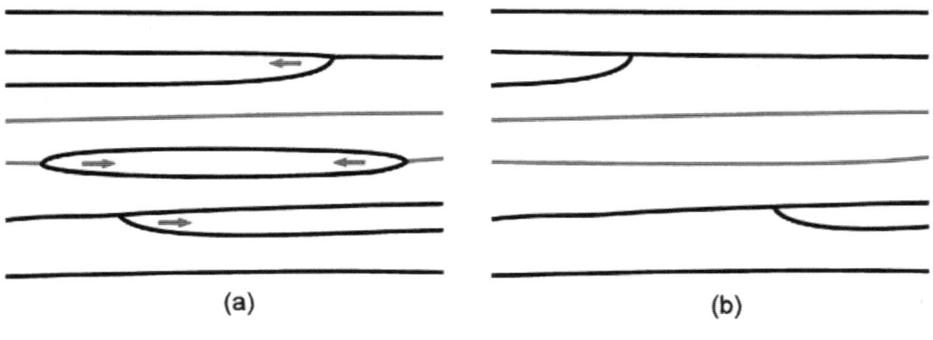

Figure 3. Schematic showing coarsening via lateral motion TJs in a deformed lamellar microstructure: (a) initial microstructure; (b) microstructure after TJ motion. Black and gray lines indicate HABs and LABs, respectively. Reprinted from [3] with permission from Elsevier.

Since in each subsurface layer there is only one dominant orientation variant, the reduction in the fraction of HABs in these layers is much more pronounced than in the center, where the dominant S component is represented by four symmetric variants [3]. Consequently, by the time pronounced recrystallization starts, the center still contains a high frequency of mobile HABs, whereas there are very few such boundaries left in the subsurface (see Fig. 2b). These differences in the frequency of mobile boundaries result in a very large difference in the frequency of recrystallization nuclei and in the recrystallization kinetics in the different layers. Accordingly, in the center layer, recrystallization was found to take place initially more rapidly and with a higher nucleation density, thus resulting in a smaller final grain size (13 μm) compared to that in the subsurface layers (~30 μm). In each layer, the nuclei were observed to form within lamellar structures and near coarse particles. The former ones typically had orientations of the rolling texture components, whereas nuclei near coarse particles had orientations of the rolling texture, or of the P {011}⟨566⟩, $Cube_{ND}$ {001}⟨310⟩ and "random" orientations. As the rolling texture in the recovered material was extremely strong (96 – 98% of all orientations), nuclei of the rolling texture components were susceptible to the effect of orientation pinning [9]. It is considered that in the subsurface layers orientation pinning experienced by nuclei of the dominant Cu component was especially pronounced, whereas the growth of nuclei with P, $Cube_{ND}$ and other orientations was mostly unhindered [3]. As a result, during recrystallization (annealing at 300 °C for 2 h) the fraction of the rolling texture components in the subsurface dropped to 23%, while the fraction of P + $Cube_{ND}$ components increased to 55%. In contrast, in the center, where orientation pinning was not as strong as in the subsurface, the fractions of the rolling texture and P + $Cube_{ND}$ components after 2 h at 300 °C were 53% and 27%, respectively. These observations emphasize the importance of considering changes during recovery when examining recrystallization in heavily deformed metals.

Conclusions

The microstructural evolution during recovery in heavily rolled aluminum is strongly influenced by the initial spatial arrangement of different rolling texture components in the deformed material. In this material, coarsening of lamellar structures via TJ motion leads to the loss of HABs. This reduction in the HAB fraction is more pronounced in the subsurface layers than in the center, and is related to the initial differences in the distribution of the rolling texture components between the layers. Recrystallization in the center, which after initial coarsening still contains a large frequency of mobile HABs, proceeds more rapidly and with a higher nucleation density, resulting in a smaller recrystallized grain size than in the subsurface layers.

Acknowledgements

The authors gratefully acknowledge the support from the Danish National Research Foundation (Grant No.DNRF86-5) and the National Natural Science Foundation of China (Grant No. 51261130091) to the Danish–Chinese Center for Nanometals.

References

[1] Yu T, Hansen N, Huang X 2011 *Proc. Roy. Soc. A* **467** 3039
[2] Yu T, Hansen N, Huang X 2013 *Acta Mater.* **61** 6577
[3] Mishin O V, Godfrey A, Juul Jensen D and Hansen N 2013 *Acta Mater.* **61** 5354
[4] Mishin O V, Bay B and Juul Jensen D 2000 *Metall. Mater. Trans.* A **31** 1653
[5] Quadir M Z, Al-Buhamad O, Bassman L, Ferry M 2007 *Acta Mater.* **55** 5438
[6] Zahid G H, Huang Y, Prangnell P B 2009 *Acta Mater.* **57** 3509
[7] Gazder A A, Hazra S S, Pereloma E V. 2011 *Mater. Sci. Eng. A* **530** 492
[8] Tian H, Suo H L, Mishin O V, Zhang Y B, Juul Jensen D 2013 *J. Mater. Sci.* **48** 4183
[9] Juul Jensen D 1995 *Acta Metall. Mater.* **43** 4117

IOP Conf. Series: Materials Science and Engineering **82** (2015) 012084 doi:10.1088/1757-899X/82/1/012084

Microstructure and texture of zinc deformed by extrusion with forward-backward rotating die (KoBo)

K Sztwiertnia[1], J Kawalko[1], M Bieda[1], M Jaskowski[2], K Pieła[2] and W Bochniak[2]

[1] Institute of Metallurgy and Materials Science, Polish Academy of Sciences, Krakow, Poland

[2] Faculty of Non-Ferrous Metals, AGH - University of Science and Technology, Krakow, Poland

Abstract. The KoBo device is a press with a cyclically rotating die enabling extrusion of ingots with the permanent destabilization of their substructure. The method involves a cyclic change of deformation path that increases the plasticity of the material and inhibits formation and propagation of cracks. The possibility of using KoBo for obtaining zinc wires with high mechanical properties was explored. A polycrystalline zinc ingot of purity of 99.995% was subjected to extrusion at room temperature. The microstructure of the material was investigated primarily via high-resolution electron backscatter diffraction using a scanning electron microscope. The microstructure is heterogeneous and consists of grains elongated slightly in the extrusion direction (ED). The material has a relatively sharp nearly axial texture with the ED scattered between the <1 0 -1 0> and <2 -1 -1 0> directions. More than 95% of high-angle grain boundaries do not correspond to any twin or low Σ CSL boundaries. The dimensions of the grains range from 10 - 20 micrometers down to the sub-micron scale. Despite the relatively large grains, the final product exhibits very good mechanical properties, which could not be explained by the Hall - Petch relation. It was found that besides grain refinement two other effects could affect the material mechanical properties. These effects are formation of nano-grains inside the large, micrometer-sized grains and formation of broad areas of high-density crystal lattice defects extended along the HAGBs.

1. Introduction

A particularly effective way to strengthen metallic materials proved to be grain refinement up to nanometric sizes, which can be realized by various methods of severe plastic deformation (SPD), e.g., [1-3]. The goal of the classical SPD methods, such as Equal Chanel Angular Pressing, Accumulative Roll Bonding, or High Pressure Torsion, is manufacturing of a material with "only" strong modification of its microstructure. In addition to the classical SPD methods, methods were developed that also change the product shape. One of these methods is the KoBo extrusion, [4, 5]. The KoBo device is a press with a forward-backward rotating die, enabling extrusion of ingots under conditions of permanent destabilization of their microstructure. The technique allows for a high-rate deformation, which favors the accumulation of lattice defects and thereby facilitates the microstructure refinement.

Polycrystalline zinc exhibits very poor mechanical properties; for example, with the grain size of approximately 40 μm, the yield strength $R_{0,2}$ reaches approximately 40 MPa and the tensile strength Rm is approximately 110 MPa. SPD methods enable a significant improvement of these properties, with R0,2 and Rm able to reach ~ 160 MPa and ~ 170 MPa, respectively. This enhancement is obtained by grain refinement to the nanometer scale, practically in the absence of dislocation and with a negligible concentration of point defects, e.g. [6, 7]. Using the KoBo device, one can obtain material with high mechanical properties [8, 9], similar to those after SPD, wherein the average grain size does not fall below of 10 - 20 μm. Thus, the grain refinement interpretation fails. This result indicates that during the

KoBo deformation of zinc, besides grain refinement, other effects that modify the microstructure strengthen the material. In this work, we seek to identify these effects.

2. Experimental procedures

The investigated material was pure zinc (99,995%) received as ingots with a diameter of 40 mm and a length of 50 mm and a typical for casting products microstructure of long columnar grains. The KoBo process was performed at room temperature. The bars were extruded to wires using different sets of KoBo process parameters. Samples for microstructure investigations were prepared from wires extruded with extrusion ratios of $\lambda=100$, $\lambda=400$ and $\lambda=490$ from longitudinal and transverse cross sections.
Microstructures were investigated using scanning electron microscopy (SEM) and transmission electron microscopy (TEM).

3. Results and discussion

The microstructure of zinc after the KoBo process was heterogeneous, with the degree of heterogeneity being dependent on selected process parameters. Nevertheless we could specify certain characteristics of the microstructure that were common for all samples, at least qualitatively. Each of the investigated microstructures was composed of slightly elongated ED grains of a varying fragmentation degree, Fig. 1. The size of grains varied from a few hundred nanometers to several tens of micrometers. The highest refinement of grains was presented in the outer, most distorted, layers of the rod. In such microstructure, there were two effects that, although occurring with varying intensity, could strengthen the material.
Fig. 1 shows the chains of recrystallized fine grains extending through the entire cross section of the extruded rod. These chains are probably effects of recrystallization in shear bands produced by extrusion with forward-backward rotating die. The change in orientation topography between the transverse cross-section taken from the region close to the band and the longitudinal cross-section exhibits characteristic differences in the orientation of the grains in the bands and in the matrix. The matrix orientation distribution function (ODF) revealed a relatively sharp nearly axial texture with the <0001> axis perpendicular to the ED which was scattered between the <1 0 -1 0> and <2 -1 -1 0> directions., as shown in Fig. 2a.

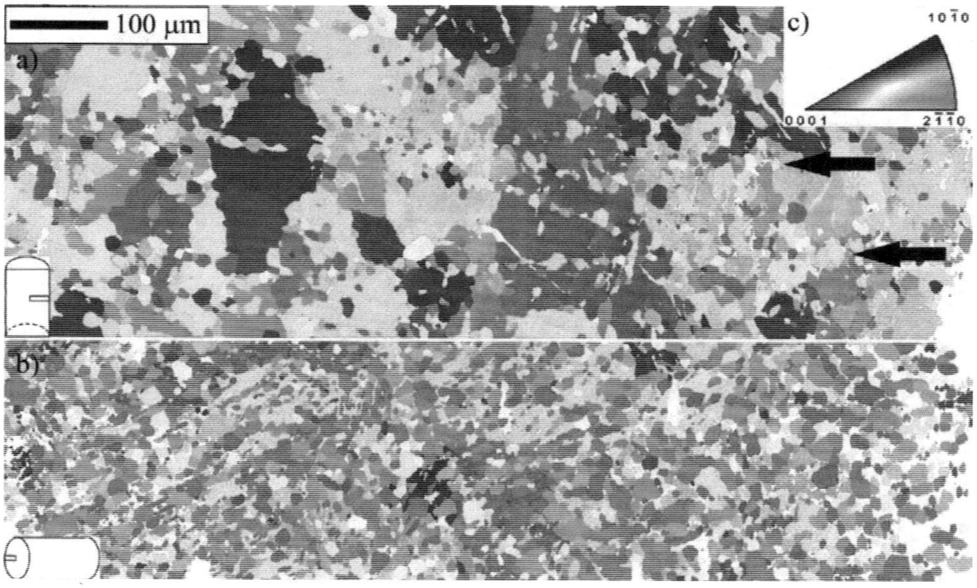

Fig. 1. Microstructure of zinc after KoBo extrusion (λ =400). Example EBSD maps in the longitudinal (a) and the transverse (b) cross-sections, with examples of recrystallized shear bands areas marked; the colors correspond to the ED in the inverse pole figure (c).

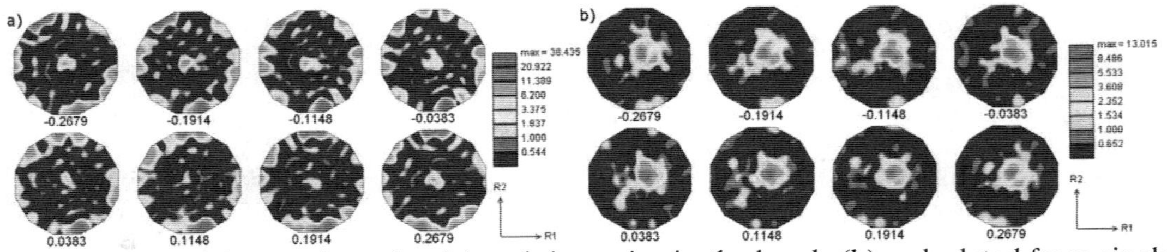

Fig. 2. ODFs of the matrix grains (a) and the grains in the bands (b); calculated from single orientations measured at transverse cross-section (fig.1. b) . Rodrigues' representation R_1 R_2 R_3, with cross-section R_3 = constant; asymmetric domain of D_6 symmetry.

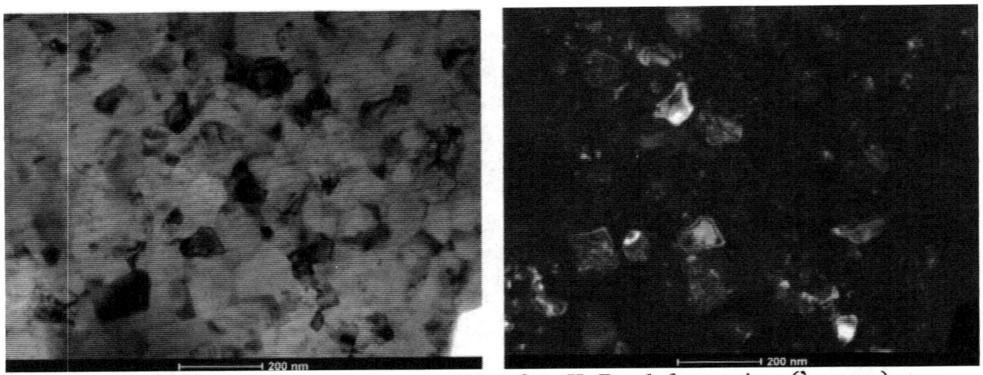

Fig. 3. An example of the zinc microstructure after KoBo deformation (λ =100), transverse cross-section; bright- and dark-field micrograph, respectively, TEM.

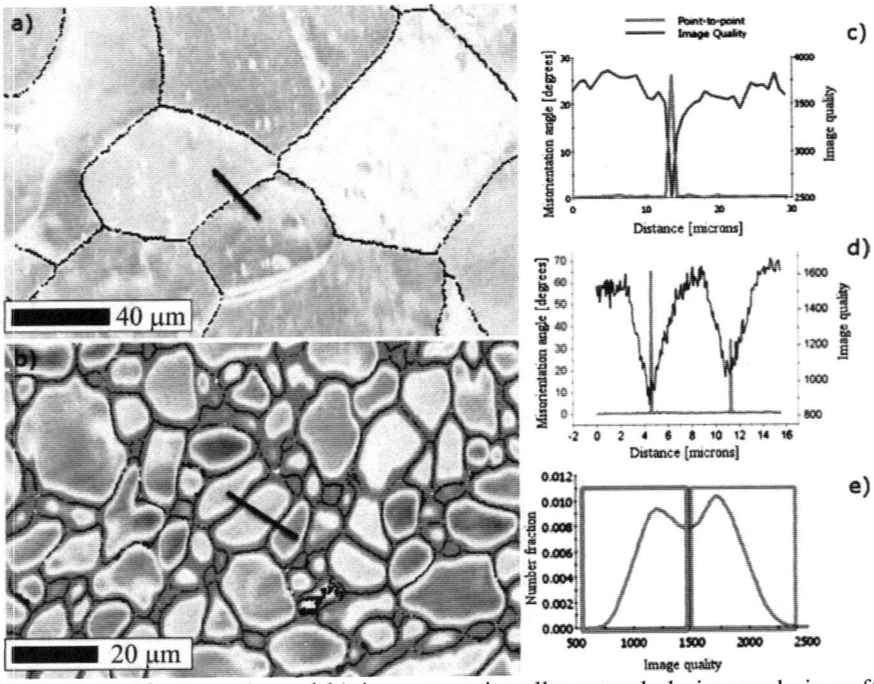

Fig. 4. Exemplary IQ maps (a and b) in conventionally extruded zinc and zinc after KoBo extrusion (λ =490); black - HAGBs, red/yellow - the areas of highly defected crystal lattice, green – areas of low lattice defect density. c) and d) misorientation angle (point to point) and IQ changes along the line segments shown in (a) and (b); number fraction with low values of

IQ – high density of lattice defects and number fraction with high values of IQ – low density of lattice defects.

Quite different situation was found in the recrystallized shear bands, where the ED was predominantly parallel to the <0001> direction, Fig. 2b. Immediately after the extrusion process is completed, nano-grains, an example of which is shown in Fig. 3 were found in microstructure. Those nano-grains are suspected to nucleate and recrystallize over time in the vicinity of shear bands and form chains of micrometer size grains observed in EBSD measurements (Fig. 1a).
The second microstructural effect that results in the strengthening of zinc after KoBo extrusion could be the formation of the unusually thick areas of the deformed crystal lattice extended mainly along the HAGBs, Fig. 4b [9]. Detection of these areas was enabled by the use of image quality (IQ) maps. The IQ of the EBSD pattern is strongly affected by the local concentration of the lattice defects. The areas with a high density of lattice defects occupied up to 30% of the volume of material within the sample and up to 60% in the surface layer.

4. Conclusions
The improvement of the mechanical characteristics in zinc after KoBo extrusion cannot be explained only on the basis of grain refinement because an average grain size does not fall below a value in the range of 10 to 20 μm, and the mechanical properties correspond to those obtained for nanocrystalline materials.
During the KoBo deformation process, besides the grain refinement, which is not as intense as that in the SPD methods, two other effects are generated:

- The first effect is the formation of nano-grains within of large, matrix grains. The nano-grains were generated during the shearing of the material due to the extrusion with forward-backward rotating die.
- The second effect is the formation of the unusually broad areas of higher density of crystal lattice defects along HAGBs.

Both of these effects require further study via High-Resolution Transmission Electron Microscopy.

Acknowledgments
This work was supported in part by the National Science Center (Poland), DEC-2011/03/B/ST8/06120 and by the European Union from resources of the European Social Fund (Project No.POKL.04.01.00-00-004/10).

References
[1] Valiev R Z, Islamgaliew R K, Alexandrov I V 2000, *Prog. Mater. Sci.* **45** 103.
[2] Meyers M A, Mishra A, Benson D J 2006 *Prog. Mater. Sci.* **51**, 427.
[3] Kumar K S, Va n Swygenhoven H, Suresh S 2003 *Acta Mater* **51**, 5743.
[4] Korbel A and Bochniak W 1998, U.S. Patent 5.737.959, 2000 European Patent 0.711.210.
[5] Korbel A, Bochniak W, Ostachowski P, Błaż L 2011, *Metall. Mater. Trans.* A **42** 2881.
[6] Zhang X, Wang H, Scattergood R O, Narayan J, Koch C C 2002 *Acta Mater.* **50** 3995.
[7] Zhang X, Wang H, Narayan J, Koch C C 2001 *Acta Mater.* **49** 1319.
[8] Korbel A, Pospiech J, Bochniak W, Tarasek A, Ostachowski P, Bonarski J 2011, *Int. J. Mat. Res.* **102** 464.
[9] Sztwiertnia K, Kawałko J, Bieda M, Berent K 2013, *Arch. Metall. Mater.* **58**, 157.

Towards effective analysis of large grain boundary data sets

K Glowinski and A Morawiec

Instiute of Metallurgy and Materials Science, Polish Academy of Sciences, Kraków, Poland

E-mail: kglowinski@ymail.com

Abstract. Grain boundaries affect properties of polycrystals. Novel experimental techniques for three-dimensional orientation mapping give new opportunities for studies of this influence. Large networks of boundaries can be analyzed based on all five 'macroscopic' boundary parameters. We demonstrate benefits of applying two methods for improving these analyses. The fractions of geometrically special boundaries in ferrite are estimated based on 'approximate' distances to the nearest special boundaries; by using these parameters, the times needed for processing boundary data sets are shortened. Moreover, grain-boundary distributions for nickel are obtained using kernel density estimation; this approach leads to distribution functions more accurate than those obtained based on partition of the space into bins.

1. Introduction

The evolution of experimental techniques for three-dimensional (3-D) microstructure imaging (e.g., automated serial sectioning, X-ray tomography) has made it possible to determine grain misorientations and boundary plane indices (i.e., all five macroscopic boundary parameters) for significant numbers of boundaries (e.g., [1, 2]). This has increased the interest in quantitative characterization of boundary networks (e.g., [3]). However, the boundary studies with five parameters taken into account turn out to be complex compared to those limited solely to grain orientations. Although progress in the development of experimental equipment has been made, acquisition of large 3-D data sets is still cumbersome. It is, therefore, important to analyze the data in a way that is both efficient and does not distort final conclusions. In this note, two practical examples of increasing effectiveness of the analysis are presented. First, the fractions of geometrically special boundaries in ferritic steel are estimated using fast-to-calculate parameters defined in [4] and [5]. These fractions supplement the characterization of the boundary network of this material given in [2]. Then, grain-boundary distributions for pure nickel, already studied in [1], are reconsidered; more accurate and reliable distributions are obtained using an approach based on multivariate kernel density estimation (KDE) [6]. Besides that, a scheme of detailed interpretation of these distributions is suggested.

2. Geometrically special grain boundaries in ferrite

Five-tuples of macroscopic boundary parameters constitute the 5-D boundary space [7]. Crystal and grain-exchange symmetries imply that different 5-tuples may represent the same physical boundary, and all these equivalent representations need to be considered. Differences in geometries of boundaries can be quantified by means of a metric defined in the space: the

Table 1. Area-weighted fractions (assuming the tolerance of 8°) of geometrically special boundaries in ferritic steel. Values in parentheses correspond to 10^7 random boundaries.

Twist	180°-tilt	Symmetric	180°-tilt & symmetric
21.2% (22.0%)	29.3% (21.9%)	2.2% (1.4%)	1.5% (0.7%)

distance between two boundaries minimized over all representations is small (large) if they are similar (distinct). Boundaries of special geometric features, e.g., tilt, twist, symmetric, 180°-tilt boundaries serve as reference points in the boundary space, and general boundaries are characterized by their distances to the nearest reference boundaries. However, calculation of the distances to the nearest special boundaries is complicated and time-consuming. In order to speed up such characterization for large numbers of boundaries, the distances can be replaced by parameters defined in [4] and [5] that are easy to calculate and strongly correlated with the true distances. These 'approximate distances' are used here for estimation of the fractions of geometrically special boundaries in a 3-D network of about 8,300 boundaries extracted from a ferritic steel sample. For the steel composition and processing history, and for the data acquisition details, see [2]. The boundaries were reconstructed in the form of a triangular mesh using *DREAM.3D* [8] program. In total, there were about 1.4×10^6 triangular boundary segments. Computation of the approximate distances for all segments took about 2 minutes on a 4-core PC whereas obtaining of accurate distances would take about 65 hours. A segment was classified as a twist, symmetric, or 180°-tilt boundary if the approximate distances to the nearest twist, symmetric, and 180°-tilt boundary, respectively, calculated for that segment were below 8°. This limit is determined by the experimental capabilities, cf. [5]. With such resolution, tilt boundaries in materials with cubic $m\bar{3}m$ crystal symmetry are indistinguishable [4]. The fractions of other special boundaries obtained for ferrite are compared to those calculated for random boundaries in Tab. 1. Only the percentages for symmetric boundaries and 180°-tilt boundaries differ significantly from those corresponding to the random data. From the distributions (Fig. 1) of the approximate distances α_S and α_I to the nearest symmetric boundaries and 180°-tilt boundaries, respectively, it follows that, indeed, near-symmetric boundaries (i.e., those with α_S close to zero) and near-180°-tilt boundaries (those with small α_I) are over-represented. It was reported in [2] that in the investigated steel, there is a high population of the $\Sigma3/(\bar{2}11)$ – simultaneously symmetric and 180°-tilt – boundaries. Therefore, it is interesting to identify the largest subgroups of all symmetric boundaries and 180°-tilt boundaries. This is done based on the distributions of α_S and α_I, respectively, obtained for the complete data set and its subsets (Fig. 1). It is inferred that the contributions of near-$\Sigma3$ boundaries to both symmetric and 180°-tilt boundaries are minor, while the dominant inputs are those of small-angle boundaries.

3. Distributions of grain boundaries in nickel

Populations of boundaries as functions of the five macroscopic parameters are the primary tool for studying 3-D boundary networks. The populations are conventionally given as multiples of random distributions (MRD) and depicted in the form of 2-D sections for fixed misorientations and varying boundary planes. For instance, sections for the $\Sigma3$, $\Sigma5$, $\Sigma7$, and $\Sigma9$ misorientations obtained for a data set collected from a pure nickel sample were shown in [1]. Those distributions were computed using a method based on partition of the boundary space into equivolume bins and counting boundary segments that fall into the bins [3]. However, as pointed out in [6], the elongated dimensions of some bins, averaging over equivalent bins, and poor correspondence between bin sizes and experimental resolutions lead to artifacts complicating interpretation of the resulting functions.

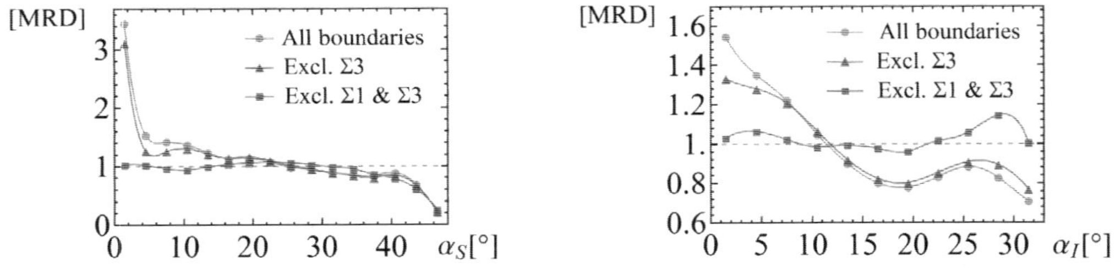

Figure 1. Distributions (in MRD – multiples of random distributions) of the approximate distances α_S and α_I to the nearest symmetric and 180°-tilt boundaries, respectively, computed for the data set of ferrite and its subsets. In the first subset, $\Sigma3$ boundaries (according to Brandon criterion) are excluded. The second subset does not contain $\Sigma1$ and $\Sigma3$ boundaries. ($\Sigma1$ boundaries have misorientation angles less than 15° and – due to data clean-up – greater than 5°.) The removed boundaries constitute 7.5% and 29.4% of the complete data, respectively.

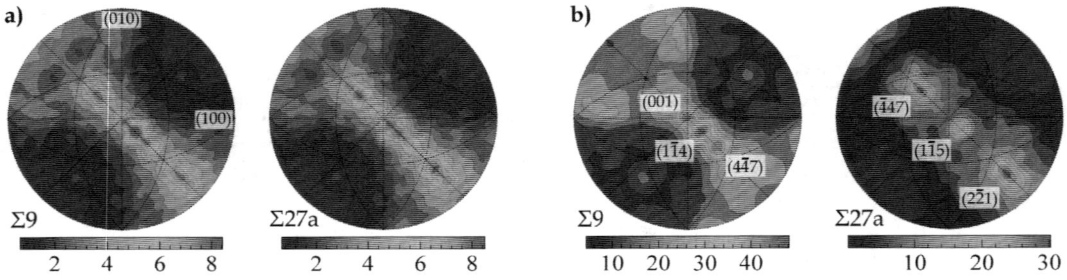

Figure 2. Populations (in multiples of random distributions) of $\Sigma9$ and $\Sigma27a$ boundaries in nickel as functions of boundary planes (plotted in stereographic projections) obtained using (a) partition into "10°-bins", and (b) the KDE-based method with $\rho_m = 3°$, $\rho_p = 7°$.

Therefore, the nickel data are reanalyzed. This time the boundary distributions were obtained using a newly developed KDE-based approach that is free from the deficiencies of the partition-based method [6]. Surfaces of about 3,500 boundaries were reconstructed in *DREAM.3D*. Formally, the boundary space is a Cartesian product of the misorientation and boundary-normal subspaces. For computational reasons and because of considerably different resolutions in determinination of misorientation and boundary-plane parameters, it is convenient to use a separate metric in each subspace. First, boundaries with misorientations located not farther from the fixed misorientation than a limiting distance ρ_m are selected. Then, the distribution is probed at evenly distributed normal directions, and boundaries whose normals deviate from a given direction by less than ρ_p are counted. The radii ρ_p and ρ_m were tailored to both resolution and amount of data and set to 3° and 7°, respectively, cf. [6]. The most significant differences between the old method and new approach were observed in the $\Sigma9$ and $\Sigma27a$ sections. The sections obtained using partition into typical "10°-bins" overlap and are barely distinguishable (Fig. 2a), whereas those computed using the new approach are clearly separated (Fig. 2b). Moreover, in the KDE-based method, statistical errors of the distributions (thus, their reliability) can be directly evaluated using the formula $\varepsilon \approx (fnv)^{1/2}$ [6]; ε is the relative error of the function at a given point, f is the value of the function at that point, n stands for the number of grain boundaries in the considered network, and v denotes the volume restricted by ρ_m and ρ_p. For instance, the heights of the peaks at the $\Sigma27a/(\bar{4}47)$ and $\Sigma27a/(2\bar{2}1)$ boundaries are

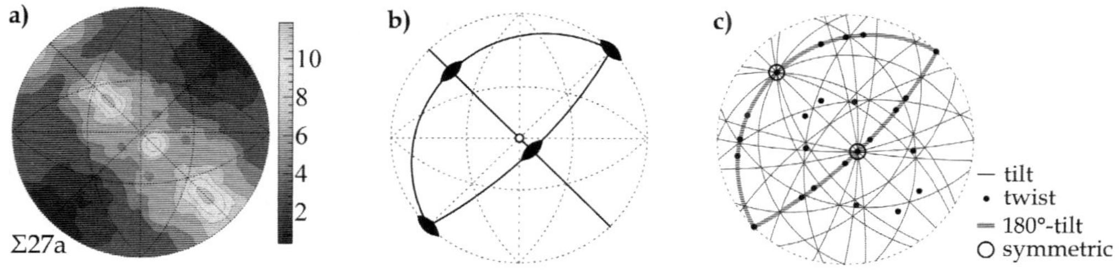

Figure 3. (a) Statistical errors of the distribution for the $\Sigma 27a$ misorientation shown in Fig. 2b. The errors are given as multiples of a random distribution. (b) Symmetries of the $\Sigma 27a$ section. (c) Locations of geometrically special boundaries for the $\Sigma 27a$ misorientation.

20 MRD and the corresponding errors are ±12 MRD. These maxima are higher than those in the sections obtained based on the bins because the volume of the space determined by $\rho_m = 3°$, $\rho_m = 7°$ (used in normalization of the distributions) is about 2 times smaller than the volume of "10°-bins". Moreover, averaging over many bins (thus, flattening of the distributions) was eliminated. Symmetry axes and mirror lines of the $\Sigma 27a$ section are shown in Fig. 3b, cf. [9]. This figure reveals, in particular, that the maxima at the $(\bar{4}47)$ and $(2\bar{2}1)$ poles are actually related to the same physical boundaries. A diagram containing the locations of all tilt, twist, symmetric, and 180°-tilt boundaries with the $\Sigma 27a$ misorientation is presented in Fig. 3c. This map indicates that the peak at the $\Sigma 27a/(2\bar{2}1)$ boundary lies at the intersection of three tilt zones – it has three distinct tilt axes.

4. Final remarks

The recently developed methods which facilitate studies of grain-boundary networks were applied to analysis of boundaries in two metals. Recognition of geometrically special boundaries is more efficient if the accurate distances to the nearest special boundaries are substituted by their approximations. Based on distributions of these parameters, it is concluded that symmetric boundaries and 180°-tilt boundaries are over-represented in the ferrite sample, but majority of them are small-angle boundaries. Most artifacts in grain-boundary distributions are reduced if KDE is used instead of the partition-based approach. In particular, the new method allows for detailed interpretation of the $\Sigma 9$ and $\Sigma 27a$ sections obtained for pure nickel as they do not overlap (as the overlap occurring in the conventional method with typical "10°-bins" is avoided).

Acknowledgments

The authors are grateful to Professor Gregory S. Rohrer (of Carnegie Mellon University) for permission to use the ferrite and nickel data. The work of K.G. was supported by the European Union under the European Social Fund within project No. POKL.04.01.00-00-004/10.

References

[1] Li J, Dillon S J and Rohrer G S 2009 *Acta Mater.* **57** 4304
[2] Beladi H and Rohrer G S 2013 *Acta Mater.* **61** 1404
[3] Saylor D M, Morawiec A and Rohrer G S 2003 *Acta Mater.* **51** 3663
[4] Morawiec A and Glowinski K 2013 *Acta Mater.* **61** 5756
[5] Glowinski K 2014 *J. Appl. Cryst.* **47** 726
[6] Glowinski K and Morawiec A 2014 *Metall. Mater. Trans.* **A45** 3189
[7] Morawiec A 2009 *J. Appl. Cryst.* **42** 783
[8] Groeber M A and Jackson M A 2014 *Integr. Mater. Manuf. Innov.* **3** 5
[9] Patala S and Schuh C A 2013 *Philos. Mag.* **93** 524

IOP Conf. Series: Materials Science and Engineering **82** (2015) 012086 doi:10.1088/1757-899X/82/1/012086

Grain boundary character distribution derived from three-dimensional microstructure reconstruction

H Pirgazi[1] and LAI Kestens[1,2]

[1] Ghent University, Department of Materials Science and Engineering, Technologiepark 903, 9052 Gent, Belgium

[2] Delft University of Technology, Department of Materials Science and Engineering, Mekelweg 2, 2628CD Delft, The Netherlands

E-mail: hadi.pirgazi@ugent.be, leo.kestens@ugent.be

Abstract. Manual serial sectioning which includes consecutive steps of sample preparation and Electron Back Scattering Diffraction (EBSD) measurement was employed to extract the two-dimensional (2D) sections of a pure nickel sample and to reconstruct the three-dimensional (3D) microstructure. A general alignment algorithm based on the minimization of misorientation between two adjacent sections has been developed to accurately align the sections. By employing this alignment algorithm, any in-plane (translational) and rotational misalignment as well as the planparallelity can be corrected. Surface triangulation technique was used to reconstruct the grain boundary surfaces. The Grain Boundary Character Distribution (GBCD) was derived from reconstructed grain boundaries. The results show that a smoother grain boundary plane can be obtained after precise translational and rotational alignment and correction of planparallelity.

The relative grain boundary energy was computed as a function of the five grain boundary parameters based on equilibrium at triple lines. The results show that the grain boundary planes carrying a $\Sigma3$ type misorientation are dominantly parallel to the {111} crystal plane, which indicates the presence of coherent twin boundaries. It was observed that coherent $\Sigma3$ type boundaries exhibit the minimum relative grain boundary energy, which is approximately 57% smaller than the average of all $\Sigma3$ boundaries, including also incoherent twin boundaries.

1. Introduction

Although many microstructural features can be characterized with conventional 2D techniques, there remain many important microstructural properties (e.g. detailed grain boundary features) that can only be measured in three-dimensions (3D). it is well-known that a full description of grain boundary planes is done with 5 parameters: 3 parameters of misorientation across the grain boundary and 2 parameters of the normal to the grain boundary plane [1]. The techniques for 3D characterization of the microstructures are based on two different approaches: (i) observation with high energy radiation and [2, 3] and (ii) by serial sectioning [4-6]. Furthermore, some stereological methods have been proposed and employed to estimate the 3D information from 2D images [7].

Serial sectioning has been the most widely used method to acquire 3D data at the macro/microscale of opaque materials. In this technique, a series of closely spaced parallel sections reveal the third dimension of the microstructure. The usual method for serial sectioning involves the cyclic removal of parallel layers of the sample, followed by imaging (e.g. by EBSD) of the planar sections [4, 8]. The removal of the material for serial sectioning can be performed with different methods: e.g. mechanical polishing, chemical polishing, Focused Ion Beam (FIB) tomography and laser or electrical discharge ablation [5]. Independent of the method employed for the preparation of the sections, the serial sectioning technique suffers from the misalignment between the sections which needs to be accurately corrected before further data processing [9].

The goal of this research is to derive complete grain boundary structural properties and to calculate the relative grain boundary energy for materials with relatively large grain size.

2. Experimental techniques

Wide field 3D Electron Back-Scattering Diffraction (EBSD) was performed on a pure nickel sample by a serial sectioning technique with mechanical polishing. The sample preparation was done by conventional mechanical polishing to remove a layer with a certain thickness (7 μm). After each polishing step, the EBSD measurements was performed in a high resolution FEG-SEM (of type FEI Quanta™450®) on a large area (3200 ×3200 μm2) with an in-plane step size of 6 μm to obtain the local crystallographic orientations. In order to meticulously fix the scan area on subsequent layers for EBSD measurements, micro-Vickers indentations were used to mark a grid on the sample. These indentations were also used to measure the thickness of each removed layer.

After collecting 50 consecutive sections the 3D volume was reconstructed. Reconstruction was carried out using two algorithms: (i) the first one for precise alignment of the sections and (ii) the second one for the surface meshing of the grain boundary planes by triangulation. An x-y alignment algorithm has been proposed by Lee et al. [10], which is based on the minimization of the average disorientation between two adjacent sections for translational alignment. In this research, this technique is improved by including rotational alignment in addition to the translational alignment. The rotation was carried out by θ° about a rotation axis which was defined by α and β as altitude and azimuth with respect to the sample coordinate system. After each rotation, the scan data were mapped into the new location and the orientations were recalculated by a transformation matrix based on θ° rotation about the axis defined by α and β angles. Figure 1 shows the results of the mentioned alignment procedure. It can be seen that any translational and rotational misalignment as well as the planparallelity (in case of 2 non-parallel consecutive sections which usually appears as a stretched image in the EBSD map) has been precisely corrected.

The second algorithm was used for grain boundary plane triangulation. To this purpose, two grain boundaries between two pairs of corresponding triple junctions were divided into equal segments and triangulation was carried out based on an algorithm proposed by Saylor et al [11].

3. Results and discussion

The results of the reconstructed microstructures are presented in Figure 2. It can be seen that by applying the full alignment procedure (translational plus rotational, cf. Figure 2c), instead of only the translational alignment (cf. Figure 2b) a smoother grain boundary plane is obtained. Together with the improvement of the grain boundary smoothness, the non-uniform orientation spread observed in Figure 2a,b does not appear in Figure 2c. It implies that the full alignment method generates more uniform orientations, which is more realistic for a fully recrystallized structure. This observation is directly related to the correction of the planparallelity and rotational misalignment, which was illustrated in Figure 1.

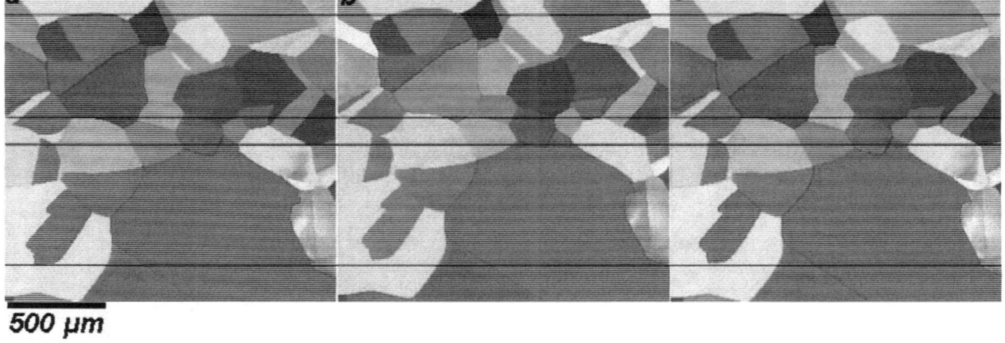

Figure 1. Results of the registration of two consecutive sections, a) the first section is considered as the fixed section, b) the second section after only the translational correction, and c) the second section after translational and rotational alignment. The black lines can be used for comparison between the results.

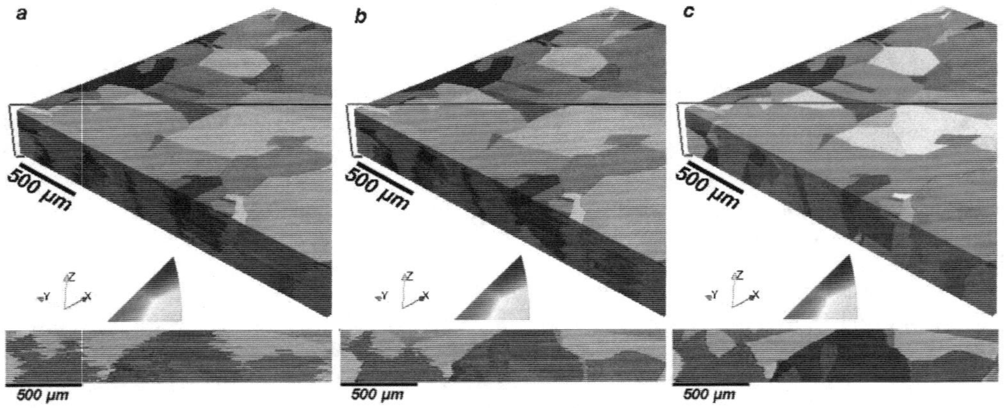

Figure 2. ND inverse pole figure volume of the reconstructed microstructure together with the cross section of the specified plane (red lines), a) the stack of the original EBSD data without any alignment, b) after translational alignment, and c) after translational and rotational alignment.

In order to calculate the grain boundary character distribution, the distribution of the normal to each triangle in the triangulated grain boundary mesh was taken into account. For each type of misorientation, the stereographic projection of the normal to the grain boundary in the crystal reference frame was plotted. Figure 3a and b shows the distribution of the grain boundary planes for Σ3 type of misorientations after translational and full alignment, respectively. It is obvious that a precise alignment of the 2D sections (cf. Figure 3b), which results in a smoother grain boundary plane, gives a more concentrated distribution with higher intensity, whereas the distribution obtained after the translational alignment is very close to the random distribution. The results also show that the grain boundary planes of Σ3 type of misorientations are dominantly terminated by {111} crystal planes, which indicates the presence of coherent twin boundaries. Similar results have been obtained from serial sectioning by FIB tomography [7].

The 3D information was also used to calculate the relative grain boundary energy based on the equilibrium at the triple junctions (Herring's relation). To this purpose, the torque term of the grain boundary energy was calculated using the normal to the grain boundary, whereas the surface tension part was obtained using the tangent to the grain boundary, cf. Equation 1.

$$\sum_{i=1}^{3}\{\sigma_i\hat{t}_i + (\partial\sigma_i / \partial\phi_i)\hat{n}_i\} = \vec{0} \qquad \text{Eq. 1}$$

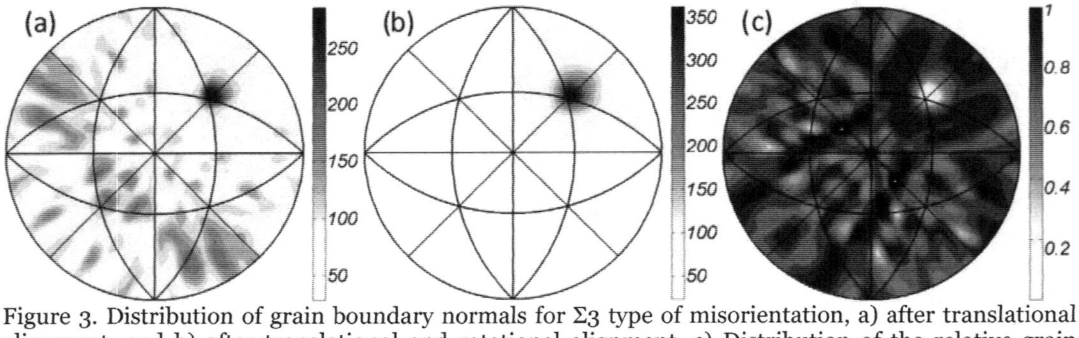

Figure 3. Distribution of grain boundary normals for Σ3 type of misorientation, a) after translational alignment, and b) after translational and rotational alignment. c) Distribution of the relative grain boundary energy for Σ3 type of misorientation.

Where σ_i is the relative grain boundary energy at triple junctions, n_i is the normal to the grain boundary, φ_i is the angle of rotation about triple line and t_i is the tangent to the grain boundary (the cross product of n and triple line).
The relative grain boundary energy was calculated using a method proposed by Adams et al. [12], which is based on discretization of the misorientation space and normal to the grain boundary space and employing multiscale statistical analysis on 3D data sets. The results of this calculation for grain boundaries carrying $\Sigma3$ type of misorientations is presented in Figure 3c. As expected, the $\Sigma3$ type of misorientations with coherent twin boundaries (terminated by {111} crystal planes) exhibit the minimum relative grain boundary energy, which is approximately 57% smaller than the average of all $\Sigma3$ boundaries, including also incoherent twin boundaries.

4. Conclusions

A general alignment algorithm was used to correct in-plane (translational) and rotational misalignment as well the planparallelity, which resulted in a better reconstruction of the 3D microstructure, illustrated by smoother grain boundary planes. Detailed grain boundary features, including the grain boundary character distribution (GBCD), were calculated from reconstructed grain boundaries. The results show that in a pure nickel sample the grain boundary planes of $\Sigma3$ type are dominantly terminated by {111} planes, which indicates the presence of coherent twin boundaries. It was also observed that for $\Sigma3$ type of grain boundaries, the relative grain boundary energy has the minimum value when the grain boundary plane is parallel to {111} plane.

References

[1] Randle V 2010 Grain boundary engineering: an overview after 25 years Labour Hist. Rev. 26 253-61
[2] Gault B, Moody M P, Cairney J M and Ringer S P 2012 Atom Probe Microscopy: Springer)
[3] Miller M K 2005 Handbook of Microscopy for Nanotechnology, ed N Yao and Z L Wang: Kluwer Academic Publisher) pp 227-46
[4] DeHoff R T 1983 Quantitative serial sectioning analysis: preview Journal of Microscopy 131 259-63
[5] Zaefferer S, Wright S I and Raabe D 2008 Three-dimensional orientation microscopy in a focused ion beam-scanning electron microscope: A new dimension of microstructure characterization Metall Mater Trans A 39A 374-89
[6] Uchic M, Groeber M, Shah M, Callahan P, Shiveley A, Scott M, Chapman M and Spowart J 2012 An Automated Multi-Modal Serial Sectioning System for Characterization of Grain-Scale Microstructures in Engineering Materials
[7] Rohrer G S, Li J, Lee S, Rollett A D, Groeber M and Uchic M D 2010 Deriving grain boundary character distributions and relative grain boundary energies from three-dimensional EBSD data Labour Hist. Rev. 26 661-9
[8] Alkemper J and Voorhees P W 2001 Quantitative serial sectioning analysis J Microsc-Oxford 201 388-94
[9] Uchic M D 2011 Computational Methods for Microstructure-Property Relationships, ed S Ghosh and D Dimiduk: Springer US) pp 31-52
[10] Lee S B, Rollett A D and Rohrer G S 2007 Recrystallization and Grain Growth III, Pts 1 and 2, ed S J L Kang, et al. pp 915-20
[11] Saylor D M, Morawiec A and Rohrer G S 2003 Distribution of grain boundaries in magnesia as a function of five macroscopic parameters Acta Materialia 51 3663-74
[12] Adams B L, Ta'asan S, Kinderlehrer D, Livshits I, Mason D E, Wu C T, Mullins W W, Rohrer G S, Rollett A D and Saylor D M 1999 Extracting grain boundary and surface energy from measurement of triple junction geometry Interface Sci 7 321-38

IOP Conf. Series: Materials Science and Engineering **82** (2015) 012087 doi:10.1088/1757-899X/82/1/012087

Microstructural evolution associated with martensitic transformation in Ni-Mn-Ga alloy

Z Li[1], Y Zhang[2,3], C Esling[2,3], X Zhao[1] and L Zuo[1]

[1] Key Laboratory for Anisotropy and Texture of Materials (Ministry of Education), Northeastern University, Shenyang 110819, People's Republic of China

[2] LEM3, CNRS UMR 7239, Université de Lorraine, 57045 Metz, France

[3] Laboratory of Excellence on Design of Alloy Metals for low-mAss Structures (DAMAS), Université de Lorraine, 57045 Metz, France

E-mail: yudong.zhang@univ-lorraine.fr; lzuo@mail.neu.edu.cn

Abstract. Based on the spatially resolved electron backscatter diffraction technique, the microstructural evolution accompanying the martensitic transformation (austenite to 7M martensite) and the intermartensitic transformation (7M martensite to NM martensite) was studied on a polycrystalline Ni53Mn22Ga25 alloy. Results show that the 7M martensite plate groups transformed from initial austenite have a diamond-shape with four twin-related variants. The 7M to NM intermartensitic transformation was accompanied by the thickening of martensite plates. With the experimental results, the characteristics of microstructural evolution during the phase transformations were further analyzed.

1. Introduction

In recent years, Ni-Mn-Ga ferromagnetic shape memory alloys have aroused great attention due to their excellent magnetic shape memory effects in martensite state. This effect, namely the magnetic-field-induced strain, is originated from the magnetic-field-driven rearrangement of martensitic variants through the variant interface movement [1]. So far, giant strains up to 10% can be achieved upon the application of a moderate magnetic field [2]. It is established that on cooling, Ni-Mn-Ga alloys undergo a first-order diffusionless thermoelastic martensitic transformation from the cubic $L2_1$ ordered austenite to a low-symmetry martensite. Depending on the alloy composition, three types of martensite, i.e. five-layered modulated (5M), seven-layered modulated (7M), and non-modulated (NM) martensite, are frequently observed [3]. In addition to the martensitic transformation, there may also exist in some alloys a first-order intermartensitic transformation from one type of martensite to another [4].

To date, many efforts have been devoted to understanding the phase transformation behaviors by taking into account the physical properties of different phases in Ni-Mn-Ga alloys [5-6]. However, there are still open issues concerning the microstructural evolutions and crystallographic correlations involved in the structural transformation. In this work, based on spatially resolved electron backscatter diffraction (EBSD) measurements, the characteristics of microstructural evolution accompanying the martensitic transformation (austenite-7M) and the intermartensitic transformation (7M-NM) were analyzed

2. Experimental

The off-stoichiometric Ni-Mn-Ga alloy with nominal composition of $Ni_{53}Mn_{22}Ga_{25}$ (at.%) was prepared by arc-melting of pure elements Ni (99.97 wt.%), Mn (99.9 wt.%) and Ga (99.99 wt.%). The as-cast ingot was sealed in a vacuum quartz tube and homogenized at 1173 K for 24h, followed by quenching in water. A part of the homogenized ingot was grinded into fine powder and annealed at 873 K for 5 h in a vacuum quartz tube. The powder X-ray diffraction (XRD) measurements were conducted on a PANalytical X'Pert diffractometer equipped with a heating and cooling stage, where the Cu Kα radiation was used. The martensitic transformation temperatures were measured by differential scanning calorimetry (DSC, TA Q100). The microstructural analyses were performed on a field emission gun scanning electron microscope (SEM, Jeol JSM 6500 F) with electron backscatter diffraction (EBSD) acquisition camera. The "beam control" mode was applied for auto-orientation mapping.

3. Results and discussion

According to the DSC measurements, the martensitic transformation start and finish temperatures of the present $Ni_{53}Mn_{22}Ga_{25}$ alloy were determined to be 286.5 K (M_s), 277.9 K (M_f), 289.3 K (A_s), and 303.1 K (A_f), respectively. Some minor peaks were also detected at lower temperatures in the DSC curves, suggesting the occurrence of an intermartensitic transformation. Powder XRD measurements at 298 K, 243 K, and 153 K confirmed that three different phases, i.e. austenite (a_A = 5.811 Å), 7M martensite (a_{7M} = 4.222 Å, b_{7M} = 5.537 Å, c_{7M} = 41.982 Å, and β = 92.5°) and NM martensite (a_{NM} = b_{NM} = 3.879 Å, and c_{NM} = 6.511 Å), may exist at certain temperatures, as displayed in Fig. 1. This clearly demonstrates the existence of a two-step structural transformation with the transformation sequence of austenite → 7M → NM on cooling.

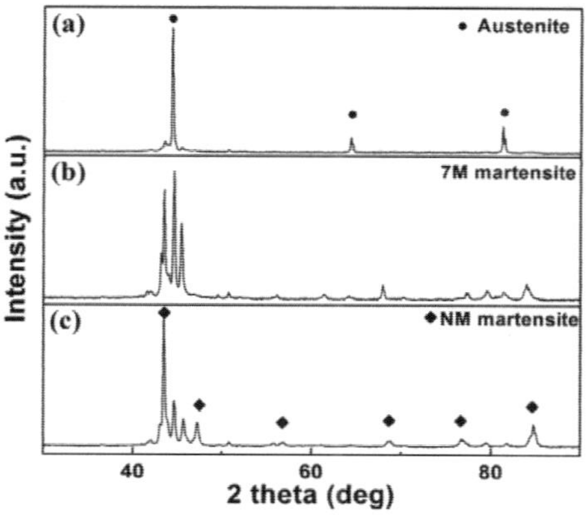

Fig. 1. Powder XRD patterns measured on cooling at (a) 298 K, (b) 243 K, and (c) 153 K.

Detailed microstructural observations revealed that the austenite co-existed with the 7M martensite at room temperature. Fig. 2a presents a typical backscatter electron (BSE) image of the co-existing austenite and 7M martensite in an original austenite grain. Notably, the initially formed 7M martensite plates had an overall diamond-like morphology, and they gradually outstretched into the entire austenite grain. Further EBSD measurements manifested that four distinct martensitic variants (A, B, C, and D) were placed back-to-back in one diamond-like martensite group, as shown in Fig. 2b. Each martensitic variant has a straight interface connecting the parent austenite.

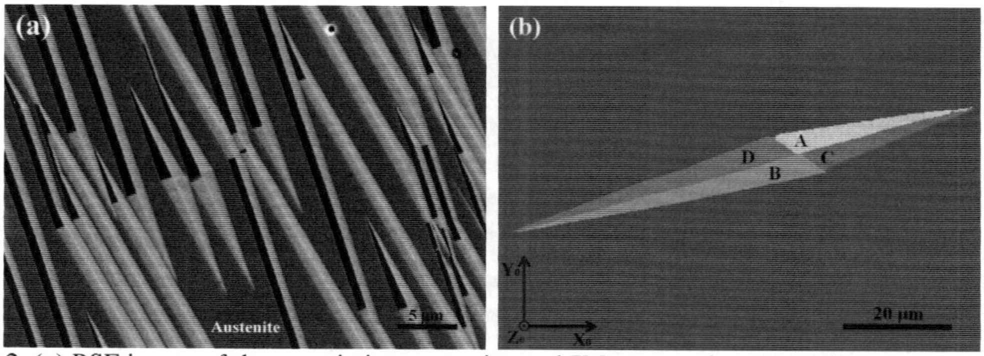

Fig. 2. (a) BSE image of the co-existing austenite and 7M martensite. (b) EBSD micrograph of the diamond-like martensite group having four variants (A, B, C, and D).

Misorientation calculations [7] showed that the four distinct martensitic variants could be twin-related one another, and the possible inter-variant twin relationships can be categorized into three types, i.e. A:C (or B:D) - type I twin, A:B (or C:D) - type II twin, and A:D (or B:C) - compound twin. Actually, only the type I and compound twin relationships exist in the diamond-like martensite group. More specifically, the long ridge of the martensite diamond represents a type I twin interface and the short ridge a compound twin interface. Using the experimentally measured orientation data, the energetically favorable orientation relationship between the austenite and the 7M martensite was derived to be the Pitsch relation, i.e. $(1\ 0\ 1)_A//(1\ \text{-}2\ \text{-}10)_{7M}$ and $[1\ 0\ \text{-}1]_A//[\text{-}10\ \text{-}10\ 1]_{7M}$, which is consistent with our previous result determined on a $Ni_{50}Mn_{30}Ga_{20}$ alloy [7].

Through the continuous nucleation and growth, the number and volume of 7M martensite plates increased at the expense of the parent austenite during the martensitic transformation. This transformation process may lead to the formation of different variant groups in the same austenite grain to reduce the transformation strains, as shown in Fig. 3.

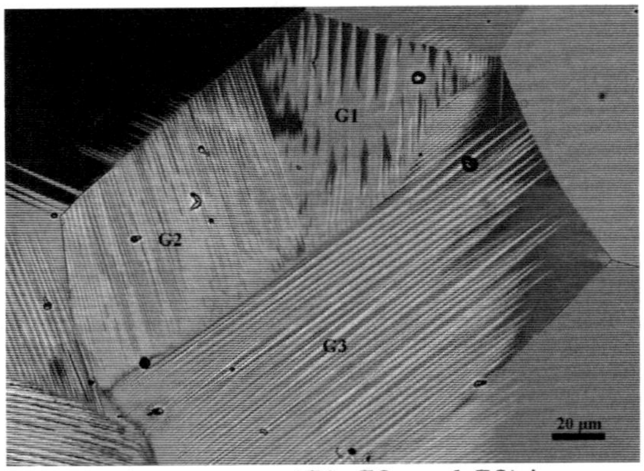

Fig. 3. Formation of different variant groups (G1, G2, and G3) in one austenite grain.

Moreover, the 7M martensite may further transform into the NM martensite in some initial austenite grains. Fig. 4a shows a selected EBSD phase-indexed map covering the co-existing 7M and NM martensite in one initial austenite grain, where the 7M martensite is colored in green and the NM martensite in gray. From Fig. 4a, it is seen that the two kinds of martensite are all in plate-shape and stretch longitudinally almost in the same direction. In general, the plate thickness of the 7M martensite is smaller than that of the NM martensite [9, 10]. When approaching the NM martensite, the 7M martensite has a tendency to thicken. Thus, the 7M to NM intermartensitic transformation is accompanied by a reduction of the specific area of martensite plate interfaces. This tendency implies that the plate interfacial energy of the NM martensite is higher than that of the 7M martensite [9]. By thickening the plates, the total interfacial area could be reduced and thus the interfacial energy is lowered.

Fig. 4b shows the EBSD orientation imaging map corresponding to Fig. 4a, where the martensite plates are colored according to their crystallographic orientations. For the 7M martensite, there exist four distinct orientation variants (A, B, C, and D) constituting one variant group. As for the NM martensite, there appear four differently oriented plates (P1, P2, P3, and P4). Each NM plate is composed of two fine twin-related lamellae alternately distributed [4], with the $\{1\ 1\ 2\}_{NM}$ plane as the twinning plane and the $<1\ 1\ \text{-}1>_{NM}$ direction as the twinning direction. One fine lamella corresponds to one orientation variant of the NM martensite and thus, one NM plate consists of two alternately distributed orientation variants. Further analyses showed that there exist specific orientation relationships between the 7M martensite and the NM martensite, i.e. $(0\ 0\ 1)_{7M}//(1\ 1\ 2)_{NM}$ and $[1\ 0\ 0]_{7M}//[1\ 1\ \text{-}1]_{NM}$ as well as $(0\ 0\ 1)_{7M}//(1\ 1\ 2)_{NM}$ and $[\text{-}1\ 0\ 0]_{7M}//[1\ 1\ \text{-}1]_{NM}$ [4]. Through further atomic reshuffling and lattice distortion, one 7M variant (plate) would finally evolve into one NM plate with two alternately distributed fine twin variants.

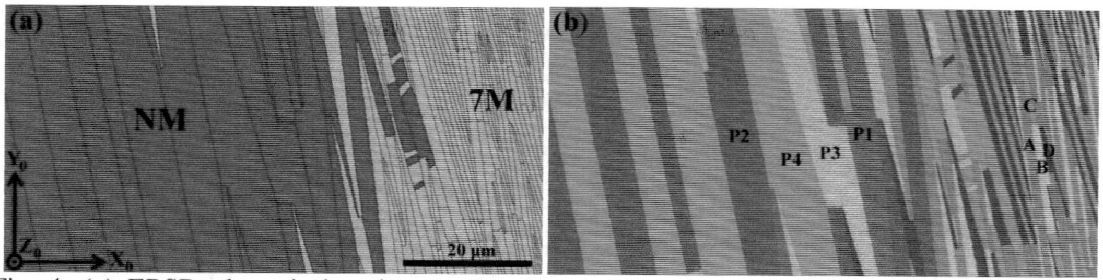

Fig. 4. (a) EBSD phase-indexed map showing the co-existence of 7M martensite and NM martensite in one original austenite grain. (b) Corresponding EBSD orientation imaging map. Four 7M variants are denoted as A, B, C, and D, and four NM plates as P1, P2, P3, and P4. The coordinate frame (X_0-Y_0-Z_0) refers to the macroscopic sample coordinate system.

4. Summary

The microstructure evolution accompanying the martensitic transformation (austenite-7M) and the intermartensitic transformation (7M-NM) was studied. The formation of the diamond-shaped 7M martensite with four twin-related variants in initial austenite grains was evidenced. Each diamond-like 7M martensite consists of type I twin (A:C or B:D pair) and compound twin (A:D or B:C pair), the long ridge of which corresponds to type I twin interface and the short ridge to compound twin interface. The 7M-NM transformation was found to be accompanied by the thickening of martensite plates, resulting in a reduction of the specific area of the NM plate interfaces. The orientation relationships between the two martensitic phases refer to $(0\ 0\ 1)_{7M}//(1\ 1\ 2)_{NM}$ and $[1\ 0\ 0]_{7M}//[1\ 1\ \text{-}1]_{NM}$ as well as $(0\ 0\ 1)_{7M}//(1\ 1\ 2)_{NM}$ and $[\text{-}1\ 0\ 0]_{7M}//[1\ 1\ \text{-}1]_{NM}$. One 7M variant (plate) finally evolves into one NM plate with two fine twin variants alternately distributed.

Acknowledgements

Financial supports from the National Natural Science Foundation of China (Grant No. 51431005), the 863 Program of China (Grant No. 2014AAQ00297), the 111 Program of China (Grant No. B07015), the Program for Liaoning Innovative Research Team in University (Grant No. LT2013007), the PhD Start-up Fund of Liaoning Province (Grant No. 20141001), the Fundamental Research Funds for the Central Universities of China (Grant No. N130110001), and the French State through the Investment in the Future Program operated by the National Research Agency and referenced by ANR-11-LABX-0008-01 (LabEx DAMAS), are highly appreciated.

References

[1] Dunand D C and Müllner P 2011 *Adv. Mater.* **23** 216
[2] Sozinov A, Likhachev A A, Lanska N and Ullakko K 2002 *Appl. Phys. Lett.* **80** 1746
[3] Pons J, Chernenko V A, Santamarta R and Cesari E 2000 *Acta Mater.* **48** 3027
[4] Li Z B, Yang B, Zhang Y D, Esling C, Zou N F, Zhao X and Zuo L 2014 *Acta Mater.* **74** 9
[5] Seguí C, Chernenko V A, Pons J, Cesari E, Khovailo V and Takagi T 2015 *Acta Mater.* **53** 111
[6] Wang W H, Liu Z H, Zhang J, Chen J L, Wu G H, Zhan W S, Chin T S, Wen G H and Zhang X X 2002 *Phys. Rev. B* **66** 052411
[7] Li Z B, Zhang Y D, Esling C, Zhao X and Zuo L 2011 *Acta Mater.* **59** 3390
[8] Li Z B, Zhang Y D, Esling C, Zhao X and Zuo L 2011 *Acta Mater.* **59** 2762
[9] Li Z B, Zhang Y D, Sánchez-Valdés C F, Sánchez Llamazares J L, Zhao X and Zuo L 2014 *Appl. Phys. Lett.* **104** 044101
[10] Li Z B, Zhang Y D, Esling C, Zhao X and Zuo L 2012 *Acta Mater.* **60** 6982

IOP Conf. Series: Materials Science and Engineering **82** (2015) 012088 doi:10.1088/1757-899X/82/1/012088

Texture evolution during hot deformation of Moly-TZM

A Chaudhuri[1], AN Behera[2], R Kapoor[2], A Sarkar[2], JKChakravartty[2], S Suwas[1]
[1]Materials Engineering, Indian Institute of Science, Bangalore 560012, India
[2]Materials Group, Bhabha Atomic Research Centre, Mumbai 400 085, India
E-mail: atanu@platinum.materials.iisc.ernet.in

Abstract. Moly-TZM was deformed at constant strain rate of 1.0 s^{-1} to investigate the high strain rate deformation behaviour by microstructural and stress response change within a temperature range of 1400-1700 ^{0}C. To correlate the deformation behaviour with orientational change, recrystallization and recovery of the material, the microstructural investigation was undertaken using scanning electron microscopy (SEM) of electron back scattered diffraction (EBSD). Depending on the grain size and orientation spread recrystallized grains were identified and texture was calculated. Change in grain boundary characteristics with increasing temperature was determined by the misorientation angle distribution for the deformed and recrystallized grains. Subgrain coalescence and increase in subgrain size with increasing temperature was observed, indicating recrystallization not only occurred from the nucleation of the dislocation free grains in grain boundaries but also from the subgrain rotation and merging of the subgrains by annihilation of the low angle grain boundaries. Detailed studies on the evolution of texture of recrystallized grains showed continuous increase in <001> fiber texture in recrystallised grains, in contrast to a mixed fiber <001> +<111> for the deformed grains.

1. Introduction

Moly-TZM is well known for its superior strength and high temperature creep resistance [1-3]. Due to its low irradiation effect at high temperature range (>800°C), Mo-TZM is a potential material for its structural application in nuclear reactors [4]. The overall composition of the alloy in wt% is 0.5–0.8%Ti, 0.08–0.1%Zr, and 0.016–0.02%C, and balance Mo. The superior strength of this alloy is attributed to the precipitation hardening from the formation of carbide precipitates of titanium (Ti) and zirconium (Zr). However in an earlier study it was revealed that the strength does not come from the pinning of the dislocation, rather the strengthening occurs by the pinning of dislocation ensembles which results in formation of substructure [1,5]. This substructure (cells/subgrains) works as dislocation barriers and hence rendering a high yield strength of Mo-TZM.

Previous studies showed that the Mo-TZM has a low strain rate sensitivity upto 900°C [3,6]. It has been predicted that at 900°C, when the material is deformed at high strain rate, some localised recrystallisation occurs [2]. A similar study of the deformation behaviour and microstructural evolution in the temperature of the order of ≈0.5Tm of Mo-TZM is lacking. In the present investigation, Mo-TZM alloy has been deformed in the temperature range of 1400-1700 °C at a constant strain rate 1.0 s^{-1}. The evolution of deformation microstructure has been investigated and analysed to elucidate the mechanism of texture formation in this material.

2. Experimental details

Cylindrical samples of dimension of 15 mm height and 10 mm diameter were prepared using EDM machining. The specimens were then annealed at 1350°C for 6 hours in vacuum (~10^{-5} millibar pressure) to obtain equiaxed microstructure. However the microstructure is not fully recovered after annealing, as deformed grains are observable in figure 1. Compression testing of the specimens was done at temperature range of 1400 to 1700°C with 100°C temperature interval at a constant strain rate of 1.0 s^{-1}. The samples were deformed up to 0.7 true strains. To retain the deformed microstructure the specimens were first cooled with high speed argon jet. The cooling rate of specimens after deformation with argon jet is typically 60-70 °C s^{-1}. The control of sample temperature was through a thermocouple is spot welded to the sample at the centre. The temperature was controlled to within ±1°C of the deformation temperature (1400, 1500, 1600 and 1700°C). Resistance heating was used with heating rate of 5°C s^{-1}. Before deformation specimens were kept at experimental temperature for

IOP Conf. Series: Materials Science and Engineering **82** (2015) 012088 doi:10.1088/1757-899X/82/1/012088

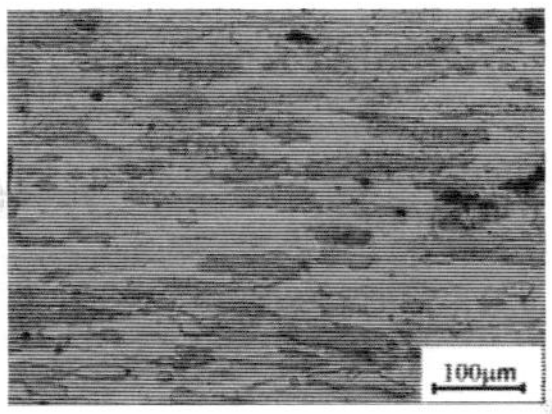

Fig.1: Optical microstructure of Moly-tzm before testing.

5 min for uniform heating. Samples were sectioned along the compression axes for metallographic characterisation. After conventional metallographic polishing, the specimens were further polished using 1μm and 0.3μm alumina slurry for mirror finishing. A solution consisting of nitric acid, hydrofluoric acid and distilled water in the ratio 2:2:5 was used for chemical etching in order to reveal optical and SEM structures. For orientation imaging, the samples were electropolished using a solution of 5% perchloric acid in methanol. Optical microstructures were taken using Carl-Zeiss microscope, while orientation imaging was done on a FEI-250 scanning electron microscope attached with TSL-OIM EBSD camera. Texture was calculated from the EBSD scans.

3. Results and discussion

Fig. 2 displays the EBSD generated microstructures represented by Inverse Pole Figure (IPF) map for the deformed materials. From the change in orientation of grains it is anticipated that apart from dynamic recovery of the grains during deformation some localised recrysallisation had occurred. In IPF image displayed in Fig 2(a-d), recovered microstructure with small dynamically recrystallised grains surrounding the deformed grains are clearly seen (subset of Figs. 2a and b). In higher temperature deformed materials, that is, the ones deformed at 1600 and 1700°C, occurrence of grain growth is very prominent, which has presumably occurred during holding the samples at deformation temperature before deformation for the uniformity of temperature throughout the specimens.

The grain sizes of the deformed and recrystallised portions of the microstructure are plotted in Fig 3(a) as a function of temperature. The recrystallised grains have been identified by the grain orientation spread (GOS) criterion. The grains with GOS<2°, have been identified as recrystallized grains. The plot indicates that the effect of temperature on the size of the recrystallised grains is minimal as the change in size of the dynamically recrystallised grains is very small. However, the size of the deformed grains increases with increasing temperature. Grain boundary character distribution with high angle boundaries (>15°), low angle boundaries (<15°) and coincide site lattice (CSL) boundaries are plotted in Fig. 2(e) after normalisation as a function of temperature. The high angle grain boundary fraction decreases for 1600 and 1700°C. The low angle grain boundary fractions in the 1600 and 1700°C deformed samples were higher than 1400 and 1500°C, possibly due to the higher fraction of subgrains within the deformed grains. The CSL boundary fraction is high for 1400 and 1500 °C. It is clear from Fig. 2 that within the temperature range of 1500- 1600 °C there is a transition in the restoration mechanism during the deformation. To observe the effect of dynamic recrystallisation on deformation, number fraction of the recrystallised grains obtained after partitioning (using grain orientation spread) are plotted in Fig. 3(b) as a function of temperature. The fraction of recrystallised grains decreases with increasing temperature.

From Fig. 3(a), it is observed that the fraction of the recrystallised grains is almost constant with temperature. However, prior grain growth is indicated for the deformed grains. Evidently at higher temperatures, grain growth is predominant in the as-received material, whereas at the lower temperatures 1400 and 1500 °C, dynamic recrystallisation is the dominating process. Possibly for the higher temperature of 1600 °C and 1700 °C the recrystallised grains were contributing to the average size growth of the pre-existing grains which were later identified as primarily deformed grains.

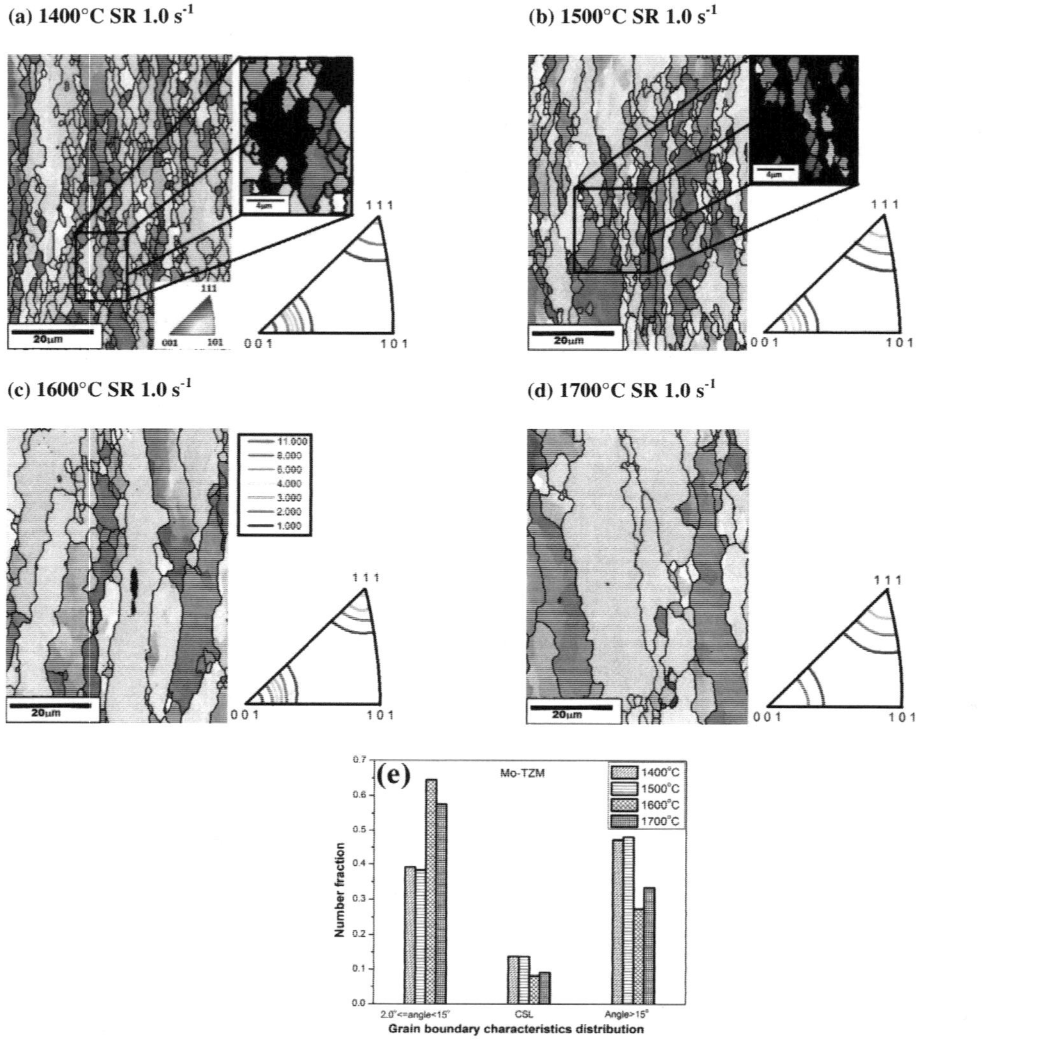

Fig. 2 (a-d): IPF maps of the Moly-TZM samples deformed at strain rate $1.0s^{-1}$ at different temperatures: (a) 1400 °C (b) 1500 °C (c) 1600 °C (d) 1700 °C. In the subset of each micrograph, at the right bottom corner, the texture is represented through inverse pole figures. Fig. 2(e): Grain boundary characteristics distribution(GBCD) for all the samples showing distinctive change in the grain boundary distribution between 1400°C, 1500°C and 1600°C, 1700°C. For Figs. (a) and (b), the magnified view of selected regions after partitioning the dynamically recrystallised grains using GOS<2°are also given at the top right corner.

Since the tests were done in the uniaxial loading condition, the textures have been depicted by inverse pole figures (IPF). The IPF are presented in the subset of Fig. 2(a-d). IPFs have been calculated on the compression plane. In all the cases, the pole density at <001> location is higher than at <111> location. Further, the pole density at <001> location is relatively higher for the samples deformed at 1400 and 1500°C. The change in ratio of pole density at <001> and <111> locations with temperature is plotted in Fig. 3(c). The ratio <001>/<111> decreases with increasing temperature, due to increase in pole density at <111>. The increase in pole density at <111> suggests the occurrence of pencil glide, where <111> is the principle direction of slip consisting of number of planes. The IPFs exclusively calculated from the recrystallised grains (partitioned using the criterion GOS<2°) are shown in Fig.4. It can be seen that most of the recrystallised grains have <001> axis parallel to the compression axis. Further, from the values of pole density, it can be observed that the texture weakens with increase in the deformation temperature.

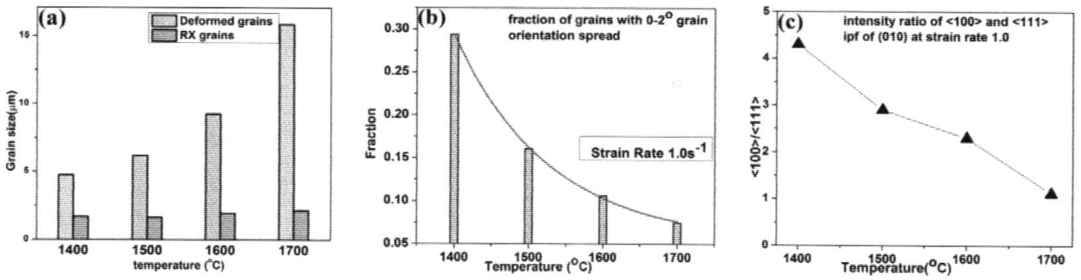

Fig.3(a): Grain size distribution for deformed and recrystallised grains showing the effect of temperature; (b): Variation of number fraction of the recrystallised grains with temperature; and (c) change in the intensity strength ratio of <100> and <111> component with increasing temperature.

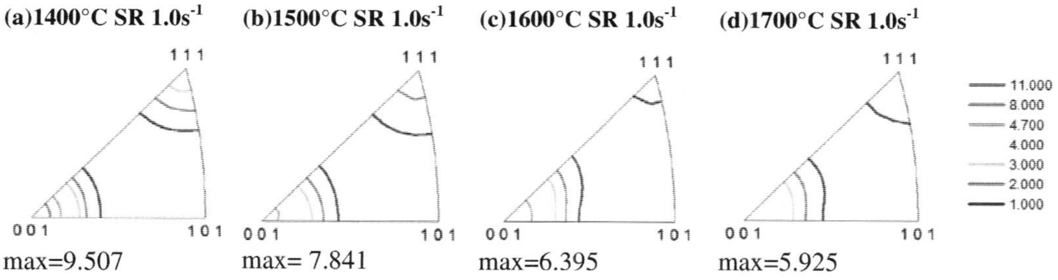

Fig.4 Inverse pole figures calculated from the recrystallised part of the samples deformed at strain rate of 1.0 s^{-1} at temperatures: (a) 1400 °C, (b)1500 °C, (c)1600 °C, and (d)1700 °C .

4. Conclusions

It is possible to carry out deformation Molybdenum–TZM alloy in the temperature range 1400-1700°C at the strain rate 1 s^{-1} without developing any instability. Microstructural analyses reveal that the dynamic restoration mechanisms (dynamic recrystallisation and recovery) are most prominent for the temperatures 1400 and 1500 °C, whereas for the temperatures 1600 and 1700°C grain growth controls the dissipation of stored energy. Texture analyses indicate that deformation mostly occurs by the activation of pencil glide with <111> direction as principle slip direction.

Acknowledgement

Authors acknowledge Board of Research in Nuclear Science (BRNS), Department of Atomic Energy (DAE), Govt. of India for the grant-in-aid for this work.

References

[1] Mrotzek T, Martin U and Hoffmann A 2010 *J. Phys. Conf. Ser.* **240** 1
[2] Calderon H, Kostorz G and Ullrich G 1993 *Mater. Sci. Eng. A*.**160** 189
[3] Majumdar S, Kapoor R, Raveendra S, Sinha H, Samajdar I, Bhargava P, Chakravartty JK, Sharma IG and Suri AK 2009 *J. Nucl. Mater.* **385** 545
[4] Cockeram BV, Smith RW, Byun TS and Snead LL 2009 *J. Nucl. Mater.* **393** 12
[5] Mrotzek T, Hoffmann A and Martin U 2006 *Int. J. Refract. Met.* H. **24** 298
[6] Filacchioni G, Casagrande E, De Angelis U, De Santis G and Ferrara D 2002 *J. Nucl. Mater.* **307–311** 705.
[7] Sarkar A,Kapoor R,Verma A,Chakravartty JK, and Suri AK 2012 *J. Nucl. Mater.* **422** 1

Hardening behavior and texture evolution of TWIP steel during strain path change

W Wen[1], M Borodachenkova[1], A Pereira[1], F Barlat[1,2] and J Gracio[1]
[1] Center of Mechanical Technology and Automation, Department of Mechanical Engineering, University of Aveiro, Campo Universitário de Santiago, 3810-193 Aveiro, Portugal
[2] Materials Mechanics Laboratory, Graduate Institute of Ferrous Technology, Pohang University of Science and Technology, San31, Hyoja-Dong, Nam-Ku, Pohang 790-784, Korea Republic
E-mail: mborodachenkova@ua.pt

Abstract. Polycrystal materials exhibit large changes in the flow stress and hardening behavior during the strain path change. Such changes are related with the crystallographic texture anisotropy and the rearrangement of dislocation structure during the pre-loading. These effects have been captured by a dislocation hardening model embedded in the visco-plastic self-consistent (VPSC) model. In this work, the texture evolution and mechanical behavior of TWIP steel during the strain path change are investigated. The experimental studies are carried out on rolled TWIP steel sheet. The mechanical responses are obtained under tensile tests along rolling direction, followed by tension along the directions with 0° and 90° from the pre-loading direction. The simulated results of strain-stress curves and the texture evolution are in good agreement with the experimental data.

1. Introduction

Twinning-induced plasticity (TWIP) steel is characterized by an excellent formability despite having an extremely high tensile strength values up to 1500MPa. The TWIP steels are typically austenitic steels, i.e. face-centered cubic (FCC) alloys, with a high manganese content in the range of 15-30 mass%. In low stacking fault energy (SFE) austenitic steel, it is well known that twinning deformation results in increasing of the strain-hardening by the creation of the twin boundaries that act as strong barriers for the dislocation glide [1,2].

A lot of studies have been carried out to explain the work hardening behavior of metals and alloys subjected to strain path change. During last years, many authors have developed the constitutive models to account the Baushinger effect, strain hardening stagnation and permanent softening after reverse loading. Recently, Beyerlein and Tomé [3] have proposed a hardening model based on the classic Voce law. This model accounts for the direction and magnitude of shear for each slip plane in each grain. This approach is able to predict the transient mechanic response due to the strain path change.

The purpose of this study is to adopt the model developed by Beyerlein and Tomé [3] to predict mechanical behavior and microstucture evolution of TWIP steel during the strain path change. A new formulation is added to account for the effect dislocation recombination. It is shown that this model captures well the mechanical response.

2. Experimental procedure

Large dog-bone samples are cut from a 1.8 mm thick sheet of FeMnAlC austenitic TWIP steel, with gauge length of 300 mm and width of 60 mm, and then subjected to uniaxial tensile test for a pre-strain of ε_p =9%. The pre-straining was carried out along the rolling direction (RD). Reloading was performed by uniaxial tensile tests done on tensile specimens cut from the pre-deformed material at 0° (RD) and 90° (TD) from the RD using as a guidance the striations left by rolling on the rolling plane. Pole figure measurements have been carried out on X-ray texture goniometer. From the processed data of the initial texture, a discrete set of 1000 grains have been weighted, extracted and used for the input of the VPSC model [3].

3. Modeling

During reloading of metals and alloys, the significant changes in the reloading yield stress and hardening evolution depending on the direction of reloading are observed [4]. Such a behavior can be explained by the texture anisotropy and the dislocaiton-based microstructure evolution in single crystals. In the present work, these mechanisms are taken into consideration and the critical resolved shear stress (CRSS) can be expressed as:

$$\tau_c^\alpha = \tau_h^\alpha + \Delta\tau_B^\alpha + \Delta\tau_{rec}^\alpha \tag{1}$$

where τ_h^α represents the strength due to the generation and patterning of dislocations. This term is calculated by the Voce law [5]. $\Delta\tau_{rec}^\alpha$ is related to the dislocation recombination. $\Delta\tau_B^\alpha$ represents the contribution of Bauschinger effect.

The dislocation structures generated during forward loading induce internal stresses, this stresses resists to the applied load and leads to immobilization of the dislocations during forward loading. During the strain path change a fraction of the dislocations (reversible dislocations) generated during pre-loading is reversed with a lower slip resistance. The reversibility of the dislocation under the applied loading leads to the yield stress drop after reloading, which is usually known as Bauschinger effect [6]. Beyerlein and Tome [3] propose the following expression for the evolution of $\Delta\tau_B^\alpha$ term with strain:

$$\Delta\tau_B^\alpha = \Delta\tau_{B,0}^\alpha \exp\left(-\frac{\upsilon_{new}^\alpha}{\upsilon_B}\right) \tag{2}$$

$\Delta\tau_{B,0}^\alpha$ represents the degree of the reversibility and the reversible dislocations created during pre-strain. We should note that the value of $\Delta\tau_{B,0}^\alpha$ is negative when the shear on the slip system is reversed. The parameter υ_B is related to how fast the backstress is nullified with the accumulated strain on slip plane α since revese (υ_{new}^α).

After reloading, the reversible dislocations generated during pre-loading will be recombined gradually. It leads to a decrease of the total dislocation density and thus a decrese of the CRSS. Notice that the dislocation recombination is also dependent on the reversibility and the previous accumulated reversible dislocations. Therefore, the $\Delta\tau_{B,0}^\alpha$ is considered in the expression of $\Delta\tau_{rec}^\alpha$:

$$\Delta\tau_{rec}^\alpha = \Delta\tau_{B,0}^\alpha\left(1-\exp\left(-\frac{\upsilon_{new}^\alpha}{\upsilon_{rec}}\right)\right) \tag{3}$$

where υ_{rec} is the "rate" of the dislocation recombination.

4. Results and discussion

The measured true strain-true stress curves are presented in Fig. 1a. The experimental monotonic true stress-true strain curve loaded until 40% along RD is also shown. The yield stress during reloading decreases as the angle with respect to the preload direction increases. Fig. 1b shows the measured initial texture which is a typical brass-type texture for rolled sheet of low stacking fault energy metal. The measured final texture of the monotonic tensile test is presented in Fig. 1c.

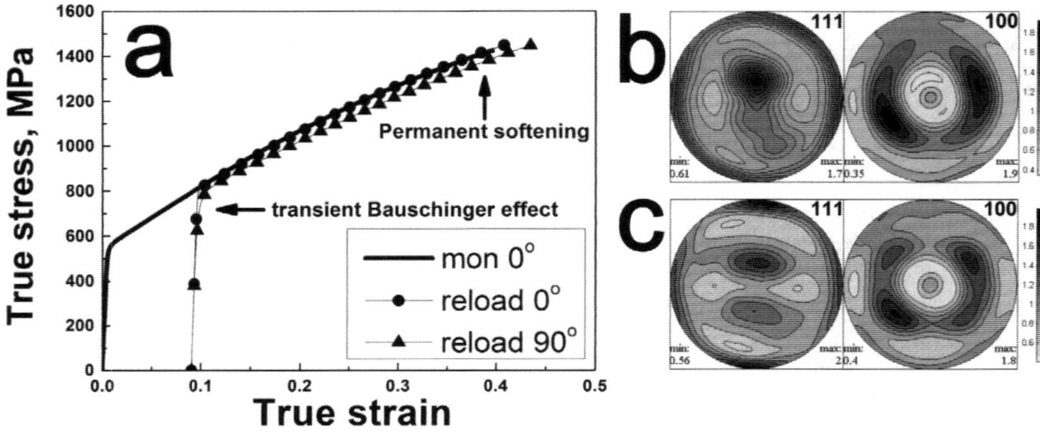

Fig.1 The experimental results (a) the measured true strain-true stress curves (b) measured initial texture (c) measured final texture after monotonic loading until 40%

Table 1. Hardening parameters

Slip				Twinning						
τ_0	τ_1	Θ_0	Θ_1	τ_0	τ_1	Θ_0	Θ_1	S^t	A^{th1}	A^{th2}
200	250	410	148	260	570	470	15	0.707	0.08	0.5

The simulation is carried out under the full strain-imposed boundary condition. The twinning reorientation is calculated by the Predominant Twin Reorientation (PTR) scheme [7]. The best results are obtained using the affine interaction. The VPSC model is not able to capture the elasto-plastic transition. In order to correctly evaluate the predicted curves, the elastic part of the experimental curves is removed. The VPSC framework associated with the new hardening law predicted successfully the deformation behavior. The values of the Voce hardening parameters (listed in Tab.1) are adjusted according to the experimental responses under the monotonic tensile test along the RD. Other hardening parameters are listed in Tab.2. The predicted true strain-true stress curves are shown in Fig. 2a and 2b. The predicted and experimental curves are in good agreement. This model captures well the effect of back-stress after reloading. The texture effect on strain path change is not evident in this work. The lower flow stress response for 90° reloading from RD is mainly contributed by the effect of dislocation recombination. The predicted texture evolution at different stages of deformation is presented in Fig. 3.

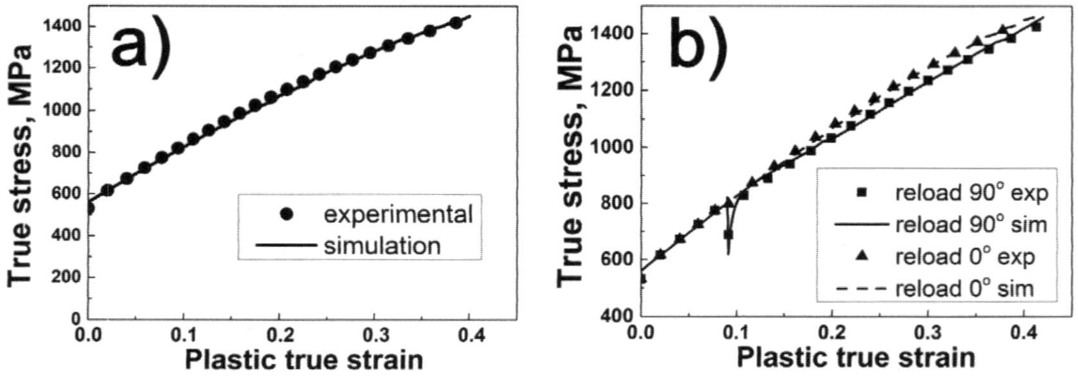

Fig.2. The simulation results for: true strain-true stress curve for monotonic tension (a) and for tension reloading at 0° and 90° from RD (b).

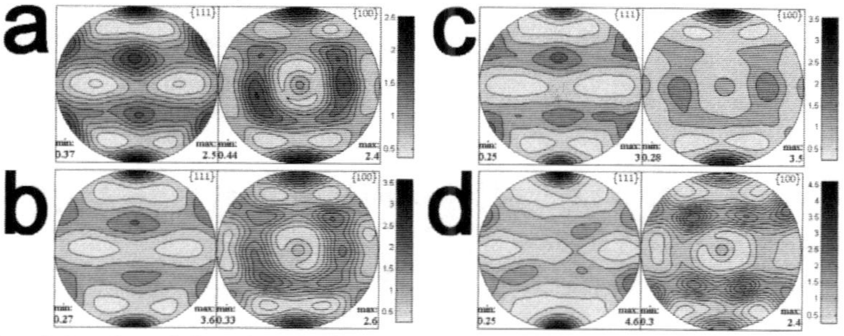

Fig.3. Predicted {111} and {100} pole figures for the texture after 9% pre-deformation (a), after monotonic loading (b), after tensile test reloading at 0° from RD (c), after tensile test reloading at 90° from RD (d).

Table 2. List of the material-dependent parameters for TWIP steel

	Value
μ (shear modulus)	60GPa
θ_B (rate of saturation for reverse shear)	15
$\upsilon_{B,sat}$ (maximum shear available for Bauschinger effect)	0.01
υ_B (related to how fast the backstress is nullified)	0.01
υ_{REC} (rate of the dislocation recombination)	0.5

5. Conclusions

In the present study, the hardening model, considering the various reversal mechanisms, is successfully applied to predict the mechanical response and texture evolution of TWIP steel subjected to two-step tension tests. The experimental results present a decrease of yielding stress during strain path change. This phenomena is well captured by presented model and contribution of the dislocation reverse-related mechanisms in the decreasing of yield stress is analyzed.

Acknowledgments

The authors acknowledge the financial support of Portuguese Foundation of Science and Technology projects PTDC/EME-TME/105688/2008, PTDC/EME-PME/116683/2010, PEST-C/EME/UI0481/2011 and PEST-C/EME/UI0481/2013.

References

[1] Karaman I, Sehitoglu H, Gall K, Chumlyakov Y. I, Maier H. J 2000 Deformation of single crystal Hadfield steel by twinning and slip *Acta Mater* **48** pp 1345–1359
[2] Grassel O, Kruger L, Frommeyer G, Meyer L.W 2000 High strength Fe±Mn±(Al, Si) TRIP/TWIP steels development -properties – application *Int J Plast* **16** pp 1391-1409
[3] Beyerlein I.J, Tomé C.N 2007 Modeling transients in the mechanical response of copper due to strain path changes *Int J Plast* **23** pp 640-664
[4] Rauch E.F, Gracio J.J, Barlat F 2007 Work-hardening model for polycrystalline metals under strain reversal at large strains *Acta Mater* **55** pp 2939–2948
[5] Voce E 1948 The relationship between stress and strain for homogeneous deformation *J Inst Metals* **74** pp 537-562
[6] Bauschinger J 1881 Ueber die Veränderung der Elasticitatagrenze und dea Elasticitatamoduls verschiadener Metalle *Zivilingenieur* **27** pp 289-348
[7] Tomé C.N., Lebensohn R.A., Kocks U.F 1991 A model for texture development dominated by deformation twinning: application to zirconium alloys *Acta Metal Mater* **39** pp 2667-2680

IOP Conf. Series: Materials Science and Engineering **82** (2015) 012090 doi:10.1088/1757-899X/82/1/012090

A *Matlab* toolbox to analyze slip transfer through grain boundaries

D Mercier[1], C Zambaldi[1] and T R Bieler[2]
[1] Max-Planck-Institut für Eisenforschung, 40237 Düsseldorf, Germany
[2] Michigan State University, East Lansing 48824 MI, USA
E-mail: d.mercier@mpie.de

Abstract. Slip transmission across grain boundaries is an essential micromechanical processes during deformation of polycrystalline materials. Slip transmission processes can be characterized based on the geometrical arrangement of active slip systems in adjacent grains and the value of the critical resolved shear stress acting on the incoming and possible outgoing slip systems. We present a Matlab toolbox which enables quantification of grain boundary slip transfer properties and comparison with experiments. Using a graphical user interface, experimental grain boundary data can be directly exported as input files for crystal plasticity finite element simulation of bicrystal experiments.

1. Introduction

The micromechanical behavior of grain boundaries (GBs) is one of the key components in understanding heterogeneous deformation of metals [1]. An increasingly important issue in constitutive modeling is to include the GB micromechanical effects in order to improve the prediction of mechanical microstructure-property relationships [2]. To investigate the nature of the strengthening effect of grain boundaries, slip transmission across interfaces has been investigated through bicrystal indentation tests performed close to grain boundaries [3-7]. To better understand the role played by the GBs and to quantify the GB micromechanics, a *Matlab* toolbox including graphical user interfaces (GUI) was developed [8]. This toolbox also links experimental results to crystal plasticity finite element (CPFE) simulations.

2. Strain transfer across grain boundaries

Luster and Morris [9] proposed a criterion to describe slip transfer across a GB based on the geometrical alignment of adjacent slip systems at the GB, see Fig. 1. To analyze the conditions for slip transfer, the crystallography of the GB and its geometry has to be known. At least five degrees of freedom are required to characterize a GB [10]: three for the rotation between the two crystals (Euler angles $(\varphi_1 \phi\, \varphi_2)$ for each crystal or misorientation axis/angle pair $([uvw]/\omega)$) and two for the orientation of the GB plane (either the trace angle and the GB inclination or the GB normal). The grain orientation and trace angle can be measured by electron backscatter diffraction (EBSD) orientation mapping and the inclination can be evaluated by serial polishing or focused ion beam sectioning. The parameter, m', and several other related parameters [1] such as the residual Burgers vector [11] and resolved shear stress based criteria are implemented in the toolbox.

3. Slip transmission analysis through the toolbox

A sample of Ti-5Al-2.5Sn (wt%) was prepared by mechanical grinding and polishing. The alpha/beta microstructure of the sample consists mainly of grains with a hexagonal structure [12]. It was characterized by EBSD and Fig. 2 displays the data in the *map GUI* component of the toolbox). Grain boundaries are color-coded to represent their maximum m' value taking into account basal and prismatic 1st order $\langle a \rangle$-slip systems, which are of importance for slip transmission in such alloys [12]. The visualization of the slip transmission parameter on EBSD maps makes the identification of interesting individual GBs convenient. A suitable bicrystal was identified, for which an intermediate potential for slip transmission was expected. The grain boundary #315 was selected and nanoindentation was performed close to

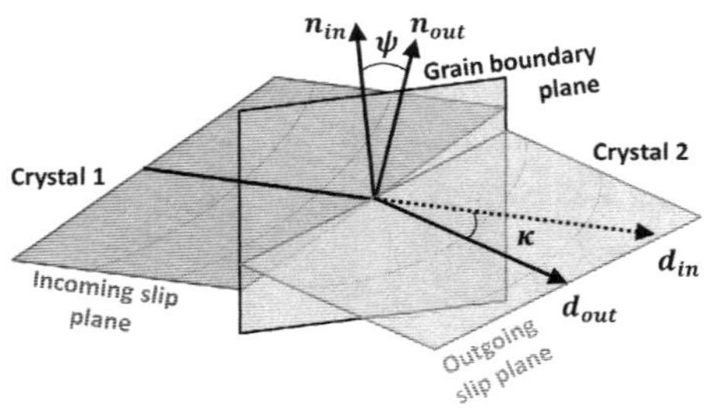

Fig. 1: Definition of m' [8], based on angle, κ, between incoming and outgoing slip directions, d_{in} and d_{out} , and angle, ψ, between slip plane normals, n_{in} and n_{out} .

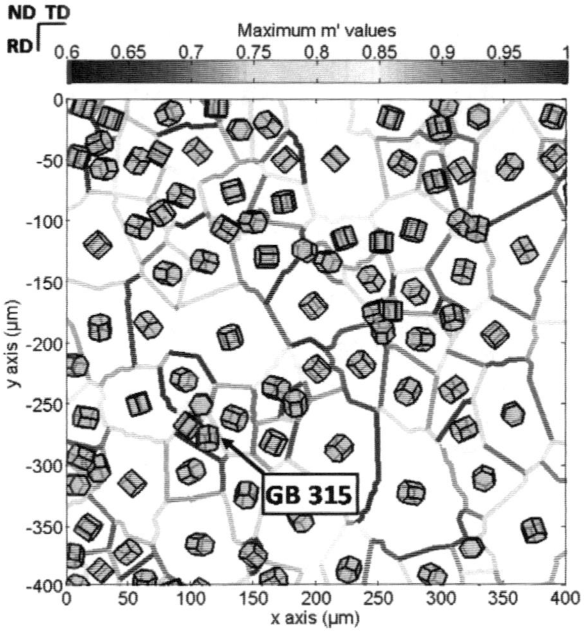

Fig. 2: EBSD orientation map of the Ti-alloy sample with grain boundaries color-coded by their maximum m' value (for basal and prismatic ⟨a⟩-slip systems).

the GB (see Fig. 3-a). A sphero-conical diamond tip with a nominal tip radius of 1 µm and a nominal cone angle of 90° was used and a maximum load of 6 mN was applied. The pile-up pattern developed around the indent is a function of the orientation of the crystal during axisymmetric indentation [13], and is in this case characteristic of basal and prismatic 1st order $\langle a \rangle$-slip activity [12]. The surface profile of the pile-up as measured by atomic force microscopy is given Fig. 3-b and the presence of a pile-up in crystal 2 probably indicates slip transmission. Analysis of slip transfer across this boundary was performed using the *bicrystal GUI* (see Fig. 3-c and 3-d). The result suggests that slip transmission occurs between the two crystals, with a prismatic 1st order $\langle a \rangle$ slip to basal slip transfer along the profile P1, with a maximum m' value of 0.77. To further analyze the spatial distribution of crystallographic shear, the *preCPFE GUI* component was used to enable correlation of the experimental results with finite elements simulations.

4. Stress state analysis using crystal plasticity finite element simulations

The *preCPFE GUI* has been created to enable rapid transfer of experimental data into simulation input files, so that a statistically significant number of indents can be assessed. A parameterized visualization of the bicrystal indentation model through the GUI allows tuning the geometry and finite element discretization and the size of the sample and the indenter, see Fig. 4-a. Once the 3D finite element model of the indentation process is set up, it is possible to generate a *Mentat procedure* file and a material configuration file using a *Python* routine [13, 14]. Theses input files are exported in order to carry out a fully automatic CPFE simulation of the bicrystal indentation using a flow rule based on Kalidindi's constitutive model available in *DAMASK* [15, 16]. Only prismatic 1st order $\langle a \rangle$, basal $\langle a \rangle$ and pyramidal 1st order $\langle c + a \rangle$ slip systems are considered and no inherent grain boundary resistance to slip is implemented. The constitutive parameters were identified from single grain indentation [13]. In this way the stress and strain fields close to the grain boundary can be rapidly quantified based on the precise experimental conditions, see Fig. 4-b and 4-c. Minor discrepancies between experimental and simulated pile-up patterns are observed. Transmission of prismatic $\langle a \rangle$-slip in crystal 1 to basal $\langle a \rangle$-slip in crystal 2 is observed, which confirms that strain transfer is mainly controlled by the geometric relationship of slip systems.

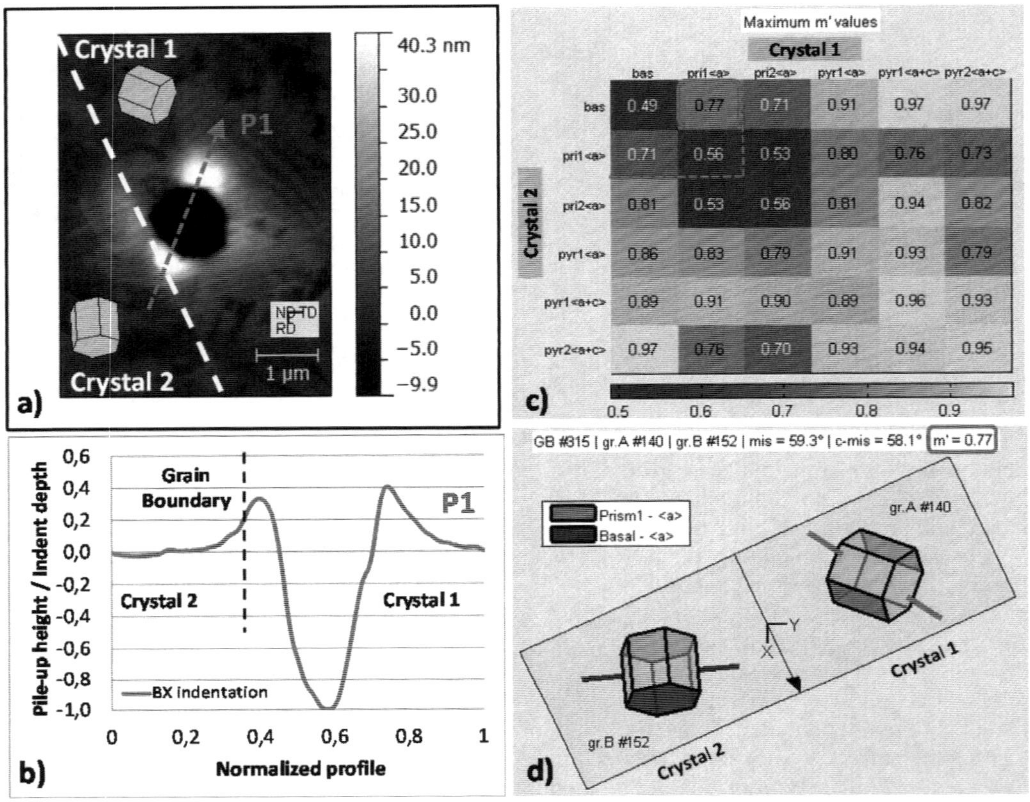

Fig. 3: a) Topography of a cono-spherical indent in Ti alloy, close to GB #315; b) line profile P1 of the pile-up; c) matrix of maximum m' values for bicrystal #315; d) Plot of the bicrystal #315 highlighting basal and prismatic $\langle a \rangle$-slip systems and corresponding surface slip traces of the systems associated with the maximum m' value.

5. Conclusion and outlook

A software toolbox and three GUIs (*map*, *bicrystal* and *preCPFE*) were developed to analyze strain transfer across grain boundaries. By a combination of spherical indentation close to GBs with EBSD mapping, grain boundary slip transmission was characterized in Ti-5Al-2.5Sn (wt%). Preliminary results suggest that slip transfer in α-titanium is mainly controlled by the geometrical alignment of adjacent slip systems at the boundary. The toolbox enables a reduction of possible sources of error in the analysis by visualization and a standardized workflow with automated data conversion. The slip systems for bcc, fcc and hcp structures and several slip transfer criterions are implemented to quantify the potential for slip transmission at GBs. The toolbox is interfaced with a *Python* package to generate CPFE simulation input files for single and bicrystal indentation testing, which is readily extendible to other experiments such as polycrystal tensile tests, micro-cantilever bending, and micropillar compression.

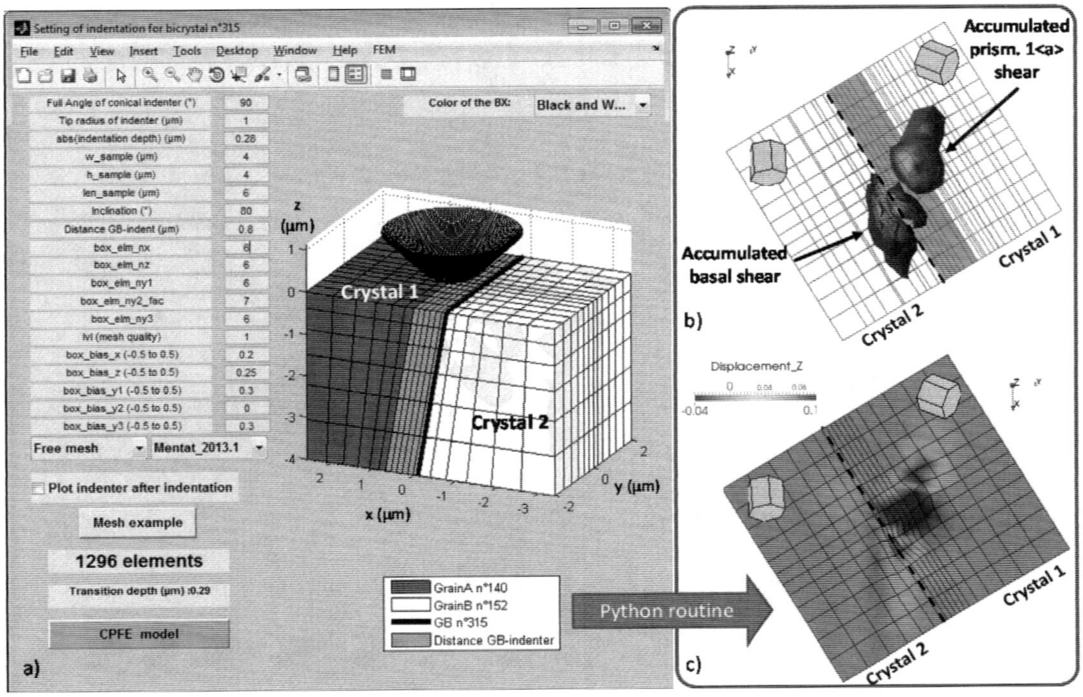

Fig. 4: a) View of the *preCPFE GUI* to optimize the mesh for bicrystal indentation experiment and generate simulation input files. b) Isosurfaces of basal shear in crystal 2 transmitted from prismatic slip activity around the indent in crystal 1. c) Topography of the simulated indent at GB #315.

Acknowledgements
Supported by the Materials World Network program (DFG ZA 523/3-1 and NSF-DMR-1108211). We acknowledge useful discussions with Y. Su, P. Eisenlohr and M. Crimp.

References

[1] Bieler T R et al 2014 *Curr. Opin. Solid State Mater. Sci.* 18(4) pp 212-226
[2] Ma A, Roters F and Raabe D 2006 *Acta Mater.* 54(8) pp 2181-2194
[3] Wo P C and Ngan A H W 2004 *J. Mater. Res.* 19(1) pp 189-201
[4] Soer W A and De Hosson J. Th. M. 2005 *Mater. Lett.* 59 pp 3192–3195
[5] Britton T B, Randman J and Wilkinson A J 2009 *J. Mater. Res.* 24(3) pp 607-615
[6] Pathak S, Michler J, Wasmer K and Kalidindi S R 2012 *J. Mater. Sci.* 47 pp 815–823
[7] Lawrence S K, Somerday B P, Moody N R, Bahr D F 2014 *JOM* 66(8) pp 1383-1389
[8] Mercier D, Zambaldi C, Bieler T R 2014 *STABiX*, http://github.com/czambaldi/stabix
[9] Luster L and Morris M A 1995 *Metall. Mater. Trans. A* 26(7) pp 1745-1756
[10] Priester L 2013 *Grain Boundaries* (Springer)
[11] Kacher J et al 2014 *Curr. Opin. Solid State Mater. Sci.* 18(4) pp 227-243
[12] Seal J R et al 2012 *Mater. Sci. Eng. A* 552 pp 61-68
[13] Zambaldi C, Yang Y, Bieler T R and Raabe D 2012 *J. Mater. Res.* 27(01) pp 356-367
[14] Zambaldi C 2014 *Python routines for CPFE preprocessing*, http://czam.de/software
[15] Roters F et al 2010 *Acta Mater.* 58(4) pp 1152-1211
[16] DAMASK—*Düsseldorf Advanced MAterial Simulation Kit*, http://damask.mpie.de

Influence of the angle between cold rolling direction and hot rolling direction on the texture evolution of non-oriented electrical steels

Y He[1], E Hilinski[2,3], M Attard[1], D Bibby[1], R Santos[1] and R Zavadil[1]
[1] CanmetMATERIALS, Natural Resources Canada, Hamilton, ON, Canada
[2] United States Steel Corporation Research and Technology Centre, Munhall, PA, USA
[3] Tempel Steel Co., Chicago, IL, USA
E-mail: yohe@nrcan.gc.ca; ehilinski@tempel.com

Abstract. In processing non-oriented electrical steel sheets using conventional rolling schemes, the most common texture components obtained after final annealing are the magnetically unfavourable <111>//ND (γ) and <110>//RD (α) fibres. A lot of researches have been carried out trying to optimize the processes to produce the favourable <001>//ND (θ) fibre. However, since the final texture is formed through a series of texture evolution steps during the solidification, hot rolling, cold rolling and annealing processes, it is quite challenging to tailor the texture of the final product. In this study, a new rolling scheme was examined, in which the cold rolling direction (CRD) was inclined to the hot rolling direction (HRD) at an angle from 0° to 90° (with a 15° increment). This was intended to alter the texture commonly produced by cold rolling along the HRD, and to optimize the final recrystallization texture. The cold rolling and recrystallization textures of two non-oriented electrical steels with 0.9% and 2.8% Si were measured. It was found that the inclination of CRD to HRD has a substantial effect on the cold rolling texture for both steels, but only in the low Si steel, does it lead to significantly different recrystallization textures. A strong cube texture was produced at an inclination angle of 60°, and the <111>//ND (γ) fibre was significantly weakened or essentially disappeared. The core losses of these steels were measured by Epstein frame method and the results showed a ~10% difference among strips cold rolled at different angles. A minimum core loss occurred at a 45° inclination angle in the low Si steel.

1. Introduction

Non-oriented electrical steels are widely employed in rotating machines and generators as the core lamination for copper windings, and the texture of the electrical steel sheets are critical to the efficiency of the electric motors or generators. The texture of the steel sheets not only depends on the chemical composition of the alloy, but it is also affected by the processing history of the material, during which the texture has gone through a series of evolution steps. In most cases, the obtained texture was dominated by the magnetically unfavourable γ fibre [1], and the ideal texture, {001}<100> (cube), is rarely produced except by some very special processing technologies such as cross rolling [2], columnar grain growth [3], phase transformation annealing [4], etc. These processes, however, are usually very difficult to be applied in industry for mass production.

In order to alter cold rolling texture and thus modify the recrystallization texture after final annealing, a new rolling scheme was tested in this study. After conventional hot rolling and hot band annealing, the steel was cold rolled at a direction different from the HRD, i.e. the cold rolling direction was inclined an angle (0° to 90°) to the HRD. In this way, the initial texture before cold rolling was intentionally changed while the microstructure remained the same since all the cold rolling samples were cut from the same hot band. Due to the change of the grain orientations, the orientation flow paths during deformation are changed [1], which would produce different cold rolling textures. After final annealing under the same conditions, various recrystallization textures would be expected. This paper investigates the influence of the inclination angle between the CRD and HRD on the texture and magnetic properties of the non-oriented electrical steels.

2. Materials and experimental procedures

The chemical compositions of the steels used in this study are listed in Table 1. The materials were processed according to the procedures shown in Figure 1a. After hot band annealing, the plates were cut to smaller samples (50 mm × 180 mm) at various angles (0°, 15°, 30°, 45°, 60°, 75°, 90°) with respect to the HRD (Figure 1b). The LowSi steel was rolled to a thickness of 0.65 mm (72% reduction) and the HighSi steel to 0.5 mm (78% reduction). After cold rolling, the steels were annealed at 800°C for 30 seconds (LowSi) or at 850°C for 12 minutes (HighSi) in argon to obtain fully recrystallized microstructures. The materials were then skin-pass rolled at room temperature with a ~2% reduction. The samples were finally annealed at 780°C for 2 hours in argon. Samples for optical microscopy were prepared using conventional metallographic procedures [5] and etched by a 2% Nital solution. Textures were measured by x-ray diffraction on the middle-thickness sheet plane. The Orientation Density Functions (ODF) were calculated from the measured incomplete pole figures, (200), (110) and (211), by using the MTEX (V3.4.2) software [6]. Magnetic properties were measured by a 200-turn Epstein frame device specifically designed for small samples and at high frequencies [7]. Testing strips (10 mm×105 mm) were cut along the rolling direction and annealed, and the testing was carried out at 400 Hz.

Table 1. Chemical compositions of the investigated steels (wt%).

Material	Si	C	Mn	Al	S	P	Fe
LowSi	0.88	0.002	0.31	0.46	0.001	0.01	Balance
HighSi	2.77	0.003	0.30	0.52	0.001	0.01	Balance

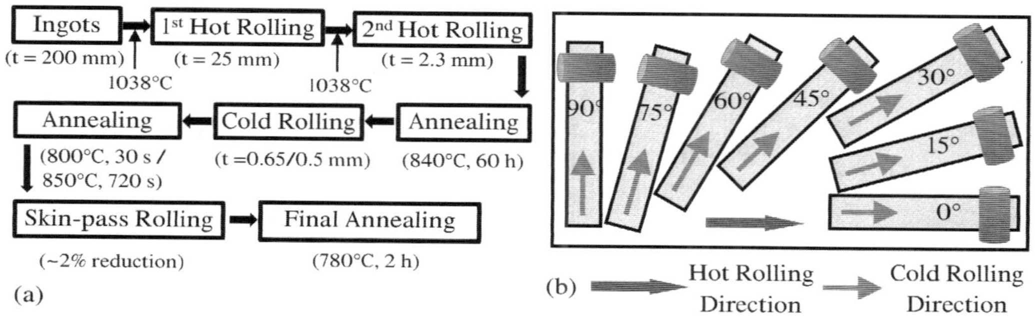

Figure 1 Steel processing procedures (a) and schematic of cold rolling samples cut from the hot band (b).

3. Results and discussion

Figure 2 shows the measured textures after cold rolling at various angles with respect to the HRD. The inclination of the CRD to HRD has a substantial effect on the cold rolling textures for both steels. For LowSi steel, the texture is generally intensified by the inclination, and at an angle of 75°, the intensity is more than 2× that of conventional rolling (0°). The continuous γ fibre observed at 0° degrades to discrete minor {111}<011> peaks when the inclination angle increases. The {001}<110> (rotated cube) component was significantly intensified at 75° and 90°. For HighSi steel, inclining the CRD to HRD only intensifies the texture at 15°, 30°, 60° and 90°. At 45° and 75°, the textures are slightly weakened. The strongest textures are observed at 15° and 90°, where a {223}<110> orientation and a {001}<110> component, respectively, dominate. Instead of having a continuous γ fibre, discrete peaks near {111}<011> (on the γ fibre) exist at all angles, and the intensities differ considerably.

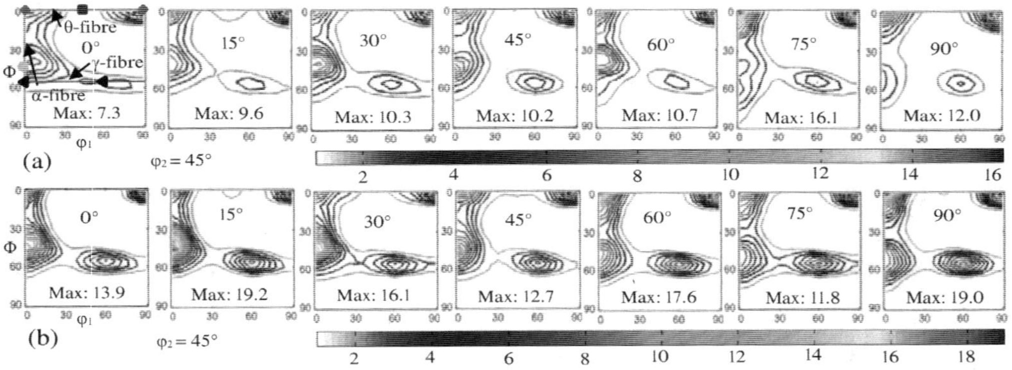

Figure 2 Textures (Bunge's notation) after cold rolling at various angles with respect to HRD: (a) LowSi steel, and (b) HighSi steel. ■{001}<100>, ◆{001}<110>, ●{223}<110>, ◀{111}<011>.

After final annealing, the textures are randomized and the intensities of the recrystallization textures drop to below 6.0 (Figure 3). For LowSi steel, the γ fibre (which was normally observed in recrystallized electrical steels) disappears. The textures are dominated by a <001>//ND fibre plus some near-Goss components (the intensities are rather low, though). At an inclination angle of 60°, a very strong cube component develops, which is much sharper than those developed at other angles. Nevertheless, this recrystallization texture does not correspond to the strongest cold rolling texture (obtained at 75°). Thus a strong cold rolling texture does not necessarily lead to a strong recrystallization texture. For HighSi steel, the recrystallization textures are very similar at all the inclination angles. The textures are dominated by a {111}<112> component (on the γ fibre) and a {001}<130> component (on the θ fibre). At all angles except 15°, the textures are weakened as compared to conventional rolling (0°), and the weakest texture is observed at an angle of 60°. This is contrary to the LowSi case, where the strongest texture was produced at 60°. There is no obvious development of a strong cube component at any of the inclination angles. It is seen that, although the inclination of CRD to HRD has a strong effect on the cold rolling textures of both steels, only on the LowSi steel, does it have a significant effect on the final recrystallization texture. The mechanism that governs the formation of a strong cube texture at an inclination angle of 60° in the LowSi steel is not known and needs to be further investigated.

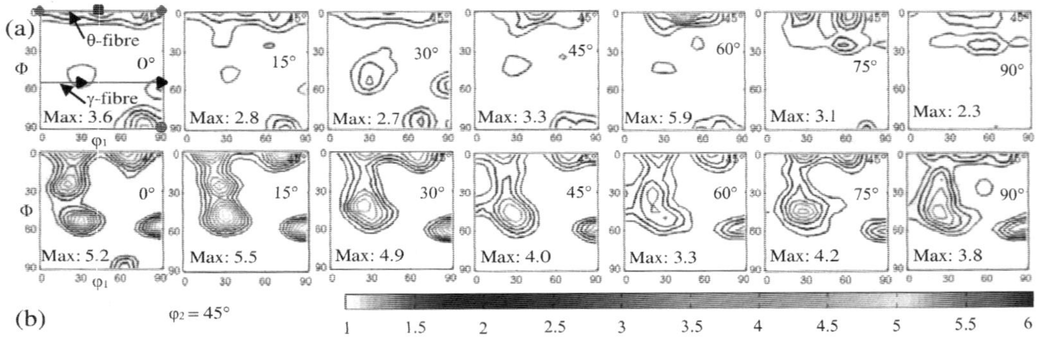

Figure 3 Textures (Bunge's notation) after final annealing at 780°C for 2 hours: (a) LowSi steel, and (b) HighSi steel. ■{001}<100>, ◆{001}<110>, ●{110}<001>, ▶{111}<112>.

Figure 4 shows the average grain sizes after final annealing as well as the core losses measured at 1.0 T, 400 Hz (RD only) for the two steels. It is seen that the inclination angle has an apparent effect on the grain size, especially for the HighSi steel. At an angle of 15°, the grain size reaches 195 μm, while at 90°, this is only 135 μm. For LowSi steel, the difference in grain size is marginal, with the largest grain size observed at 30° (150 μm) and the smallest at 0° and 45° (~130 μm).

Although it has been shown that at an inclination angle of 60°, a strong cube texture (intensity 5.9) was produced in the LowSi steel, the measured core loss (Figure 4b) at this angle is not the lowest. Instead, at a 45° inclination angle (which only has a moderate cube component with an intensity of 3.3), the core loss has the lowest value. It should be noted that, a {110}<001> texture (Goss) also develops at this angle, which contributes to the lowering of the core loss in the rolling direction. On the other hand, an examination of the microstructure showed that, at 60°, some extra-large grains >1 mm in diameter (data not shown) were produced, which may have some adverse effect on the core loss, since in low Si steels, the eddy current loss is a major part of the core loss, and the eddy current loss increases with grain size [8]. For HighSi steel, the core loss seems to correspond more closely to the grain size than to the texture, i.e. a smaller grain size seems to give rise to a lower core loss.

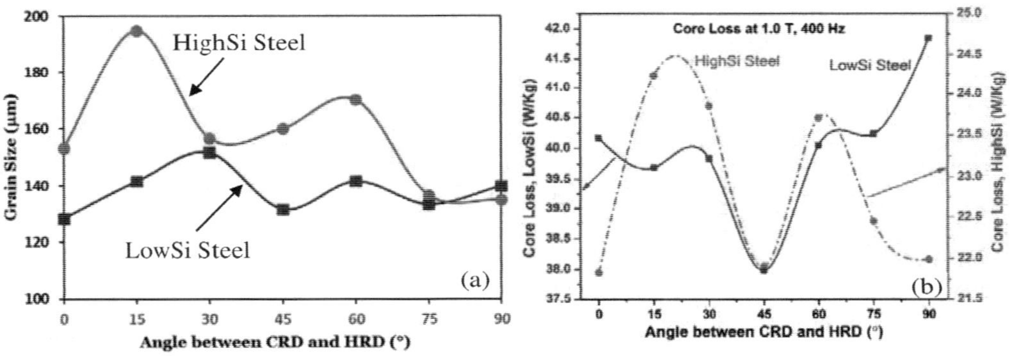

Figure 4 Average grain sizes (a) and core losses (b) of the two steels after final annealing.

4. Conclusions

By inclining the CRD to HRD, the cold rolling texture was substantially altered, which leads to significantly different recrystallization textures in the low Si steel. A strong cube texture develops when the angle is 60°. However, the lowest core loss is observed at an inclination angle of 45°, not 60°. For high Si steel, the effect of the angle between CRD and HRD on the recrystallization texture is not obvious, but it has a significant effect on the grain size.

References

[1] Kestens L and Jacobs S 2008 *Texture, Stress, and Microstructure* Article ID 173083
[2] Hayakawa Y and Kurosawa M 2002 *Acta Mater.* **50** 4527
[3] Tomida T, Uenoya S and Sano N 2003 *Mater. Trans.* **44** 1106
[4] Tomida T 1996 *J. Appl. Phys.* **79** 5443
[5] Vander Voort G 1999 *Metallography Principles and Practice* (Materials Park, OH: ASM)
[6] Hielscher R and Schaeben H 2008 *J. Appl. Crystal.* **41** 1024.
[7] Mthombeni T, Pillay P and Strnat R 2007 *IEEE Trans. Energy Conv.* **22** 614
[8] Matsumura K and Fukuda R 1984 *IEEE Trans. Magn.* **MAG-20** (5) 1533

IOP Conf. Series: Materials Science and Engineering **82** (2015) 012092 doi:10.1088/1757-899X/82/1/012092

Texture effect on mechanical properties anisotropy of products from Zr-based alloys

M Isaenkova, Yu Perlovich, O Krymskaya and S Pakhomov
National Research Nuclear University MEPhI (Moscow Engineering Physics Institute), Moscow, Russia, E-mail: isamarg@mail.ru, yuperl@mail.ru

Abstract. Technique of the macroscopic yield stress anisotropy calculation of Zr alloys based on texture *f*-parameters was validated in the frame of this work. In order to compare experimental and calculated anisotropy of products, samples cut out from rolled slabs of Zr-1%Nb alloy along three directions (rolling, transversal and normal) were subjected to uniaxial tension and compression up to different deformation degrees. Investigation of texture evolution under loading was implemented by measuring of several direct pole figures by means of X-ray diffractometric methods for samples in the initial state as well as after their deformation. Using obtained texture data three normalized yield stresses for orthogonal directions of studied samples were estimated and compared with experimental values measured by initial loading curves.

1. Introduction

Construction elements of the nuclear reactor core from Zr-based alloys are characterized with anisotropy of different mechanical properties. Degree of such anisotropy is mainly determined in the initial state of products and is based on static tensile or compressive tests along different directions, what requires labor-intensive sample preparation. At the same time various authors have developed mathematical algorithms, based on crystallographic texture data, which allow to calculate anisotropy of properties as well as modeling of its evolution under external loading [1-3]. Along with complicated models for calculation of the texture evolution there exists a simple technique for the express evaluation of the products anisotropy for materials having HCP crystalline structure based on well known integral Kearns *f*-parameters [4]. In this context, the aim of presented study is to validate such technique for products from Zr-based alloys.

2. Estimation of the yield stress anisotropy by the texture data

Simple method of the physical-mechanical properties averaging is known for materials having HCP crystalline structure. The method is based on integral Kearns *f*-parameters being the sum of volume fraction projection of basal normals [0001] on different directions of sample [4, 5]. In case of HCP materials with orthotropic sample symmetry the Mises yield criterion for the triaxial stress state, taking texture *f*-parameters into account, has the following form [4, 6]:

$$\sigma_e{}^2 = \frac{3}{2}\left[f_1(\sigma_2-\sigma_3)^2 + f_2(\sigma_3-\sigma_1)^2 + f_3(\sigma_1-\sigma_2)^2\right],$$

where σ_e – effective stress, 1, 2, 3 – orthotropic axes, i.e. RD, TD and ND for rolled products. The initiation of the plastic flow during uniaxial loading along one of the directions (i) occures under condition $\sigma_e{}^2 = \frac{3}{2}(1-f_i)(\sigma_{0.2}^i)^2$, where $\sigma_{0.2}^i$ – material's yield stress measured in the i-th direction in course of uniaxial test. Thereby for determination of the yield anisotropy one can reduce the general equation to the dimensionless form by normalization to the $\sigma_{0.2}^i$ in any direction. Thus construction of the yield locus for any type of stress state can be conducted by determination of *f*-parameters and the yield stress for one of product's directions.

3. Experimental results

Samples, cut out from rolled products of Zr-1%Nb alloys along different directions, were subjected to uniaxial compression and tension up to different deformation degrees in order to study texture evolution under simple loadings. Tension tests were conducted using plane samples, cut out from the sheet of 2 mm thickness along rolling (RD) and transversal directions (TD), with test portion length equal to 20 mm and cross section area equal to 20 mm². Samples were strained prior to the moment of the neck initiation and after that probes from the central part along the sample's length were cut for the texture investigation. Compression tests were conducted on cylindrical samples cut out from hot-rolled slab along RD, TD (∅10×15mm) and normal direction (ND) (∅8×12mm). After the deformation samples were cut parallel to the base of the cylinder at the mid-height for further studies. Test surfaces of all samples were subjected to grinding and fine polishing for removing of the work-hardened layer.

Investigation of the texture evolution under loading was carried out by means of X-ray diffractometric measurement of several incomplete pole figures (PF) for samples in the initial condition as well as after their deformation [5]. These PFs were used for ODF restoration followed by the complete PFs construction and calculation of *f*-parameters by the LaboTex program [7]. Using obtained *f*-parameters, normalized yield stresses for three orthogonal directions of samples were calculated. Anisotropy of these yield stresses was compared with experimental one measured by uniaxial loading curves.

Tension test

Obtained tensile loading curves are presented in fig. 1 together with PF (0001) and {11$\bar{2}$0} for initaial state as well as after tension along RD and TD. Moreover variation of normalized to $\sigma_{0.2}^{RD}$ yield locus of the sheet during its tension along different directions is presented in fig. 1-*b* while calculated anisotropy coefficient $R=\sigma_{0.2}^{TD}/\sigma_{0.2}^{RD}$ in table 1.

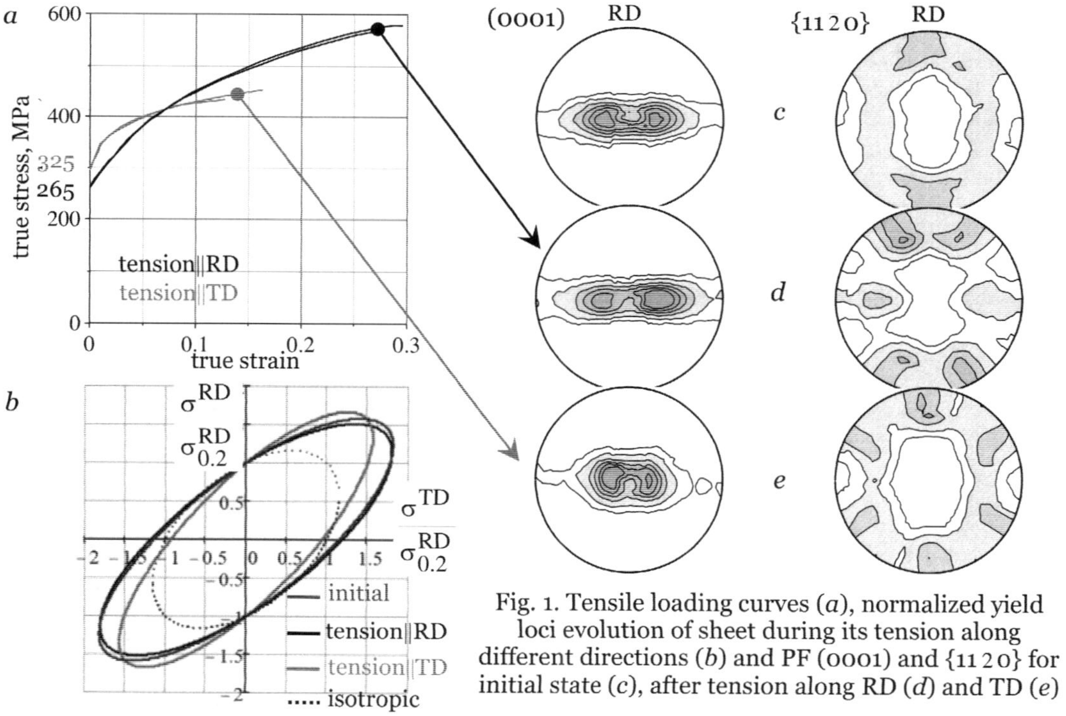

Fig. 1. Tensile loading curves (*a*), normalized yield loci evolution of sheet during its tension along different directions (*b*) and PF (0001) and {11$\bar{2}$0} for initial state (*c*), after tension along RD (*d*) and TD (*e*)

It's seen from fig. 1-*a* that material's yield stress for tension along TD is higher than for tension along RD while elongation, which characterizes plasticity, is lower. In case of uniaxial tension along RD basal normals position is stable what is seen on PF (0001) as well as by *f*-parameters values (table 1). Rotation of prismatic normals observed by PF {11$\bar{2}$0} for this

case (fig. 1-*d*) testifies that deformation takes place at the expense of the prismatic slip operation as confirmed by modeling in [8].
In case of uniaxial tension along TD the basal normals distribution changes (fig. 1-*e*), *f*-parameter for tensile direction (f_{TD}) decreases and f_{RD} increases at the expense of the twinning activization by planes {10$\bar{1}$2} in grains having the TD-orientation [9]. This fact explains the higher yield stress under tension along TD because critical resolved shear stress is higher for twinning than for prismatic slip. There is no prismatic normals rotation in this case and basal normals reorientate towards the ND due to the basal slip activization.

Table 1. Tension anisotropy of sheet from Zr-based alloy.

$R_{exp}=\sigma_{0.2}^{TD}/\sigma_{0.2}^{RD}=1.23$			
Sample	f_{RD}	f_{TD}	R_{calc}
initial state	0.08	0.29	1.15
tension || RD	0.07	0.28	1.19
tension || TD	0.11	0.22	0.94

Texture evolution under loading along RD results in the increasing of the calculated yield anisotropy coefficient R_{calc} in the sheet plane and under tension along TD anisotropy changes to the inverse value (table 1, fig. 1-*b*). On the basis of the obtained data one can see that, if strain direction during tensile test coincides with strain direction under preceding mechanical treatment, then product anisotropy intensifies and, if strain direction reverses, then degree of anisotropy decreases or changes to the inverse value.
Difference between the experimental and calculated anisotropy coefficient of the initial sheet takes place because the used calculation algorithm doesn't involve any information about various operating deformation mechanisms and difference of their critical resolved shear stresses. Despite this fact it quite applicable for express evaluation of anisotropy degree and its evolution under subsequent loading.

Compression test

Results for samples after compression tests are presented in figure 2. Due to the intermediate behavior of the deformation diagramm for compression along TD it will not be considered further in detail. Thus anisotropy coefficient $R=\sigma_{0.2}^{ND}/\sigma_{0.2}^{RD}$ is used in this case. Two stages of material's hardening can be found on compression curves for samples cut out along RD. This is connected with the drastic reorientation due to activization of twinnig by {10$\bar{1}$2} planes at the initial stage of the compression along to the direction, which was the tension direction under preceding mechanical treatment – RD. Basal slip having higher critical resolved shear stresses is involved during further deformation, what causes an increase of flow stress. Compression along the initial compressive direction under rolling (ND) doesn't have such features. This can be explained by lower critical resolved shear stress for twinning activated during compression along RD, than for basal slip activated during compression along ND. Thus the texture under compression along ND of slab is more stable and under compression along RD drastic reorientation of basal normals takes place.
Initail experimental anisotropy coefficient of the slab R=1.31. Calculated coefficients obtained by measured PF and their variation under uniaxial loading are presented in the table 2. On the basis of the obtained data one can see that, as in the case of uniaxial tension, coincidence of the strain direction during test with the strain direction under preceding mechanical treatment (RD is tensile direction and ND is compressive during rolling) results in anisotropy intensification and reversing of strain direction results in anisotropy decreasing or changing to the inverse value.

Table 2. Compression anisotropy of Zr-slab.

$R_{exp}=\sigma_{0.2}^{ND}/\sigma_{0.2}^{RD}=1.31$							
compression || ND				compression || RD			
true strain	*f*-parameters RD	*f*-parameters ND	R_{calc}	true strain	*f*-parameters RD	*f*-parameters ND	R_{calc}
0	0.16	0.48	1.27	0	0.14	0.38	1.29
0.11	0.15	0.54	1.36	0.11	0.28	0.27	1.14
0.18	0.18	0.57	1.39	0.18	0.31	0.29	1.07
0.38	0.14	0.65	1.58	0.50	0.49	0.21	0.86

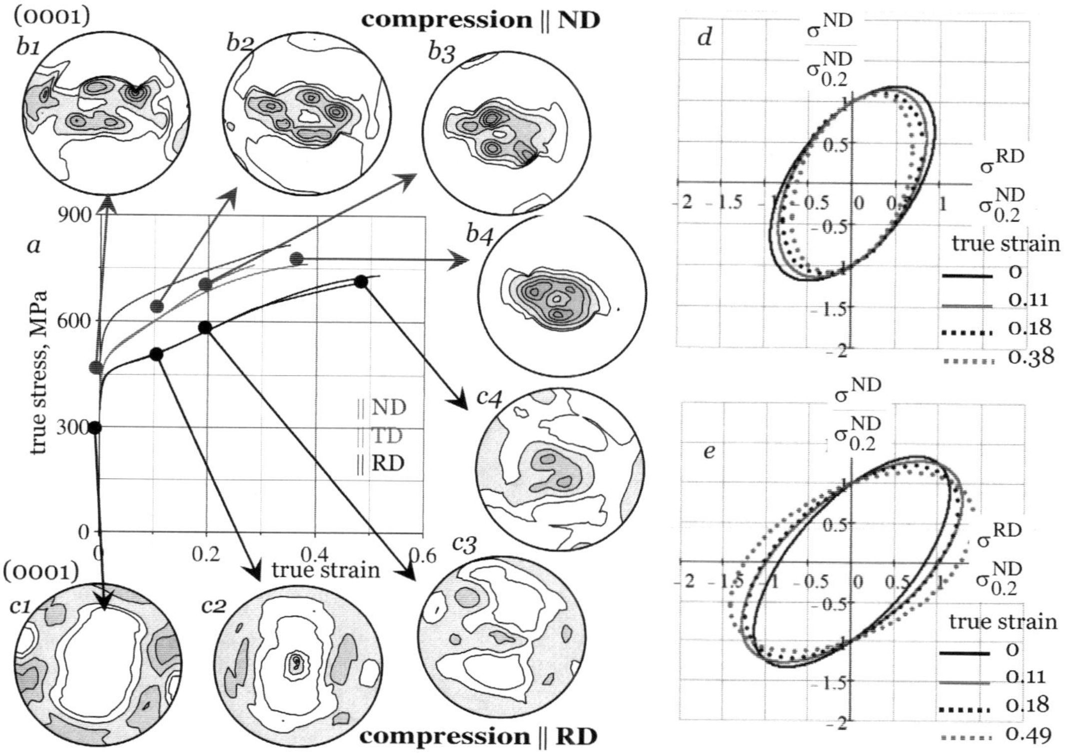

Fig. 2. Compression loading curves (*a*), PF (0001) for samples deformed along ND (*b1-b4*) and RD (*c1-c4*) up to different deformation degrees and yield loci evolution of slab after its compression along ND (*d*) and RD (*e*).

4. Conclusions

Interrelation between the crystallographic texture evolution of products from Zr-based alloys during uniaxial loading along different directions and their yield stress anisotropy is established.

Contributions of different plastic deformation mechanisms, activated during uniaxial tension and compression along different directions, are compared.

It was shown that technique of calculation mechanical properties anisotropy,based on integral texture parameters, allows to evaluate degree of anisotropy of products from Zr-based alloys.

References

[1] Bunge H J 1982 *Texture Analysis in Materials Science* (Butterworth, London)
[2] Kocks U F et al 1998 *Texture and anisotropy* (Cambridge, University Press)
[3] Wenk H R and Van Houtte P 2004 Texture and anisotropy *Rep. Prog. Phys* N**67** pp 1367–1428
[4] Tempest P A 1980 *J. Nucl. Mater.* vol **92** pp 191-200
[5] Perlovich Y, Isaenkova M, Fesenko V 2013 *Zavodsckaya laboratoriya* **79** № 7 (1) p 25
[6] Ablogin A L and Matsegorin I V 1988 *Industrial laboratory* vol **54** (1) p 27
[7] LaboTex v. 3.0 by LaboSoft (Krakow, Poland) http://www.labosoft.com.pl
[8] Xu F, Holt R A, Daymond M R 2009 *J. Nucl. Mater.* vol **394** pp 9-19
[9] Isaenkova M and Perlovich Y 1991 *Fizika metallov i metallovedenie* N **5** pp 87-92

Bulk rock elastic moduli at high pressures, derived from the mineral textures and from extrapolated laboratory data

K Ullemeyer[1], T Lokajíček[2], R Keppler[1], R N Vasin[3] and J H Behrmann[4]

[1] Institute of Geosciences, University of Kiel, Kiel, Germany

[2] Institute of Geology AS CR, Prague, Czech Republic

[3] Frank Laboratory of Neutron Physics, Joint Institute of Nuclear Research, Dubna, Russia

[4] GEOMAR Helmholtz Centre for Ocean Research, Kiel, Germany

E-mail: kullemeyer@geomar.de

Abstract. The elastic anisotropy of bulk rock depends on the mineral textures, the crack fabric and external parameters like, e.g., confining pressure. The texture-related contribution to elastic anisotropy can be predicted from the mineral textures, the largely sample-dependent contribution of the other parameters must be determined experimentally. Laboratory measurements of the elastic wave velocities are mostly limited to pressures of the intermediate crust. We describe a method, how the elastic wave velocity trends and, by this means, the elastic constants can be extrapolated to the pressure conditions of the lower crust. The extrapolated elastic constants are compared to the texture-derived ones. Pronounced elastic anisotropy is evident for phyllosilicate minerals, hence, the approach is demonstrated for two phyllosilicate-rich gneisses with approximately identical volume fractions of the phyllosilicates but different texture types.

1. Introduction

Many studies have shown that the 'extrinsic' crack effects on bulk rock elastic anisotropy are largely sample-dependent and must be determined experimentally, whereas the texture-related 'intrinsic' contribution to bulk rock elastic anisotropy can be predicted from the mineral textures [1]. Existing equipment for the measurement of elastic wave velocity trends as a function of confining pressure P_c is often limited to pressures of the intermediate crust (400 – 600 MPa) [1, 2]. In order to get information on the elastic behavior of rocks in the lower crust, extrapolation of the elastic wave velocities obtained at intermediate pressures to pressures of the lower crust (≈ 1000 MPa) is desirable.

Several model functions were proposed to describe velocity – pressure trends in rocks, for an overview refer to [3]. The relation recommended in [2] appears to be most suitable for the purpose of extrapolation, because it provides the 'intrinsic' velocity trend, which, at high pressures, is assumed to describe the overall velocity trend sufficiently well. It has been shown that the 'intrinsic' velocity trend still can be refined [3], leading to a better estimate of the extrapolated velocities. We used the relations suggested in [2] and [3] to evaluate refined 'intrinsic' velocity trends from cube sample measurements performed at Institute of Geosciences (Kiel) [1]. Complete *P*- wave velocity distributions were determined on spherical samples at Geological Institute AS CR (Prague) [2] and extrapolated using the concept described below. From combined cube and spherical sample data the elastic constants were calculated for pressures P_c = 200 MPa (based on experimental data) and 1000 MPa (based on extrapolated data) [4, 5]. The latter were compared to the texture-derived elastic constants.

Two gneisses (G1 and G2 in the following) bearing approx. 30% phyllosilicates were selected for demonstration of the approach. The main difference between the samples is the texture type of the phyllosilicates, this offers good opportunity to quantify the influence of texture variation of the highly anisotropic phyllosilicates [6, 7] on bulk rock elastic behavior.

2. Experimental details and data evaluation procedures

Initially, complete (132 sample directions) V_P distributions at $P_c = 200$ MPa were measured on spherical sample. From the velocity distributions the best fit orthogonal coordinate system (X, Y, Z) for cube sample preparation was determined by means of eigenvalue analyses. The cube sample data set comprises V_P, V_{S_1} and V_{S_2} measurements at 16 pressure stages (12 – 600 MPa) for these sample directions. For the common pressure of $P_c = 200$ MPa, the elastic constants were calculated from the V_P distributions and the six cube sample shear wave velocities [4]. The cube sample 'intrinsic' velocity trends were estimated using the relation described in [2] and refined according to the procedure explained in [3]. For extrapolation of the V_P distributions to the desired pressure of $P_c = 1000$ MPa we assumed that the 'intrinsic' pressure gradients derived for the cube sample apply to the spherical sample as well. By this means, the V_P distributions as a whole can be extrapolated, provided that the outlines of the V_P patterns and orientations of the principal directions X, Y and Z do not change significantly. At least, the elastic constants for $P_c = 1000$ MPa were calculated from the extrapolated data sets.

For the texture measurements the neutron diffraction was applied to ensure sufficient grain statistics, because large sample volumes had to be investigated. The measurements were performed at the SKAT texture diffractometer at the pulsed reactor IBR-2M in Dubna, Russia [8]. Recording TOF spectra offered the opportunity to evaluate the textures by means of 'Rietveld Texture Analysis' (RTA) [9, 10], which is the most promising method to extract the mineral textures from diffraction spectra with many peak overlaps. We used the MAUD software for the texture evaluation [11, 12]. Subsequently, the elastic constants of bulk rock were modeled from the textures applying the Voigt averaging scheme.

3. Results and discussion

In addition to the composition of the samples (G1: muscovite–17%, chlorite–13%, quartz–38%, albite–28%, microcline–3%, calcite–1%; G2: muscovite–15%, biotite–14%, chlorite–5%, quartz–47%, albite–19%), RTA revealed girdle-like pole density distributions for the muscovite and chlorite (001) pole figures of sample G1. Due to this characteristic, the texture strength is weak, as indicated by texture indices f2 of 1.28 and 1.11 and maximum pole densities of 2.01 and 1.36 [m.r.d], respectively (Fig. 1). In contrast, the (001) pole figures of muscovite, biotite and chlorite of sample G2 display axial symmetric pole density distributions and mostly stronger textures, with texture strengths f2 = 4.25/1.40/1.02 and pole density maxima of 12.4/4.09/1.25 [m.r.d.], respectively (Fig. 1).

The textures of the other rock constituents are not presented, because they are weak or particular volume fractions are minor. Nevertheless, all textures were included in the averaging procedure.

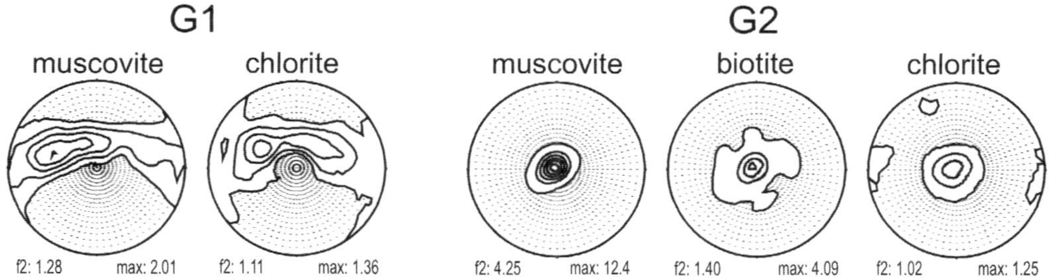

Figure 1. Phyllosilicate basal plane pole figures of samples G1 and G2. f2: texture index; max: maximum pole density [m.r.d.]. Dots in the pole figures indicate pole densities < 1.0 [m.r.d.].

The cube sample *P*-wave velocity trends for sample G1 suggest orthorhombic symmetry of the V_P distribution, this is confirmed by the spherical sample measurements at 200 MPa. The refined 'intrinsic' trends for directions Y and Z converge, leading to an axial symmetric V_P distribution at 1000 MPa. The V_P distributions (200/1000 MPa) of sample G2 are axial symmetric (Fig. 2A). Both samples show shear wave splitting for directions X and Y, which is more pronounced for sample G2 (Fig. 2B).

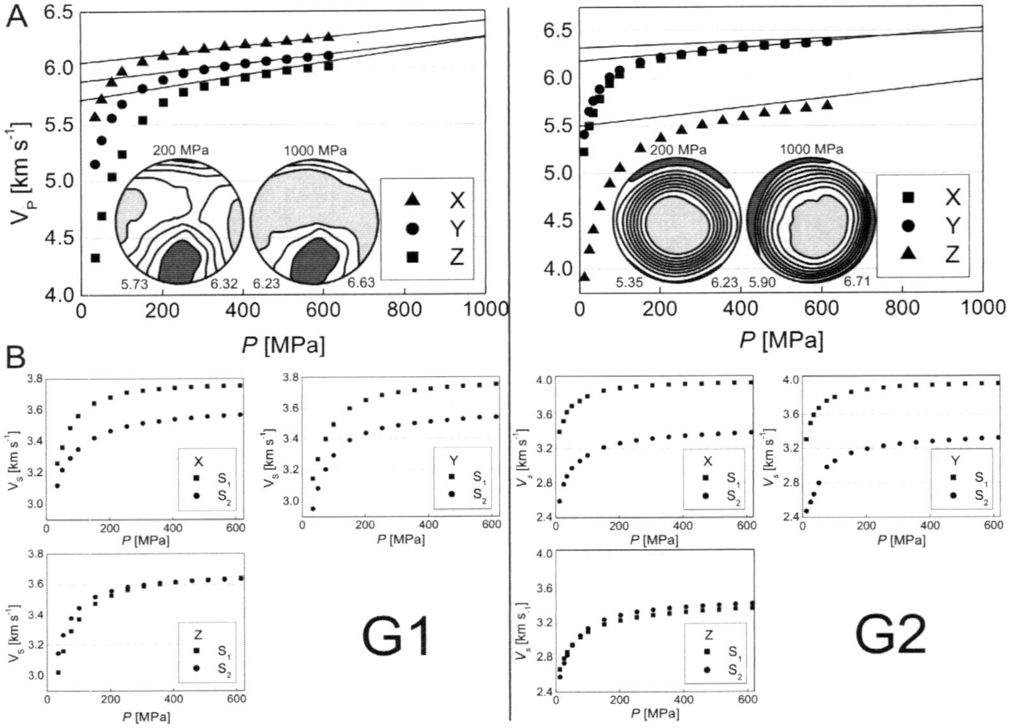

Figure 2. Compilation of velocity measurements. **A:** Cube sample *P*-wave trends. Straight lines: refined 'intrinsic' velocity trends. Inset diagrams: experimental (200 MPa) and extrapolated (1000 MPa) *P*- wave velocity distributions [km s^{-1}] from spherical sample measurements. Dark-shaded/light-shaded regions indicate regions of maximum/minimum velocity. **B:** shear wave splitting.

For both samples, recalculated *P*- and *S*-wave distributions from particular sets of elastic constants (Fig. 3) demonstrate the diminishing effect of the crack fabric. Anisotropies of the V_P distributions at P_c = 200/1000 MPa decrease only slightly with increasing pressure, which can be also inferred from consistent increase of the minimum and maximum velocities. The degree of anisotropy is larger for the well-textured sample G2. In contrast, a large drop of anisotropy is observed for the texture-derived V_P distributions, mainly due to the much larger difference of the minimum velocities. The maximum velocities of the '1000 MPa' and 'texture-derived' V_P distributions are comparable within the expected error ranges, *i.e.*, the crack influence vanishes for sample directions perpendicular to the phyllosilicate basal plane pole density maxima but persists for directions parallel to the maxima. Hence, there is some correlation between the phyllosilicate basal plane pole figures and the crack fabric. Due to these observations, we expect differences of the experimental and texture-derived *S*- wave distributions as well. The shapes of the 'G1' *S*- wave patterns differ largely, but the texture-derived velocity variations are in the order of some tenths of meters and, therefore, observed differences may be erroneous and should not be judged. In contrast, the 'G2' *S*- wave patterns are considered to be more reliable due to larger velocity variation of the texture-derived *S*- wave distributions. The general layout is axial symmetric, however, the texture-derived *S*- wave distributions show distinct maxima on the girdle-like velocity distributions. Such differences support the above assumption that some crack influence persists even at confining pressures of 1000 MPa.

IOP Conf. Series: Materials Science and Engineering **82** (2015) 012093 doi:10.1088/1757-899X/82/1/012093

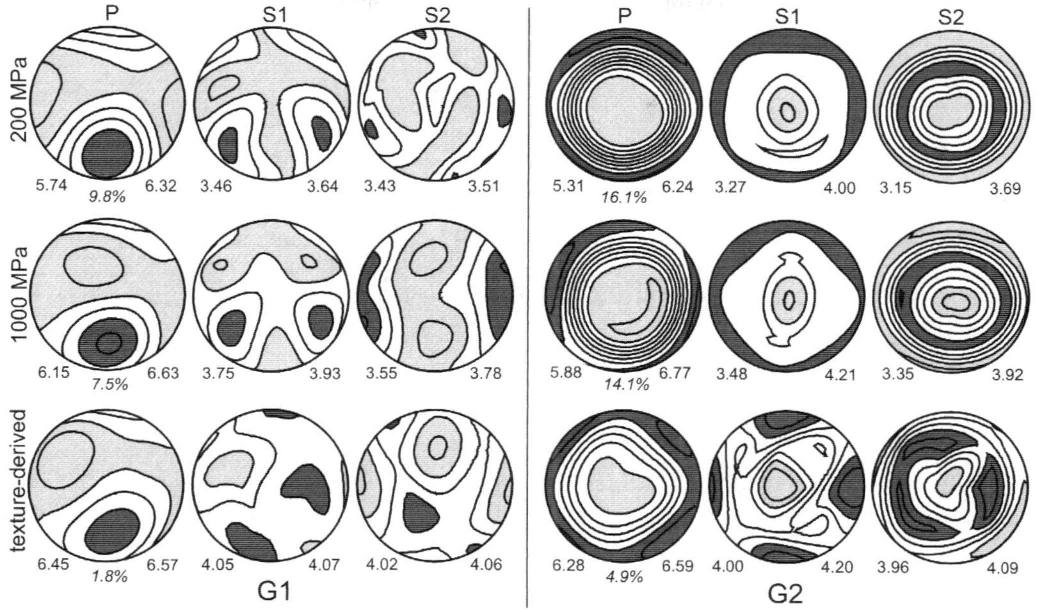

Fig. 3. Recalculated *P*- and *S*- wave distributions [km s^{-1}]. The minimum and maximum velocities are indicated, percentages refer to anisotropy $(Vp_{max}-Vp_{min})/Vp_{mean}$. Dark-shaded/light-shaded regions indicate regions of maximum/minimum velocity.

4. Conclusions

We conclude that elastic anisotropy of rocks containing large volume fractions of phyllosilicates is mainly controlled by the phyllosilicate textures. This holds true for strong axial symmetric textures and for weak girdle-like textures as well. The contribution of the other mineral phases is neglectible due to weak textures and/or small volume fractions. Differences between the recalculated '1000 MPa' and 'texture-derived' *P*- and *S*- wave distributions indicate persisting crack influence even in the lower crust. The correlation between the phyllosilicate basal plane pole figures and the crack fabric offers the opportunity to model the influence of the crack fabric from particular textures [12].

Acknowledgement Financial support of the German Federal Ministry of Education and Research through grants 05K09FKA and 05K13FKB is gratefully acknowledged.

References

[1] Kern H, Ivankina T I, Nikitin A N, Lokajíček T, and Pros Z 2008 *Tectonophysics* **457** 143
[2] Pros Z, Lokajíček T, and Klima K 1998 *Pure and Applied Geophysics* **151** 619
[3] Ullemeyer K, Nikolayev D I, Christensen N I, and Behrmann J H 2011 *Geophys. J. Int.* **185** 1312
[4] Klíma K 1973 *Studia geophys. geod.* **17** 115
[5] Jech J 1991 *Phys. Earth Planet. Inter.* **66** 153
[6] Vaughan M T, and Guggenheim S 1986 *J. Geophys. Res.* **91** 4657
[7] Militzer B, Wenk H R, Stackhouse S, and Stixrude L 2011 *Amer. Min.* **96** 125
[8] Keppler R, Ullemeyer K, Behrmann J H, and Stipp M 2014 *J. Appl. Cryst.* **47**, doi: 10.1107/S1600576714015830
[9] Matthies S, Lutterotti L, and Wenk H R 1997 *J. Appl. Cryst.* **30** 31
[10] Von Dreele R B 1997 *J. Appl. Cryst.* **30** 517
[11] Wenk H R, Lutterotti L, and Vogel S C 2010 *Powder Diffraction* **25** 283
[12] Wenk H R, Vasin R N, Kern H, et al. 2012 *Tectonophysics* **570-571** 123

IOP Conf. Series: Materials Science and Engineering **82** (2015) 012094 doi:10.1088/1757-899X/82/1/012094

On the correlation of meso-scale local texture and roping profile in AA6xxx sheet alloys

L Qin, M Seefeldt and P Van Houtte

Department of Materials Engineering, KU Leuven, Kasteelpark Arenberg 44, Mailbox-2450, BE-3001 Leuven, Belgium

E-mail: qinling.whu@gmail.com, marc.seefeldt@mtm.kuleuven.be

Abstract. Roping as a heterogeneous plastic deformation is generally attributed to the occurrence of the meso-scale clustering of grains with similar orientations. Large-scale electron backscattered diffraction (EBSD) orientation maps are customarily used to correlate the orientation topography with the roping profile. The most common way of investigating this phenomenon is to extract the predominant texture components and then to correlate them with the roping profile, since grains belong to different texture components lead to different plastic responses. Instead of using a microscopic representative volume element in the length scale of the grain size, the present work proposes a moving window mechanical model to use a representative volume element of the meso-scale, corresponding to a grain cluster, to simulate roping. For a tensile test in the transverse direction, a quantitative prediction of surface roping profile can be obtained. For an artificial EBSD orientation map, the proposed model can yield both roping wavelength and amplitude.

1. Introduction

Aluminium alloys for automobile body panel application often show a specific type of band-shaped surface roughening upon stretching, called "roping" or "ridging". Roping, manifesting itself as a series of ridges and valleys along the rolling direction (RD), has attracted much attention within the communities of crystallographic texture and crystal plasticity. Thanks to the rapid development of automated EBSD technique, clustering of similarly orientated grains, i.e., orientation banding, has been found by several research groups [1–5]. Furthermore, much effort has been made to correlate the spatial distribution of grains, which belong to a specific texture component, with roping. Thus, the band-like distribution of grains with Cube orientation [3, 6] as well as Goss [7, 8], R [9], X [10] and combinations of Goss-Cube [11] and X-Cube-Goss orientations [10] was proposed as predominant cause of roping. All the aforementioned research has indicated that the evolving surface roughness profiles can be related to the surface texture patterning.

An earlier work by the present authors [12] showed that roping could not be understood simply by banding of grains belonging to only specific orientations. Therefore, the present work will simulate roping based on a representative volume element (RVE) of the meso-scale with contrasting texture. To understand roping, two aspects, namely the roping wavelength and amplitude, need to be analysed together. The wavelength has already been predicted by different authors through various methods [8, 13]. Further understanding of the amplitude formation can provide a deeper insight into the origin of roping. However, no satisfactory roping profile with

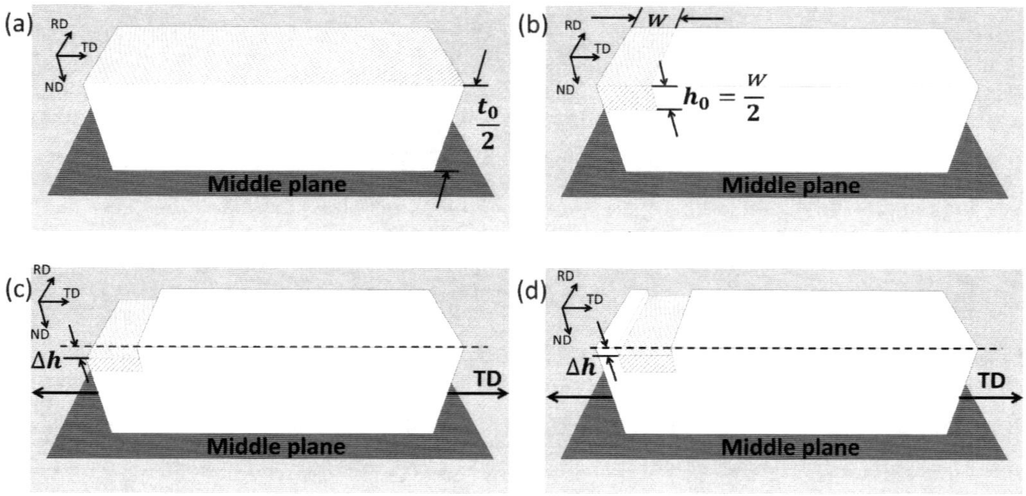

Figure 1. Schematic illustration of the meso-scale moving window mechanical model: (a) An EBSD map illustrated by the hatching lines is measured on the sheet surface; (b) A RVE of the meso-scale illustrated by the hatched volume is selected; (c) A plastic tensile strain is exerted to the selected RVE, resulting in its thickness change of Δh; (d) Scanning the RVE through the entire surface EBSD map yields a hypothetical surface profile under this plastic tensile strain.

substantial roping amplitude has been predicted based on the surface texture patterning for AA6xxx sheets to the knowledge of the authors.

2. Methodology

A large-scale EBSD orientation map is used as input to simulate surface roughening, in particular roping, with the help of the moving window method [12, 14]. Figure 1 schematically illustrates the simulation procedure of the meso-scale mechanical model. Firstly, the surface EBSD orientation map is measured and fed into this model. Secondly, a RVE of the meso-scale is selected by a window, of which the width is w and the length extends over the entire RD dimension of the EBSD map. Additionally, the height of this RVE of the meso-scale, h_0, is assumed to be $w/2$ (see Figure 1(b)). Due to lack of information on spatial distribution of texture through the thickness, the spatial distribution of the surface texture is used to represent that of the RVE. Then, the thickness change of the RVE of the meso-scale, Δh, is simulated by the Full Constraints (FC) Taylor model assuming that each RVE of the meso-scale behaviors like a standard free-standing tensile test sample [15–18]. Lastly, Δh is given as a function of the position of the window, when the window scans through the entire EBSD map in the transverse diection (TD).

In order to study the effect of window width, w, on the predicted roping profile, artificial EBSD orientation distribution maps are made using MTEX [19] and the MTM-FHM software system [20]. Figure 2(a) shows such an EBSD map with alternating equal bands of grains with Cube and R orientation. The banded pattern of Cube and R is chosen because Cube and R are typical texture components in AA6016-T4 [5, 10, 21]. Additionally, the development of the surface roping profile is attributed to the spatial distribution of grains with R orientation in the form of RD-elongated colonies embedded in the Cube orientation dominant matrix, according to Wittridge and Knutsen [9]. For the EBSD map shown in Figure 2, the band width is 504 μm and the step size is 8 μm.

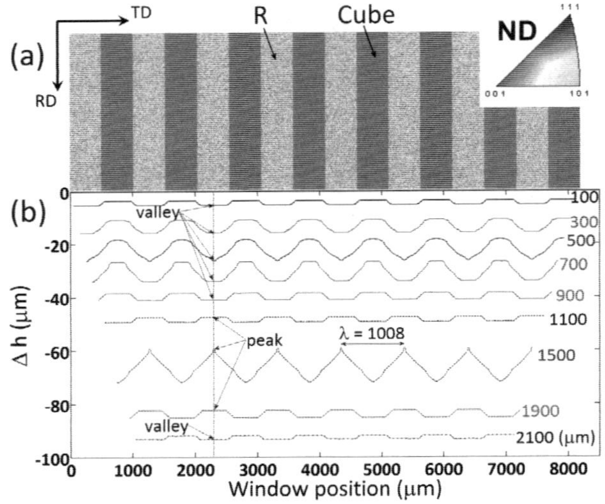

Figure 2. Effect of the window width on the hypothetical surface profiles simulated based on the artificial EBSD orientation map: (a) the EBSD IPF map of alternating equal bands with R and Cube orientation. The color coding scheme, which is illustrated by the unit triangle on the top right of the EBSD map, is defined according to the crystallographic direction along the sample's normal direction (ND) (b) the hypothetical surface profiles simulated by using different window widths which are indicated on the right of the profiles

3. Results and Discussions

Figure 2 shows the effect of the window width on the hypothetical surface profile simulated based on the banded R/Cube EBSD map. Since the period of a pair of R and Cube band along TD is 1008 μm, this value will be the wavelength of the roping profile upon deformation. Crystal plasticity simulations using the FC Taylor model give r-values of 0.35 and 1.0 for the ideal R and Cube orientation, respectively, if the tensile axis is parallel to TD. Therefore, the simulated profile is characterized as the positive tendency if the R and Cube bands form valleys and peaks, respectively. It is negative in case of the opposite. As shown in Figure 2, the positive and negative tendency profile will repeatedly alternate with increasing window width by the multiple of the wavelength. Consequently, the effect of the window width on the roping profile morphology can be summarized as:

$$\begin{array}{ll} positive\ tendency, & if\ 2n < \frac{w}{\lambda} < 2n+1; \\ negative\ tendency, & if\ 2n+1 < \frac{w}{\lambda} < 2n+2;\ , \\ & n = 0, 1, 2\cdots, \end{array} \quad (1)$$

where, w is the window width and λ the wavelength.

No matter which window width is used, the wavelength can be identified as shown in Figure 2(b). Therefore, this meso-scale model is insensitive to the choice of the window width for the purpose of identifying the wavelength. The amplitude changes with the choice of the window width due to the correlation of the volume height with the window width as shown in Figure 1(b). When the window width satisfies $w = \left(n + \frac{1}{2}\right)\lambda$ with $n = 0, 1, 2\cdots$, the amplitude approximately reaches the maximum. It is unfortunate that the study does not include simulation results on experimental EBSD maps. Nevertheless, the proposed meso-scale moving window model extends the scope of roping analysis towards the formation of roping amplitude.

4. Conclusions

The proposed meso-scale moving window mechanical model can simulate the roping wavelength and amplitude in a objective way based on the selected the meso-scale local texture. This model developed based on the MTM-FHM software system interprets roping in terms of the existence of mesoscopic RVEs with contrasting local textures. A quantitative criterion for estimating the roping wavelength using the positive and negative tendency property (Eq. 1) is proposed. The prediction of roping wavelength is insensitive to the choice of window width. The results of the application of this meso-scale model to the experimental EBSD maps will be reported elsewhere.

Acknowledgments

The authors gratefully acknowledge the financial support of the Interuniversity Attraction Poles Program (P7/21) from the Federal Government of Belgium and the M2i Institute in the Nederlands.

References

[1] Beaudoin A, Bryant J and Korzekwa D A 1998 *Metallurgical and Materials Transactions A* **29** 2323–2332

[2] Raabe D, Sachtleber M, Weiland H, Scheele G and Zhao Z 2003 *Acta Materialia* **51** 1539–1560

[3] Bennett T, Petrov R and Kestens L 2009 *Scripta Materialia* **61** 733–736

[4] Jin H, Wu P D, Ball M D and Lloyd D J 2005 *Materials Science and Technology* **21** 419–428

[5] Engler O and Hirsch J 2002 *Materials Science and Engineering: A* **336** 249 – 262

[6] Engler O and Brünger E 2002 *Materials Science Forum* **396-402** 345–350

[7] Baczynski G, Guzzo R, Ball M and Lloyd D 2000 *Acta Materialia* **48** 3361–3376

[8] Wu P and Lloyd D 2004 *Acta Materialia* **52** 1785 – 1798

[9] Wittridge N and Knutsen R 1999 *Materials Science and Engineering: A* **269** 205–216

[10] Wu P, Lloyd D, Bosland A, Jin H and MacEwen S 2003 *Acta Materialia* **51** 1945 – 1957

[11] Jin H and Gupta A 2012 *Materials Science Forum* **702 - 703** 273 – 278

[12] Qin L, Seefeldt M, Bennett T A, Petrov R H and Van Houtte P 2011 *Materials Science Forum* **702-703** 955–958

[13] Engler O, Schäfer C and Brinkman H J 2012 *Acta Materialia* **60** 5217–5232

[14] Qin L, Seefeldt M and Van Houtte P 2013 *Proceedings of Materials Science and Technology 2013* ed Isac M, Chiesa F and Guthrie R pp 1247–1283

[15] Hosford W F and Backofen W A 1964 *Fundamentals of Deformation Processing* ed Backofen W A (Syracuse, NY, USA: Syracuse University Press) pp 259–292

[16] Bunge H J 1970 *Kristall und Technik* **5**(1) 145–175

[17] Van Houtte P 1998 *Materials Science Forum* **273 - 275** 67–76

[18] Skrotzki W, Tamm R, Oertel C G, Beckers B, Brokmeier H G and Rybacki E 2001 *Materials Science and Engineering: A* **319321** 364 – 367

[19] Bachmann F, Hielscher R and Schaeben H 2010 *Texture and Anisotropy of Polycrystals III* vol 160 ed Klein H and Schwarzer R (Trans Tech Publ) pp 63–68

[20] Van Houtte P 2004 *The "MTM-FHM" software system* MTM, KU Leuven

[21] Hirsch J and O E 2001 *Recrystallization and Grain Growth Proceedings of the First Joint International Conference* ed Gottstein G and Molodov D (RWTH Aachen, Germany: Springer-Verlag) pp 731–740

IOP Conf. Series: Materials Science and Engineering **82** (2015) 012095 doi:10.1088/1757-899X/82/1/012095

Tensile failure of thin aluminium sheet observed by in-situ EBSD

S Kahl[1,2], R L Peng[1] and S Johansson[1]

[1] Division of Engineering Materials, Linköping university, Linköping, Sweden
[2] Gränges Technology, Gränges AB, Finspong, Sweden
E-mail: soren.kahl@liu.se

Abstract. Tensile tests on two similar 75-μm-thick aluminium sheet materials were carried out inside a scanning electron microscope equipped with an electron backscatter detector. The materials were subjected to simulated brazing prior to the test because this type of material is used for fins in automotive heat exchangers. Grain sizes were large relative to sheet thickness and ND-rotated cube and P texture components dominated the recrystallization textures; their volume fractions differed strongly in the two different materials, though. Strains over the microscope image fields were determined from positions of constituent particles or from grain sizes; the two methods gave consistent results. Grains with high Schmid factors accumulated significantly more deformation than grains with low Schmid factors. Cracks nucleated in high-Schmid factor grains, or in groups of such grains, at the specimen edges. When only low-Schmid factor grains were present at the specimen edges, the crack nucleated inside the specimen. The subsequent crack growth was intragranular and occurred at approximately 90° relative to the load direction.

1. Introduction

The onset of plastic deformation in single crystals of aluminium is influenced by the orientation of the crystal lattice relative to the load direction according to Schmid's law [1]. In polycrystalline materials, grain-to-grain interactions complicate the situation, especially for fine-grained materials where the grain size is significantly smaller than the smallest sample dimension.

Aluminium materials for aluminium heat exchangers have large grains after brazing [2]. This is especially significant for the thin fin materials where grain lengths in the rolling direction often reach a few hundred micrometers while thicknesses are typically between 50 and 100 μm. Sometimes, grains also extend over the complete thickness. Therefore, grain interactions are expected to be smaller in these materials as compared to most other engineering materials.

Fin materials are responsible for the major part of the heat exchange between air and heat exchanger. They are typically formed into wavy shapes and placed between tubes or plates of the respective heat exchanger. Due to this geometry, the mechanical loading usually occurs in the rolling direction of the material.

On a polished tensile test specimen, the evolution of surface morphology and changes in the crystallographic orientation of the surface layer can be investigated in a scanning electron microscope (SEM) equipped with a detector for analysis of electron backscatter diffraction (EBSD). In our contribution, we present details of tensile plastic deformation and fracture of two aluminium fin materials that differ mainly with respect to their texture. Tensile tests with small specimens were performed both inside and outside of the SEM.

2. Experimental procedure

75-μm-thick aluminium materials produced by Gränges AB, Sweden, were selected for this study. The most important production steps of these materials are direct-chill casting, hot rolling and cold rolling. However, we did not sample material from standard commercial materials. Instead, we chose materials from development stages which exhibited texture variations that made them interesting for our present study. Compositions were determined from melt charges by optical emission spectroscopy and are given in Table 1.

Before any testing, the material was subjected to a simulated brazing procedure that included heating to 600 °C, soaking for 3 min at 600 °C and subsequent fast cooling in forced air.

During brazing of aluminium heat exchangers, the material is heated to 600 °C and kept there for a few minutes. Since the materials of this investigation were heat exchanger materials, we intended them to be in the same state as in the product.

Table 1: Composition determined by optical emission spectroscopy.

Material	Si (wt.%)	Fe (wt.%)	Mn (wt.%)	Zn (wt.%)	Zr (wt.%)	Ti (wt.%)
A	0.77	0.23	1.61	1.46	0.12	0.04
B	0.84	0.26	1.62	0.03	0.12	0.03

Regular tensile tests followed the ISO6892-1:2009 standard. Tensile tests with small specimens were carried out on a Gatan-Deben microtest stage that could be inserted into a Hitachi SU-70 field emission gun SEM. The parallel section of the small specimen covers an area of 5 mm x 5 mm. More details of this setup and of the test procedure were provided in reference [3].

3. Results and discussion

Results from regular tensile tests are provided in Table 2. Material B has slightly higher strength values but a smaller elongation to fracture.

Table 2: Tensile test results in rolling direction on material after simulated brazing, determined from tensile tests according to ISO6892-1:2009.

Material	Thickness (μm)	Yield strength (MPa)	Tensile strength (MPa)	Elongation to fracture (%)
A	75 μm	54	135	8.5
B	75 μm	60	138	5.7

Material B has larger grains than Material A as shown in Table 3. Most grains of the investigated specimen of Material B have a <111> crystal direction parallel to the rolling direction (RD). For Material A, the fraction of grains with <111>||RD is similar to the fraction of grains with <120>||RD. It is interesting to investigate materials that contain these two types of grain orientations because the Schmid factor for loading parallel to RD is only 0.27 for <111>||RD but has a high value of 0.49 for <120>||RD. Most grains with <111>||RD are represented by the P-texture component while the cube texture component with a rotation around the normal direction (ND) of the rolled sheet stands for most grains with <120>||RD.

Table 3: Grains sizes and volume fractions of texture components, allowing a maximum of 15° deviation from the ideal orientation.

Material	Grain size	<111> || RD	{111}<110> P-texture	<120> || RD	{001}<120> ND-rot. cube
A, specimen 1	54 μm (256 grains)	39%	33%	36%	24%
specimen 2	47 μm (305 grains)	37%	31%	36%	19%
B	113 μm (661 grains)	82%	82%	4%	4%

We tested two different simple methods to determine strain over the SEM field of view. The first one is based on the change in distances between particles, as described in detail in reference [3]. The second method calculates strains from the change in sizes of bounding boxes that surround individual grains. Both methods require manual identification of the features (particles or grains) that should be included into the strain calculation. The particle method gave consistent results: In the case where the strain ε could be determined directly from two images or as the sum of several strain steps ε_1, ε_2, ε_3, we obtained $\varepsilon = \varepsilon_1+\varepsilon_2+\varepsilon_3$ within better than 0.1% strain. Resolution and accuracy of the grain method depends strongly on the step size of the EBSD measurement. In the present investigation, we only reached a resolution of approximately 1% strain. However, accuracy is improved by averaging over many grains; when we averaged over 60 grains we obtained the same result as from the particle method, within 0.1% strain.

The strains measured with the particle method in neighboring fields of view in the central region of specimen 1 of Material A as well as corresponding orientation maps are given in Figure 1. Fields of view numbered 2 and 3 contained a larger region of connected "blue grains" and thus exhibited smaller strains than fields of view numbered 1 and 4. "Blue grains" with low Schmid factors accumulate less deformation than "yellow grains".

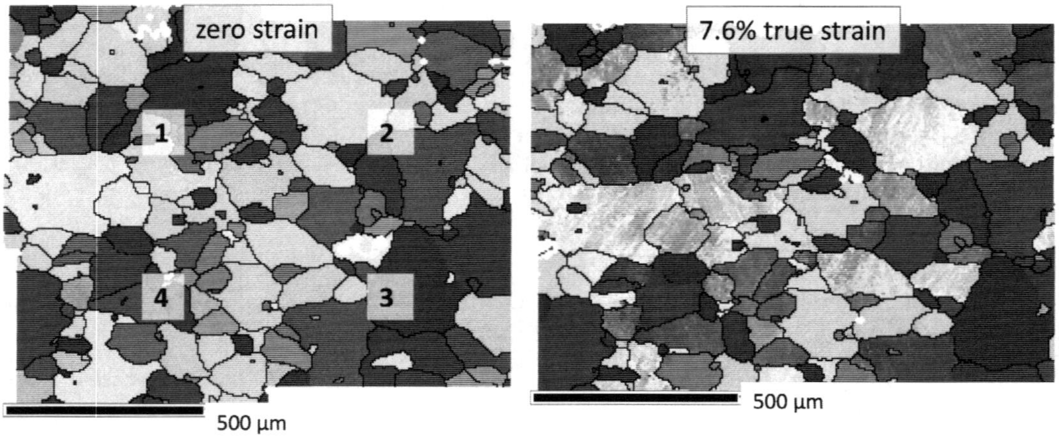

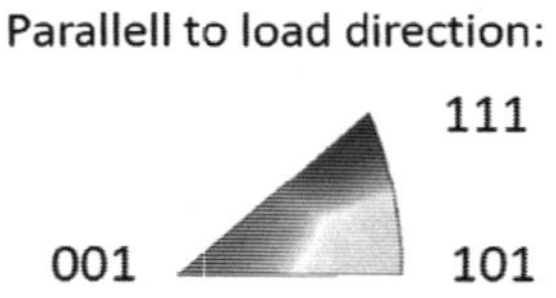

Extension	Field 1 True strain	Field 2 True strain	Field 3 True strain	Field 4 True strain
505 µm	4.1%	3.4%	3.1%	3.9%
814 µm	8.2%	7.1%	6.8%	8.3%

Figure 1: Orientation map in inverse pole figure coloring of the central region of specimen 1 of Material A.

The nucleation and growth of a crack in specimen 2 of Material A is depicted in Figure 2. The crack nucleated in a group of "yellow grains" near the upper specimen edge and then propagated at right angle to the load direction. Prior to rupture, the main crack made a sharp turn in order to meet a secondary crack that started at the lower edge, presumably in another group of "yellow grains". At 9% strain, the indication of a localized through-thickness necking at an angle of 50°-60° to the load direction is visible and indicated by the white arrows. Such a local neck may be expected for isotropic, thin, homogeneous sheet specimen [4]. However, in the presented case as well as for all other specimens tested in the project, the fracture was dominated by crack initiation and growth. We did not observe indications that growth was influenced by crystallography.

From Material B, we prepared a specimen that contained only "blue grains" with the exception of two large central "yellow grains", as shown in Figure 3, and two small "yellow grains". Deformation was strongly localized by the presence of the two large "yellow grains" and the crack started at a grain boundary triple junction at the upper grain boundary of the upper "yellow grain". Except for a very short distance above the crack nucleation point, crack growth was exclusively intragranular. Nevertheless, crack growth in this specimen contained a crystallographic aspect because the crack changed direction each time a grain boundary of a misorientation angle above 10° was crossed.

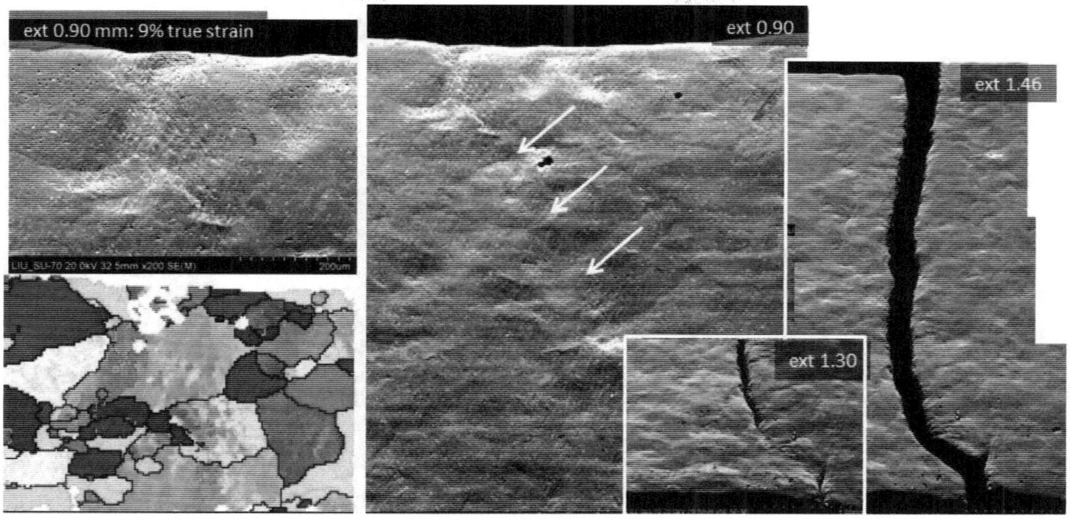

Figure 2: Nucleation and propagation of the tensile crack in specimen 2 of Material A. Pole figure coloring is the same as in Figure 1. Extensions are in millimeters. White arrows mark the indication of a local through-thickness necking.

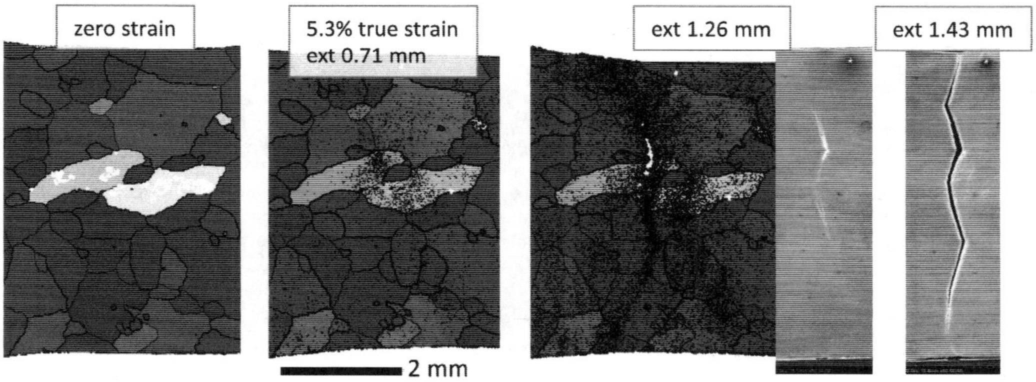

Figure 3: Nucleation and propagation of the tensile crack in small tensile specimen of material B. The complete specimen width is shown. Pole figure coloring is the same as in Figure 1. Thin black lines represent grain broundaries above 2° and thick black lines above 10° misorientation angle.

6. Conclusions

We prepared thin aluminium sheet materials that mainly consisted of large grains with <111>||RD and <120>||RD. The deformation and fracture characteristics of tensile tests parallel to RD were determined by the spatial distribution of these two types of grains: Grains with <120>||RD accumulated more strain than grains with <111>||RD, fracture nucleated in groups of <120>||RD grains, and cracks grew perpendicular to the load direction.

References

[1] Dieter GE 1988 *Mechanical metallurgy* McCraw-Hill Book Company (Singapore)
[2] Hirsch J, ed. 2006. *Virtual fabrication of aluminum products*. Weinheim: WILEY-VCH. 19-26 pp.
[3] Kahl S, Peng RL, Calmunger M, Olsson B, Johansson S 2014 *Micron* **58** 15-24
[4] Backofen WA 1972 *Deformation processing* Addison-Wesley (Reading, Mass., U.S.A.)

IOP Conf. Series: Materials Science and Engineering **82** (2015) 012096 doi:10.1088/1757-899X/82/1/012096

Effect of the grain shape on the q-value evolution of steel sheets*

Q Xie[1], Y Wang[1] and A Van Bael[2]
[1] The State Key Laboratory for Advanced Metals and Materials, the University of Science and Technology Beijing, 10083, China
[2] Department of Metallurgy and Materials Engineering, KU Leuven, B-3000 Leuven, Belgium
E-mail: Qingge.Xie@ustb.edu.cn; YdWang@mail.neu.edu.cn; Albert.VanBael@mtm.kuleuven.be

Abstract. Two polycrystalline plasticity models, namely the ALAMEL model and the visco-plastic self-consistent (VPSC) model were used to study the effect of the grain shape on the evolution of plastic anisotropy. The latter was characterized by the q-value in this paper. The two models both predict the evolution of grain shape and texture and then can predict the effect of the initial grain shape on the texture. The final anisotropy of the materials was also assessed with the help of these models. Data of tensile tests in three directions for three low carbon steel sheets with either equiaxed or a pancake grain shape were used to evaluate the model predictions. Both models give the same trend of the grain shape effect on the evolution of the q-values. Results demonstrate that the effect of the grain shape can be clearly reflected in the *x*-value distribution as defined and discussed in this paper. But it is not so obvious in the predicted textures for the investigated three materials according to both models.

1. Introduction

For steel and aluminium sheets used in automotive and packaging applications, the evolution of the q-value (the contraction ratio proposed by Bunge [1]) profile is used to characterize the anisotropic plastic property of a sheet. The q-value is defined as the ratio of plastic strain rate in the width direction to the plastic strain rate in the tensile direction during a uniaxial tensile test. It also corresponds to the slope of a tangent line of the yield locus for a given material. The importance of an accurate determination of the q-value in metal forming is that it is closely related with the deep drawing property of a material. A high q-value retards necking in sheet metal forming processes whereas its variation with the tensile test direction in the sheet plane correlates well with the earing tendency of a drawn cup.

Crystallographic texture is known as the main source of this plastic anisotropy. Other sources of anisotropy, such as the grain shape, directional substructure and uneven distribution or a non-equiaxed shape of the second phase may also play a role. In this paper, we employed two crystal plasticity models: ALAMEL [2] and VPSC [3], to study the initial grain shape effect on the q-value evolutions for three low carbon steel sheets. Similar studies have been given by Chung and Cho [4, 5] and by Delannay [6-8]. However, the former mainly focused on the experimental evidence and the emphasis of the latter was on the initial q-value affected by

* Most of the work reported in this paper was done when Qingge Xie did his PhD research at the department of MTM KU Leuven in Belgium.

the grain shape. This paper did investigations on not only the effect of the grain shape on the initial q-value, but also the effect of the grain shape on the evolution of the q-values.

The ALAMEL model assumes that each grain 'sees' its single neighboring grain. Through compensating relaxations along the common grain boundary between two neighboring grains, the shear stress equilibrium at the grain boundary is achieved by adopting the generalized Schmid law. These additional relaxations (induced by grain interactions) along the grain boundary satisfy the continuity of the velocity at the grain boundary and they vanish somewhere in the grain (cf. [9]). The grain shape in the ALAMEL model is indirectly reflected by the orientation distribution of the grain boundary. The VPSC model assumes that each grain is an inclusion embedded in an equivalent medium, the property of which is the weighted average of that of all consisting grains in the aggregate. The VPSC model employs the Eshelby tensor to take the aspect ratio of each grain into account. These two models use two different methods to calculate the q-values (cf. [10]), which were proved to be equivalent by Van Houtte (cf. the appendix in [10]) for materials with orthorhombic symmetry. The textures in the studied materials in this paper indeed feature orthorhombic symmetry.

2. Results and discussion

The weighted average of the aspect ratios of the three low carbon steel sheets are 1.53:1.19:0.55 (RD: TD: ND) for T52 and 1.28:1.19:0.65 for both T57 and T61. In the model, all grains are considered to have the same shape and the same size but a different lattice orientation. To understand the effect of pancake grain shape on the q-value evolution of tensile tests, for both ALAMEL and VPSC, two types of simulations were carried out: one that uses a (hypothetical) equiaxed microstructure as initial material, and one that uses the measured non-equiaxed ('pancake') structure as initial material. Each time the experimentally measured textures were used as initial texture of the simulations, with a representative volume element (RVE) consisting of 5000 grains. The present study assumes isotropic hardening at the level of individual grains. Fig. 1 gives the evolution of the q-values calculated by the two models. Besides that, the model proposed by Delannay [8] which captures the grain shape/size effect in each grain by introducing a back stress τ_b^α on the critical resolved shear stress (CRSS) of potential slip system is also adopted here (cf. equation 3 in [8]). The mean-free path of mobile dislocations within the grain is a physically sound reason of this model by considering the fact that grain shape may cause individual slip systems to strengthen differently. Following the method described in [10], the CRSS of slip system is calculated by equations (1)-(3).

$$\tau_c^\alpha = \tau_{sat} - (\tau_{sat} - \tau_b^\alpha)\exp(-\Gamma) \tag{1}$$

$$\Gamma = \int \sum \dot{\gamma}^\alpha \, dt \tag{2}$$

$$\tau_b^\alpha = \tau_0 \left(1 + \frac{0.02}{d_l^\alpha}\right) \tag{3}$$

τ_b^α is the back stress term of slip system α. τ_0 is a reference parameter. d_l^α is related to the grain shape effect and it can be calculated following equations (8)-(10) in [8]. τ_c^α is the CRSS of slip system α. τ_{sat} is the saturated value of the CRSS of the slip system in the studied material. Γ is the total slip in a given grain on all potential slip systems. This model was introduced into the VPSC model in this paper. It was named as 'VPSC-pancake adapted' in Fig.1 and only the case of initially pancake grain shape is given (the effect of this new approach is only obvious when the grain shape deviates from the equiaxed one before the work hardening saturates).

The q-value of the real material declines with the strain in all cases in Fig.1. Clearly, according to both models, this declining trend is induced by the initial pancake grain shape. Interestingly, the 'VPSC-pancake adapted' gives the best predictions for T52. In the meantime, the predictions for T57 and T61 did not become worse. This new approach almost does not affect the quality of the texture evolution and the evolution of the *x*-value distribution (as discussed in the next paragraph). It is easy to be understood that by introducing a back stress term on the CRSS of each slip system, the yield locus of each grain is changed. This further on affects the shape of the yield locus of the whole aggregate. Fortunately, by using suitable model parameters as given in equations (1)-(3), the q-value prediction seems to be slightly improved.

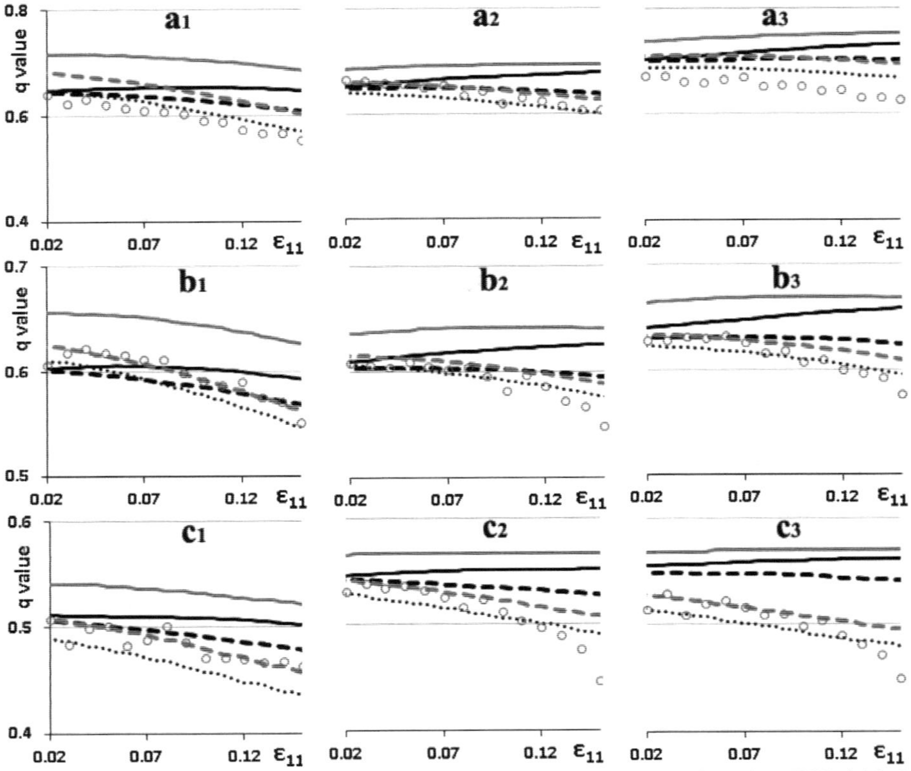

Fig.1 The q-value evolution for tensile tests of a_1 T52-0°, a_2 T52-45° and a_3 T52-90°; b_1 T57-0°, b_2 T57-45° and b_3 T57-90°; c_1 T61-0°, c_2 T61-45° and c_3 T61-90°. Experiments (open circles), ALAMEL-equiaxed (black full lines), ALAMEL-pancake (black dashed lines), VPSC-equiaxed (grey full lines), VPSC-pancake (grey dashed lines), VPSC-pancake-adapted (black dotted lines).

An interesting difference between these two models is the strain rate distribution calculated for each grain in the aggregate. Hereby, we plot the distribution of the ratio x defined in equation (4).

$$x = \mathrm{sign}(\mathbf{d}_{||} : \mathbf{D}) \, ||\mathbf{d}_{||}|| / ||\mathbf{D}|| \qquad (4)$$

$\mathbf{d}_{||}$ is the component of the tensor **d** (the local strain rate in a given grain in a polycrystal) which is parallel with the imposed strain rate **D**. The similar method was already used in [11].

'sign' indicates the sign of the double dot product between $\mathbf{d}_{||}$ and **D**. The x-value of a given grain can indicate the local strain rate deviation from **D** in the direction of **D**. The distribution of the x-value can reflect the strain rate distribution in a polycrystal (i.e. deformation homogeneity in a polycrystal in the direction of **D**) and can also indicate the size of the relaxation for both soft (x>1) and hard (x<1) grains in the direction of **D**. The corresponding figures were given in Fig.7 and Fig.8 in [10]. We found that the x-value distribution is sensitive to the initial grain shape. The pancake grain shape can make the x-value distribution narrower than the case of an equiaxed structure (i.e. materials with pancake grain shape feature more homogeneous strain rate distribution in the direction of **D** than that with initially equiaxed structure.). Throughout the tensile deformation, the change of the x-value distribution is small.

The texture prediction by using the VPSC model and the ALAMEL model with and without considering the initial grain shape was studied too. We found that the effect of the pancake grain shape on the deformation texture is small according to both models.

Currently employed models can already give very good q-value predictions. In reality, the dislocation substructure developing within the grains hardens the slip systems differently. Accounting for this additional source of anisotropy may improve predictions further.

3. Conclusions

The comparison between experimentally observed q-value evolutions during the tensile tests for three materials with the predicted q-value evolutions by two models suggests that the grain shape has a clear effect on the q-value. The pancake grain shape affect the strain rate distribution in the polycrystal (reflected by the x-value distribution), which further on affect the evolution of the q-values. Both the ALAMEL model and the VPSC model can give a good estimation of the initial q-value and its evolution in most cases. The model proposed by Delannay [8] was also used in this paper and it seems to be able to improve the q-value predictions further.

The effect of the initial grain shape on the texture evolution was also studied. This effect is quite small for the studied three materials according to both models.

Acknowledgements

This research was carried out under the project number M41.2.08307/M41.10.08307 in the framework of the Research Program of the Materials innovation institute M2i (www.m2i.nl). Y Wang and Q Xie also thank the financial support from National Science Foundation of China (NSFC) (Grant No. 51231002), the Fundamental Research Funds for the Central Universities (Grant No. 06111020), and the fundamental research fund at the State Key Laboratory for Advanced Metals and Materials (2014Z-01).

References

[1] Bunge H J 1970 *Kristall und Technik* **5** 145
[2] Van Houtte P, Li S, Seefeldt M, Delannay L 2005 *Int J Plasticity* **21** 589
[3] Lebensohn R A and Tomé C N 1993 *Acta metal Mater* **41** 2611
[4] Chung Y H, Cho K K, Han J H and Shin M C 2000 *Scripta Mater* **43** 759
[5] Cho K K, Chung Y H et al 1999 *Scripta Mater* **40** 651
[6] Delannay L and Barnett M R 2012 *Int J Plasticity* **32–33** 70
[7] Delannay L, Melchior M A, Signorelli J W, Remacle J F, Kuwabara T 2009 *Comp Mater Sci* **45** 739
[8] Delannay L 2012 Mater Sci Forum **702-703** 182
[9] Xie Q 2014 Ph. D. Thesis KU Leuven
[10] Xie Q, Eyckens P, Vegter H, Moerman J, Van Bael A, Van Houtte P 2013 *Mat Sci Eng A-Struct* **581** 66
[11] Sarma G B, Dawson P R 1996 *Int J Plasticity* ***12*** *1023*

IOP Conf. Series: Materials Science and Engineering **82** (2015) 012097 doi:10.1088/1757-899X/82/1/012097

Reduction in mechanical anisotropy through high temperature heat treatment of Hastelloy X processed by Selective Laser Melting (SLM)

T Etter[1], K Kunze[2], F Geiger[1] and H Meidani[1]
[1] ALSTOM (Switzerland) Ltd, Baden, Switzerland
[2] Scientific Centre for Optical and Electron Microscopy, ETH Zurich, Switzerland
E-mail: thomas.etter@power.alstom.com; karsten.kunze@scopem.ethz.ch; fabian.geiger@power.alstom.com; hossein.meidani@power.alstom.com

Abstract. Selective Laser Melting (SLM) is an additive manufacturing technology used to directly produce metallic parts from thin powder layers. To evaluate the anisotropic mechanical properties, tensile test specimens of the Ni-base alloy Hastelloy X were built with the loading direction oriented either parallel (z-specimens) or perpendicular to the build-up direction (xy-specimens). Specimens were investigated in the "as-built" condition and after high temperature heat treatment. Tensile tests at room temperature and at 850°C of "as-built" material have shown different mechanical properties for z- and xy-specimens. The anisotropy is reflected in the Young's modulus, with lower values measured parallel to the build-up direction. It is shown that the anisotropy is significantly reduced by a subsequent recrystallization heat treatment. The characterization of microstructural and textural anisotropy was done by Electron Back Scatter Diffraction (EBSD) analysis. Predictions of Young's modulus calculated from the measured textures compare well with the data from tensile tests.

1. Introduction

Ni-base superalloys are extensively used for components in aero- and industrial gas turbine engines [1, 2]. The solid-solution strengthened Ni-base alloy Hastelloy X has been widely used in today's heavy-duty gas turbines for hot gas path components such as combustors, transition pieces and exhaust-end components [3]. Progress in Ni-base alloy performance would not be possible without the parallel improvement of processing technology. Additive manufacturing processes such as selective laser melting (SLM) [4, 5] are prospective candidates to complement or replace conventional machining/production processes such as cutting and casting. Due to powder based article production and the inherently high cooling rates in these processes, the resulting material is very homogeneous with respect to chemical composition and principally free of segregations and residual eutectic phases [6, 7].

The high thermal gradients in SLM processing cause crystals to grow preferentially in well-defined directions, resulting in specific microstructures and textures [8]. Suitable scanning strategies favor either sharp single component textures or more uniformly distributed crystal orientations [9]. The anisotropy of mechanical properties is largely controlled by the texture and can be modeled based on the texture and the single crystal mechanical properties [10, 11]. In the present study, texture and microstructural anisotropy in SLM made Hastelloy X specimens are analyzed using EBSD (electron backscatter diffraction) and correlated with anisotropic material behavior observed during mechanical testing (Young's modulus). We report on processing routes to reduce the amount of anisotropy by (i) optimized laser scanning strategy, and (ii) suitable heat treatment.

2. Material and methods

2.1. Sample processing and testing

In the SLM process, thin powder layers are deposited on a metallic base plate or on the already produced skeleton of a designed object [12]. The cross-sections of a sliced CAD file are scanned subsequently using a high power laser beam in order to locally melt the powder material. The Hastelloy X alloy powder (www.praxairsurfacetechnologies.com) has a nominal weight percentage composition of 22% Cr, 18% Fe, 9% Mo, 1.5% Co, 0.7% W, <1% Si, <1% Mn, 0.10% C, <0.008% B, Ni (balance). A SLM machine 280HL (www.stage.slm-solutions.com) was used to build specimens with different orientations relative to build-up (z-axis) and laser scanning directions (in the xy-plane).

Three different laser scanning strategies were performed for subsequent layers with nominal thickness of 30 μm deposited in the xy-plane (one fixed scanning direction per layer; see also Fig. 1, top):

A. lines rotate 90° (parallel/perpendicular to the testing direction),
B. lines rotate 90° (+45°/-45° to the testing direction),
C. lines rotate 67°.

The SLM built pieces were cut from the substrate plate by electro-discharge machining. Samples were investigated either in the "as-built" condition or after heat treatment at 1200°C for 4h (Table 1). Tensile tests of M10 specimens were carried out according to ISO 6892 at ambient and elevated temperature (850°C) with a 50 kN load cell, at a strain rate of ~4 %/min, with a hold time of ~30 min after reaching the testing temperature in a resistance furnace.

2.2. Characterization
Metallographic sections were cut from the thread areas of the tensile test specimens for microstructural characterization. Orientation mappings were performed on a FEG-SEM MIRA3 XMU (www.tescan.com) equipped with an AZtec EBSD system (www.oxford-instruments.com), or on a FEG-SEM Quanta 200F (www.fei.com) equipped with a Hikari EBSD module and software OIM 6.2 (www.edax.com) enabling large area combo scans. EBSD data acquisition was carried out at 20 kV acceleration voltage, 15-25 mm working distance, and a nominal beam current of 2-7 nA. Only reliable data (confidence index > 0.1) were used for texture representation and subsequent modeling.

2.3. Texture based property modelling
The elastic tensor for single crystal Hastelloy X (c_{11} = 230.40 GPa, c_{12} = 156.12 GPa, c_{44} = 121.77 GPa at room temperature, derived from [13]) possesses a marked anisotropy, with minimum of Young's modulus of 104 GPa parallel to <100> and maximum of 294 GPa parallel to <111>. Using the MATLAB toolbox Mtex [14] (http://code.google.com/p/mtex/), we derived texture based lower (Reuss) and upper (Voigt) bounds of the polycrystal elastic property, as well as their arithmetic average (Hill estimate) [15].

Table 1: Young's modulus measured on SLM specimens processed by three scanning strategies.

Specimen testing direction	**Heat treatment**	**Test Temperature [°C]**	**Scanning strategy**		
			A (0°/90°)	B (-45°/45°)	C (random)
			Young's Modulus [GPa]		
XY	as-built	23	146	190	171
XY	as-built	850	99	129	118
XY	1200°C/4h	23	195	198	192
XY	1200°C/4h	850	136	132	139
Z	as-built	23	148	-	148
Z	as-built	850	101	-	106
Z	1200°C/4h	23	182	-	187
Z	1200°C/4h	850	121	-	130

3. Results
3.1. Mechanical testing
The experimental results (Table 1) indicate a major impact of the scanning strategy on the mechanical anisotropy measured in the as-built condition at room and elevated temperature. Young's modulus hardly differs for z-specimens (parallel to the build-up direction) between strategies A and C, which also matches the value in the xy-plane for strategy A (i.e. parallel to the laser scanning direction). After heat treatment, the differences between the three scanning strategies are nearly extinct, and the anisotropy of the modulus is largely reduced [16].

3.2. Microstructure and texture analysis
Metallographic observations as well as orientation mapping (by EBSD) on as-built specimens reveal a columnar grain morphology, with the longer grain axes aligned preferentially parallel to the build-up direction z. A grain boundary pattern roughly parallel to the two perpendicular

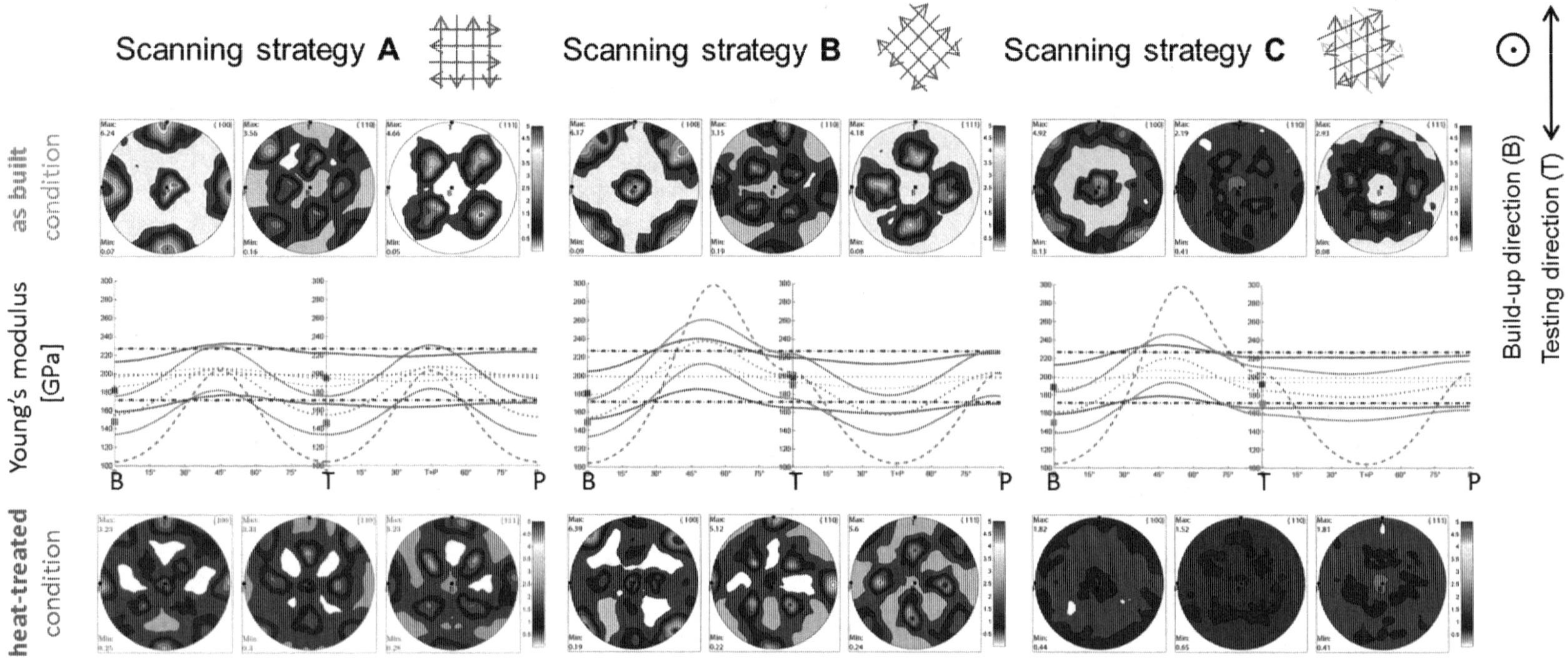

Fig. 1
Texture (pole figures) and derived Young's modulus (profiles B-T-P) for Hastelloy X specimens, processed by SLM with scanning strategies A, B and C. The profiles show Young's modulus for a single crystal (red dashed), for the textured material in as-built (green) and in heat-treated (blue) condition, and for the isotropic case of a uniform texture (black dash-dot). Upper (Voigt) and lower (Reuss) bounds by solid lines, Hill average by dashed lines.

laser scanning directions was observed in the xy-plane of samples processed by strategies A and B. All samples possess a pronounced texture with a concentration of <001> parallel to the build-up direction z (Fig. 1, top row). For A and B strategies, the <100> axes align preferentially parallel to the respective laser scanning directions, thus the textures are basically single components <100>{001} for A and <110>{001} for B with some scatter, when referred to testing and build-up direction. For scanning strategy C, a rather random distribution of <100> within the xy-plane was observed, thus a fiber texture with <001> parallel to z.

After heat treatment, the microstructure is largely modified by recrystallization, resulting in significant grain coarsening and more equiaxed grain shapes. No indications of the as-built columnar grain shapes are remaining. On the other hand, a larger part of the materials has retained the as-built texture, which is complemented by additional texture components and a higher uniform component (Fig. 1, bottom row), resulting generally in a reduction in texture strength from as-built to heat-treated conditions.

3.3. Texture based property modelling

Young's modulus is predicted from the texture for any desired sample direction and plotted as profile from B (build-up direction z) to T (testing direction) and further to P (perpendicular to both) (Fig. 1, center row). For comparison, the corresponding profiles are also shown for an ideally oriented single crystal (red dashed lines). The anisotropy for the single crystal is maintained in type, but reduced in extent for the textured samples in as-built condition for scanning strategies A and B, while there is only little in-plane (xy) anisotropy for strategy C. The broadening of textures after heat treatment result in further reduced anisotropy of Young's modulus, which approaches the isotropic bounds in the xy-plane for strategy C. For all three strategies, Young's modulus still maintains a minimum parallel to the build-up direction z, caused by a maximum in the center of the (001) pole figure.

4. Discussion and conclusions

It was shown that suitable scanning strategies allow tailoring the texture and thus anisotropy of mechanical properties. Rotation of the laser scanning direction by 90° between consecutive layers results in single texture components and minima of Young's modulus along the build-up and laser scanning directions. Rotation of the laser scanning direction by 67° leads to <001> fiber textures and to a nearly transverse isotropic elastic tensor, both with respect to the build-up direction. Additional texture components develop by recrystallization through suitable heat treatment, which generally reduce the overall elastic anisotropy. The SLM technology enables to control the crystallographic textures with respect to two distinct sample directions (build-up and laser scanning), which offers new possibilities optimizing the design of components, either for minimized anisotropy or taking advantage of a suitably maximized anisotropy.

References

1. Durand-Charre M and Davidson JH 1997 *The microstructure of superalloys* (Amsterdam: Gordon and Breach Science Publ.) 124 S.
2. Sims CT 1987 *Superalloys II [high-temperature materials for aerospace and industrial power]* (New York, N.Y.: Wiley) 615 S.
3. Zhao JC et al 2000 Mat Sci Eng A **293** 112-119.
4. Das S et al 2000 Mater Design **21**(2) 63-73.
5. Marquis FDS et al 2002 *Rapid prototyping of materials proceedings of symposium* (Warrendale, Pa.: TMS) 224 S.
6. Rickenbacher L et al 2013 Rapid Prototyping J **19**(4) 282-290.
7. Wang F 2012 Int J Adv Manuf Technol **58**(5-8) 545-551.
8. Kunze K et al 2014 Mat Sci Eng A **620** 213-222.
9. Thijs L et al 2013 Acta Mater **61**(5) 1809-1819.
10. Kocks UF et al 1998 *Texture and anisotropy preferred orientations in polycrystals and their effect on materials properties* (Cambridge: Cambridge University Press) XII, 676 S.
11. Niehuesbernd J et al 2013 Mat Sci Eng a-Struct **560** 273-277.
12. Rombouts M et al 2006 Cirp Ann-Manuf Techn **55**(1) 187-192.
13. Canistraro HA et al 1998 J Eng Mater-T Asme **120**(3) 242-247.
14. Bachmann F et al 2010 Solid State Phenomena **160** 63-68.
15. Mainprice D et al 2011 in *Deformation Mechanisms, Rheology and Tectonics: Micro-structures, Mechanics and Anisotropy* ed Prior DJ et al (Geological Society: London). 175-192.
16. Etter T et al 2014 TMS2014 143rd annual meeting & exhibition, 230.

IOP Conf. Series: Materials Science and Engineering **82** (2015) 012098 doi:10.1088/1757-899X/82/1/012098

Origin of magnetic fabrics in ultramafic rocks

AR Biedermann[1], K Kunze[2], AS Zappone[3] and AM Hirt[1]
[1] Institute of Geophysics, ETH Zurich, Switzerland
[2] Scientific Centre for Optical and Electron Microscopy, ETH Zurich, Switzerland
[3] Swiss Seismological Service, Zurich, Switzerland
E-mail: andrea.biedermann@erdw.ethz.ch; karsten.kunze@scopem.ethz.ch; alba.zappone@sed.ethz.ch; ann.hirt@erdw.ethz.ch

Abstract. The magnetic fabric of a rock, defined by the anisotropy of magnetic susceptibility (AMS), is often used as a tectonic indicator. In order to establish a quantitative relationship between AMS and mineral texture, it is important to understand the single crystal intrinsic AMS of each mineral that contributes to the AMS of the rock. The AMS and crystallographic preferred orientation (CPO) of amphiboles, olivine and pyroxenes has been analyzed in a series of amphibolites, peridotites and pyroxenites that do show preferred mineral alignment. The CPO of each mineral phase was determined based on electron backscatter diffraction (EBSD). Whole-rock AMS was computed based on the CPO and single crystal AMS of the respective minerals. A comparison between measured and modelled magnetic anisotropy shows that the directions of the principal susceptibility axes agree well in amphibolite and peridotite. Pyroxenite is a good example for competing AMS fabrics in polyphase rocks.

1. Introduction

Anisotropy of magnetic susceptibility (AMS) is often used as a proxy for mineral texture in geologic applications, e.g. to gain information on deformation processes in metamorphic rocks or emplacement and flow processes in igneous rocks [1-3]. Magnetic susceptibility is a second-order symmetric tensor and typically relates to the arrangement of iron in the crystal lattice. Therefore, AMS in a rock can yield information on the preferred mineral alignment, given that (1) the minerals that carry the AMS can be identified, and (2) their intrinsic AMS is known. Rock texture is usually carried by the abundant paramagnetic minerals, e.g. silicates, while even small amounts of ferromagnetic grains, e.g. iron oxides, often dominate the magnetic susceptibility. Hence it is important to isolate the paramagnetic component of the AMS [4]. Ultramafic minerals, such as amphiboles, pyroxenes and olivine, often control the paramagnetic AMS and texture in mafic and ultramafic rocks [5-8].

Electron backscatter diffraction (EBSD) is a powerful technique that determines mineral phases and their CPO on large areas of a sample [9]. The CPO can then be correlated with the anisotropy of physical properties, most often for seismic (elastic) anisotropy. EBSD has also served to better understand what minerals carry the AMS in polyphase mafic rocks and how CPO and AMS relate qualitatively. For example, in the La Palisse lava flow in the French Massif Central, a correlation was found between plagioclase and clinopyroxene preferred orientation and the orientation of the principal susceptibility axes [10].

The aim of this study is to test how mathematical models that combine the EBSD-derived CPO with the single-crystal AMS of the constituent minerals compare to measured AMS in ultramafic rocks. For this, a series of deformed mono- and polyphase rocks like amphibolite, peridotite and pyroxenite are investigated through EBSD and AMS measurements. The results from this study allow using AMS as qualitative and possibly quantitative proxy for mono- or polyphase mineral texture, and as such as a tectonic marker.

2. Material and methods

2.1. Sample description

After carefully performed pre-inspection by standard petrographic techniques, four samples each were selected from three different rock types: amphibolite, peridotite and pyroxenite. Two amphibolite samples were drilled in the Ivrea Zone, Northern Italy, and consist of hornblende, quartz and feldspar. The other two amphibolite samples were taken from the Møre-Trøndelag Fault Zone, Norway, and contain hornblende, biotite, quartz and feldspar. The four peridotite samples come from the Ivrea Zone and are made up mainly of olivine, with minor components of hornblende, clinopyroxene and an opaque mineral, probably magnetite. The pyroxenite samples are from Oman, consist of clinopyroxene, orthopyroxene and hornblende and are strongly serpentinized. One pyroxenite sample contains large grains, so that is was not possible to obtain a representative CPO. Deformed rocks were chosen in order to ensure that they possess a CPO and magnetic fabric. The lineation and foliation were determined if possible from macroscopic and thin section evidence.

2.2 Determination of AMS

Cores of 2.5 cm diameter and 2.2 cm length or 1.7 cm diameter and 1.6 cm length were drilled from each rock in its geographic reference system. AMS was measured on a torque magnetometer. Measurements were performed in high fields between 1.0 and 1.5 T in order to isolate the paramagnetic component of the AMS. The measured AMS is described in Fig. 1 by the orientation of the principal axes, i.e. the eigenvectors, of the susceptibility tensor.

2.2 Determination of CPO

The core surfaces were ground, polished, and lapped with colloidal silica in order to obtain a smooth surface low in defect content and thus suitable for EBSD measurements. The sides of each core were covered with silver paint to reduce charging problems, but no surface coating was applied. EBSD data were acquired on a SEM EOscan (Tescan, Brno CZ), equipped with a Pegasus OIM version 6.2 EBSD+EDS system by Ametek-Edax (Mahway, NJ, USA). EBSD patterns and EDS counts of specific elements were mapped simultaneously with a beam current of 3-5 nA and 20 kV acceleration voltage. Only one high-symmetry phase was indexed online to increase acquisition speed. The data were later reprocessed using the ChiScan routine of OIM Data Collection to re-index all relevant mineral phases filtered by user-defined windows for EDS counts. The measured CPO is presented in Fig. 1 by contoured pole figures of three mutually perpendicular crystal axes.

2.3 Modelling of rock AMS

AMS was modelled based on the CPO and single crystal susceptibility tensors [11]. Model calculations of Hill averages [12] were performed using MTex [13]. Separate models were made for (1) each relevant phase, and (2) a combination of all phases according to their modal composition for each rock. The modelled susceptibility ellipsoid is displayed in Fig. 1 by colored isolevel stereoplots and by the orientation of the corresponding principal axes.

3. Results

A summary of CPOs and comparison between measured and modelled AMS of selected cores is given in Fig. 1. Modelled principal susceptibility directions agree well with observations for all amphibolites, both from the Ivrea Zone and the Møre-Trøndelag Fault Zone. These samples are clearly dominated by the hornblende fabric, as quartz and plagioclase (1) possess weak CPOs and (2) have weak susceptibilities. The biotite fraction in the amphibolite from the Møre-Trøndelag Fault Zone is less than 5 vol%. Thus, the magnetic fabric of the amphibolites can be considered as monomineralic.

For the peridotite, AMS was modelled for each phase – olivine, clinopyroxene and hornblende – and the combination of phases, weighted according to the modal composition. As can be seen in Fig. 1c, olivine predominates the total AMS signal. The hornblende AMS competes with the olivine AMS. However, because olivine composes >93% of these samples, both hornblende and clinopyroxene AMS only add minor contributions to the total AMS. The principal directions match well between observed and modelled AMS both if (1) the AMS is calculated exclusively for olivine, and (2) the AMS is computed for the combination of phases.

In the pyroxenite samples, the measured maximum susceptibility coincides with that modelled for clinopyroxene, which is the main phase (Fig. 1d). The orientation of the minimum axes of measured and whole-rock modelled AMS agree. The latter is strongly influenced by the orthopyroxene component, which makes up only 14% of the rock.

4. Discussion

The orientation of principal susceptibility axes has been successfully modelled in rocks whose AMS is dominated by a single mineral phase, e.g. by phyllosilicates [14]. This study confirms these results for ultramafic and mafic rocks. In the case of amphibolite, the orientations of the maximum and minimum axes of the modelled AMS based on hornblende

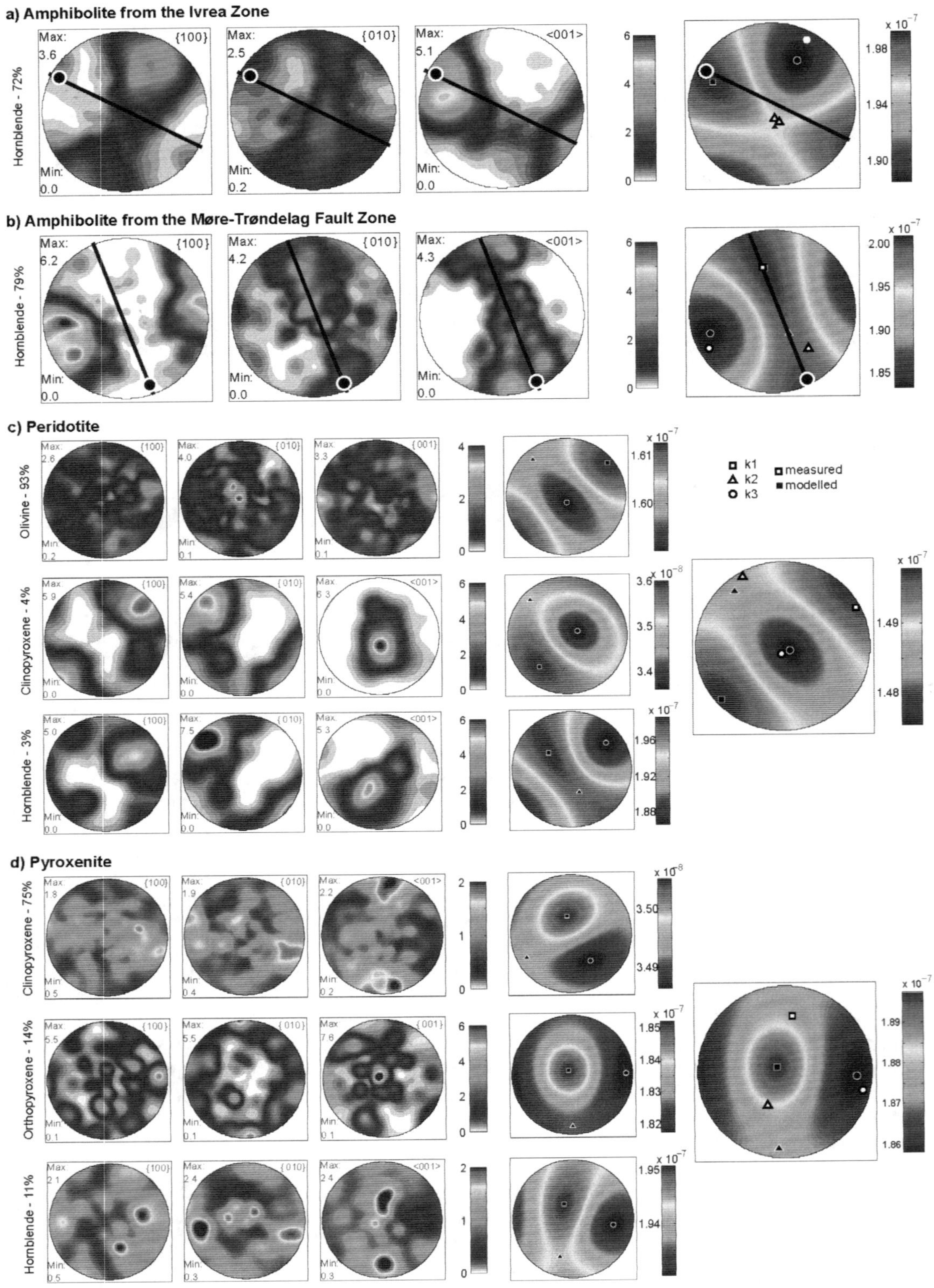

Figure 1: CPO of relevant phases, modelled and measured AMS for amphibolite (a,b), peridotite (c), and pyroxenite (d). The AMS contribution of each phase and the combination is shown separately for the peridotite and pyroxenite. Black lines and circles indicate foliation and lineation directions if visible.

CPO are in good agreement with the measured AMS. The orientation of the principal susceptibility axes with respect to macroscopic lineation will depend on the hornblende texture. The intrinsic maximum susceptibility of hornblende is parallel to the crystallographic *b*-axis, the minimum is parallel to a^*, and the intermediate susceptibility, which is parallel to *c*, has a value close to the maximum susceptibility. The girdle distribution of the a^* and *b*-axes of hornblende in the Ivrea samples forces the maximum susceptibility to align with the macroscopic lineation. In contrast, the crystallographic axes in the samples from the Møre-Trøndelag Fault Zone show a point distribution and the maximum susceptibility is sub-parallel the concentration of crystallographic *b*-axes. Therefore, the maximum principal susceptibility, i.e. the magnetic lineation, coincides with the mineral lineation for a fiber texture with *c*-axes parallel lineation. If the *b*-axes are grouped, it will flip and follow the *b*-axes maximum, which is typically formed perpendicular to the rock lineation. This transition in AMS related to variation of CPO was verified by modelling (not shown here) using the Bingham distribution [15] implemented in MTex [13].

Olivine dominates the magnetic fabric of the peridotite samples. There is a good agreement between the modelled and measured AMS. The intrinsic AMS for olivine has the maximum susceptibility along the *c*-axis, and the minimum along the *b*-axis for Fe concentrations >6 wt% FeO. A comparison between olivine CPO and measured AMS shows that the olivine in the peridotites contains >6 wt% FeO.

The pyroxenite samples display a complex relationship between the measured and modelled AMS, which is related to the competing magnetic fabrics of clinopyroxene, orthopyroxene and hornblende. The measured AMS does not agree with any modelled single crystal anisotropy; the measured maximum susceptibility is near the predicted orientation for clinopyroxene, but the minimum axis is closer to what is expected for orthopyroxene or hornblende.

5. Conclusions

This study is a further demonstration of modelling magnetic fabrics based on the modal composition and CPO of relevant phases in a rock [cf. 14]. A good agreement has been found between measured and modelled AMS in particular for amphibolite and peridotite, where the AMS is dominated by a single mineral. The magnetic fabric, even though clearly related to the CPO, is not necessarily coaxial with the mineral fabric. Pyroxenite is a good example for a composite fabric from minerals whose individual magnetic fabrics compete with one another. When modelling AMS from CPO it is important: (1) to have enough grains to allow for reliable orientation statistics in the EBSD data; (2) to use a sample surface that is representative for the whole sample; and (3) to know the intrinsic AMS of minerals responsible for the magnetic fabric. This study demonstrates, newly also for mafic and ultramafic rocks, that AMS can be applied as proxy for rock texture in geologic applications, in particular when it is dominated by a single mineral.

References

[1] Borradaile GJ and Henry B 1997 *Earth-Science Reviews* **42** 49-93
[2] Borradaile GJ and Jackson M 2010 *Journal of Structural Geology* **32** 1519-1551
[3] Hrouda, F 1982 *Geophysical Surveys* **5** 37-82
[4] Martín-Hernández F and Ferré EC 2007 *Journal of Geophysical Research* **112** B03105
[5] Archanjo CJ et al 1994 *Journal of Structural Geology* **16** 323-335
[6] Borradaile GJ, Steward RA and Werner T 1993 *Tectonophysics* **227** 1-15
[7] Schulmann K and Ježek J 2011 *International Journal of Earth Sciences* **101** 705-714
[8] Zak J, Verner K and Tycova P 2008 *Geological Magazine* **145** 537-551
[9] Prior DJ et al 1999 *American Mineralogist* **84** 1741-1759
[10] Boiron T et al 2013 *Geophysical Journal International* **193** 585-602
[11] Biedermann AR 2014 *PhD Thesis 21989* ETH Zurich (Zurich, Switzerland)
[12] Hill R 1952 *Proceedings of the Physical Society A* **65** 349-354
[13] Mainprice D, Hielscher R and Schaeben H 2011 *The Geological Society Special Publications* **360** 175-192
[14] Martín-Hernández F et al 2005 *Journal of Geophysical Research* **110** B06102
[15] Kunze K and Schaeben H 2004 *Mathematical Geology* **36** 917-943

IOP Conf. Series: Materials Science and Engineering **82** (2015) 012099 doi:10.1088/1757-899X/82/1/012099

Texture and anisotropy of yield strength in multistep isothermally forged Mg-5.8Zn-0.65Zr alloy

D R Nugmanov [1], O Sh Sitdikov [2] and M V Markushev [3]
[1] Researcher, IMSP RAS, Ufa, Russia
[2] Senior researcher, IMSP RAS, Ufa, Russia
[3] Head of laboratory, IMSP RAS, Ufa, Russia
E-mail: Dnugmanov@imsp.ru

Abstract. The effect of multistep isothermal forging (MIF) on microstructure, texture and anisotropy of ambient temperature yield strength of MA14 (Mg-5,8Mg-0,65Zr (%, wt)) magnesium alloy hot-pressed rod was analyzed. It has been found that the initial axial texture is quite stable under 1st MIF step up to strain e~4.2 at 400°C and has been gradually transformed into much weaker single-peak one at further processing at 300 and 200°C to total strain of 10.2. Such texture changes were accompanied by strong grain refinement, along with significant reduction of the alloy strength anisotropy.

1. Introduction

It is well known, that magnesium-based alloys are quite attractive materials for industrial applications owing to extraordinary high specific strength (balance of high strength and low density) and damping properties. However, because of rather strong anisotropy of service and technological properties, they are less competitive than cubic light materials, such as aluminum alloys [1]. Therefore, development of effective and cost-affordable commercial method to decrease anisotropy of their properties, especially of tensile strength parameters, is quite essential. Such method should be conditioned by simplicity in mass-productions. Besides, it should result in deep structure-texture transformations via deformation grain refinement up to fine and even ultrafine sizes (grain size less than 10 and 1 μm, consequently) to process evenly distributed grain orientations in bulk billets. Judging to literature data, multistep isothermal forging (MIF) [2,3] corresponds to these requirements, even despite its very few applications for magnesium alloys.

The purpose of the present investigation is the study of structure, texture and strength evolutions in the hot-pressed rod of MA14 (ZK60) alloy during MIF.

2. Material and procedure

Billets with sizes of 70×70×170 mm cut out from a commercial hot-pressed 90 mm rod of MA14 (Mg-5,8Mg-0,65Zr (%, wt)) alloy were processed to MIF in three sequential steps at 400, 300 and 200 °C using hydraulic press with isothermal die set. Each step has been consisted of several cycles, involving few setting with axis changing (see more details in [4,5]). The orientations of the axes (directions of deformation) in the initial rod and in the MIF billet are shown in Fig. 1. Total number of cycles was 18 with corresponding true strain of a billet e=10.2 (i.e., 4.2 at 400, 3.0 at 300 and 3.0 at 200°C). The alloy structure and texture were analyzed by means of electron backscattered diffraction (EBSD) method using Tescan Mira LM scanning electron microscope and HKL CHANNEL 5 software. Pole figures were processed using the data of 5-6 EBSD maps 100×100 μm² from different areas, being enough for correct texture analysis [6]. Texture was also characterized by maximum intensity (I_{max}) of the {0001} pole figures and the integral texture index (F), estimated as [7]:

$$F_i = \int_0^{\frac{\pi}{2}} I(\varphi)\sin(\varphi)\cos^2(\varphi) * d(\varphi) \qquad (1)$$

where φ is the angle of the pole orientation, $I(\varphi)$ – texture intensity at φ.

Tensile tests were carried out on Instron 5982 using flat samples with 10x2x2 mm gage, cut from initial rod and MIF billet with orientations indicated in Figure 1. The alloy yield strength (YS) anisotropy was characterized by the coefficient ΔYS, determined as:

$$\Delta\, YS = 2\,\frac{YS_{max} - YS_{min}}{YS_{max} + YS_{min}} \tag{2}$$

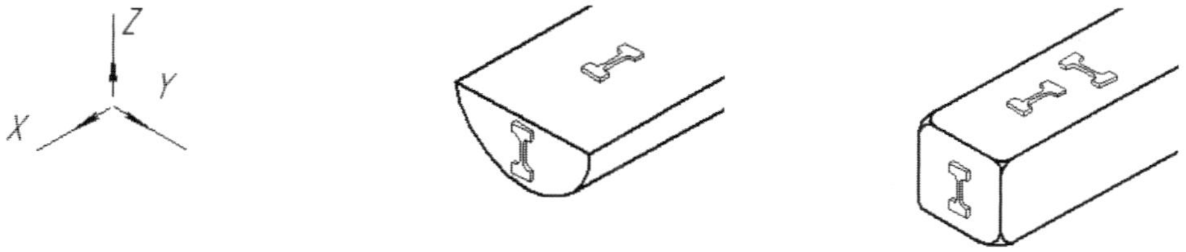

Figure 1. Rod and MIF billet with axis designation and corresponding tensile samples orientations.

3. Results and discussion

It was found that MIF led to the alloy deep grain refinement: initial coarse fibered structure, containing ~40% of fine 9 μm grains, was transformed into predominantly recrystallized one with 7 μm grains even after first MIF step, and under subsequent processing - into near homogeneous one with 2-4 μm in size grains [4].

The {0001} pole figures and inverse pole figures (IPF) plotted for X direction in the initial rod (a) and in the MIF billets processed at various steps (b trough d) are represented in Fig. 2. Initial rod had the typical for a hot-pressed semi-product axial texture with an axis of the symmetry parallel to the pressing direction (X) and with maximum intensity more than 12 (Fig. 2a). Hence, the X directions in IPF fill in all ranges along the line from <$01\bar{1}0$> to <$\bar{1}2\bar{1}0$>. Surprisingly, but after the 1st step of MIF, the texture similar to initial one was still observed, even though it was characterized by high intensity and additional sharpening of X orientation density in IPF at near <$\bar{1}3\bar{2}0$> and <$\bar{1}3\bar{2}1$> as well as its spreading toward the pyramidal components <$\bar{1}2\bar{1}2$> and <$01\bar{1}2$> (Fig. 2b).

2nd step of MIF led to significant texture changes (Fig. 2c). The maximum intensity of the practically single component texture {$11\bar{2}2$} was only a bit higher than 5, that is more than twice weaker, than after the first MIF step. According to IPF data the deformation microstructure is reoriented to provide the higher orientation density in a wide range between <$\bar{1}2\bar{1}2$>, <$\bar{1}2\bar{1}1$>, <$01\bar{1}2$> and <$02\bar{2}1$> directions with three relatively weak maximums of orientations (Fig. 2c). It is necessary to note that spreading of the initial texture was not so expected, as the temperature of the second step was still in single-phase region. However, in our case, the alloy was processed to severe plastic deformation, under which homogeneous recrystallized fine-grained structure has been formed.

Under the 3rd MIF step, near 45° rotation of a texture maximum was found in the {0001} pole figure, resulting in a single component having the same intensity (Fig. 2d). The final structure is characterized by a fiber texture, the maximum of orientation density of X axes of which roughly occurs between <$01\bar{1}2$> and <$\bar{1}2\bar{1}2$>.

Evolution of the alloy yield strength and its anisotropy under MIF is shown in Table 1 and Figure 3 along with I_{max} and F changes vs strain. As seen, the initial alloy demonstrates strong strength anisotropy: the YS in the pressing direction is nearly 30% higher, than in the transverse one. After the 1st cycle of MIF, the alloy strength in longitudinal direction has not been changed, but it decreased on ~30 MPa in transverse directions, increasing ΔYS value.

Further forging at 300 °C led to significant YS decrease along the billet and some rise in other directions, resulting in anisotropy minimum. After the last MIF step the alloy strength decreased in all directions, lightly increasing the YS anisotropy.

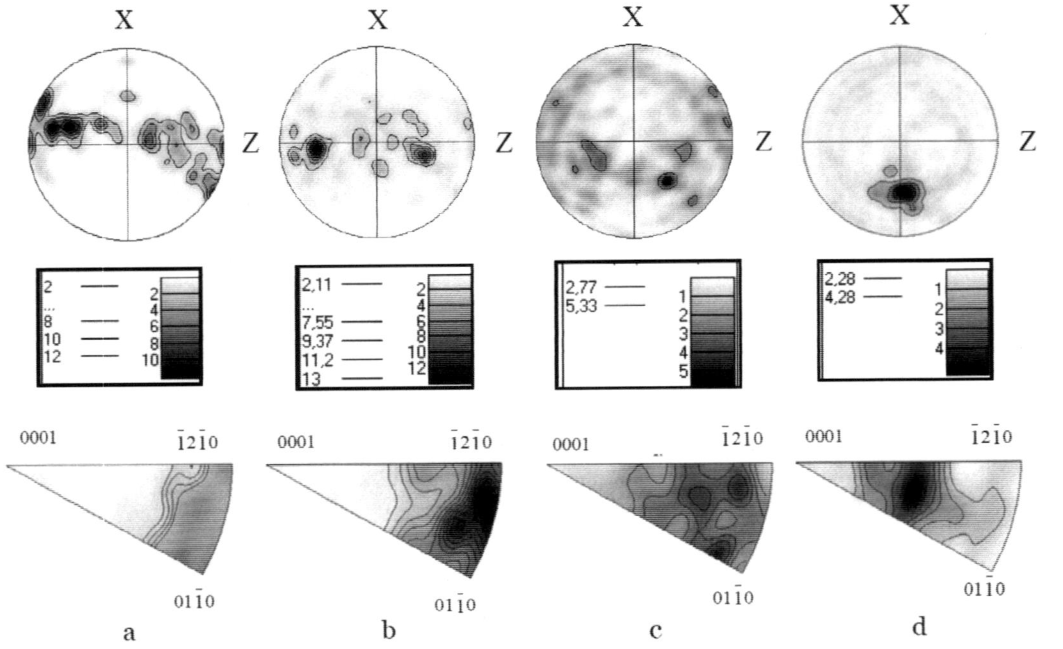

Figure 2. Pole figures {0001} and inverse pole figure for X direction in the initial rod (a) and in the MIF billet after 1st (b), 2nd (c) and 3rd (d) steps.

Table 1. MA14 alloy yield strength in different conditions

Condition	YS, MPa		
	X*	Y	Z
Hot-pressed rod	200	-	130
MIF 1st step	200	105	95
MIF 2nd step	145	135	150
MIF 3rd step	135	115	100

*- X, Y, Z correspond to samples tensile directions (cutting) in Figure 1.

It is well known that except texture, yield strength of a material depends on a number of structural factors [8]. For the current alloy they are: the lattice friction, solutes, precipitates, grain size, dislocation density, etc., which along with the corresponding parameters can influence the yield strength. Although in the present case the average alloy strength is decreased in general, testifying prevailing influence of some softening structural factors, owing to recrystallization, decomposition of magnesium solid solution, formation and coarsening of precipitates under processing at temperatures below solvus, etc. [4-6], it is evident that its anisotropy is strongly governed by the texture.

Due to the data at Figure 3 there could be found some correlations between the changes in the YS anisotropy, maximum texture intensity and integral texture index under the alloy processing. Thus, it can be concluded, that the alloy ΔYS is clearly depends on texture parameters. These data had also clearly shown that due to sequential changes in loading directions, taking place during MIF performed under decreasing temperature conditions, the texture of the Mg alloy significantly weakens, thereby resulting in decrease in its strength anisotropy.

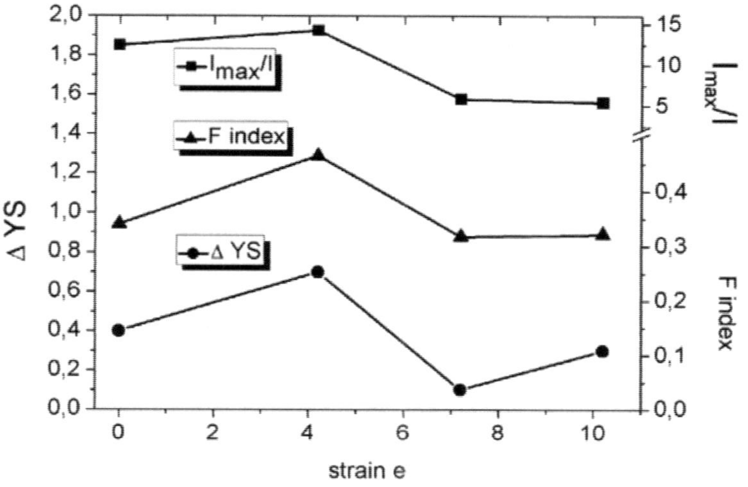

Figure 3. Dependences of the alloy coefficient of yield strength anisotropy (Δ YS), maximum texture intensity (I_{max}/I) and texture index F on MIF strain.

4. Conclusions

MIF, performed under decreasing temperature conditions in the range from 400 to 200°C, results in significant grain refinement and weakening the texture of the hot-pressed Mg alloy MA14, thereby promoting decrease in its yield strength anisotropy. MIF at 400°C has a minor effect on the alloy yield strength and its anisotropy, corresponding to minimum changes in type of texture in spite of strong microstructural changes. Further MIF at 300°C gives the main effect on YS anisotropy due to changes in type of texture and its intensity decrease.
There is a strong correlation between the maximum texture intensity, the integral texture index and strength anisotropy. In particular, high texture intensity results in high yield strength anisotropy and, to the contrary, texture weakening promotes a less anisotropy of strength. The latter becomes more pronounced in case of texture type changes.

Acknowledgment

The study was partially supported by RFBR; project № 14-08-31344.

References

[1] Mordike BL, Lukac P 2007. Physical Metallurgy. In: Magnesium Technology. Metall., Des. Data, Appl., ed by Friedrich H.E., Mordike B.L., New York.: Springer, Chapter 5, 667.
[2] Mulyukov RR 2008 *J. Mater. Sci.* **43**. 257.
[3] Markusehv MV 2011 *Letters on Materials*. **1**. 36.
[4] Nugmanov D 2012 *Bas. Probl. Mat. Sci.* **9** 230.
[5] Nugmanov DR 2014 *Letters on Materials* **4** 209.
[6] Nugmanov DR 2013 *Perspect. Mater.* **15** 101.
[7] Czerwinski F *Magnesium Alloys* 2011 *Design, Processing and Properties* (Rijeka: InTech)
[8] Estrin Y, Vinogradov A 2013. *Acta Mater.* **61** 782.

IOP Conf. Series: Materials Science and Engineering **82** (2015) 012100 doi:10.1088/1757-899X/82/1/012100

Texture dependent lattice strains and texture gradient in Al7020

Z Y Zhong[1,2], H-G Brokmeier[1,3], E Maawad[2] and N Schell[2]
[1] Technische Universität Clausthal, IWW-TEXMAT, Agricolstraße 6, 38678 Clausthal-Zellerfeld, Germany
[2] Helmholtz-Zentrum Geesthacht, GEMS Outstation at DESY, Notkestraße 85, 22607 Hamburg, Germany
[3] Helmholtz-Zentrum Geesthacht, Max-Planck-Straße 1, 21502 Geesthacht, Germany
E-mail: zhengye.zhong@hzg.de

Abstract. Firstly, an Al7020 block was characterized by the texture gradient, which was remarkably strong. Texture sharpness in the center with about 30 mrd (multiple random distribution) shows typical plain strain texture components. On the surface and close to the surface the texture sharpness is much weaker showing also shear components. Strongest shear was not directly at the surface but 4 mm deeper. The texture analysis at HEMS Beamline (Petra III/DESY-Hamburg) was done with continuous scanning mode to include all grains to improve the grain statistics. Secondly, with an in situ synchrotron experiment the texture dependent lattice strain behavior was investigated using flat tensile samples oriented 0°, 45° and 90° to the rolling direction (RD). Texture induced anisotropy influenced on the lattice dependent yield strength and the lattice dependent stress-strain behavior, which will be discussed in detail. Due to the high energy synchrotron beam complete Debye-Scherrer rings were obtained so that the patterns, parallel and perpendicular to loading direction (LD), were obtained simultaneously.

1. Introduction

The rolling texture of fcc metals with high stacking fault energy, such as aluminum alloys, exhibits orientations along β fiber at large degrees of rolling under plane strain state, where the plastic deformation is realized by slips on the {111}<110> slip systems, without presence of twining. However, due to the friction between rolls and workpiece and roll gap geometry, texture gradient through the thickness will be formed. Investigations revealed that the friction influenced on the thin surface layer, whereas the roll gap geometry may cause shear deformation through the thickness [1, 2]. Besides, in the present study, the texture induced anisotropic behaviors of the highly textured center layer are characterized by the lattice strain evolution during uniaxial tension along 0°, 45° and 90° to RD , respectively.

2. Experimental procedure

The as-received Al7020 block has a thickness of 29.7 mm, from which the texture-gradient sample was cut along the normal direction with dimensions of 7.09 × 6.36 × 29.7 mm^3. The investigated layers are described by the parameter $s = 2t/t_0$ (t is the distance from the center and t_0 is the plate thickness). Five layers were investigated, corresponding to the s values of 0.97, 0.73, 0.49, 0.26 and 0, respectively. Pole figure measurements were carried out at the high energy X-ray beamline HEMS P07B at PETRA III (DESY, Hamburg), with X-ray wavelength of 0.1422 Å, beam size of 0.5 × 0.5 mm^2, sample detector distance of 1111 mm and mar345 detector. The ω range for pole figure measurement was from -92.5° to 92.5° with 5° per step using continuous scanning mode. Complete orientation distribution function (ODF) was calculated from the measured (111), (200) and (220) pole figures by iterative series expansion method with an expansion degree of $L_{max} = 35$.
The flat tensile samples for lattice strain evolution were cut from the center layer (s = 0) along 0°, 45° and 90° to RD, respectively, with gauge volume of 3 × 3.4 × 18 mm^3 (thickness × width × length). In situ tensile test was carried out at HEMS P07B, with X-ray wavelength of 0.1420 Å, beam size of 0.5 × 0.5 mm^2, sample detector distance of 1119 mm and Perkin Elmer XRD 1622 detector, on a universal tensile machine with loading speed of 5 × 10^{-4}

mm/s till UTS at room temperature. In order to get lattice spacing d_{hkl} in both parallel and perpendicular to LD, the intensity-2θ spectra were integrated from complete Debye-Scherrer ring within 10° segment in γ along the two directions. Lattice strains under different loads were obtained by $\varepsilon = (d_{hkl} - d_{o,hkl})/ d_{o,hkl}$, where $d_{o,hkl}$ is the starting lattice spacing of stable elastic deformation. The misalignment of detector was calibrated by LaB_6 powder.

3. Results and discussion

3.1. Texture gradient along the normal direction of the Al7020 block

The complete ODF analysis results at φ_2 of 45°, 65° and 90° sections are shown in Fig. 1 for the above mentioned 5 layers. Table 1 shows the main ideal orientations observed in Fig. 1. In the surface layer (s = 0.97), the texture is consisted of the orientations along the β fiber and weak shear components. But the Bs component scatters in the vertical direction in the φ_2 = 90° section, while the Copper component shifts to the Dillamore component which is at the lower Φ angle in the φ_2 = 45° section and S component shifts to the lower Φ angle in the φ_2 = 65° section. Besides, the rotated cube component scatters towards {112}<110> shear orientation. In the subsurface layer (s = 0.73), rotated cube component becomes stronger and the orientations along the β fiber get weaker than the surface layer. Furthermore, the scattering of the β fiber orientations grow larger. In the quarter layer (s = 0.49), the orientations along the β fiber become stronger and tend to shift towards their ideal positions. The scattering of the Copper, Bs and S components also get smaller. The rotated cube component gets very weak. Moreover, there is a weak Goss component. Further towards the center (s = 0.26 and s = 0), the texture is composed of the typical plain strain rolling components which concentrate along the β fiber, and the shear components totally disappear.

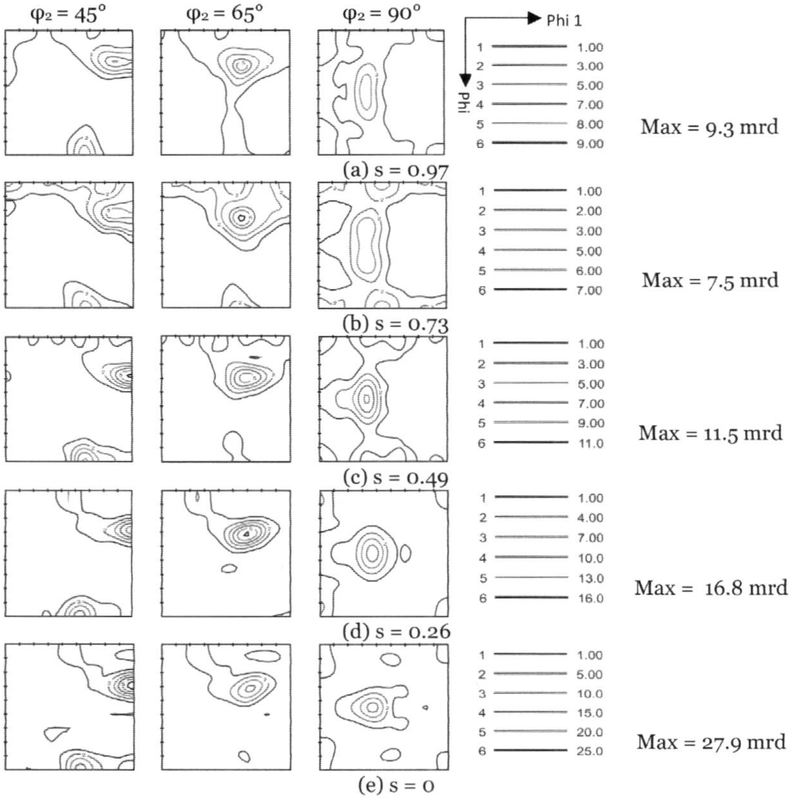

Fig. 1: ODF results of different layers of the Al 7020 plate

Table 1: Miller indices and Euler angles of the main ideal orientations observed in Fig.1.

Designation	Miller indices	Euler angles			Designation	Miller indices	Euler angles		
	{hkl}<uvw>	φ_1/°	Φ/°	φ_2/°		{hkl}<uvw>	φ_1/°	Φ/°	φ_2/°
Copper	{112}<111>	90	30	45	Goss	{011}<100>	0	45	0/90
Dillamore	{4 4 11}<11 11 8>	90	27	45	Cube	{001}<100>	0	0	0/90
S	{123}<634>	59	34	65	Rotated cube	{001}<110>	45	0	0/90
Bs	{011}<211>	35	45	0/90	Shear	{112}<110>	0	35	45

Deviation from plain strain deformation state where $\varepsilon_{11} = -\varepsilon_{33}$ and $\varepsilon_{12} = \varepsilon_{13} = \varepsilon_{22} = \varepsilon_{23} = 0$ leads to shear texture (1, 2, 3 stand for RD, TD and ND of the as-received block, respectively) [3]. Relaxation of any of the shear components (ε_{12}, ε_{13}, ε_{23}) could be achieved by rolling, resulting in different texture changes with respect to the plain strain deformation [4]. In the s=0.97 and s = 0.73 layers, at φ_2 = 45° section, the scattering of Dillamore towards the rotated cube component, especially at s = 0.73 layer, can be attributed to the relaxation of ε_{13}. The increase of rotated cube component, Dillamore component scattering to the lower and lower Φ angle, and Bs and S components at φ_2 = 65° and φ_2 = 90° sections scattering further towards the rotated cube component, indicate ε_{13} value increasing from surface to subsurface. In the s = 0.49 layer, Dillamore component shifts to the higher Φ angle where it corresponds to the Copper component at φ_2 = 45° section, indicating the ε_{13} value gets smaller. In φ_2 = 90° section, besides the scattering towards the rotated cube component, Bs component also scatters towards the Goss component owning to the relaxation of ε_{12}. From s = 0.26 to s = 0 layer, the texture sharpness gets stronger and stronger and plain strain deformation prevails, making the Copper, Bs and S components tend to locate at their ideal positions and the scattering of these components get smaller.

3.2. Lattice strain evolution during uniaxial tension

The lattice strain evolution along 0°, 45° and 90° to RD during uniaxial tension is shown in Fig. 2. The differences in lattice strain evolution along these three directions directly show the anisotropic behavior. In the elastic region, all results show the expected behavior related to less single crystal elastic anisotropy, which is expressed as $2C_{44}/(C_{11}-C_{12})$ with a value equal to 1.22 [5]. Means, the texture dependent lattice strain evolution in elastic region is not evident. The lattice dependent yield strengths of individual planes are shown in Table 2. As one can see the lattice dependent yield strength varies due to the orientation of the tensile direction with respect to the rolling geometry, because texture induced anisotropy makes the shape of yield surface deviate from the isotropic state. Therefore, different yield strengths can be obtained through tensile loading in different directions. In the plastic region, much information can be obtained, but in the present article, only the orientation dependent work hardening behaviors of individual lattice planes and the abnormal behaviors of {200} planes are discussed. When plastic deformation starts, the grain oriented work hardening behavior (see the single crystal investigations of Kocks [6]) dominates the lattice behavior, as shown in Fig. 2. For all the three samples, the work hardening behaviors of individual lattice planes can be divided into two groups. The first group shows a flat work hardening curve after yield while the second group shows an increasing work hardening curve. The different work hardening behaviors are due to the different stress states. The stress state of the first group is dominant by the [100] tension, whereas that of the second group is dominant by [111] tension (see Bishop [7] for the different stress states for polyslip). However, the lattice strain of {200} planes exhibits a dramatic change after yield in both parallel and perpendicular to the LD of the 45° and 90° to RD samples. The change is mainly related to the stress state of the <110> directions of the {200} planes [5, 8, 9]. When the <110> of the {200} is stressed, the active slip systems are (111)[10-1], (111)[01-1],(11-1)[101] and (11-1)[011]. If the amount of slip on the

4 active slip systems is the same, the deformation is asymmetric and it will cause large contraction in the [001] direction. When the contraction is opposite to the LD, it will cause a compressive stress. If the contraction is parallel to LD, a tensile stress will be induced. For the {200} lattice strain perpendicular to the LD, the contraction is much larger than the matrix, leading to tensile stress in this direction, as shown in Fig. 2 (b).

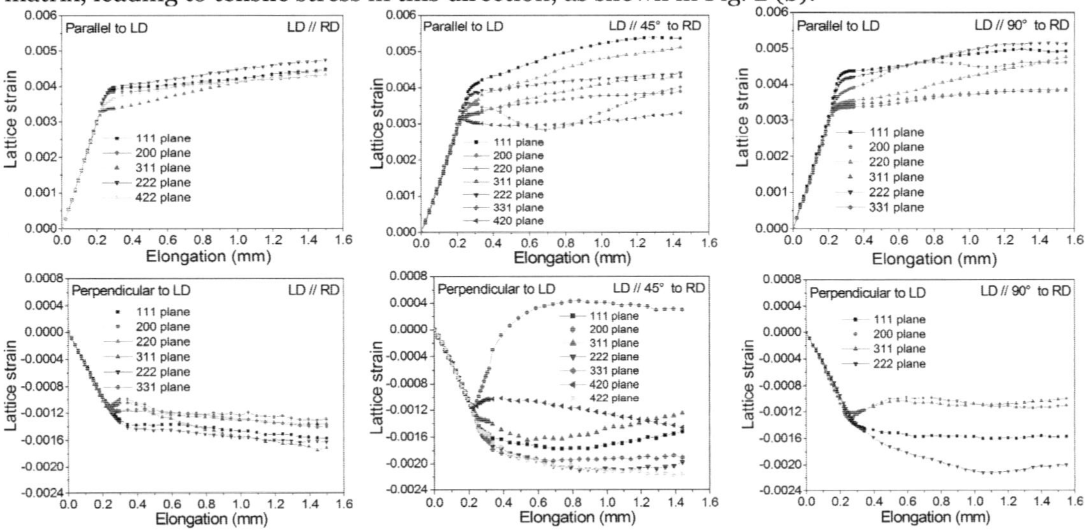

(a) RD (b) 45° to RD (c) 90° to RD

Fig. 2: Lattice evolution along different directions during uniaxial tension

Table 2: Lattice dependent yield strength (MPa)

Lattice plane	111	200	220	311	222	331	420	422
RD	315.40	314.07	---	312.37	314.63	---	---	322.24
45° to RD	293.65	290.25	298.62	296.13	291.17	304.77	299.81	---
90° to RD	298.49	295.86	312.75	311.48	295.51	308.86	---	---

4. Conclusions

Texture gradient is found in the as-received Al7020 block. From the surface to subsurface, the deformation mode is shear plus plain strain. From subsurface to center, plain strain deformation prevails. Besides, texture induced anisotropy is characterized by lattice strain evolution during uniaxial tension along the 0°, 45° and 90° to RD directions. The results show that lattice dependent stress-strain behaviors are strongly related to the stress states, which are caused by both external load and the interactions between neighboring grains.

References

[1] Trusykowski W, Krol J and Major B 1980 *Metall. Trans.* **11A** 749
[2] Asbeck H O and Mecking H 1978 *Mat. Sci. Eng.* **34** 111
[3] Hirsch J and Lücke K 1988 *Acta Metall.* **36** 2883
[4] Engler O, Huh M Y and Tomé C N 2000 *Metall. Mater. Trans. A* **31A** 2299
[5] Clausen B, Lorentzen T and Leffers T 1998 *Acta Mater.* **46** 3087
[6] Kocks U F 1960 *Acta Metall.* **8** 345
[7] Bishop J F W 1953 *Phil. Mag.* **44** 51
[8] Hosford W F 1964 *Trans. Met. Soc. AIME* **230** 12
[9] Daymond M R, Tomé C N and Bourke M A M 2000 *Acta Mater.* **48** 553

IOP Conf. Series: Materials Science and Engineering **82** (2015) 012101 doi:10.1088/1757-899X/82/1/012101

Texture evolution of warm-rolled and annealed 304L and 316L austenitic stainless steels

D Lindell[1]
[1] Researcher, Swerea KIMAB AB, Box 7047, 165 07 Kista, SWEDEN
E-mail: david.lindell@swerea.se

Abstract. The brass-to-copper rolling texture transition is observed during warm rolling austenitic stainless steels. In the current paper austenitic stainless steels 304L and 316L have been subjected to warm rolling at 700°C to 90 % reduction. The evolution of microstructure and texture during subsequent annealing has been studied using dilatometry and electron backscatter diffraction. Recrystallisation texture for 304L was primarily cube with some retained rolling texture while 316L only had retained rolling texture. The different behaviour between the two steels is believed to originate from differences in molybdenum content.

1. Introduction

Stainless steels have attractive combinations of mechanical and chemical properties that can be tailored to fit many different type of applications. Highly fluctuating cost of alloying elements has increased the interest for ferritic grades and lean duplex grades. Lean duplex grades are usually benchmarked against austenitic grades where they have an advantage when it comes to specific weight. This trend towards a more diversified stainless steel portfolio has reinforced the motivation to understand the delicate interplay between microstructure and various types of corrosion. This in turn can lead to new grades that are optimised not only in terms of chemistry but also by the microstructure. One example is certain high temperature grades for which the grain size is a balance requirement in terms of creep strength and corrosion resistance. A more futuristic view includes the concept of grain boundary engineering. Steels with over 80 % coincidence lattice boundaries (CSL) and a consequent impressive resistance against intergranular corrosion have been demonstrated [1, 2]. Further, crystallographic effects on localised corrosion [3, 4] and corrosion in sulphuric acid [5] have been observed. The aim of the current paper was to use a processing rather than an application point of view to explore the possibilities of texture and microstructure control of warm rolled material. Stainless steels have low-to-medium stacking fault energy so the rolling texture is usually of the brass type [6-8]. However, Hu [8, 9] has demonstrated that AISI 304 undergoes the brass-to-copper texture transition during warm rolling. The warm rolling approach is appealing because recrystallisation textures of materials with copper-type rolling texture can be very sensitive to processing history and chemistry.

2. Experimental

Two 26 mm plates of 304L and 316L were received from Outokumpu Stainless in Degerfors, Sweden. The chemical composition of the steels is given in Table 1. In order to refine the grain size prior to warm rolling, the 26 mm thick plates were cold rolled to 50 % reduction and annealed at 850°C for 1 h. The now 12 mm thick plates were cut to pieces of 20 x 8 x 12 mm^3 thereafter reheated to 600, 700 and 800°C for 15 min and immediately warm rolled in six passes down to 90 % nominal reduction. The surface temperature after these six passes was about 50-80°C lower than after reheating. For 304L, the copper rolling texture was pronounced in all three strips. However the traces of brass texture were largest for 600°C and some degree of recrystallisation was observed for the 800°C specimen so the 700°C specimen was selected for further studies. For comparison, the 700°C version of 316L was also selected. The actual rolling reductions of these two strips were 91 and 88 % for 304L and 316L respectively. The evolution of microstructure and texture during annealing was studied using dilatometry and EBSD. The annealing temperatures were 800, 900 and 1140°C and the annealing times (time at annealing temperature) were 3, 10, 30 and 120 s. The heating and cooling rate was in all cases 50°Cs⁻¹. EBSD maps comprising at least 30k points with coarse step size were collected for texture analysis and higher magnification maps were collected for grain size determination. Because no standard lubrication could be used during rolling the surfaces consists of a sheared zone, which was excluded from further analysis. Orientation distribution functions were calculated using the serial expansion method using the software Salsa within the HKL software package using L_{max} = 22 and 3° Gaussian spread. Grain sizes extracted from the EBSD were determined using the linear intercept method along the rolling direction with a critical disorientation of 10°.

Table 1. Heat analysis of the steels investigated (wt%).

	Cr	Ni	Mn	Mo	Cu	Si	C	N
304L	18,3	8.1	1.5	0.5	0.4	0.3	0.021	0.07
316L	17.5	11.1	1.2	2.3	0.5	0.4	0.014	0.07

3. Results and discussion

3.1 Evolution of microstructure

The recrystallisation fraction and average grain size as a function of annealing cycle is given in Table 2. 304L recrystallises at 800°C, even though not fully at the longest time investigated, while 316L does not recrytallise at all within this time frame. Both steels recrystallise at 900°C with a grain size of ~4 μm when fully recrytallised. It is often stated that the relative high recrystallisation temperature of the stainless steels make it impossible to achieve grain sizes below about 10 μm [10, 11] without special treatments such as martensite reversion. Alloying for this kind of purpose is a balancing act between the effect of alloying on recrystallisation temperature and on grain boundary mobility. The chemistries used here are by no means optimised in this respect, however it demonstrates that at least half of the typically stated 10 μm can be achieved in practise with simple means. The effect of alloying on grain boundary mobility is most clearly seen at the highest annealing temperature, 1140°C. The Mo alloyed 316L had a grain size of 20 μm after the annealing for 120 s at 1140°C while 304L had a grain size of 54 μm after the same treatment.

Table 2. Average grain size and recrystallisation fraction as a function of annealing cycle.

		304L			316L		
			Average grain size (μm)**			Average grain size (μm)**	
Annealing temperature (°C)	Annealing time (s)	Degree of recrystallisation* (%)	Σ3 included	Σ3 excluded	Degree of recrystallisation* (%)	Σ3 included	Σ3 excluded
800	3	0	-	-	0	-	-
800	10	5	1	1	0	-	-
800	30	18	1	1	0	-	-
800	120	65	2	3	0	-	-
900	3	87	2	3	5	1	2
900	10	100	3	5	55	2	3
900	30	100	5	10	81	2	3
900	120	100	7	12	100	3	4
1140	3	100	7	12	100	5	8
1140	10	100	9	16	100	6	11
1140	30	100	13	22	100	8	13
1140	120	100	28	54	100	11	20

*Assumed that the deviation from 99 % hit rate originate from deformed material only.
**Critical disorientation = 10°.

3.2 Evolution of recrystallisation texture

ODF sections for 304L with constant $\varphi_2 = 0^o$ and $\varphi_2 = 45^o$ are given in Figure 1. The as-rolled texture (a) shows typical copper rolling texture features with a pronounced fraction of the copper orientation {112}<111>^{C}, along with S {123}<634>^{S} (not shown in these sections) and brass {110}<112>^{B}, which shows some spread towards the Goss orientation {110}<001>^{G}. At the early stage of recrystallisation (b) the copper and brass components are weakened while the Goss orientation and all other orientations along $\varphi_1 = \varphi_2 = 0^o$ are strengthened. Towards the end of recrystallisation and during the early grain growth stage (c-e), two maxima along $\varphi_1 = \varphi_2 = 0^o$ start to form at $\Phi = 0^o$ (cube orientation) and $\Phi \approx 30^o$. Further grain growth (f-g) results in strengthening of the cube orientation, but there are still remains of the rolling texture even after excessive grain growth. Here the evolution of texture is described on basis of the grain size, however there are also indications that a higher temperature for a given final grain size facilitates cube texture formation. 900°C-120s and 1140°C-3s resulted in the same average grain size but the latter had a sharper cube component, see Figure 1e-f.

A comparison between the recrystallisation texture of 304L and 316L is given in Figure 2, where the ODFs are plotted for $\varphi_1 = \varphi_2 = 0^o$. 316L had, compared to 304L, a stronger Goss component in the deformation texture. During grain growth in 316L, the Goss component weakens and a maximum at $\Phi \approx 30^o$ is formed. In contrast to 304L further grain growth does not result in the formation of cube. The difference between the two steels, and the fact that rolling texture was retained in 304L, is believed to be due to differences in grain boundary mobility by solute drag effects, mainly from molybdenum. Even the 304L used in this work had 0.5 wt% Mo, which may explain why components in the rolling texture are retained. The work by Hu [8, 9] shows that this can be avoided, unfortunately no information regarding the Mo content in their work could be found. Another explanation that cube was not formed is that insufficient number of nuclei is retained from earlier processing steps. This explanation is however not likely because the number of cube oriented grains was very similar for 304L and 316L.

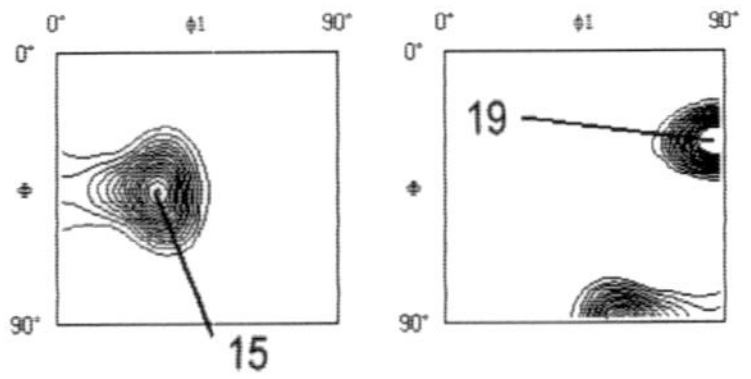

a) As warm rolled at 700°C to 91% reduction.

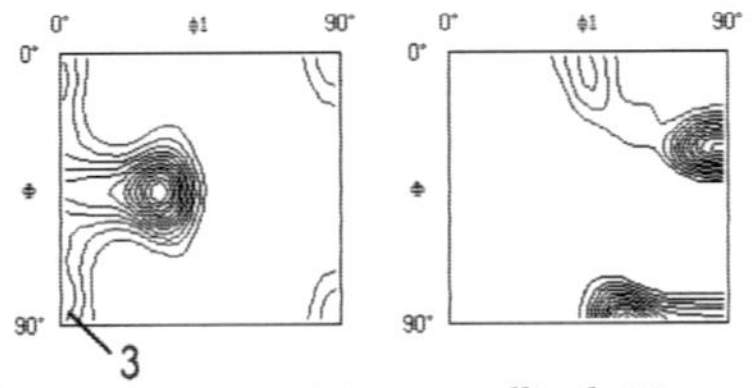

b) 800°C ; 30 s – 18 % recrystallised, GS = 1 µm. The ODF is calculated from the recrystallised grains i.e. not the full dataset.

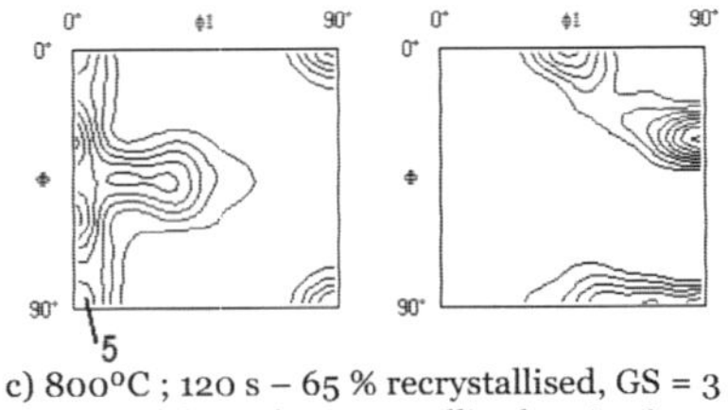

c) 800°C ; 120 s – 65 % recrystallised, GS = 3 µm. The ODF is calculated from the recrystallised grains i.e. not the full dataset.

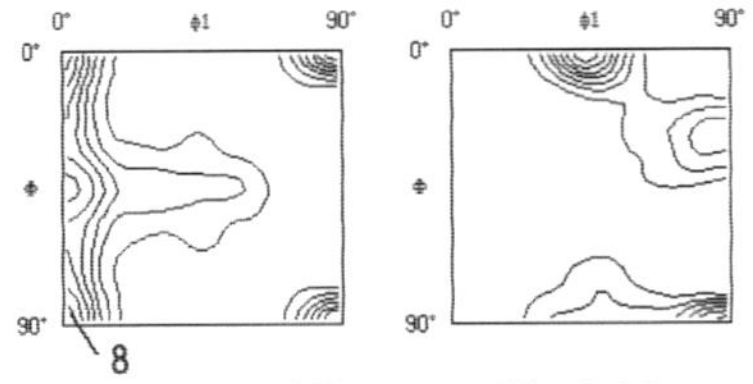

d) 900°C ; 10 s – fully recrystallised, GS = 5 µm.

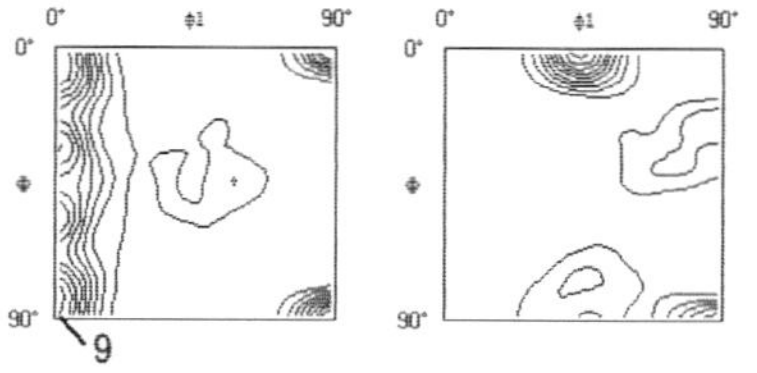

e) 900°C ; 120 s – fully recrystallised, GS = 12 μm.

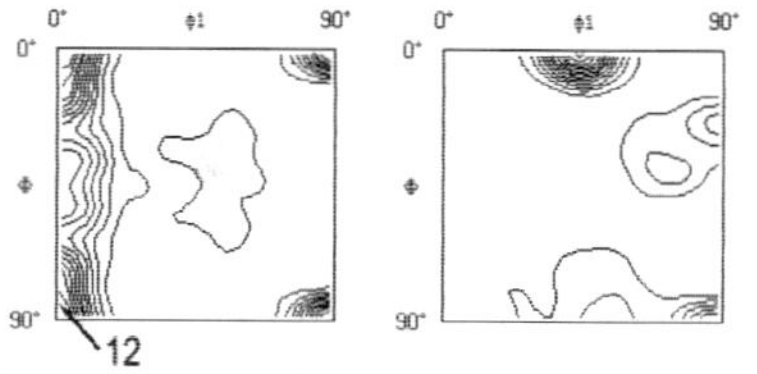

f) 1140°C ; 3 s – fully recrystallised, GS = 12 μm.

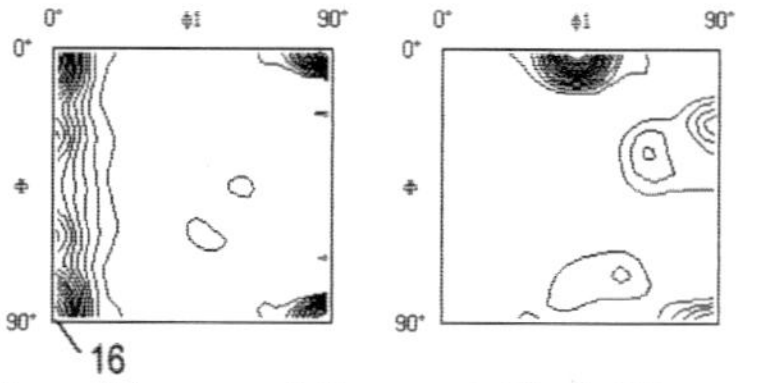

g) 1140°C ; 30 s – fully recrystallised, GS = 22 μm.

Figure 1. Examples of ODF sections for $\varphi_2 = 0^o$ (left) and 45° (right) for 304L. For the partially recrystallised material given in a and b the orientation functions are calculated on basis of the recrystallised fraction only. The contour steps are all unity.

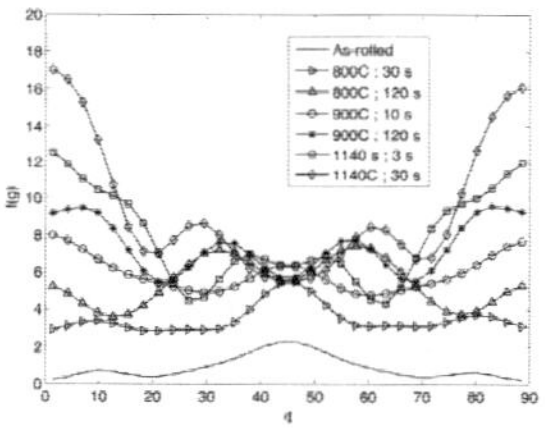

a) 304L.

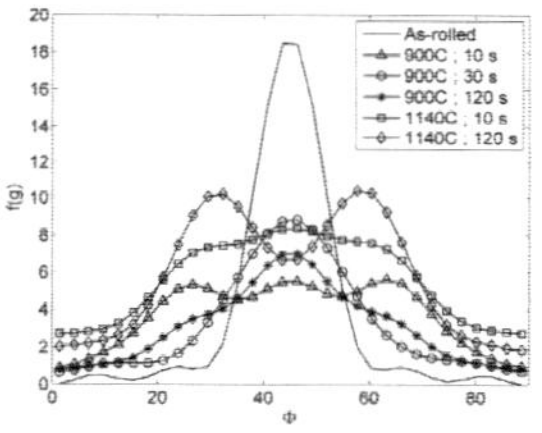

b) 316L.

Figure 2. Plot of the ODF for constant $\varphi_1 = \varphi_2 = 0$.

5. Conclusions

Microstructure and texture evolution of 304L and 316L during annealing after warm rolling have been studied. Recrystallisation proceeds with a strengthening of the Goss component that spreads along the $\varphi_1 = \varphi_2 = 0^o$ section (<100> // RD). During growth, 304L develops cube texture but with some rolling texture retained, while for 316L the rolling texture is retained without any formation of cube texture. To increase the strength of the cube texture the chemistry has to be further optimised, the current belief is that molybdenum should be avoided in order to do so.

Acknowledgments

The Hugo Carlsson foundation is greatly acknowledged for the financial support of this work. Prof. Rachel Pettersson and Prof. Bevis Hutchison are greatly acknowledged for reading the script and coming up with ideas.

References

[1] Michiuchi M, Kokawa H, Wang Z J, Sato K and Sakai K 2006 *Acta Materialia* **54** p 5179
[2] Shimada M, Kokawa H, Wang Z J, Sato Y S and Karibe I 2002 *Acta Materialia* **50** p 2331
[3] Shahryari A, Szpunar J and Omanovic S 2009 *Corrosion Science* **51** p 677
[4] Krishnan S, Dumre J, Bhatt S, Esther T, Akinlabi T and Ramalingam R 2013 *International Journal of Mechanical, Industrial Science and Engineering* **7** p 246
[5] Lindell D and Pettersson R 2014 *Submitted to Materials and Corrosion*
[6] Chowdhury S G, Das S and De P K 2005 *Acta Materiala* **53** p 3951
[7] Kumar B R, Mahato B, Bandyopadhyay N R and Bhattacharya D K 2005 *Materials Science and Engineering A* **394** p 296
[8] Hu H 1974 *Texture* **1** p 233
[9] Hu H and Goodman S R 1984 *Material Science and Engineering* **67** p 143
[10] Schino A D and Kenny J M 2003 *Materials Letters* **57** p 3182
[11] Tomimura K, Takaki S, Tanimoto S and Tokunaga Y 1991 *ISIJ International* **31** p 721

IOP Conf. Series: Materials Science and Engineering **82** (2015) 012102 doi:10.1088/1757-899X/82/1/012102

Texture gradient in a copper tube at maximum and minimum wall thickness

H.-G. Brokmeier[1,2], A. Carradó[3], N. Al-hamdany[1],T. Pirling[5], R. Wimpory[6], N. Schell[2] and H.Palkowski[4]

[1] Institut für Werkstoffkunde und Werkstofftechnik, TU Clausthal, Agricolastr. 6, 38678 Clausthal-Zellerfeld, Germany

[2] Helmholtz-Zentrum Geesthacht, Max-Planck-Straße 1, 21502 Geesthacht, Germany

[3] Institut de Physique et Chemie des Matériaux de Strasbourg, UMR7504 CNRS-UDS 23 rue du Loess, 67034 Strasbourg, France

[4] Institut für Metallurgie, TU Clausthal, Robert Koch Strasse 42, 38678 Clausthal-Zellerfeld, Germany

[5] Institute Laue Langevin, 71 avenue de Martyrs, 38000 Grenoble, France

[6] Helmholtz-Zentrum Berlin, Hahn-Meitner-Platz 1, 14109 Berlin, Germany

E-mail: heinz-guenter.brokmeier.@tu-clausthal.de

Abstract. Tube drawing is an inhomogeneous process with anisotropic flow. Quantitative texture analysis was used to get inside in eccentricity and ovality of a Cu-tube drawn without a plug. To obtain the texture gradient over the wall thickness high resolution synchrotron radiation was used. One obtains on one hand different textures at the surfaces than in the central region and the other hand a variation of the texture gradient at the maximum wall thickness of the tube compared to the minimum wall thickness.

1. Introduction

Seamless tubes are used for applications where homogeneity of strength, microstructure and extended product life are important design parameters. Cold drawing is widely applied in the industrial production of seamless tubes, employed for various mechanical applications. In order to achieve the final diameter and wall thickness, the extruded pre-tubes are reduced successively in several cold drawing steps. This can be done by just drawing the tube through a die or by adding a plug, which results in better defined wall thickness and inner surface quality. Eccentricity (E), the deviation of the wall thickness over the circumference of the tube from its average value, of pre-tubes causes non homogeneous deformation over the circumference in the single drawing steps.

Texture investigations were carried out at the high energy materials science beamline HEMS P07B at Petra III/DESY-Hamburg. Synchrotron radiation was used because of the high penetration power and the high brilliance. From the texture side it is known that tube drawing produces monoclinic sample symmetry due to a complex deformation process. Based on the process parameters reduction is a combination of reduction in wall thickness and in tube diameter. Most tube investigations were carried out at zircaloy tubes, which are based on the importance of zircaloy tubes in nuclear industries [1,2,3]. A good introduction in this topic is given by Tenckhoff [4]. Investigations on Cu tubes become recently of more interest [5]. But there is a lack of texture gradients investigations on tubes.

2. Sample description

The test sample, made of SF-Cu: 99.90 % min Cu, 0.015-0.040 % P, was a seamless tube produced by drawing without a plug as a final deformation step, see figure 1a. The pre-tube obtained by extrusion was reduced in several cold drawing steps to its final dimension. In a first step a ring of 4mm length was cut from the drawn Cu-tube. Due to the variation of the wall thickness three small cubes were prepared with cross section of about 4x4 mm^2, one at maximum thickness, one at minimum thickness and one at medium thickness position, see

IOP Conf. Series: Materials Science and Engineering **82** (2015) 012102 doi:10.1088/1757-899X/82/1/012102

figure 1b. It should be noticed that a sample cut will not change the crystallographic texture but the stress state.

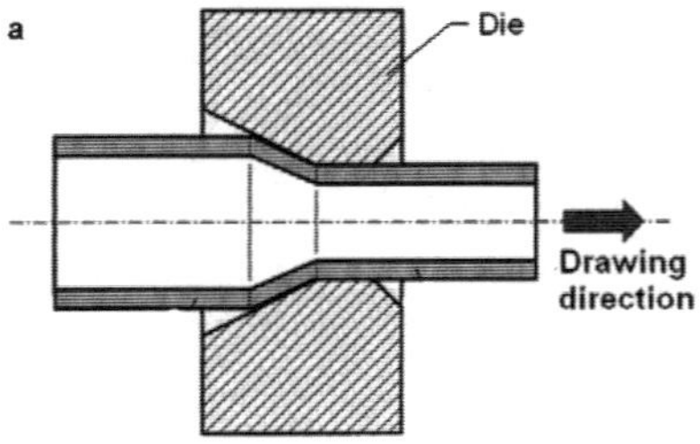

b

Parameter		Pre-tube	Final tube
Outer Ø d	(mm)	50.0	40.9
Average thickness t	(mm)	4.2	4.37
Δ thickness Δt	(mm)	0.8	0.65
Eccentricity E	(%)	9.5	7.4

Fig. 1: Plug-less drawing (a) and geometrical data of the final tube (b) over tube wall thickness

3. Experiment

Pole figure measurements were carried out at the high energy beamline HEMS-PO7B@Petra III/Hasylab - Hamburg with an energy of 86,76 keV. The wavelength of 0.1430 Å, equivalent to 86,76 keV, was much lower than standard CuKα X-rays with 1.5418 Å of laboratory X-rays. Basic advantages of the lower wavelength are on one side a much higher penetration power and on the other side very low scattering angles to obtain complete Debye-Scherrer rings. For this purpose a set of complete pole figures were obtained without sample tilt [6]. A typical set up to optimize the counting time for overnight measurements is shown in figure 2, in which three small samples have been glued together. Based on prior experiments with only three positions over the wall thickness nine positions were scanned to obtain the texture gradient more precize. A beam size of 300x500μm² (figure 2), gives a sufficient local resolution. For one sample position the total pole figure measurement consists of 37 area detector images, each of them lasted 20 sec.

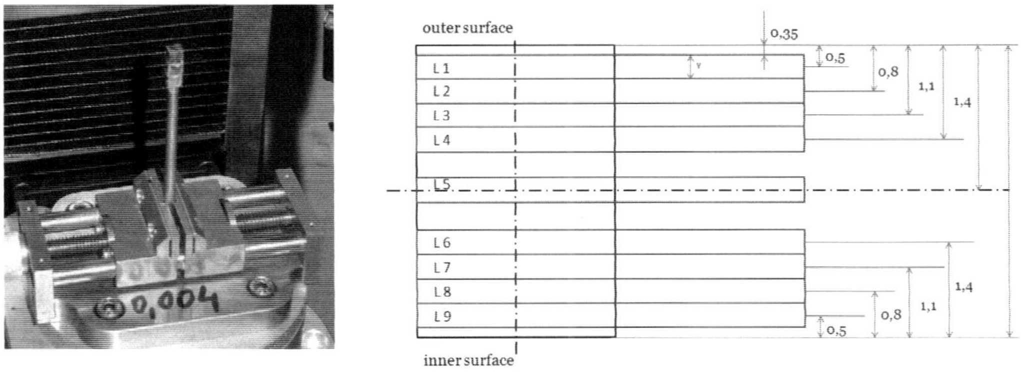

Fig. 2: Samples mounted on a pin and technical drawing of measured positions

4. Results

Quantitative texture was calculated firstly by extracting the pole figure data from area detector data by the software package SABO for three Cu reflections (111), (200) and (220). Thereafter the iterative series expansion method after Bunge was used to obtain the orientation distribution function (ODF). All steps were done with triclinic sample symmetry. As an example, figure 3 presents the φ2=45° ODF sections for the outer surface (L1), the

IOP Conf. Series: Materials Science and Engineering **82** (2015) 012102 doi:10.1088/1757-899X/82/1/012102

inner surface (L9) and the central position (L5) for two samples. As expected the texture gradient at maximum wall thickness is similar to the texture gradient at minimum wall thickness.

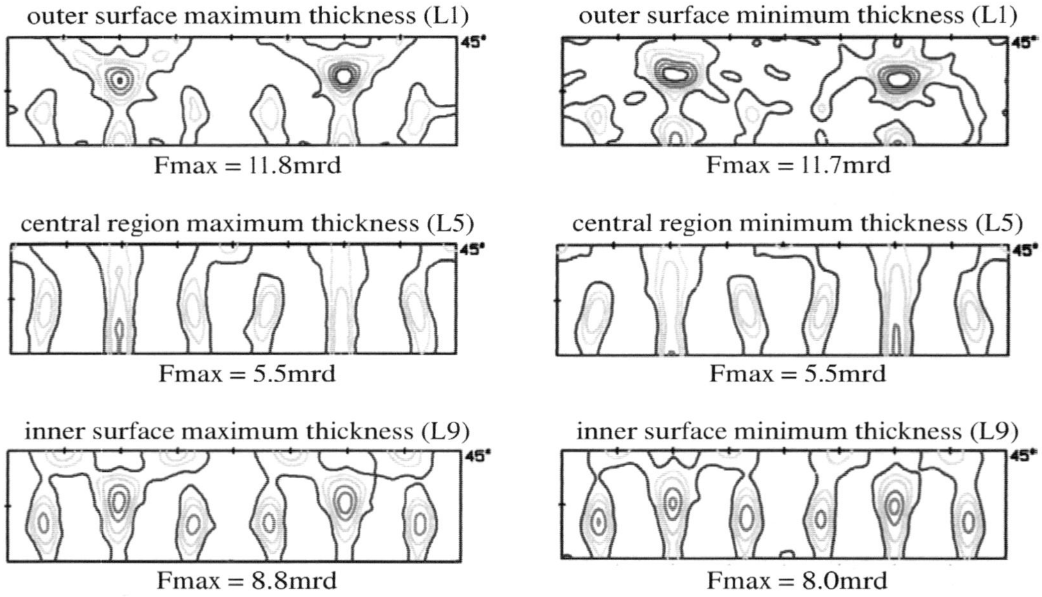

Fig. 3: ODF sections φ2=45° at outer surface, central position and inner surface

Comparing with ideal texture components and texture fibers in the orientation space (figure 4) one can see that the τ-fiber <110>//TD dominates the texture evolution with a strong variation from outer region to inner region for both samples. Particular at the inner surface (L9) a weak cube component is present, while at central position (L5) shows a minor rotated cube. Another texture component at φ1=30° to 35°,Φ= 55° to 60°, φ2=45° which is close to the γ-fiber shows a remarkable gradient from outer to inner surface.

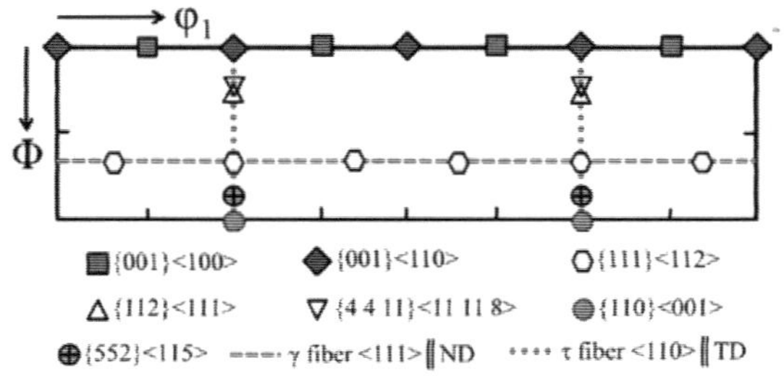

Fig. 4: Ideal texture components and texture fibers in ODF sections φ2=45°

The comparison of the two texture components at φ1=30°, Φ=60°, φ2=45° ({111}<112>) and φ1=45°, Φ=0°, φ2=45° ({001}<100>) shows for maximum thickness and for minimum thickness the same tendency (figure 5). One difference is a stronger cube component at both surfaces (L1 and L9) for the sample at maximum wall thickness. The ({111}<112> is little stronger at the inner surface (L9) for minimum wall thickness with 7.2 mrd against 5.8mrd.

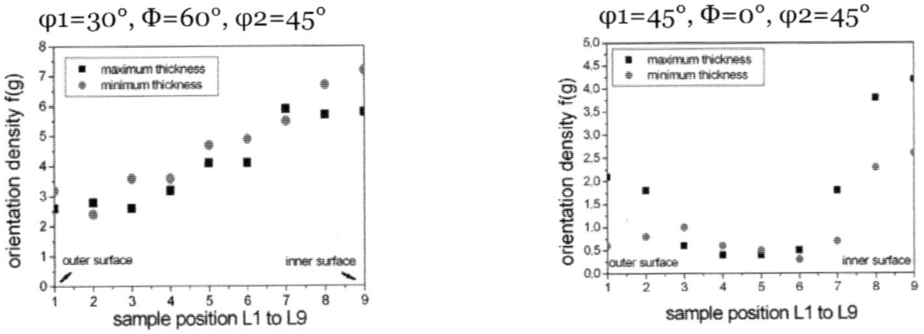

Fig. 5: Texture gradient of components $\varphi1=30°$, $\Phi=60°$, $\varphi2=45°$and $\varphi1=45°$, $\Phi=0°$, $\varphi2=45°$

The τ-fiber evolution has been discussed intensively by Hirsch et al. [7] for Cu30%Zn with different rolling degrees. There is a shift of the Cu-component to higher Φ-values which goes faster at maximum thickness than on minimum thickness. Strongest texture is at the outer surface while in the central region a much weaker texture exists.

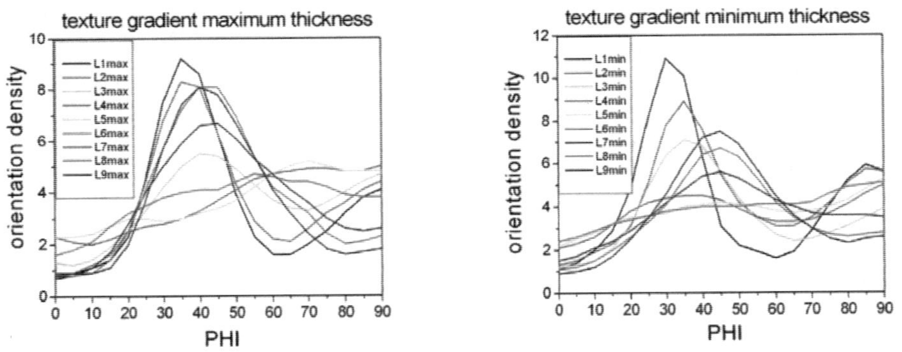

Fig. 6: τ-fiber evolution

It should be noticed that a correlation between texture gradient and tube drawing parameters needs more experiments including the initial texture.

References

[1] Vicente Alvarez M A, Santisteban J R, Domizzi G and Almer J 2011 *Acta Mat.* **59** 2210
[2] Mareau C and Daymond M R 2011 Mater. Sci. Eng. A **528** 8676
[3] Li Y, Rogge R and Holt R A 2006 Mater. Sci. Eng. A **437** 10
[4] Tenckhoff E 1980 *Verformungsmechanismen, Textur und Anisotropie in Zirkonium und Zirkaloy* (Berlin Gebrüder Bornträger Verlag)
[5] Carradò A, Brokmeier H-G, Pirling T, Wimpory R, Schell N and Palkowski H 2013 *Advanced Eng. Materials* **15** 469
[6] Brokmeier H-G and Yi S B 2008 *Texture and Texture Analysis in Engineering Materials* (Neutrons and Synchrotron Radiation in Engineering Materials Science) eds W. Reimers, A. Pyzalla, A. Schreyer, H. Clemens (Weinheim Wiley VCH Verlag) chapter 10 pp.57-77
[7] Hirsch J, Lücke K and Hatherly M 1988 *Acta metal* **36** 2905

IOP Conf. Series: Materials Science and Engineering **82** (2015) 012103 doi:10.1088/1757-899X/82/1/012103

Local strain distributions in partially recrystallized copper determined by *in situ* tensile investigation

F X Lin[1], H S Ubhi[2], M Petrenec[3], Y B Zhang[1] and D Juul Jensen[1]

[1] Danish-Chinese Center for Nanometals, Section for Materials Science and Advanced Characterization, Department of Wind Energy, Technical University of Denmark, Risø Campus, 4000 Roskilde, Denmark

[2] Oxford Instruments NanoAnalysis, Halifax Road, High Wycombe, HP12 3SE, Great Britain, harvinderubhi@gmail.com

[3] Tescan Orsay Holding, a.s., Libušina tř. 21, 623 00 Brno, Czech Republic
E-mail: lnfe@dtu.dk

Abstract. A partially recrystallized copper sample produced by cold-rolling and annealing was deformed *in situ* by uniaxial tension in a scanning electron microscope, and electron backscatter diffraction data were collected before and after deformation to certain strains. The local strain distributions are quantified using digital image correlation. Distributions of the normal strain along the tensile direction (ε_{xx}) are shown in this paper. The recrystallized grains are found to deform more than the remaining unrecrystallized matrix. When ε_{xx} is averaged along lines perpendicular to the tensile direction, significant variation are observed, which may be related to the local recrystallized volume fraction.

1. Introduction

Nanostructured and ultra-fine-grained metals produced by plastic deformations to high strains often have high strength but poor ductility, which limits their practical application. Introduction of recrystallized grains into the deformed matrix to create a 'composite' structure has been suggested to improve ductility of such materials. However, to achieve certain ductility, the recrystallized volume fraction (V_V) significantly varies for different samples. For example, for copper deformed by dynamic plastic deformation (DPD), the ductility does not improve much until V_V is larger than 80% [1], whereas, if the sample was rolled after DPD, the ductility is observed to improve when V_V is ~20% [2]. The different "efficiency" of the recrystallized material for improving the ductility may be related to heterogeneities within the sample [3], which leads to different local strain distributions. Digital image correlation (DIC) has recently been used to map local strains [4]. The correlation can be based on painted or deposited patterns [e.g. 5], or on samples' own microstructure [e.g. 6,7]. The latter has the advantage of further linking the strains to the microstructure. In this work, we investigated a ~45% partially recrystallized copper produced by cold-rolling and annealing. The microstructural evolution was followed by electron backscatter diffraction (EBSD) during *in situ* tensile tests, based on which local strain distributions were determined using DIC. The strains in the recrystallized grains and in the remaining unrecrystallized matrix are compared and discussed for understanding the effects of heterogeneities on the local strain distribution.

2. Experimental

Oxygen free high conductivity copper (purity 99.95%) was rolled at room temperature to a thickness reduction of 90%, corresponding to a von Mises strain of 2.7. A tensile specimen (shape and dimensions shown in Fig. 1a) was cut from the rolled plate with the tensile direction parallel to the rolling direction (RD). The specimen was annealed at 230 °C for 30 min, after which it was 45% recrystallized according to previous EBSD measurements [8]. The specimen was ground and electro-polished before tensile testing, and the sample thickness after electro-polishing was 1.58 mm.

The *in situ* tensile test was conducted in a TESCAN MIRA 3 XM scanning electron microscope (SEM) using a Gatan Microtest 2000EW tension stage. The cross head velocity was 0.4 mm/min. Before the tensile test, an area of 120×150 μm² was mapped by EBSD on the rolling section (defined by RD and the transverse direction (TD)) using a step size of 0.2 μm. During the tensile test, the deformation was interrupted at several strains, and the same region was characterized by EBSD repeatedly using the same step size.

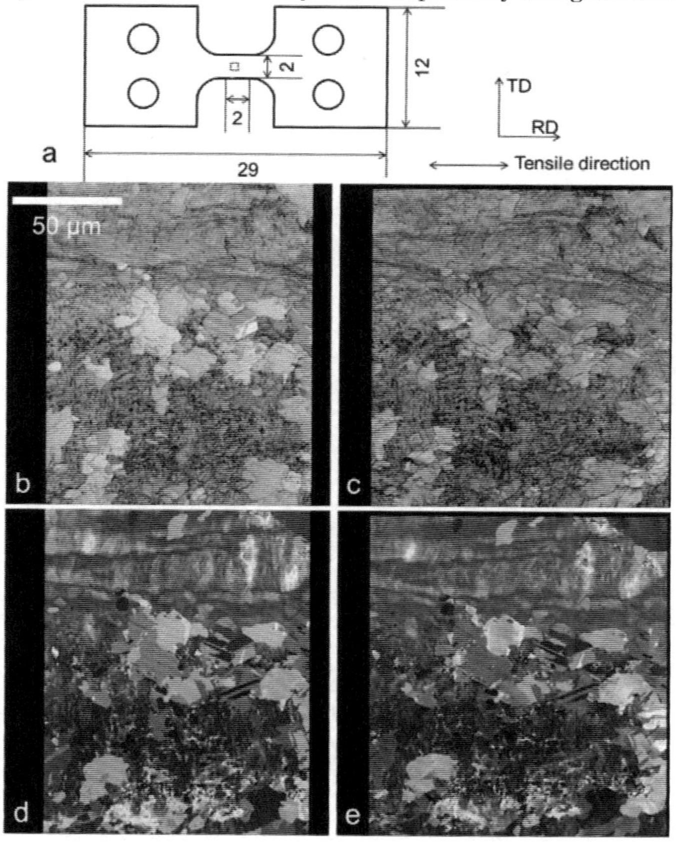

Fig. 1 Specimen shape (a) and EBSD maps (b-e) used for DIC. The unit of the specimen dimensions in (a) is mm. The small window in (a) illustrates the position of the area mapped by EBSD. b) and d) are maps before loading, while c) and e) are those at the maximum load. b) and c) are band contrast (BC) maps, and d) and e) are colored according to inverse pole figure of RD (IPF-RD). The IPF-RD maps have been converted to a gray scale for DIC.

EBSD maps before and after straining were compared using a MATLAB-based DIC algorithm [9], from which local strains were calculated. Various maps can be generated based on the EBSD data, and we used four types of maps for DIC: band contrast (BC) maps (Figs. 1 b and c), maps colored according to inverse pole figure of RD (IPF-RD maps, Figs. 1 d and e), of TD (IPF-TD maps) and of the normal direction (IPF-ND maps). For each type of maps, displacements at a grid of 59×71 points with a spacing of 10 pixels were determined using DIC, and for each grid point a subset size of 40×40 pixels (8×8 μm²) was used. In this paper, we only compare the map at the maximum load with the map before loading. At the maximum load, the engineering stress is about 330 MPa. The engineering strain is unknown for the *in situ* experiment, but according to previous conventional tensile tests of a similar sample, the engineering strain is about 10% [8].

3. Results and discussion

3.1. Image correlation and displacements

Fig.1 shows the microstructures at no load and the maximum load. The recrystallized volume fraction in the mapped area is 29%, smaller than the overall V_V (45%), which is related to the sample heterogeneity. After straining, individual recrystallized grains and subgrains of the unrecrystallized matrix can still be recognized, with only small changes of shapes. The microstructure can thus be used directly for DIC. The intercept length (boundaries >2°) is 2.2 μm for the recrystallized grains when twin boundaries are considered as normal boundary, and is less than 1 μm for the subgrains in the unrecrystallized matrix. Therefore, a subset of 8×8 μm^2 contains enough features for DIC.

The DIC analysis determines the displacements for 81%, 86%, 78% and 87% of the grid points using BC, IPF-RD, IPF-TD and IPF-ND maps, respectively (see Figs. 2 a and b for the displacement field determined from BC and IPF-RD maps). It is found that: 1) except a few incorrectly indexed points, the displacements calculated from different maps are similar; 2) the unindexed points are not in the same locations. Therefore, it is possible to optimize the displacement field by combining the calculations from the four different maps (i.e. BC/IPF-RD/IPF-TD/IPF-ND maps). For each grid point, if the displacement is indexed only in one type of maps, use it directly, otherwise, use the average value from the two nearest calculations. The combined maps are shown in Fig. 2c, and displacements of 95% grid points are indexed. The displacements are then smoothed using a Gaussian kernel smoother of a size of 5×5 points (Fig. 2d).

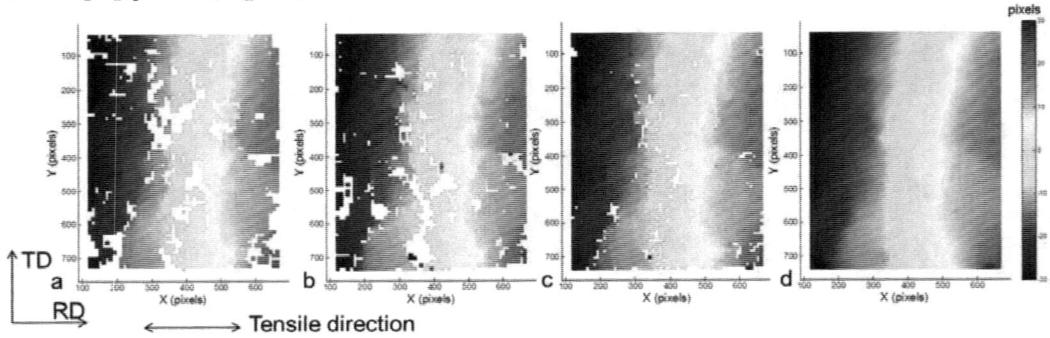

Fig. 2 Maps showing displacements along the tensile direction (i.e. RD) at the maximum load. (a) DIC using BC maps, (b) DIC using IPF-RD maps, (c) Combining displacements from all the four types of maps (BC/IPF-RD/IPF-TD/IPF-ND). d) Smoothed displacements. The white parts in a)-c) are unindexed grid points. In b), a few isolated points, which are incorrectly indexed, have totally different colors from their neighboring points. The incorrectly indexed points were removed during the smooth process.

3.2. Strain distribution in the recrystallized grains and unrecrystallized matrix

Strains were calculated by interpolating displacements using bi-cubic finite element shape functions [10]. In this work, we only discuss the normal strains along the tensile direction ε_{xx}, which are shown in Fig. 3. The average ε_{xx} is 8.8%, close to the macroscopic elongation reported in [8] for a similar sample. This value is also in good agreement with the manual calculations based on the distance change between two points (usually two triple junctions), which confirms that the strain calculated with the automatic DIC method is reliable. The average ε_{xx} of the recrystallized grains and of the unrecrystallized matrix is 10.9% and 8.0%, respectively, i.e. the recrystallized part has been deformed slightly more. "Hot spots" with strains larger than 15% are observed in the strain map (red parts in Fig. 3). Most of the "hot spots" are associated with recrystallized grains, but a few "hot spots" appear in the unrecrystallized matrix, such as that marked by an arrow in Fig. 3.

The "hot spots", i.e. the highly deformed regions in Fig. 3, tend to be elongated along TD. The ε_{xx} is therefore averaged along lines parallel to TD. As shown in Fig. 4, the average ε_{xx} significantly varies along RD. These variations are at some positions closely related to V_V, as lines with large average ε_{xx} also have large V_V. However, this relationship is far from complete, and it is not yet clear why some lines with high V_V in Fig. 4 have relatively small local strains (may be related to the fact that V_V is only measured at the surface, while the ε_{xx} reflects bulk behavior). It is interesting to point out that the "hot spots" in the unrecrystallized matrix (marked in Fig. 3) locate at lines with high V_V, which suggests that the variation of V_V may have strong effects on the local strain distribution, and the "efficiency" of the recrystallized grains may be related to their spatial distribution.

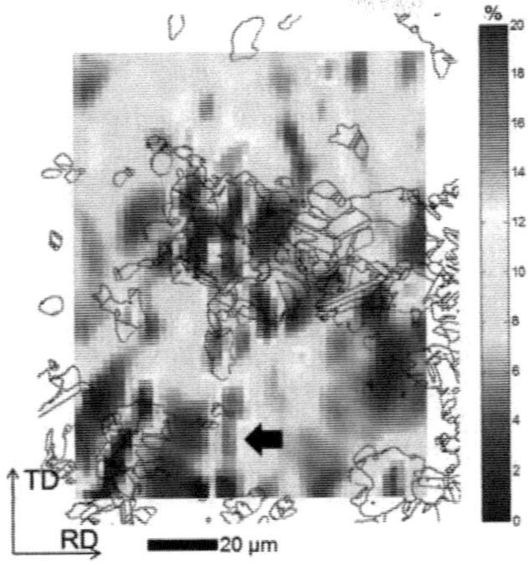

Fig. 3 Maps showing the normal strains parallel to the tensile direction ε_{xx}. Boundaries of the recrystallized grains are shown by black lines on top of the strain map. The arrow marks a region where large local deformation occurs in the unrecrystallized matrix.

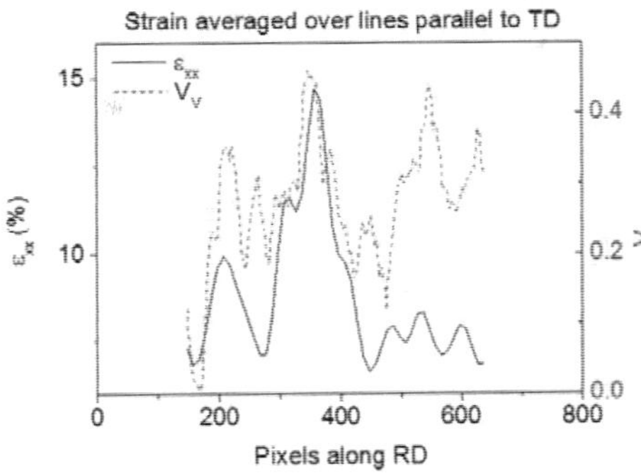

Fig. 4 ε_{xx} averaged over lines parallel to TD, i.e. perpendicular to the tensile direction. V_V of the lines are also plotted.

4. Conclusions

We investigated the local strain distributions of a partially recrystallized copper sample after *in situ* tensile deformation. It is demonstrated that microstructures from EBSD maps can be used as input for DIC and for determination of the local strains. The recrystallized grains deform on average more than the unrecrystallized matrix, but some unrecrystallized matrix is also deformed to relatively high strains, which may be related to the recrystallized grains in the vicinity. Local "hot spots" of highly deformed unrecrystallized matrix may be important for the mechanical properties of the sample. Detailed analysis of the link between the microstructure, texture and local strains is planned for further work.

Acknowledgements

FXL, YBZ and DJJ gratefully acknowledge the support from the Danish National Research Foundation (Grant No. DNRF86-5) and the National Natural Science Foundation of China (Grant No. 51261130091) for the Danish-Chinese Center for Nanometals, within which this work was performed. The authors also acknowledge GATAN for the loan of the stage.

References

[1] Li Y S et al 2008 *Scr Mater* **59** 475
[2] Zhang Y et al in preparation
[3] Lin F X et al 2014 *Acta Mater* **72** 252
[4] Withers P 2008 *Strain* **44** 421
[5] Kidmose J et al 2012 *J Mater Sci* **47** 7901
[6] Quinta da Fonseca J et al 2004 *J Micros* **218** 9
[7] Keskar N et al 2014 *Acta Mater* **69** 265
[8] Lin F X et al 2014 *Proceedings of the 35th Risø International Symposium on Materials Science* 373
[9] Jones E 2013 "Improved digital image correlation (DIC)," http://www.mathworks.com/matlabcentral/fileexchange/43073-improved-digital-image-correlation-dic
[10] Jones E et al 2014 *Exp Mech* **54** 971

IOP Conf. Series: Materials Science and Engineering **82** (2015) 012104 doi:10.1088/1757-899X/82/1/012104

Combined materials characterization by area detector investigations using hard X-rays

H.-G. Brokmeier[1,2], E. Maawad[2], R.E. Bolmaro[3], Z.Y. Zhong[1] and N. Schell[2]
[1] Institut für Werkstoffkunde und Werkstofftechnik, TU Clausthal, Agricolastr. 6, D-38678 Clausthal-Zellerfeld, Germany
[2] Helmholtz-Zentrum Geesthacht, Max-Planck-Straße 1, 21502 Geesthacht, Germany
[3] Instituto de Física Rosario, IFIR-CONICET, Bv. 27 de Febrero 210 bis, S2000 EZP, Rosario, Argentina.
E-mail: heinz-guenter.brokmeier.@tu-clausthal.de

Abstract. Materials characterization includes a long list of information's which are on one hand important for process optimizations and on the other hand key knowledge for application, lifetime and failure predictions. With the help of area detector investigations one is able to get simultaneously all information related to line position, line intensity and line broadening of Bragg reflections. Examples are residual stress profiles, quantitative texture mapping, diffraction elastic constants, lattice dependent stress-strain curves, thermal expansion coefficients and defect densities. A very special case is high energetic synchrotron radiation which allows very fast and complete detection of generalized pole figures, namely intensity pole figures (texture), line broadening pole figures (micro-strain) and line position pole figures (macro-strain) using complete Debye-Scherrer rings.

1. Introduction

It is known that complete diffraction patterns contain a lot of information's. New technologies in experimental facilities and data analysis open many fields for combined materials characterization. Materials characterization in terms of engineering science and applications means all kinds of structure information needed to predict or explain materials behaviour, critical limits in processing and application and failure analysis [1,2]. Materials characterization by diffraction methods is intensively used since many years. Standard is the measurement of phase composition and phase transition as well as texture and residual stress analysis by X-rays. Less often microstrain and domain size investigations have been reported [3,4] as well as the determination of materials constants such as the diffraction elastic constants, thermal expansion coefficients and lattice dependent yield points. But for all investigations one needs a specific instrumental set-up which optimized for some of the investigations and longtime one uses different instruments for texture and residual stress analysis. The availability of area detectors opens the field to get much information simultaneously. Together with high brilliant synchrotron sources a dynamic materials characterization or a mapping of information becomes possible in reasonable time. Among laboratory X-rays and neutrons, hard X-rays with energies > 50 keV are in use.

2. Hard X-rays

Great advantage of hard X-rays with energies up to 200keV is, among others, the low scattering angle, which means comparably small area detectors can be used to get a set of complete Debye-Scherrer cones without sample movement. Figure 1 shows a typical area detector image with a set of complete Debye-Scherrer rings. As one can see, two different types of information can be extracted. First is the intensity distribution along the 2-θ scale, so that one gets a sum diffraction pattern as integration over complete Debye-Scherrer rings or a set of individual cake diffraction pattern. The sum diffraction pattern is mainly used to identify the quantitative phase distribution and determine crystallographic structures. Cake diffraction pattern give much more details related to specific sample orientations. This may come from sample processing such as rolling with typical mechanical test directions at 0°, 45° and 90° to RD (RD – rolling direction) or by in situ loading tests with cake diffraction

IOP Conf. Series: Materials Science and Engineering **82** (2015) 012104 doi:10.1088/1757-899X/82/1/012104

patterns with LD || RD or LD ⊥ RD (LD – loading direction. The hkl-dependent information contain in much better way the anisotropic bevahior of polycrystalline materials. A second kind of data extraction is done to obtain the intensity distribution along Debye-Scherrer rings. This intensity distribution is used to calculate generalized pole figures. Mostly known is the peak intensity pole figure named as crystallographic texture pole figure. With line broadening and peak position pole figures, which can be obtained simultaneously, additional informations are available directly, comparing macro- and micro-strains with texture components.

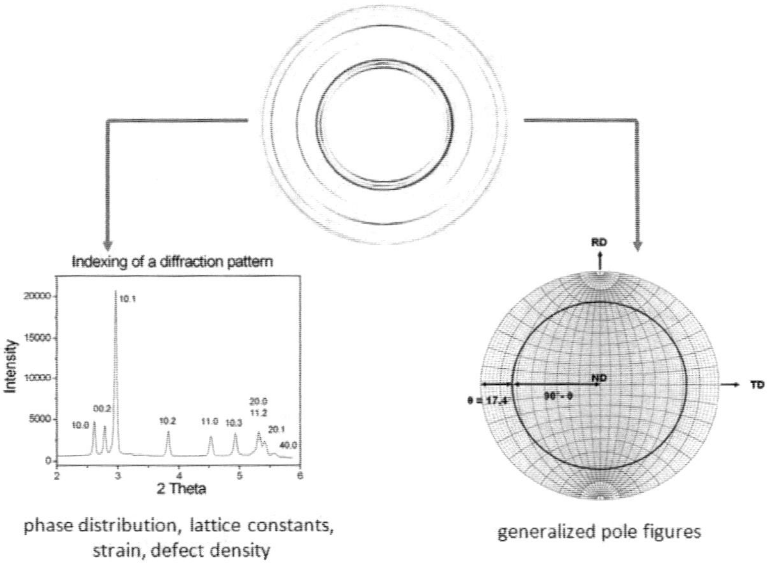

Fig. 1: Various information obtained simultaneously from area detector images

For low scattering angles less than 5° in 2-θ the intensity distribution of the Debye-Scherrer rings are very close to the outer great circle. That has the great advantage that for generalized pole figure scanning only sample rotation is needed but no sample tilt. Another advantage is that one gets from area detector images diffraction patterns to use the $\sin^2\psi$ – method for in-plane residual stress analysis. Figure 2 shows an area detector image of a rolled Al sample.

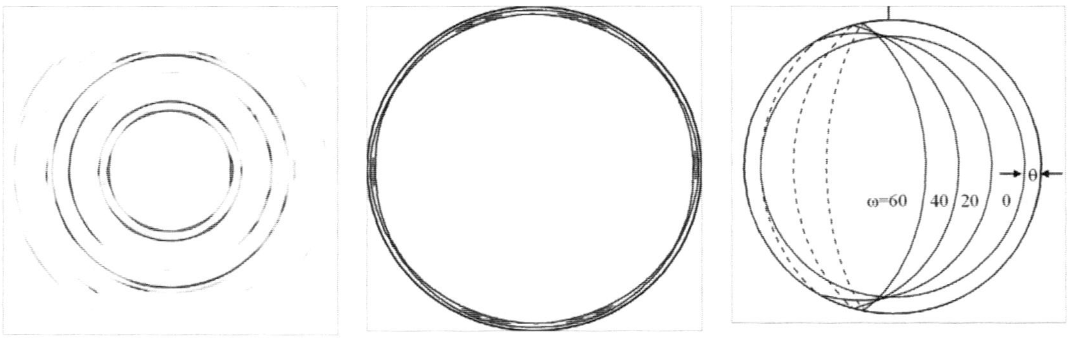

Fig. 2: Area detector image of Al (left), intensity distribution of the first Debye-Scherrer ring in pole figure plot (Al (111) – middle), ω-rotation to cover complete pole figures

For FCC materials, the first Debye-Scherrer ring with the lowest 2-θ value is the (111) reflection. The intensity distribution of the (111) reflection in the pole figure plot indicates already same typical information of the texture. In principle this sole ring on the different images is enough to use $\sin^2\psi$ – method for residual strain analysis. In this special case the texture is to strong to get sufficient results. In the right figure one can see that an ω-rotation is enough for complete pole figure coverage. Step size in ω must be related to the accuracy of the pole figure. Due to the high brilliance of synchrotron radiation, measurements with ω-rotations of 1° or less can be carried out very fast without problem.

3. Examples

All Experiments were carried out at the high energy beamline HEMS P07B at Petra III/DESY-Hamburg. With energies >80keV sample from mm to cm thickness can be measured in transmission mode using the pin hole method in 1 to 5 sec.

3.1. In situ tensile test on of Ti2.5Cu

In situ experiments on Ti2.5Cu have been used to measure on one hand the texture evolution during tensile deformation and to determine on the other hand polycrystalline X-ray elastic constants, which are microstructure and texture dependent [5]. A loading device, see figure 3 left, has been used to measure continuously diffraction patterns during tensile load. Result was a set of 193 diffraction patterns. The colours in figure 3 middle represent 4 different areas of the stress-strain curve. A typical MAUD-fit of one cake diffraction (±5°‖ to tensile direction) is shown in figure 3 right.

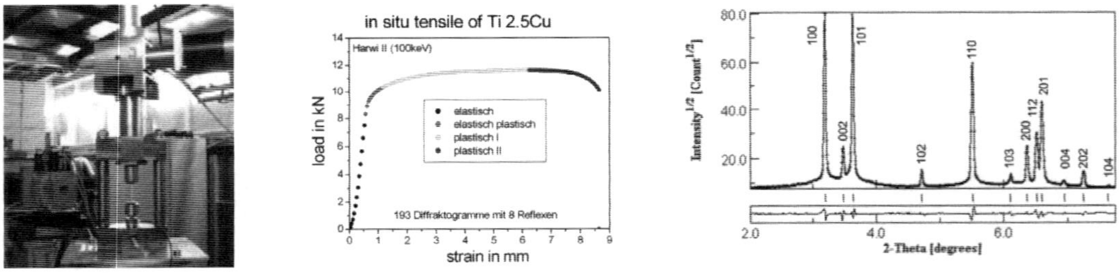

Fig. 3: In-situ loading (left), loading curve (middle), MAUD-fit of one pattern (right)

The analysis of the elastic area (black points) was carried out to the get lattice dependent diffraction elastic constants (DECs), hkl dependent yield points and Poisson ratios as well as load dependent change of the c/a ratio. As expected the polycrystalline DECs differ from pure Ti and influence the residual stress calculation. A variation of the lattice constants and the c/a ratio between LD and TD indicate the pre-residual stress of the sample originated by processing. The texture undergoes a strong texture change which is discussed in more detail by Maawad et al. [5], see figure 4.

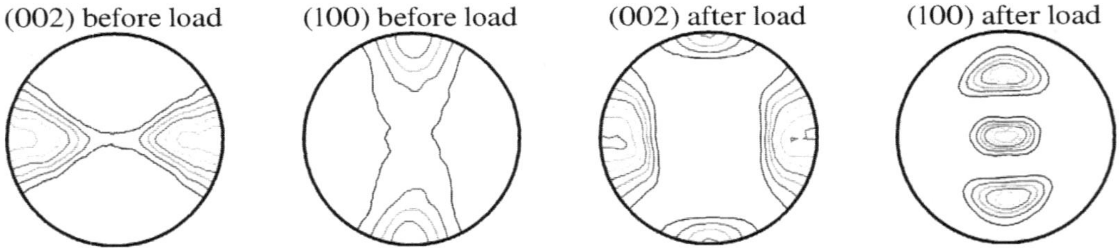

Fig. 4: Ti pole figures before and after load

3.1. Intensity and line broadening pole figure of F138 steel.

The material is F138 austenitic stainless steel, a kind of AISI316LVM, with a moderate strong preferred orientation [6]. For the first test the standard set up for pole figure measurement was used. As shown in figure 1 left from each area detector image one gets easily 360 diffraction patterns, totaling 13320 diffraction patterns in case of 5° rotation in ω. To extract intensity pole figures and line broadening or full width of maximum pole figures (FWHM) we have used STECA [7]. STECA is a software package looking on individual reflections. A key problem is the intensity statistics for strongly textured samples. Due to pole figure areas with less scattered intensities the error of FWHM determination increases. One way to overcome this problem is to improve the counting statistics and another way is to use only those pole figure areas having high enough intensities.

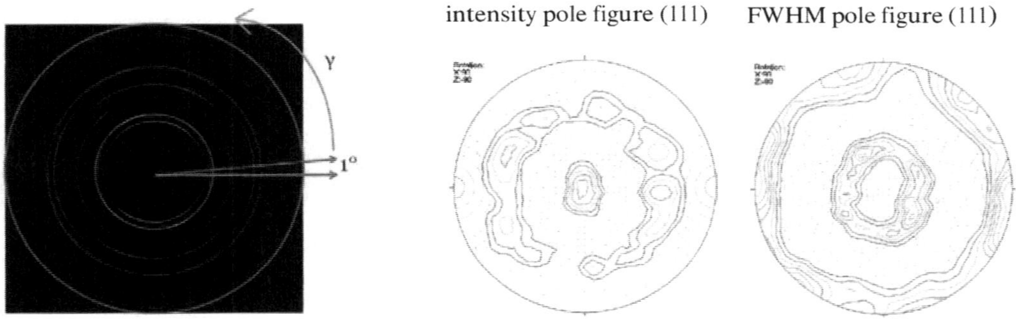

Fig. 5: Area detector image of F138, intensity pole figure, FWHM pole figure

4. Conclusion

Area detector experiments with all kinds of radiation show a high potential to make combined materials characterization [8,9,10]. Key point for these studies is firstly a perfect characterization of the used instrument, which is only roughly needed for texture research only. Second is that combined interpretations require well known gauge volume and sufficient statistics in intensity as well as in detector resolution. In two examples it has been shown that the anisotropy of the materials dominated by texture and microstructure can be understood much better if a combined analyses is done.

References

[1] Van Swygenhoven H and Van Petegem S *Materials Characterization* 2013 **78** 47
[2] Brokmeier H-G *Procedia Engineering* 2011 **10** 1657
[3] Mittemeijer E J and Welzel U *Z. Kristallogr.* 2008 **223** 552
[4] Alkhazraji H, Salih M Z, Zhong Z Y, Mhaede M, Brokmeier H-G, Wagner L and Schell N 2014 *Metallurgical and Materials Transaction* B **45** 1579
[5] Maawad E, Brokmeier H-G, Zhong Z, Al-hamdany N, Salih M, Wagner L and Schell N 2014 *Mat. Sci. Eng.* A **594** 62
[6] Bolmaro R E, De Vincentis N S, Benatti E, M Kliauga A M, Avalos M C, Schell N and Brokmeier H-G 2014 Mat. Sci. Eng. : IOP Conference Series: **63** 012148
[7] Randau C, Garbe U and Brokmeier H-G 2011 *J. Appl. Cryst.* **44** 2011 641
[8] Gan W M, Randau C, Hofmann M, Brokmeier H-G, Mueller M and Schreyer 2012 *J. Phys.: Conf. Series* **340** 012100
[9] Bernier J and Miller M 2006 *J. Appl. Cryst.* **39** 358
[10] Van Swygenhoven H, Schmitt B, Derlet P M, Van Petegem S, Cervellino A, Budrovic Z, Brandstetter S, Bollhalder A and Schild M *Rev. Sci. Instruments* 2006 **77** 0139002

IOP Conf. Series: Materials Science and Engineering **82** (2015) 012105 doi:10.1088/1757-899X/82/1/012105

Diffraction on heavy samples at STRESS-SPEC using a robot system

N Al-Hamdany[1], WM Gan[2], C Randau[3], H-G Brokmeier[1,2] and M Hofmann[4]

[1] Institut für Werkstoffkunde und Werkstofftechnik, TU Clausthal, Agricolastr. 6, D-38678 Clausthal-Zellerfeld, Germany

[2] Helmholtz-Zentrum Geesthacht, Max-Planck-Straße 1, 21502 Geesthacht, Germany

[3] Georg-August Universität Göttingen, Geowissenschaftliches Zentrum, 37077 Göttingen, Germany

[4] Heinz Maier-Leibnitz Zentrum (MLZ), Lichtenbergstr. 1, 85747 Garching, Germany
E-mail: nowfal.alhamdany@tu-clausthal.de

Abstract. The material science diffractometer STRESS-SPEC has high flux and a high flexible monochromator arrangement to optimize the needed wavelength. Many specific sample handling stages and sample environments are available. One of them is a Stäubli RX 160 robot with nominal load capacity of 20 kg and more freedom for texture mapping than the Huber 512 Eulerian type cradle. Demonstration experiments of non-destructive pole figures and strain measurements of Cu-tube segments weighing 12 kg weight and 250 mm in length and 140 mm diameter have been carried out. The residual strains measured by the robot and by the XYZ-stage fit quite well, that means the robot is reliable for strain measurements. The texture of the Cu-tube has dominant recrystallization texture components represented by the cube and the rotated cube.

1. Introduction
Neutron diffraction is well established for many kinds of investigations in materials science. For crystallographic texture investigation neutrons are favored for bulk information [1], but local studies are needed for residual stress profiles [2]. The beam port STRESS-SPEC at Heinz Maier-Leibnitz Zentrum (MLZ) at Garching/Germany is designed as a materials science diffractometer, which is able to offer strain [3] and texture analysis [4]. Both investigations can be done because of a comparably high neutron flux and a large flexibility in choosing the optimized wavelength for strain (2θ=90° arrangement) and texture (high intensity for low (*hkl*)) research.
STRESS-SPEC is equipped with a large sample stage, composed of XYZ-stage and an ω-rotary table for strain scanning. This stage can handle samples up to 300 kg. For pole figure measurements a Huber 512 type Eulerian cradle with an asymmetric φ-table must be installed on top due to the need to tilt and rotate the sample. Texture mapping is strongly limited by sample weight as well as by sample movement. Some investigations have been done with complex geometries, such as large tubes, shaft, turbine blade, etc. [5] with manual sample positioning and automatic pole figure scanning. To overcome these limitations a Stäubli RX160 robot was installed at STRESS-SPEC. The working range for strain scanning and texture mapping is within a circle of 500 mm (safety radius) and the present limitation is given by a nominal load of 20 kg. The big brother Stäubli RX170 with a nominal load of 30kg can be used only for low Bragg angles due to restricted space at STRESS-SPEC, and is not suitable for residual strain measurements. The aim of this work is to illustrate the capabilities and advantages of the RX160 robot in diffraction analysis of heavy samples, giving an example of texture and strain analysis of a 12 kg Cu-tube segment.

2. Experimental procedure
2.1. Sample description
The test sample was an industrial produced seamless SF-copper tube of 140 mm outer diameter, 10 mm wall thickness and 1m length. The chemical composition of the SF-Cu tube

is: 0.0168% P, 0.0041% Pb, 0.005% Al, 0.0016 Ni and the rest is Cu. The average grain size obtained by optical microscopy was about 76 μm. A tube segment of 250 mm length was cut

2.2. Three dimensional laser scanner

For both pole figure and residual strain measurements, the sample within the sample holder must be scanned by a 3D laser scanner as shown in fig.1 a,d. The scanner captures the profiles generated by the interaction of a projected laser and the target topography by using triangulation. In the next step the coordinates for all sample positions for texture mapping and strain scanning were defined. An in house developed mathematical library was used to extract the numerical values for the selected points (X, Y, Z, RX, RY and RZ with respect to robot plate) using Geomagic Studio software and then implemented in the robot software. Sample including sample holder were mounted on the robot arm (fig. 1b).

2.3. Pole figure measurement by the Robot

The pole figure measurements were carried out using a dual wavelength of $\lambda_{PG(004)}$ =1.55 Å and $\lambda_{PG(006)}$ =1.03 Å to save beam time. STRESS-SPEC is equipped with an area detector of 300x300 mm², so that the two wavelengths result in three Cu pole figures, namely Cu (111) and Cu (200) with 1.55 Å, and Cu (220) with 1.03 Å, see fig 1c. Depending on the 2θ angle (2θ= 47°) and the sample detector distance (1065 mm) the sample was tilted five times in χ. For each χ-position a continuous scanning along φ followed and an automatic data collection every 5° was performed. Continuous scanning, which is very easy with the robot, has the advantages of reducing counting time and collecting information from all crystallites in the gauge volume. The disadvantage is the averaging over 5° in φ, not useful for sharp textures. The results obtained are incomplete pole figures, due to heavy absorption along the length of the sample (250 mm). As an example, the texture from the central part of the tube wall at the 90° position was obtained using a primary slit of 5 mm in diameter and a radial collimator (5 x 5 mm²) as secondary slit. The exposure time for each detector image (5° rotation in φ) was 2 min. The software package StressTextureCalculator [6] has been used to extract pole figure data from area detector data using the mathematical formalism of Bunge and Klein [7].

2.4. Strain measurement by the robot

The strain measurements were carried out using a neutron wavelength of 1.53 Å obtained from a Si(400) monochromator. A rectangular primary slit 2 x 2 mm² made of cadmium and a radial collimator (2 x 2 mm²), acting as a secondary slit, were used. The sample to detector distance was 1065 mm and the primary slit to sample distance was 50 mm. To measure the peak position of Cu(311) for strain analysis the detector position was 2θ=89.6°. The strains at sample positions 0°, 90° and 180° across the thickness (at 6 positions) for every direction (axial, radial and hoop) were measured. The exposure time for every measuring point was 5 min. The strains obtained by the robot were compared with those obtained using the XYZ-stage at the identical sample positions. The same experimental conditions have been applied except that 8 positions across the thickness were measured when using the XYZ-stage. The exposure time for measurement was 2 min (because of limited beam time) for every step. It should be mentioned that the configuration for the axial strain was adjusted manually. Figure 1e and f shows the axial configuration for strain measurement using the robot and the XYZ-stage.

IOP Conf. Series: Materials Science and Engineering **82** (2015) 012105 doi:10.1088/1757-899X/82/1/012105

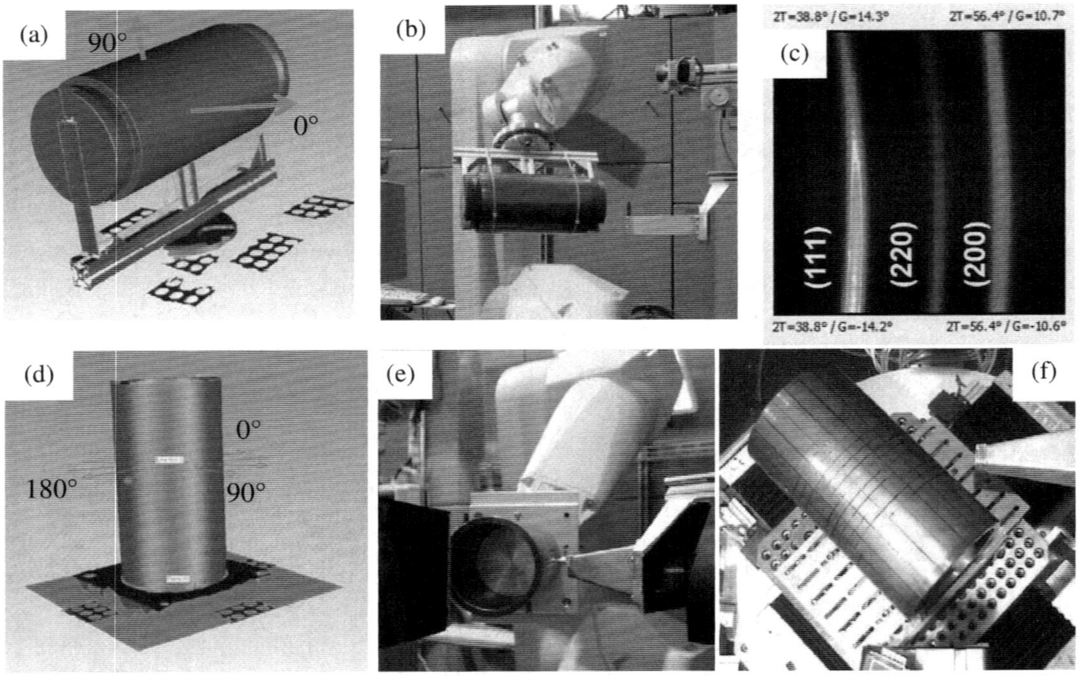

Fig. 1 (a) 3D laser scan for Cu tube mounted on sample holder for pole figure measurement(b) Cu tube with the sample holder mounted on the robot arm (c) 2D diffraction image (d) 3D laser scan for the Cu tube mounted on the sample holder for strain measurement (e) axial configuration for strain measurement using the robot (f) axial configuration for strain measurement using the XYZ-stage

3. Result

The Orientation Distribution Function (ODF) was calculated by the iterative series expansion method using incomplete (111), (200) and (220) pole figures. The degree of series expansion was Lmax=22. The sample symmetry of the ODF-calculation was triclinic. For the ODF results of face centered cubic (fcc) materials sections φ_2=0° and φ_2=45° have been used to describe the texture components because most of them are visible as shown in fig. 2. The texture in this Cu tube is characterized by 6 components. The recrystallization components [8, 11] are cube {001}<100>, rotated cube C_{RD} {013}<100>, Goss{110}<001>. Rot C_{RD} was described as recrystallization twins, which have been observed in Cu during static and dynamic recrystallization. The deformation components [8, 11] are copper {112}<111>, brass {110}<112>, and S {123}<643> also known as β-fiber. The dominant components are the rotated cube C_{RD} {013}<100> with an intensity of 11 multiples of random distribution (mrd) and the cube {001}<100> with 8.7 mrd. This copper tube was made by hot deformation which results in recrystallization during and after extrusion.

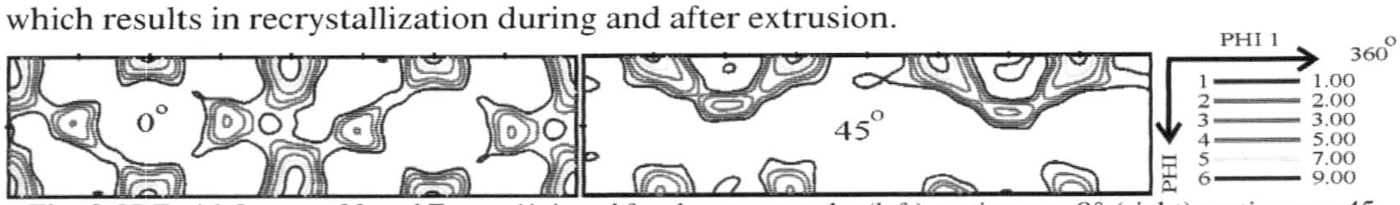

Fig. 2 ODF with Lmax = 22 and Fmax=11.4 mrd for the copper tube (left) section φ_2=0° (right) section φ_2=45

Fig. 3 shows comparisons for the axial and radial strains at the 90° position for both the robot and the XYZ-stage measurements. They are in good agreement. The axial strain varies from tension in outer region to compression in inner region and the radial strain profile goes from compression in the outer region to tension in inner region.

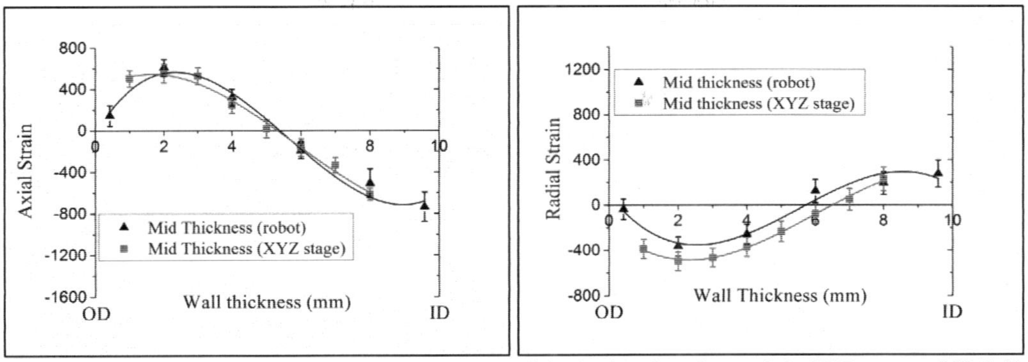

Fig. 3 Comparision between the measurements accomplished using the robot and the XYZ-stage at 90° (left) axial strain and (right) radial strain (OD: outside diameter and ID: inside diameter)

4. Conclusion
The high penetration power of neutrons and sample manipulation by the Stäubli RX160 robot combine well for local and non-destructive texture and residual strain measurements. The texture of the investigated tube shows a combination of deformation represented by the β fiber and recrystallization textures represented by the rotated cube C_{RD} {013}<100> and cube {001}<100>. The robot can handle larger samples than the Eulerian cradle. The residual strain measurements using the robot show a good agreement with the strains determined using the XYZ-stage. The positioning *of* the robot was sufficiently good for strain analysis. A great advantage was that all the three strain directions can be measured with identical set up, being not necessary to change the sample alignment manually.

Acknowledgements
We would like to thank the DAAD for their support. The robot project was supported by the German BMBF under number of 05KN7MCA. We would like to thank Mr. Karl Braun for his help in manufacturing the sample holders for the copper tube.

References
[1] Brokmeier H.-G. 2006 Physica B: Condensed Matter **385**, 623
[2] Lorentzen T., Hutchings M.T., Withers P.J. 2005 Introduction to Characterization of Residual Stress by neutron Diffraction (Taylor and Francis, London)
[3] Hofmann M., Robelo-Kornmeier J., Garbe U., Wimpory R.C., Repper J., Seidl G.A., Brokmeier H.-G., Schneider R. 2007 Neutron News **18** , 27
[4] Brokmeier H.-G., Gan W. M., Randau C., Rebelo-Kornmeier J., Hofmann M. 2011 Nuclear Instruments and Methods in Physics A **642** , 87.
[5] Brokmeier H.-G., Randau C., Gan W.M., Hofmann M., Lippmann T., Schell N.2012 Mater. Sci. Forum. **702-703**, 499.
[6] Randau C., Garbe U., Brokmeier H.-G. 2011 J. Appl. Crystallogr. **44**, pp. 641-646.
[7] Bunge H. J., Klein H. 1996 Z. Metallkunde .**87**, 465.
[8] Lücke K. 1984 Proc. of 7th International Conference on Textures of Materials, Netherlands Society for Materials Science, 195.
[9] Wasserman G. 1963 Z. Metallkd. **54**, 61
[10] O. Engler: Acta mater. Vol. 48, P4827 (2000).
[11] Humphreys F. J. and Hatherly M. 2004 Recrystallization and related annealing phenomena (Amsterdam, Elsevier), chap.3, p. 67 and chap.12, p.379.

IOP Conf. Series: Materials Science and Engineering **82** (2015) 012106 doi:10.1088/1757-899X/82/1/012106

Effect of interlayer macrostress on hydride orientation in shell tubes from Zr-based alloys

M. Isaenkova, Yu. Perlovich[1], O. Krymskaya and P. Medvedev
National Research Nuclear University MEPhI (Moscow Engineering Physics Institute), Moscow, Russia
E-mail: [1]yuperl@mail.ru

Abstract. Methods of texture analysis allowed to show that inhomogeneity of the crystallographic texture in shell tube wall is responsible for rise of tangential macrostress in the tube. This inhomogeneity predetermines a difference of thermal expansion between outside and inside layers of tube at its exploitation temperature. The indicator of such stress is formation of radially-oriented hydrides in the transverse section of tube, subjected to treatment in autoclave.

1. Introduction

Shell tubes from Zr-based alloys are responsible construction elements of atomic reactors. During exploitation these tubes undergo to oxidation and hydrogen saturation, which can become a reason for their fracture. Crystallographic texture is among the factors that have a very significant and sometimes decisive influence on the corrosion behavior of zirconium alloy products. In particular, the unfavorable hydride orientation in shell tubes from Zr-based alloys under operating conditions of the reactor dramatically reduces their service life. When considering the influence of texture on the hydride orientation in tube wall, one should take into account, that in most cases this effect is detected indirectly, in particular - by initiating a tensile stress due to non-uniformity of texture. At the same time, crystallographic texture is the primordial cause responsible for that or another parameters of corrosion process. In this paper, authors are considering the mechanisms of texture layering in tubes from Zr-based alloy on its corrosion and hydride orientation, irrelative of chemical or phase composition of these alloys.

2. Studied samples and methods of investigation

Studied samples were cut out from several series of shell tubes, submitted to regular technological treatment. These samples were measured by X-ray texture method of direct pole figures (PF) [1]. At that electro-erosion cutting of tubes into separate segments and assembling of these segments into a square sample were used. For composite samples $PF(0001)_{\alpha\text{-}Zr}$ were constructed and integral Kearns parameters were calculated. The similar fragments of the same tubes were saturated with hydrogen in the vapor-water atmosphere within an autoclave during 42 hours at 350°C for. Then cross-sections of tubes were studied using optical microscopy in order to reveal hydride particles [2].

3. Experimental results and discussion

Application of precise contemporary X-ray methods by study of shell tubes from Zr-based alloys permitted to reveal a number of structure features, able to quicken diffusion of oxygen and hydrogen into the matrix. First of all this relates to inhomogeneity of the crystallographic texture through the tube wall, arising due to different values of Q-factor, calculated for layers, which adjoin to inside and outside surfaces of tube. It is equal to the ratio of reduction by tube wall $\Delta t/t$ to reduction by its diameter $\Delta D/D$, where t – thickness of wall and D – diameter of tube. As Q-factor increases, texture maxima in PF(0001) shift to the center of stereographic projection, moving away from the transversal tube direction T to the tube radial direction R. Q-factor is the major technological parameter of the tube rolling and

determines a stress state in tube wall during its deformation. Therefore, Q-factor influences on the crystallographic texture, forming at these conditions.
Usually Q-factor is calculated, when starting from the thickness of whole tube wall. However if we distinguish in the tube at least two layers and determine for them values of Q-factor by assumption, that deformation through these layers is homogeneous, we would see that obtained values of Q-factor inevitably differ from each other. Though at the first approximation textures of these layers are visually similar, their corresponding integral Kearns parameters f_R, f_T and f_L, used for quantitative estimation of texture in tubes from Zr-based alloys and calculated as projections of the basal axes distribution on radial, tangential and axial directions of tube [3], prove to be somewhat different. Influence of this effect on hydride orientation in shell tube wall is considered lower.
Results of tube texture analysis are presented in Fig. 1. At the left part of Fig. 1 a fragment of shell tube and ifs flat sections, studied by X-ray texture measurements, are depicted, whereas at its right part – results of measurements. From above the partial PF(0001) (with angular radius 80°) for the outside layer of tube with calculated Kearns parameters are shown, while from below – the same for the inside layer of tube. Further equatorial sections R-T of PF(0001) for both layers are constructed (Fig 1-c). The difference of Kearns parameters for studied layers corresponds to the limiting sensitivity of the method for their determination [3], but the following example in Fig. 2 visually shows that this difference is sufficiently big for rise of essential tangential macrostress.

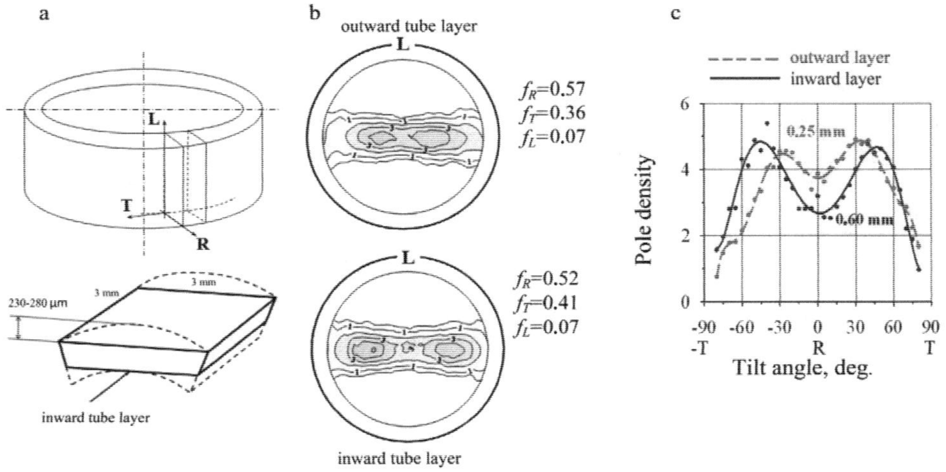

Figure 1. Scheme of cutting and preparing studied inside and outside surfaces of tube (a), PF (0001) of different tube layers (b) and distributions of basal normals in R-T section (c)

Coefficient of thermal expansion for any direction in tube, for example – for tangential direction T, is connected with texture Kearns parameter f_T by the following way:

$$\alpha_T = \alpha_c f_T + \alpha_a(1 - f_T),$$

where α_c and α_a – coefficients of thermal expansion for α-Zr single crystal along axes c and a in its crystalline lattice. At that, Kearns parameter is calculated namely for the direction T. Mean values of coefficients α_c и α_a at 350°C, found for a number of literature data [4], are equal to 5.5 deg^{-1} and 8.8 deg^{-1}, respectively.
Differences in thermal expansion of outside and inside layers along the tangential direction of tube at its exploitation temperature is a reason for rise in it of tangential macrostress.

When the coefficient of thermal expansion α_T of inside layer exceeds the analogous coefficient of outside layer, tangential tensile stresses operate at the outside layer of tube.

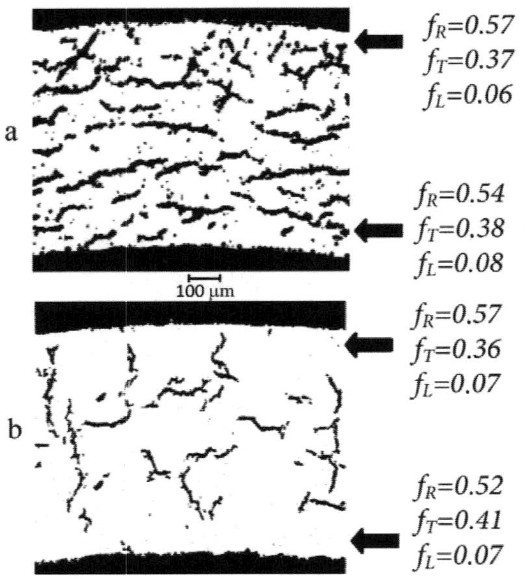

Figure 2. Radially-oriented hydrides in the cross-section of tubes, submitted to heating up to 350°C in a autoclave

A reliable indicator of such stresses is rise of radially-oriented hydrides in the cross-section of tube, submitted to keeping in autoclave (Fig. 2). By heating of tube up to 350°C solubility of hydrogen in α-Zr increases more than by 10 times, and by next cooling this hydrogen enters in the composition of hydride phase, precipitating in tube in accordance with direction of operating macrostresses, when taking into account, that precipitation of hydrides is a mechanism of these macrostresses relaxation.

Cross-sections of two shell tubes, presented in Fig. 2, differ in arrangement of hydride phase precipitations. At that, in the first tube hydrides are situated approximately along tangential direction, while in the second tube – along radial direction. It is known, that Hydride precipitates in α-Zr matrix have habit planes of numerous types, including basal planes and grain boundaries. However, by presence of tensile stresses they prove to be the strongest factor among those, determining orientations of hydride precipitates, which line up by perpendicular to operating stress. Namely because of this circumstance hydride precipitates reorient in the course of delayed hydride cracking in the field of stresses. Thus, the distribution of hydrides in Fig. 2-a responds to their arrangement by habit planes, whereas the hydride distribution in Fig. 2-b confirms existence of tangential macrostress in the second tube before precipitation in it of the hydride phase. This macrostress promotes penetration of oxygen inside the tube wall.

It is seen in Fig. 2-b, that under the influence of tangential macrostresses lamellar hydride particles take up radial positions, perpendicular to outside surface of tube, and coincide apparently with grain boundaries, having analogous orientations and being ways of active hydrogen diffusion under operation of tangential macrostresses. On the contrary, in the first tube (Fig. 2-a) in the presence of tangentially-oriented hydrides penetration of hydrogen inside the tube wall is difficult because of absence of tangential stresses as well as transparent radial ways of hydrogen diffusion, generated by these stresses.

Calculated deformations of shell tubes in the tangential direction as a result of their heating up to 350°C are presented in Table 1 in order to illustrate, how insignificant can be layer-by-layer texture inhomogeneity in tube, causing nevertheless a rise of essential macrostresses, responsible for precipitation of radially-oriented hydrides and promoting diffusion of oxygen inside the tube wall. Here ε_T – strain relative tangential deformation of the corresponding tube layer, and $\varepsilon_T^{in}/\varepsilon_T^{out}$ – the ratio of these values for inside and outside tube layers.

Table 1. Thermal expansion of different layers of shell tube by its heating up to 350°C

Sample number	Depth of layer, t, μm	f_T	α_T 10^{-6}, deg.	$\varepsilon_{T,}$ 10^{-3}	$\varepsilon_T^{in}/\varepsilon_T^{out}$
1	250	0.37	6.721	2.2179	1.005
	600	0.38	6.724	2.2288	
2	240	0.36	6.688	2.2070	1.025
	620	0.41	6.853	2.2615	

Comparison of deformation textures for inside and outside layers of tubes testifies, that in the second tube the relative tangential deformation of inside layer by $\Delta f_T = 0.05$ already exceeds the deformation, responding to the yield stress. When taking into account values $\varepsilon_T^{in}/\varepsilon_T^{out}$, presented in the last column, it becomes understandable, that by control of production the first tube was accepted as satisfying to technical conditions, whereas the second tube was rejected as spoilt.

4. Summary

As obtained experimental data show, the layer-by-layer texture gradient is sufficient for noticeable variations of dimensional thermal changes in successive layers. In the case, when the coefficient of thermal expansion for the inside layer exceeds that for the outside layer, tangential tensile stresses operate at the outside tube surface. As a result, intensification of corrosion processes takes place, including diffusion of oxygen and hydrogen.

Presented work is performed within the framework of Nuclear Systems&Materials Center under the government support of Improving Competitiveness Program of National Research Nuclear University MEPhI (Moscow Engineering Physics Institute).

References

[1] Perlovich Y, Isaenkova M 2002 *Metallurgical and materials transactions A* **33A** (3) p. 867
[2] Zaimovskiy A S, Nikulina A V, Reshetnikov N G 1981 *Zirconium alloys in atomic energy*, Energoizdat, Moscow, (in Russian).
[3] Perlovich Y, Isaenkova M, Fesenko V 2013 *Zavodsckaya laboratoriya. Diagnostica materialov* **79** № 7 (1) p. 25 (in Russian).
[4] Douglass D L 1971 *The Metallurgy of Zirconium*, Int. Atomic Energy, Vienna.

Investigation of rail wheel steel crystallographic texture changes due to modification and thermo-mechanical treatment

T Lychagina [1] D Nikolayev [1] A Sanin [2] J Tatarko [2] and K Ullemeyer [3]
[1] Joint Institute for Nuclear Research, Dubna, Russia
[2] Dnepropetrovsk National University, Dnepropetrovsk, Ukraine
[3] Institute of Geosciences, University of Kiel, Kiel, Germany
E-mail: lychagina@jinr.ru

Abstract. In this work crystallographic texture for a set of rail wheel steel samples with different regimes of thermo-mechanical treatment and with and without modification by system Al-Mg-Si-Fe-C-Ca-Ti-Ce was measured by neutron diffraction. The texture measurements were carried out by using time-of-flight technique at SKAT diffractometer situated at IBR-2 reactor (Dubna, JINR, Russia). The three complete pole figures (110), (200), (211) of α-Fe phase in $5^{\circ}\times 5^{\circ}$grid were extracted from a set of 1368 spectra measured for each sample. The samples were cut from rail wheel rim and from transitional zone (between rail wheel hub and wheel disk). It was concluded that the steel modification and some changes in the heat treatment modes of the rail wheels from the experimental (modified) and the conventional (non-modified) steel lead to reorientation of texture component.

1. Introduction

Solid-rolled wheels for railway rolling stock are being currently produced by about fifteen industrial enterprises in different countries (Belgium, the Czech Republic, France, Russia, Ukraine, etc.). According to Railway Wheels and Wheelsets Committee (ERWA), global demand for wheels is about 4.8 million units per year. Passenger rolling stock market in Europe increases annually by 3% (according to the International Union of Railways). It is need about 500 kg of steel to make one rail wheel. The price of a steel increased since 2004 to 2010 by 100 - 120%. The rising of a steel price leads to increasing of wheels price. Therefore, the important task for operating companies becomes to prolong life of wheels, and for producers of wheels is to use the most efficient ways of high-quality metal production. That is why the study of a rail wheel steel quality is urgent from both scientific and practical sides.

One way to improve the quality of the wheels is a modification of the molten steel. The modification is admixture of small amounts of additives to modify the properties of alloy through a change its microstructure (grains size, grains shape) and prefered orientations, i.e. texture. The modifiers admixture result in grain refinement. Besides, the additional atoms incorporated into crystal lattice leads to its distortion. This process could result in the blocking of dislocations and consequently to alloy hardening.

The final operational properties of the wheels depend on several factors, namely the chemical composition, the macro-and microstructure and also the crystallographic texture of the steel. The production technology of the rail wheels includes such operations as forging, press forming and rolling that can result in texture formation. The subsequent temperature treatment (annealing, quenching and tempering for rim) can also influence texture changing. The crystallographic texture can result in brittle cracking in steel products [1-3], so it can be one of the rail wheels damage reasons. The aim of this work is to study the impact of the modification and the thermo-mechanical treatment on crystallographic texture of wheels made from steel R7. So we tested whether modification and technological operations lead to the formation of undesirable texture in the rail wheels.

IOP Conf. Series: Materials Science and Engineering **82** (2015) 012107 doi:10.1088/1757-899X/82/1/012107

2. Material and samples

We investigated the samples that were cut from the rail wheels produced from the steel R7. This type of steel is used for all freight wagon wheels and for the most passenger coaches exploited by Deutsche Bahn. The samples were cut as from both the conventional (non-modified) and the experimental (modified) R7 steel. The system Al-Mg-Si-Fe-C-Ca-Ti-Ce was admixed in steel for modification. This modifier was developed in Dnepropetrovsk National University [4]. The developed modifier has a discrete structure, high solubility in the melt and the intensive multi-functional interaction with steel. R7 grade steel is multicomponent and incorporates six carbide-forming elements (C, Mn, Cr, Ti, V, Mo), two harmful impurities (S, P) and micro-alloying components (Si, Ni, Cu, Al). The chemical composition of the conventional and the modified R7 steel is within limits of the European standard 13262EN. It was found out that the modification of the steel R7 improves a set of properties. It results in grain refinement, reduces the concentration of harmful impurities, reduces the amount of non-metallic inclusions and changes theirs shape from oblong to globular, stabilizes the chemical composition and mechanical properties [5-6]. The amount of oxygen and hydrogen impurities is also below in the modified steel comparing with the conventional one. Sulfur and oxygen reduction results in decreasing of sulfides and oxides in the modified steel. The reduction of non-metallic inclusions and change of their shape from oblong to globular decreases the risk of cracking (see Fig. 1).

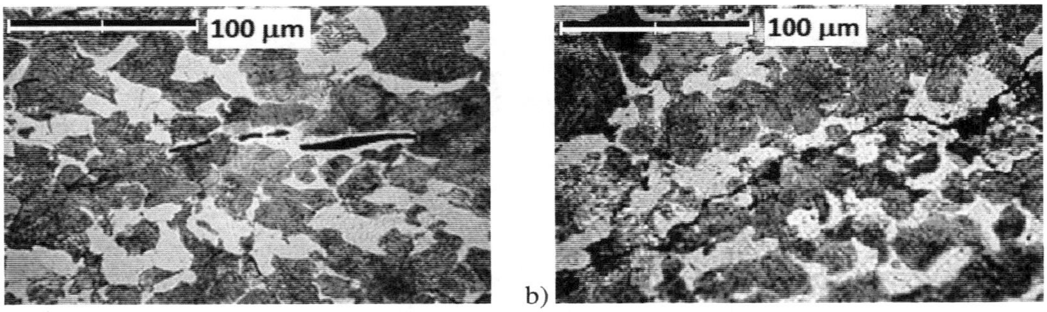

a) b)

Fig. 1 : a) Non-metallic inclusions in ferrite in the conventional steel, b) the cracking in the conventional steel, metalgraph microscopy (NEOPHOT-21), x 500

We had for texture investigation four samples from the wheel rim and four samples from the transitional zone between the wheel hub and disk. Namely in this transitional zone cracks sometimes appear under fatigue cycling tests. The scheme for samples cutting from the rail wheel is presented in Fig.2.

Fig. 2: The scheme for samples cutting from the rail wheel.

IOP Conf. Series: Materials Science and Engineering **82** (2015) 012107 doi:10.1088/1757-899X/82/1/012107

The axis Z for the samples cutting from the transition zone coincides with the wheel radius. The axis Z for the samples cutting from the rim is along the tangential direction of the rim. The samples 1 and 3 were cut from the modified steel wheel, whereas the samples 2 and 4 were cut from the conventional steel wheel. The samples 3 and 4 were undergone deformation according to technological operations of rail wheels production, and the samples 1 and 2 were undergone temperature treatment following this deformation. The samples from the modified steel were annealed at 880° C during 2h 20min. The samples from the conventional steel were annealed at 850° C during 1h 40 min. Then rims were quenched at 860° C during 2h and tempered at 560° C during 3h 30min for the modified steel and at 540° C during 2h 40min for the conventional steel.

3. Results and discussion

The texture measurements were carried out using neutron diffraction time-of-flight technique at SKAT diffractometer [7-8] situated at pulsed IBR-2 reactor (Dubna, JINR, Russia). The samples volume was about 10cm³. We had good grain statistics because the grain size in the both types of steel is about 25μm. The three complete pole figures (PFs) (110), (200), (211) for α-Fe phase were extracted from a set of 1368 measured spectra for each sample. 1368 were obtained by 19 detectors arranged at diffraction angle $2\theta = 90^\circ$. That corresponds to a 5-degree grid for the polar angle and azimuth as well (19 x 72). The resolution of SKAT texture spectrometer is rather high $\Delta d / d = 5\times10^{-3}$ at $d = 2.5\,\text{Å}$ and $2\theta = 90^\circ$ [8]. This resolution allows to have non-overlapped diffraction peaks for α-Fe phase of the steel. In Fig.3 the spectrum for sample 1 obtained as sum over 1368 individual measured spectra is presented to demonstrate SKAT resolution. The measuring time was 22 hours. It was used the local peak fit procedure for the PFs extraction [9-10].

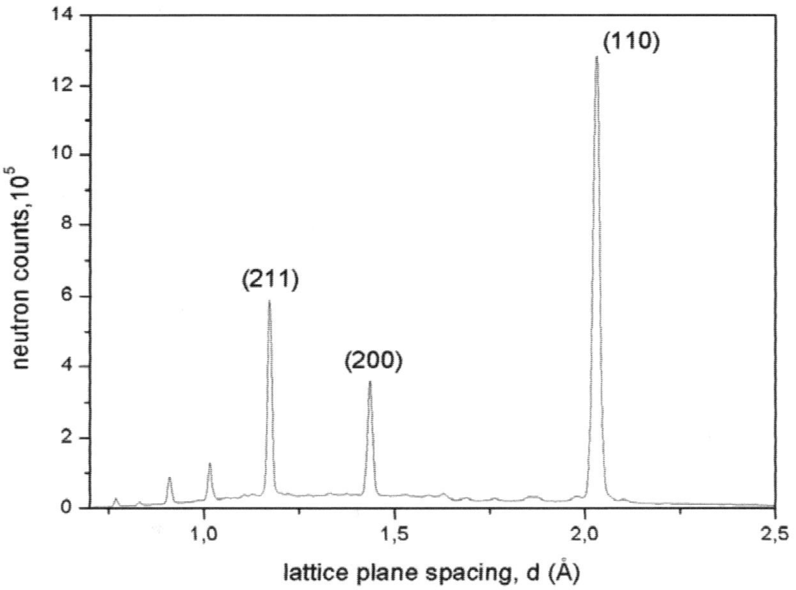

Fig. 3: The sum spectrum measured at SKAT for sample 1 from modified steel.

Figure 4 displays the three measured complete PFs for four samples from the transition zone of the rail wheel. The texture in the transition zone of the rail wheel is rather weak for all investigated samples. The modification results in the reorientation of the texture component. This follows from PFs comparison of the samples 1 and 2. The modification results in slightly

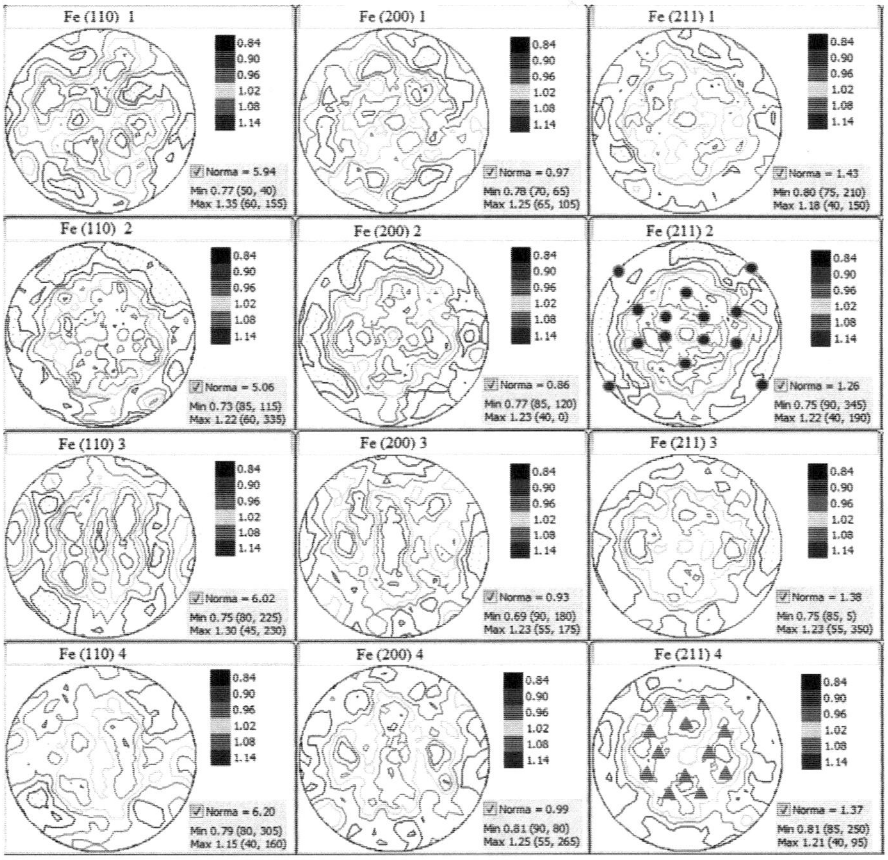

Fig. 4: The smoothed pole figures (110), (200), (211) of α-Fe phase for four samples from the transition zone of rail wheel. The red triangles mark the deformation texture component whereas the blue points mark the recrystallization texture component. The top raw is for the sample 1, the next one is for the sample 2, etc. The axis Z of the sample coordinate system is in the center of the PFs.

enhancing of the texture component on the PF (110) for the sample 3 in comparison with the sample 4. The annealing does not weaken the texture. The pole density maximum is 1.35 mrd in PF (110) and 1.25 in PF (200) for the sample 1 (after annealing) whereas 1.3 mrd in PF (110) and 1.23 mrd in PF (200) for the sample 3 (after deformation without annealing). It can be noticed the deformation component (001)<110> which is typical for body-centered cubic materials for the samples 3 and 4 whereas for the samples 1 and 2 there is the recrystallization component (011)<100> which is also typical for bcc materials. The deformation component is marked by the red triangles and the recrystallization component is marked by the blue points in PF (211) on the Fig.3. It was shown for only one PF to avoid the covering of texture details and the picture overflow. The position of these texture components are in accordance with calculations presented in [11]. We would like to underline that our measurements were able to show properly of the texture components corresponding to the technological operations even for so weak pole density values.

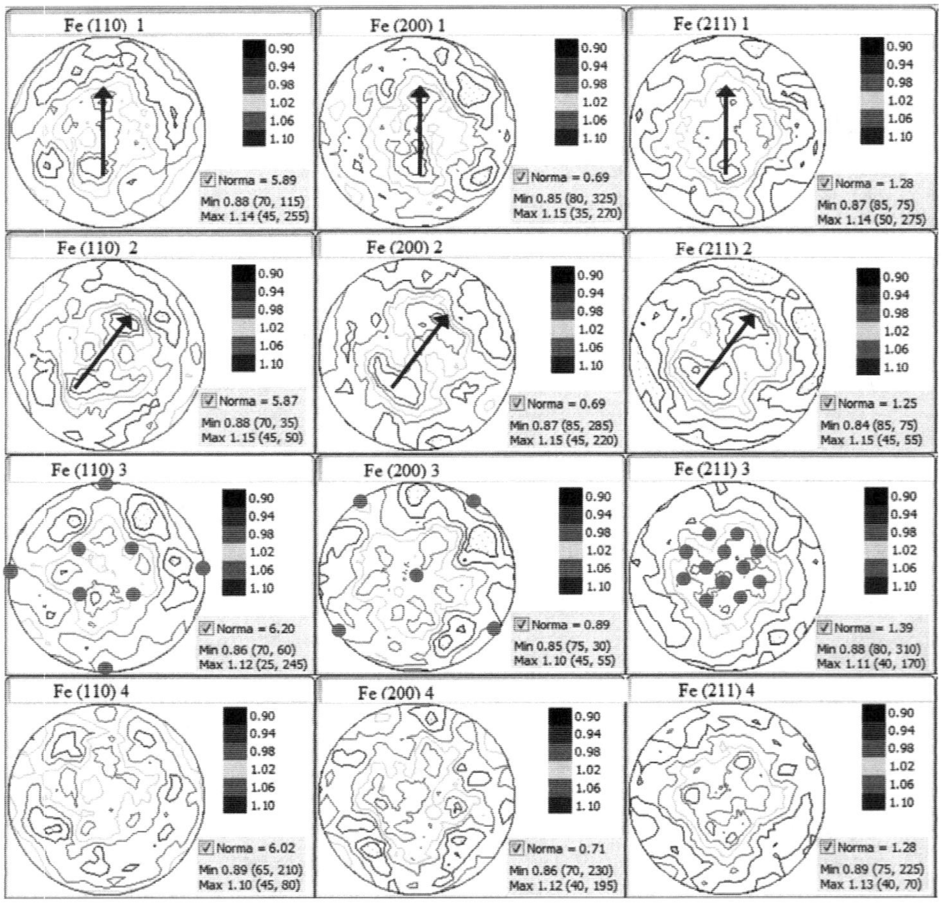

Fig. 5: The smoothed pole figures (110), (200), (211) of α-Fe phase for four samples from the rail wheel rims. The red points mark the deformation texture component. The blue arrows demonstrate the texture component rotation in the sample 1 (from the modified steel) according to the sample 2 (from the conventional steel). The top raw is for the sample 1; the next one is for the sample 2, etc. The axis Z of the sample coordinate system is in the center of the PFs.

The texture component reorientation during heat treatment (annealing) can be explained by the secondary recrystallization processes and phase transition in steel. The changes in scattering of texture components in the modified samples are associated with the introduction of the modifier as well as several differences in the heat treatment modes that were applied to the experimental and the conventional steels.
The three measured complete PFs for four samples from the rail wheel rims are presented in Fig.5. Texture in the rims is very weak in all measured samples and it almost does not enhance after the heat treatment. It should be noticed that texture in the samples from the

rail wheel rim is weaker than in the samples from the transition zone. The modifier admixture together with some changes in the modes of heat treatment (annealing, quenching, tempering) rotates the texture component at approximately 45°. This rotation is shown by the arrows in Fig.4. The tendency to alignment of α-Fe (110), (200) and (211) planes near the tangential direction of the rim in the sample from the modified steel after the heat treatment can be observed.

4. Summary

The texture in all investigated samples from the transition zone (between the rail wheel hub and the wheel disk) and from the rail wheel rim is appeared to be very weak. The modification influences texture changes in the steel R7. This is valid for both transition zone and rim. The modifier admixture in the experimental steel results in reorientation of texture component. Besides, the modification inhibits the process of the deformation texture destruction and formation of recrystallization texture during annealing in the investigated transition zone of the rail wheel.

The measurements at SKAT spectrometer were able to show properly the texture changes in spite of the texture weakness. So the texture development during rail wheels operation and after fatigue cycling tests can be examined by using this instrument and will be in the focus of our interest in future.

References

[1] Shkatulyak N M and Tkachuk E N 2012 Influence of internal cycling pressure on texture and fracture of pipes from low carbon steel *Fizika i texnika vysokix davlenii* **22-1** 107 (in Russian)

[2] Usov V V et al 1990 *Russian Metallurgy (Metally)* **1** 120

[3] Lyakishev N P et al 2000 *Russian Metallurgy (Metally)* **2** 68

[4] Shapovalova O M et al 2010 Deoxidizer -modifier for steel and alloy treatment *Patent* №92955 92955 C22C 35/00 in Ukraine

[5] Markova I A et al 2013 *Metallurgical and Mining Industry* **6** 73

[6] Tatarko J V and Sanin A F 2012 Influence of modification on quality of rail wheel steel *Metalloznavstvo ta obrobka metalliv* **4** 35 (in Ukranian)

[7] Ullemeyer K et al 1998 The SKAT texture diffractometer at the pulsed reactor IBR-2 at Dubna: experimental layout and first measurements *Nuclear Instruments and Methods in Physics Research A* **412-1** 80

[8] Keppler R et al 2014 Potential of full pattern fit methods for the texture analysis of geological materials: implications from texture measurements at the recently upgraded neutron time-of-flight diffractometer SKAT *Journal of Applied Crystallography* **47** 1520

[9] Lychagina T and Nikolayev D 2007 *Crystallography Reports* **52-5** 774

[10] Lychagina T, Nikolayev D and Wagner F *Texture, Stress and Microstructure* 2009 237485

[11] Matthias S, Vinel G and Helming K 1987 *Standard distributions in Texture Analysis* (Berlin: Akademie-Verlag)

IOP Conf. Series: Materials Science and Engineering **82** (2015) 012108 doi:10.1088/1757-899X/82/1/012108

Texture analysis of tinplate steel and its application in production of double reduced high strength tinplate grades with controlled earing properties

M Černík[1], R Gburík[2], L Hrabčáková[3] and P Vranec[4]

[1] Specialist, U. S. Steel Košice, s.r.o., Slovakia
[2] CTS manager, U. S. Steel Košice, s.r.o., Slovakia
[3] Specialist, U. S. Steel Košice, s.r.o., Slovakia
[4] Researcher, U. S. Steel Košice, s.r.o., Slovakia
E-mail: mcernik@sk.uss.com

Abstract. Several years of texture investigation via X-ray Diffraction and EBSD techniques clearly highlight the effect of steel chemistry and processing on the final texture development in individual tinplate grades produced by U.S. Steel Košice. The influence of chemical composition, degree of deformation, and the annealing process on the texture development in tinplate steel is presented for selected grades processed in production and using the material controlled laboratory experiments. "Non-earing and Ultra-low Earing" tinplate materials provide a good case study where the desired final texture, which contains a strong γ fiber along with other texture component, is achieved with precise control of key processing variables. Development and production of high strength double reduced material with isotropic properties can be achieved by detailed study of crystallographic texture. As-measured Inverse Pole Figure (IPF) maps and calculated Orientation Distribution Functions (ODFs) were used to characterize the strength of the γ fiber and selected texture components. Calculation of the normal anisotropy index, r, from the acquired texture information, was accomplished for these steels using available crystal plasticity models. Achieved results of detailed texture analysis of tinplate in the manufacture of double reduced high strength tinplate steel grades have been successfully applied.

1. Introduction

For double reduction tinplate steel sheets made from low carbon steel, besides the required mechanical properties, the texture is important in view of the deep drawing and earing. Nowadays the materials are annealed in the batch or continuous annealing lines before double reduction. It is relatively difficult to produce material with non-earing properties when the desired mechanical values of double reduction tinplate sheets of low carbon steel are obtained. The tendency of producers is to make the different grades of tinplate steels using one chemical composition with the adjustment of technological process. The double reduction is a part of this technology.

There is a change of microstructure and texture during the cold rolling process depending on the degree of material reduction. Crystallographic texture of double reduced tinplate steels using X-ray and electron diffraction was studied. The aim of the standard X-ray and EBSD methods was to determine the relation between the microstructure and texture of tinplate steels on the double reduction. Values of orientation density function F(g), volume fractions of selected texture components and γ fibers (ND//<111>)[1,2] were evaluated on the examined specimens. R(α) – value of normal anisotropy (Lankford value) and ΔR(α) – planar anisotropy coefficient were calculated from the series expansion coefficients using the popLA software [3]. The values of R(α) are directly dependent on a strong γ fiber and low values of α (RD//<110>) a θ (ND//<100>) fibers. However, it is not possible to measure values of R(α) mechanically as the double reduction tinplate steels are very thin.

2. Materials and experimental procedures

The commonly produced samples of double reduced tinplate grades from continuous annealing line were taken for the texture investigation. Typically, double cold reduction of commonly produced tinplate grades up to 40 % is applied. The examined samples were rolled in the laboratory pilot line with up to 50% reduction. All samples have similar chemical composition given in Table 1.

Table 1. Chemical composition (in wt. % alloying elements) of low carbon steel

C [%]	Mn [%]	Si [%]	P [%]	S [%]	Al [%]	N [%]
0.04-0.08	0.18-0.35	Max. 0.03	Max. 0.02	Max. 0.02	0.02-0.08	Max. 0.08

Three types of tinplate steels with different initial texture before double reduction were investigated: Type A – standard, B – non-earring and C – ultra low earring. X-ray diffraction equipment with texture goniometer was used to measure textures. Four pole figures of (110), (200), (112) and (103) planes were measured, which were used for the calculation of ODF (orientation distribution function) by the SH (Bunge), WIMV (Wiliams, Imhof, Matthies, Vinel) [4] and ADC (Pawlik) [5] methods. LaboTex and popLA programs were used for the calculation of ODF. Analyses of microstructures and textures were carried out using TSL-OIM software [6]. IQ, IPF maps, grain size, GAM (grain average misorientation) etc. were evaluated qualitatively and quantitatively. Due to clarity, only selected results of EBSD are documented.

3. Results

Samples of A, B and C types were subsequently reduced up to 50%. Such reduction is not used in practice as the material would be too thin and thus useless. The aim of experimental rolling was to obtain the knowledge about textures after double reduction related to initial texture. The results of X-ray texture analysis are briefly shown in Table 2, where the calculated ODFs for 10%, 30% and 50% reduction are depicted. The initial weak γ fiber in sample of A type gradually disappears and is practically not present at all in the 50% reduction, only textural component close to (100)<0-11> remains. In the case of type B sample the γ fiber is significantly reduced for the reduction at 50%, but is still readable along with the rotated cube texture component (100)<0-11>. For the 30% reduction, the γ fiber is even stronger than for the 10% reduction. Texture evaluated by ODF for C type sample is very stable; strong γ fiber remains even at 50% reduction. This result was also confirmed by the volume fraction of texture γ fiber calculated from the ODF using the LaboTex software with ACD method, figure 1. On the other side, the volume fraction of rotated cube texture component (100)<0-11> for the sample C increases with the degree of deformation from value 1% up to 2.5%. The initial values of (100)<0-11> for the samples A and B are significantly higher and decrease with the degree of reduction to value of 2.5% too, figure 2.

Table 2. The X-ray deformation texture for samples A, B and C. ODF for φ(2)=45°.

Type	Deformation 10%	Deformation 30%	Deformation 50%
A	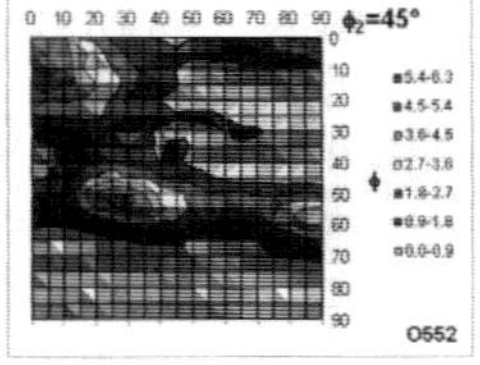	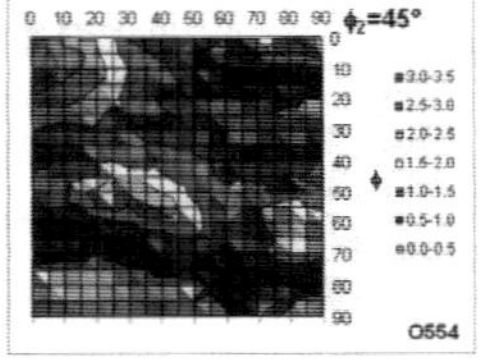	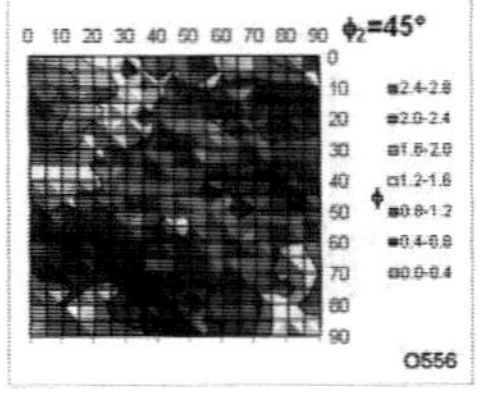

IOP Conf. Series: Materials Science and Engineering **82** (2015) 012108 doi:10.1088/1757-899X/82/1/012108

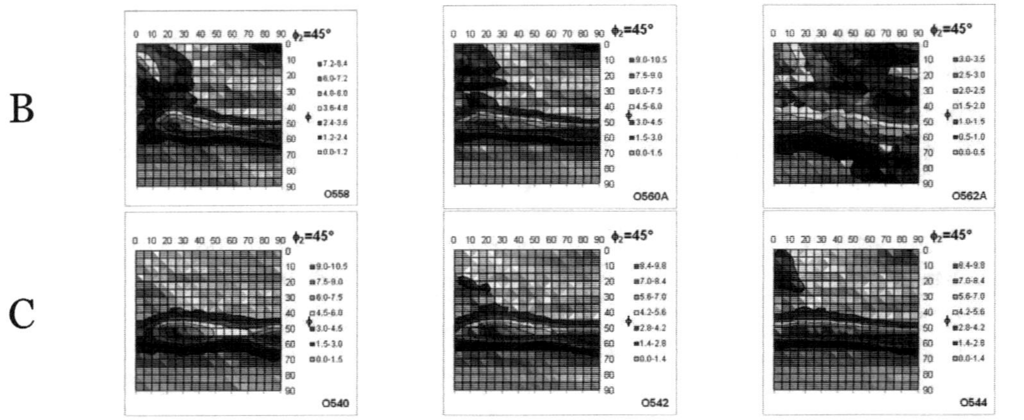

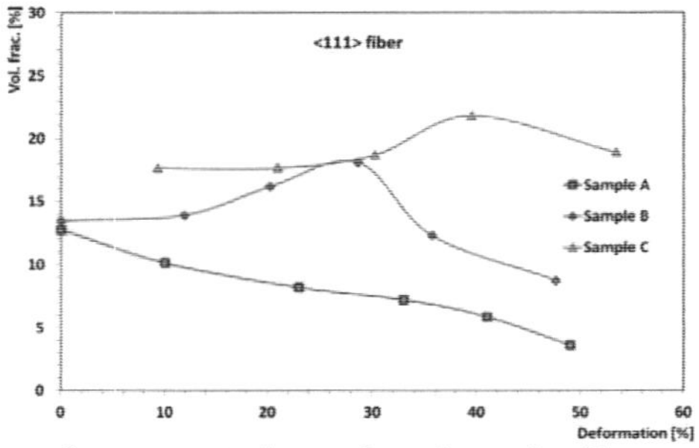

Figure 1 Volume fraction of γ fiber samples A, B and C.

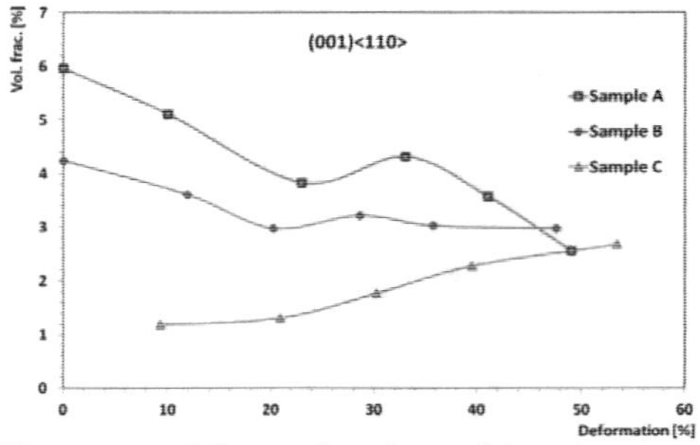

Figure 2 Volume fraction of (100)<0-11> texture component - samples A, B and C.

Lankford value R(α) for samples A, B and C (α degree from RD=0° to TD=90°), which defines ratio between the in-plane and through thickness strain were calculated using popLA software and are graphically described in the figure 3. Graph in the figure 3 clearly shows that the R(α) value is highest for sample C and lowest for sample A. The character of B curve is parabola with the maximum value at 25%. Curves of ΔR(α), which have even negative values, are shown in the figure 4. Because the ΔR(α) values were calculated from ODF, a small uncertainty from the experimental values should be considered, which cannot be measured due to thin thickness of double reduction tinplate steels.

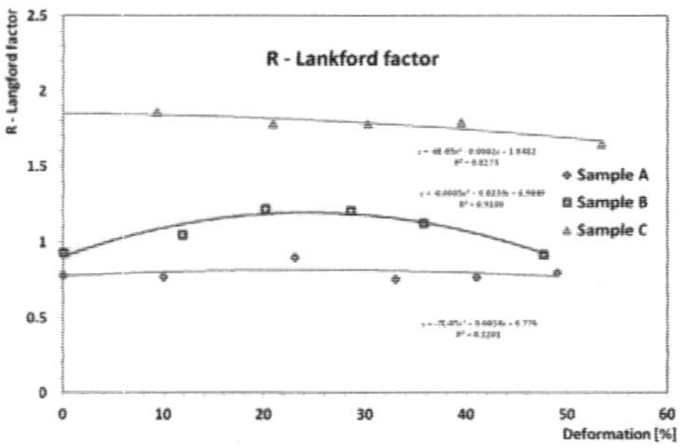

Figure 3 Lankford parameter R(α) for samples A, B and C.

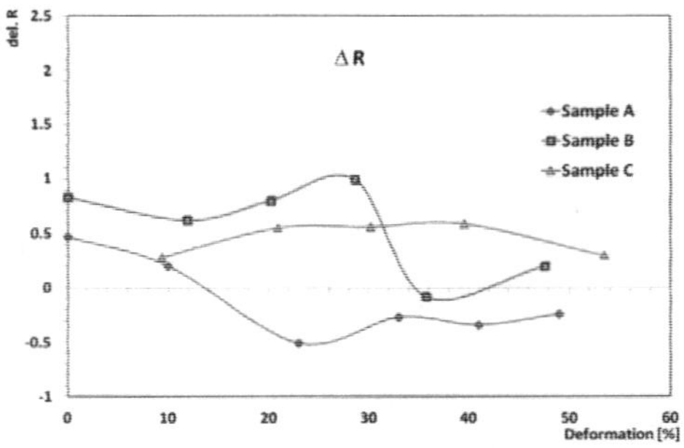

Figure 4 Delta R(α) for samples A, B and C.

EBSD texture analysis also allows calculating the ODF, but only for a limited number of measured data; therefore the result is less accurate than the one from the X-ray analysis. On the other side, EBSD method is powerful tool for microstructural analysis of grains, grain boundaries, angular boundaries, twins, GAM, etc. According to many authors [7, 8] the GAM parameter best describes the microstructural internal stress of deformed textural samples. The distribution of misorientation for deformed samples C calculated by OIM software are shown in figure 5. The graph in figure 5 clearly shows that the average value of misorientation increases with the degree of deformation. The average GAM for B and C samples are shown in the figure 6. Both two curves have character of parabola with the maximum at B=31.5% a C=57.2%. The values of C curve are significantly lower than values of B curve. The internal stress in the deformed samples C is lower; the steel sheet is more resistant to deformation.

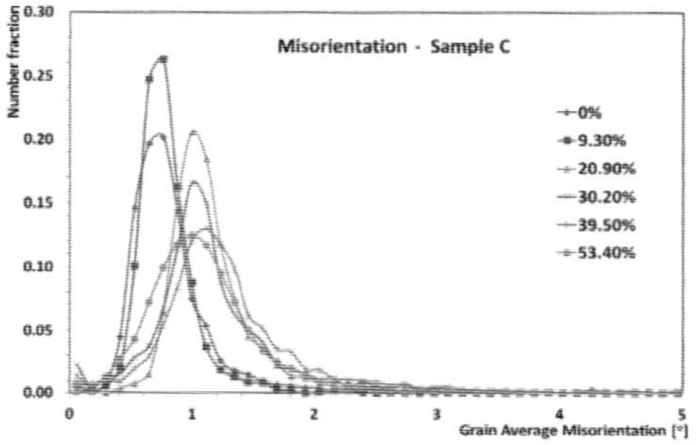

Figure 5 GAM-grain average misorientation sample C.

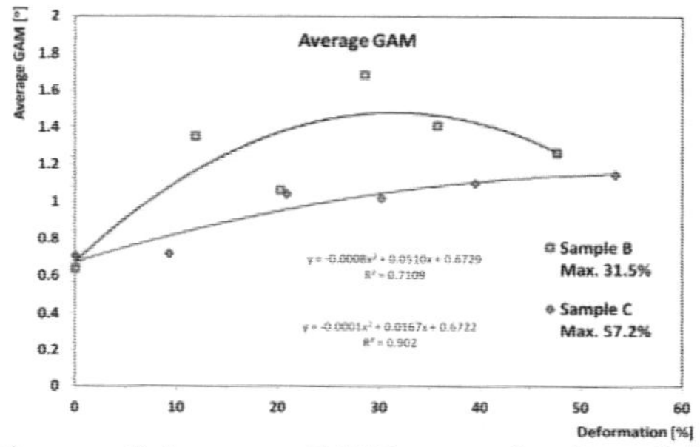

Figure 6 Average GAM - samples B and C.

4. Summary

ODFs, volume fractions of selected texture components and γ fibers were calculated from the measured data on X-ray equipment for reduced A, B and C samples. The best values of γ fiber were obtained for the sample C – ultra low earing even at 50%.

The values of normal anisotropy coefficient R(α) and planar aniosotropy ΔR(α) perfectly correlate with the results of texture fibers and selected texture components. The sample of C type has the highest values of R(α). The values of ΔR(α) remain positive and relatively low.

Internal stresses for reduced samples B and C were evaluated by the parameters of GAM using EBSD analysis. The curves of average GAM dependence on deformation are of parabola character with the maximum at B=31.5% and C=57.2%. The internal stress in the deformed C samples is lower; the sheet steel is more resistant to deformation.

References

[1] Bunge H J 1982 Texture Analysis in Materials Science (Butteworth-Heineman, London)
[2] Engler O and Randle V 2010 Introduction to texture Analysis: Macrotexture, Microtexture, Orientation mapping (RCD Press, Boca Raton)
[3] Koks U F, Tome C N and Wenk H R 1998 Texture and Anisotropy (Cambridge University Press UK)
[4] Kallend J S, Koks U F, Rollet A D and Wenk H R 1991 Operational texture analysis, *Materials Science and Engineering*, **A132** pp 1-11
[5] Pawlik K, Pospiech J, Lucke K 1991 Textures and Microstructures, Vol.14-18, pp 25-30
[6] OIM Analysis 4.4 Manual
[7] Wright S I, Nowell M M, Field D P 2011 Microsc. Microanal., Vol. 17, pp 316-329
[8] Khatirkar R, Krishna K V, Kestens L A I Petrov R, Pant P, Samajdar I 2011 ISIJ International, Vol. 51 No. 5, pp 849-856

IOP Conf. Series: Materials Science and Engineering **82** (2015) 012109 doi:10.1088/1757-899X/82/1/012109

Local texture of microstructural inhomogeneities in rolled microalloyed steel

N Zolotorevsky[1], S Panpurin[1], A Kazakov[2], O Pakhomova[2] and S Petrov[3]

[1] Department of Mechanics and Control Processes, Polytechnical University, Saint-Petersburg, Russia

[2] Department of Metallurgical Technologies, Polytechnical University, Saint-Petersburg, Russia

[3] Central Research Institute of Structural Materials "Prometey", Saint-Petersburg, Russia
E-mail: zolotorevsky@phmf.spbstu.ru

Abstract. Specific inhomogeneities consisting of coarse-grained bainite are observed in the microstructure of low carbon microalloyed steels after hot rolling. Earlier a special etching method has been developed allowing to reveal that these inhomogeneities markedly affect a fracture toughness of steels. In the present work their crystal geometry was studied using EBSD technique, and orientations of former austenite grains were reconstructed. The austenite, from which the coarse-grained bainite regions have been produced, is shown to have orientations concentrated predominantly within the brass component of austenite rolling texture. The inhomogeneities of steel microstructure are promoted by orientation dependency of the deformation substructure of heavily deformed austenite grains.

1. Introduction

Mechanical properties of low carbon steels for pipeline productions were shown to depend on an appearance of specific structural inhomogeneities [1]. When using optical metallography, these inhomogeneities manifest themselves as regions with non-polygonal or even lath-like bainite structure [1]. As it was shown by EBSD analysis [2], crystallites within these regions have relatively low-angle misorientations and, as a result, form very coarse "effective grains". Large size of these effective grains promotes propagation of trans-granular cleavage cracks from one high-angle boundary to another, leading to degradation of toughness [3].

It was suggested that the regions with lath-like structure were formed by a transformation of abnormally coarse austenite grains [2,4]. At the same time, even coarse austenite grains are known to be subdivided by highly misoriented bainite blocks on a rather fine scale due to a great number of variants of orientation relationship between α and γ phases [5,6]. Therefore, strong variant selection is needed to form a coarse effective grain. Actually, a nonrandom variant selection takes place in the case of plastically deformed austenite [7] due to anisotropy of its deformation substructure, which is known to be dependent on grain orientation [8]. One can thus suggest that not only large size but also specific orientation of a former austenite grain plays a role in formation of structural inhomogeneities considered. The aim of the present work is to verify this hypothesis.

2. Experimental

Industrially processed low-carbon microalloyed steel was investigated. Its chemical composition is 0.06C-1.7Mn-0.2Si-0.2Mo-0.1(V+Nb+Ti) (wt pct). Steel plate was hot rolled from 300 mm slab to 27.7 mm under industrial conditions. Seven initial passes were conducted in the temperature range from 1010 to 930°C. Then a plate was hold at the temperature about 930°C during 20 minutes that is sufficient for recrystallization. Subsequent passes were conducted at temperatures 750±40°C. For the steel investigated a recrystallization does not occur within this temperature range. True strain accumulated at this later stage is about 1.6. After rolling, the steel was cooled with a mean rate about 10 K/s.

The samples were prepared from a central region of steel plate. The longitudinal section was examined. EBSD analysis was carried out by SEM Quanta 3D FEG using the EDAX Pegasus system with a step size 0.5 μm. The data were treated using MTEX software. Evaluation of orientation relationship (OR) between α and γ phases was performed, as distinct from the earlier works [9,10], on a base of misorientations between α-phase crystallites only without reconstruction of austenite. After that, statistics of inter-variant misorientations was determined and orientations of former austenite were reconstructed. The methods of OR evaluation and austenite reconstruction will be described elsewhere.

3. Results and discussion

3.1. General characterization of microstructure

As it is seen in Fig. 1, the steel examined has non-uniform microstructure: one can observe very coarse-grained regions along with fine-grained ones. The regions of different microstructure are separated by much extended high-angle boundaries. Supposedly these are boundaries of former austenite grains subjected by severe deformation. Acicular morphology is predominant in the coarse-grained regions, but polygonal – in some fine-grained ones. It is to be noted that fields 1 and 2 do not represent a typical microstructure: they were chosen for examination of the structural inhomogeneities.

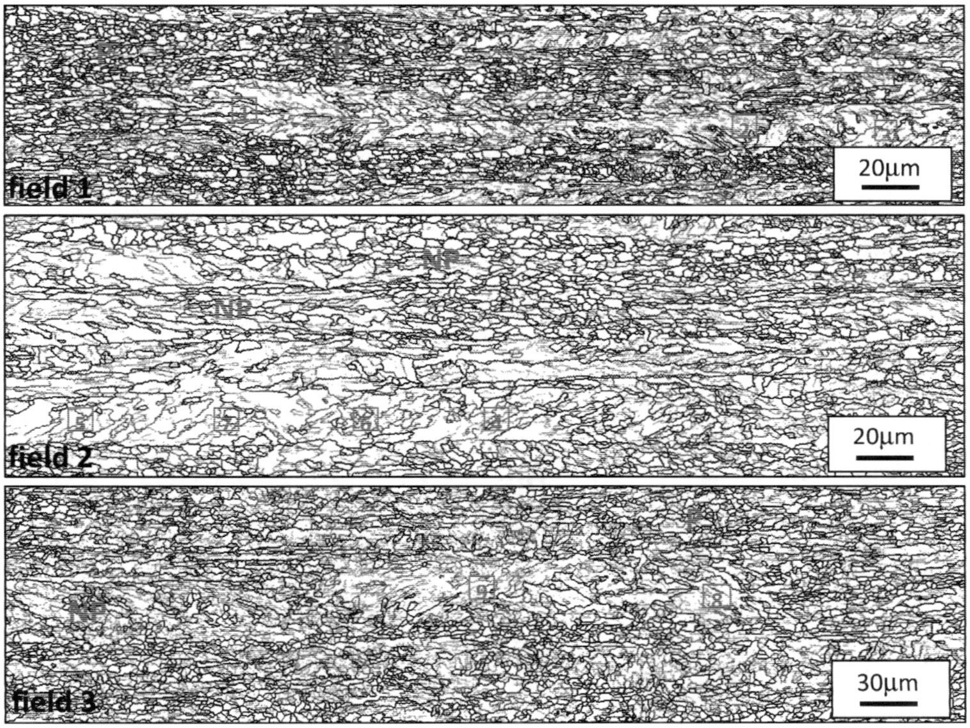

Fig. 1. Boundary maps of specimen examined. The boundaries are drawn by green (2°<θ<5°), blue (5°<θ<15°) and black(θ>15°) lines.

Usually the structure of industrially processed hot-rolled microalloyed steels is characterized as ferrite-bainite [1]. Since it is complicated to differentiate bainite and quasi-polygonal ferrite of microalloyed steel by their morphology only, crystallographic analysis can be used for this purpose. OR determined for above three fields are presented in Fig. 2a by help of deviation angles of close-packed planes and directions. Here the Greninger-Troiano (G-T) OR and the ORs reported by Takayama et all [11] are shown for comparison also. The latter ORs are obtained for a bainite of low carbon steel formed during isothermal holding (the holding temperatures are indicated at the plot). Assuming similarity of bainite crystallography in low carbon steels [6,10], one can conclude from the comparison that we deal with upper bainite formed at relatively high temperatures in the steel examined. As judging from morphological features, certain fraction of α-phase in the fine-grained regions may relate to a polygonal ferrite, but this fraction seems to be small since the OR fulfills much less strictly for a ferrite than for a bainite [12]. Above conclusion is supported by the spectrum of inter-variant boundaries (Fig. 2b), where the low-angle boundaries, V1/V4, V1/V8 and V1/V11&V13 (nomenclature of variants corresponds to generally accepted one [11]), are predominant in accordance with the results obtained for bainite formed both isothermally [11] and by continuous cooling [13].

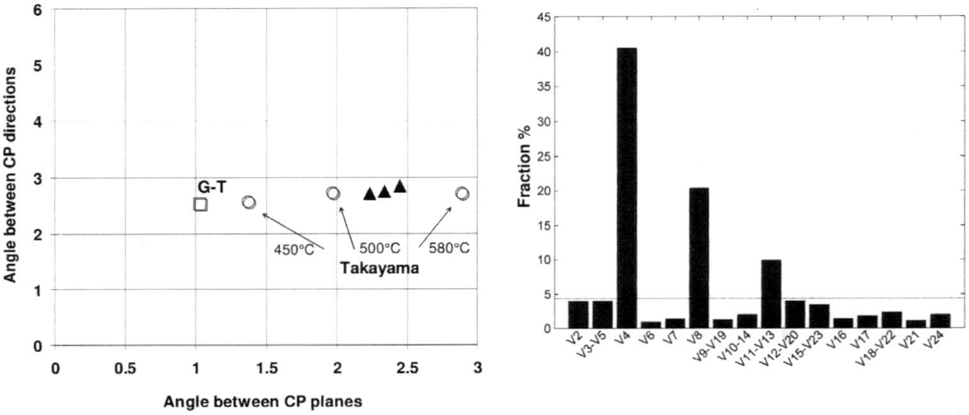

Fig. 2. Crystallographic characterization of specimens shown in Fig. 1: ORs presented by angles between planes $\{111\}_\gamma$ and $\{110\}_\alpha$ as a function of angles between directions $<110>_\gamma$ and $<111>_\alpha$ (black triangles) together with Greninger-Troiano OR and the ORs reported by Takayama et all [11] (a); length fractions of inter-variant boundaries (b).

Based on the orientation data obtained, local textures were calculated. It turned out that the local texture of field 3, which has quite typical microstructure, has the features inherent for the transformation texture developed from the plastically deformed austenite [16]. The texture of fields 1 and 2, where the coarse structure is predominant, has less usual character.

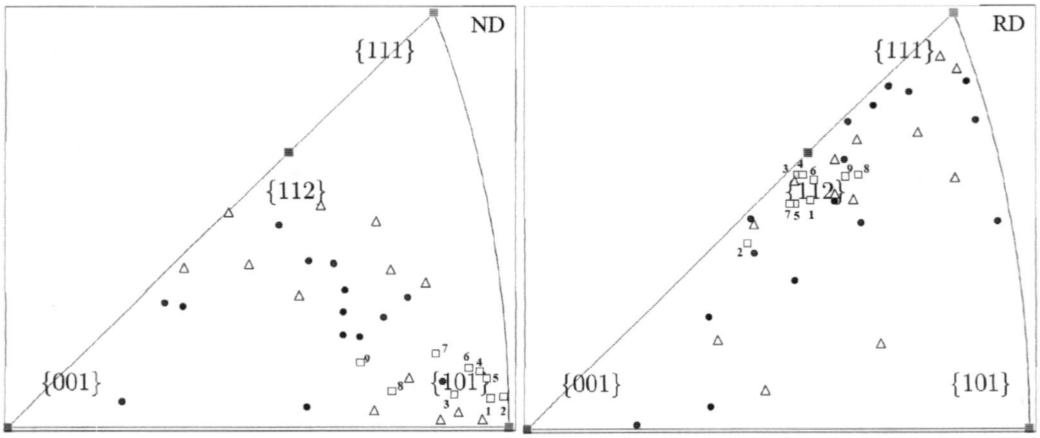

Fig.3. IPF for normal direction (ND) and rolling direction (RD) obtained from former austenite orientations determined in the regions with fine polygonal (circles), nonpolygonal (triangles) and coarse-grained structure (squares). Numbering corresponds to Fig. 1.

3.2. Reconstruction of former austenite microtexture

Orientations of former austenite were determined for the regions of three types: (1) the regions of fine-grained polygonal structure (marked by **P** in Fig. 1; corresponding poles are marked by circles in Fig. 3); (2) the regions of more coarse nonpolygonal structure with visible extended boundaries of former austenite grains (marked by **NP** in Fig. 1; the poles are marked by triangles in Fig. 3); (3) the coarse-grained regions associated with the structural inhomogeneities (regions are numbered in Fig. 1; the poles marked by squares).

As one can see from Fig. 3, the orientations of austenite correspond well to a rolling texture of FCC metals; the most of them belong to the β-fiber. The austenite, from which the regions of first and second type were produced, has orientations distributed rather uniformly within the β-fiber, mainly around the S-component. However, the austenite producing the regions of third type differs considerably: its orientations are concentrated around the brass component.

One can suggest that this observation relates to an orientation dependency of former austenite deformation behavior. Actually, during plastic deformation crystals are known to be subdivided (fragmented) to misoriented cells, cell blocks or subgrains. The misoriented regions therewith form bands, often extended along active slip planes [8]. Additionally to this regular band substructure, the bands of localized shear appear at medium to high strains. In the study of cold rolled Al it was shown that grains of near-brass orientations have minimal values of misorientations among other orientations belonging to β-fiber [14]. Additionally, it was revealed that single crystals having the brass orientation show less shear localization compared to crystals of the *S* and copper orientations [15].

Note that occurrence of regular band substructure promotes variant selection at nucleation and growth of bainitic ferrite, and this variant selection results in an increase of bainite packet size [7]. On the other hand, both high-angle misorientations and localized shear nonuniformities promote nucleation of different variants of bainitic ferrite. Therefore, peculiarities of the brass orientation – lower level of misorientations and greater homogeneity – lead to a more strict variant selection during phase transformation and, as a result, promote formation of larger bainite packets as well as formation of bainite blocks with low misorientations. Since the packets of upper bainite formed at a relatively high temperature contain mostly low-angle inter-variant boundaries [11,13], the bainite characterized by extraordinary coarse effective grain size arises from the deformed austenite grains of near-brass orientation.

4. Conclusion

Structural inhomogeneities observed in the industrially processed low carbon steels appear to consist of upper bainite containing blocks and packets with low-angle misorientations. The coarse-grained constitution of these regions is predetermined by specific deformation substructure of heavily deformed austenite grains having near-brass, (110)[$1\bar{1}2$],orientation.

References

[1] Kazakov A, Kiselev D and Pakhomova O 2012 *CIS Iron & Steel Review* 4
[2] Uranga P, Lopez B and Rodrigues-Ibabe J 2009 *Ironmaking and Steelmaking* **36** 162
[3] Hanamura T, Yin F and Nagai K 2012 *ISIJ Int.* **44** 610
[4] Roy S, Chakrabarti D and Dey G 2013 *Metall. and Mater. Trans.* A **44A** 717
[5] Lambert-Perlade A, Gourgues A and Pineau A 2004 *Acta Mater.* **52** 2337
[6] Furuhara T, Kawata H, Morito S and Maki T 2006 *Mater. Sci. and Eng.* A **431** 228
[7] Malet L, Barnett M, Jacques P and Godet S 2009 Scripta Mater. **61** 520
[8] Liu Q, Juul Jensen D and Hansen N 1998 Acta Mater. **46** 5819
[9] Miyamoto G, Takayama N and Furuhara T *Scripta Materialia* 2009 **60** 1113
[10] Humbert M, Blaineau P, Germain L and Gey N *Scripta Mater.* 2011 **64** 114
[11] Takayama N, Miyamoto G and Furuhara T 2012 *Acta Mater.* **60** 2387
[12] Novillo E et all. 2004 *Mater. Sci. and Eng.* A **385** 83
[13] Panpurin S et all. 2013 *Materials Science Forum* **762** 110
[14] Delannay L, Mishin O, Juul Jensen and Van Houtte P 2001 *Acta mater.* **49** 2441
[15] Godfrey A, Jull Jensen D and Hansen N 1998 *Acta Mater.* **46** 823
[16] Ray R, Jonas J, Butron-Guillen M and Savoie J 1994 *ISIJ Int.* **34** 927

IOP Conf. Series: Materials Science and Engineering **82** (2015) 012110 doi:10.1088/1757-899X/82/1/012110

Evolution of texture and its influence on the failure of components in some aluminium alloys

Ramesh Narayanan Parameswaran[1,a], Suwas Satyam[2,b], Sinha Parmeshwar Prasad[1,c] and Ranganathan Srinivasa[2,d]

[1] Vikram Sarabhai Space Centre, ISRO, Thiruvananthapuram 695022, India
[2] Department of Materials Engineering, Indian Institute of Science, Bangalore 560012, India

[a] rameshnarayananp@yahoo.co.in, [b]satyamsuwas@materials.iisc.ernet.in,
[c] pp_sinha@vssc.gov.in, [d]rangu2001@yahoo.com

Abstract. This paper describes the evolution of crystallographic texture in three of the most important high strength aluminium alloys, viz., AA2219, AA7075 and AFNOR7020 in the cold rolled and artificially aged condition. Bulk texture results were obtained by plotting pole figures from X-ray diffraction results followed by Orientation Distribution Function (ODF) analysis and micro-textures were measured using EBSD. The results indicate that the deformation texture components Cu, Bs and S, which were also present in the starting materials, strengthen with increase in amount of deformation. On the other hand, recrystallization texture components Goss and Cube weaken. The Bs component is stronger in the deformation texture. This is attributed to the shear banding.

In-service applications indicate that the as-processed AFNOR7020 alloy fails more frequently compared to the other high strength Al alloys used in the aerospace industry. Detailed study of deformation texture revealed that strong Brass (Bs) component could be associated to shear banding, which in turn could explain the frequent failures in AFNOR7020 alloy. The alloying elements in this alloy that could possibly influence the stacking fault energy of the material could be accounted for the strong Bs component in the texture.

1. Introduction

The aerospace industry uses a variety of aluminium alloys in different forms like sheets, forgings, extruded rods, welded components and machined components for launch vehicle and satellite applications. The aluminum alloys used in the aerospace industry are subjected to a variety of processing operations to realize the final product. These processes invariably introduce crystallographic texture in the material which in turn influences the mechanical properties. In this context, it is important to characterize the texture formed, both at macro and micro level.

This paper deals with the evolution of crystallographic texture three most important aerospace aluminium alloys AA2219, AA7075 and AFNOR7020 in a cold rolled and artificially aged condition. In-service applications indicate that the as-processed AFNOR7020 alloy fails more frequently compared to the other high strength Al alloys used in the aerospace industry [9-12]. An attempt has been made to correlate the texture to the failure pattern in the three aluminium alloys used in the aerospace industry.

Texture was determined by pole figure measurement and by calculating the Orientation Distribution Function (ODF). The ODFs were further analyzed for the estimation of the texture fibres and the volume fraction of the texture components. Micro-texture measurements were performed using Electron Back-Scatter Diffraction (EBSD).

2. Experimental

2.1 Starting materials

AA2219 alloy was received in the form of sheets and plates. These alloys are used in the T87 temper, wherein a small amount of cold working (7%) is given to the material after solution treatment before artificially aging it. The alloy AA7075 was received in the form of a thick plate in the T7351 condition (solution treatment + stabilized), while the alloy AFNOR 7020 was received in the form of sheets in

T651 condition (solution treatment plus artificial aging). The chemical composition and temper designations of the Al alloys used for the study are given in Table 1.

Table 1 : Chemical composition and temper designations of the Al alloys studied

Alloy Designation	Initial condition	Alloy forms	Zn	Mg	Cu	Cr	Mn	Zr	Fe	Si	Al
AA2219	T87 (ST*+CW**+AA***)	Plates (7.2 mm)	0.1 max	0.02 max	5.8-6.8	0.18	0.2-0.4	0.1-0.25	0.3 max	0.2 Max	Bal
AA7075	T7351 (ST+SB#)	Plates (15 mm)	5.1-6.1	2.1-2.9	1.2-2.0	0.2-0.3	0.3 max	0.05 max	0.5 max	0.40 Max	Bal
AFNOR 7020	T651 (ST+AA)	Sheets (1.8mm)	4.0-5.0	1.0-1.4	0.20	0.1-0.35	0.05-0.5	0.08-0.2	0.4 max	0.35	Bal

*ST = Solution treated
**CW = Cold worked
***AA = Artificially aged
#SB = Stabilized

2.2 Cold Rolling and Aging
The as-received AA2219 sheets and plates were solutionized at 535±5 ^{0}C followed by water quenching and then subjected to multi-pass rolling at room temperature without any intermediate annealing. After the final pass, the samples were artificially aged at 163 ^{0}C for 24 hours and air cooled. The as-received AA7075 and AFNOR7020 alloy sheets and plates were solutionized at the respective solutionizing temperatures (465±5 ^{0}C) followed by water quenching and then subjected to rolling at room temperature in multiple passes. The alloy AA7075, after cold rolling, was artificially aged at 163 ^{0}C for 24 hours and then air cooled. For the AFNOR 7020 alloys, a two step aging cycle was followed. In the first cycle, samples were subjected to 100 ^{0}C for 6-8 h followed by air cooling. This was immediately followed by the second step aging at 163 ^{0}C for 24 h and air cooling.

2.3 Bulk texture measurements
Bulk texture measurements were carried out using a PANanlytical MRD X-ray texture goniometer with Cu K_α radiation. Four incomplete pole figures, namely, {111}, {200}, {220} and {311} were measured at the mid-thickness sections of the specimens. The measured incomplete pole figure data was used to calculate the orientation distribution function (ODF). The complete pole figures were re-calculated from the ODFs. The volume fraction of the individual texture components was determined from the ODFs considering a Gaussian spread of ±10^0.

2.4 Electron back-scatter diffraction (EBSD) analysis
EBSD measurements were carried out on the FEI QUANTA FEG-SEM, with TSL data acquisition software. An area 100 μm X 200 μm was scanned with a step size of 0.2 μm. The criterion for identification of low and high angle grain boundaries was set as misorientation angle range 2-15° and 15-180° respectively.

3. Results and discussion

3.1 Bulk texture
The (111) pole figures for the material AA2219 in the starting condition as well as the cold rolled and aged materials is given in Figs. 1(a) and (b). The (111) pole figure of the starting material shows some near cube orientations. The overall texture, however, appears weak. As the material is cold rolled up to 95% thickness reduction, the deformation texture components appear and the overall texture becomes stronger. The (111) pole figures for the starting material AA7075 along with the cold rolled samples are presented in Figs. 1(c) and (d) and that for the AFNOR7020 alloy are given in Figs. 1(e) and (f). It is clear that the texture of both these starting materials has strong Cu and weak Bs and S components.

With increase in rolling deformation up to 70%, the Cu component gradually weakens and the Bs and S texture components become stronger.

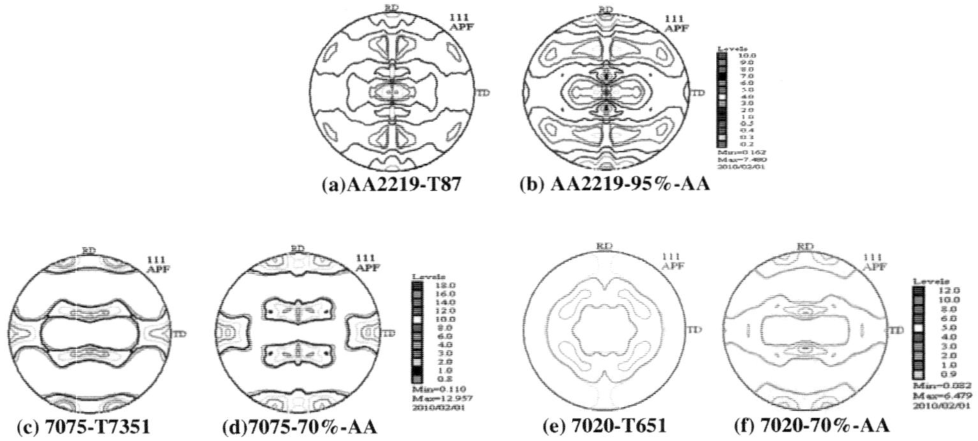

Fig. 1: (111) pole figures for (a,b) AA229, (c,d) AFNOR7020, and (e,f) AA7075 alloy samples.

The constant $\phi_2 = 0^o$, 45° and 65° sections of the ODFs for the starting material in the solution treated condition and after 70% rolling deformation are shown in Fig.2. The major texture components are observed to be Cu, Bs and S. In the solution treated and naturally aged condition, weak cube component is also observed. On increasing the amount of cold deformation, the cube component disappears. As the amount of cold rolling is increased to 70%, the Cu component gradually weakens from the 50% cold rolled to the 70% cold rolled condition.

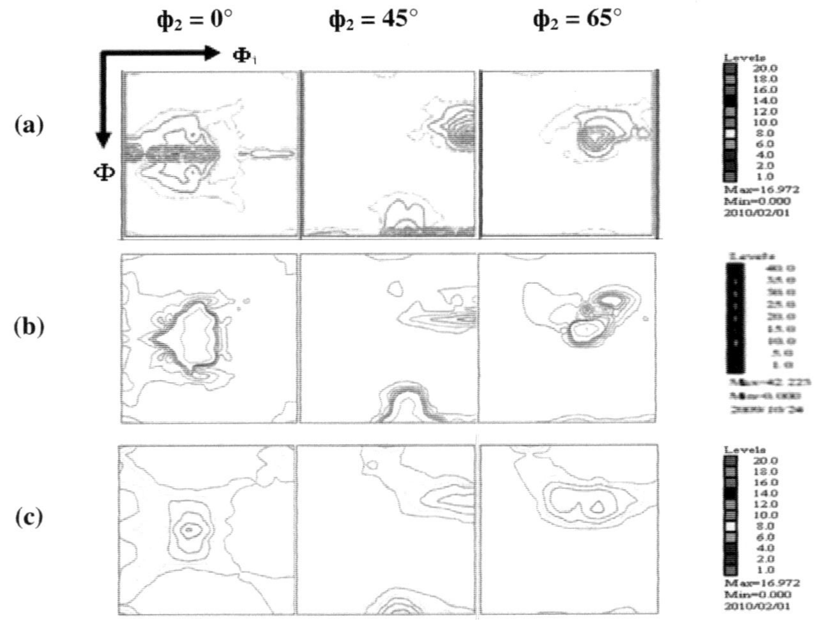

Fig 2. $\phi_2 = 0^o$, 45° and 65° sections of the ODFs for 95 % cold rolled (a) AA2219 and for 70% cold rolled (b) AA7075 and (b) AFNOR7020 alloy samples obtained from the mid thickness.

The ODFs also reveal that Bs and S components of texture become stronger, as the amount of cold rolling is increased from 50% to 70%. The texture of the starting material is characterized by discrete [001]||ND fibre with the intensity maxima at rotated cube position, more precisely at $(001)[1\bar{3}0]$ location. A sharp increase in the Bs component with increase in rolling deformation is clearly discernable in the ODF sections. The strengths of Cu and S components, however, do not change with cold rolling.

3.2 Microstructural analyses

Image Quality (IQ) maps for cold rolled AA2219 alloy samples are shown in Fig. 3 and that for the AA7075 and AFNOR7020 alloy samples are shown in Fig. 4. It is clear that the materials are highly cold worked. The IQ map also becomes more diffuse and darker, as the amount of cold work in the materials increase. Black spots in the IQ maps are due to cavities where the second phase precipitates were present and have come out during electro-polishing. The microstructure of the AFNOR7020 alloy consists of extensive shear bands predominantly aligned at angles from the rolling direction. The shear banding could be either due to the higher volume fraction of the precipitates and the associated strain fields around them or it could be attributed to reduced stacking fault energy of aluminium due to the alloying elements.

In order to understand the texture development in the three types of alloys, ϕ_2 sections of the ODFs for similar cold reduction (70%) were compared. It is clear that AFNOR7020 alloy develops a stronger texture compared to the AA2219 and AA7075 alloys. A comparison of the texture development for the three alloys as a function of cold rolling deformation indicates that all the three deformation texture components, viz., Cu, Bs and S become stronger with increase in cold rolling. Amongst the deformation texture components, the Bs and S components emerge stronger in comparison with the Cu component as the amount of the cold rolling increases [1-6]. Cu component is observed to be weak in AA7075 alloy and relatively stronger in AFNOR7020 alloy samples.

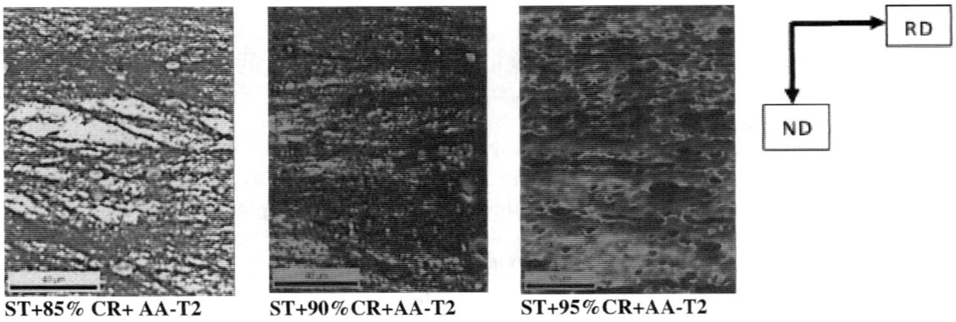

Fig 3: Image Quality Maps for cold rolled AA2219 Al alloy samples

The strongest Bs component is observed in the AFNOR7020 alloy. This is attributed to the shear banding. Shear bands are clearly observed in the cold rolled samples of AFNOR7020 alloy. This is in agreement with Engler [16], who suggested that such a texture transition in aluminium alloys could be due to shear banding. Humphreys and Hatherly [14] have observed shear bands in Al-Zn-Mg alloy cold rolled to 90% similar to the one observed in the AFNOR7020 alloy cold rolled to 70% used in this study. Previous studies on Al-Zn-Mg alloy [15, 16] have indicated that colonies of shear bands are usually several grains thick and the bands in alternate colonies are in opposite senses so that a herringbone pattern develops. This is exactly similar to the shear bands observed in the AFNOR7020 alloy cold rolled to 70% used in this study as given by the IQ maps in Fig. 4. The shear bands marked

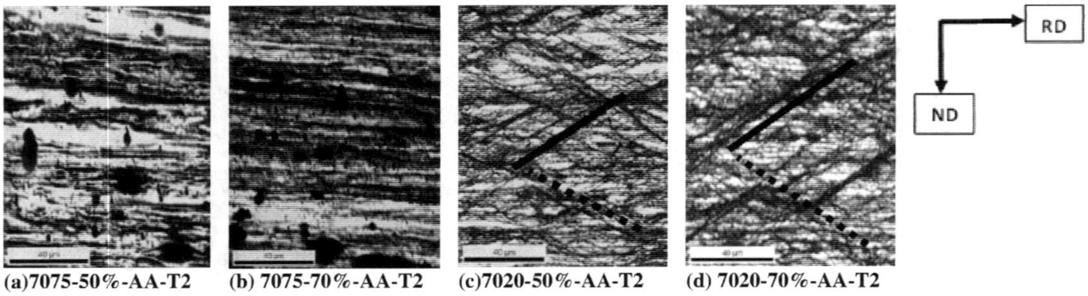

Fig 4: Image Quality Maps for cold rolled samples of (a,b) AA7075 alloy, and (c,d) AFRNOR7020 alloy.

adjacent to the thick black line have an orientation of 40° to the rolling plane whereas those marked parallel to the dotted black lines have an orientation of 35° to the rolling plane. The former is known as 'Brass type' shear band while the latter is known as 'Copper type' shear band. Any grain boundaries in the colony are crossed without deviation. The above mentioned "copper type" shear bands occur in a variety of materials and their occurrence depends on a number of factors like grain size and the addition of magnesium to aluminium which promotes shear banding.

Usually, Brass type texture is favored in low stacking fault energy (SFE) materials. The presence of Mg and Zn in the Al-Zn-Mg alloy lowers the SFE, hence favoring Bs type of texture [8]. Amongst the alloys studied, AFNOR7020 shows pronounced shear bands compared to AA7075 probably due to the higher Zn: Mg ratio in this alloy. This could explain the more frequent failures in AFNOR7020 alloy as shear bands are known to cut across the thickness of the material and cause catastrophic failures.

4. Conclusions

1. The already present deformation texture components in the starting material, Cu, Bs and S strengthen with increase in rolling deformation, while the recrystallization texture components, Goss and Cube, weaken.
2. The examination of the ODFs reveals that the alloy AFNOR 7020 developed a stronger texture compared to the other two alloys.
3. In AA7075 alloy, the S component is stronger than the other two deformation texture components. No shear bands are observed even at high percentages of rolling in the AA2219 and AA7075 alloy samples.
4. The Bs component is stronger in AFNOR 7020 alloy, which is attributed to shear banding. Cu and Bs type shear bands are observed in the cold rolled samples. This is probably due to the higher Zn: Mg ratio in this alloy.
5. The presence of shear bands could account for the frequent failures in AFNOR7020 alloy as shear bands are known to cut across the thickness of the material and cause catastrophic failures.

Acknowledgements
The authors wish to express their gratitude to Director, VSSC for permitting to publish this work, and to the Deputy Director, VSSC (MME) for the encouragement during various stages of the work.

References
[1] Dillamore, I.L. and Roberts, W.T., 1964, *Acta Metall.*, **12**, 281
[2] Pospiech,J. and Lücke, K., 1978, *Acta Metall.*, **26**, 1709
[3] Suwas, S., Singh, A.K., Rao, K.N. and Singh, T., 2002 , *Z. Metallkde.*,**93**, 918.
[4] Suwas, S., Singh, A.K., Rao, K.N. and Singh, T., 2002 , *Z. Metallkde.*,**93**, 928.
[5] Suwas, S., Singh, A.K., Rao, K.N. and Singh, T., 2003, *Z. Metallkde.*,**94**, 1313.

IOP Conf. Series: Materials Science and Engineering **82** (2015) 012110 doi:10.1088/1757-899X/82/1/012110

[6] Suwas, S., and Singh, A.K., 2003, *Mater. Sci. Engg.*, **356A**, 368
[7] Leffers, T. and Ray, R.K., 2009, *Prog. Mater. Sci.*, **54**, 351
[8] Smallman, R.E. and Green, D., 1964, *Acta Metall.*, **12**,145
[9] Jha, A.K., Ramesh Narayanan P., Sreekumar. K. and Sinha, P.P., 2010, *Engg. Failure Anlys.*, **17**, 562
[10] Jha, A.K., Murty, S.V.S.N and Jacob E., 2002, *Engg. Failure Anlys.*, **9**,709
[11] Jha, A.K., Naga Sirisha G., Ramesh Narayanan P. and Sreekumar. K., 2009, *J. Failure Anlys. Prevn.*, **9**, 414
[12] Jha, A.K., Ramesh Narayanan, P., Diwakar, V., Sree Kumar, K., and Mittal, M.C., 2004, *Engg. Failure Anlys.*, **11**, 463
[13] Ramesh Narayanan, P., Suwas, S., Sree Kumar, K., Sinha, P.P., and Ranganathan, S., 2012, *Mater. Sci. Forum* **702-703**, 315
[14] Humpherys, F.J., and Hatherly, M., 2004, *Recrystallization and Related Annealing Phenomena*, Second Edition, (Elsevier, U.K.)
[15] Duckham, A., Knutsen, R.D., and Engler, O., 2001, *Acta Mater.*,**49**, 2739
[16] Engler, O., 2000, *Acta Mater.*,**48**, 4827

IOP Conf. Series: Materials Science and Engineering **82** (2015) 012111 doi:10.1088/1757-899X/82/1/012111

Correlation of texture and intergranular corrosion in Al-Mg 5xxx series alloys

O Engler, T Hentschel and H-J Brinkman
Hydro Aluminium Rolled Products, R&D Bonn, PO Box 2468, D-53014 Bonn, Germany
E-mail: olaf.engler@hydro.com

Abstract. Aluminium-alloys of the AA 5xxx series with Mg contents in excess of 3% may suffer from intergranular corrosion (IGC) when exposed to temperatures in the range 60 to 200°C. At these temperatures Al-Mg alloys are rendered susceptible to IGC by precipitation of β-Al_8Mg_5 phases along the grain boundaries. Accordingly, susceptibility to IGC will depend on grain size as well as type and orientation of the grain boundaries present in the material, that is, on the crystallographic texture of the material at final gauge.
Therefore, it is of great interest to study the correlation of texture and precipitation of β-AlMg phases and, therewith, susceptibility to IGC. For this purpose, different AA 5182 samples were processed so as to produce different crystallographic textures and characterized with respect to microstructure and resistance against IGC. EBSD local texture analysis was applied to provide information about the grain boundary character distribution. Eventually, this may enable Al industry to reduce the susceptibility of Al-Mg alloys to IGC by proper control of the final gauge texture, such that higher Mg-contents may be used in IGC-critical applications.

1. Introduction

Aluminium alloys of the AA 5xxx series with Mg contents in excess of 3% may suffer from intergranular corrosion (IGC) when they are exposed for a long time to temperatures in the range 60 to 200°C [1,2]. At these temperatures Al-Mg alloys are rendered susceptible to IGC by precipitation of β-Al_8Mg_5 phases along the grain boundaries. Fig. 1(a) shows an example of alloy AA 5182 with about 4.5% Mg, which was heat treated for 17 h at 130°C. The sample was etched for 30 min in 6% phosphoric acid (BSI-test) so as to reveal the spatial arrangement of β-Al_8Mg_5 phase particles. Obviously, in this heavily sensitized condition the β-Al_8Mg_5 particles have precipitated along the grain boundaries which, in the presence of a corrosive medium, renders the material susceptible to IGC (Fig. 1(b)). For these reasons, 5xxx alloys for IGC-critical automotive applications are generally limited to Mg-contents below 3%, which prevents industry from utilizing the full strength potential and, in turn, the full potential for weight savings of 5xxx series alloys.
Volume and rate of precipitation of β-Al_8Mg_5 phases along the grain boundaries will depend on density, type and orientation of the grain boundaries present in the material, that is, on the grain size and on the crystallographic texture of the material at final gauge. Regarding texture, it has recently been demonstrated that there is a clear correlation between texture and resistance against IGC in 5xxx series alloys [3]. For two differently processed samples of alloy AA 5182 a decrease in IGC mass losses from 40 mg/cm² down to 30 mg/cm² could unequivocally be ascribed to different texture, more specifically, to differences in the grain boundary character distribution of the two sheets. More specifically, a batch-annealed sample with a weak cube recrystallization texture plus retained rolling texture had better IGC stability than a flash annealed sample with a more pronounced cube texture. The texture of the batch annealed had a higher density of special CSL-grain boundaries and, most notably, low-angle grain boundaries which are known to play a key role in the resistance to IGC [4,5].
In the present study we have applied the above information on correlation of texture and IGC so as to improve the corrosion behaviour of alloy AA 5182. For this purpose AA 5182 hot strips were taken from two different hot mills, which are known to produce different hot strip texture. Then, the hot strips were lab-processed with and without inter-annealing to final gauge in order to explore the texture differences that can be produced in an industrial setting. The materials were characterized with respect to microstructure, texture and resistance against IGC. Eventually, this may enable us to reduce the susceptibility of Al-Mg alloys to IGC by proper control of the final gauge texture, such that higher Mg-contents may be used for IGC-critical applications.

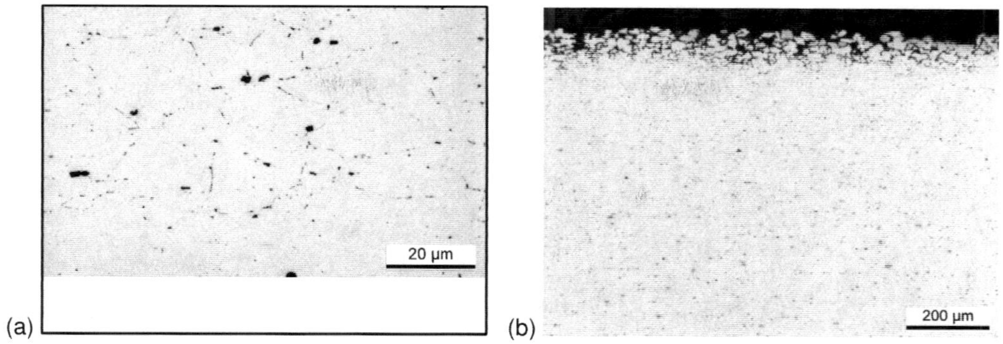

Fig. 1. Micrographs of AA 5182 after sensitization annealing for 17 h at 130°C. (a) BSI-etched, showing grain boundary precipitation; (b) microstructure after NAMLT test, showing heavy corrosion attack by IGC (longitudinal section).

2. Experimental

In the present study two hot strip samples of AA 5182 (EN AW-5182 / ISO: AlMg4.5 Mn0.4) were sourced from different hot mills. Both samples, referred to as I and II, revealed a fine grained, fully recrystallized microstructure, but were differentiated by their recrystallization texture. Samples from the two hot strips were processed on the lab facilities of the Hydro R&D Center in Bonn so as to ensure maximum comparability of the down-stream processing. One set of samples was cold rolled by a total thickness reduction of 50% and soft annealed with a simulated batch annealing practice (peak temperature 350°C) with a controlled cooling cycle. The second set of samples was cold rolled 50%, inter-annealed, again rolled 50% and then soft annealed as described before. The third series of samples was cold rolled in several passes to a total of 75% without inter-annealing, before being soft annealed.

Susceptibility to IGC was determined in the standard nitric acid mass loss test (NAMLT) according to ASTM G67. Samples were tested both in the as-received condition ("none") and following an artificial sensitization treatment for 17 h at 130°C. For each material and state two specimens with a size of approximately 50 × 60 mm^2 were prepared and the weight loss was determined after submersion of the samples in concentrated nitric acid for 24 h.

Grain size was determined by quantitative image analysis in optical micrographs, obtained by anodical oxidation in longitudinal sections of the sheet samples. Crystallographic textures of the sheets were determined at about quarter thickness by standard X-ray diffraction techniques at RWTH Aachen [6]. From several pole figures the 3-dimensional orientation distribution functions (ODFs) $f(g)$ were calculated by the series expansion method [7]. The ODFs were corrected with respect to the so-called ghost error following the positivity method of Dahms and Bunge [8] and represented by plotting iso-intensity lines in three characteristic sections with φ_2 =45°, 65° and 90° through the Euler orientation space. Preliminary EBSD analysis performed close to the sheet surface proved that the textures of the sub-surface layers were quite similar as the bulk textures, which is characteristic of a homogeneous plane-strain state upon cold rolling of thin sheet.

3. Results

It is known that the susceptibility to IGC depends on grain size, with small grain size being more prone to IGC than larger grains. Therefore, for all samples the grain sizes were determined quantitatively; two examples with noticeably different grain sizes are presented in Fig. 2. The hot strip samples showed some variation in average grain size, viz. 15 µm (I) and 19 µm (II), respectively. After cold rolling and soft annealing the differences between the two variants largely vanished, however. For the samples with a rolling degree of 50% the final grain size varied between 15 and 17 µm (e.g. Fig. 2(a)); two passes of 50% with inter-annealing led to very similar results. After two consecutive 50% passes without inter-annealing (i.e. a total of 75%) the final grain size was smaller, some 11-12 µm (Fig. 2(b)), but again there were no differences among the two different hot strip variants.

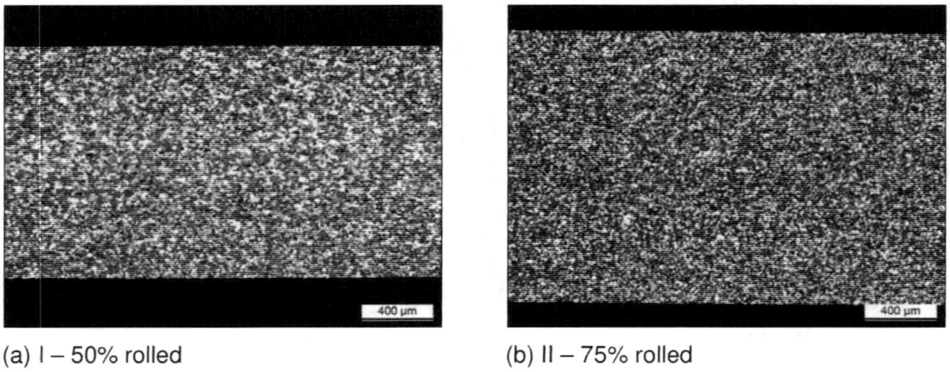

Fig. 2. Microstructure of two specimens with different grain size (a) hot strip I – 50% cold rolled, (b) hot strip II – 75% cold rolled.

Fig. 3 summarizes the results from the NAMLT test in terms of the IGC mass losses for the various samples. It appears that in all cases the additional heat treatment of 17 h at 130°C significantly amplified the mass losses, yet without changing the relative ratio between different samples. For the hot strip I (labelled "0%" in Fig. 3) mass losses were larger than for hot strip II. This tendency exactly followed the grain size data of the two hot strips.

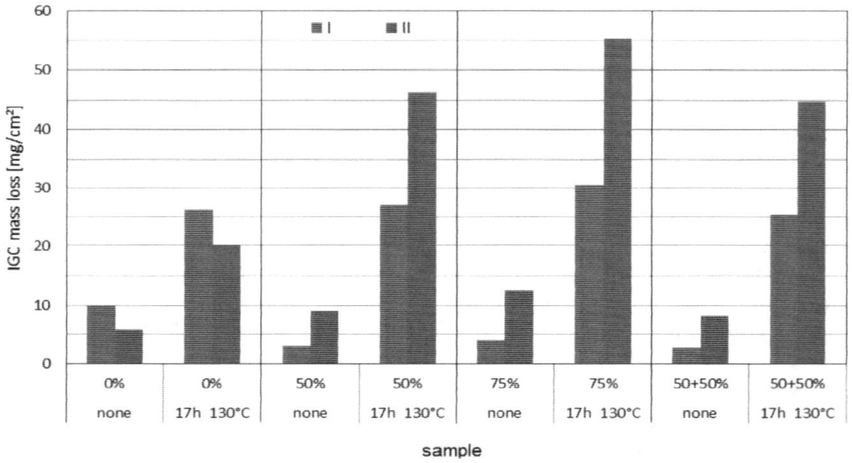

Fig. 3. IGC mass losses during NAMLT testing for the various samples.

After down-stream processing these findings were reversed, however. In all cases material II revealed larger mass losses than material I. With a view to the three different processing routes it turned out that the 75% cold rolled samples had highest mass losses. 75% cold rolling with inter-annealing (i.e. 50% + 50%) or a single pass of 50% both gave noticeably better resistance to IGC, in contrast (Fig. 3). These differences are again ascribed to the significantly lower grain size in the 75% rolled material (Fig. 2). Note that both the quantitative mass loss data and, especially, the differences between the differently processed sheets are comparable to the results obtained in the previous study [3].
Results of the texture measurements are presented in Figs. 4 – 7 in the form of the ODFs obtained for the various hot strips and the three different processing routes, respectively. As already alluded to earlier, the textures are all comprised of the cube recrystallization texture as well as orientations of the retained rolling texture with varying ratios. Hot strip I had a quite sharp cube texture with a maximum intensity $f(g)$ = 18 together with some retained rolling texture (Fig. 4(a)). Such textures which are characteristic of multi-pass tandem lines are typically encountered in AA 3104 can body stock (e.g. [9]).

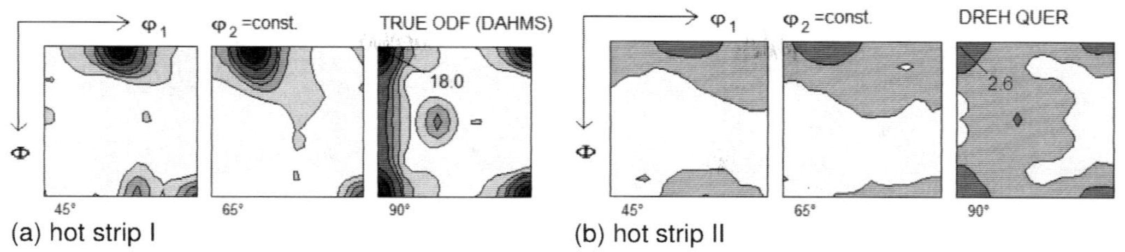

Fig. 4. ODFs of the two different hot strip samples.

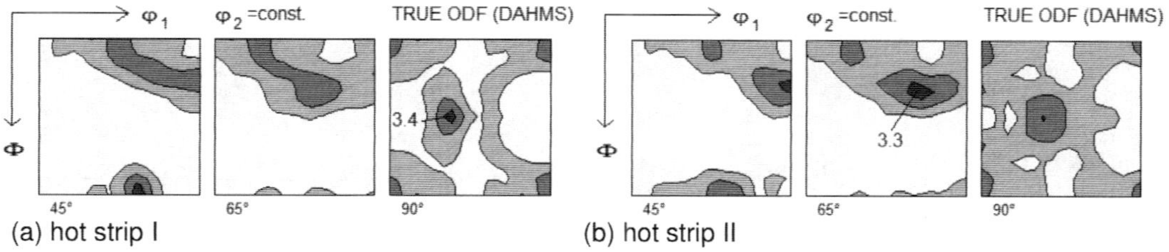

Fig. 5. ODFs of the two hot strips after 50% cold rolling and recrystallization.

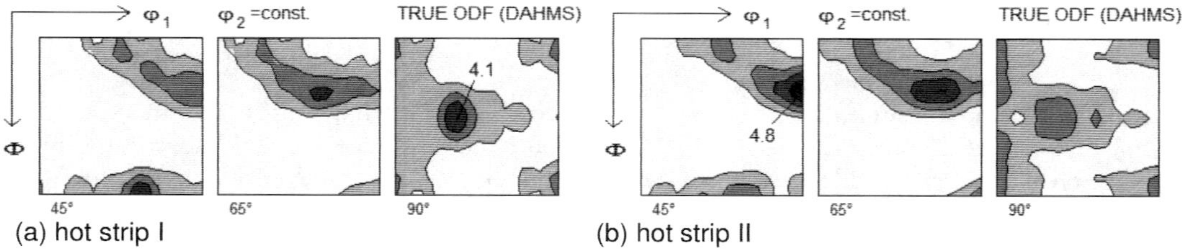

Fig. 6. ODFs of the two hot strips after 75% cold rolling and recrystallization.

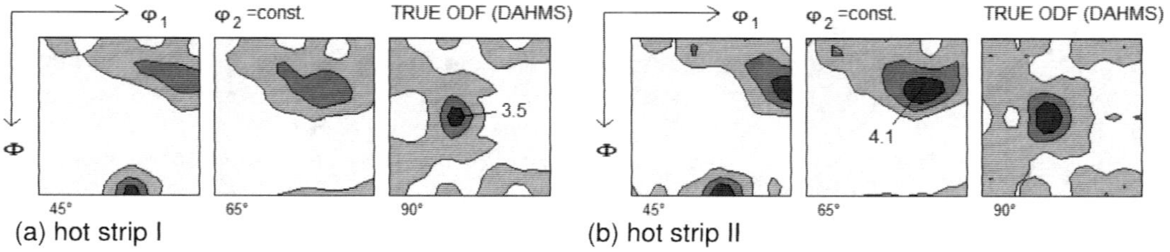

Fig. 7. ODFs of the two hot strips after 75% cold rolling with inter-annealing (50%+50%).

All other textures – hot strip II as well as all cold rolled and soft annealed samples – were quite weak, with maximum intensities ranging from 3 to 5. The texture results of the 75% rolled samples are further summarized in Fig. 8 in form of their main texture fibres, viz. (i) the cube$_{RD}$-fibre comprising the cube-orientation with its RD-scatter (open symbols) and (ii) the β-fibre with the retained rolling texture orientations (filled symbols). Evidently, in both variants the intensities of the retained rolling texture orientations notably exceed those of the cube$_{RD}$-fibre. It is noted that in all cases the samples of series I revealed lower intensities, for both the cube$_{RD}$-fibre and, especially, the retained rolling texture orientations (Figs. 5 – 8).

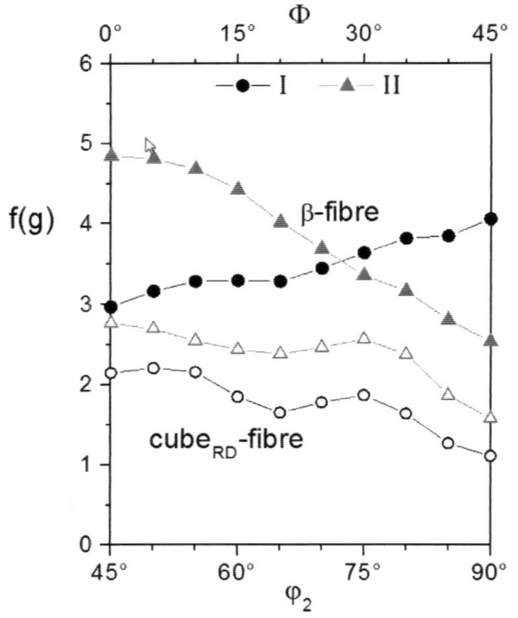

Fig. 8. Texture intensities along the cube$_{RD}$-fibre (open symbols) and β-fibre (filled symbols) for the two 75% rolled samples.

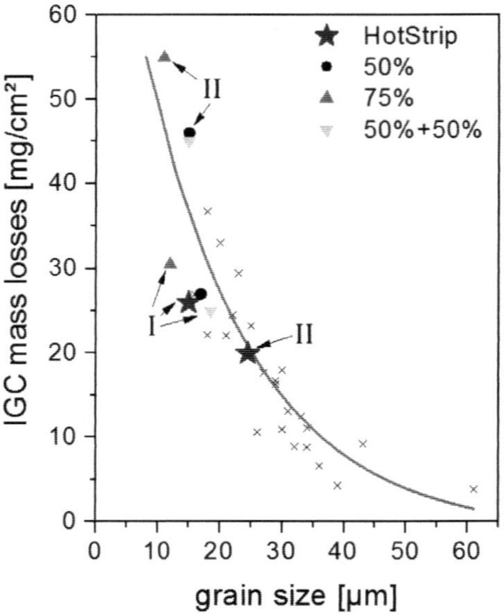

Fig. 9. Mass losses due to IGC in alloy AA 5182 as a function of grain size (sensitized for 17 h at 130°C).

Within the individual series I and II it appears that the 75% rolled sample revealed strongest β-fibre intensities, followed by 50%+50% and, finally, 50%. For the cube$_{RD}$-fibre orientations the sequence is slightly different: 75% is again strongest, but followed by 50% and then 50%+50% (Figs. 5 – 7).
It is worthwhile emphasizing that in all cases the texture evolution followed the well-known patterns of texture formation during cold rolling and recrystallization in Al alloys, see, e.g., Refs. [9-11].

4. Discussion and conclusions

As already stated above, intergranular corrosion (IGC) in Al-Mg alloys depends on grain size and on the crystallographic texture of the material at final gauge. Fig. 9 shows a plot of the mass losses due to IGC in alloy AA 5182 as a function of grain size, revealing an appreciable reduction in mass losses with increasing grain size. Comparison of the dependency on grain size summarized in Fig. 9 with the different mass losses obtained in the previous study [3] suggests that the influence of grain size is much bigger than that of texture. Accordingly, the differences in IGC mass losses of the two different hot strip samples can readily be attributed to the different grain sizes of the same, while a possible impact of texture may be screened.
In order to study this more deeply, Fig. 9 also contains the IGC data for the two hot strips, revealing that the results for hot strip II come to lie rather close to the red best-fit curve. Sample I shows slightly lower mass losses than what would be expected for the actual grain size, however, which, with all due care, seems to imply that the actual texture is beneficial in improving IGC resistance in this very material. This material I with its pronounced cube recrystallization texture (Fig. 4(a)) will comprise a rather high fraction of low-angle grain boundaries, whereas hot strip II with a very weak hot strip texture (Fig. 4(b)) will presumably comprise a much higher fraction of high-angle grain boundaries, similar to the well-known Mackenzie-distribution for randomly oriented microstructures [6]. Low-angle boundaries are

known to play a key role in the resistance to IGC, in that these boundaries were previously found to be most resistant to corrosion in high-purity Al [4].

After down-stream processing (cold rolling and soft annealing) it was observed that the 75% cold rolled samples had higher mass losses than the 50% cold rolled samples (both one pass or two passes with inter-annealing, i.e. 50%+50%; Fig. 3). Again, these differences can readily be attributed to the significantly lower grain size in the 75% rolled material (Fig. 2).

Vice versa, a direct impact of texture on IGC can only be derived for samples with identical grain size. In the present study this applies to the three series with identical down-stream processing, i.e. 50°, 50%+50% and 75%, respectively. Within these three processing routes it turned out that alloy II always revealed higher mass losses due to IGC than material I (Fig. 3). The IGC results are also marked in Fig. 8, indicating that despite some subtle differences in grain size the mass losses of material II tend to lie closer to the red best-fit curve than those of material I.

Thus, it may be concluded that the weak mixed textures composed of the cube recrystallization texture and some retained rolling texture orientations (e.g. Figs. 5 – 7) reveal optimum conditions for exhibiting resistance against IGC. The reason for the differences between the current materials I and II with rather similar mixed textures could not be clarified yet, however. In the previous study it was shown that preferable textures were characterized by a high density of 'special' CSL-grain boundaries and, most notably, low-angle grain boundaries [3]. Both low-angle grain boundaries and special CSL-boundaries are commonly assumed to have lower energy than regular high-angle grain boundaries (e.g. [12]). As precipitation of β-Al_8Mg_5 phases preferably occurs along un-ordered high-energy boundaries, an increased fraction of special boundaries will reduce precipitate formation, thus reducing IGC intensity [4,5]. A detailed classification of the grain boundary character distribution of the two materials is currently being performed by means of EBSD local texture analysis.

References

[1] Searles JL, Gouma PI and Buchheit RG 2001 *Metall. Mater. Trans.* **32A** 2859-2867
[2] Tan L and Allen TR 2010 *Corrosion Sci.* **52** 548-554
[3] Hill L, Randle V and Engler O 2013 *Mater. Sci. Tech.* **29** 1006-1011
[4] Kim SH, Erb U, Aust KT and Palumbo G 2001 *Scripta Mater.* **44** 835-839
[5] Aust KT, Erb U and Palumbo G 1994 *Mater. Sci. Eng.* **A176** 329-334
[6] Engler O and Randle V 2010 *Introduction to Texture Analysis: Macrotexture, Microtexture and Orientation Mapping* (Boca Raton, FL: CRC Press)
[7] Bunge HJ 1982 *Texture Analysis in Materials Science* (London: Butterworths)
[8] Dahms M and Bunge 1989 *J. Appl. Cryst.* **22** 439-447
[9] Engler O, Löchte L and Hirsch J 2007 *Acta Mater.* **55** 5449-5463
[10] Engler O, Crumbach M and Li S 2005 *Acta Mater.* **53** 2241-2257
[11] Engler O and Hirsch J 2009 *Inter. J. Mater. Res.* **100** 564-575
[12] Lin H and Pope DP 1995 *Mater. Sci. Eng.* **A394** 192-193

IOP Conf. Series: Materials Science and Engineering **82** (2015) 012112 doi:10.1088/1757-899X/82/1/012112

Residual stresses determination in textured substrates for plasma sprayed coatings

J Capek[1], Z Pala[1,2] and O Kovarik[1]
[1] Faculty of Nuclear Sciences and Physical Engineering, Czech Technical University in Prague, Trojanova 13, 12000 Prague, Czech Republic
[2] Institute of Plasma Physics, Za Slovankou 1782/3, 18200 Prague, Czech Republic
E-mail: capekjir@fjfi.cvut.cz

Abstract. In this contribution, we have striven to respond to the desire of obtaining the residual stress tensor in the both cold-rolled and hot-rolled substrates designated for deposition of thermal coatings by plasma spraying. Residual stresses play an important role in the coating adhesion to the substrate and, as such, it is a good practice to analyse them. Prior to spraying, the substrate is often being grit blasted. Residual stresses and texture were quantitatively assessed in both virgin and grit blasted sample employing three attitudes. Firstly without taking preferred orientation into account, secondly from measurements of interplanar lattice spacings of planes with high Miller indices using Mo*Kα* radiation. And eventually, by calculating anisotropic elastic constants as a weighted average between single-crystal and X-ray elastic constants with weighting being done according to the amount of textured and isotropic material in the irradiated volume. In the ensuing verification analyses, it was established that the latter approach is suitable for materials with either very strong or very weak presence of texture.

1. Introduction

Joint occurrence of texture and residual stresses in metals is no rarity and it is well known that determination of the stress tensor by X-ray diffraction is not, in this case, a straightforward task and calls for both experimental procedures and data processing algorithms beyond the standard ones employed for isotropic and fine-grained polycrystalline materials. For now, a universal approach with the potential to correctly and properly evaluate the residual stresses in textured materials is, sadly, still lacking and this issue is tackled either, in the worst scenario, by neglecting the texture or by choosing one of the hitherto proposed methods.

The value and homogeneity of residual stresses and surface roughness of materials used as substrates for thermal coatings have also significant influence [1] on their adhesion. Grit blasting not only leads to rough surface and redistribution of residual stresses, but also considerably changes the preferred orientation in the affected surface layer. In this particular study, we analysed hot-rolled and cold-rolled steel substrates dedicated for ceramic coating deposition by WSP (water-stabilized plasma) spraying [2]. Particles of ceramic material are melted in the plasma jet and solidify upon impacting the substrate; hence, the overall properties of substrate-coating interface, including residual stresses, play in important role in the fatigue resistance [3] of the sprayed component.

2. Theory

Except for the so called $\sin^2\psi$ *method*, the most widely used approach for macroscopic residual stresses determination, there also exists an algorithm proposed by Winholtz and Cohen [4] which is distinguished by lower calculation error. This method does not require any linear approximation but determines the unknown stress tensor components by least-squares minimization of the difference X^2, see equation (1), between calculated ε^{calc} and measured ε^{meas} lattice strain, where the unknown stress components σ^S are refined.

$$X^2 = \sum_i [\varepsilon_i^{calc}(\sigma^S) - \varepsilon_i^{meas}]\,. \qquad (1)$$

For stress determination in textured materials, Hauk and Nikolin [5] suggested analysing diffraction planes *{hkl}* with high values of Miller indices. In this case, the diffraction profiles

correspond to X-rays diffracted by higher number of differently oriented grains and the deviation from the linearity in the $\sin^2\psi$ plot is practically non-existent even with presence of strong texture. For iron-based materials, it is suitable to use Mo*Ka* radiation and measure *{732+651}* diffraction profile. The main advantage of this method is suppression of texture influence on residual stress analysis, but overlapping and relatively low diffracted intensity count among its disadvantages.
More appropriate methods are those which do not use X-ray elastic constants (XEC), but the so called X-ray stress factors (XSF). These parameters are functions of *(i)* Miller indices, *(ii)* angles φ, ψ, representing rotation of specimen, and *(iii)* texture. Viewing textured material as a state of grains' ordering somewhere between ideal polycrystal and single-crystal, it should be possible to propose an algorithm for residual stress determination by combining textured and non-textured state. On the basis of a work by Dölle [6], the program was developed for calculation of the residual stress state in specimens with preferred orientation. Equation (2) represents determination of XSF (R_{ij}):

$$R_{ij}\ (hkl, \varphi, \psi) = \frac{\lambda^i\, r_{ij}\,(hkl) + \lambda^a\, S_{ij}}{\lambda^i + \lambda^a}. \qquad (2)$$

Constants λ^i and λ^a denote the volume ratio of untextured and textured parts in the specimen, respectively, dependent on the individual inclination during measurement and corresponding to relative intensities of PF (pole figure). Parameters r_{ij} represent the combination of XEC calculated on the basis of the selected linear elastic model and S_{ij} represent single-crystal elastic constants in the laboratory reference frame. Residual stresses are calculated following *Winholtz and Cohen method* using XSF. For this particular approach, it is necessary to comply with specific conditions; otherwise the calculation error grows rapidly. In particular, this method is applicable in the case of very weak, or virtually non-existent, texture, and also when strong and sharp texture is present.

3. Experiment

The analysed samples were made of *ISO 630:1980* steel. One sample was hot-rolled (RH) and the other cold-rolled (RC). After this procedure, each sample was grit blasted by coarse corundum particles with an incidence angle of 75° on one side of both samples (RH-blasted and RC-blasted). Strains were measured in three directions on the rolling plane. Stresses determined in the directions parallel with the rolling direction (RD) are termed stresses σ_{RD}, perpendicular (transversal direction) stresses σ_{TD}. Stresses termed σ_{45} halve RD and TD and measurements in this direction are carried out in order to compute σ_{12} stress tensor component. The shape of primary beam had a rectangular shape with dimensions 10×1 mm^2. Generally with tilt increasing, the irradiated surface area enlarges and the penetration depth decreases. We assumed biaxial state of stress and used the *Winholtz and Cohen method* to compute residual stresses. Diffraction profiles of *{211}* and *{732+651}* of bcc iron were analysed by Cr*Ka* and Mo*Ka* radiation, respectively. Profiles corresponding to $K_{\alpha 1}$ radiation were fitted by *Pearson VII* function. In the generalised Hooke's law, we used XEC obtained in accordance with *Eshelby-Kröner model* [7] using single crystal elastic constants for iron S_{11} = 7.6 TPa^{-1}, S_{12} = -2.8 TPa^{-1}, S_{44} = 8.6 TPa^{-1}. Very important parameter to calculate residual stresses correctly, especially the component σ_{33}, is stress-free lattice plane distance d_0. This parameter was determined on the basis of finding the so called stress-free direction ψ^* of the RH sample according to the algorithm that can be found [8] in chapter 2.112c, which yielded the value of 0.117024 nm for *{211}* planes.
The texture was qualitatively determined using Schultz reflection geometry and radiation from X-ray tube with cobalt anode. The Co*Ka* radiation was used, because 1D detector for Cr*Ka* was not at our disposal; the effective penetration depth T^{ef} of Co*Ka* for *{110}*, *{200}* and *{211}* profile is 5 µm, 7 µm and 9 µm, respectively which is approximately 1.5 higher values

than 3 μm, 5 μm and 6 μm for Cr*Kα*. The shape of primary beam was rectangular with dimensions 0.25×1 mm². Pole figures were re-calculated from the orientation distribution function which was calculated on the basis of *WIMV* method [9] from measured PF.

4. Results

Components of macroscopic residual stress tensors relevant for surface layers can be seen in Tab. 1, where Cr*Kα* (XSF) means that the method using XSF was employed. Next, the values listed for Cr*Kα* and Mo*Kα* are based on the XEC for random texture. Pole figures of *{211}* planes are seen for cold-rolled sheet in Fig. 1.

Table 1. Surface distributions of macroscopic residual stresses.

Stress, MPa	Radiation	RC	RH	RC-blasted	RH-blasted
σ_{RD}	Cr*Kα*	-70 ± 5	-6 ± 8	-141 ± 3	-201 ± 3
	Cr*Kα* (XSF)			-147 ± 6	-187 ± 10
	Mo*Kα*	-61 ± 7	-24 ± 6	-205 ± 14	-260 ± 13
σ_{45}	Cr*Kα*	-85 ± 5	-14 ± 6	-132 ± 5	-200 ± 5
	Cr*Kα* (XSF)			-130 ± 7	-170 ± 12
	Mo*Kα*	-37 ± 4	-11 ± 4	-200 ± 15	-252 ± 9
σ_{TD}	Cr*Kα*	-125 ± 3	2 ± 10	-142 ± 3	-194 ± 6
	Cr*Kα* (XSF)			-130 ± 4	-174 ± 13
	Mo*Kα*	-33 ± 8	-16 ± 3	-188 ± 16	-212 ± 16
σ_{33}	Cr*Kα*	-15 ± 8	10 ± 10	-19 ± 2	-29 ± 3
	Cr*Kα* (XSF)			-19 ± 2	-14 ± 4
	Mo*Kα*	33 ± 4	73 ± 2	165 ± 7	149 ± 6

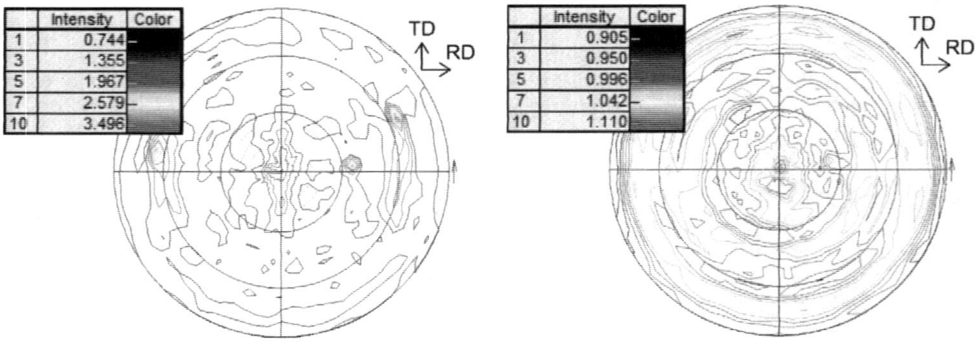

Figure 1. PF of *{211}* planes for RC (left) and RC-blasted (right) samples.

5. Discussion

Firstly, in the used experimental lay-out it was impossible to calculate stresses in virgin samples employing the method with XSF, because of a too coarse-grained material in the irradiated volume. Sadly we were not able to increase the irradiated surface area without significant defocusation errors. Hence, in virgin samples only calculation with texture omission and with Mo*Kα* can be compared which leads us to observation of fairly large discrepancies in stress values. The responsible factor for this is most likely twofold. Firstly different gauge volumes correspond to these values (with penetration depth for Mo*Kα* being approximately triple of T^{ef} for Cr*Kα*) and since the presence of stress gradients in virgin

IOP Conf. Series: Materials Science and Engineering **82** (2015) 012112 doi:10.1088/1757-899X/82/1/012112

samples cannot be ruled out, it is not surprising that the presented values are not the same. Secondly, the first method uses algorithm for stress calculation in isotropic material while the investigated material was textured. As for the large discrepancy in σ_{33} values in grit-blasted samples, the effect of stress depth gradient and difference in T^{ef} should be considered.
The pole figures in Fig. 1 clearly show that grit-blasting leads to much lower degree of preferred orientation in surface layer. The main reason is most likely an increasing orientation spread in the grains due to plastic deformation. The stress tensor was evaluated in three manners for the grit blasted samples. Thus on the basis of the above mentioned model, XSF were calculated and afterwards residual stresses were determined. Because the level of preferred orientation in the blasted samples is very low, relative intensities are nearly one and resulting XSF almost equal XEC. Differences between values of stress components calculated according the first two approaches are very small as seen in Tab. 1. The reason for this observed difference stems from difference in relative intensities, i.e. even the very low level of texture has some influence on stress values. Hence, the applied method according to equation (2) leads only to small alterations in stress values for weak texture, which is in accordance with expectations. On the other hand, this algorithm remains to be verified on strongly textured samples.

6. Conclusion

Even though the values of residual stresses calculated with neglecting the texture in non-blasted samples show sound experimental error, they should not be trusted because the applied algorithm does not correspond to assumptions of the method. On the other hand, it was not possible to use the algorithm with XSF for determination of residual stresses because of coarse grained nature of the irradiated volume. Thus, to us it seems the most appropriate to use Mo*Kα* radiation for residual stress determination is such case.
From pole figures, it is obvious that grit blasting of both cold and hot rolled steel surface leads to quantitative change in texture with very low level of preferred orientation in grit blasted surfaces. When we compare the two approaches without and with XSF, they lead to very similar values of stress. However, employing Mo*Kα* radiation correspond to different values of stress and one has to bear in mind a big difference in penetration depth which is almost threefold in comparison with Cr*Kα*. It should also be noted that grit-blasted hot-rolled sheet has larger values of compressive residual stresses when compared to cold-rolled.

Acknowledgements

J. C. acknowledges support of the Grant Agency of the Czech Technical University in Prague, grant No. SGS13/219/OHK4/3T/14. Z. P. was supported by GAP108/12/1872, a Czech Science Foundation grant.

References

[1] Harris A F and Beevers A 1999 *Int. J. Adhes. Adhes.* **19** 445–452
[2] Chraska P and Hrabovsky M 1992 *Therm. Spray: Int. Adv. Coat. Technol.* **1** 81–85
[3] Kovarik O et al. 2014 *Surf. Coat. Technol.* **251** 143–150
[4] Winholtz R A and Cohen J B 1988 *J. Phys.* **41** 189–199
[5] Hauk V and Nikolin H -J 1988 *Texture Microstruct.* **8/9** 693–716
[6] Dolle H 1979 *J. Appl. Cryst.* **12** 489–501
[7] Eshelby J D 1957 *Proc. Roy. Soc.* A **241** 376–396
[8] Hauk V, ed. 1997 *Structural and Residual Stress Analysis by Nondestructive Methods: Evaluation-Application-Assessment* (Elsevier) 640
[9] Kocks U F et al. 2005 *Texture and Anisotropy: Preferred Orientation in Polycrystals and their Effect on Materials Properties* (London, Cambridge University Press) 115–118

Grain boundary engineering in a thermo-mechanically processed Nb-stabilized austenitic stainless steel

A Yunquera [1], D Jorge-Badiola [1], I Gutiérrez [1] and A Iza-Mendia [1]
[1] CEIT and Tecnun (University of Navarra), Donostia-San Sebastián, Basque Country, Spain
E-mail: djbadiola@ceit.es

Abstract. Three different thermo-mechanical strategies—annealing, strain recrystallization and strain annealing—were applied to a Nb-stabilized 304H austenitic stainless steel in order to study their effects on grain boundary character distribution (GBCD). An Electron Backscatter Diffraction (EBSD) analysis revealed specific combinations of cold reduction-temperature-time that favor annealing twinning. A uniform increase in microstructural size and special boundaries (particularly for Σ3, Σ9 and Σ27 boundaries) was achieved under strain annealing conditions (low cold reductions) and long times at high temperatures (≥ 990°C). These conditions provide a high fraction of special boundaries (about 80%), which replace the random grain boundary network and thus optimize the GBCD. The profuse presence of $\Sigma 3^n$ boundaries is attributed to the geometric interaction of twin-related variants during grain boundary migration. In addition to all this, precipitation takes place at the temperature range where optimum GBCD is achieved. The significance of precipitation in the different strategies was also tackled.

1. Introduction

Grain boundaries are crystallographic discontinuities where phenomena such as migration, sliding, solute segregation and precipitation can occur. However, some boundaries present special properties that improve material performance under the above mentioned degradation phenomena. The boundaries are usually described through the coincidence site lattice (CSL) model [1], which defines boundaries by the density of common sites (Σ) between grains sharing a boundary. Among all Σ boundaries, $\Sigma 3^n$ type boundaries are assumed to have a higher specialness, particularly Σ3 boundaries. Within Σ3 boundaries, many analyses have concluded that only {111} plane-based Σ3 boundaries, also known as coherent Σ3 annealing twins, are special [2]. Thus, increasing the proportion of coherent Σ3 boundaries by decreasing random High Angle Grain Boundaries (HAGB) is critical for obtaining grain boundary character distribution (GBCD) control and for improving intergranular properties. In that regard, Grain Boundary Engineering (GBE) aims to increase the density of special grain boundaries in materials with low SFE (stacking fault energy) through various thermo-mechanical routes that basically consist of applying a cold strain and following it with an annealing treatment [3,4]. Two types of GBE can be distinguished: strain recrystallization treatments that increase the proportion of special boundaries by forming new grains during recrystallization, and strain annealing treatments that promote high fractions of $\Sigma 3^n$ without forming new HAGB boundaries.

The excellent properties of austenitic stainless steels (ASS) under severe conditions make them an interesting objective for studying their behavior under supercritical conditions [5]. Over the last decade several studies have addressed GBE research on austenitic stainless steels [6,7]. More than 80% of special boundaries are obtained with strain annealing treatments [8,9]. Unlike previous studies, this study is focused on a Nb-stabilized 304H stainless steel, one of the materials likely to be employed under supercritical conditions, in which Nb(C,N) precipitation could play a role either as a microstructural controlling factor or as a matrix strengthener . The present study is intended to explore and analyze the microstructures obtained when a Nb-stabilized 304H stainless steel is subjected to different final processing strategies that seek to achieve an optimum grain boundary network.

2. Experimental

The material used in this study is an 18Cr9Ni3CuNbN austenitic stainless steel. Various series of thermo-mechanical processes, whose experimental details are shown in **Table 1**, were applied to the material and can be categorized as:

- Additional annealing treatments on heavily cold reduced material (*strain recrystallization*).
- Low cold reduction (applied in tensile mode, 5-15%) plus annealing treatments (*strain annealing*). This treatment type was applied to two different as-received states of the material, solution annealed SA1 and SA2, which exhibited a fully recrystallized microstructure with differences in grain size as well as in precipitation.
- Additional annealing treatments on fully recrystallized material so as to induce *grain growth*.

Table 1 Thermo-mechanical treatments carried out on 18Cr9Ni3CuNbN ASS

Thermo-mechanical treatments	Reduction (%)	Temperature (°C)	Time (h)
Heavily cold rolled + annealing treatment	~65	1050-1150	0.5-2
Solution annealed + cold reduction + annealing treatment	5-15	950-1200	0.16-24
Solution annealed + additional annealing treatment	-	990-1200	0.5-1

After each annealing treatment, the samples were immediately cooled down in water. Following that, EBSD and FEG-SEM were employed to characterize the samples. The analysis of the HAGB character was analyzed by looking at the low-Σ type boundaries on areas larger than 600x600μm^2, with a 1μm step size. In particular, for the measurements of the coherent Σ3 annealing twins, the trace criterion {111} –with tolerance lower than 8°– and the Σ3 relationship were adopted [1,10]. Moreover, the Kernel Average Misorientation (KAM) parameter was used to define the strain state of the material.The samples were assumed to be strain free material when KAM (third neighbor, 3°) was below 0.5°.

3. Results and discussion

3.1. Effect of the thermo-mechanical treatments on low-Σ boundaries

As a result of applying different thermo-mechanical treatments to this austenitic stainless steel, a wide range of microstructures were generated. The strain annealed samples showed a convoluted grain boundary network, whereas in the samples where only recrystallization and/or grain growth had taken place a more equiaxed HAGB network was attained. The analysis was based on measuring the total CSL boundaries from Σ3 to Σ27 as the length fraction of the total HAGB (>15°). Some trends can be detected when plotting this fraction as a function of the specific grain boundary area S_v, **Figure 1.a**. All the treatments involve a uniform increase in the grain size, and therefore the decrease in S_v, along with the amount of special grain boundaries, except for grain growth treatments at low temperatures and strain annealing treatments under the lowest cold reduction-temperature-time conditions (5%-950/990°C-<3h). Under such conditions, no effect on the grain boundaries and the microstructural unit size was detected. The highest effectiveness of the strain annealing processes with regard to the other processing routes is fairly evident in terms of the CSL boundary population. In fact, the length fraction of low-Σ boundaries for 5%-990/1030°C-24h is 85%. This means that the grain boundary network almost reaches the so called twin-limited microstructure in terms of Σ3 boundaries [11], since most of the triple points contain two Σ3 boundaries. Other austenitic stainless steels for similar combinations of cold reduction-annealing temperatures are able to develop the same low-Σ boundary fraction [6].

Additionally, the microstructural evolution for a fixed strain annealing treatment depends on the as-received recrystallized state, due to the differences in the initial S_v and in precipitation.

As far as the coherent Σ3 boundaries are concerned, two readily distinct behaviors are observed when their length fraction is plotted against the fraction of low-Σ boundaries, **Figure 1.b.** Strain recrystallization and pure grain growth treatments at high temperatures (above 1150ºC) slightly enhance the CSL fraction despite the significant increase in coherent Σ3 twin boundaries. Conversely, the strain annealing treatments involve the simultaneous increase of CSL and coherent Σ3 length fractions. High amounts of Σ9 and Σ27 were achieved in strain annealed samples -exceeding 9% in most cases- whereas the other samples hardly reached 3%. Regardless of the strategy, the coherent Σ3 fraction levels off at the same fraction (~42%) for low S_v. Therefore, the mechanisms responsible for coherent Σ3 boundary formation and for $\Sigma 3^n$ type boundary formation are operative when strain annealing treatments are applied. Instead, the mechanism for coherent Σ3 boundary formation predominates under recrystallization strain and grain growth treatments at high temperatures.

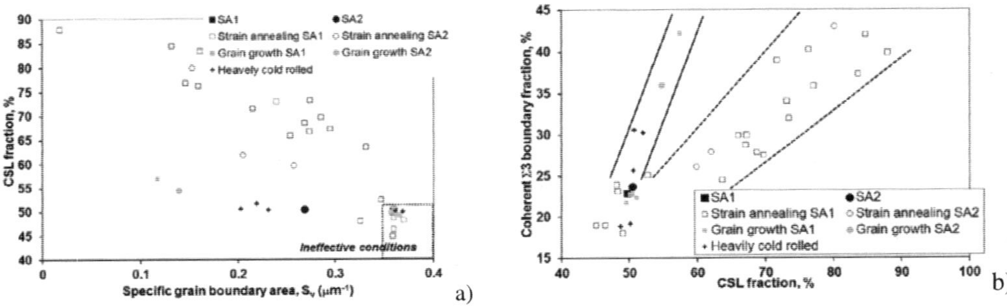

Figure 1 a) Fraction of low-Σ boundaries as a function of the specific grain boundary area. b) Fraction of coherent Σ3 annealing twins vs. low-Σ boundary fraction.

3.2. Evolution over time of low-Σ boundaries for strain annealing treatments

In order to analyze the development of the special grain boundaries over time, different annealing treatments after a 5% reduction at 990ºC and 1030ºC were carried out for various holding times, **Figure 2.a**. CSL boundary formation kinetics are faster at the highest temperature, 1030ºC, but over time both behaviors converge. The main increase in the amount of special grain boundaries for strain-annealed samples is limited to a time interval in which all the strain energy is largely consumed. After that interval, the amount of Σ boundaries also increases due to subsequent grain growth, but at a much slower rate. A zoom of a region in which strain annealing is effective, after 0.5h at 1030ºC, shows some clues about the generation of the GBCD, **Figure 2.b**. Several strain free regions (black regions, low kernel) are made of a highly twinned microstructure. A few general high angle boundaries in this region appear to be related to the impingement of different highly twinned regions that meet together (arrow between regions A and B).

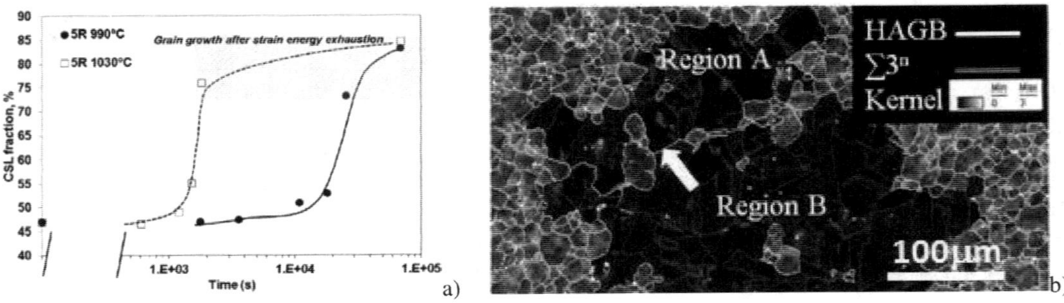

Figure 2 a) Evolution of the CSL fraction over time for a 5% reduction and further annealing (SA1) and **b)** Kernel map with overlapped HAGB network for the 5R-1030ºC-0.5h annealing treatment

3.3. Effect of the precipitation on the low-Σ boundary formation

Precipitation of Nb(C,N) is present in the as-received solution annealed states and during additional annealing treatments. The existence of these precipitates substantially impinges the grain boundary migration, **Figure 3.a**. However, the delaying effect of the precipitation on kinetics does not bring about a noticeable change in the GBCD, as shown by the high fraction of special grain boundaries. This conclusion is also supported by the experimental observations reported for other ASS, where no precipitation was observed at the annealing temperatures but very close fractions of low-Σ boundaries resulted [78-9], **Figure 3.b**, and by a previous theoretical study on particles as a way to control the grain boundary character [12].

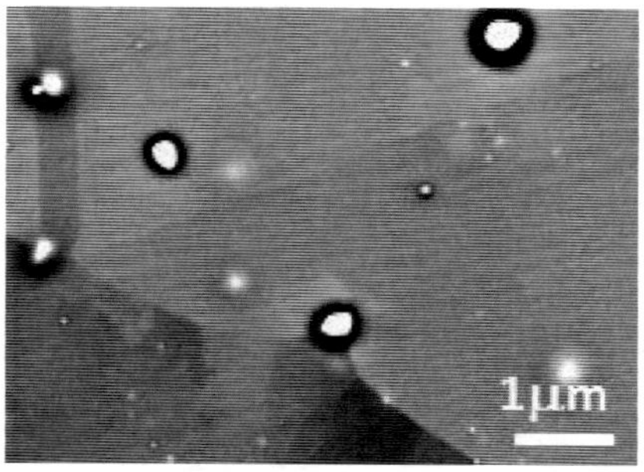

a)

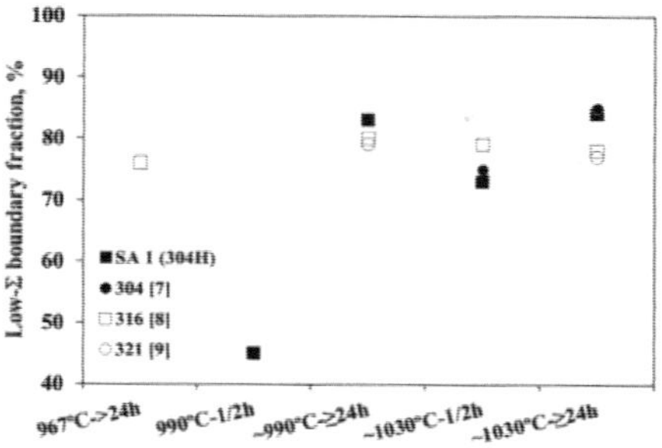

b)

Figure 3 a) Grain boundary migration impingement by Nb(C,N) precipitation and b) low-Σ boundary fraction for different ASS and thermo-mechanical sequences (cold reduction 5%).

4. Conclusions

The evolution of GBCD depending on the thermo-mechanical sequence applied has been tackled in a Nb-stabilized austenitic stainless steel. Recrystallization and/or pure grain grow at temperatures above 1150ºC enhance the formation of annealing twins, whereas strain annealing treatments augment the overall length fraction of low-Σ boundaries. In general, both types of sequences bring about an increase in the grain size. Nb(C,N) precipitation delays the kinetics of the grain boundary migration and its effect on the character of the final grain boundaries appears to be limited. For the time intervals analyzed, further grain growth after strain energy exhaustion involved an increase in the low-Σ boundary fraction in strain annealing treatments.

References

[1] Brandon D G 1966 *Acta Metall.* **14** 1479
[2] Randle V 2006 *Scripta Mater.* **54** 1011
[3] Randle V 2004 *Acta Mater.* **52** 4067
[4] Kumar M, King W E and Schwartz A J *Acta Mater.* 2000 **48** 2081
[5] Beddoes J and Parr J G 1999 *Introduction to Stainless Steels*, ASM International, OH (USA)
[6] Fang X, Zhang K, Guo H, Wang W and Zhou B *Mater. Sci. Eng. A* 2008 **487** 7
[7] Shimada M, Kokawa H, Wang Z J, Sato Y S and Karibe I 2002 *Acta Mater.* **50** 2331
[8] Michiuchi M, Kokawa H, Wang Z J, Sato Y S and Sakai K 2006 *Acta Mater.* **54** 5179
[9] Kurihara K et al. 2011 *J. Mater. Sci.* **46** 4270
[10] Randle V, Davies H 2002 *Ultramicroscopy* **90** 153
[11] Palumbo G et al. *Phys. Stat. Sol.* 1992 **131** 425
[12] Miodownik M A 2006 *Scripta Mater.* **54** 993

IOP Conf. Series: Materials Science and Engineering **82** (2015) 012114 doi:10.1088/1757-899X/82/1/012114

Effect of hot working on the damping capacity and mechanical properties of AZ31 magnesium alloy

K Lee[1,a], C Kang[1,b] and K Kim[1,c]
[1] Department of Metallurgical Engineering, Pukyong National University, Busan 608-739, Republic of Korea
E-mail: [a]leekh@swhitech.com, [b]metkcy@pknu.ac.kr, [c]mrppeng@pknu.ac.kr

Abstract. Magnesium alloys have received much attention for their lightweight and other excellent properties, such as low density, high specific strength, and good castability, for use in several industrial and commercial applications. However, both magnesium and its alloys show limited room-temperature formability owing to the limited number of slip systems associated with their hexagonal close-packed crystal structure. It is well known that crystallographic texture plays an important role in both plastic deformation and macroscopic anisotropy of magnesium alloys. Many authors have concentrated on improving the room-temperature formability of Mg alloys. However, despite having a lot of excellent properties in magnesium alloy, the study for various properties of magnesium alloy have not been clarified enough yet.

Mg alloys are known to have a good damping capacity compared to other known metals and their alloys. Also, the damping properties of metals are generally recognized to be dependent on microstructural factors such as grain size and texture. However, there are very few studies on the relationship between the damping capacity and texture of Magnesium alloys. Therefore, in this study, specimens of the AZ31 magnesium alloy, were processed by hot working, and their texture and damping property investigated. A 60 mm × 60 mm × 40 mm rectangular plate was cut out by machining an ingot of AZ31 magnesium alloy (Mg-3Al-1Zn in mass%), and rolling was carried out at 673 K to a rolling reduction of 30%. Then, heat treatment was carried out at temperatures in the range of 573–723 K for durations in the range of 30–180 min. The samples were immediately quenched in oil after heat treatment to prevent any change in the microstructure. Texture was evaluated on the compression planes by the Schulz reflection method using nickel-filtered Cu Kα radiation. Electron backscatter diffraction measurements were conducted to observe the spatial distribution of various orientations. Specimens for damping capacity measurements were machined from the rolled specimen, to have a length of 120 mm, width of 20 mm, and thickness of 1 mm. The damping capacity was measured with a flexural internal friction measurement machine at room temperature. It was found that the damping capacity increases with both increasing heat-treatment temperature and time, due to grain growth and the increased pole densities of textures.

1. Introduction

There has been a growing interest in using magnesium alloys for various industrial and commercial applications, including electronics and automotive components parts, owing to their low density and high specific strength. However, magnesium alloys have poor formability at room temperature because of the limited number of slip systems available, given their hexagonal close-packed structure [1]. To extend their applicability, their room-temperature formability obviously needs to be improved, and ways to achieve which, including microstructure refinement and texture control, have been studied by various researchers [2]. However, excluding study for such formability improvement, other characteristics of magnesium alloys have not been clarified enough yet

Magnesium alloys are known to have higher damping capacities compared to other metals and their alloys [3]. The damping capacity of commercial magnesium alloys is lower than that of pure magnesium [4]. In general, the mechanical properties and damping capacity of materials are known to depend on grain size, presence of precipitates, and texture among a host of attributes. In a previous study, the authors studied the effect of heat treatment on the damping capacity of ductile cast iron and found that the mechanical properties and damping capacity varied depending on the heat treatment conditions such as temperature and time [5]. It is therefore believed that similar trends can be expected of magnesium alloys.

In this study, the magnesium alloy, AZ31, subjected to hot-rolling was experimentally investigated under various heat-treatment conditions. Especially, the effect of texture on the mechanical properties and damping capacity of AZ31 was investigated.

2. Experimental procedure

Specimens for evaluation were prepared from commercial cast AZ31 magnesium alloy ingots. A rectangular plate of dimensions 60mm (width) × 60mm (length) × 40mm (thickness) was machined out from an ingot and was hot-rolled at 673 K to a rolling reduction of 30%. Figure 1 is the inverse pole figure showing the axis density distribution. The mean axis density is used as unit. The maximum axis density and its position are shown at the bottom of the figure. The position is expressed as (α, β), with α being the angle between (0001) and the position of the maximum axis density, and β being the rotation angle around [0001] from the [0001]-[10$\bar{1}$0] line to the position of the maximum axis density. The maximum axis density is 4.8 times the mean axis density, and the position of the maximum axis density is (0, 0); (0, 0) implies [0001] (basal) texture [6].

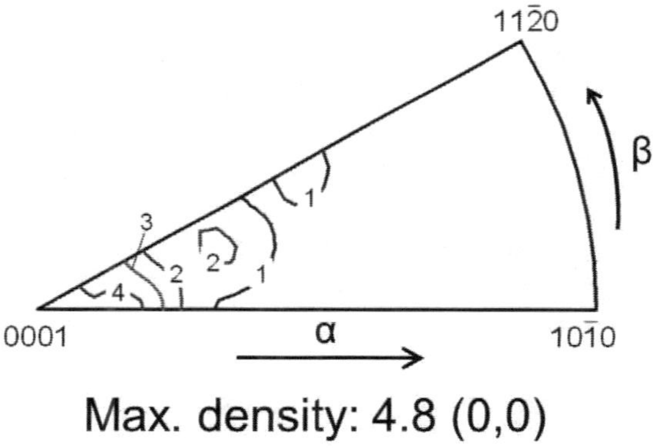

Fig. 1 Inverse pole figure showing the density distribution of axis. Mean axis density is used as a unit [6].

In order to investigate the effect of heat treatment on damping capacity, the specimens were annealed at temperatures between 573 K and 773 K for durations in the range of 30–180 min. The specimens were quenched in oil immediately after annealing to prevent any change in microstructure. The mid-plane section of specimens was prepared by mechanical polishing for microstructure observation.

Tensile tests were conducted at room temperature with a cross head speed of 2 mm/min, and the damping capacity was measured with a flexural internal friction measurement machine at room temperature. The vibration amplitude of the specimens was measured by means of an optical sensor throughout the vibration attenuation process. The damping, denoted by the logarithmic decrement (δ), is defined as follows:

$$\delta = 1/n \ \mathrm{In}(A_o/A_n), \quad (1)$$

where n is the vibration cycle, while Ao and An represent the vibration amplitudes of the initial and n^{th} cycles, respectively. The damping capacity was calculated from the decay of the vibration [5].

Texture measurement was carried out by the Schulz reflection method using nickel-filtered Cu Kα radiation. Based on five pole figures, the crystal orientation distribution function (ODF) was determined by the Dahms and Bunge method [7]. Electron backscatter diffraction (EBSD) measurements were performed after electrolytic polishing of the mechanically polished samples.

3. Results and discussion

Figure 2 shows the optical micrographs of the specimens before and after heat treatment. As shown in Fig. 1(a), dendrites were observed along the grain boundaries, these were composed of $Mg_{17}Al_{12}$ [8]. Upon heat treatment, this intermetallic, which had formed before heat treatment disappeared, as shown in Fig.2 (b).

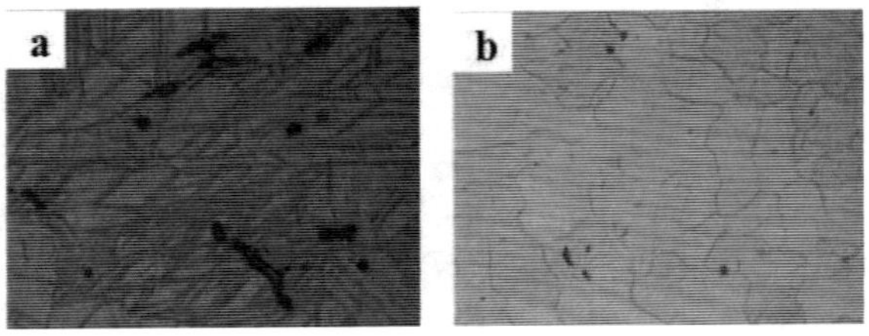

Fig. 2 Optical microstructure showing before and after the heat treatment. (a) before deformation, (d) after heat treatment (623 K, 180 min).

Figure 3 shows the effect of heat treatment time on the tensile strength and hardness of Mg alloys heat-treated at 673 K. As can be seen, both the hardness and tensile strength are inversely proportional to heat treatment time. Additionally, both these properties exhibited a similar relationship with the heat treatment temperature.

Grain size is known to increase with increasing heat-treatment temperature and time; the same was observed in this study. It is believed that the changes in tensile strength are effected by grain growth, and a reduction in the number of defects such as dislocations.

Figure 4 shows the relationship between the heat treatment time and the maximum axis density in the inverse pole figure for heat treatment at 623 K. It is known that a basal

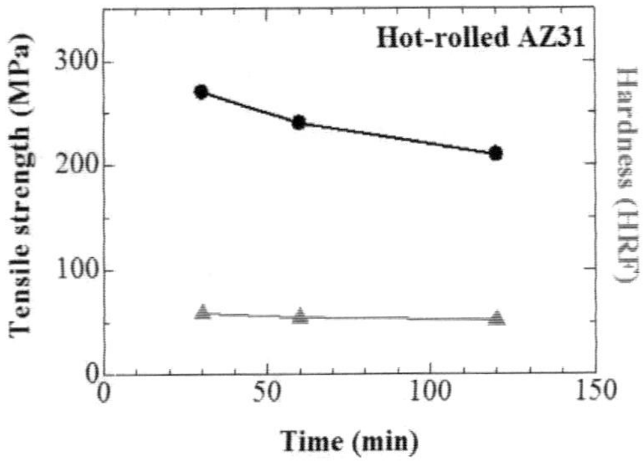

Fig. 3 Effect of heat treatment time on tensile strength and hardness for heat treatment at 673K.

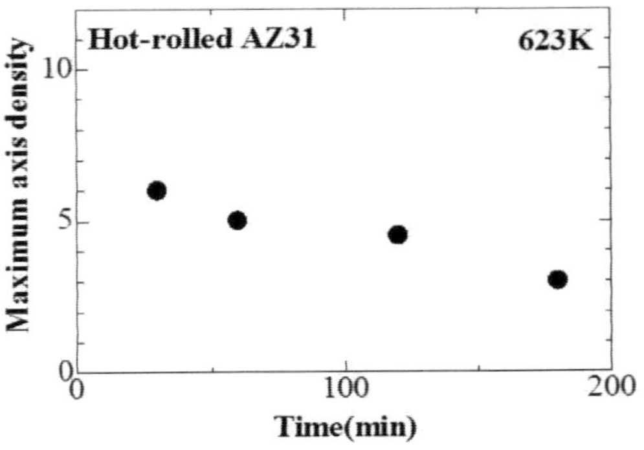

Fig. 4 The relationship between heat treatment time and maximum axis density of (0001) for heat treatment at 623 K.

texture is formed in magnesium alloys upon hot-rolling [9]. In this study, a basal texture was formed in all specimens despite annealing at different times and temperatures, and the maximum axis density of the basal texture was found to decrease with increasing heat-treatment time and temperature, as seen in Fig. 4.

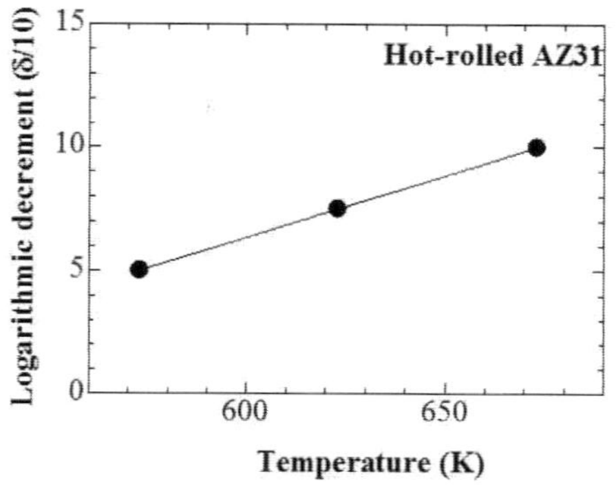

Fig. 5 Effect of heat treatment temperature on logarithmic decrement of hot-rolled AZ31.

At higher temperatures and longer durations, the damping capacity was gradually increased by the reduction in the number of defects such as dislocations in Fig. 5. Consequently, the grain size and texture are affected by the hot-working conditions, and that the changes in grain size and texture influence mechanical the properties (strength and hardness) and damping capacity of magnesium and its alloys.

4. Summary

In order to clarify effect of texture on the mechanical properties and damping capacity of magnesium and its alloys, specimens of AZ31 magnesium alloy were hot-rolled and experimentally investigated after being subjected to various heat-treatment conditions. It was found that with increasing heat-treatment temperature and time, the grain size of the specimens increased and the basal texture progressively weakened. Consequently, the mechanical properties (strength and hardness) deteriorated and the damping capacity decreased.

Acknowledgement

This work was supported by a Research Grant of Pukyong National University (2014).

References

[1] Barnett M R, Nave M D and Bettles C J 2004 Mater. Sci. Eng. A **386** 205

[2] Agnew S R and Duygulu 2003 Mat. Sci. Forum **419-422** 177

[3] Sugimoto K, Matsui K, Okamoto T and Kishitake K 1975 JIM **16** 647

[4] Watanabe H, Mukai T, Sugioka M and Ishikawa K 2004 Scr. Mater. **51** 291

[5] Kang C, Sung J, Kim G, Kim B, Kim I 2009 Mater. Trans. **50** 1390

[6] Kim J, Okayasu K and Fukutomi H 2013 Mater. Trans. **64** 192

[7] Dahms M and Bunge H J 1989 J. Appl. Cryst. **22** 439

[8] Celotto S 2000 Acta Mater. **48** 1775

[9] Ohtoshi K and Katsuta M 2001 J. JILM **51** 534

IOP Conf. Series: Materials Science and Engineering **82** (2015) 012115 doi:10.1088/1757-899X/82/1/012115

In-situ observation of dislocation and analysis of residual stresses by FEM/DDM modeling in water cavitation peening of pure titanium

DY Ju[1,2] and B Han[2]
[1] Department of Material Science and Engineering, Saitama Institute of Technology, Japan
[2] Department of Mechanical Engineering, University of Science and Technology Liaoning, Qianshan load 184, Anshan, China
E-mail: dyju@sit.ac.jp

Abstract. In this paper, in order to approach this problem, specimens of pure titanium were treated with WCP, and the subsequent changes in microstructure, residual stress, and surface morphologies were investigated as a function of WCP duration. The influence of water cavitation peening (WCP) treatment on the microstructure of pure titanium was investigated. A novel combined finite element and dislocation density method (FEM/DDM), proposed for predicting macro and micro residual stresses induced on the material subsurface treated with water cavitation peening, is also presented. A bilinear elastic-plastic finite element method was conducted to predict macro-residual stresses and a dislocation density method was conducted to predict micro-residual stresses. These approaches made possible the prediction of the magnitude and depth of residual stress fields in pure titanium. The effect of applied impact pressures on the residual stresses was also presented. The results of the FEM/DDM modeling were in good agreement with those of the experimental measurements.

1. Introduction

Cavitation impact has historically attracted attention due to its costly damage to hydraulic mechanical parts, such as hydrofoil surfaces, turbopump impellers, pumps, and valves. As such, many researchers [1-5] have traditionally focused their investigations into the damage mechanism of cavitation.

More recently, cavitation impact has been successfully developed as a means to improve the fatigue performance of mechanical components by introducing residual stress into the superficial layer of metallic components, in a manner similar to the more conventional process of shot peening. Ju and Qin et al. [16] designed a new ventilation nozzle, through which suitable air can be aerated into the extra high-velocity flow in the nozzle throat, thereby forming a tremendous pressure gradient between the upstream and downstream flows. This specific method is referred to as water cavitation peening (WCP) [6, 7]. Sahaya Grinspan and Gnanamoorthy successfully employed another cavitation process by injecting a high-speed oil jet into an oil-filled tank. This method, referred to as oil jet peening (OJP), effectively reduces the erosion of mechanical components [8-10].

Several recent investigations have revealed that WCP improves fatigue performance by introducing compressive residual stresses in the surface of metallic components [7- 9, 11, 12. The method has also been demonstrated to induce high uniform compressive residual stresses in gear tooth surfaces, since complicated and narrow surfaces can be more easily peened [16]. Compared with conventional shot peening, WCP can obtain the smoother surfaces. The distributions of impact pressure are isotropic, therefore, process capability of WCP is uniform at different incidence angle [16]; however, it is unknown if WCP can induce changes in the microstructures within the strengthened layers of metallic components, as shot peening does [18- 21].

Residual stresses in materials are often the result of metallurgical processes, such as casting, forging, welding, and quenching. Residual stresses from metallurgical processes usually depend on changes in thermal sources and volume accompanying the phase transformation. Generally, two types of residual stresses should be considered, that is, macro-residual stresses and micro-residual stresses [23]. Macro-residual stresses depend on the plastic deformation of solid materials due to rapid non-uniform cooling, while micro-residual stresses are caused by strain and deformation due to phase transformation and changes in microstructure. We also know that distortions due to thermal and elastic-plastic deformation and strain, as well changes in phase transformation and texture in manufactured materials, are important regardless of the type residual stresses. Unfortunately, since the post-WCP distortion process is minimal, it is not sufficiently clear why the WCP process creates such large compressive residual stresses, similar to traditional peenings. Therefore, it is important to clarify the cause of residual stress generation through basic material properties research.

In order to approach this problem, specimens of pure titanium were treated with WCP, and the subsequent changes in microstructure, residual stress, and surface morphologies were investigated as a function of WCP duration. The influence of water cavitation peening (WCP) treatment on the microstructure of pure titanium was investigated. The microstructural evolution in the near-surface of pure titanium as a function of WCP time was characterized by X-ray diffraction (XRD), optical microscopy (OM), scanning electron microscopy (SEM), and transmission electron microscopy (TEM). After WCP treatment, changes in the microstructure, as well as residual stress and surface morphologies as functions of WCP time, were recorded using a novel experimental design involving an in-situ observation function. The obtained results indicate that twinning plays an important role in the plastic deformation and residual stresses of hexagonal close-packed (HCP) structured metal materials, and therein, that the deformation twinning and twinning interaction were induced by WCP in the strengthening layer. A stable compressive residual stress layer was found in the near-surface of the investigated pure titanium.

In this paper, a novel combined finite element and dislocation density method (FEM/DDM), proposed for predicting macro and micro residual stresses induced on the material subsurface treated with water cavitation peening, is also presented. A bilinear elastic-plastic finite element method was conducted to predict macro-residual stresses and a dislocation density method was conducted to predict micro-residual stresses. These approaches made possible the prediction of the magnitude and depth of residual stress fields in pure titanium. The effect of applied impact pressures on the residual stresses was also presented. The results of the FEM/DDM modeling were in good agreement with those of the experimental measurements.

2. Experimental procedures

2.1. Test specimen and WCP conditions

Fig. 1 depicts a schematic of the test specimen and clamp. The test specimens consisted of pure titanium in as-annealed process state. The size of each test specimen was 4 mm × 6mm × 32 mm. The special clamp depicted in Fig. 2 was used to ideally combine the specimens (A) and (B) into one part.

The key process conditions of WCP include aeration flux, standoff distances (SOD), and WCP duration. All of these attributes significantly impact the degree of enhancement of the process, and therein, it is necessary to determine ideal process conditions prior to treating the test specimens. According to recent observations[15] regarding impact pressure distribution measurements, an aeration flux of 0.4 L/min and a SOD of 85 mm were demonstrated to induce a tremendous impact pressure. As shown in Fig. 2, the specimens were treated by WCP in the bottom center of the test water tank. WCP durations of 15 min, 30 min, 45 min, and 60 min were investigated in this study.

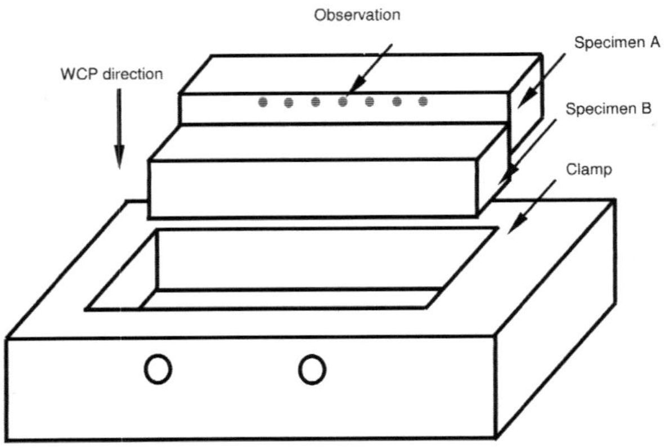

Fig. 1: Schematic of the test specimen and the clamp.

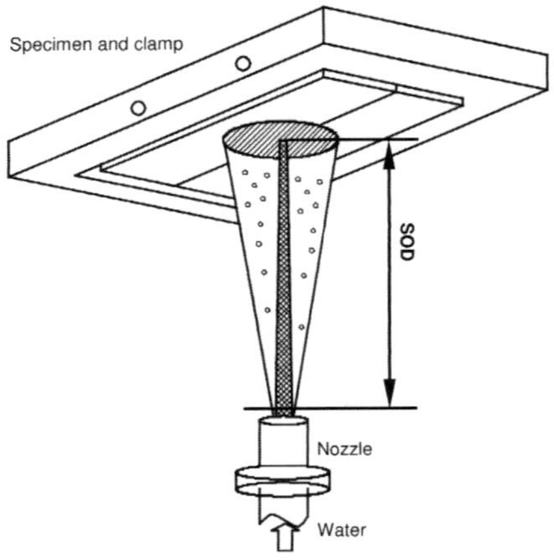

Fig. 2: Schematic of the WCP process

2.2. *OM observation of microstructures*

The specimens A and B were first ground and polished, and the surfaces of the microstructure observation points were eroded with a solution of 10% HF + 5% HNO_3 + 85% H_2O. The microstructure photos showing the different optimal points were recorded by OM. As depicted in Figs. 1 and 2, the same two specimens were joined by the special clamp in the WCP process. Finally, the microstructures at the same points were again recorded by OM after the WCP treatment. In the preceding method, microstructural changes in the superficial layer were recorded at each step for each WCP duration.

2.3. *TEM observation of microstructures*

The TEM (JEOL2100) was used to examine specimen microstructures at a depth of 30 to 40 μm from the surface in order to further investigate any microstructural changes due to the WCP treatment. In order to obtain a clear image of the microstructure, samples were observed using SEM after etching in a solution containing 10%HF + 5%HNO3 +85%H2O. A Twin-Jet Electropolisher (FISCHIONE-140 Digital Power Control) and a solution of 5% HClO4 + 95% CH3OH was used to etch the specimens to an optimal thickness for TEM analysis.

2.4. *Residual stress measurement*

The surface residual stress and depth distribution of residual stress in the near-surface layer were measured by X-ray diffraction stress analysis using the conventional $\sin^2\Psi$ method. A special optical system jointly extracted precise information concerning the residual stresses in areas as small as 0.15 mm in diameter. In this study, a side inclination method was also used for stress measurement. The X-ray tube (Co-Kα type) was operated at 30 kV and 10 mA, with a slit diameter of 2 mm. The shift of the α-Ti (114) diffraction profile was detected at angles φ = 10°, 20°, 30°, 40°, and 45°. The diffractive angle $2\theta_0$ was 154.4°. The stress constant of the X-ray diffraction analysis was –180 MPa/°. Moreover, Vickers hardness was measured on the peened surface (peened for 45 min) and unpeened specimens with a load of 9.8 N. In order to investigate the depth distribution of the residual stress, the near-surface layer of the test point was removed by step-by-step electrolytical polishing using a Proto Electrolytic Polisher-Model 8818 (Proto Manufacturing Ltd.). The depth of the electrolytic polishing was approximately 10 to 25 μm at each step, and was adjustable via the voltage control, timer control, and flow rate of the electrolyte control. The depth distribution of the residual stress was obtained by measuring the residual stress at each step.

3. Analysis Model by FEM/DDM

Considering not only macro-plastic strain but also micro-plastic strain, the total residual stress $\sigma_{residual}$ induced by WCP is given by

$$\sigma_{residual} = \sigma_{FEM} + \sigma_{DDM} \quad (1)$$

where, σ_{FEM} is the resolved residual stress induced by macro-plastic deformation, which can be obtained from the finite element method; and σ_{DDM} is the resolved residual stress induced by macro-plastic deformation, which can be obtained from the dislocation density method. Under uniaxial strain conditions, the highest elastic stress level in the impact wave propagation is defined as the Hugoniot Elastic Limit (HEL). According to Hook's law [27], the surface plastic strain ε_p can be expressed as

$$\varepsilon_p = \frac{-2HEL}{3\lambda + 2\mu}\left(\frac{P}{HEL} - 1\right) \quad (2)$$

where, P is the impact pressure and a function of pressure pulse duration, λ is Lame's constants, and μ is Lame's constants. The residual stress σ_{FEM} induced by WCP can be written as

$$\sigma_{\text{FEM}} = \sigma_0 - \left[\frac{\mu\varepsilon_p(1+\nu)}{(1-\nu)+\sigma_0}\right]\left[1 - \frac{4L_p(1+\nu)}{\pi r_p}\right] \quad (3)$$

where, r_p is the radius of circular impact zone on the target, σ_0 is initial residual stress, and L_p is the depth of plastic affection.

$$\sigma_{DDM} = -A\mu b\sqrt{N_m} + \sigma_0 \quad (4)$$

where, σ_{DDM} is the residual stress induced by dislocations; A is the constant caused by unknown sources; μ is the shear modulus; N_m is the density of mobile dislocation ; b is the magnitude of Burgers vector, and σ_0 is the original stress. According to Eq. 2 and Eq. 1, the total residual stress $\sigma_{residual}$ can be written as

$$\sigma_{residual} = -\left[\frac{\mu\varepsilon_p(1+\nu)}{(1-\nu)+\sigma_0}\right]\left[1 - \frac{4L_p(1+\nu)}{\pi r_p}\right] - A\mu b\sqrt{N_m} + \sigma_0 \quad (5)$$

Based on Eq. 1, the total residual stress $\sigma_{residual}$ induced by WCP can be obtained by the combined finite element method and dislocation density method (FEM/DDM) approach.

4. Results and discussion

4.1. Microstructures

The microstructural changes of the superficial layers as a function of WCP duration, at the same observation points, are shown in Fig .3. Figure 3 (a) depicts typical equiaxial α grains of annealed pure titanium, with grain sizes in the range of 20 to 100 μm. From Fig. 3 (b-d), the microstructural evolution of the titanium specimens at WCP durations of 15 min, 30 min and 45 min can be clearly observed. The deformation twinning induced by WCP at a depth of 30 to 40 μm from the surface can be observed via the OM micrographs in Fig. 4 (a,b). The density and quantity of deformation twinning increase gradually with increasing WCP duration, and decrease gradually with increasing layer depth from the treated surface. The density and quantity of deformation twinning no longer significantly increase when the WCP time exceeds 45 min and when the depth of deformation twinning reaches approximately 150 μm. The trend in microstructural change is similar to that of the compressive residual stress. At the same time, the integrity of the treated surface is maintained, proving that WCP can produce a smoother surface with minimal structure losses.

The TEM microstructures at a depth of 30 to 40 μm are depicted in Fig. 5. The interaction of deformation twinning in two different directions, as well as the high density dislocations among the bands of twinning, can be seen in Fig. 5 (a). Three types of twinning systems cross at the same point, as shown in Fig. 5 (b).

4.2. *Residual stress*

The typical residual stress distributions in the superficial layers, as functions of WCP duration, are depicted in Fig.6. From Figure 6, it can be observed that the maxima of the compressive residual stresses due to WCP are at the surfaces of the specimens for all of the investigated WCP treatment durations. This observation result differs from that of shot-peened. The maximum residual stress is usually observed below the surface during the shot-peening process. The distributions of compressive residual stresses depend on material characteristics and peening conditions. Furthermore, the load mode of WCP is an impact wave pressure due to cavitation collapse. The obvious macro-non-uniform plastic deformation is not induced by WCP, as compared to the deformation one would observe in other traditional methods. Crystalline defects, such as vacancies, interstitials, dislocations, and twinning, are the primary causes of compressive residual stress distributions during WCP process. The maximum of compressive residual stress observed was –620 MPa at a WCP duration of 60 min. As the depth from the surface was increased to 150 μm, the compressive residual stress was observed to gradually decrease. In comparison to residual stress distributions of WCP-treated specimens, those of original specimens essentially show zero residual stress throughout the entire near-surface. Since the compressive residual stress was observed to slowly increase after 45 minutes, this can be considered as the saturation time of compressive residual stress. This result indicates that the compressive residual stress can be induced in the near-surface of the specimen through WCP. Vickers hardness measurements were conducted on the peened (45 min) and unpeened specimen surfaces to evaluate their respective hardnesses. Table 2 depicts the changes in surface hardness of the peened and unpeened specimens, and therein, the average hardness of the peened specimens was about 17 HV higher than that of the unpeened specimens, indicating strengthening due to WCP.

4.3. *Analysis results of residual stress by FEM/DDM model*

According to the predicting method proposed in this paper, the residual stress distributions were predicted after 45 min with the different applied impact pressures. Figure 7 shows the predicted results of residual stress after 45 min with the different applied impact pressures. The black line indicates the predicted results based on the EFM model, while the red line represents the prediction results based on the DDM model. The green line shows the total residual stresses based on the FEM/DDM model. In Fig. 7, the predicted results with different applied impact pressures after 45 min and the experimental results measured by XRD method are compared. The predicted results considering the macro-residual stress and the micro-residual stress with applied impact pressure around 800 MPa are in good agreement with the experimental results by XRD method. These results indicate that the new model could potentially be a useful tool in predicting the residual stresses and investigating the effect of processing materials parameters on the distribution of residual stress.

4.4 *Strengthening modeling and mechanism of WCP*

Fig. 8 illustrates the proposed mechanism of WCP, and depicts the strengthening model of the surface microstructure for a pure titanium plate during the WCP process. The modeling is based on the residual stress concentration generated by a pile-up caused by a dislocation activation of the impact source [21-26], such as that shown in Fig. 8. The amount of dislocation-induced pile-up is determined by the distance L between the dislocation lines created by the source and barrier, and the applied stress. The local stress in front of the barrier is equal to the product of the initial residual stress and the number of piled-up dislocations. The unique twinning that is created will strongly depend on micro-residual stresses and dislocation-induced pile-up. The initiation and propagation of the twinning in a neighboring grain are also shown in Fig.8. The crystal structure of pure titanium is a typical HCP structure at room temperature. The basal planes (0001) are the close-packed planes, and there are three close-packed directions <1120>; hence, in HCP, there are three slip systems. This low number of slip systems means it is difficult to plastically deform an HCP metal. It can be shown that five slip systems must operate in each grain of a polycrystal if it is to deform in a manner compatible with the deformations of its neighbors. In light of these constraints, the deformation of an HCP polycrystal must occur via mechanical twinning to avoid cracking at the grain boundaries.

The dislocation sources in crystal grains are continuously induced in the near-surface of specimens due to the shear stresses. The dislocation density increases gradually, and many dislocation pile-ups form. When plastic deformation occurs, deformation twinning is induced at the location of the dislocation block due to the low number of slip systems. Impact waves caused by the collapse of a cavitation bubble can produce significant strain, a high strain rate, and numerous direction cycle loads, which all induce the overlapping and interaction of deformation twinning. This indicates that the enhancement mechanism of WCP might be related to the activity of the deformation twinning and the high density of dislocations. At the same time, the evolution of microstructures can also induce compressive residual stress in the near-surface of the specimens.

From the results in this paper, we conclude that the mechanism corresponding to the model proposed in this paper is appropriate. Since it accurately describes the systems investigated here, we feel it may be useful for designing WCP process conditions, and perhaps for harnessing the possible reinforcing effect observed in the WCP-processed specimens.

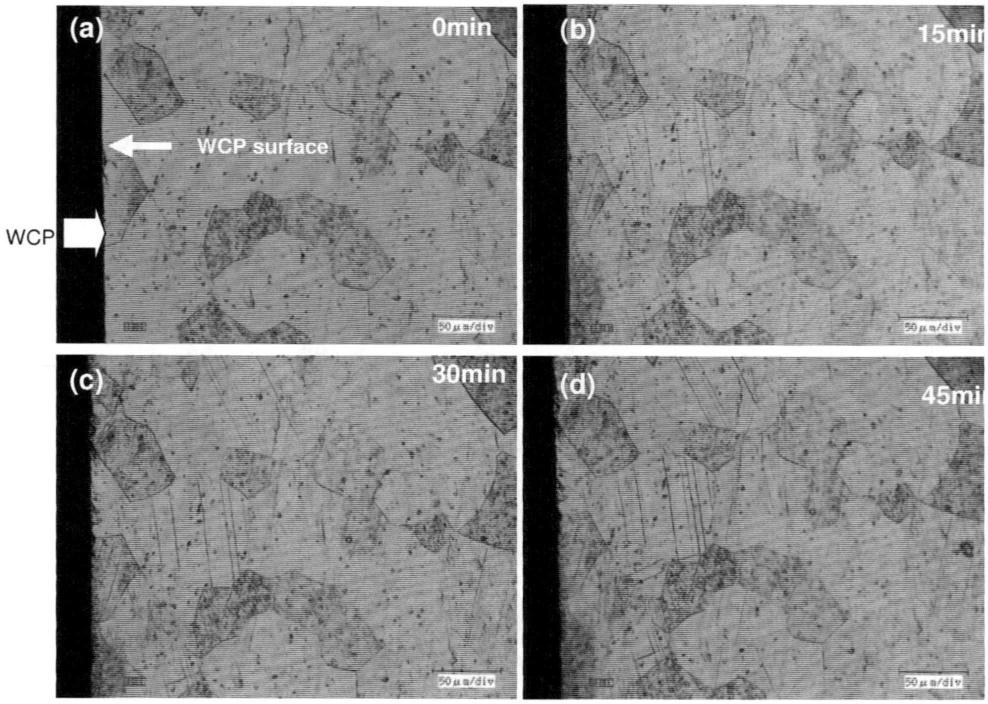

Fig. 3: The microstructural evolution process of pure titanium with WCP duration, at the same observation point.

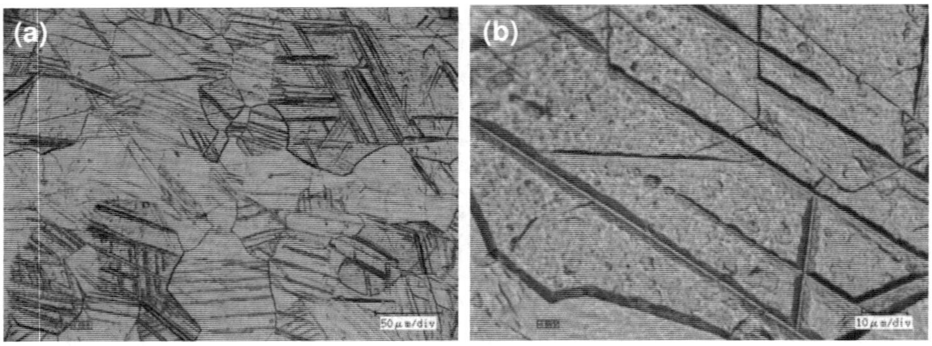

Fig. 4: OM microstructures at a depth of 30-40 μm from the surface.

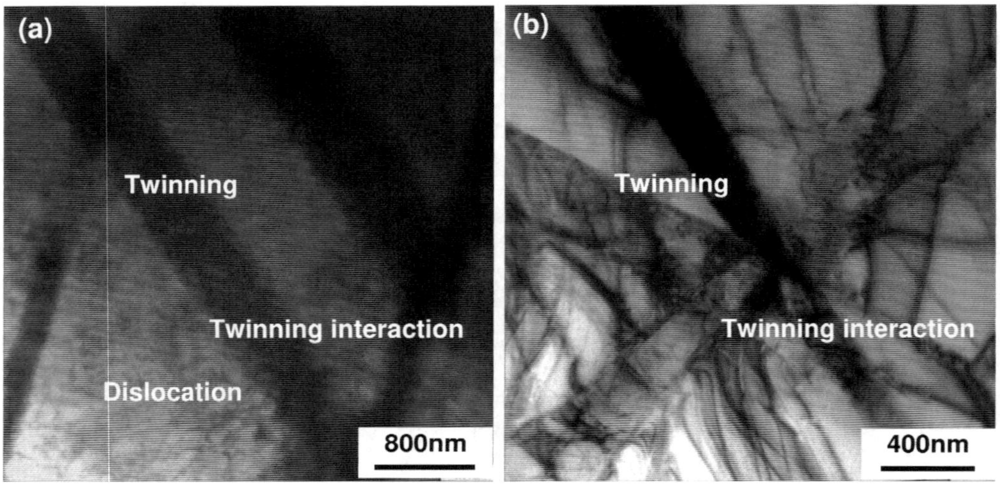

Fig. 5: TEM microstructures at a depth of 30-40 μm from the surface. (a) The interaction of deformation twinning in two different directions, as well as the high density dislocations among the bands of twinning ; (b) Three types of twinning systems cross at the same point.

Table 2 : The hardness changes induced by WCP on the surface

Process	Vickers hardness (HV)					Average (HV)
Unpeened	148	147	160	147	156	151.6
Peened	165	160	175	170	173	168.7

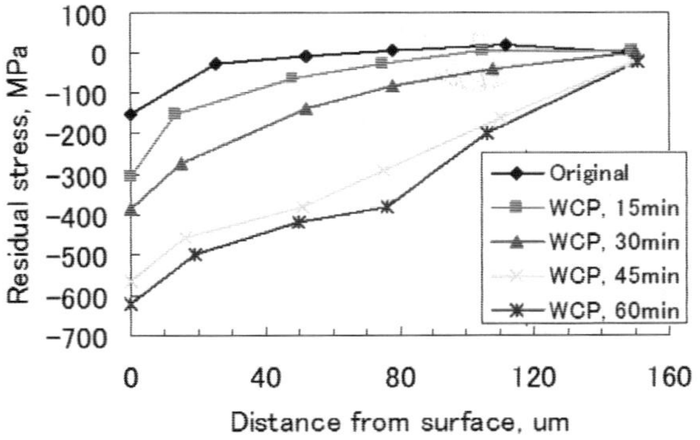

Fig .6: Distributions of residual stress in the strengthening layer.

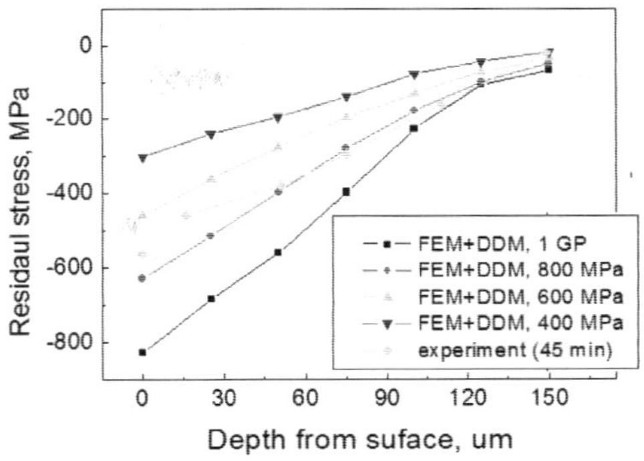

Fig. 7: Predicted results of residual stress versus experimental data by FEM/DDM.

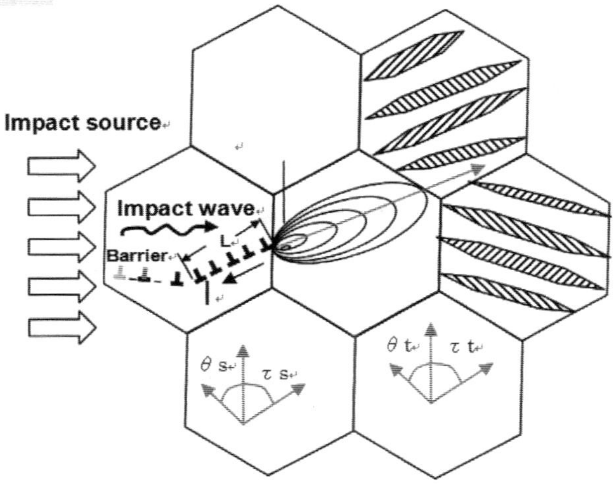

Fig. 8: Modeling the strengthening of surface microstructures during the WCP process.

4. Conclusions

The results obtained in this investigation show that WCP can induce a stable compressive residual stress in the superficial layer of titanium specimens. The depth of the compressive residual stress zone can be up to approximately 150 μm. Deformation twinning and dislocations play an important role in the micro-plastic deformation of HCP-structured metal materials.

The following conclusions can be drawn from this paper concerning the residual stress generation mechanism. The dislocation density of the α-phase of pure titanium increases when the impact energy caused in the WCP process exceeds the internal energy of the pure titanium crystal in the initial WCP process stages. This was confirmed to generate twinning deformation via the dislocation-induced formation and development of a slipping zone. The deformation due to twinning, the interaction between these deformation twins and dislocations, and local micro-plastic deformation are all induced by WCP in the strengthening layer. These findings indicate that the enhancement mechanism of WCP might be related to both the activity of the deformation twinning, and to the dislocations. The predicted results of residual stress in the near-surface of pure titanium were in good agreement with those of the experimental measurements

Acknowledgements
This research receives ongoing support from the High-Tech Research Center and Open Research Center at the Saitama Institute of Technology.

References

[1] Hammitt FG and De MK. 1979 *Wear* **52** 243-262
[2] Tomita Y and Shima A 1986 *J. Fluid Mech.* **169** 535-564
[3] Chen YL, Kuhl T and Israelachvili J 1992 *Wear* 153 31-51
[4] Sun Z, Kang XQ, Wang XH 2005 *Mater. Des.* **26** 59-63
[5] Tang CH, Cheng FT and Man HC 2006 *J Surf. Coat. Tech.* **200** 2602-2609
[6] Qin M, Ju DY and Oba R 2006 *Surf. Coat. Tech.* **200** 5364-5369
[7] Han B, Ju DY and Jia WP 2007 *Appl. Surf. Sci.* **253** 9342-9346
[8] Sahaya Grinspan A and Gnanamoorthy, R., 2006 Appl. Surf. Sci. 253, 989-996.
[9] Sahaya Grinspan A and Gnanamoorthy R 2006 *Appl. Surf. Sci.* **253** 997-1006
[10] Sahaya Grinspan A and Gnanamoorthy R 2006 *J Surf. Coat. Tech.* **201** 1768-1775
[11] Ramulu M, Kunaporn S, Jenkins M, Hashish M and Hopkins J.2002 *J. Press. Vess. Technol.* **124** 118-123
[12] Kunaporn S, Ramulu M, Jenkins M and Hashish M 2005 *J. Press. Vess. Technol.* **127** 186-192
[13] Sahaya Grinspan A and Gnanamoorthy R 2007 *J. Manuf. Sci. Eng.* **129** 601-606
[14] Sahaya Grinspan A and Gnanamoorthy R 2007 *J. Eng. Mater. Technol.* **129** 609-613
[15] Han B, Ju DY and Nemoto T 2007 *Mater. Sci. Forum* **561-565** 2485-2488
[16] Ju DY, Qin M, Koubayashi T and Oba R 2006 *Surf. Eng.* **22** 219-223
[17] Qin M, Ju DY and Oba R 2006 *Surf. Coat. Tech.* **201** 1409-1413
[18] Martin U, Altenberger I, Scholtes B, Kremmer K and Oettel H 1998 *Mater. Sci. Eng.* A **246** 69-80
[19] Altenberger I, Scholtes B, Martin U and Oettel H 1999 *Mater. Sci. Eng.* A **264** 1-16
[20] Wu X, Tao N, Hong Y, Xu B, Lu J and Lu K 2002 *Acta Mater.* **50** 2075-2084
[21] Harada Y, Fukaura K and Haga S 2007 *J. Mater. Process. Tech.* **191** 297-301
[22] Wang TS, Lu B, Zhang M, Hou RJ and Zhang FC 2007 *Mater. Sci. Eng.* A **458** 249-252
[23] Noyan C and Cohen JB, 1987 *Residual Stress Measurement by Diffraction and Interpretation*, Springer-Verlag, New York, pp. 2-101
[24] Ju DY 2002 *Handbook of Residual Stress and Deformation of Steel.* In: Totten G, Howes M and Inoue T (Eds.), ASM International, Ohio, pp. 372-398
[25] Meyers MA, Vöhringer O and Lubarda VA 2001 *Acta Mater.* **49** 4025-4039
[26] Gilman JJ 1997 *Phil. Mag.* **76** 329-336
[27] Jordan AS, Caruso R and von Neida AR. 1980 *Bell Syst. Tech* **59** 593-637

AUTHOR INDEX

AUTHOR INDEX

AUTHOR INDEX

AUTHOR INDEX